The Early Life History of Fish

The Proceedings of an International Symposium Held at the Dunstaffnage Marine Research Laboratory of the Scottish Marine Biological Association at Oban, Scotland, from May 17–23, 1973

Edited by J. H. S. Blaxter

With 299 Figures

Springer-Verlag Berlin Heidelberg New York 1974

John H. S. Blaxter, M. A., D. Sc.
Scottish Marine Biological Association
Dunstaffnage Marine Research Laboratory
P. O. B. 3
Oban, Argyll PA 34 4AD/Great Britain

ISBN 3-540-06719-1 Springer-Verlag Berlin · Heidelberg · New York
ISBN 0-387-06719-1 Springer-Verlag New York · Heidelberg · Berlin

Offsetprinting: Julius Beltz, Hemsbach/Bergstr. Bookbinding: Brühlsche Universitätsdruckerei, Gießen.

Introduction

Plans for an International Symposium on the Early Life History of Fish were first discussed in 1967 at the F.A.O. Advisory Committee on Marine Resources Research (A.C.M.R.R.). It was considered that studies on fish eggs and larvae were of value in estimating the size of fish stocks, in appraising the stock-recruitment relationship, and in helping to answer questions on the systematics and taxonomy of fish.

An A.C.M.R.R. Working Party recommended in 1969 that a Symposium should be held and after discussion with the International Association of Biological Oceanography (I.A.B.O.) Dr. J.H.S. Blaxter of the Scottish Marine Biological Association (S.M.B.A.) agreed to convene the meeting with the help of a steering committee. Various international agencies were interested in the meeting and agreed to sponsor it in various ways. These agencies included F.A.O., I.A.B.O., the International Council for the Exploration of the Sea (I.C.E.S.) and the Scientific Committee for Oceanic Research (S.C.O.R.). In the final event F.A.O. supported the meeting by publishing the abstracts beforehand, providing travel funds for participants and publishing a report after the meeting. I.A.B.O. and S.C.O.R. provided travel funds and the Royal Society of London and S.M.B.A. finance for local expenses. Without this financial help the Symposium would have been very much less successful.

A Steering Committee was set up in 1972 which consisted of the following members:

Convenors: Dr. J.H.S. Blaxter (U.K.), Prof. G. Hempel (F.G.R., I.A.B.O. representative), Prof. S. Tanaka (Japan, A.C.M.R.R. representative).
Committee: Dr. E.H. Ahlstrom (U.S.A.), Dr. D.F. Alderdice (Canada), Prof. O. Dragesund (Norway, I.C.E.S. representative), Dr. E. Fagetti (F.A.O.), Prof. C. Hubbs (U.S.A.), Dr. R. Lasker (U.S.A.), Mr. O. Ledoux (France), Prof. K. Lillelund (F.G.R.), Prof. T. Rass (U.S.S.R.), Dr. W.J. Richards (U.S.A.), Mr. A. Saville (U.K.).

With their help the Symposium was organized and then held from 17-23 May 1973 at the Dunstaffnage Marine Research Laboratory in Oban, Scotland. About 160 participants from 26 countries attended to hear 62 papers arranged in plenary and parallel sessions.

The dominant theme of the papers was the use of egg and larval surveys in fish population dynamics. It seems to be generally accepted that high mortalities in these early stages are an important factor in determining the future brood strength and success of a year class, and thus the recruitment to the fishery. High mortalities of eggs were reported in a number of species which seem to be explained by imperfections in development rather than by predation or current systems carrying the eggs out of the area. After hatching, losses due to predation and starvation are heavy and many larvae are small, requiring food of very small size. Attempts have been made to quantify the nutritional status of larvae to see whether sea-caught larvae are viable and whether the weaker ones are selectively sampled by plankton nets. Additional mortality due to the delicate integument and general fragility of larvae may follow from unsuccessful strikes by predators.

The Symposium showed that our knowledge of micro-distribution is also inadequate. Many plankton nets filter large quantities of water and are not equipped with closing devices to limit the depth of sampling. Larvae make small-scale horizontal and vertical migrations, and some initial evidence is also available that larvae may aggregate on small food particles. A re-shaping of ideas on larval distribution may emerge from the Symposium.

Much more is now known about the tolerance of eggs and larvae to such harmful factors as high temperature, low oxygen, H_2S and heavy metals. These data are helpful in assessing losses due to entrainment in power station cooling systems and the possibility of exploiting heated effluents in aquaculture. The improvement in culture techniques of late has been remarkable. New sources of small food organisms, the possibility of micro-encapsulation of food and the advances in husbandry generally mean that it will soon be possible to rear most species from the egg, given adequate finance. It is still not clear, however, whether larvae can utilize dissolved organic matter (Pütter's Theory).

One of the main existing needs is the production of mature parent fish on demand. The use of appropriate regimes of temperature and photoperiod and of hormone injections may help to solve this problem. In the field of aquaculture, a new form of research is being developed, on hybridization and other types of genetic manipulation, which may lead to improved growth and viability.

Taxonomic studies also occupied a part of the Symposium and it is clear that even now the larvae of many fish are not well-known. This is partly due to the multiplicity of species but also to the tendency for larvae to show smaller morphological distinctions than adults. Perhaps this is the most neglected of the fields covered by the Symposium.

These published proceedings of the Symposium should be useful in representing the current status of research into marine fish eggs and larvae. The coverage of the various branches of the research is wide and the presence together at the Symposium of so many of the active research workers provided an excellent opportunity to crystallize opinion and plan for the future.

Acknowledgement is due to some members of the Committee, especially Dr. Ahlstrom and Mr. Saville, who provided editorial help.

February 1974 J.H.S. Blaxter

Contents

POPULATION STUDIES

Chairmen of sessions: Dr. D.H. Cushing, Mr. A. Saville, Prof. G. Hempel, and Dr. J.J. Zijlstra

Rapporteur: Mr. R. Jones

DISTRIBUTION

Chairman of session: Prof. O. Dragesund

Rapporteur: Dr. J.D.M. Gordon

FEEDING AND METABOLISM

Chairman of session: Prof. K. Lillelund

Rapporteur: Dr. K.F. Ehrlich

PHYSIOLOGICAL ECOLOGY

Chairmen of sessions: Dr. R. Lasker and Dr. D.F. Alderdice

Rapporteur: Dr. D.N. Outram

DEVELOPMENTAL EVENTS

Chairman of session: Prof. C. Hubbs

Rapporteur: Dr. C.C. Lindsey

BEHAVIOUR

Chairman of session: Prof. G. Hempel

Rapporteur: Dr. R.N. Gibson

TAXONOMY

Chairmen of sessions: Dr. E.H. Ahlstrom and Dr. W.J. Richards

Rapporteurs: Dr. E.H. Ahlstrom and Dr. W.J. Richards

REARING

Chairman of session: Mr. O. Ledoux

Rapporteur: Mr. A.B. Bowers

Population Studies

Larval Mortality in Marine Fishes and the Critical Period Concept[1]

R. C. May

INTRODUCTION

The high fecundity of many marine fishes implies that an extremely high rate of mortality must be experienced by each year-class. Most of this mortality occurs during the pelagic larval stage, and for this reason the characteristics of larval mortality are intimately related to basic problems in the population dynamics of fishes, including density-dependent regulatory mechanisms, the relation between stock and recruitment, and the determination of year-class strength (Beverton, 1962; Gulland, 1965; Hempel, 1965; Cushing, 1969). The critical period concept may be defined as the proposal that most larval mortality is concentrated during a relatively short period in early development. This review will first consider several variants of this broad definition and then examine evidence for a critical period (in its most widely accepted sense) in marine fishes.

DEFINITIONS OF THE TERM "CRITICAL PERIOD"

The term "critical period" was first applied to fish by two early French fish culturists, Fabre-Domergue and Biétrix (1897), who used it to describe the time of complete yolk absorption, when they had observed high mortality among marine fish larvae in laboratory rearing attempts. Fish culturists still use the term critical period to refer to stages when mortality among captive larvae is high (e.g. Liao et al., 1970).

In the fisheries literature the term "critical period" has been used in a different sense. Thus Gulland (1965) defines the "critical phase" as "the phase during which the strength of a year-class is determined", without specifying a developmental stage when this phase occurs or a cause of mortality during the phase.

The most generally understood definition of the term "critical period" was put forth by the Norwegian fishery biologist Johan Hjort. In a pioneering study, Hjort established that the year-class strength of Norwegian herring and cod stocks varied widely and that the strength of a year-class was determined early in its history (Hjort, 1914, 1926). He went on to say, "This ... leads us to the question, at which stage of development the most critical period is to be sought. Nothing is known with certainty as to this; such data as are available, however, appear to indicate _the very earliest larval and young fry stages_ as most important (Hjort, 1914, p. 204; original emphasis)".

---

[1] Contribution no. 437, Hawaii Institute of Marine Biology.

Hjort suggested that survival through these stages might be affected in two ways: 1. by a lack of food at the time the larvae begin to feed, which could cause catastrophic mortality; 2. by currents that could transport larvae to areas unfavorable to their further development, resulting in high mortality. Hjort emphasized the first of the above mechanisms and cited the experiences of fish culturists as evidence that the time of yolk exhaustion was a particularly sensitive period. Moreover, the stock with which Hjort was most concerned, the Norwegian spring-spawning herring, spawn at about the same time that the spring plankton bloom commences. Hjort thus reasoned that a slight retardation of the bloom or hastening of the spawning could leave early larvae without food and result in high mortality and low recruitment. The extremely successful 1904 year-class of herring, which resulted from eggs spawned late in the season, seemed to support this theory.

The first mechanism proposed by Hjort, referred to here as "Hjort's critical period concept", has been the subject of much comment and speculation. Restated, the concept maintains that the strength of a year-class is determined by the availability of planktonic food shortly after the larval yolk supply has been exhausted. Despite consistent references in the literature to Hjort's idea of a critical period, the only published review of the concept is Marr's (1956) critical evaluation. Much relevant information has accumulated in the 17 years since Marr's review appeared, including laboratory studies of larval behavior and food requirements and field studies of larval survival and condition. This review will assess Hjort's critical period concept in the light of recent findings by considering: 1. survival curves for natural populations of larval fish, 2. evidence for the starvation of larvae at sea, and 3. the sensitivity of larval fish to food deprivation.

SURVIVAL CURVES FOR NATURAL POPULATIONS OF LARVAL FISH

Is larval mortality at sea concentrated at the end of the yolk-sac stage, as Hjort's critical period concept predicts? The evidence most closely related to this question is found in survival curves for natural populations of larval fish; but unfortunately there are few such curves in the literature, and those that do exist are difficult to interpret.

All estimates of larval survival in nature are based on collections made with plankton nets. To yield meaningful results, the water strained must contain a representative sample of the larval population and the sampling gear must efficiently capture the larvae in that water. The "patchiness" which characterizes plankton (Cassie, 1963; Wiebe, 1970) and the vertical distribution patterns of larvae (Colton et al., 1961; Wood, 1971) give little assurance that the larvae in the sampled water truly represent the population under study. This fact is reflected in the wide confidence limits applied to plankton samples (Taft, 1960; Ida, 1972); and it has long been agreed that towed plankton nets are selective (Tranter, 1968). The early larval stages (when the postulated critical period would occur) are frequently undersampled because of mesh selection (Ahlstrom, 1954; Saville, 1959; Lenarz, 1972); and later larvae are undersampled because of net avoidance (Ahlstrom, 1954; Murphy and Clutter, 1972; Barkley, 1972). Larvae obtained in field collections are routinely counted and measured. In order to generate a survival curve from these data, the larval growth rate is needed. Usually growth rates are not known but are estimated from the progression of length-frequency modes (Sette, 1943) or from the assumption of a particular form of growth curve (Ahlstrom, 1954). Occasionally growth rates have been obtained by very dubious means, such as measuring the "growth" of starved larvae in aquaria (Farris, 1961). The effects of temperature are generally ignored in these estimates of larval growth. Furthermore, since food availability affects larval growth rate (O'Connell and Raymond, 1970), mortality resulting from starvation would be confounded with a decreased growth rate, and a reliable survival curve would be nearly impossible to derive. A further complication arises when samples from different areas and different seasons or years are pooled to give a single curve (e.g. Nakai et al., 1955). This procedure will mask any spatial or temporal variations in the shape of the survival curve.

Without dismissing these reservations, the available survival curves have been surveyed and classed as to whether or not they indicate greatly increased mortality at, or shortly after, complete yolk absorption (Table 1). Most of the curves either show no increased mortality at yolk absorption or permit varying interpretations. Karlovac (1967) describes the only clear-cut instance of increased mortality at yolk absorption, although Dragesund and Nakken (1971), while not presenting a complete survival curve, report 94% mortality among Norwegian herring at the transition to active feeding. Several instances of high mortality during embryonic and yolk-sac stages (Farris, 1961; Dekhnik, 1963, 1964) could be due to predation, sampling errors, or poor egg quality. Certain studies (e.g. Saville, 1956; Lenarz, 1972) that present plots of larval numbers against size give no indication of a critical period. It should be noted, however, that the semi-logarithmic plots commonly used for survival and length-frequency curves tend to under-emphasize the magnitude of mortality in the early larval stages. To summarize, the available data do not allow one to answer the question of whether or not mortality is concentrated at the end of the yolk-sac stage in natural populations. This is essentially the conclusion which Marr (1956) reached, with only two studies (Sette, 1943; Ahlstrom, 1954) available to him. The studies that have appeared since 1956 have by no means settled the matter, largely because of the variability and uncertainty of the data.

It is not the purpose of this section to discredit information based on plankton collections. Indeed, collections of fish eggs and larvae have yielded invaluable data on distribution and relative abundances and are a powerful tool for resource assessment (Ahlstrom, 1965, 1966, 1968). However, it should simply be stated that with present sampling techniques such collections are generally not precise or accurate enough to disclose whether or not larval fish mortality is concentrated at the end of the yolk-sac stage. Even with perfect sampling, a unique answer to this question will probably never emerge. One should expect variations between fish species, and between years and areas within species, and in the shapes of survival curves.

Table 1. Studies which provide survival curves for larval fish at sea. Question marks indicate that interpretations of the curves may vary

Fish studied	Location	"Critical period"	Reference
Scomber scombrus	N.W. Atlantic	no	Sette, 1943
Sardinops caerulea	California Current	no(?)	Ahlstrom, 1954
Sardinops melanosticta	Japan	yes(?)	Nakai et al., 1955
Engraulis japonica	Japan	yes(?)	Nakai et al., 1955
Trachurus symmetricus	California Current	no	Farris, 1961
Pseudopleuronectes americanus	Mystic River estuary (Connecticut, U.S.A.)	no(?)	Pearcy, 1962
Clupea pallasii	British Columbia	no	Stevenson, 1962
Engraulis encrasicholus	Black Sea	no	Dekhnik, 1963
Trachurus mediterraneus	Black Sea	no	Dekhnik, 1964
Clupea pallasii	Japan	no	Iizuka, 1966
Sardina pilchardus	Adriatic Sea	yes	Karlovac, 1967

EVIDENCE FOR LARVAL STARVATION AT SEA

Regardless of the true shape of larval survival curves at sea, is there evidence that starvation at sea is a major cause of larval mortality at the end of the yolk-sac stage, as predicted by Hjort's critical period concept? Several lines of evidence refer to this question.

Incidence of Feeding. This term refers to the percentage of sea-caught larvae which have food in their guts; a low incidence of feeding might be taken as evidence that larvae were starving. Values reported for larval fishes are highly variable, probably reflecting sampling problems as well as real variations. In early larvae, especially clupeoids, the incidence of feeding is sometimes very low (e.g. Lebour, 1920; Bowers and Williamson, 1951; Arthur, 1956; Berner, 1959). Several explanations of the low feeding incidence in clupeoids have been given, including: a rapid rate of digestion (Lebour, 1921), nutrition from dissolved organic matter (Morris, 1955), and a low food requirement (Arthur, 1956). There is no evidence to support any of these explanations and it is now clear that low feeding incidences are not an inviolable rule among early larval clupeoids; for example, Ciechomski (1967) found that 52% of first-feeding Argentine anchovy. *Engraulis anchoita* had food in their guts, and Burdick (1969) found feeding incidences of 80 to 100% among first-feeding Hawaiian anchovy, *Stolephorus purpureus*. Furthermore, some reported cases of low feeding incidences may be results of diurnal feeding patterns (Blaxter, 1965) or defecation upon capture (Hardy, 1924). It has also been suggested that larvae that have been feeding and are in good condition might have a better chance of avoiding the plankton net, thus increasing the percentage of larvae taken with empty guts (Arthur, 1956). However, Burdick (1969) found similar feeding incidences in larval *S. purpureus* collected during the day with a standard 1 m net and those taken with a plankton purse seine, a device which largely eliminates the bias of net avoidance (Murphy and Clutter, 1972). From these findings, he concluded that day-caught larvae were no less healthy than those caught at night (at least in Kaneohe Bay, Hawaii, where his samples were taken). This contradicts Isaacs' (1965) theory that day-caught larvae represent the fraction of the population removed by natural mortality (Ahlstrom [1965] gives further evidence refuting this theory). Duka (1961, 1969) suggests that the low feeding incidence among day-caught anchovy (*Engraulis encrasicholus*) larvae in the Black Sea and Azov Sea is caused by a peculiar mode of feeding, in which food is taken at widely spaced intervals and assimilated with high efficiency as an adaptation to limited food supply. This idea is obviously not applicable to engraulid species such as *S. purpureus*, which have high feeding incidences, or *Engraulis mordax*, which have high feeding rates when food is abundant (Hunter, 1972).

Zaika and Ostrovskaya (1972) derive equations which relate feeding incidence to food concentration, incorporating terms for digestive rate and the searching and capturing abilities of the larvae. Using behavioral data for herring, taken from Rosenthal and Hempel (1970), Zaika and Ostrovskaya calculate that the low feeding incidence (5%) reported by Arthur (1956) for *Sardinops caerulea* larvae is close to theoretical expectation. Using Hunter's (1972) behavioral data for *E. mordax*, and assuming a digestive rate comparable to that of sardine larvae, Arthur's (1956) feeding incidence for this species (15%) is also close to expectation at the ambient food densities of 1-3/l. The high feeding incidences (80-100%) reported by Burdick (1969) for *S. purpureus* are to be expected at the high food densities (up to 200 copepod nauplii/l) in Kaneohe Bay, Hawaii (Burdick, 1969; Bartholomew, 1973). The contribution of Zaika and Ostrovskaya is important, since it indicates that many values for feeding incidence in the literature may truly reflect food availability.

Several papers have correlated feeding incidence with food availability. Berner (1959) found *E. mordax* larvae with food in their guts only in areas where food was concentrated, and Bainbridge and Forsyth (1971) found that the biomass of gut contents in Clyde herring (*Clupea harengus*) larvae was higher when the biomass of available food was high. Lisivnenko (1963) found a higher feeding incidence among Baltic herring (*C. harengus*) larvae in years of high food availability. Nakai et al. (1966) reported significant correlations between the density of copepod nauplii and the feeding incidence of Japanese anchovy (*Engraulis japonica*) larvae, between naupliar density and the number of food items in the larval guts, and between the monthly landings of immature anchovies and the naupliar density of adjacent regions in corresponding months.

<u>Emaciated or Dead Larvae Associated with Poor Plankton</u>. Soleim (1942) found "great quantities of dead or semi-dissolved larvae" (*C. harengus*) in areas of the sea where there were exceptionally few copepod nauplii, the major food of these larvae, and most such larvae had just completed yolk-absorption. Marr (1956), however, cast doubt on this evidence by suggesting that the dead larvae were a result of inadequate washing of the plankton nets between hauls. In a more convincing study, Shelbourne (1957) assessed the physical condition of plaice (*Pleuronectes platessa*) larvae from good and bad plankton patches. In samples taken in January, a scarcity of planktonic food was reflected in the deteriorating condition of larvae caught at the end of the yolk-sac stage. In March the plankton was far richer and plaice larvae at yolk-absorption were in a much improved condition;

also, more late-stage larvae were found than in January. Shelbourne concluded that a truly critical stage, dependent on food conditions, may occur at the end of the yolk-sac stage. Nakai (1962) found that postlarval sardines (*Sardinops melanosticta*) with thick guts were more common over the continental shelf of Japan than offshore, where copepod nauplii (the larvae's major food) were scarce. He concluded that gut diameter might reflect conditions - namely, food availability - influencing the growth and survival of larvae. Nakai et al. (1969) found a correlation of gut diameter with body depth in larval anchovies and assumed that the gut size reflected the feeding history, and thus the survival, of the larvae.

Blaxter (1965, 1971) measured condition factors (weight/length^3) and body heights of Clyde herring (*C. harengus*) larvae from the sea and compared them with values for larvae dying of starvation in the laboratory. Condition factor was the more sensitive indicator of larval condition; paradoxically, the condition factors of sea-caught larvae greater than 12 mm were well below those of larvae starving in the laboratory. This was attributed either to different growth characteristics in rearing tanks and in the sea or to selective capture of weak or moribund larvae by the plankton net. Another surprising outcome of this study was a negative correlation of condition factors with the biomass of planktonic food at sea; it was suggested that this may have reflected grazing. However, as expected, there seemed to be a positive relation between condition factor and resulting year-class strength. Chenoweth (1970) found that very low condition factors of *C. harengus* larvae along the Maine coast in the winter of 1965 were correlated with a high rate of mortality. Villela and Zijlstra (1971), however, found no clear relation between condition factors of larval *C. harengus* in the central and southern North Sea and the resulting year-class strength, although the year with the highest larval condition factor produced the largest year-class. Thus food supply, larval survival, and year-class strength may be more closely related in some areas than in others.

THE SENSITIVITY OF LARVAL FISH TO FOOD DEPRIVATION

A major assumption of Hjort's critical period concept is that first-feeding larvae are extremely sensitive to food deprivation. Whether larvae are sensitive or not in this respect will depend on their ability to withstand lack of food for certain periods and their ability to capture adequate amounts of food at food densities prevailing in the ocean. A number of experimental studies exploring these abilities have been made in recent years.

The Effect of Delayed Initial Feeding. Blaxter and Hempel (1963) noted that starved herring larvae kept in aquaria became progressively weaker and gradually reached a point where they would not show feeding behavior even if food were made available. Since it is obviously unlikely that larvae could recover after passing this point, these authors referred to it as the "point of no return". They listed values for the point of no return, ranging from 5 days after yolk absorption (Kiel herring at 12°C) to 9 days after yolk absorption (Norway herring at 8°C). The point of no return was positively correlated with egg size and the amount of yolk present at hatching, which Blaxter and Hempel (1963) have shown to be population characteristics. Lasker et al. (1970) investigated the effect of delayed feeding on the survival of newly hatched anchovy (*E. mordax*) larvae. Feeding could be delayed up to 1.5 days after yolk absorption without increasing mortality, but a delay of 2.5 days resulted in catastrophic mortality and a survival curve almost identical to that of larvae which received no food at all. This pattern of mortality is essentially what Hjort's critical period concept predicts. Yet, as Fabre-Domergue and Biétrix (1898) pointed out in the last century, laboratory-reared larvae that are provided with suitable food when they first require it do not show increased mortality at the time of complete yolk absorption. The morphological and physiological transitions which occur at this time do not in themselves appear to make the period more critical, as Morris (1956) suggested.

Wyatt (1972) found that plaice larvae passed a "point of no return" after 8 days of starvation (at 10°C); older plaice larvae were more resistant to starvation, not passing a point of no return even after 25 days without food. May (1971) investigated the effects of delayed initial feeding on larvae of the grunion (*Leuresthes tenuis*) and found that regardless of how long feeding was delayed (at 18°C), as long as some starved larvae were still alive, 80% or more of them began feeding when food was made available, and at least 40% of them were able to survive and grow until the end of a 20-day experiment. Grunion larvae thus never passed a point of no return; if they were deprived of food in nature, mortality extending over a number of days resulted (other factors being optimal) and hence these larvae did not show a critical period in the sense of Hjort.

Effect of Food Density. O'Connell and Raymond (1970) assessed the mortality and growth of anchovy (*E. mordax*) larvae over the first 12 days of life at various food densities and found that containers of larvae receiving 1000 copepod nauplii/l/day or less showed drastic mortality on the 6th or 7th days after hatching, while containers receiving 4000 nauplii/l/day or more showed no such trend. The average lengths of survivors at the end of the experiment varied in proportion to the feeding levels, 4000 and 8000 nauplii/l/day producing lengths 2.5 to 3 times greater than 1000 nauplii/l/day. O'Connell and Raymond concluded that food concentration could affect survival and ultimately year-class strength in the anchovy. Saksena and Houde (1972) found that laboratory-reared larvae of bay anchovy (*Anchoa mitchilli*) survived better and grew faster at medium (1330-1688 organisms/l) and high (2811-3323 organisms/l) concentrations of wild plankton than at low concentrations (621-692 organisms/l), whereas the larger larvae of the scaled sardine (*Harengula pensacolae*) grew equally well at low and high food concentrations (444 and 1324 organisms/l, respectively) and survived better at low concentrations. Despite rather low survival rates, this experiment points again to the importance of interspecific differences in larval responses to food availability.

Riley (1966) and Wyatt (1972) showed that food (*Artemia* nauplii) availability could influence the growth and survival of plaice larvae in laboratory experiments, and Wyatt (1972) demonstrated an effect of food density on the condition (height : length ratio) of these larvae. From Wyatt's work it appears that food densities of 1000 nauplii/l or greater are required for optimal growth and condition among captive plaice larvae. In laboratory rearing efforts it is common to use food concentrations of this general order of magnitude (e.g. Houde and Palko, 1970; Lasker et al., 1970).

Food concentrations in the ocean are usually orders of magnitude below those shown to be optimal for larval fish in laboratory studies. Arthur (1956) reported that in the California Current the density of copepod nauplii (the principal larval fish food in this region) was 1/l or greater in 70%, and 3/l or greater in 50% of his stations. Beers and Stewart (1967), in a transect extending 600 miles offshore from San Diego, found densities of copepod nauplii (between 35 and 103 μ in size) ranging from about 2 to 7/l, and in the eastern tropical Pacific they found (Beers and Stewart, 1971) concentrations of nauplii ranging from 10 to 44/l. Reliable data from other areas are inadequate, and the values of Beers and Stewart are perhaps the best available estimates of food levels in the open ocean.

However, the situation may be quite different in more enclosed areas. The 14-year mean concentration of larval food organisms for the month of May in the brackish Gulf of Taganrog, Azov Sea, was 223/l (range, 39-546/l; Mikhman 1969), and in the eutrophic southern sector of Kaneohe Bay, Hawaii, concentrations of copepod nauplii as high as 200/l have been recorded (Burdick, 1969); values of 50-100/l are not uncommon at several times of the year (Bartholomew, 1973). These relatively high figures are still far below densities used in laboratory rearing.

Certain field data provide indirect evidence that food density affects larval survival. Correlations of food abundance with larval condition and feeding incidence have been referred to above. Lisivnenko (1961) found that for the years 1955-1959 the concentration of larval food organisms in the Baltic Sea ranged between approximately 5 and 20/l and was positively correlated with larval herring abundance. Pavlovskaya (1963) found for anchovy in the Black Sea a similar correlation of year-class strength with larval food concentration over the range 4-16/l for the period 1953-1960. Mikhman (1969) noted increased yields of young *Clupeonella delicatula* in the Azov Sea in years when the food density exceeded 250-300/l. Other correlations have been made between larval abundance and plankton biomass (Johansen, 1927; Pinus, 1970). Correlations, of course, do not prove causal relationships, and Gulland (1965) has rightly emphasized the dangers of correlations in fisheries work, both because of the multiplicity of possible correlations and because of the unreliability of the data on which the correlations are based.

Various theoretical estimates of food concentrations necessary for larval survival have been made. Nishimura (1957) assumed values for larval swimming speeds and perception ranges and calculated that a sardine (*Sardinops melanosticta*) larva would require a minimum food density of 22/l to maintain itself and grow. Zaika and Ostrovskaya (1972) calculated that at a food concentration of 20/l, about 50% of Azov anchovy (*E. encrasicholus*) larvae would have food in their guts, while 100% of horse mackerel (*Trachurus mediterraneus*) larvae would contain food at the same concentration, indicating the different feeding behavior of the two species.

It is obvious from the above information that food density can affect larval survival; however, food is not necessarily a decisive factor in all situations. Dekhnik et al. (1970), summarizing their extensive work on fish larvae in the Black Sea, arranged the following evidence to support their contention that food was not a limiting factor for larvae in that body of water: 1. larval mortality was highest during the yolk-sac stage, 2. few larvae had empty guts, 3. there was no direct relation between food abundance and the rate of food consumption of larval fishes, 4. larvae consumed only a small fraction of the food available, and 5. similar preferred foods were found in different species of larvae and under different conditions of plankton abundance and composition.

Searching Ability and Feeding Efficiency. Related to the problem of the food density necessary for larval survival, what is the efficiency of larvae in locating and capturing food items? If a larva is successful in only a small percentage of its strikes at food, it will require a relatively high density of food particles to satisfy its nutritional requirements.

The feeding efficiency of fish larvae was first studied by Braum (summarized in Braum, 1967). Working in freshwater with *Coregonus* larvae, Braum found that larvae were successful in only about 3% of their feeding strikes during their first 8 days of feeding. Similar work with marine fish larvae (Table 2) has indicated a similarly low rate of success for clupeoids at first feeding but a high feeding efficiency for plaice and belonids. Feeding efficiency depends to some extent on the prey, *Artemia* nauplii being captured by herring and belonids more easily than wild plankton (Rosenthal and Hempel, 1970; Rosenthal, 1970). With additional estimates of swimming speed, perceptive range, and food requirements, the food density necessary to satisfy these requirements has been calculated for herring and anchovy larvae (Table 2); these estimates tend to be higher than typical densities at sea (see above). Clupeoids show a rapid increase in feeding efficiency with larval growth, accompanied by an increased search capacity, so that these larvae are undoubtedly most vulnerable to starvation just after yolk absorption. When larvae go without food their swimming abilities decline rapidly (Laurence, 1972), making contact with food progressively less likely. Behavioral studies thus emphasize the importance (at least for clupeoid larvae) of finding food as soon as possible after the yolk is absorbed.

Laboratory investigations of feeding behavior and of survival at different food densities indicate that the first-feeding larvae of some species require food densities much higher than they would be likely to encounter in the ocean. This disparity is puzzling and suggests that the true availability of food to larvae at sea is perhaps not being measured. As Hunter (1972) has suggested, small-scale patchiness of food and the ability of larvae to locate and remain in a patch may be crucial to early larval survival. Ivlev (1965) calculated that *Trachurus* larvae in the Black Sea would search in a day a volume of water containing only 4.6 food organisms, even at maximum plankton concentrations, whereas these larvae were known from stomach content analyses to consume an average of 70-80 food organisms per day. This discrepancy he interpreted as an indication that the food organisms must be moving, thereby increasing their contacts with the larvae. Ivlev calculated that 38.2 to 46.4% of the *Trachurus* larvae would die of starvation; he concluded that the food supply of early larvae had "an important but not decisive" effect on survival.

Table 2. Feeding efficiency of marine fish larvae at the end of the yolk-sac stage, and estimates of required food concentrations. (Feeding efficiency = % successful feeding strikes)

Fish studied	Length at first feeding (mm)	Feeding efficiency (%)	Required food density (no./l)	Reference
Clupea harengus (Baltic)	8	1	--	Rosenthal and Hempel, 1970
Clupea harengus (Downs)	10-11	6- 40[1]	4-42[5]	Rosenthal and Hempel, 1970
Clupea harengus (Clyde)	8-11	<5	--	Blaxter and Staines, 1971
Sardina pilchardus	4- 5	<5	--	Blaxter and Staines, 1971
Pleuronectes platessa	6- 7	40- 70[2]	--	Blaxter and Staines, 1971
Engraulis mordax	3.5	10- 30[3]	105[6]	Hunter, 1972
Belone belone	12	60-100[4]	--	Rosenthal, 1970

[1]Variation due to different assumptions about accuracy of predation.

[2]Variation between experiments.

[3]Variation between larvae.

[4]Depends on food type.

[5]Density of *Artemia* nauplii required to keep larval digestive tract full.

[6]Density of rotifers required to meet metabolic demand.

FACTORS OTHER THAN FOOD AVAILABILITY

Space limitations do not permit a detailed treatment here of factors other than food availability that may influence larval survival. However, a few words must be said to place the above information in perspective: Predation is widely considered to be one of the greatest causes of larval fish mortality (e.g. Murphy, 1961; Hempel, 1965; Dekhnik et al., 1970), and there is good experimental (e.g. Lebour, 1923, 1925; Lillelund and Lasker, 1971) and field (e.g. Stevenson, 1962; Fraser, 1969) evidence to support this; but quantitative estimates of the magnitude of mortality due to predation are totally lacking, as are similar estimates for mortality due to starvation. Vladimirov (1970) argues cogently that inherent qualitative differences among sexual products have a bearing on embryonic and larval survival, a subject which is particularly well developed in the Russian literature (see also Nikolskii, 1969). Abiotic factors certainly influence early survival (Lillelund, 1965; Blaxter, 1969) and may interact with food availability. Currents, for instance, may transport larvae to areas of low food abundance (note that this was one of Hjort's original suggestions - see above, p. 4), and Ivlev (1961) has demonstrated that starvation increases the susceptibility of fishes to various deleterious influences, including: toxic substances, pH, low O_2 tension, infection, and predation. The effect of inadequate food on survival may therefore be indirect as well as direct.

DISCUSSION

In considering the validity of Hjort's critical period concept, one must bear in mind the multiplicity of ecological and species-specific variables that affect larval survival. It would be impossible for any single mechanism to explain larval survival in all cases, and the critical period concept does not do so. Yet, considering the field and laboratory evidence together, it is apparent that in many cases the availability of food at the time of complete yolk absorption can and does affect larval survival, as Hjort surmised. The magnitude of natural mortality from starvation remains uncertain, as does its influence on year-class strength. In some instances it is likely that year-class strength is influenced, or even largely determined, by food availability at the critical period; but because of the difficulties of investigating natural mortality at sea, the case is built largely on circumstantial evidence. The most necessary evidence - reliable comparative data on rates of survival at yolk absorption for good and bad year-classes - is lacking.

Although information is available for only a few species, recent laboratory work has shown that the ability to survive under conditions of limited food availability varies greatly between species. Those larvae, such as atherinids, which are relatively large in size and hatch late in development, are much more able to cope with food deprivation than larvae, such as engraulids, which hatch in a poorly developed state after a short incubation period. Resistance to starvation may also vary on an intraspecific level. Winter - and spring - spawning herring (*C. harengus*) produce larger eggs with a later "point of no return" than summer - and autumn - spawners (Blaxter and Hempel, 1963), an apparent adaptation to reduced food availability in winter and spring (Hempel, 1965; Cushing, 1967). Egg size apparently varies, adapting to the time of spawning, in other species also (Bagenal, 1971).

Susceptibility to starvation, and whether or not a particular species will be likely to pass through a critical period after yolk exhaustion, are facets of its general strategy of survival. Considering these aspects from this comparative perspective, the important question becomes not, "does a critical period exist?" but rather, "how do the characteristics of reproduction and early development reflect the overall adaptation of the species to its environment?"

Recent work (Dragesund, 1969) indicates that Hjort may have been essentially correct in his explanation of factors determining year-class strength in Norwegian spring-spawning herring. The considerable amount of work which has been done since 1914 does not support generalization of Hjort's concept to all, or even a major portion, of other fish stocks. There are still many unknowns. Hjort himself, discussing studies devoted to working out the conditions which determine year-class strength, said:

> As a matter of fact, the object can never be fully attained; new questions will constantly arise, as the knowledge obtained creates the demand for new, and it will always be possible to increase and intensify our comprehension of the vital conditions affecting the organisms in question (Hjort, 1914, p. 209; original emphasis).

SUMMARY

1. The critical period concept put forth by Johan Hjort maintains that the strength of a year-class is determined by the availability of planktonic food shortly after the larval yolk supply has been exhausted. This concept has been reviewed in the light of ecological and experimental data.

2. Survival curves for larval fishes collected in the sea are difficult to interpret and do not justify firm rejection or acceptance of Hjort's concept.

3. Although other factors undoubtedly also influence larval survival at sea, field and laboratory data suggest that starvation may be an important cause of larval mortality at the end of the yolk-sac stage.

4. The relation between starvation-induced mortality at yolk-absorption and year-class strength remains unclear. It seems likely that year-class strength is influenced in many cases by food availability for first-feeding larvae, as Hjort hypothesized, but the case is based largely on circumstantial evidence.

5. Different species react differently to food deprivation, and the feeding conditions facing a given species will vary both spatially and temporally. Whether a critical period in Hjort's sense of the term exists in a given instance will depend on a number of environmental and species-specific factors.

ACKNOWLEDGEMENTS

I wish to thank the Institute of Marine Resources, State of California, who supported me while an early version of this review was being prepared, and the NOAA Office of Sea Grant who provided support during the writing of the final version. A National Science Foundation travel grant enabled me to attend this Symposium.

REFERENCES

Ahlstrom, E.H., 1954. Distribution and abundance of egg and larval populations of the Pacific sardine. U.S. Fish Wildl. Serv., Fish. Bull. 56, 83-140.

Ahlstrom, E.H., 1965. A review of the effects of the environment of the Pacific sardine. Spec. Publ. ICNAF 6, 53-74.

Ahlstrom, E.H., 1966. Distribution and abundance of sardine and anchovy larvae in the California current region off California and Baja California, 1951-1964: A summary. U.S. Fish. Wildl. Serv., Spec. Sci. Rept. - Fish. (534), 71 pp.

Ahlstrom, E.H., 1968. What might be gained from an oceanwide survey of fish eggs and larvae in various seasons. Calif. Coop. Oceanic Fish. Invest., Rep. 12, 64-67.

Arthur, D.K., 1956. The particulate food and the food resources of the larvae of three pelagic fishes, especially the Pacific sardine, *Sardinops caerulea* (Girard). Ph.D. Thesis, Univ. Calif., Scripps Inst. Oceanogr., 231 pp.

Bagenal, T.B., 1971. The interrelations of the size of fish eggs, the date of spawning and the production cycle. J. Fish Biol. 3, 207-219.

Bainbridge, V. and Forsyth, D.C.T., 1971. The feeding of herring larvae in the Clyde. Rapport Process-Verbaux Réunions Conseil Perm. Intern. Exploration Mer 160, 104-113.

Barkley, R.A., 1972. Selectivity of towed-net samplers. Fish. Bull., U.S. 70, 799-820.

Bartholomew, E.F., 1973. The production of microcopepods in Kaneohe Bay, Oahu, Hawaii. M.S. Thesis, Univ. of Hawaii, 43 pp.

Beers, J.R. and Stewart, G.L., 1967. Micro-zooplankton in the euphotic zone at five locations in the California current. J. Fish. Res. Bd. Can. 24, 2053-2068.

Beers, J.R. and Stewart, G.L., 1971. Micro-zooplankters in the plankton communities of the upper waters of the eastern tropical Pacific. Deep-Sea Res. 18, 861-883.

Berner, L., 1959. The food of the larvae of the northern anchovy *Engraulis mordax*. Bull. Inter-Amer. Trop. Tuna Comm. 4, 3-22.

Beverton, R.J.H., 1962. Long-term dynamics of certain North Sea fish populations. In: E.D. Le Cren and M.W. Holdgate (Eds.): The exploitation of natural animal populations, pp. 242-264. London: Blackwell.

Blaxter, J.H.S., 1965. The feeding of herring larvae and their ecology in relation to feeding. Calif. Coop. Oceanic Fish. Invest., Rep. 10, 79-88.

Blaxter, J.H.S., 1969. Development: Eggs and larvae. In: Fish Physiology, Vol. 3 (Eds. W.S. Hoar and D.J. Randall), p. 485. New York: Academic Press.

Blaxter, J.H.S., 1971. Feeding and condition of Clyde herring larvae. Rapport Process-Verbaux Réunions Conseil Perm. Intern. Exploration Mer 160, 128-136.

Blaxter, J.H.S. and Hempel, G., 1963. The influence of egg size on herring larvae. J. Conseil Intern. Exploration Mer 28, 211-240.

Blaxter, J.H.S. and Staines, M.E., 1971. Food searching potential in marine fish larvae. Proc. 4th European Marine Biol. Symp., 467-485.
Bowers, A.B. and Williamson, D.I., 1951. Food of larval and early post-larval stages of autumn spawned herring in Manx waters. Rep. mar. biol. Stat. Pt. Erin 63, 17-26.
Braum, E., 1967. The survival of fish larvae with reference to their feeding behaviour and food supply. In: S.D. Gerking (Ed.): The Biological Basis of Freshwater Fish Production, p. 113-131. Oxford and Edinburgh: Blackwell.
Burdick, J.E., 1969. The feeding habits of nehu (Hawaiian anchovy) larvae. M.S. Thesis, Univ. of Hawaii, 54 pp.
Cassie, R.M., 1963. Microdistribution of plankton. Oceanogr. Mar. Biol. Ann. Rev. 1, 223-252.
Chenoweth, S.B., 1970. Seasonal variations in condition of larval herring in Boothbay area of the Main coast. J. Fish. Res. Bd. Can. 27, 1875-1879.
de Ciechomski, J. Dz., 1967. Investigations of food and feeding habits of larvae and juveniles of the Argentine anchovy *Engraulis anchoita*. Calif. Coop. Oceanic Fish. Invest., Rep. 11, 72-81.
Colton, J.B., Jr., Honey, K.A. and Temple, R.F., 1961. The effectiveness of sampling methods used to study the distribution of larval herring in the Gulf of Maine. J. Conseil Intern. Exploration Mer 26, 180-190.
Cushing, D.H., 1967. The grouping of herring populations. J. Marine Biol. Assoc. U.K. 47, 193-208.
Cushing, D.H., 1969. The fluctuation of year-classes and the regulation of fisheries. Fiskeridir. Skr. Ser. Havunders. 15, 368-379.
Dekhnik, T.V., 1963. Patterns of variation in abundance and mortality of *Engraulis encrasicholus ponticus* Alex eggs and larvae in the Black Sea. Tr. Sevastopolsk. Biól. Sta. Akad. Nauk SSR 16, 340-358.
Dekhnik, T.V., 1964. Changes in the abundance of Black Sea mackerel eggs and larvae during the development period. Tr. Sevastopolsk. Biól. Sta. Akad. Nauk SSR 15, 292-301.
Dekhnik, T.V., Duka, L.A. and Sinukova, V.I., 1970. Food supply and the causes of mortality among the larvae of some common Black Sea fishes. Prob. Ichthyol. 10, 304-310. Translation from Vopr. Ikhtiol.
Dragesund, O., 1969. Factors influencing year-class strength of Norwegian spring spawning herring (*Clupea harengus* Linne). Fiskeridir. Skr. Ser. Havunders. 15, 381-450.
Dragesund, O. and Nakken, O., 1971. Mortality of herring during the early larval stage in 1967. Rapport Process-Verbaux Réunions Conseil Perm. Intern. Exploration Mer 160, 142-146.
Duka, L.A., 1961. Food of anchovy larvae in the Black Sea. Tr. Sevastopolsk. Biól. Sta. Akad. Nauk SSR 14, 242-256.
Duka, L.A., 1969. Feeding of larvae of the anchovy (*Engraulis encrasicholus maeoticus* Pusanov) in the Azov Sea. Prob. Ichthyol. 9, 223-230. Translation from Vopr. Ikhtiol.
Fabre-Domergue and Biétrix, E., 1897. La période critique post-larvaire des poissons marins. Bull. Mus. Nat. Hist. Nat. Paris 3, 57-58.
Fabre-Domergue and Biétrix, E., 1898. Role de la vésicule vitelline dans la nutrition larvaire des poissons marins. C. R. Mem. Soc. Biol. (Paris) 10 Sér., Tome 5, 466-468.
Farris, D.A., 1961. Abundance and distribution of eggs and larvae and survival of larvae of jack mackerel (*Trachurus symmetricus*). U.S. Fish. Wildl. Serv., Fish. Bull. 61, 247-279.
Fraser, J.H., 1969. Experimental feeding of some medusae and chaetognatha. J. Fish. Res. Bd. Can. 26, 1743-1762.
Gulland, J.A., 1965. Survival of the youngest stages of fish, and its relation to year-class strength. Spec. Publ. ICNAF 6, 363-371.

Hardy, A.C., 1924. The herring in relation to its animate environment. Pt. I. The food and feeding habits of the herring with special reference to the east coast of England. Fish. Invest. (Lond), Ser. II, 7, 53 pp.
Hempel, G., 1965. On the importance of larval survival for the population dynamics of marine food fish. Calif. Coop. Oceanic Fish. Invest., Rep. 10, 13-23.
Hjort, J., 1914. Fluctuations in the great fisheries of northern Europe viewed in the light of biological research. Rapport Process-Verbaux Réunions Conseil Perm. Intern. Exploration Mer 20, 1-228.
Hjort, J., 1926. Fluctuations in the year classes of important food fishes. J. Conseil Intern. Exploration Mer 1, 5-38.
Houde, E.D. and Palko, B.J., 1970. Laboratory rearing of the clupeid fish *Harengula pensacolae* from fertilized eggs. Mar. Biol. 5, 354-358.
Hunter, J.R., 1972. Swimming and feeding behavior of larval anchovy *Engraulis mordax*. Fish. Bull., U.S. 70, 821-838.
Ida, H., 1972. Variability in the number of fish taken by larva nets. Bull. Jap. Soc. Sci. Fish. 38, 965-980.
Iizuka, A., 1966. Studies on the early life history of herring (*Clupea pallasii* C. et V.) in Akkeshi Bay and Lake Akkeshi, Hokkaido. Bull. Hokkaido Reg. Fish. Res. Lab. 31, 18-63 (in Jap., Eng. summ.).
Isaacs, J.D., 1965. Night-caught and day-caught larvae of the California sardine. Science 144, 1132-1133.
Ivlev, V.S., 1961. Experimental ecology of the feeding of fishes. Yale Univ. Press, New Haven, 302 pp.
Ivlev, V.S., 1965. On the quantitative relationship between survival rate of larvae and their food supply. Bull. Math. Biophys. 27 (spec. issue), 215-222.
Johansen, A.C., 1927. On the fluctuations in the quantity of young fry among plaice and certain other species of fish and causes of same. Rep. Danish Biol. Sta. 33, 1-16.
Karlovac, J., 1967. Etude de l'écologie de la sardine *Sardina pilchardus* Walb. dans la phase planctonique de sa vie en Adriatique moyenne. Acta Adriat. 13, 1-109.
Lasker, R., Feder, H.M., Theilacker, G.H. and May, R.C., 1970. Feeding, growth, and survival of *Engraulis mordax* larvae reared in the laboratory. Mar. Biol. 5, 345-353.
Laurence, G.C., 1972. Comparative swimming abilities of fed and starved larval largemouth bass (*Micropterus salmoides*). J. Fish. Biol. 4, 73-78.
Lebour, M.V., 1920. The food of young fish. No. III (1919). J. Marine Biol. Assoc. U.K. 12, 261-324.
Lebour, M.V., 1921. The food of young clupeoids. J. Marine Biol. Assoc. U.K. 12, 458-467.
Lebour, M.V., 1923. The food of plankton organisms, II. J. Marine Biol. Assoc. U.K. 13, 70-92.
Lebour, M.V., 1925. Young anglers in captivity and some of their enemies. A study in plunger jar. J. Marine Biol. Assoc. U.K. 13, 721-734.
Lenarz, W.H., 1972. Mesh retention of larvae of *Sardinops caerulea* and *Engraulis mordax* by plankton nets. Fish. Bull, U.S. 70, 839-848.
Liao, I.C., Lu, Y.J., Huang, T.L. and Lin, M.C., 1970. Experiments on induced breeding of the grey mullet *Mugil cephalus* Linnaeus. Indo-Pac. Fish. Counc. 14th sess., Bangkok, Thailand, Nov. 18-27, 1970. (mimeo) 26 pp.
Lillelund, K., 1965. Effect of abiotic factors in young stages of marine fish. Spec. Publ. ICNAF 6, 673-686.
Lillelund, K. and Lasker, R., 1971. Laboratory studies of predation by marine copepods on fish larvae. Fish. Bull., U.S. 69, 655-667.
Lisivnenko, L.N., 1961. Plankton and the food of larval Baltic herring, in the Gulf of Riga. Trud. Nauch.-Issled. Inst. Ryb. Khoz., Latvian SSR 3, 105-138. Fish. Res. Bd. Can., Transla. Ser. No. 444.

Marr, J.C., 1956. The "critical period" in the early life history of marine fishes. J. Conseil Intern. Exploration Mer 21, 160-170.
May, R.C., 1971. Effects of delayed initial feeding on larvae of the grunion, *Leuresthes tenuis* (Ayres). Fish. Bull., U.S. 69, 411-425.
Mikhman, A.S., 1969. Some new data on the larval feeding of the Azov tyul'ka *Clupeonella delicatula* (Nordm.) and on the role of the nutritional factor in fluctuations in its abundance. Prob. Ichthyol. 9, 666-673. Translation from Vopr. Ikhtiol.
Morris, R.W., 1955. Some considerations regarding the nutrition of marine fish larvae. J. Conseil Intern. Exploration Mer 20, 255-265.
Morris, R.W., 1956. Some aspects of the problem of rearing marine fishes. Bull. Inst. Oceanogr. Monaco, 1082, 61 pp.
Murphy, G.I., 1961. Oceanography and variations in the Pacific sardine populations. Calif. Coop. Oceanic Fish. Invest., Rep. 8, 55-64.
Murphy, G.I. and Clutter, R.I., 1972. Sampling anchovy larvae with a plankton purse seine. Fish. Bull., U.S. 70, 789-798.
Nakai, Z., 1962. Studies relevant to mechanisms underlying the fluctuation in the catch of the Japanese sardine, *Sardinops melanosticta* (Temminck and Schlegel). Jap. J. Ichthyol. 9, 1-115.
Nakai, Z., Koji, H., Shigemasa, H., Takashi, K. and Hideya, S., 1966. Further examples of Hjort's hunger theory. 2nd Int. Oceanogr. Cong., Abstr., p. 263.
Nakai, Z., Kosaka, M., Ogura, M., Hayashida, C. and Shimozono, H., 1969. Feeding habit, and depth of body and diameter of digestive tract of shirasu, in relation with nutritious condition. J. Coll. Mar. Sci. Tech. Tokai Univ. 3, 23-34.
Nakai, Z., Usami, S., Hattori, S., Honjo, K. and Hayashi, S., 1955. Progress report of the cooperative Iwashi resources investigations, April 1949 - December 1951. Fish. Agency Tokai Reg. Fish. Res. Lab. Tokyo, 116 pp.
Nikolskii, G.V., 1969. Theory of fish population dynamics. Edinburgh: Oliver and Boyd, 323 pp.
Nishimura, S., 1957. Some considerations regarding the amount of foods daily taken by an early post larva of sardine. Ann. Rept. Jap. Sea Reg. Fish. Res. Lab. 3, 78-84.
O'Connell, C.P. and Raymond, L.P., 1970. The effect of food density on survival and growth of early post yolk-sac larvae of the northern anchovy (*Engraulis mordax* Girard) in the laboratory. J. Exp. Mar. Biol. Ecol. 5, 187-197.
Pavlovskaya, R.M., 1963. The main causes of fecundity variation in year classes of Black Sea anchovy. Nauch- Tekh. Inf. VNIRO, No. 9, cited by Nikolskii 1969.
Pearcy, W.G., 1962. Ecology of an estuarine population of winter flounder, *Pseudopleuronectes americanus* (Walbaum). Bull. Bingham Oceanogr. Collect. Yale Univ. 18, 1-78.
Pinus, G.N., 1970. The food of Azov tyul'ka larvae (*Clupeonella d. delicatula*) and the effect of feeding conditions on the abundance of the progeny. J. Ichthyol. 10, 519-527. Translation from Vopr. Ikhtiol.
Riley, J.D., 1966. Marine fish culture in Britain. VII. Plaice (*Pleuronectes platessa* L.) post larval feeding on *Artemia salina* L. nauplii and the effects of varying feeding levels. J. Conseil Intern. Exploration Mer 30, 204-221.
Rosenthal, H., 1970. Anfütterung und Wachstum der Larven und Jungfische des Hornhechts *Belone belone*. Helgolaender Wiss. Meeresuntersuch. 21, 320-332.
Rosenthal, H. and Hempel, G., 1970. Experimental studies in feeding and food requirements of herring larvae (*Clupea harengus* L.). In: J.H. Steele (Ed.): Marine food chains, 344-364. Berkeley: Univ. Calif. Press.

Saksena, V.P. and Houde, E.D., 1972. Effect of food level on the growth and survival of laboratory-reared larvae of bay anchovy (*Anchoa mitchilli* Valenciennes) and scaled sardine (*Harengula pensacolae* Goode and Bean). J. Exp. Mar. Biol. Ecol. 8, 249-258.
Saville, A., 1956. Eggs and larvae of haddock (*Gadus aeglefinus* L.) at Faroe. Mar. Res. Scot. 4, 1-27.
Saville, A., 1959. Mesh selection in plankton nets. J. Conseil Intern. Exploration Mer 23, 192-201.
Sette, O.E., 1943. Biology of the atlantic mackerel (*Scomber scombrus*) of North America. Part I: Early life history, including the growth, drift, and mortality of the egg and larval populations. U.S. Fish. Wildl. Serv. Fish. Bull. 38, 149-237.
Shelbourne, J.E., 1957. The feeding and condition of plaice larvae in good and bad plankton patches. J. Marine Biol. Assoc. U.K. 36, 539-552.
Soleim, P.A., 1942. Årsaker til rike og fattige årganger av sild. Fiskeridir. Skr. Ser. Havunders. 7, 1-39.
Stevenson, J.C., 1962. Distribution and survival of herring larvae (*Clupea pallasii* Valenciennes) in British Columbia waters. J. Fish. Res. Bd. Can. 19, 735-810.
Taft, Bruce A., 1960. A statistical study of the estimation of abundance of sardine (*Sardinops caerulea*) eggs. Limnol. Oceanogr. 5, 245-264.
Tranter, D.J. (Ed.), 1968. Zooplankton sampling. UNESCO, Paris, 174 pp.
Vilella, M.H. and Zijlstra, J.J., 1971. On the condition of herring larvae in the central and southern North Sea. Rapport Process-Verbaux Réunions Conseil Perm. Intern. Exploration Mer 160, 137-141.
Vladimirov, V.I., 1970. Ontogenetic qualitative differences as one factor in the dynamics of a fish population (research tasks). Hydrobiol. J. 6, 7-18. Translation from Gidrobiol. Zh.
Wiebe, P.H., 1970. Small-scale spatial distribution in oceanic zooplankton. Limnol. Oceanogr. 15, 205-217.
Wood, R.J., 1971. Some observations on the vertical distribution of herring larvae. Rapport Process-Verbaux Réunions Conseil Perm. Intern. Exploration Mer 160, 60-64.
Wyatt, T., 1972. Some effects of food density on the growth and behaviour of plaice larvae. Mar. Biol. 14, 210-216.
Zaika, V.Y. and Ostrovskaya, N.A., 1972. Indicators of the availability of food to fish larvae. 1. The presence of food in the intestines as an indicator of feeding conditions. J. Ichthyol. 12, 94-103. Translation from Vopr. Ikhtiol.

R.C. May
Hawaii Institute of Marine Biology
University of Hawaii
P.O. Box 1346
Kaneohe, Hawaii 96744 / USA

Larval Mortality and Subsequent Year-Class Strength in the Plaice (*Pleuronectes platessa* L.)

R. C. A. Bannister, D. Harding, and S. J. Lockwood

INTRODUCTION

The ability of a population to replace itself is determined by the age structure, growth and fecundity of the mature stock. Where this comprises a large number of mature age groups, with slow growth to a relatively large size, and where fecundity is high, the correspondingly high egg production, greatly in excess of the ultimate recruitment to the adult stock, implies the occurrence of appropriately high egg and larval mortality rates. Since egg production is potentially variable over the possible range of equilibrium stock size between the virgin state and the heavily exploited state, it is the precise nature of the pre-recruit mortality which, through the corresponding stock and recruitment relation, or lack of it, determines the stability of the stock and influences its resistance to the effects of exploitation.

For a number of years it was a convenient assumption of fishery management that the recruitment to demersal stocks was independent of parent-stock size, and the North Sea plaice stock was identified as a prime example of this concept. In such a situation, where recruitment is not simply proportional to stock, it has become axiomatic to look upon the abundance of the pre-recruit phase as the determining element, the pre-recruit mortality being expressed by a combination of density-independent and density-dependent terms in mathematical models which are associated with appropriate, but largely untested, behavioural hypotheses. Ricker (1954, 1958) introduced the concept of replacement stock and formulated the density-independent and density-dependent terms in such a way as to suggest that there would commonly be a maximum in the recruitment curve. Beverton and Holt (1957) described one expression which generated a Ricker "dome", and another which allowed recruitment to reach an asymptote. More recently Harris (in prep.) has distinguished between density-dependent mortality (due to densities during the pre-recruit phase) and what he calls stock-dependent mortality (arising from the total number of eggs spawned). His purpose is to show that the occurrence of a domed or asymptotic curve could depend on whichever of the two factors has the dominant effect.

For a variety of reasons the development of the models has tended to exceed those studies of the egg and larval stages necessary to test their validity. The object of this paper is to redress the balance by presenting the results of egg and larval surveys conducted on the plaice spawning grounds of the southern and central North Sea during the last decade. Whatever the nature of the pre-recruit mortality process, the plaice stock is notable for achieving replacement with a minimum of variation in recruitment ($109.0-565.8 \times 10^6$) over a range of stock size. However, stock-record data from the commercial fishery show that during this period the plaice stock actually benefited from the recruitment of a year-class, spawned in the cold winter of 1963, which was substantially more abundant than the average expected year-class. The pattern of egg mortality observed in that year provides an important comparison with the patterns observed in other years and thereby allows some discussion of the nature of the controlling processes in the planktonic phase and the possible effect of the potential increase in egg production originating from the 1963 cohort.

METHODS

Stock Record Data. Quarterly and annual stock-record data derived from the routine length and age sampling of male and female plaice landed by the commercial fleets of the United Kingdom and the Netherlands (ICES Statistical Newsletters) are available for all or part of the post-war period from 1947 to 1971. They provide indices of the relative abundance of the different age groups in the landings. Between them the Netherlands and United Kingdom fleets fish most of the available part of the North Sea plaice stock. By raising their combined age-composition data to the total international catch and applying the now well established technique of virtual population analysis (Fry, 1957; Bishop, 1959; Gulland, 1965; Garrod, 1967; Schumacher, 1970), it is possible to estimate the total number of plaice in each age group for those year-classes present in the exploited stock. Recruitment is therefore accessible as the number of 2-year-old plaice estimated by this method. Potential egg production at spawning is calculable from the stock-record data and estimates of the mean fecundity of each age group in the mature stock. Fecundity has been estimated from a total sample of 230 mature female gonads collected from the southern and central North Sea during sampling by research and commercial vessels between 1969 and 1972. It had been established by covariance analysis that the log length-log fecundity relationships for different years and different grounds were not significantly different (Bannister, unpublished data). For the period 1962 to 1971 fecundity at different ages can be applied directly to the virtual population estimates of spawning stock to give 'absolute' potential egg production. For the years prior to 1962 the available virtual population data are insufficient and the fecundity has to be applied to the catch per effort of each age group (represented by the number of fish per 100 h fishing in the Lowestoft landings in the first quarter of the year), to give an index of egg production. The mid-point of the maturation ogive has been taken as 35 cm, which corresponds to the mean length of 5-year-old female plaice. The spawning stock has accordingly been estimated from age groups 5 to 16, the upper cut-off reflecting the fact that data for fish older than 16 are not available for the whole series of years.

Egg and Larval Surveys. Extensive surveys of the known spawning grounds of the plaice in the English Channel, Southern Bight and central North Sea were made between 1962 and 1971 to sample the pelagic eggs and larvae quantitatively. The Lowestoft multipurpose high-speed plankton sampler described by Beverton and Tungate (1967) was used for these collections, which were part of a larger study of the whole plankton community and the environmental changes occurring during the spawning season. To support the work at sea, laboratory studies were made on the development rates of eggs and larvae in relation to temperature (Ryland and Thacker, unpublished data); in both experimental and sea material, the eggs and larvae were sorted into the same morphological stages as those described for eggs by Simpson (1969) and for larvae by Ryland (1966).

The counts of ten different categories of eggs and larvae at each station on each cruise were charted, contoured at selected levels and the total abundance in the cruise area was calculated by planimetry (see Simpson, 1959). These values were then corrected for the stage duration, using the temperatures observed at sea, in order to estimate numbers produced per day for each development state on each cruise. Annual production curves for each development stage in each season were then plotted, using the mid-point of the cruise as the mean sampling date. The final estimates of seasonal production for each stage were made by measuring the areas under these curves. The values determined in this way were used to estimate the losses between stages in each season. By using the mean sea temperature for each production period, the average duration for each stage was calculated. By plotting seasonal abundance against cumulative stage duration (age) generalized loss-rate curves were obtained for each season. The instantaneous coefficient of loss per day, Z, can then be derived from these seasonal data either as $N_2/N_1 = e^{-Zt}$ (where N_1 is the initial number of eggs or larvae and N_2 the number surviving to the next stage in the period t days, t being the mean stage duration), or from fitted regressions of $\log_e$ numbers of eggs or larvae produced in each stage, against their cumulative stage durations (Harding and Talbot, 1970). Estimates of variability of the average numbers of eggs and larvae were made by analysis of variance of the logarithms of the numbers produced per metre squared per day, following a technique described for pilchard eggs (Cushing, 1957), and the residual variances were used to calculate fiducial limits for station, cruise and seasonal means (Bagenal, 1955) which were then applied to the total abundance estimates (Harding and Talbot, 1970).

O- and 1-Group Surveys. Surveys with standard beam trawls (Riley and Corlett, 1966) and push nets (Riley, 1971) have been carried out in plaice nursery areas around the British Isles. Of particular relevance were those conducted in Filey Bay on the northeast coast of England in 1968 and 1969 (Lockwood, in press). Densities of O- and 1-group plaice were estimated from the trawl catches, expressed as numbers per 100 m^2, charted, contoured, and the total abundance estimated by planimetry for each survey. The estimate of population size was checked in 1969 by a tagging experiment. The variation in numbers between surveys, and the growth of juvenile plaice, estimated from length frequency analyses on the trawl-caught fish, were then used to estimate seasonal mortality rates amongst O- and 1-group fish. These methods compare with those used in the Scottish surveys in Lock Ewe (Edwards and Steele, 1968) and Dutch surveys in the Waddensee (Zijlstra, 1972).

RESULTS

Recruitment to the Fishery. For the 1947 to 1968 year-classes Fig. 1 shows the virtual population estimate of 2-year-old recruits, expressed as units of the 1947 to 1968 year-class mean of 337.4 x 10^6 plaice. The figure illustrates the low variability of recruitment, the coefficient of variation being 48.8%, which can be compared, for example, with a value of 87% for the recruitment of the North Sea

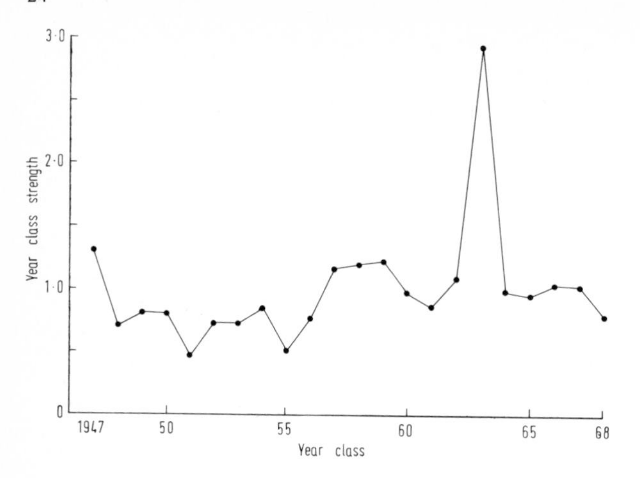

Fig. 1. North Sea plaice. Post-war recruitment of 2-year-old plaice (virtual population estimate), as units of the 1947-68 year-class mean

haddock broods spawned between 1919 and 1924 (Raitt, 1936). Since steady recruitment alone is not notable, Fig. 2 indicates the range of stock over which recruitment has been maintained. Recruitment is shown as the virtual population estimate of 2-year-old plaice, and stock is expressed as the potential egg production emanating from the number of 5- to 16-year-old females landed per 100 h fishing at Lowestoft in the first quarter of the year. In this figure the points for the 1957 to 1968 year-classes lie above and to the right of their predecessors. To some extent the change in the level of recruitment may reflect problems with the virtual population estimates for more recent years, during which the Lowestoft and Netherlands fleets have fished more efficiently on a wider part of the stock than in the immediate post-war years. On the other hand, Gulland (1968) shows how the reduced amount of fishing on the small plaice grounds during the 1950s has generated a recognizable increase in the stock, as revealed by increases in both the catch per effort and the average size of plaice in the landings. For the two periods 1950-1956 and 1957-1964 the average annual mean length of Lowestoft plaice landings and the average plaice catch per 100 h fishing by English motor trawlers for three typical statistical squares in the North Sea, i.e. G7, L8 and J8, are:

Period	Mean length of Lowestoft landings (cm)		English catch per effort (metric tons/100 hours)		
	♂	♀	G7	L8	J8
1950-56	31.0	33.5	36.9	62.2	55.2
1957-64	34.0	38.0	49.2	99.8	77.9

Both Fig. 1 and Fig. 2 demonstrate clearly the unique nature of the well-above-average recruitment of the 1963 year-class, which was estimated to comprise 990 x 10^6 2-year-olds, compared with, for example, 371 x 10^6 for the 1962 year-class and 268 x 10^6 for the 1968 year-class. Fig. 3 shows the importance of the 1963 year-class to the United Kingdom fishery. Between 1958 and 1962 the stock achieved an approximately steady state, with UK landings showing the mean age composition indicated by Fig. 3i. The 1963 spawning stock was of similar age composition (Fig. 3ii). Between 1966 and 1971 the in-

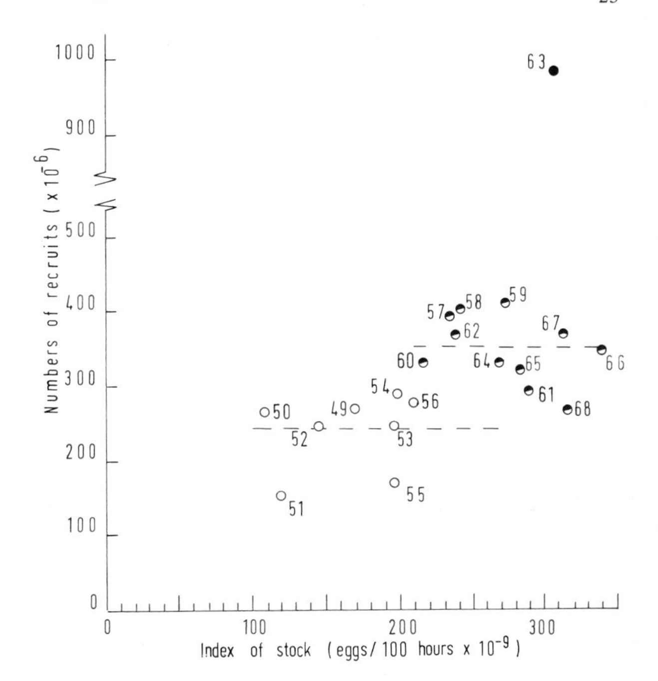

Fig. 2. North Sea plaice. Recruitment and the spawning stock; recruits at two years of age (virtual population estimate); stock as eggs/100 hours from Lowestoft first quarter age composition

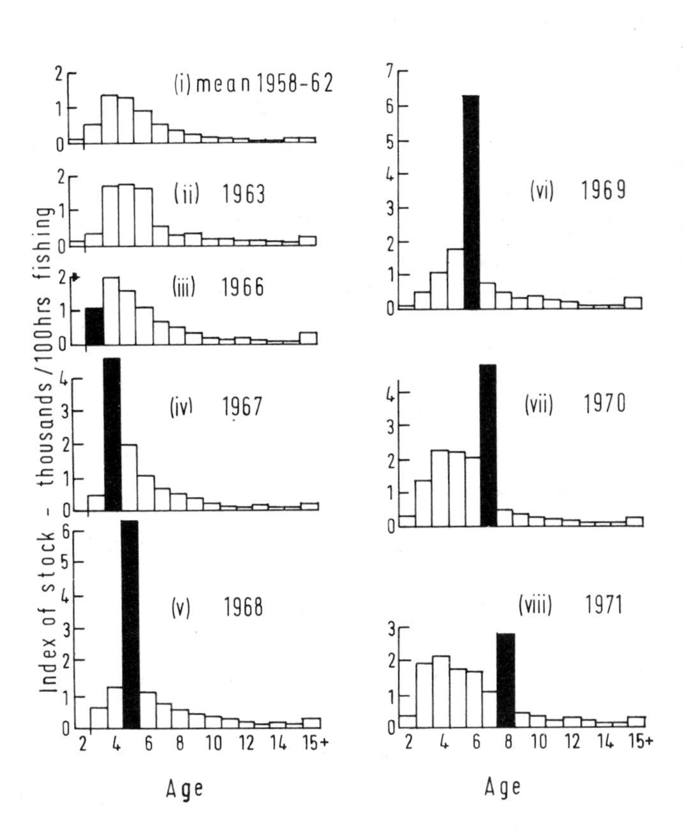

Fig. 3. North Sea plaice. Age composition data for the United Kingdom fishery, 1958-71

fluence of the 1963 year-class recruitment is clearly indicated (Fig. 3, iii-viii). A further consequence of the high abundance of the 1963 year-class is the density-dependent decrease in growth. The differences between the mean lengths (cm) of 5-, 6-, and 7-year-old male and female plaice of this year-class and the 1965-1971 period means are:

	♂	♀
Age 5		
1963 year-class	32.47	34.23
1965-71 mean	34.08	36.38
Age 6		
1963 year-class	34.04	36.07
1965-71 mean	35.93	39.71
Age 7		
1963 year-class	35.57	38.46
1965-71 mean	37.13	40.68

The foregoing suggests that, either the spawning of the 1963 year-class was very successful, or the survival of pelagic eggs and larvae, and of juvenile plaice must have been higher, or that a combination of both factors operated.

The following is a description of the egg and larval mortality patterns giving rise to the normal and abnormal recruitments to the stock in the period 1962 to 1971.

The Egg and Larval Stages. The main spawning grounds of the North Sea plaice are very extensive, ranging from the Eastern Channel through the Southern Bight into the central North Sea, as may be seen from the charted distribution of stage I eggs collected during the joint cruises of RV CORELLA and RV TRIDENS near the peak of spawning in 1971 (Fig. 4). The relative intensities of the spawning on the different grounds may change from year to year, but it is apparent from recent surveys that the spawning in the central North Sea is now much more extensive than was reported for the period up to 1948 (Simpson, 1959). In 1968 the total number of stage I eggs produced in the Channel and North Sea spawnings up to 56°30'N was estimated at approximately 15 billion (15×10^{12}), of which more than 10 billion were produced in the central North Sea (Fig. 5) as calculated from the area beneath the curve in Fig. 5.

In the past decade the Southern Bight has been surveyed five times, during the winters of 1962, 1963, 1968, 1969 and 1971. It has been possible to measure the seasonal abundance of the eggs and larvae of the plaice and to obtain estimates of the mortalities which occurred between development stages in each year (Table 1 and Fig. 6). Examination of the data for these years shows that egg production in the Southern Bight ranged between 4.56 and 2.50 billion stage I eggs and that the total losses were generally in excess of 90% between stage I eggs and stage 4 larvae (just prior to metamorphosis). Further examination shows that the results fall into two groups. Those for 1962, 1965, and 1971, which may be classified as normal years, show that

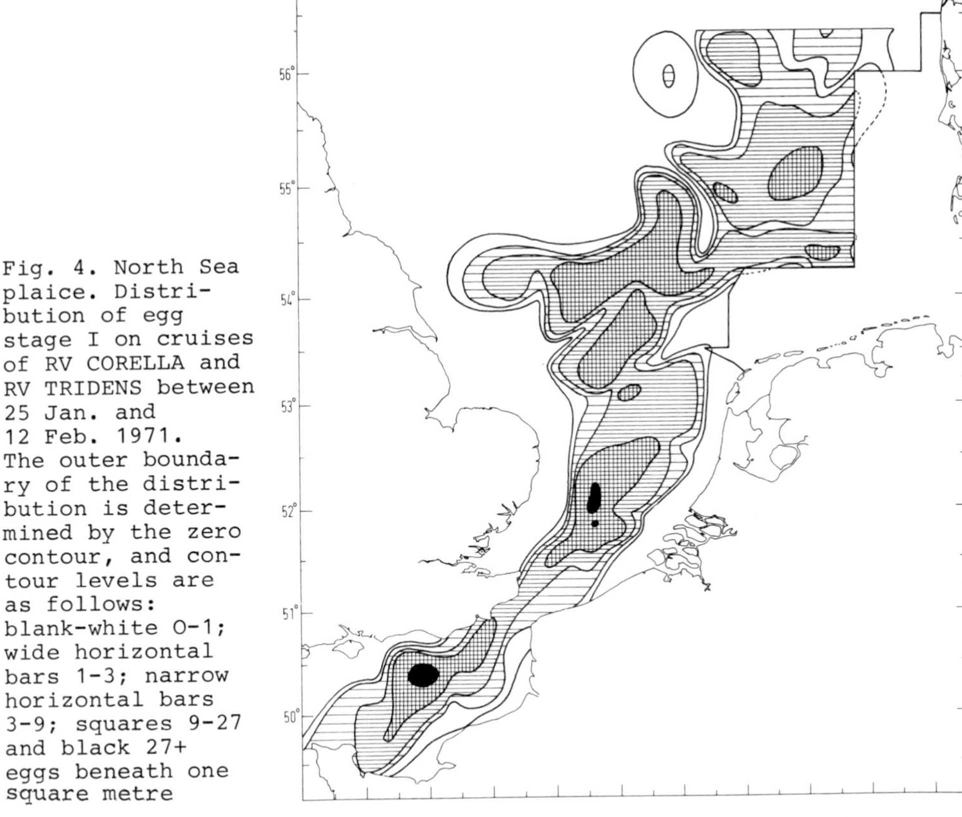

Fig. 4. North Sea plaice. Distribution of egg stage I on cruises of RV CORELLA and RV TRIDENS between 25 Jan. and 12 Feb. 1971. The outer boundary of the distribution is determined by the zero contour, and contour levels are as follows: blank-white 0-1; wide horizontal bars 1-3; narrow horizontal bars 3-9; squares 9-27 and black 27+ eggs beneath one square metre

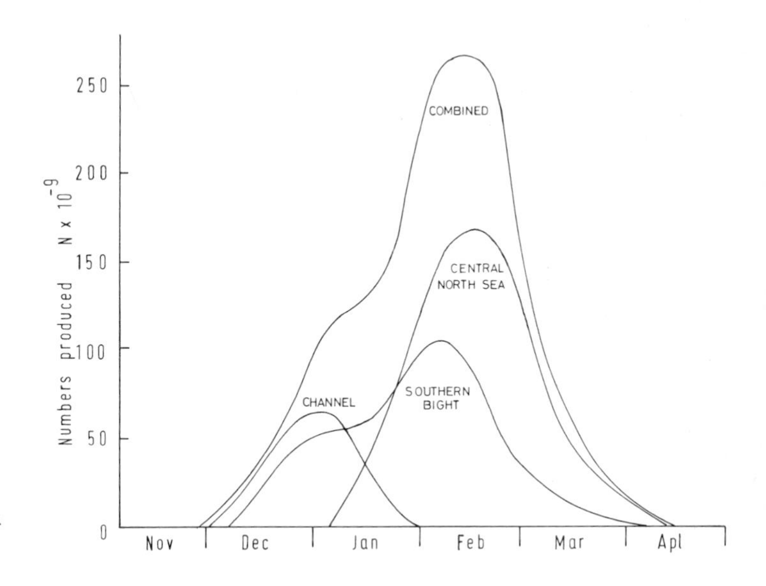

Fig. 5. North Sea plaice. Production of stage I eggs of plaice in the English Channel and North Sea in 1968

Table 1. Abundance, mortality and mortality rate estimates for planktonic plaice eggs and larvae in the Southern Bight of the North Sea

Development stage	1962			1968		
	$N \times 10^{-12}$	Mortality		$N \times 10^{-12}$	Mortality	
		%	Z		%	Z
Eggs						
I	3.802			3.113		
		12.94	0.040			
II	3.310			3.411		
		8.87	0.024		21.52	0.072
III	3.017			2.677		
		16.97	0.036		15.54	0.036
IV	2.624			2.261		
		61.72	0.017		58.65	0.017
V	0.959			0.935		
Larvae		79.17	0.825		71.34	0.014
1	0.199			0.268		
		58.28	0.033		88.06	0.080
2	0.083			0.032		
		94.60	0.099		79.69	0.054
3	0.0045			0.0065		
		73.53	0.037		84.92	0.026
4	0.0012			0.00098		
Eggs I to V		74.78	0.081		69.97	0.073
Larvae 1 to 4		94.75	0.082		99.63	0.062
Eggs I to larvae 4		99.97	0.064		99.97	0.065

Abundance estimates (as numbers produced x 10^{-12} in the spawning season) and mortality estimates (Z) for plaice eggs and larvae throughout the seasons 1962, 1963, 1968, 1969 and 1971 in the North Sea. Z is calculated from $N_t/N_o = e^{-Zt}$, where Z is the instantaneous mortality rate per day, N_o the number of eggs or larvae produced at time 0 and N_t the numbers of eggs or larvae surviving at time t, t being the duration in days between stages.

Table 1 (continued)

1971			1963			1969		
$N \times 10^{-12}$	Mortality		$N \times 10^{-12}$	Mortality		$N \times 10^{-12}$	Mortality	
	%	Z		%	Z		%	Z
2.530			4.556			2.969		
	27.19	0.107		26.05	0.046		23.38	0.081
1.842			3.369			2.275		
	23.07	0.076		1.28	0.002		6.59	0.018
1.417			3.331			2.125		
	30.70	0.105		10.60	0.017		34.12	0.111
0.982			2.978			1.400		
	51.73	0.159		23.41	0.028		44.64	0.086
0.474			2.281			0.873		
	60.13	0.056					36.08	0.025
0.189						0.558		
	67.73	0.052					44.80	0.024
0.061						0.308		
	70.49	0.048					3.25	0.0011
0.018						0.298		
	85.00	0.054					26.85	0.0093
0.0027						0.218		
	81.27	0.116		49.93	0.019		70.60	0.075
	98.57	0.052					60.93	0.017
	99.89	0.061					92.66	0.021

Preliminary estimates of the errors involved in these abundance assessments were made for the 1962 and 1963 data in an analysis of variance. The 95% confidence limits calculated as a percentage of the logarithmic means were then applied to the seasonal abundance estimates to give the fiducial limits. On the average these estimates were of the order of half to one-third of the abundance estimates in the egg stages and one-third to a quarter for the larval estimates of abundance. These are lower than the limits of half to double the estimated number often quoted in the literature (Saville, 1964).

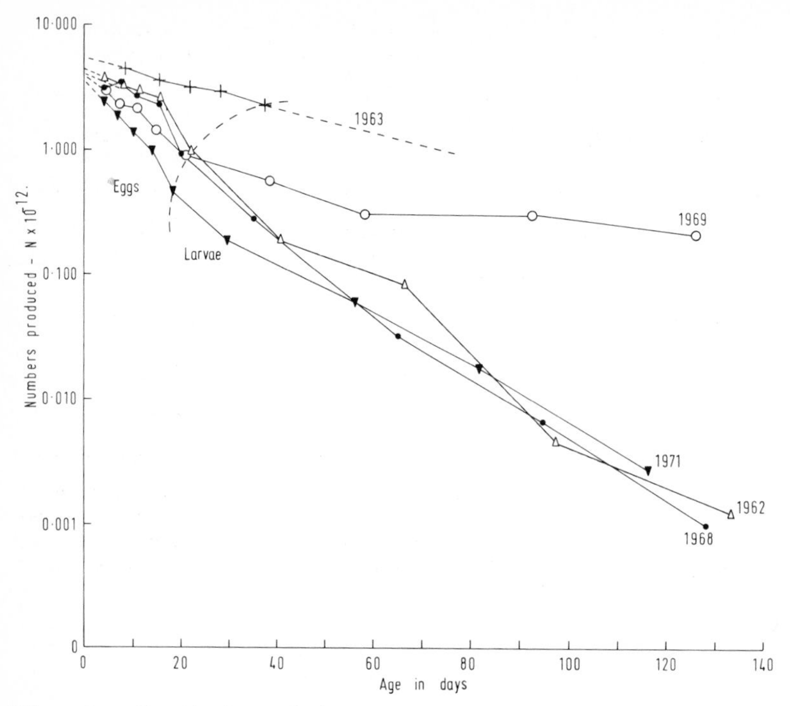

Fig. 6. North Sea plaice. Survival curves for eggs and larvae for various years between 1962 and 1971

losses were higher in the egg stages than in the larval phase, the overall seasonal loss being in excess of 99%. The slopes of the mortality curves are steep, corresponding to values of the instantaneous coefficient of mortality, Z, of 0.061 to 0.065 per day, which are equivalent to an average mortality rate of approximately 84% per month. For these years the recruitment to the adult stock was close to average. The results for 1963 and 1969 may be termed abnormal. In 1963, for example, mortality in the egg stages was lower, being only 49.9% overall. In 1969 the distributions were fully sampled and the survival of larvae was even better than in the egg stage, resulting in a total larval abundance at stage 4 which was more than 200 times greater than that of 1968, starting from similar values of stage I egg abundance in both years. The overall seasonal loss from egg stage I to larval stage 4 was 93%. Unfortunately the 1963 surveys were only completed for the egg distributions and no data were obtained for larvae. However, on the basis both of the 1969 data and the obviously high recruitment of the 1963 year-class as 2-year-olds there is little doubt that the curve of egg survival would have continued into the larval stages at approximately the same rate in that year too.

Fitted regressions to each year's data give the average slopes for the mortality curves shown in Fig. 6 (Table 2). The regressions show that the slopes are similar in 'normal' years and significantly less in 1963 and 1969, which were therefore good years for survival in the pelagic phase. Hence the results demonstrate that in 1963 good survival in the egg phase led to a demonstrably large recruitment. We can thus predict with some confidence that the good larval survival in 1969 will also lead to an above-average recruitment, as indeed the results from the 1971 stock record data seem to indicate.

Table 2. Mortality rates* for eggs and larvae of plaice in the Southern Bight of the North Sea

Year	Egg stages I to V	Larval stages 1 to 4	Egg stage I to larval stage 4
1962	-0.0744	-0.0589	-0.0659
1963	-0.0209	-	-
1968	-0.0714	-0.0611	-0.0685
1969	-0.0740	-0.0095	-0.0209
1971	-0.1185	-0.0576	-0.0665

*Regressions of $\log_e$ numbers produced per spawning season for each year plotted against the average age in days calculated from the seasonal mean temperature for each development stage. The slope of the regression gives the daily mortality rate.

Table 3. Estimates of abundance of plaice eggs (numbers x 10^{-12})

Year	Stage I	Fertile eggs*
1962	3.802	6.171
1963	4.556	5.104
1968	3.113	5.258
1969	2.969	4.178
1971	2.530	4.058

*Estimated from the intercept of the regression of log numbers of eggs produced in development Stages I to V against average age for each season.

The intercept of the regression for the egg development stages I to V gives a rough estimate of the number of fertilized eggs at zero time, immediately after spawning (Table 3); this must be a closer estimate of the number of eggs actually spawned. If this value is applied proportionately to the total stage I egg abundance estimated as 15 billion for the North Sea in 1968, the estimate of total eggs spawned would be approximately 25 billion. The corresponding estimates of potential egg production calculated by applying fecundity-at-age data to the virtual population estimates of the total abundance of age groups 5 to 16 in the stock are shown below, as numbers x 10^{-12}.

1962	1963	1964	1965	1966	1967	1968	1969	1970
42.4	43.4	41.5	40.8	40.7	40.8	55.6	53.9	52.8

On this basis, egg production should have been between 40 and 42 billion for the period 1962 to 1967, and as high as 50+ billion for the succeeding three years, during which the 1963 year-class is included in the mature stock. Though the 1968 figures are of the same order of magnitude, a discrepancy exists. Several factors relate to this: partly the difference between the sea area in which the individual egg-patch abundances have been raised and the overall area of the stock referred to by the virtual population data, partly the assumption that all of the nominal virtual population mature stock does in fact take part in spawning each year, and partly the assumption that all the eggs produced from this stock will be fertilized. If we set aside these factors, the between-year comparison in the calculated figures should still be valid. It is also notable that, for 1969, when egg production should have been enhanced by the contribution of the 1963 cohort, the ultimate survival should have been so good.

O- and 1-Group Stages. At Filey Bay the abundance of O-group plaice at peak of recruitment in July 1968 was estimated as 0.263×10^6. In June 1969 the estimate was 15×10^6, which is an increase of 56 times over the 1968 figure. This shows that the increased strength of the 1969 year-class already evident in the pelagic larval phase was maintained into the juvenile population in this nursery area.

Observations of mortality in O-group plaice nursery areas are generally calculated from the slope of the regression of $\log_e$ numbers against time from peak of recruitment to the onset of winter, and may be extended into the next year to include 1-group fish. This rate of decrease in O-group numbers is an overestimate, since there is typically a drop in catches in winter followed by an apparent recruitment of 1-group plaice in the spring (Riley and Corlett, 1966; Corlett, 1966; Macer, 1967; Lockwood, 1972). This pattern is also typical of juvenile winter flounder (Pearcey, 1962) and turbot (Jones, 1973). The tagging experiments carried out in Filey Bay suggest that there is no seaward migration of juvenile plaice followed by 1-group recruitment in spring (Lockwood, 1972 and in press), so that the overestimation in the observed rate of decrease must reflect changes in the availability of the young fish to the fishing gear during the winter.

The observed mean rates of population decline for the O- and 1-group plaice of the Irish Sea (Riley and Corlett, 1966; Macer, 1967), Loch Ewe (Edwards and Steele, 1968; Steele and Edwards, 1970) and the North Sea (Lockwood, 1972; Zijlstra, 1972) are summarized in Table 4. The mean mortality rates for O-group plaice were generally in the range 30-50% per month. In Filey Bay the mortality rate was 46% in 1968, a normal year in the North Sea; but in 1969, when the density was shown to be higher, the mortality rate was 70% per month. O- to 1-group mortality rates on the other hand were lower (in the range 9-19% per month) and were of the same order to the 1968 and 1969 year-classes, although the observations are in this case from two different sources, Filey Bay and the Waddensee. These results suggest that the density-dependent processes regulating the numbers in the early development stages are still operative in the O- and 1-group populations.

Table 4. Observed mean monthly rates of young plaice population decrease on beaches

Region	Years	Percentage mortality (± 2 S.D.)	Sampling period
a <u>0-group plaice</u>			
Irish Sea			
Port Erin Bay (Riley and Corlett,1966)	1963	30	
	1964	35	
	1965	50	
Red Wharf Bay (Macer,1967)	1964	40	
West coast of Scotland			
Loch Ewe (Edwards and Steele,1968; Steele and Edwards,1970)	1965	46 (39-51)	
	1966	46 (36-54)	
	1967	49 (23-66)	
	1968	39 (32-45)	
North Sea			
Filey Bay (Lockwood,1972)	1968	46 (34-56)	
	1969	70 (54-81)	
	1972	43 (22-58)	
b <u>0- to 1-group plaice</u>			
Irish Sea			
Port Erin Bay (Riley and Corlett,1966)	1963-65	15	Dec-Apr
North Sea			
Waddensee	1969-70	9	Sep-May
Zeeland	1969-70	14	Sep-May
Open coast (calculated from Zijlstra,1972)	1969-70	19	Sep-May
Filey Bay (Lockwood,1972)	1968	13	Jul-Sep

Fig. 7 summarizes these results in the form of survival curves representing the 1962 and 1963 cohorts. Real abundance estimates were used for the egg, larval, and adult stages, while for 0- and 1-group fish abundances were estimated from average mortality rates. The monthly natural mortality rate used for 0- to 1-group fish was the exponential decrease from 0.71 to 0.14. A curve fitted using these rates results in an annual instantaneous mortality rate of 0.1 by the time the fish are two years old. The figure further illustrates both the observed abundance following exploitation and the expected abundance in an unexploited stock for a natural instantaneous mortality rate per annum of 0.1. The assumption is that the sets of data from the different sources are compatible with one another. The curves express the density-dependent nature of the pre-recruit mortality process, in which heavy early mortality reduces as abundance declines. The curves also illustrate the difference between the early mortality rates for the two year-classes in question.

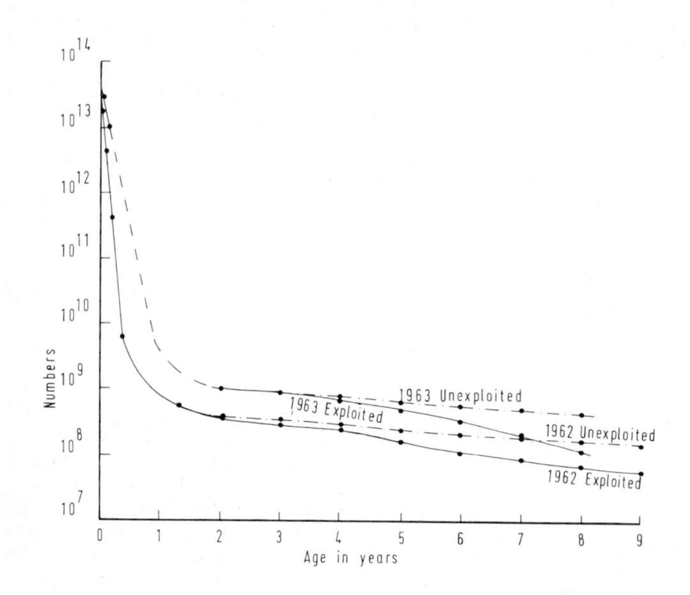

Fig. 7. North Sea plaice. Survival curves for the 1962 and 1963 cohorts

DISCUSSION

For the period between egg production and the end of the 1-group stage the data demonstrate the high early mortality that we regard as being characteristic of fecund populations. The data are also in line with Hjort's notion that year-class strength may be largely determined in the planktonic phase (Hjort, 1914). With reference to the average 1962 year-class, the 1963 egg survey results could have been used to give a qualitative prediction of the very good recruitment to the fishery two or three years later. Similarly, with reference to the larval mortality patterns shown in 1962, 1968, and 1971, the 1969 larval abundance data might presage good recruitment by the 1969 year-class.

Conceptually the principal point of interest in these data is whether or not the pre-recruit mortality is observed to vary with population density or recognizable environmental features. Within the life of an average year-class there are, of course, obvious differences between the mortality observed in each phase, so that whereas 80% of eggs and larvae may die per month, the 0-group die at 40% per month and the 1-group at 10 to 20% per month. In the first years of adult life (ages 5-15) the death rate is approximately 10% per year. Mortality thus declines as the cohort abundance declines with age.

Looking more specifically at egg and larval stages, we can see within-year and between-year differences in the mortality patterns. Fig. 6 shows that within a cohort the mortality rate from egg stage I to egg stage IV is constant. After larval stage 1 mortality rates are either constant or decrease slightly. In the 1962, 1968, and 1971 year-classes there is a step down in numbers between egg stages IV and V coincident with hatching. Constant egg mortality within a year-class could occur by development failure or predation. The constant or slightly decreasing larval mortality through the morphological stages is probably due to predation alone. For the 1962, 1968 and 1971 year-classes the cause of the increased mortality at hatching is unknown.

The between-year effects are compared by ranking the abundance of egg stages I and V and larval stages 1 and 4 as shown below:

Rank	Eggs I	Eggs V	Larvae 1	Larvae 4
High	63	63	63?	?
	62	62	69	69
	68	68	68	71
	69	69	62	62
Low	71	71	71	68

Despite the changes in abundance in 1962, 1968, and 1971 there is no change in the ranking order up to egg stage V but there are some changes thereafter. Compared to the original egg abundance the order of year-classes 1962, 1968 and 1971 is reversed and the 1969 larval abundance outranks all except the 1963 year-class. Table 1 and Fig. 6 show that in 1962, 1968 and 1971 such changes could be consistent with a density-dependent effect in the larval phase. However, the 1963 year-class originated from the highest abundance of eggs and the 1969 year-class from a stage V egg abundance similar to that for the average 1962 and 1968 year-classes. For these two broods density-independent effects are postulated. However, this is not to say that year-classes can necessarily be classed as being of density-dependent or density-independent origin, for both processes must contribute to the abundance of any year-class.

In a previous discussion Harding and Talbot (1973) referred to abnormally low temperature conditions in 1963, which also produced good year-classes of cod and sole, and give us some indication of the probable course of events in 1963. There are possibly two effects here: Reduced temperature will prolong development and increase the time of potential exposure to predation. We could thus assume that at the very low 1963 temperatures the predators were perhaps low in number, abnormally distributed, or were not feeding as heavily as at average temperatures. This could reduce both egg and larval mortality rates. In addition, if the delayed hatching allowed the larvae to synchronize more closely with the production of their food organisms (Cushing, in press), we would expect improved larval survival even if the level of predation was normal. It is not yet known whether events in 1969 were consistent with a similar temperature effect.

REFERENCES

Bagenal, M., 1955. A note on the relations of certain parameters following a logarithmic transformation. J. Mar. biol. Ass. U.K. 34, 289-296.
Beverton, R.J.H. and Holt, S.J., 1957. On the dynamics of exploited fish populations. Fishery Invest. (Lond.), Ser. 2, 19, 533 pp.
Beverton, R.J.H. and Tungate, D.S., 1967. A multi-purpose plankton sampler. J. Cons. perm. int. Explor. Mer 31, 145-157.
Bishop, Y.M.M., 1959. Errors in estimates of mortality obtained from virtual populations. J. Fish. Res. Bd. Can. 16, 73-90.
Corlett, J., 1966. Mortality of O-group plaice in the northern Irish Sea. ICES C.M. 1966/C:6, 4 pp., figs. (mimeo).
Cushing, D.H., 1957. The number of pilchards in the Channel. Fishery Invest. (Lond.), Ser. 2, 21 (5), 27 pp.
Dickson, R. and Lee, A., 1972. Recent hydro-meteorological trends on the North Atlantic fishing ground. Fish. Ind. Rev. 2 (2), 4-11.
Edwards, R. and Steele, J.H., 1968. The ecology of O-group plaice and common dabs in Loch Ewe. 1. Population and food. J. Exp. Mar. Biol. Ecol. 2, 215-238.
Fry, F.E.J., 1957. Assessment of mortalities by use of the virtual population. Proc. Jt Scient. Meet. of ICNAF, ICES and FAO, on "Fishing Effort, the Effect of Fishing on Resources and the Selectivity of Fishing Gear", Lisbon, 1957. (Paper P. 15, 8 pp., mimeo).
Garrod, D.J., 1967. Population dynamics of the Arcto-Norwegian cod. J. Fish. Res. Bd. Can. 24, 145-190.
Gulland, J.A., 1965. Estimation of mortality rates. Annex to Arctic Fisheries Working Group (Gadoid Fish. Comm.). ICES C.M. 1965, Doc. 3, mimeo.
Gulland, J.A., 1968. Recent changes in the North Sea plaice fishery. J. Cons. perm. int. Explor. Mer 31, 305-322.
Harding, D. and Talbot, J.W., 1973. Recent studies on the eggs and larvae of the plaice (*Pleuronectes platessa* L.) in the Southern Bight. Rapp. P.-v. Réun. Cons. perm. int. Explor. Mer 164, 261-269.
Hjort, J., 1914. Fluctuations in the great fisheries of Northern Europe viewed in the light of biological research. Rapp. P.-v. Réun. Cons. perm. int. Explor. Mer 21, 160-170.
Jones, A., 1973. The ecology of young turbot, *Scophthalmus maximus* (L.), at Borth, Cardiganshire, Wales. J. Fish. Biol. 5, 367-383.
Lockwood, S.J., 1972. An ecological survey of an O-group plaice (*Pleuronectes platessa* L.) population, Filey Bay, Yorkshire. Ph.D. Thesis, University of East Anglia, 169 pp.
Lockwood, S.J., (in press). The movements of juvenile plaice off the north Yorkshire coast. Proc. Challenger Soc.
Macer, C.T., 1967. The food web in Red Wharf Bay (North Wales) with particular reference to young plaice (*Pleuronectes platessa*). Helgolaender Wiss. Meeresuntersuch. 15, 560-573.
Pearcey, W.G., 1962. The ecology of an estuarine population of winter flounder, *Pseudopleuronectes americanus* (Walbaum), Pts 1-4. Bull. Bingham oceanogr. Coll. 18 (1), 1-78.
Raitt, D.S., 1936. Stock replenishment and fishing intensity in the haddock of the North Sea. J. Cons. perm. int. Explor. Mer 11, 211-218.
Ricker, W.E., 1954. Stock and recruitment. J. Fish. Res. Bd. Can. 11, 559-623.
Ricker, W.E., 1958. Handbook of computations for biological statistics of fish populations. Bull. Fish. Res. Bd. Can. (119), 300 pp.
Riley, J.D., 1971. The Riley push-net, Appendix I. In: Methods for the Study of Marine Benthos (Eds. N.A. Holme and A.D. McIntyre), p. 286-290. Oxford and Edinb.: Blackwell, 334 pp. IBP Handbook, No. 16.

Riley, J.D. and Corlett, J., 1966. The number of O-group plaice in Port Erin Bay, 1964-1966. Rep. Mar. Biol. Stn Port Erin 78, 51-56.
Ryland, J.S., 1966. Observations on the development of larvae of the plaice, *Pleuronectes platessa* L., in aquaria. J. Cons. perm. int. Explor. Mer 30, 177-195.
Saville, A., 1964. Estimation of the abundance of a fish stock from egg and larval surveys. Rapp. P.-v. Réun. Cons. perm. int. Explor. Mer 155, 164-170.
Schumacher, A., 1970. Bestimmung der fischereilichen Sterblichkeit beim Kabeljaubestand vor Westgrönland. (Estimation of fishing mortality in the stock of cod off West Greenland). Ber. dt. wiss. Kommn. Meeresforsch. 21, 248-259 (Eng. Abstract).
Simpson, A.C., 1959. The spawning of the plaice (*Pleuronectes platessa*) in the North Sea. Fishery Invest. (Lond.), Ser. 2, 22 (7), 111 pp.
Steele, J.H. and Edwards, R.R.C. The ecology of O-group plaice and common dabs in Loch Ewe. IV. Dynamics of the plaice and dab populations. J. Exp. Mar. Biol. Ecol. 4, 174-187.
Zijlstra, J.J., 1972. On the importance of the Waddensee as a nursery area in relation to the conservation of the southern North Sea fishery resources. Symp. zool. Soc. Lond. (29), p. 233-258.

R.C.A. Bannister
Fisheries Laboratory
Lowestoft / GREAT BRITAIN

D. Harding
Fisheries Laboratory
Lowestoft / GREAT BRITAIN

S.J. Lockwood
Fisheries Laboratory
Lowestoft / GREAT BRITAIN

The Distribution and Mortality of Sole Eggs [*Solea solea* (L.)] in Inshore Areas

J. D. Riley

INTRODUCTION

Studies of the ecology of flatfish during their first year revealed that very little was known of the association of spawning grounds with 'nursery areas' inshore, and that virtually no recent data were available on the relative sizes of these inshore spawning grounds and nursery areas. A survey programme was started in a selected area for sole in 1969, and in 1970 the first part of a complete survey of the inshore area of the coast of England and Wales was undertaken when the east coast of England was examined for the first time. The main purpose of these inshore and estuarine surveys has been to gather quantitative data on the abundance of planktonic eggs, larvae and O- and 1-group fish.

The results described here on the distribution and mortality of sole eggs have been obtained from an intensive survey of the plankton of the River Blackwater in 1969, over the main spawning period of the sole there, and from an extensive survey of the east coast in 1970.

MATERIALS AND METHODS

The timing of both surveys coincided with the expected peak spawning of the sole in a normal year. In the Blackwater Estuary sole move inshore during April when the water temperature is less than 9°C. Spawning usually starts about the third week of April and is virtually over by early June, by which time the water temperature is about 15°C.

Temperature data were available for the coastal region from a series of ten coastal stations situated from Whitby in the north to Leigh-on-Sea in the south (Fig. 1). These stations provide 8 to 10 'high water' seawater temperatures per month to the hydrographic section of the Lowestoft laboratory; they are taken by local harbour masters, lifeboat-station staff, etc.

In 1969 sampling began in the Blackwater Estuary on 6 May, when the seawater temperature was about 11°C and spawning had already started; the survey was completed on 5 June, when the water temperature had risen to 15°C and spawning was virtually over. Temperature data for the neighbouring coastal stations of Harwich and Leigh-on-Sea show that the 1969 May values were about 1°C higher than the average of the years 1966-1972. In spite of this variation, spawning took place at about the usual time for that area.

The 1970 survey of the east coast took place between 30 April and 19 May, and the sea temperatures during tow-netting ranged between 7.5°C and 13.8°C (mean 10.02°C). A few sole eggs were taken in

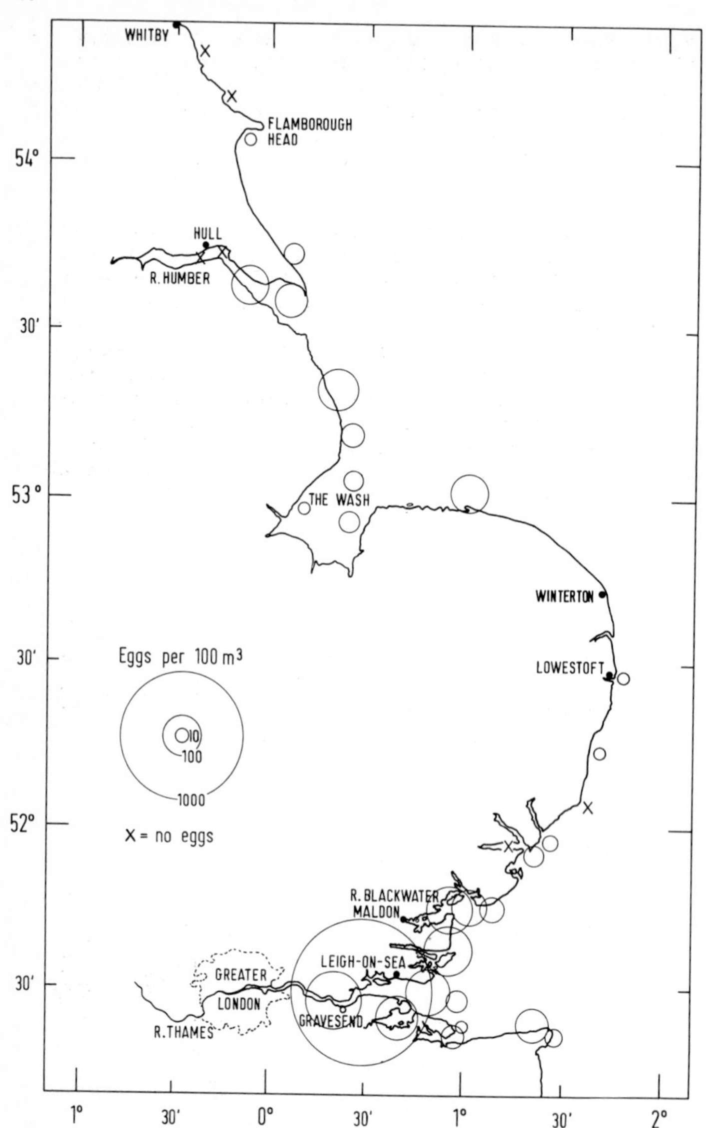

Fig. 1. Distribution of sole eggs - English east coast, April-May 1970. For clarity in the Thames region, mean values for one set of three or one set of two stations have been shown

water at the lowest temperature (7.5°C) but egg densities greater than 1 per m^3 were only found in water at temperatures of 10.0°C or more.

In 1970 the May average water temperature of the ten coastal stations was 10.6°C. This was the lowest May value for the period 1966-72, when the mean May value was 11.0°C. From the available information on the times of spawning, however, we can conclude that the sampling period in 1970 spanned the spawning period of the sole as well as could any 20-day period.

All the plankton samples were collected by an encased 30 cm diameter high-speed tow-net fitted with a conical net of 24 meshes/cm. The flow of water through the net was controlled by the size of the circular aperture cut in the moulded fibreglass nose-cone, and the volume filtered was measured by a flowmeter mounted immediately behind the nose-cone opening. The area of the nose-cone opening was 100 cm^2 in 1969 and 250 cm^2 in 1970.

Stability of the net through the water was achieved by the tubular shape of the moulded fibreglass casing and the single vertical fin fitted on the tail (Fig. 2). Diving was facilitated by the use of a 'Scripps' depressor and the net was used to complete a series of multiple oblique hauls between the surface and the bottom. In deeper water only two complete dives might be required per 10-min. haul, but at shallow stations 7 or 8 would be completed at the standard speed of diving and hauling, for the same time period. All the sampling was done from the Laboratory's inshore vessel RV TELLINA (17 m length overall), keeping the engine revolutions and pitch of the variable pitch propeller constant so as to give a strictly comparable speed at all stations of 1 m/sec (approximately 2 kn).

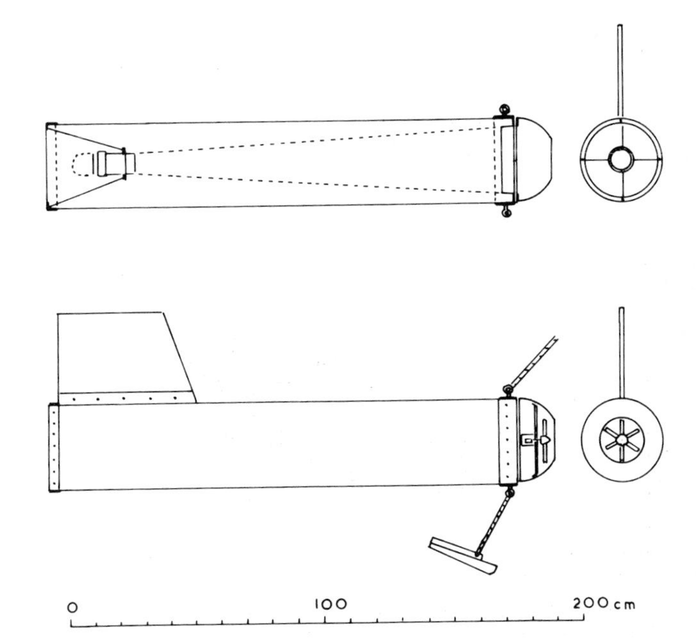

Fig. 2. High-speed tow-net, Lowestoft pattern, 30 cm diameter. Upper left, diagrammatic section of sampler showing position of nylon net; upper right, mounting of flowmeter; lower right, front view; lower left, in fishing position showing flowmeter

Immediately after collection the plankton was preserved in 4% neutral formalin. In the laboratory all the fish eggs and larvae were sorted and identified and the sole eggs were also sorted into developmental stages. The stages used were similar to those of Simpson (1959) for plaice eggs, which are a modification of H.J. Buchanan Wollaston's original stages. In the development of sole eggs, however, there is no equivalent of Simpson's stage V, because hatching occurs at late stage IV.

The sole egg developmental stages used were (Fig. 3):

stage IA - blastula stages from fertilization up to the point at which the blastoderm diameter is 4/5 of the egg diameter;

IB - gastrulation stages up to the appearance of a distinct primitive streak on the embryonic shield;

II - gastrulation from the end of stage IB up to the closure of the blastopore;

III - growth and organogenesis up to the growth of the tail (free from yolk-sac attachment) to the equator of the egg, with the anterior body axis running from pole to pole;

IV - growth and organogenesis tail-tip growth through the equator to a position between 60° and 40° from nose, when hatching occurs.

These stages are somewhat arbitrary, but are recognizable with some precision in preserved material taken at sea. In laboratory experiments it was shown that one embryonic feature described above - the end of stage IB - appeared slightly differently in live and preserved material (see discussion).

Plankton Sampling Schedules

A. River Blackwater, 1969. The estuary (see Fig. 1) runs from Maldon in Essex to the estuary Bar, a distance of about 25 km, and has a tidal exchange rate of 2-5% with the neighbouring sea area, for although between 50 and 70% of the water leaves the estuary on the ebb, a high proportion subsequently returns with the succeeding flood (Talbot, 1967). The estuary has therefore a high degree of mixing, a factor confirmed by the absence of any appreciable or persistent vertical salinity stratification at high water over the lower 90% of the estuary. The salinity difference between the Bar and the upper reaches was usually $<1.5^{\circ}/oo$ at high water. The estuary is therefore one which does not lose much water or many sole eggs to the sea to the east and is landlocked on all other sides. In this confined situation it was assumed that there was a thoroughly mixed population of sole eggs which was sampled over a 30-day period and whose incubation period was relatively short, i.e. 6 days at $12.0^{\circ}C$.

Sampling was done in two groups: in the first, samples were taken over either a complete ebb or flood at a fixed position in the lower reaches of the estuary, known as the 'Bench Head Buoy'. Each half-tidal series at this position included samples taken at high and low water, at mid-tide and at intermediate times between high tide and mid-tide and mid-tide and low tide (5 in all). These samples were taken every 2 days during daylight hours; consequently, some samples were taken on a falling tide and others on a rising tide. The tidal flow of the estuary would cause between 50 and 70% of the water in the estuary at high tide to flow past a line drawn across the river through the sampling position at the Bench Head Buoy. The second group of samples was taken on a transect of the estuary from the Bar Buoy up the estuary to within 4 km of Maldon. The duration of each haul was between 10 and 12 min.

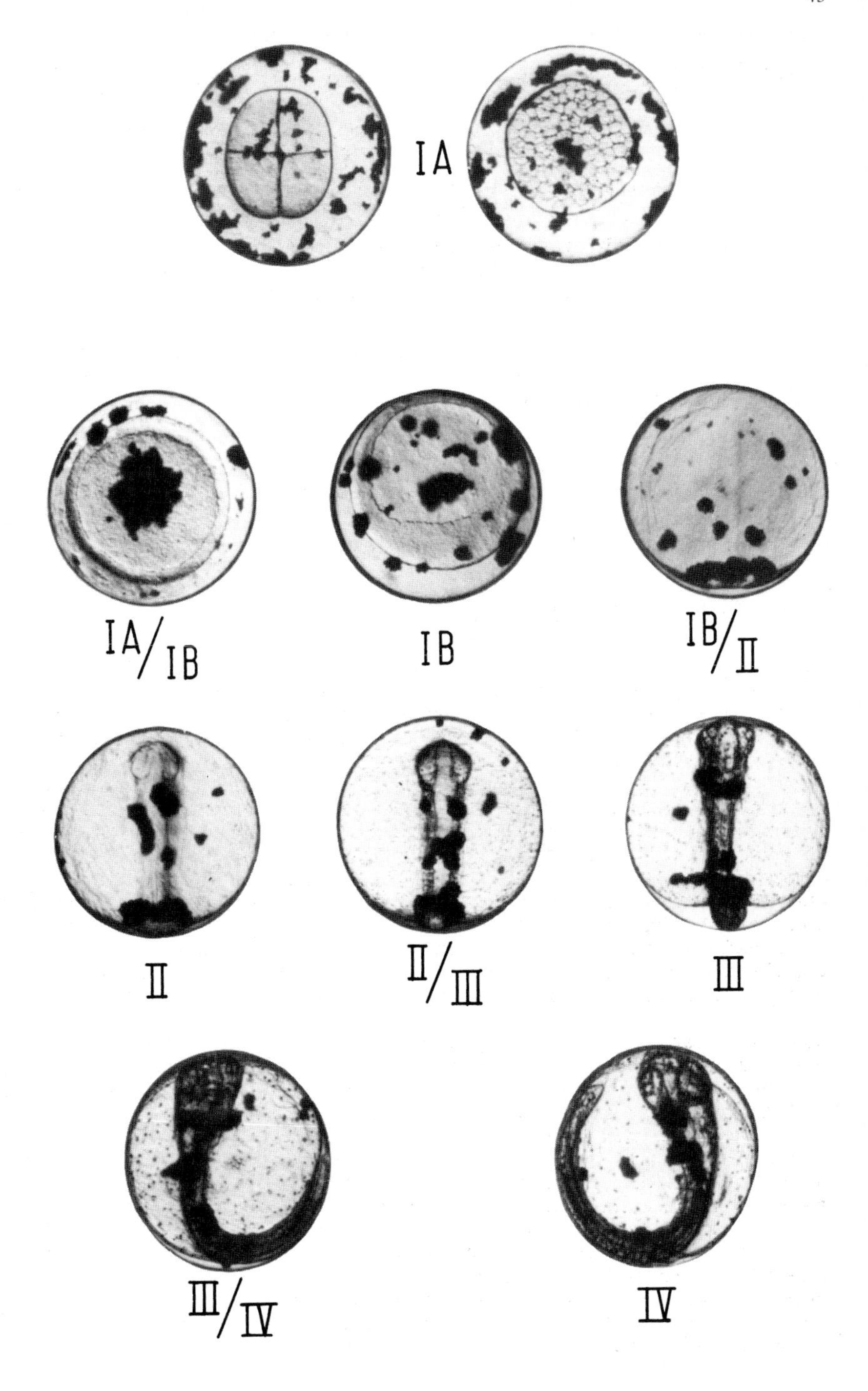

Fig. 3. Incubation stages of sole eggs

B. English East Coast Inshore Survey, 1970. The planktonic fish eggs and larvae were sampled at 42 stations from Broadstairs to Berwick between 30 April and 19 May. Whenever possible the sample was collected in a bay or estuary where the sea bed was likely to be soft and relatively level. This was done in order to avoid mechanical damage of the sampler. Stations were usually between 15 and 25 km apart and within 3 km of the coast. Water depths varied from 4 to 50 m. As a result of experience gained in 1969, the time of each haul was increased to about 30 min in order to sample adequately the relatively low densities of fish eggs and larvae expected at some stations.

Laboratory Experimental Measurement of a Stage Duration in Sole Eggs

The surveys had been planned in the anticipation that the major, and perhaps the only, factors which would affect the proportion of the various stages of incubation in the sole eggs in the plankton would be mortality and stage duration. If the stage duration effect could be allowed for, mortality could be measured.

In order to measure the stage duration precisely over the whole biotic range a special incubator was constructed in the laboratory workshops, using the principle of Halldal and French (1956)(Fig. 4). A range of temperatures was achieved by standing the small glass incubator vessels on a block of aluminium which was cooled at one end and heated at the other by circulating water from thermostatically-controlled and insulated reservoirs. Thermostatic control of the refrigeration compressor and the immersion heater was by capillary thermostats and relays; each circulation was maintained by small continuously rated centrifugal pumps constructed of polypropylene and delivering about 20 l/min. By varying the temperature of the two circulating waters the temperature levels and gradient along the aluminium block could provide the desired range of incubation temperatures. For the sole, incubation temperatures of 6-20°C were chosen. Adult sole caught in the River Blackwater were used to provide artificially-fertilized eggs for the experiment (Riley and Thacker, 1969) and these were transferred within 1 h to the incubator vessels into water already at the correct temperatures.

Artificial fertilization makes precise timing of development possible. Subsequent development was monitored both visually and photographically every hour for the first 6 h, then every 3 h for the next 2 1/2 days, and finally every 6 h thereafter, until hatching.

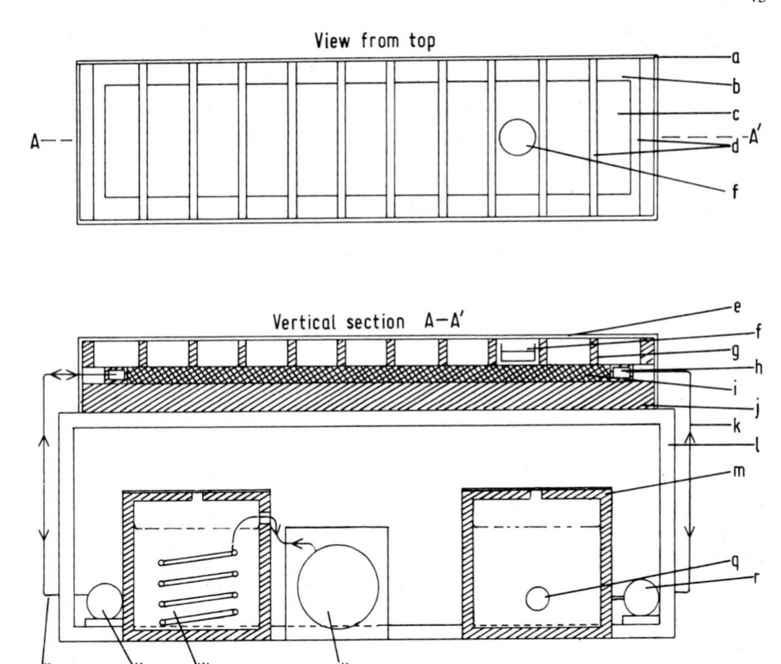

Fig. 4. Incubator apparatus for pelagic fish eggs: (a) marine plywood (12 mm) sides to incubator chamber; (b) marine plywood (12 mm) surround to aluminium block; (c) aluminium block 5 cm thick; (d) expanded polystyrene insulation; (e) acrylic sheet, hinged cover; (f) glass incubation vessel; (g) expanded polystyrene partition; (h) circulation cavity in aluminium block; (i) aluminium block; (j) expanded polystyrene insulators; (k) flow and return pipes to either side of cavity in block (heating); (l) 35 cm equal angle steel frame; (m) insulation of warm reservoir contained in box of 12 mm resin-bonded plywood, 70 l fibreglass reservoir; (q) immersion heater (250 W); (r) circulating pump; (u) refrigerator unit; (w) cooling coil 1 cm O.D. copper coil; (x) cooling flow and return pipes to block; (y) cooling circulating pump

RESULTS

Laboratory Measurement of Stage Duration. Sole eggs failed to gastrulate at temperatures higher than 16°C, although if eggs were taken through gastrulation into stage III at temperatures between 7 and 15°C and then held at temperatures up to 20°C subsequent development and hatching was normal. Similarly, at temperatures below 6°C early cleavage stages proceeded slowly and abnormally and the eggs died after about 2 days; but when eggs were taken through gastrulation at temperatures between 7 and 15°C subsequent incubation at 6°C through stages III and IV produced normal larvae.

Observations of the incubation stages were recorded against time, and the maximum and minimum durations of each of the 5 stages were obtained at each of the 5 temperatures at which incubation was completed. These time periods for the five stages IA to IV gave calculated regressions on a log temperature x log duration plot (Fig. 5) from which the duration of any stage at any temperature can be estimated. The range for successful incubation was between 7 and 15°C and was determined by the temperature tolerance during early gastrulation. The temperature limits for successful feeding by the hatched larvae are generally higher (13-19°C) in the laboratory.

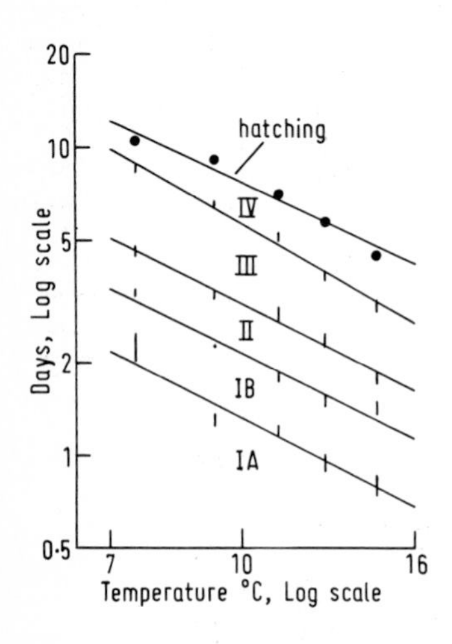

Fig. 5. Sole egg incubation stage durations - laboratory observations. Regressions as follows:

Stage		
IA	$\log_e y$	$= -1.4121 \log_e x + 3.5341$
IB		$= -1.3327 \log_e x + 3.8447$
II		$= -1.3638 \log_e x + 4.2821$
III		$= -1.5888 \log_e x + 5.4022$
IV		$= -1.2852 \log_e x + 5.0072$

River Blackwater, 1969. A total of 12 tidal station sample series was taken between 6 and 28 May; the tenth and twelfth series contained no sole eggs and have been disregarded. Three transects up the river were completed on 9, 15, and 21 May. The five most abundant fish species, in descending order of egg density, were *Sprattus sprattus* (L.), *Solea solea* (L.), 'rockling' spp., *Platichthys flesus* (L.) and *Engraulis encrasicholus* Cuvier. The five most abundant fish larvae, in descending order, were *Clupea harengus* (L.), *Gobius* sp., *Sprattus sprattus*, *Platichthys flesus* and *Solea solea*, over the sampling period as a whole.

The sole egg densities for each development stage in the tidal series at the Bench Head Buoy (Table 1) were divided by the individual stage durations at the mean temperature of each sampling series to give a daily production for each stage. These values were plotted against time, and the area below the production curve gave the total production value for each stage. These total production estimates were then plotted against the mid-stage duration, calculated from the mean sea temperature of the sampling stations and the laboratory results. The regression line fitted to these data gave calculated intercepts when x = 0 (fertilization) and 5.45 days (hatching at 13.09°C) of 30.1 and 0.797, indicating a survival through incubation of 2.65% and a mortality rate of 49% per day (Fig. 6).

Table 1. River Blackwater sole eggs, 1969: corrected catch and stage production - tidal series of samples

Date	Mean temp. (°C)	Stage					Total
		IA	IB	II	III	IV	
		(a) Catch per 5 m^3					
6 May	11.90	15.4	2.5	4.2	1.7	0	23.8
8	12.33	23.3	1.4	2.9	2.1	0	29.7
10	12.19	11.3	1.2	2.8	2.3	1.0	18.6
12	13.30	8.2	1.9	3.6	0.8	0	14.5
14	13.97	7.0	2.1	3.0	5.1	0	17.2
16	13.46	10.8	0	3.1	2.3	4.3	20.5
18	13.08	1.7	0.8	1.7	1.7	0	5.9
20	13.02	2.7	0	0.9	0	0.9	4.5
22	13.32	1.0	0	0	2.1	0	3.1
26	14.30	0	0	0	1.2	0	1.2
Total		81.4	9.9	22.2	19.3	6.2	139.0
		(b) Days duration					
6 May	11.90	1.05	0.65	0.80	1.80	1.85	
8	12.33	1.00	0.65	0.70	1.80	1.75	
10	12.19	1.00	0.65	0.75	1.80	1.80	
12	13.30	0.90	0.60	0.65	1.55	1.65	
14	13.97	0.83	0.57	0.60	1.35	1.60	
16	13.46	0.88	0.57	0.65	1.45	1.70	
18	13.08	0.91	0.64	0.65	1.55	1.70	
20	13.02	0.92	0.63	0.65	1.60	1.70	
22	13.32	0.90	0.60	0.65	1.55	1.65	
26	14.30	0.81	0.54	0.60	1.35	1.55	
		(c) Production per 5 m^3 per day					
6 May	11.90	14.7	3.8	5.3	0.9	0	
8	12.33	23.3	2.2	4.1	1.2	0	
10	12.19	11.3	1.8	3.7	1.3	0.6	
12	13.30	9.1	3.2	5.5	0.5	0	
14	13.97	8.4	3.7	5.0	3.8	0	
16	13.46	12.3	0	4.8	1.6	2.5	
18	13.08	1.9	1.3	2.6	1.1	0	
20	13.02	2.9	0	1.4	0	0.5	
22	13.32	1.1	0	0	1.4	0	
26	14.30	0	0	0	0.9	0	
		(d) Production (eggs per m^3) by planimetry					
		31.03	6.62	11.9	4.56	1.37	
		(e) Midstage time (days) at 13.09°C					
		0.46	1.22	1.85	2.96	4.60	

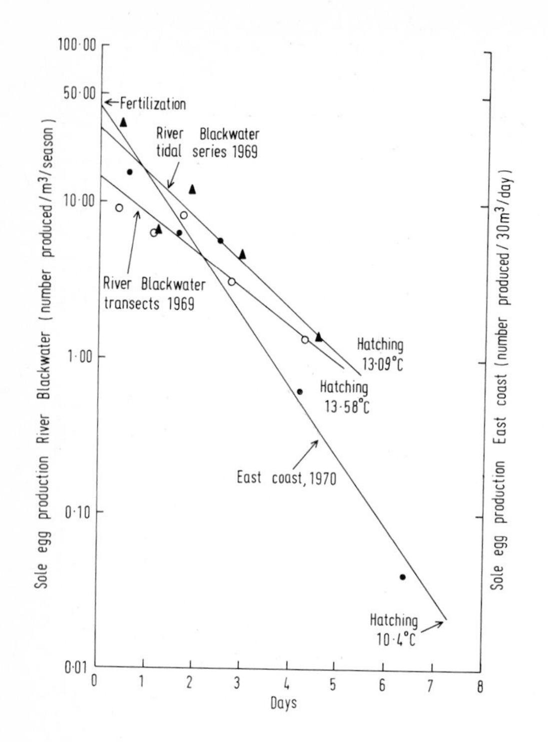

Fig. 6. Planktonic sole egg mortality, 1969 and 1970. Plot of stage production indices against time. Calculated regression lines

The catches of sole eggs from the three transects up the estuary were treated in a similar way (Table 2), and the production estimates were plotted against the mid-stage duration at the mean sea temperature of the sampling stations, i.e. 13.58°C (Fig. 6). The regression line fitted to these data gave calculated intercepts when x = 0 (fertilization) and x = 5.15 (hatching at 13.58°C) of 14.8 and 0.895, indicating a survival through incubation of 6.05%, while the slope of the regression gave a mortality rate of 42% per day.

The different values obtained from the two sampling methods is probably due to the low numbers of eggs obtained on the transects which extended beyond the centre of the spawning area. For the purposes of comparison with 1970, the two survival percentages of 2.65 and 6.05 have been meaned (4.35%).

English East Coast, 1970. Sole eggs were found in 30 of the 42 stations of the survey and all of these were south of Flamborough Head (see Fig. 1). The greatest density of eggs was found in the Thames Estuary and a minor concentration was associated with the Humber/Wash area. The open coastlines of Kent, East Anglia and north of the Humber showed concentrations all less than 1 egg/m^3.

In the samples the sole eggs were outnumbered by eggs of the sprat (*Sprattus sprattus* (L.)) which were found off the entire coast, being absent only from the samples taken in the Rivers Humber, Thames, Swale and Stour. Eggs of the dab (*Limanda limanda* (L.)) and 'rockling' spp. were predominant north of the Humber. Eggs of nine other species were present at lower densities. No sole larvae were caught during the survey. The sole egg samples were considered to be representative

Table 2. River Blackwater sole eggs, 1969: corrected catch and stage production - transects series of samples

Date	Temp. (°C)	Stage				
		IA	IB	II	III	IV
		(a) Catch per 5 m^3				
9 May	12.5	11.25	3.45	5.75	5.75	1.9
15 May	14.84	0	0.85	1.15	1.15	1.15
21 May	13.40	2.25	1.00	1.10	0	0
		(b) Days duration				
9 May	12.5	0.97	0.63	0.70	1.70	1.7
15 May	14.84	0.76	0.52	0.54	1.23	1.55
21 May	13.40	0.88	0.58	0.64	1.50	1.70
		(c) Production per 5 m^3 per day				
9 May	12.5	11.6	5.5	8.2	3.4	1.1
15 May	14.84	0	1.6	2.1	0.9	0.7
21 May	13.40	2.6	1.7	1.7	0	0
		(d) Production (eggs per m^3 per season) by planimetry				
		8.86	6.26	8.32	3.01	1.32
		(e) Midstage time (days) at 13.58°C				
		0.86	1.15	1.76	2.82	4.35

of the whole coast south of Flamborough Head and of a large proportion of the spawning season. After sorting and staging, production values were obtained for each stage at individual stations, depending on the duration of that stage at the temperature of the water on sampling. The stations north of Flamborough Head which contained no sole eggs were disregarded.

The mean production values of the stations for each stage (Table 3) were plotted against the midpoint of incubation time at the mean temperature of the stations, 10.4°C (Fig. 6). The regression of this plot gave calculated intercepts, when $x = 0$ (fertilization) and 7.30 days (hatching), of 42.32 and 0.0212 respectively, a survival through incubation of 0.05% with the slope indicating a mortality rate of 65% per day.

For the purposes of comparison, the coastline south of Flamborough Head was divided at Winterton (see Fig. 1) into two spawning regions, Thames and Humber/Wash. When the mean sole egg stage production values for these two regions were considered separately the values for the Thames were about three times greater than for the Humber/Wash region (Table 3).

Table 3. Sole egg stage production, Flamborough Head southwards, 1970

Region	Number of stations	Stage IA	IB	II	III	IV
		(a) Total production				
All	34	517.41	210.10	196.45	20.58	1.29
Humber/Wash	12	75.98	24.55	36.73	2.76	0
Thames	22	441.43	185.55	159.72	17.82	1.29
		(b) Mean production per 30 m^3				
All		15.22	6.18	5.78	0.61	0.04
Humber/Wash		6.33	2.05	3.06	0.23	0
Thames		20.07	8.43	7.26	0.81	0.06
		(c) Midstage time at 10.4°C				
		0.63	1.68	2.55	4.25	6.40

A χ^2 test on the mean egg stage production values indicated no significant difference in the proportions of the stages in the two regions,

$$\chi^2 = 0.25 < \chi_{0.05(2)} = 5.99,$$

and as the mean station temperatures for the two regions were similar, 10.9°C and 10.1°C, the regression line fitted to data from both the spawning regions represents the egg survival of the whole area sampled (Table 3 and Fig. 6).

DISCUSSION

The points on Fig. 6 have one common pattern about the regression; the production estimates for stage IB appear to be too low and those for stage II too high. This is due to the difficulty in distinguishing between the two stages in preserved material. The onset of stage II is the development of the primitive streak on the embryonic shield, and in preserved and stored material the whole gastrula becomes opaque and the thickening streak shows up much more distinctly than in live material. Results indicate that a proportion of stage II eggs had been mis-identified from the Blackwater and a smaller proportion of stage II eggs from the east coast. The higher percentage from the 1969 material was probably due to it being in preservative for 16 months before analysis, compared with 1 to 2 months for the 1970 material.

The percentage survival figures of 4.35 for 1969 and 0.05 for 1970 are in a ratio of 87:1, indicating wide differences in survival for the 2 years. During the 1970 plankton survey, which subsequently indicated a very high egg mortality of 99.95%, no larvae of the sole were caught, whereas in 1969, with an egg mortality of 95.65%, sole larvae were caught at densities averaging $0.1/m^3$ over the survey period.

No data are yet available, from the fishery statistics, on the relative strengths of the two year-classes of sole on the English east coast grounds. Until they are available no indication can be given as to whether year-class strengths may be decided as early as the incubation stages.

Although survival through incubation differed greatly, the percentage survival per day differed much less markedly and the highest survival occurred in warmer water. During the spawning season of the sole (late April to early June), tidal estuaries are usually considerably warmer than the open coast and offshore areas. There may be a survival advantage for the sole in spawning early, and on the English east coast they do so, spawning first in the estuaries where temperatures are highest.

Young sole in their first year are also found predominantly in the same areas as the sole eggs, and a beam-trawl survey of the east coast in 1970 showed that some sole had survived from the 1970 spawning in spite of the high egg mortality. However, the results are not sufficient on their own to indicate comparative year-class strength. During their first winter young sole are not caught inshore and it is assumed that they move off to deeper water. The east coast stock survival may be largely dependent on the restricted estuarine spawnings in areas which unfortunately receive some of the highest effluent loads in Britain.

SUMMARY

1. The distribution of sole eggs and larvae off the English east coast was surveyed using a Lowestoft 30 cm encased sampler. This provided samples over the whole coastal spawning area in 1970, and over an almost complete spawning season in one particular area in 1969 (the River Blackwater). Samples of sole eggs were sorted into developmental stages, and the production of the stages was compared. Proportions of egg stages were found not to be significantly different between different regions (1970).

2. The abundance of each developmental stage in the samples is dependent upon the numbers of eggs present and the duration of each stage. Stage duration was measured in the laboratory and the abundance of each egg stage in the sea samples was divided by the stage duration to give an estimate of the numbers produced per day. These estimates were plotted against the midpoints of the five developmental stages, and the slopes of the fitted regression lines indicated an egg mortality of 95.65% during 1969 in the River Blackwater and 99.95% on the English east coast in 1970.

3. Spawning on the east coast in 1970 was restricted to the regions south of Flamborough Head and was more concentrated in the Humber/ Wash and Thames regions.

REFERENCES

Halldal, P. and French, C.S., 1956. The growth of algae in crossed gradients of light intensity and temperature. Yb. Carnegie Instn (Wash.), No. 55, p. 261-265.

Riley, J.D. and Thacker, G.T., 1969. New intergeneric cross within the Pleuronectidae, dab x flounder. Nature (Lond.) 221, 484-486.
Simpson, A.C., 1959. The spawning of the plaice (*Pleuronectes platessa*) in the North Sea. Fishery Invest., Lond., Ser. 2, 22 (7), 111 pp.
Talbot, J.W., 1967. The hydrography of the River Blackwater. Fish. Invest. (Lond.), Ser. 2, 25 (6), 92 pp.

J.D. Riley
Fisheries Laboratory
Lowestoft / GREAT BRITAIN

Seasonal Changes in Dimensions and Viability of the Developing Eggs of the Cornish Pilchard (*Sardina pilchardus* Walbaum) off Plymouth

A. J. Southward and N. Demir

INTRODUCTION

The fishery for the Cornish Pilchard is generally regarded as a northern extension of the much greater fishery for the Atlantic Sardine in the Bay of Biscay. Most of the pilchards taken there are four years old or more, and younger specimens are rare in commercially caught samples. It has been suggested, therefore, that the population is maintained by migration of older fish from the northern Bay of Biscay, where there are large numbers of 2-year olds. This area is presumed to act as a 'nursery ground' for any young fish surviving from the more northerly spawnings (Furnestin, 1944; Hickling, 1945; Le Gall, 1950).

Considerable spawning does take place outside the main part of the Bay of Biscay, north of 48°. Eggs and post-larvae can be found in the western approaches to the English Channel in spring and early summer, off the south and west Irish coasts in summer and autumn, and in the Eastern Channel and Southern North Sea in late summer; off South Devon and Cornwall some eggs and post-larvae occur in the plankton almost all year round (Corbin, 1947, 1950; Cushing, 1957; Southward, 1962, 1963; Arbault and Boutin, 1968; Arbault and Lacroix Boutin, 1969; Fives, 1970; Arbault and Lacroix, 1971; Kennedy, Fitzmaurice and Champ, 1973).

We have recently tried to assess the seasonal pattern of breeding of the species in the inshore waters off the coasts of South Devon and Cornwall by means of a series of closely-spaced plankton samples taken at approximately monthly intervals in 1969 and 1970. The general methods of the survey and the hydrographic changes observed have already been described (Southward and Demir, 1972): this contribution deals with egg size and mortality in relation to environmental factors.

We would like to express our thanks to Sir Frederick Russell for his advice and encouragement during the course of this work, to Dr. J.H.S. Blaxter for useful discussions on the problem, and to Mrs. Helen Egginton for help with the statistical computations.

METHODS

The samples were taken over a grid of 30 stations extending 40 miles along the south coasts of Devon and Cornwall and out to the 40 fathom (74 m) depth contour 20 miles out from the coast (Fig. 1). Nearly all samples were collected with a modified Plymouth 1 m-diameter net, towed obliquely through the water column at a speed of 4 kn, using a Scripps pattern depressor. The net had a terylene mesh with a hole size of 0.7 x 0.9 mm (Southward, 1970), and each 10-min haul filtered an average volume of 750 m^3 (Southward and Demir, 1972). The formalin-fixed eggs were graded into the series of stages described by Gamulin and Hure (1955) for the Adriatic Sardine, corresponding essentially to the stages used by Ahlstrom (1950) for the Pacific Sardine. The 11 stages of this scheme were found to be more easily identifiable than the 20 stages used by Cushing (1957), and also allow comparison to be made with other races and species of sardine. The eggs were examined for viability on a subjective basis, using criteria similar to those discussed by Lee (1961); eggs with a cloudy perivitelline fluid, with obviously damaged yolk-masses and embryos, and those with signs of autolysis/bacterial action were regarded as dead. We must note that in this report mortality means the percentage of these observed dead eggs in the total sample, and has no direct relation to the concept of theoretical mortality deduced from observed changes in proportions of the successive developmental stages. All but a small fraction of the dead eggs could be allocated to the various stages. Subsamples of the viable eggs were measured for diameter of the capsule, of the yolk mass, and of the oil globule, using a calibrated micrometer eyepiece in a low power stereomicroscope. In sardine eggs the size and relative proportions of the yolk mass vary with the stage of development (Demir, 1969). The diameter of the yolk mass was therefore measured along the transverse axis at stages 5 and 6, when there is least difference between the vertical and transverse axes.

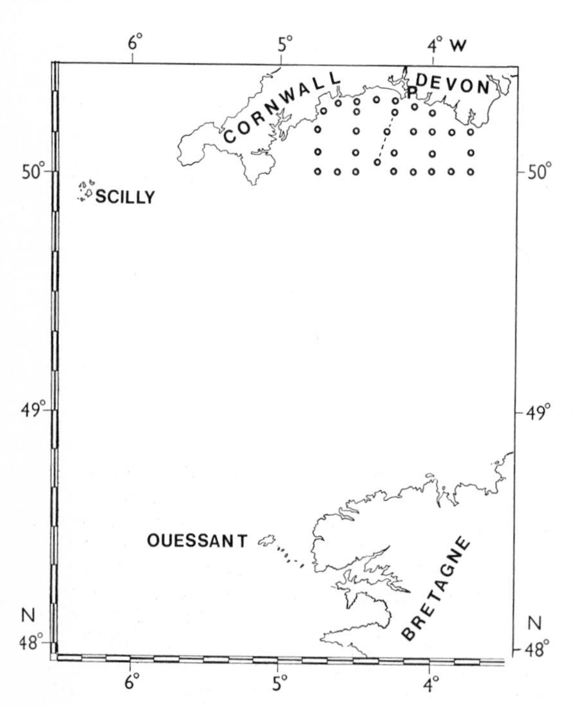

Fig. 1. The grid of stations sampled in 1969 and 1970 in relation to the mouth of the English Channel. The position of Plymouth is indicated by P, and the stations sampled more frequently, at weekly intervals, are joined by a dotted line. Note that the fishery for the Cornish Pilchard covers a wider area than the sampling grid, extending to the western tip of Cornwall. The nearest fishery for the Atlantic Sardine is just south of Ouessant

The samples taken each month have been pooled to study the seasonal changes in egg size and mortality. It should be noted that there was much seasonal variation in intensity and location of spawning. Full details will be given in a later contribution, but as an example of seasonal changes we show (Fig. 2) the monthly means of total egg numbers found at stations worked at weekly intervals during 1969 and 1970. In both years most of the spawning occurred in June and July, with a second, slightly smaller peak in October or November. It should also be noted that no stage 1 eggs (just fertilized) could be found in any of the 300 samples examined: this will also be discussed in a later contribution.

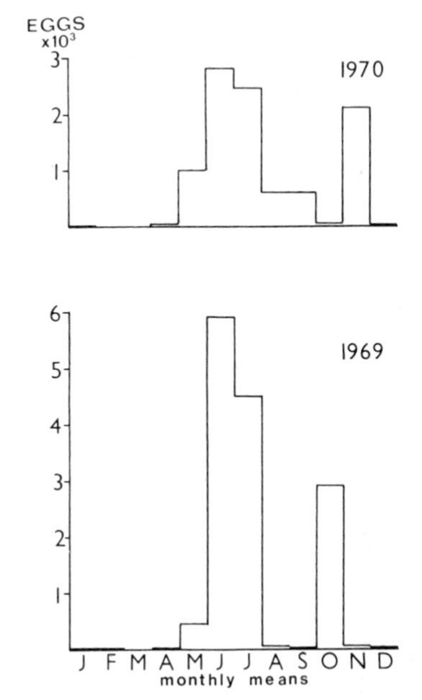

Fig. 2. Seasonal variation in spawning intensity of pilchard off Plymouth. Monthly means of total numbers of eggs found at the stations sampled at weekly intervals in 1969 and 1970. These samples were taken with the 2 m net hauled obliquely for 20 mins at 4 kn

RESULTS

Viability of the Eggs. Pilchard egg samples collected off Plymouth and farther afield in the Western Channel have always contained a large number of dead or damaged eggs, but this fact has not hitherto been commented upon (e.g. Corbin, 1947; Cushing, 1957; Southward, 1962), nor have the exact proportions always been recorded. The overall mortality was very high in our samples, varying from 38 to 92%, with an average of approximately 50%. This is higher than the mortality found in the Northern Adriatic (Gamulin and Hure, 1955) where the proportion judged dead varied from 3 to 50%. Somewhat lower mortalities than in our samples have also been reported from Rousillon and the Golfe du Lion, varying from 10 to 46% with an average of 29% (Lee et al., 1967; Aldebert and Tournier, 1971), and from Marseille and Sète, ranging from 46 to 52% (Lee, 1961). Arbault et al. (1972) have recently reported a mortality of only 13% in the northern Bay of Biscay in May 1970, at temperatures between 13 and 15°C.

Our samples show a well-marked seasonal variation in mortality, with generally better survival of eggs in the warmer months of the year. Fig. 3a shows the regression of mortality at stages 2 and 6 on sea surface temperature at the time of sampling, and Fig. 3b shows a similar regression for total mortality of all stages. There is clearly some relationship with temperature at the time of spawning, but the correlation coefficient is barely significant at the 5% level. A slightly better correlation (r = 0.57, p = 5%) can be obtained for total mortality by omitting the sample for January 1970, which showed an abnormally high proportion of older stages. However, it is obvious that there is much variance due to factors other than temperature.

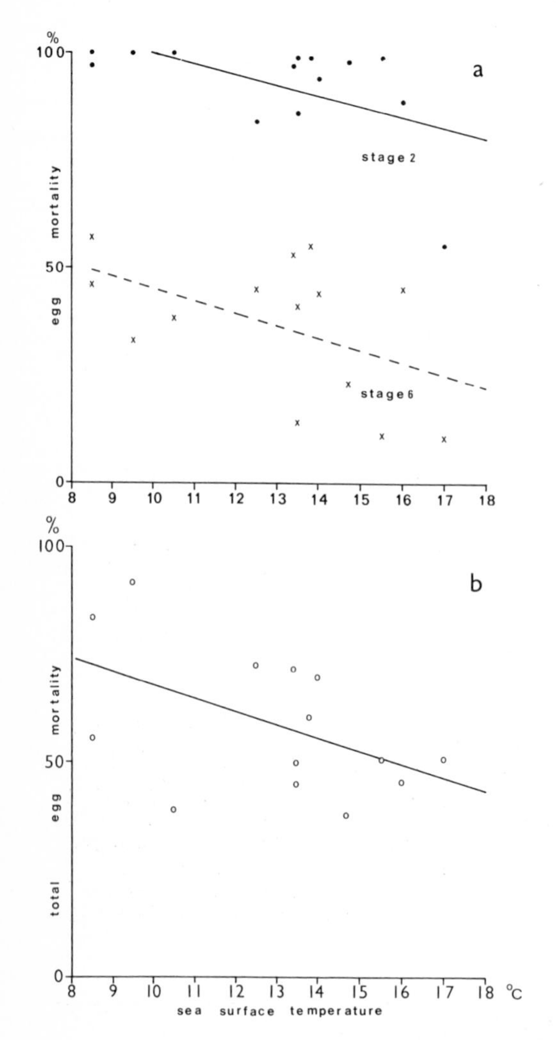

Fig. 3. Regression of egg mortality on sea surface temperature at the time of sampling: (a) plotted separately, stage 2 (r = -.52) and stage 6 (r = -.49); (b) all stages pooled (r = -.49). Sea temperature measured at Plymouth hydrographic station E 1, corresponding to one of the stations on the grid

In our samples most of the dead eggs belong to the earlier stages; this is shown clearly if we pool all the samples, as in Fig. 4. The pronounced sigmoid-form of the mortality curve illustrates the very substantial improvement in viability of the eggs between stages 5 and 6. Similar sigmoid curves are produced if we plot the results from individual months for which there are large enough samples of eggs (e.g. June, July, October and November). However, the samples collected in the two colder months of this group, June and November, show greater mortality of the later stages.

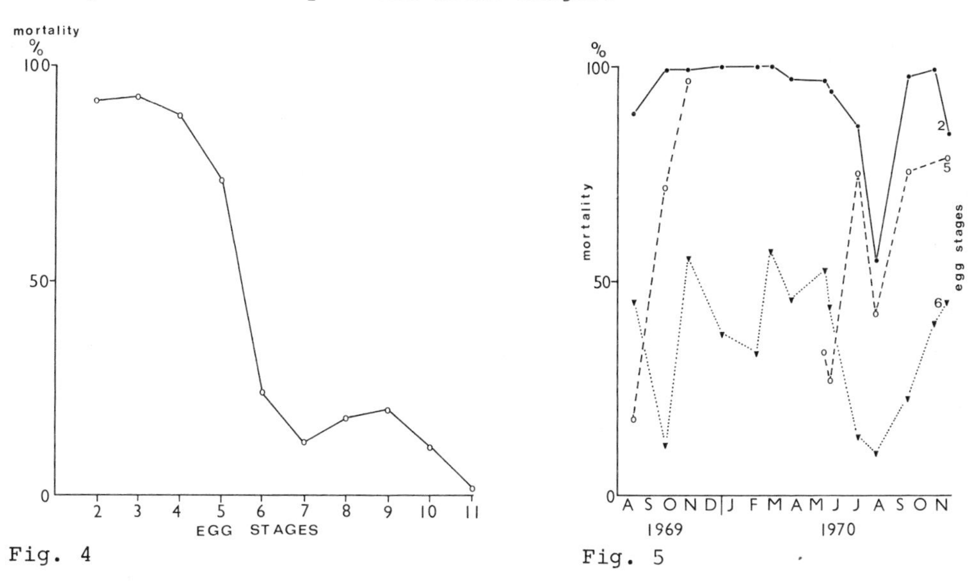

Fig. 4. Distribution of egg mortality at the different stages of development. All samples from 1969 and 1970 pooled

Fig. 5. Seasonal variation in the mortality of the early stages of development (stages 2, 5 and 6). The mortality at stages 3 and 4 was in between that at stages 2 and 5, but has been omitted from the graph for reasons of clarity. Note that some samples contained too few eggs to allow reliable calculation of percentage mortality of certain developmental stages

The seasonal changes in relative mortality of the different developmental stages are best seen if we consider the stages separately (Fig. 5). At stage 2 there was a very high mortality throughout the year, except in the warmest months, July and August; stages 3 and 4, not shown in Fig. 5, have a similar trend in mortality. At stage 5 the mortality is more variable, though there is still a well-marked period, from May to October, when mortality is least. In stage 6 and later stages there is much better survival of eggs all through the year, though there is still a period of somewhat reduced mortality from May to November.

Seasonal Changes in Egg Dimensions. The results of our measurements of egg diameter and diameter of the yolk mass and oil globule are given in Table 1. The size distribution found each month has been compared with that which would have been expected if egg dimensions were distributed normally. The mean values for diameter of the capsule (Fig. 6a) suggest a steady cyclic change through the year. Most of the samples did not differ significantly from normal when tested by the chi-square method, indicating that the samples are probably homogenous; in October 1969 and May, 1970, however, the distribution appeared to differ significantly from normal ($p = 0.1$ or less), possibly indicating the presence of two different classes of eggs. When all the samples from 1969 and 1970 are pooled and tested in the same way, the result is more significantly different from normal ($p = 0.01$).

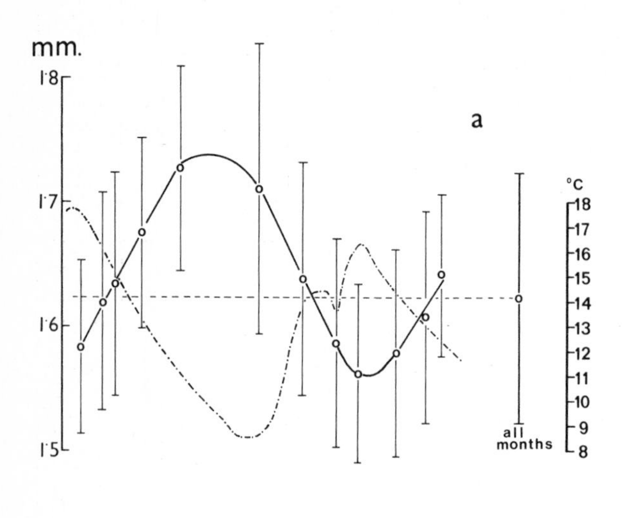

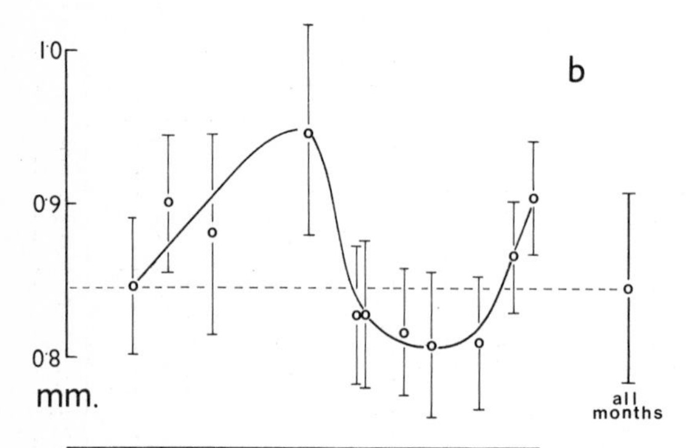

Fig. 6 (a) Seasonal changes in diameter of the egg capsule, shown as means and one standard deviation either side of the mean (circles and horizontal bars). The broken line indicates the mean of all samples pooled for the period; the broken and dotted line gives the trend in sea-surface temperature at International Station E 1, which corresponds to one of the grid stations. (b) Seasonal changes in diameter of the yolk mass. Symbols as in (a)

Table 1a. Diameter of the capsule of eggs of *Sardina pilchardus* collected off Plymouth in 1969 and 1970. The size is given as eyepiece units (20 units = 1 mm) with the means converted to mm

Date	micrometer units																		Mean (units)	Mean (mm)	Num-ber
	23	24	25	26	27	28	29	30	31	32	33	34	35	36	37	38	39	40			
25-26 VIII 1969	-	-	-	-	1	4	1	13	27	37	23	5	2	-	-	-	-	-	31.664	1.583	113
23 IX 1969	-	-	-	-	1	0	5	9	15	24	17	17	11	1	1	-	-	-	32.396	1.620	101
9-10 X 1969	-	-	-	2	6	5	5	36	60	93	132	86	57	15	3	-	-	-	32.682	1.634	500
13-14 XI 1969	-	-	-	-	-	-	1	3	8	24	40	28	25	11	2	1	-	-	33.497	1.675	143
1-2 I 1970	-	-	-	-	-	-	-	-	4	4	5	12	18	10	5	0	1	-	34.559	1.728	59
8-9 IV 1970	-	-	-	1	1	1	1	3	10	11	19	21	13	20	13	4	3	1	34.164	1.708	122
27 V 1970	-	-	-	2	0	1	0	1	6	4	4	6	2	5	1	0	0	1	32.939	1.647	33
2-3 VI 1970	-	1	1	0	0	1	3	3	13	23	27	12	12	4	3	-	-	-	32.689	1.634	103
14-15 VII 1970	1	0	1	1	1	4	13	24	49	61	34	24	7	2	-	-	-	-	31.694	1.585	222
11-12 VIII 1970	-	-	-	-	1	3	7	15	27	19	17	2	1	-	-	-	-	-	31.228	1.561	92
29-30 IX 1970	-	-	-	2	1	9	14	40	54	60	49	20	8	-	-	-	-	-	31.549	1.577	257
4-5 XI 1970	-	1	2	1	0	0	6	12	26	40	26	18	10	2	1	-	-	-	32.062	1.603	145
24-25 XI 1970	-	-	-	-	-	-	-	3	9	14	23	15	5	0	1	-	-	-	32.829	1.641	70
Total	1	2	4	9	12	28	56	162	308	414	416	266	171	70	30	5	4	2	32.452	1.623	1960

Table 1b. Diameter of the yolk mass of eggs of *Sardina pilchardus* collected off Plymouth in 1969 and 1970. The dimensions are given in eyepiece units (20 units = 1 mm), as measured across the transverse axes of eggs in stages 5 and 6 (see p. 54)

Date	micrometer units												Mean (Units)	Mean (mm)	Number
	13	14	15	16	17	18	19	20	21	22	23	24			
9-10 X 1969	-	-	1	26	60	15	1	-	-	-	-	-	16.893	0.845	103
13-15 XI 1969	-	-	-	2	31	51	23	7	-	-	-	-	18.018	0.901	114
1-2 I 1970	-	-	3	2	7	13	5	2	-	-	-	-	17.656	0.883	32
8-9 IV 1970	-	-	-	4	13	8	22	20	6	0	1	1	18.933	0.947	75
27 V 1970	-	-	5	9	14	3	1	-	-	-	-	-	16.563	0.828	32
2-3 VI 1970	-	1	7	41	41	12	-	-	-	-	-	-	16.549	0.827	102
14-15 VII 1970	-	-	30	106	65	18	1	-	-	-	-	-	16.336	0.817	220
11-12 VIII 1970	-	2	23	34	26	6	1	-	-	-	-	-	16.152	0.808	92
29-30 IX 1970	1	5	48	110	79	14	-	-	-	-	-	-	16.179	0.809	257
4-5 XI 1970	1	0	0	14	75	49	6	-	-	-	-	-	17.297	0.865	145
24-25 XI 1970	-	-	-	1	11	42	13	3	-	-	-	-	18.086	0.904	70
Total	2	8	117	349	422	231	73	32	6	0	1	1	16.915	0.846	1242

Table 1c. Diameter of the oil-globule of eggs of *Sardina pilchardus* collected at Plymouth in 1969 and 1970. The dimensions are given in eyepiece units (40 units = 1 mm)

Date	micrometer units							Mean (units)	Mean (mm)	Number
	4	5	6	7	8	9	10			
25-26 VIII 1969	1	28	21	-	-	-	-	5.400	0.135	50
23 IX 1969	-	12	52	5	-	-	-	5.899	0.147	69
9-10 X 1969	1	5	77	21	2	-	-	6.170	0.154	106
1-2 I 1970	-	-	1	20	32	3	-	7.661	0.192	56
8-9 IV 1970	-	-	4	44	20	-	-	7.235	0.181	68
27 V 1970	-	-	7	22	3	-	-	6.875	0.172	32
2-3 VI 1970	-	1	66	33	-	-	-	6.320	0.158	100
14-15 VII 1970	-	111	94	3	-	-	-	5.481	0.137	208
11-12 VIII 1970	-	39	49	3	-	-	-	5.604	0.140	91
29-30 IX 1970	9	20	206	7	4	-	-	5.907	0.148	246
24-25 XI 1970	-	2	44	9	13	-	-	6.485	0.162	68
Total	11	218	621	167	74	3	-	6.077	0.152	1094

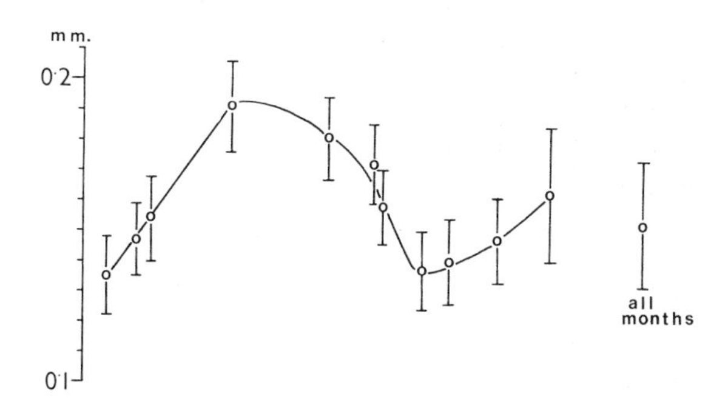

Fig. 7. Seasonal changes in diameter of the oil globule. Symbols as in Fig. 6

The seasonal trend in yolk mass size is not quite as smooth as egg diameter (cf. Fig. 6a and 6b), and several of the samples, notably those collected in April, July and November deviate significantly from normal ($p = <0.05$). Fat globule size (Fig. 7) appears to show a smoother seasonal trend, with a maximum in January and a minimum in July, but only one of the sets of samples approaches agreement with normality. This may be due more to the difficulty in measuring the small oil globule at the magnification we used rather than to any apparent inhomogeneity of the samples.

Fig. 6a compares the cycle of egg size with the trend in sea-surface temperature, and it can be seen that the largest eggs are found in the coldest months and the smallest in the warmest months. The general inverse correlation is brought out more clearly in Fig. 8, and there is a fairly high statistical significance ($r = -.76$, $p = <.01$). However, the existence of such a relationship does not necessarily imply a direct effect of temperature, or an effect of temperature alone, on egg size.

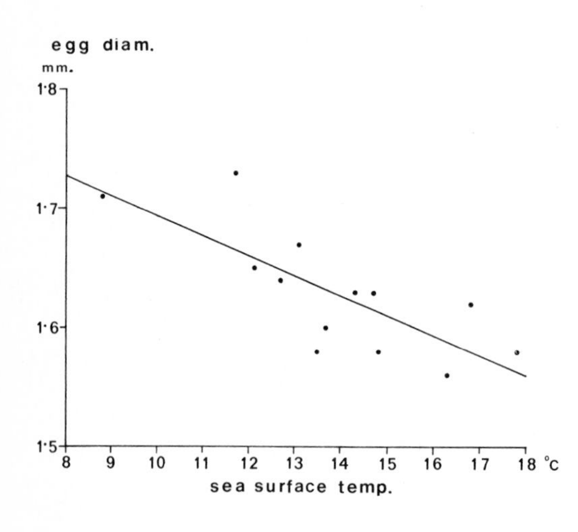

Fig. 8. Regression of egg size (capsule diameter) against mid-monthly sea surface temperature at Station E 1

DISCUSSION

Causes of Mortality. Before discussing the significance of the observed seasonal changes in mortality, we have to determine to what extent such mortality would be expected in the sea and how much is due to the methods of capture. This question is impossible to answer fully, as it is not feasible to collect naturally spawned eggs without using a net; yet, as Rollefsen (1933) has shown, fish eggs are very susceptible to mechanical damage (e.g. being dropped onto silk netting in air) especially during the early stages. This susceptibility has recently been studied in more detail by Pommeranz (this symposium). It has been suggested that mortality is increased by the use of high towing speeds, and we have therefore taken a series of samples with an identical net hauled vertically at a reduced speed of about 2 kn (a typical towing speed of 1 m/sec as used for large coarse meshed vertical nets). The samples were taken between 1800 and 0600 h in early June during a period of calm weather. In addition to pilchard eggs the samples also contained large numbers of delicate medusae and postlarval flatfish, all of which appeared to be fully alive on capture and which showed no signs of damage when examined after fixation. The mortality curve for all these vertical samples pooled is shown in Fig. 9; alongside it is the mortality for a series of high speed samples taken two years earlier, also in a period of calm weather in June. Clearly, in spite of the apparent lack of damage to other plankters, there was still considerable pilchard-egg mortality in the vertical hauls, and the figures for stage 2 are practically the same; at all the other stages however, the mortality was less in the vertical nets. A like comparison needs to be repeated under different circumstances to be relied upon, nevertheless we can tentatively suggest that mortality is still high in vertically-hauled nets, but that at least some of the mortality of the later stages in high speed tows might be related to the speed of the tow and greater velocity of impact of the eggs with the meshes. As both sets of samples were taken in very calm weather it would seem that the effect of wave action in causing egg mortality may well have been overrated in the past (see also Pommeranz).

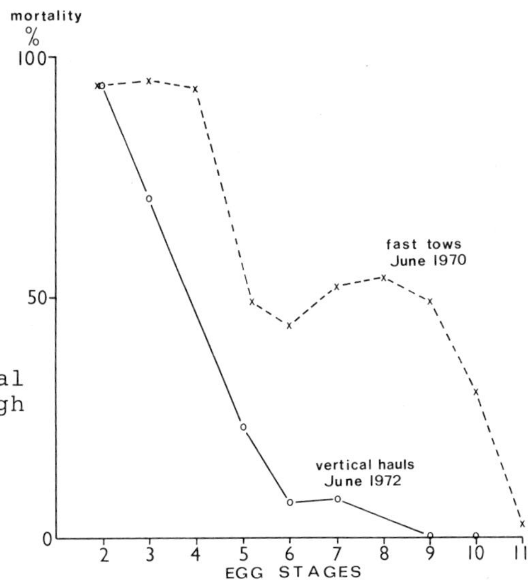

Fig. 9. Comparison of egg mortality at the different developmental stages in samples taken by the high speed method in June 1970 and by vertical tows of the same net in June 1972. Both series in very calm weather. Insufficient stage 4 eggs were collected in 1972 to allow calculation of percentage

Ignoring for the moment any seasonal variation, it is clear that the early stages of pilchard eggs are very delicate, but that after stage 5 has been passed there is a considerable improvement, coinciding with closure of the blastopore. Rollefsen (1933) has noted that an increasing resistance to mechanical damage in cod eggs corresponds to closure of the blastopore, the yolk being completely invested in embryonal tissue by stage 6. Bonnet (1939) came to a similar conclusion regarding the mortality of cod eggs maintained at different temperatures, though he noted a further increase in mortality in the late stages just before hatching. A distribution of egg mortality similar to that of the pilchard seems to occur in the Venezuelan anchovy studied by Simpson and Gonzalez (1967). These authors report a high mortality in the early stages, mostly "those that had not closed the blastopore, which occurs during stage 6". It is possible, therefore, that the extremely high early mortality of pilchard and other sardine eggs, might be related to the permeability of the egg membrane in stages I and II when the perivitelline space is enlarging and water must be entering (cf. Miller, 1952).

Seasonal Variation in Mortality. Although, as we have discussed, some of the egg mortality could be due to the method of capture, the seasonal differences in mortality are still valid. It has already been noted that there appears to be a general relationship between temperature and egg mortality, as all developmental stages show the greatest mortality in the colder part of the year, from November to May. It is possible that there is a similar general relationship to temperature in the N. Adriatic, since Gamulin and Hure (1955) found the highest mortality in March. In other parts of the Mediterranean, however, temperature would seem to be less important, and Aldebert and Tournier (1971) suggest instead a relationship to salinity, the eggs being more able to withstand decreases than increases. As was noted above, low temperatures or changes in temperature might also be expected to accentuate any natural susceptibility of the eggs to changes in salinity or mechanical damage, and hence we need look no further for an explanation of the very high mortality observed in the Western Channel in winter. The inhibiting effect of low temperature seems confirmed by the distribution data given by l'Herrou (1971), and by the experimental results of Cushing (1957). As very few eggs at all are found in winter there seems little chance of significant recruitment to the stock of fish, whether or not the overall mortality is high. However, comparatively few eggs are present in August, the warmest month, even though egg mortality is then lowest, and it would seem that June-July and September-October, when peak spawning coincides with relatively low mortality, are the only periods in which some recruitment could be expected. It is therefore difficult to explain the persistent, though low-level, spawning that is found through the rest of the year, since this could not be expected to have much effect on the genetic make-up of the population. Perhaps the winter spawners found off Plymouth are fish originating from successful winter spawning farther south in the warmer waters of the Bay of Biscay.

Egg Dimensions. The close correspondence in seasonal trends of egg diameter, yolk mass, and oil globule indicates that the cause must be sought in changes influencing the oocyte stage in the ovary, and not in an immediate response of the egg to conditions after spawning. The problem of egg size in relation to season has recently been reviewed by Bagenal (1971). In northern waters most fishes show a gradual decrease in egg diameter as the spawning season progresses. This seasonal change is often larger than differences in mean sizes found in different geographical areas of northern seas (excluding the Baltic) at any one season, and Bagenal does not believe there is any direct connection between seasonal variation in egg size and that in temperature or salinity. Nevertheless, it is well established that salinity does influence the size of fish eggs (see, for example, Hiemstra, 1962; Solemdal, 1967) and there can be considerable differences in capsule shape and size and yolk size in eggs collected from waters of different salinity (see, for example, Demir, 1968). Furthermore, the evidence we have collated (Table 2) shows that in *Sardina* the eggs tend to get smaller in the more southern and warmer water localities.

Table 2. Extent of geographical variation in size of eggs of *Sardina pilchardus*: diameter of capsule in mm

Locality	Range	Mean (winter)	Overall mean	Mean (summer)
Plymouth	1.15-2.00	1.70	-	1.56
Golfe de Gascogne[1]	1.50-1.80	-	-	-
Marseille[2]	1.33-1.79	-	1.52	-
Marseille[3]	-	1.65	-	1.32
Sète[2]	1.27-1.75	-	1.50	-
Rousillon[4]	1.38-1.67	-	-	-
N. Adriatic[5]	1.37-1.75	-	1.57	-
Golfo di Napoli[6]	1.50-1.70	-	-	-
Castellon[7]	1.31-1.68	-	1.52	-
Mer d'Alboran[8]	1.29-1.75	-	-	-
Baie d'Alger[9]	1.35-1.80	-	-	-
Black Sea[10]	2.00-2.10	-	-	-
Sea of Marmora[11]	1.20-1.85	-	1.52	-
NE Aegean Sea[11]	1.30-1.65	-	1.47	-
Medit. Turkish Coast[11]	1.30-1.65	-	1.50	-

[1]Arbault and Boutin (1968)
[2]Lee (1961)
[3]Aboussouan (1964)
[4]Ruivo and Wirz (1953)
[5]Gamulin and Hure (1955)
[6]Raffaele (1888)
[7]Andreu and Rodriguez-Roda (1951)
[8]Massuti (1955)
[9]Marinaro (1971)
[10]Vodyanitsky and Kazanova (1954)
[11]Demir (1969)

It seems possible, therefore, that the seasonal cycle of egg size in the Cornish Pilchard could be related to changes in environmental conditions, especially temperature, as already suggested by the comparison made in Figs. 6 and 8. There could be direct influence of temperature on the rate of maturation of the oocytes, so that a larger number of smaller eggs would be produced in the warmer part of the year, and a reduced number of larger eggs in the colder part of the year.

At present we cannot relate this theory to the gonad cycle. The last previous report on gonad development in the Cornish Pilchard (Hickling, 1945) covers the period 1935-1938, when there was only one peak of spawning, in June-July. The cyclic changes reported at that time would agree with the present data in leading to a smaller egg size in August, but otherwise do not fit in with the occurrence of a second peak of spawning in October-November.

In the herring, with demersal eggs, there is a similar seasonal trend in egg size so that winter spawners tend to have larger eggs and lower fecundity than summer spawners. Blaxter and Hempel (1963) have shown that this seasonal difference in size results in the production of larger larvae which can survive longer in winter, when conditions are less favourable for feeding; they suggest that this size differenc could be regarded as an adaptation to this end. At present we cannot say if there is any corresponding difference in size of pilchard larvae hatched at different seasons, and it is not clear if the summer and autumn spawning peaks are produced by the same fish or by differer groups, as in the herring. However, the number of viable pilchard eggs present in the area in winter is so small that their larger size is less likely to be of any adaptive significance.

SUMMARY

A grid of 30 stations was sampled monthly from August 1969 to November 1970, using a modified 1 m net hauled obliquely at 4 kn. The eggs were staged and their mortality and dimensions assessed.

1. The overall mortality averaged 50%, varying from 38% in August to 92% in February, most of the dead eggs belonging to stages 2, 3 and 4.

2. Some of the mortality after stage 2 may be due to the high towing speed.

3. The reduction in mortality after stage 5 appears to be connected with the closure of the blastopore, as may be the case with eggs of other species of fish.

4. The high mortality in winter appears to be related to sea temperature.

5. Any spawning outside the periods June-July and October-November, when peak numbers of eggs coincide with relatively low mortality, is unlikely to have much effect on recruitment to the fishery, or on the genetic composition of the shoals.

6. There was considerable seasonal change in capsule diameter and in diameter of the yolk mass and oil globule, apparently inversely related to sea temperature, the smallest eggs occurring in summer, the largest in winter.

7. The small summer eggs are closer in size to those found in more southern localities.

REFERENCES

Aboussouan, A., 1964. Contribution a l'étude des oeufs et larves pelagiques des poissons téléostéens dans le golfe de Marseille. Rec. Trav. St. mar. Endoume 32 (48), 87-173.

Ahlstrom, E.H., 1950. Influence of temperature on the rate of development of pilchard eggs in nature. Special Sci. Repts. U.S. Fish. Wildl. Serv. 15, 132-167. Reissue of report No. 23, originally printed in 1943.

Aldebert, Y. and Tournier, H., 1971. La reproduction de la sardine et de l'anchois dans le golfe du Lion. Rev. Trav. Inst. scient. techn. Pêch. marit. 35, 57-75.

Andreu, B. and Rodriguez-Roda, J., 1951. Estudio comparativo del ciclo sexual, engrasamiento y repleción estomacal de la sardina, alacha y anchoa del mar catalán, acompanado de relacion de pescas de huevos planctónicos de estas especies. Publ. Inst. Biol. Applicada Barcelona 9, 193-232.

Arbault, S. and Boutin, N., 1968. Oeufs et larves de poissons téléostéens dans le golfe de Gascogne en 1964. Rev. Trav. Inst. scient. techn. Pêch. marit. 32, 413-476.
Arbault, S. and Lacroix Boutin, N., 1969. Epoques et aires de ponte des poissons téléostéens du golfe de Gascogne en 1965-1966 (oeufs et larves). Rev. Trav. Inst. scient. techn. Pêch. marit. 33, 181-202.
Arbault, S. and Lacroix, N., 1971. Aires de ponte de la sardine, du sprat et de l'anchois dans le golfe de Gascogne et sur la plateau celtique. Resultats de 6 années d'étude. Rev. Trav. Inst. scient. techn. Pêch. marit. 35, 36-56.
Arbault, S., Beaudoin, J. and Lacroix, N., 1972. Zones-test dans le golfe de Gascogne en 1970. Icthyoplankton-zooplankton. Ann. Biol. (Copenh.) 27, 71-72.
Bagenal, T.B., 1971. The inter-relation of the size of fish eggs, the data of spawning and the production cycle. J. Fish. Biol. 3, 207-219.
Blaxter, J.H.S. and Hempel, G., 1963. The influence of egg size on herring larvae. J. Conseil Perm. Intern. Exploration Mer 28, 211-240.
Bonnet, D.D., 1939. Mortality of the cod egg in relation to temperature. Biol. Bull. Mar. Biol. Lab. Woods Hole 76, 428-441.
Corbin, P.G., 1947. The spawning of the mackerel, *Scomber scombrus* L., and pilchard, *Clupea pilchardus* Walbaum, in the Celtic Sea in 1937-39, with observations on the zooplankton indicator species, *Sagitta* and *Muggiaea*. J. Marine Biol. Assoc. U.K. 27, 65-132.
Corbin, P.G., 1950. Records of pilchard spawning in the English Channel. J. Marine Biol. Assoc. U.K. 29, 91-95.
Cushing, D.H., 1957. The number of pilchards in the Channel. Fish. Invest., Ser. 2, 21 (5), 1-27.
Demir, N., 1968. Analysis of local populations of the anchovy *Engraulis encrasicolus* (L.) in Turkish waters based on meristic characters. Instanbul Üniv. Fen. Fak. Mecm. B 33, 25-57.
Demir, N., 1969. The pelagic eggs and larvae of teleostean fishes in Turkish waters. I. Clupeidae. Istanbul Üniv. Fen. Fak. Mecm. B 34, 43-74.
Fives, J.M., 1970. Investigations of the plankton of the west coast of Ireland. IV. Larval and postlarval stages of fishes taken from the plankton of the west coast in surveys during the years 1958-1966. Proc. Roy. Irish Acad. B 70, 15-93.
Furnestin, J., 1944. Contribution à l'étude biologique de la sardine atlantique (*Sardina pilchardus* Walbaum). Rev. Trav. Inst. scient. techn. Pêch. marit. 13, 221-386.
Gamulin, T. and Hure, J., 1955. Contribution à la connaissance de l'ecologie de la ponte de la sardine (*Sardina pilchardus* Walb.) dans l'Adriatique. Acta Adriatica 7, (8), 1-23.
Hickling, C.F., 1945. The seasonal cycle in the Cornish Pilchard, *Sardina pilchardus* Walbaum. J. Marine Biol. Assoc. U.K. 26, 115-138.
Hiemstra, W.H., 1962. A correlation table as an aid for identifying pelagic fish eggs in plankton samples. J. Conseil Perm. Intern. Exploration Mer 27, 100-108.
Kennedy, M., Fitzmaurice, P. and Champ, T., 1973. Pelagic eggs of fishes taken on the Irish Coast. Irish Fish. Invest. (B) 8, 1-23.
Lee, J.Y., 1961. La sardine du golfe du Lion (*Sardina pilchardus sardina* Regan). Rev. Trav. Inst. scient. techn. Pêch. marit. 25, 417-511.
Lee, J.Y., Park, J.S., Tournier, H. and Aldebert, Y., 1967. Répartition des principales aires de ponte de la sardine en fonction des conditions de milieu dans le golfe du Lion. Rev. Trav. Inst. scient. techn. Pêch. marit. 31, 343-350.
Le Gall, J., 1950. Biologie des Clupéidés (le hareng excepté). Mem. Inst. scient. techn. Pêch. marit., ser. spec. 14, 1-126.

L'Herrou, R., 1971. Etude biologique de la sardine du golfe de Gascogne et du plateau celtique. Rev. Trav. Inst. scient. techn. Pêch. marit. 35, 455-473.

Marinaro, J.Y., 1971. Contribution a l'étude des oeufs et larves pelagiques de poissons mediterranées. V. Oeufs pelagiques de la Baie d'Alger. Pelagus 3 (1), 1-115.

Massuti, M.O., 1955. La ponte de la sardine (*Sardina pilchardus* Walb.) dans le dètroit de Gibraltar, la mer d'Alboran, les eaux du levant Espagnol et des Iles Baléares. Proc. Techn. Pap. gen. Fish. Council Medit. 3, 103-129.

Miller, D.J., 1952. Development through the prolarval stage of artificially fertilized eggs of the Pacific Sardine (*Sardinops caerulea*) Calif. Fish Game 38, 587-591.

Raffaele, F., 1888. Le uova galleganti e le larve dei Teleostei nel golfo di Napoli. Mitt. Zool. Stat. Neapel 8 (1), 1-84.

Rollefsen, G., 1933. The susceptibility of cod eggs to external influences. J. Conseil Perm. Intern. Exploration Mer 7, 367-373.

Ruivo, M. and Wirz, K., 1953. Biologie et ecologie de la sardine (*Sardina pilchardus* Walb.) des eaux de Banyuls. I. Observations sur la ponte en Automne-Hiver, 1951. Vie et Milieu 3, 151-189.

Simpson, J.G. and Gonzalez, G.G., 1967. Some aspects of the early life-history and environment of the sardine *Sardinella anchovia* in eastern Venezuela. Ser. Recursos y Explotacion Pesqueros 1 (2), 1-93.

Solemdal, P., 1967. The effect of salinity on buoyancy, size and development of flounder eggs. Sarsia 29, 431-442.

Southward, A.J., 1962. The distribution of some plankton animals in the English Channel and approaches. II. Surveys with the Gulf III. high-speed sampler, 1958-60. J. Marine Biol. Assoc. U.K. 42, 275-375.

Southward, A.J., 1963. The distribution of some plankton animals in the English Channel and approaches. III. Theories about long-term biological changes, including fish. J. Marine Biol. Assoc. U.K. 43, 1-29.

Southward, A.J., 1970. Improved methods of sampling post-larval young fish and macroplankton. J. Marine Biol. Assoc. U.K. 50, 689-712.

Southward, A.J. and Demir, N., 1972. The abundance and distribution of eggs and larvae of some teleost fishes off Plymouth in 1969 and 1970. I. Methods and hydrography. J. Marine Biol. Assoc. U.K. 52, 787-796.

Vodyanitsky, V.A. and Kazanova, I.I., 1954. Opredelitel pelagicheskii lichinok ryb Chernogomorya. Trudy vses. naucho-issled. Inst. morsk ryb Khoz. 28, 240-323.

A.J. Southward
Marine Biological Association
The Laboratory, Citadel Hill
Plymouth, PL 1 2 PB / GREAT BRITAIN

N. Demir
Department of Zoology (II)
University of Istanbul
Vezneciler - Istanbul / TURKEY

Environmental Influences on the Survival of North Sea Cod

R. R. Dickson, J. G. Pope, and M. J. Holden

INTRODUCTION

The problem of relating indices of recruitment to the physical environment of a species is, as Larkin (1970, p. 9) has pointed out, that "virtually any set of stock-recruit data is sufficiently variable to inspire untestable hypotheses about the effects of trends in environments, especially with the wealth of meteorological and oceanographic data that can be mined for real and fortuitous correlations". Consequently, any observed relationship between recruitment and the environment should be examined critically, and any conclusions reached should be considered as tentative, at least until they can be tested against predicted results.

In general terms, however, it has been shown rather clearly in the literature that the year-class strength of cod appears to show a positive relationship to temperature at the poleward limits of its range (Hermann, 1953; Hermann et al., 1965; Elizarov, 1965; Kislyakov, 1961), and a negative relationship at its equatorward range limits (Martin and Kohler, 1965; Dickson and Lamb, 1972).

In keeping with this generalization, the recent cooling tendency observed in the North Sea (Dickson and Lamb, 1972) has been accompanied by an increase in the abundance of cod which, in this area, is tending towards its southern limit of range. During the period 1950-1963, the total international landings of cod from the whole of the North Sea averaged 94 138 metric tons (whole weight) per annum, with a minimum of 65 440 metric tons in 1951 and a maximum of 115 647 metric tons in 1959 (ICES Bulletins Statistiques des Pêches Maritimes, 1952-1960). Since 1963 annual landings have risen, almost without interruption, to 320 031 metric tons in 1971 which represents the highest recorded landing from the North Sea. The average annual landing over the period 1964-1971 was 247 882 metric tons. Holden (1972) has shown that this increase in landings has resulted almost entirely from an increase in recruitment, with increased fishing intensity playing only a minor part. Geographically, the greatest increase in abundance has occurred to the south of the Dogger Bank (latitude 54°N approximately). In this southern area, recruitment has also been observed to be more variable than in the northern part of the North Sea, with year-class size (estimated from the numbers of 2 year-old cod) varying by a factor of 5.5 for the former area and 2.8 for the latter, in the case of the year classes between 1961 and 1968 (Holden and Flatman, 1972).

Within the North Sea, tagging returns have shown the existence of several populations of cod. These stocks are not geographically isolated and their boundaries may overlap or may change seasonally with migration, so that a given area may be occupied by different stocks at different times of year. Spawning occurs from the beginning of January until April and takes place on several relatively distinct spawning grounds (Fig. 1); it is not yet possible to link individual stocks with spawning grounds.

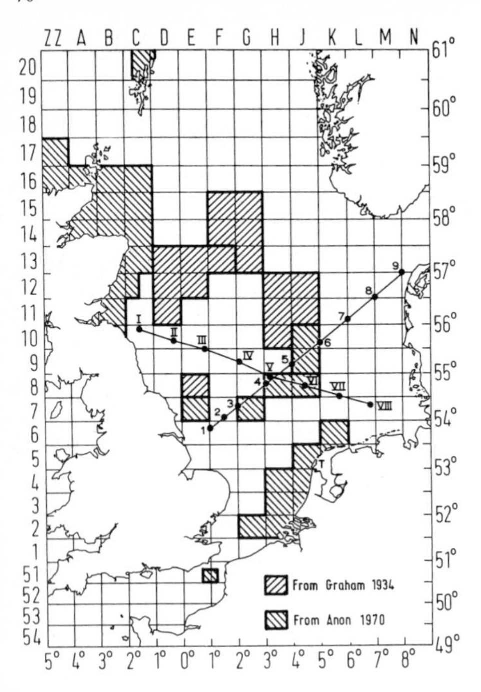

Fig. 1. Main spawning grounds of cod in the North Sea based on Graham (1934) and Anon (1970) showing routes on which surface temperatures were taken

METHODS

In analysing the environmental influences on the survival of North Sea cod, a number of possible indices of recruitment and an almost embarrassing wealth of hydrographic and meteorological data are available. In this case, the indices of recruitment chosen were the catches of 2 year-old cod per 100 h effort in the Grimsby trawl and seine fishery (data from Holden, 1972, 1973), since these two fleets have shown only minor changes in fishing power over the period considered here. The annual distribution of catches from the two fleets in 1968 is shown in Figs. 2 and 3 together with heavy black lines representing suggested discriminants between the stocks (ICES North Sea Roundfish Working Group, 1973 Meeting), and from this it can be seen that neither fleet fishes entirely within the boundaries of any one cod stock. Table 1 lists the available indices of recruitment for the two vessel types over the period 1954-1968, and it is apparent that the two are closely related, with a correlation coefficient of 0.86. The differences which do occur are attributed partly to the differing proportions of the various sub-stocks that are exploited by the two fleets, and partly to random errors.

The temperature data used here were chosen to cover the principal spawning grounds of the central North Sea from Flamborough Head to the Great Fisher Bank; they also cover the principal areas from which the Grimsby fleets derive their catch, see Fig. 4.

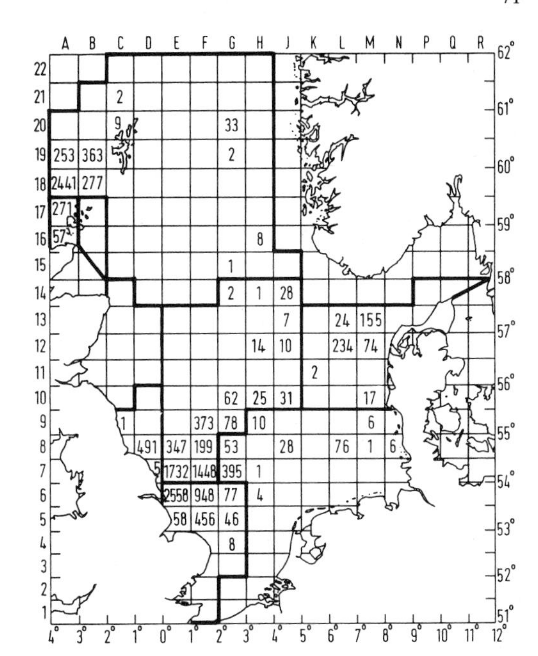

Fig. 2. Distribution of cod catches by Grimsby trawlers from the North Sea in 1968; discriminants between stocks shown by heavy lines

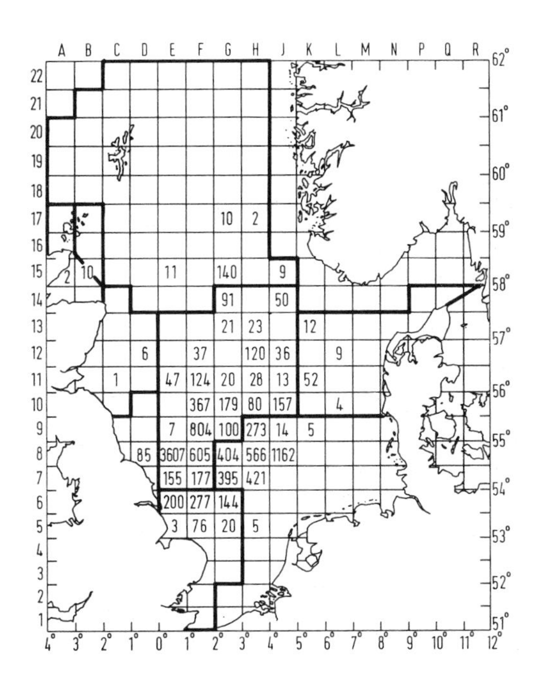

Fig. 3. Distribution of cod catches by Grimsby Danish seiners from the North Sea in 1968; discriminants between stocks shown by heavy lines

Table 1. Relative year-class strength

Year Class	Number of 2 year-old cod caught per 100 h fishing	
	Grimsby seiners	Grimsby trawlers
1954	75	376
1955	151	1160
1956	55	316
1957	198	794
1958	287	1347
1959	217	298
1960	194	299
1961	213	349
1962	327	508
1963	1741	1860
1964	963	1703
1965	548	1756
1966	1219	2626
1967	187	617
1968	223	659

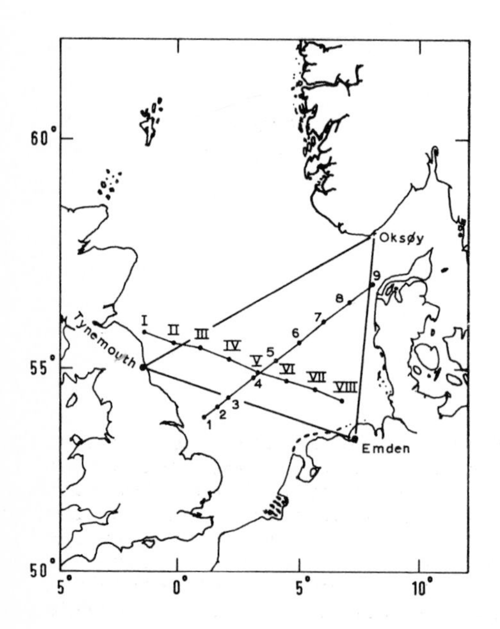

Fig. 4. Sources of data referred to in the text. Sea temperature data were obtained from the Hull-Hanstholm route (stations 1-9) and the Leith-Bremen route (stations I-VIII) and from a single station near Texel at 53°01'N 4°22'E. Wind data were computed from pressure gradients in the triangle Oksöy-Emden-Tynemouth (Schott, 1970)

RESULTS

Fig. 5 (a and b) shows the relationship between the estimates of cod year-class strength (as measured by the subsequent catch of 2 year-olds per 100 h effort) and surface temperature for the 9 stations along the Hull-Hanstholm route in each month of the year. At each station the 1954-1968 time series of temperature for each month has been compared with the two series of year-class strength estimates (trawl and seine) over the same period to produce two matrices of correlation coefficients. The resulting 108 correlation coefficients in each matrix have then been contoured. A significant year-class strength/temperature relationship is apparently shown in those station/months with a high negative correlation coefficient. On average these are highest in January, at the time of first spawning, and in April-May during the period of larval drift, but are unexpectedly low in February and March. In June, the relationship between year-class strength and temperature breaks down completely, and with the exception of a slight reappearance of higher negative correlation coefficients in October at the north-eastern end of the section, it appears that no year-class strength/temperature relation-

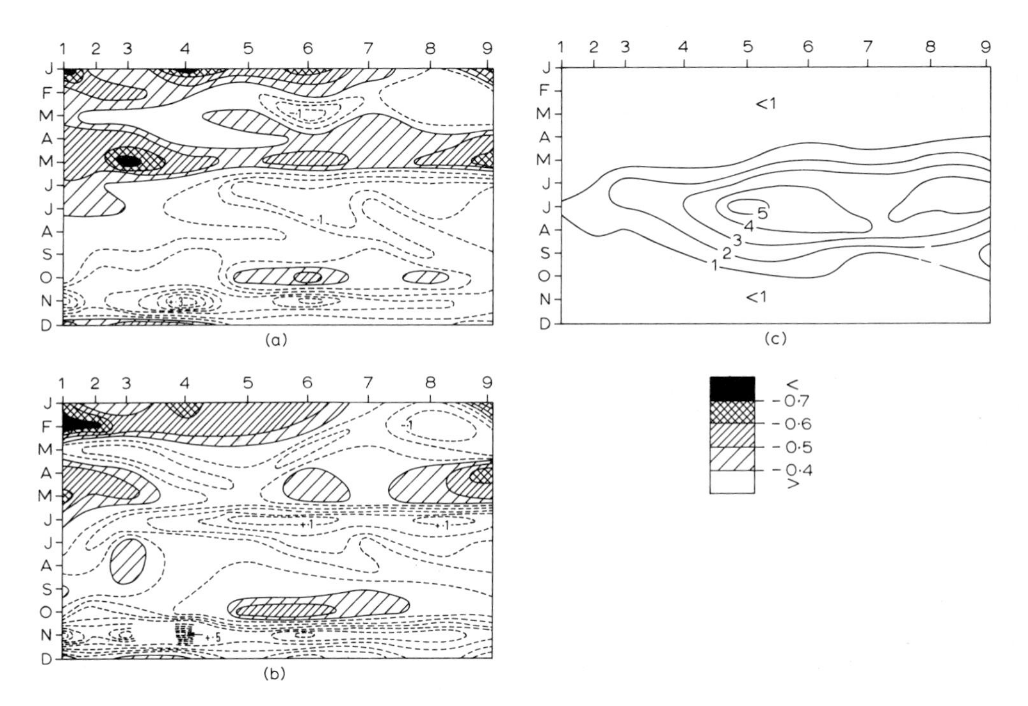

Fig. 5a-c. The relationship of cod year-class strength to temperature (Hull-Hanstholm section). In (a) and (b) the 1954-1968 time series of surface temperature for each month at each station has been correlated with the time-series of cod year-class strengths derived from the Grimsby trawl and seine catch respectively. The resulting correlation coefficients have then been contoured. In (c) the average timing and intensity of the seasonal thermocline on the Hull-Hanstholm section is represented by the average temperature difference between 7.5 m and 30 m depth at each station, in each month. (Data from Tomczak and Goedecke, 1962)

ship exists in the latter half of the year. Fig. 5c shows the average temperature difference (in ^{O}C) between 7.5 m and 30 m depth at each station on the Hull-Hanstholm section (data from Tomczak and Goedecke, 1962) and thus illustrates the average occurrence, intensity, and geographical distribution of the seasonal thermocline. Comparing Figs. 5a and 5c it is clear that the breakdown of the year-class strength/temperature relationship in June is associated with the establishment of the thermocline. Even in detail, the least significant correlation coefficients are found at those points on the section where the thermocline is most intense. Thus at first sight it would appear that along the whole length of the Hull-Hanstholm section, the link between temperature and year-class strength becomes established in the months of January, April, and May, during the period of early spawning and larval drift, and breaks down in the surface layers in June with the establishment of the thermocline. The fact that the vast majority of correlation coefficients are negative is in itself an indication that the relationships between recruitment and surface temperature are real and not merely fortuitous correlations resulting from the examination of 108 separate series.

Fig. 6 (a, b and c) provides identical data for the Leith-Bremen section. In this case, temperature data are only available for the

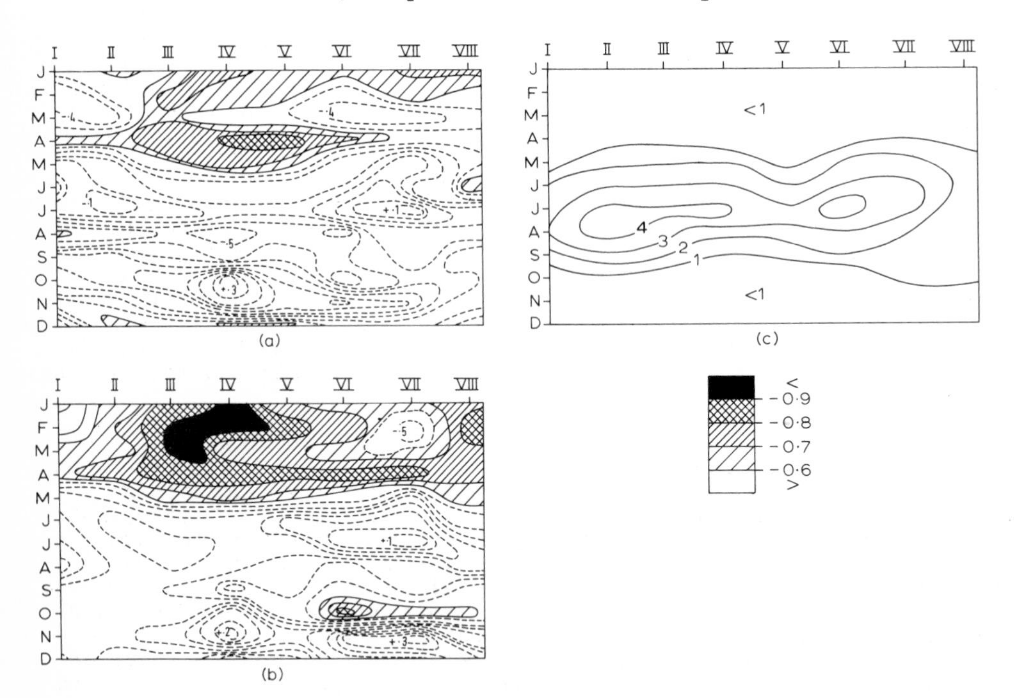

Fig. 6a-c. The relationship of cod year-class strength to temperature (Leith-Bremen section). In (a) and (b) the 1960-1968 time series of surface temperature for each month at each station has been correlated with the time series of cod year-class strengths derived from the Grimsby trawl and seine catch, respectively. The resulting correlation coefficients have then been contoured. In (c) the average timing and intensity of the seasonal thermocline on the Leith-Bremen section is represented by the average temperature difference between 7.5 m and 30 m depth at each station, in each month. (Data from Tomczak and Goedecke, 1962)

shorter period 1960-1968 so that, although correlation coefficients appear to be higher than on the Hull-Hanstholm section, they have a lower significance. The Leith-Bremen data show an essentially similar situation to that just described; high negative correlation coefficients are found in January, but these decrease greatly in February and March. The year-class strength/temperature relationship increases in significance again during April and May (mainly in the central part of the section between stations III and VI where the Hull-Hanstholm and Leith-Bremen sections cross) and breaks down rapidly in June with the establishment of the seasonal thermocline. Again the link between temperature and year-class strength is minimal where the intensity of the thermocline is strongest. The fact that the correlation coefficients in April/May are significant along the whole length of the Hull-Hanstholm section, but only along the central portion of the Leith-Bremen section, provides some slight indication that the main effect of temperature on year-class strength occurs within an elliptical area of the central North Sea with its major axis trending NE-SW, though the dangers of relying on this indication are clear in view of our lack of comparable data from other North Sea sections.

Further grounds for caution are inherent also in the fact that only sea surface temperatures were used in the year-class strength-temperature correlations. Although on both sections the relationship between the two parameters breaks down in the surface layers in June with the establishment of the thermocline, it is entirely possible that temperatures below the thermocline will remain relatively unchanged compared with those in the surface layers. Thus if a proportion of young cod occupy the near-bottom layer during the summer it is possible for a year-class strength/temperature relationship to persist undetected throughout the summer below the thermocline. We have some indication that this occurs in the weak reappearance of higher negative correlation coefficients in October on the Hull-Hanstholm section, when the thermocline breaks down. It is not thought likely, however, that a strong year-class strength/temperature relationship can persist in the deeper layers much beyond August, since at the extreme south-western end of the Hull-Hanstholm section, where the thermocline is never strongly established, the high negative correlation coefficients disappear at this time (Fig. 5).

To examine the relationships just described more critically, a two-way analysis of variance was performed on normalised transformations of the correlation coefficients. "Station averages" and "month averages" of correlation coefficients were calculated for the first 6 months of the year only, covering the period in which the most significant correlations had been found. In the case of the Hull-Hanstholm seine data (Table 2a) this analysis showed that the average correlation coefficient for each month considered, except June, was significantly greater than zero and that the average correlation coefficients for January, February, April and May differed significantly from those of March and June. Similar analyses were performed on the Leith-Bremen seine correlation coefficients (Table 2b) and the Hull-Hanstholm and Leith-Bremen trawl data (Tables 3a and 3b). All these analyses showed significant differences in the average monthly correlation coefficients and in all cases the averages are significantly greater than zero. The two trawl analyses (Tables 3a and 3b) also show a significantly lower regression in March compared with January and April. For all four analyses the various station averages proved to differ significantly, although these inter-station differences are

Table 2. Analyses of variance of normalized correlation coefficients

(a) Hull-Hanstholm (stations-months) with year-class strength (Seine)

Cause	Degrees of Freedom	Sum of Squares	Mean Squares	F. Ratios
Months	5	1.794	0.3588	10.622
Stations	8	0.795	0.0993	2.940
Residuals	40	1.352	0.0338	
TOTAL	53	3.940		

(b) Leith-Bremen (stations-months) with year-class strength (Seine)

Cause	Degrees of Freedom	Sum of Squares	Mean Squares	F. Ratios
Months	5	5.057	1.0114	27.6238
Stations	7	2.201	0.3144	8.5889
Residuals	35	1.281	0.0366	
TOTAL	47	8.539		

less than the inter-month differences just described. For the Hull-Hanstholm section seine data, stations 1 and 2 had significantly higher correlations than stations 3 to 8, while stations 1 and 3 had significantly higher correlations when the trawl data were considered. Amongst the stations of the Leith-Bremen section, stations III, IV, V and VIII had significantly higher correlations with both trawl and seine data than did other stations on the section.

Table 3. Analyses of variance of normalized correlation coefficients

(a) Hull-Hanstholm (stations-months) with year-class strength (Trawl)

Cause	Degrees of Freedom	Sum of Squares	Mean Squares	F. Ratios
Months	5	1.3218	0.2644	14.61
Stations	8	0.3494	0.0437	2.41
Residuals	40	0.7241	0.0181	
TOTAL	53	2.3953		

(b) Leith-Bremen (stations-months) with year-class strength (Trawl)

Cause	Degrees of Freedom	Sum of Squares	Mean Squares	F. Ratios
Months	5	1.526	0.3050	10.336
Stations	7	0.723	0.1030	3.50
Residuals	35	1.033	0.0295	
TOTAL	47	3.282		

In summary, there is strong evidence that a general relationship exists between the spring sea-surface temperature and the subsequent recruitment of that year-class of cod throughout the central North Sea; this relationship seems in general to be most consistently high in January and April and to a lesser extent in May also. The relationship appears to be strongest at the southwestern end of the Hull-Hanstholm section, and at the central and extreme eastern parts of the Leith-Bremen section. A simple correlation between surface temperatures off Texel (53° 01'N 4° 22'E) and the two recruitment estimates show that to some extent the relationship may also be found in this area of the Southern Bight (Table 4).

Table 4. Correlation coefficients for regressions between the number of 2 year-old cod per 100 h effort in the Grimsby Trawl and Seine fishery, and the sea-surface temperature off Texel (at 53°01'N, 04° 22'E) two years previously

Type	Jan	Feb	Mar	Apr	May	June
Grimsby Trawl	-.42	-.23	-.73	-.44	-.24	-.13
Grimsby Seine	-.75	-.55	-.51	-.53	-.35	-.11

In this case, the months of maximum correlation are not the same for the two categories of fishing effort, but the slight indication that the maximum correlation occurs earlier than in the central North Sea is perhaps in keeping with Daan's finding that cod spawning occurs earliest in the Southern Bight of the North Sea (Daan, 1973).

DISCUSSION

Thus far, the relationship between temperature and cod year-class strength has been examined empirically with no discussion as to the possible mechanisms involved. Two possible temperature-operated mechanisms have recently been suggested in the literature. Alderdice and Forrester (1971) have shown that for the Pacific cod, *Gadus macrocephalus*, there is an optimum temperature range of 3-5°C for egg hatching within which variations in salinity and oxygen concentration have little effect over a wide range of values. Within this optimum temperature range the cod larvae at hatching are larger than those incubated outside this range. During those years in which the temperature range is at or near the optimum, a greater proportion of eggs will hatch and the larvae will be large, giving a greater chance of survival. The metabolism of a poikilothermic species is temperature-related and, at higher temperatures, this necessitates a higher food intake than at lower temperatures. The total food intake of a population of cod larvae will be higher at high temperatures than low and the chance of density-dependent mortality, resulting from the population consuming its food supply faster than it is produced, is therefore greater. On the other hand, at very low temperatures the larvae will become so inactive that they cannot catch and consume food, and so die. It is not suggested that this latter situation ever exists in the North Sea, but it does appear to occur off West Greenland.
A second possible mechanism has been suggested by Cushing (1969, 1972), i.e. that although the spawning of many fish species is relatively fixed in time, variations in their larval survival and subsequent recruitment are controlled by inter-annual variations in the match or mismatch between the production of larvae and the production of

their food. He also suggests that the larval and food production cycles are varied in time by different and largely unconnected factors, with larval production varying according to the temperature effect on egg and larval development, and with food production being varied by changes in windspeed and direction (and hence in solar radiation and in the depth of the wind-mixed layer).

In order to examine Cushing's theory more closely, the analysis described above was extended to include an examination of the relationship between the speed and direction of the surface wind and the seine and trawl indices of recruitment for the North Sea cod.

The wind data used were those computed by Schott (1970) from the air pressure gradients in the triangle Oksöy-Emden-Tynemouth (Fig. 4) over the period 1954-1967. The correlation coefficients of the regression of this data with the seine and trawl data are shown in Tables 5 and 6. Several of these correlation coefficients are nominally significant at the 5% level, but in the context of 18 data series two significant results are not particularly surprising and it is probably safer to regard the relationship between the wind data and the trawl and seine data as being unproven at present. Nevertheless, the levels of correlation coefficients in April, May, and June are suggestive of some meaningful relationship.

Table 5. Correlation coefficients for regressions between year-class strength (Seine) and Schott's components of wind strength in the central North Sea

Type	Jan	Feb	Mar	Apr	May	June
North Component[a]	+.23	+.02	-.22	-.57	-.42	-.40
East Component[b]	+.46	+.28	+.03	+.64	+.04	+.48
Velocity	+.06	-.09	+.17	+.14	-.14	-.45

[a]A northerly component is recorded as positive.
[b]An easterly component is recorded as positive.

Table 6. Correlation coefficients for regressions between year-class strength (Trawl) and Schott's components of wind strength in the central North Sea

Type	Jan	Feb	Mar	Apr	May	June
North Component[a]	+.07	+.17	+.08	-.30	-.55	-.42
East Component[b]	+.44	+.19	-.11	+.58	-.10	+.64
Velocity	+.08	-.12	+.05	+.11	-.01	-.58

[a]A northerly component is recorded as positive.
[b]An easterly component is recorded as positive.

The two temperature-operated mechanisms just described are not mutually exclusive. Since North Sea cod lie towards the southern limits of the species range, a cooling of the marine climate may well (as Alderdice and Forrester, 1971 suggest) represent a shift towards an optimum temperature range for the survival of cod in the egg and larval stages, leading to a greater percentage hatching of eggs, larger larvae at hatching, and a larval metabolic rate that is more in keeping with the local rate of food production.

If this cooling also induced a shift in the timing of larval production, which tended towards a greater match between the optimum larval food requirement and the local production of food, then this influence on survival would be additional to those just described; it is possible that the abundant North Sea year classes associated with cold winters arise because all these conditions for survival are optimized under these extreme climatic situations. Generally stated, the more variable recruitment to the southern North Sea cod population compared with the northern is perhaps a reflection of the more extreme variability of temperature in the shallower southern area.

SUMMARY

We concluded that sea temperature is one of the main factors which has a direct relationship with year-class size of North Sea cod and that the increase in the average size of cod from 1963 onwards has resulted from sea temperatures in the first half of the year falling to a level which has optimized conditions for both hatching and larval survival.

1. Average annual landings of cod from the North Sea more than doubled between 1950-1963 and 1964-1971. Over this period there has been a general fall in sea temperatures in this area.

2. The relationship between year-class size of cod, based on two indices of abundance, and monthly sea temperatures was examined over two sea-surface temperature sections. Negative correlations, significantly greater than zero, were obtained for the first five months of the year in all cases with, in general, highest correlation coefficients in January and April.

3. With the available data there was no provable relationship between year-class size of cod and either wind strength or direction.

4. It is suggested that low sea temperatures during both the time of spawning and larval drift act by maximizing the proportion of eggs which hatch, by enhancing the chance of a match between hatching and production, and by optimizing the metabolic range of the species.

REFERENCES

Alderdice, D.F. and Forrester, C.R., 1971. Effects of salinity, temperature and dissolved oxygen on the early development of the Pacific cod (*Gadus macrocephalus*). J. Fish. Res. Bd. Can. 28, 883-902.

Anon, 1970. Interim report of the North Sea Cod Working Group ICES Doc. CM 1970/F 15, 6 pp., figs. (mimeo).

Cushing, D.H., 1969. The regularity of the spawning season of some fishes. J. Cons. perm. int. Explor. Mer 33, 81-92.
Cushing, D.H., 1972. The production cycle and the numbers of marine fish. Symp. Zool. Soc. Lond. (29), 213-232.
Daan, N., 1973. Dutch cod egg and larvae surveys in 1971. Ann. Biol. (Copenh.) 28, 98-99.
Dickson, R.R. and Lamb, H.H., 1972. A review of recent hydrometeorological events in the North Atlantic sector. Spec. Publs. int. Commn NW. Atlant. Fish. (8), 35-62.
Elizarov, A.A., 1965. Long-term variations of oceanographic conditions and stocks of cod observed in the areas of West Greenland, Labrador and Newfoundland. Spec. Publs. int. Commn NW. Atlant. Fish. (6), 827-831.
Graham, M., 1934. Report on the North Sea cod. Fish. Invest. (Lond.), Ser. II, 13 (4), 160 pp.
Hermann, F., 1953. Influence of temperature on strength of cod year classes. Ann. Biol. (Copenh.) 9, 31-32.
Hermann, F., Hansen, P.M. and Horsted, Sv.A., 1965. The effect of temperature and currents on the distribution and survival of cod larvae at West Greenland. Spec. Publs. int. Commn NW. Atlant. Fish. (6), 389-395.
Holden, M.J., 1972. Variations in year-class strengths of cod in the English North Sea fisheries from 1954 to 1967 and their relation to sea temperature. Ann. Biol. (Copenh.) 27, 86.
Holden, M.J., 1973. Year-class strengths of cod in the English North Sea fisheries in 1971. Ann. Biol. (Copenh.) 28, 100.
Holden, M.J. and Flatman, S., 1972. An assessment of North Sea cod stocks using virtual population analysis. ICES Doc. CM 1972/F 21, 4 pp., tables, fig. (mimeo).
Kislyakov, A.G., 1961. The relationship between hydrological conditions and variations of cod year-class abundance. Trudy Soveshch. ikhtiol. Kom. 13, 260-264.
Larkin, P.A., 1970. Some observations on models of stock and recruitment relationships for fishes. ICES Symposium on Stock and Recruitment. Doc. No. 17, 14 pp. (mimeo).
Martin, W.R. and Kohler, A.C., 1965. Variation in recruitment of cod (*Gadus morhua* L.) in southern ICNAF waters, as related to environmental changes. Spec. Publs int. Commn NW. Atlant. Fish. (6), 833-846.
Schott, F., 1970. Monthly mean winds over sea areas around Britain during 1950-1967. ICES Service Hydrographique, Charlottenlund Slot, Denmark, 17 pp. (mimeo).
Tomczak, G. and Goedecke, E., 1962. Monatskarten der Temperatur der Nordsee, dargestellt für verschiedene Tiefenhorizonte. Ergänz. Hft. Dt. Hydrograph. Z., Reiche B (4°), Nr. 7, 16 pp., charts.

R.R. Dickson
Fisheries Laboratory
Lowestoft / GREAT BRITAIN

J.G. Pope
Fisheries Laboratory
Lowestoft / GREAT BRITAIN

M.J. Holden
Fisheries Laboratory
Lowestoft / GREAT BRITAIN

Some Factors Influencing Early Survival and Abundance of *Clupeonella* in the Sea of Azov

G. N. Pinus

INTRODUCTION

Tiulka (*Clupeonella delicatula delicatula* N.) is the most abundant fish of the Sea of Azov, with catches amounting to 40-50% of the total fish landed from this sea. These fish fatten and winter in the Sea of Azov and spawn mainly in the Bay of Taganrog.

To find out the causes of fluctuations in abundance the environmental conditions for reproduction from 1950-1960 were analyzed. The material was collected in the Bay of Taganrog and partly in the Sea of Azov from April to July during complex standard cruises of the AzcherNIRO ships, exploratory cruises of the Black and Azov Seas Fishing Reconnaissance Department (AzCherPromrazvedka) and by ships of the Zdanov regional inspection of the Azov Fish Cultural Section (Azovrybvod).

METHODS

Ichthyoplankton samples were taken with large nets (80 cm-diameter mouth) towed for 10 min with a speed of 0.18-0.20 m/sec, which allowed good survival of embryos. Ichthyoplankton sampling was accompanied by water temperature observations, salinity and dissolved oxygen determinations, sea condition recordings, and zooplankton sampling. During the investigations 486 ichthyoplankton samples were processed and 231 salinity and 147 dissolved oxygen determinations were made. Of the 5243 larvae which were measured, 1129 were examined for food.

Experiments were also conducted to observe the development of eggs in natural conditions, to determine the survival rate of embryos at optimal temperatures and to study their response to light. The materials used for the experiments were eggs and milt obtained from live spawners caught with pound nets. The influence of salinity, sea state, light intensity, oxygen, and temperature on survival of eggs was assessed by measuring the mortality of eggs in natural situations where one factor, e.g. salinity, varied and the other factors were optimal. For this purpose both naturally and experimentally spawned eggs were used. Most of the materials collected were processed in the laboratory. This processing consisted mainly of determining the total number of eggs in the samples, their stages of development, and counting living and dead eggs. Live and dead tiulka eggs were distinguished by Dementieva's (1958) method for the eggs of the Azov anchovy, with an additional subdivision of the dead eggs into two groups. The first group included the eggs with an opaque chorion and yolk sacs, and embryos appearing as diffuse clots at the initial stage and more developed but deformed at later stages. The second group consisted of semi-decayed eggs, with stages of development impossible to determine. Due to their large oil globules dead tiulka eggs usually float on the surface, where they decay.

RESULTS

Salinity. While investigators generally believe that salinity has a pronounced effect on the strength of tiulka year classes (Monastyrsky, 1952; Kostyuchenko, 1955; Karpevich, 1960) there is no agreement on the optimum salinities for reproduction. Kostyuchenko (1955), Maisky (1960) and Karpevich (1960) considered that mass spawning is successful in water with salinities up to 9°/oo. According to Logvinovich (1955) eggs and larvae at early stages in experimental conditions develop in a regular way and mortality is not above normal in water with salinities up to 13.1°/oo. Smirnov's data (1969) suggest salinities from 0.5 to 5°/oo to be most favourable for tiulka larvae.

To find out the effect of salinity on the survival of tiulka eggs, situations were selected when the water temperature was at an optimum (15-18°C), sea state was 1 to 2, while salinity varied. At any one salinity the lowest survival was observed among eggs at earlier stages and the highest among those at the final stage of development. The highest number of eggs developed in water with salinity below 9°/oo. At a salinity of 9-10°/oo there was a sharp decrease in the number of surviving eggs. This confirms the opinion of those investigators who suggest that the spawning of tiulka is successful in water with salinity below 9°/oo.

The comparison of average salinities in the western, central, and eastern parts of the Bay of Taganrog and the abundance of underyearling tiulka in 1952-1960 did not reveal any relationship (the correlation coefficients were -0.15, -0.21 and -0.18, respectively). Apparently, salinities observed in the Bay of Taganrog during the period under investigation did not produce a decisive effect on the strength of recruitment to the tiulka stock.

Sea State. The mechanical action of waves adversely affects the survival of eggs of most species (Rollefsen, 1930; Molander, 1949; Pavlovskaya, 1955, 1958; Grudinin, 1961, and others). This is also true of tiulka eggs when wave disturbance increases the mortality of embryos. The first and second stages are most sensitive to wave disturbance. Only 15% of eggs at the first stage of development were found to survive at sea state 4. With calmer seas at sea state 1-3 the number of surviving first- and second-stage eggs increased to 32-48%. At later development stages the number of live eggs increased considerably, with all stage 4 embryos surviving at sea state 0 to 4.

The comparison of data on the abundance of underyearling tiulka with wind data for the Bay of Taganrog in 1950-1960 revealed no agreement between the fluctuations in the abundance and the variations in wind force during the spawning period. For example, the lowest number of underyearlings (73) was observed in 1959 when wind conditions were favourable. A relatively low abundance of underyearlings in 1950-1952, 1957, 1958, and 1960 was observed at different wind conditions in May. The correlation coefficient between the number of days with wind force 3-4 and the abundance of underyearlings for the period of 1950-1960 is -0.22 ± 0.08, which is not significant. It can be said, therefore, that variations in sea state observed in the Bay of Taganrog seem to be of minor importance and do not determine the results of spawning.

Light Intensity. There are no data on the effect of light on the development of tiulka eggs. For some species, however, this effect is quite appreciable. Our observations on the survival of tiulka eggs were conducted at different degrees of cloudiness, other environmental conditions (temperature, sea state, salinity) being favourable. At the gastrula stage (the first stage of development) and at later stages the number of developing eggs was not found to depend on light intensity. To check this, experiments were performed on the development of tiulka eggs in light and in the dark. Second- and fourth-stage eggs were caught in the Bay of Taganrog at an optimal temperature (17°C). The experiments showed that the survival of these eggs did not depend on light conditions. The data obtained during the 24 h station indicated that the maximum spawning took place around midnight. This suggests that bright light does not contribute to better survival of eggs at the initial stage of development. First-stage eggs develop at night; survival at later stages does not depend on variations in light intensity. In general, it seems that light does not limit the abundance of tiulka.

Oxygen. Dissolved oxygen content plays an important role in the development of planktonic organisms, including pelagic fish eggs. Pelagic eggs are usually free of any respiratory pigment and under conditions of low oxygen they die earlier than demersal eggs (Kryzanovsky et al., 1953). Tiulka eggs differ from other fish eggs in the presence of rather large amethyst-coloured crystals of pigment in their yolk sacs. According to Kryzanovsky (1956) the biological importance of this pigment for tiulka is unknown, though it is believed to contribute to better assimilation of oxygen by the eggs.

The analysis of our data on the content of dissolved oxygen in the surface layer of the Bay of Taganrog during the spawning period revealed that the development of eggs was regular and the mortality not above normal. The lowest dissolved oxygen content observed, 6 ml/l (80% saturation), was not found to affect mortality of tiulka embryos. Variations in the dissolved oxygen content in the surface waters of the Bay of Taganrog over 11 years (1950-1960) were not found to be related to the strength of tiulka year classes (correlation coefficient $+ 0.19 \pm 0.44$), which seems to indicate that the oxygen regime in the Bay of Taganrog during the reproduction period is usually favourable.

Temperature. Many investigators believe temperature to be the main factor determining the survival of pelagic eggs of fishes (Revina, 1958; Pavlovskaya, 1958; Dementieva, 1958, and others). Temperature plays an important role in the reproduction of tiulka. Our studies revealed temperature-dependent changes in the duration of development stages. This varied from 92 h at an average temperature of 11.3°C to 21 h at the temperature of 23°C, i.e. a more than fourfold difference.

The effect of temperature on the survival of tiulka embryos is also great, as shown from Table 1 based on data collected at optimum salinities (1 to 7°/oo) and at sea state 2 or less and with temperatures varying between 6 and 20°C.

It is seen from the table that survival varied according to the stage of development and to water temperature. Spawning began at the temperature of 6 to 8°C but at these temperatures only few eggs survived to stage 4. The highest mortality at different water temperatures was recorded at early development stages and the lowest was observed in eggs just about to hatch. This conclusion is corroborated by experimental results.

Table 1. Percentage survival of eggs at different temperatures

Development stage	Beginning of spawning		Peak of spawning		End of spawning		
	6-8°C :	11-13°C	13-15°C :	15-18°C	18-20°C :	16-18°C :	18-20°C
First	15	23	37	48	37	32	31
Second	19	31	44	82	66	36	36
Third	34	75	73	98	83	63	63
Fourth	41	71	81	100	100	100	82
Number of samples	12	6	9	34	15	16	21

Water temperature (column group heading)

Optimum conditions for the survival of tiulka eggs are found when the temperature reaches 15-18°C. It is at these temperatures that peak spawning occurs. A correlation was found between the number of days in May with a temperature of 15-18°C and the number of underyearlings, which showed that in years when sharp water temperature gradients were observed in spring (1952 and 1960), or when the Bay of Taganrog warmed up rapidly (1955, 1958 and 1961), and the water temperature of 15-18°C in May continued for about 10 days the abundance of underyearlings was low. In other years (1950, 1953, 1954, 1956, 1962 and 1963), when the optimum temperature continued for 18 to 22 days, the abundance was high.

There is a distinct relationship between the temperature conditions during peak spawning and the strength of the year class. When there were fewer days with the temperature favourable for the development of eggs a lower abundance of underyearlings resulted. The correlation between the number of days in May with a temperature of 15 to 18°C and the abundance of underyearlings for the period of 1950 to 1963 is + 0.89 $\pm$ 0.15, which is highly significant.

Feeding. Having considered the effect of abiotic factors on the embryonic development of tiulka we now turn to the analysis of feeding conditions of larvae which are believed to be of great importance. In papers dealing with predictions of the productivity of the Sea of Azov following the regulation of the flow of the Don it was believed that a salinization of the Bay of Taganrog might result in the disappearance of some freshwater zooplankton forms. Bokova (1955) believed that a higher salinity of the Bay water would lead to the disappearance of rotifers, which would affect the feeding of tiulka larvae and result in a decline of the stock.

A detailed analysis of the feeding conditions of larvae showed that the composition of food varied in different areas of the Bay, those planktonic organisms which occurred in higher quantities predominating in the larval diet. The biomass of rotifers was low and they were not part of the tiulka diet, their place being taken by the copepodite stages of copepods; there was no decline in the intensity of feeding. The investigations showed that in May the relative amount of food (copepods) was fairly high and, as a rule, was not found to be a factor limiting the abundance of year classes.

SUMMARY

The comprehensive study of the conditions for reproduction of the Azov tiulka revealed some regularities in abundance fluctuations. The consideration of factors which can affect the strength of tiulka year classes, directly or indirectly, showed the critical importance of temperature conditions during the period of embryonic development. In years with optimum temperatures (15 to 18°C) continuing for only about 10 days poor year classes result, while in those years with optimum temperatures continuing for an average of 20 days a sharp increase in the abundance of recruits is observed. Hence, the number of days with the temperature of 15 to 18°C in May, with proper consideration for the hydrological regime and the characteristics of the development of zooplankton in the Bay of Taganrog, may be used as a rough indication of the strength of tiulka year classes.

REFERENCES

Bokova, E.N., 1955. Food availability of juvenile Azov tiulka under conditions of the regulated discharge. Voprosy Ikhtiol. 4, 137-151.

Dementyeva, T.F., 1958. Methods of studying the relationship between environmental factors and the abundance of the Azov khamsa. Trudy VNIRO 34, 30-62.

Grudinin, P.I., 1961. The effect of food supply on the survival of the Azov khamsa larvae. Proc. of Conference on fish population dynamics, 13, 454-456. Izdatelstvo Akademii Nauk SSSR.

Karpevich, A.F., 1960. The effect of varying river discharge and the regime of the Sea of Azov on its commercial fauna. Trudy AzNIIRKH, 1, 3-113.

Kostyuchenko, R.A., 1955. The change in the stock of the Azov tiulka (*Clupeonella delicatula delicatula* N.) following the regulation of river discharge. Trudy VNIRO, 31, 188-195.

Kryzhanovsky, S.G., 1956. Material on development of clupeids. Tr. Instituta Morfol. Zhivotn. 17.

Kryzhanovsky, S.G., Disler, N.N. and Smirnova, E.I., 1953. Ecologo-morphological regularities of development of percids. Tr. Instituta Morfol. Zhivotn. 10, 3-138.

Logvinovich, D.N., 1955. On some factors governing the abundance of the juvenile Azov tiulka. Trudy AzCherNIRO, 16, 241-251.

Maisky, V.N., 1960. The state of stock of gobiids, khamsa and tiulka from the Sea of Azov in the period from 1931 to 1958. Trudy AzNIIRKH 1, 381-412.

Molander, 1949. Sprat and milieu conditions. Ann. Biol. (Copenh.) 1, 165-174.

Monastiyrsky, G.N., 1952. Dynamics of commercial fish populations. Trudy VNIRO, 21, 3-162.

Pavlovskaya, R.M., 1955. Early survival of Black Sea khamsa. Trudy AcCherNIRO 16, 99-120.

Pavlovskaya, R.M., 1958. Some aspects of the Black Sea khamsa biology reproduction and development related to the problems of population dynamics. Trudy AzCherNIRO 17, 75-109.

Revina, N.I., 1958. A contribution to reproduction and survival of eggs and juveniles of "large sized" horse mackerel from the Black Sea. Trudy AzCherNIRO 17, 9-30.

Rollefsen, G., 1930. Observation on the cod eggs. P.-v Réun. Cons. perm. int. Explor. Mer 65, 31-35.

Smirnov, A.N., 1969. The effect of the ecological factors on the spawning results of some fish species in the Azov Sea Bay of Taganrog. Voprosy Ikhtiol. 9, 651-656.

G.N. Pinus
All Union Research Institute for Marine Fisheries and Oceanograph
(VNIRO)
Moscow B - 140 / USSR

Nutrition de la larve de turbot [*Scophthalmus maximus* (L.)] avant la métamorphose[1]

M. Girin

INTRODUCTION

Un pas important dans l'élevage du Turbot a été franchi en 1972 avec les premières métamorphoses de larves obtenues par Jones (1972b), Alderson et Howell (Comm. pers.) et Girin (1972). Il n'a malheureusement pas été possible de cerner avec précision les facteurs de ces succès. En particulier, ni le régime alimentaire, ni les conditions de vie offerts durant la métamorphose n'ont semblé déterminants.

Ceci nous a incité à tenter d'analyser plus finement les relations entre la survie et le régime alimentaire proposé dès le début de la nutrition. En effet, si la métamorphose représente une crise dans la vie de la larve, elle a vraisemblablement d'autant plus de chances d'être surmontée que celle-ci a pu se développer auparavant dans de meilleures conditions. Il ne serait donc pas surprenant que des régimes fournissant des pourcentages de survie élevés et une bonne croissance pendant le développement symétrique se traduisent par des pourcentages élevés de métamorphoses.

Sur le plan qualitatif, Jones (1972a) a montré que l'adjonction de plancton naturel (nauplii de Copépodes) au Rotifère *Brachionus plicatilis* O.F. Müller, habituellement employé comme première nourriture, améliorait sensiblement les pourcentages de survie au début de la vie larvaire. Nous avons cherché à aborder l'aspect quantitatif de ce problème sur un régime simple composé seulement de *Brachionus* et de nauplii du Crustacé Branchiopode *Artemia salina* L.

METHODES

Les oeufs ont été fournis, exceptionnellement tard dans la saison (5 juillet), par une femelle sauvage pêchée en état de maturité, et fécondés artificiellement. L'incubation a débuté en bac cylindrique de 30 l, en eau stagnante brassée par une forte aération, à 19°C, près du lieu de pêche. Les oeufs fécondés ont été transportés jusqu'au laboratoire à la fin de la gastrulation, dans des bocaux non aérés où la température est montée accidentellement à 21°C. Devenus pratiquement tous benthiques à l'arrivée, ils ont été partagés entre un aquarium légèrement aéré, un panier d'incubation en circuit ouvert, et un bac cylindrique fortement aéré identique au premier, de nouveau à 19°C. Les oeufs placés dans l'aquarium ont tous avorté.

La veille de l'éclosion, 12 000 oeufs ont été transférés dans 7 aquariums parallélépipèdiques transparents ("altuglas") de 60 l, contenant

[1]Contribution n° 172 du Département Scientifique du Centre Océanologique de Bretagne.

chacun 35 l d'eau de mer à 21°C, filtrée sur un filtre à sable. Le tableau 1 présente leur schéma de répartition et le nombre de larves nées, les oeufs avortés ayant été enlevés 24 heures après le début des éclosions.

Tableau 1. Répartition des oeufs et nombre de larves vivantes 24 heures après le début des éclosions

N° de l'aquarium	Incubation achevée en:	Nombre d'oeufs	Larves nées	Pourcentage d'éclosions
1	panier	1 000	974	97%
2	panier	2 000	1 700	85%
3	bac	2 000	1 802	90%
4	bac	2 000	1 460	73%
5	bac	2 000	1 746	87%
6	bac	2 000	1 740	87%
7	panier	1 000	509	51%

Les figures 1a et 1b montrent les régimes alimentaires employés. 48 heures après le début des éclosions, 40 000 *Brachionus* ont été placés dans l'aquarium 6, 120 000 dans les aquariums 1, 2, 3, 5 et 7, et 200 000 dans l'aquarium 4; puis les volumes ont été complétés à 40 litres, afin de mettre à la disposition des larves respectivement 1, 3 et 5 Rotifères par ml. Les proies consommées furent remplacées quotidiennement jusqu'au 11ème jour selon le même régime dans tous les lots, sauf dans le lot 3 où le transfert des *Brachionus* aux nauplii d'*Artemia* a été fait plus rapidement. L'offre en nauplii d'*Artemia* a été ensuite adaptée au nombre de larves survivantes et à la consommation (les très faibles quantités offertes le 13ème jour sont dues à des problèmes d'incubation d'oeufs d'*Artemia* de mauvaise qualité).

La dose quotidienne de nourriture a été diluée dans 2 l d'eau filtrée du 3ème au 12ème jour inclus, ce qui a progressivement porté les volumes des élevages à 60 l. A partir du 13ème jour, 6 l (10%) ont été renouvelés chaque jour; sauf dans les aquariums 4 à partir du 17ème jour, et 1 à partir du 18ème jour, où le renouvellement a été porté à 50% devant une mortalité exceptionnelle paraissant due à une accumulation de déchets dans le milieu.

La température a été maintenue à 21°C pendant toute la durée de l'expérience, à un demi-degré près.

Les larves mortes ont été dénombrées et éliminées chaque jour, tandis que les déchets étaient siphonnés.

Les courbes de survie journalières (fig. 1c) permettent de tester l'influence sur les larves de quatre paramètres différents: le mode d'incubation (tous aquariums avant la nutrition, puis aquariums 2 et 5), la concentration des proies au début de l'alimentation (aquariums 4, 5 et 6), la quantité de proies offerte à chaque larve pendant les deux premières semaines de la nutrition (aquariums 1, 2 et 7) et la période de passage d'un régime de *Brachionus*, à un régime de nauplii d'*Artemia* (aquariums 3 et 5).

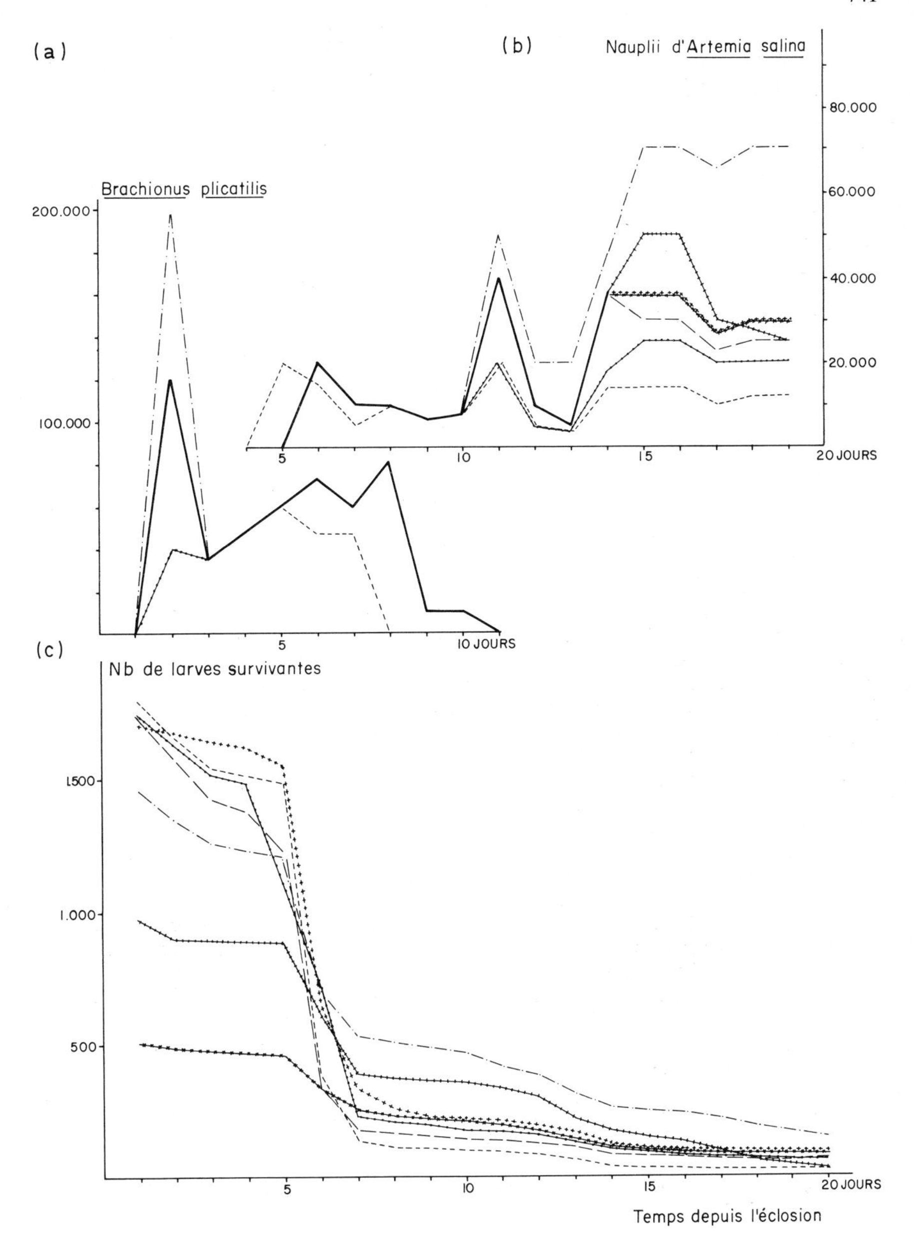

Fig. 1a-c. Ration et survie journalières globales. Aquarium n° 1 ; n° 2 ++++; n° 3 ----; n° 4 -·-·-·; n° 5 ; n° 6 ———; n° 7 *****. (a) Ration de *Brachionus plicatilis*. (b) Ration de nauplii d'*Artemia salina*. (c) Survie

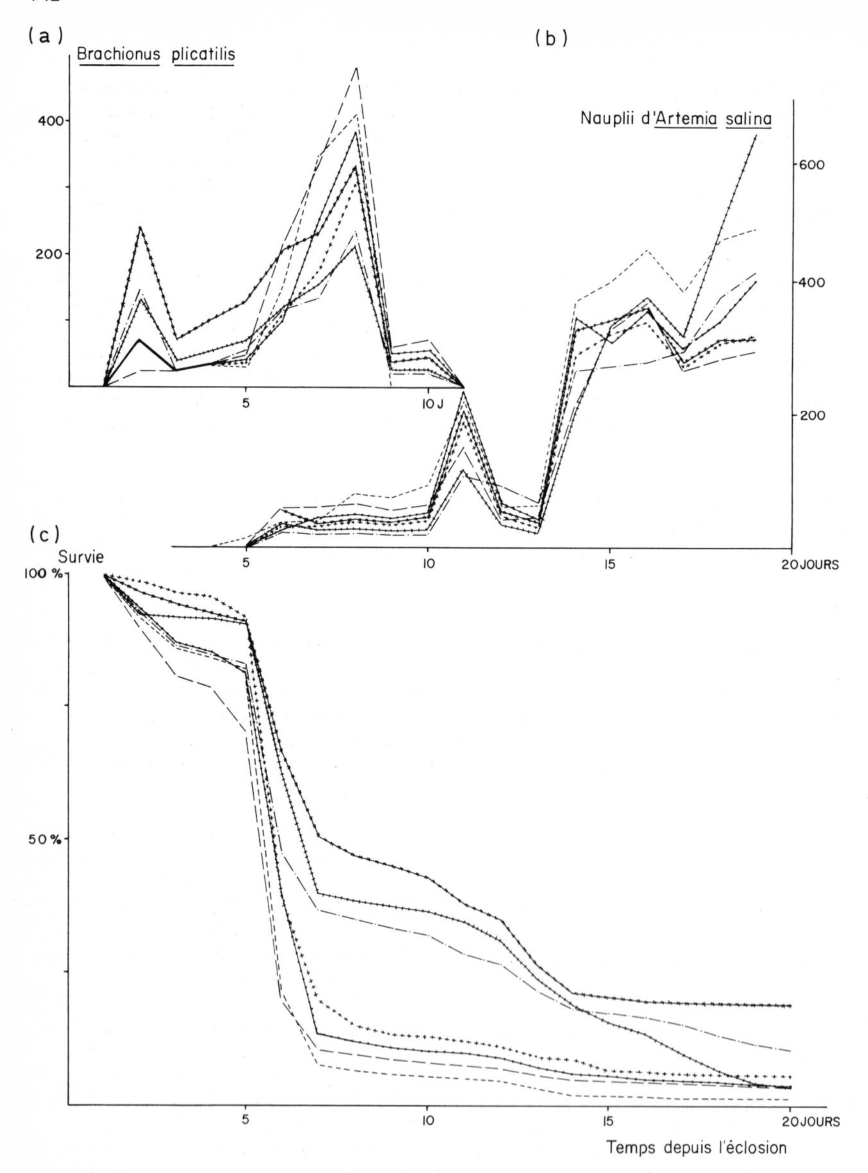

Fig. 2a-c. Ration journalière individuelle et pourcentages de survie. La ration est exprimée par rapport au nombre de larves vivantes au moment de sa distribution; la survie, en pourcentage du nombre de larves vivantes 24 heures après le début des éclosions. Aquarium n° 1 ┼┼┼┼┼; n° 2 ++++; n° 3 ----; n° 4 -·-·-; n° 5 —•—•—; n° 6 — — —; n° 7 ✶✶✶✶✶. (a) Ration individuelle de *Brachionus plicatilis*. (b) Ration individuelle de nauplii d'*Artemia salina*. (c) Pourcentages de survie

RESULTATS

Mode d'Incubation. Il n'est pas possible de tirer des pourcentages d'éclosion (tableau 1) une conclusion sur l'efficacité relative des deux types d'incubation employés avec succès: les chiffres obtenus après l'incubation en panier sont très étalés. Ceci est vraisemblablement dû au fait qu'on a mis en suspension un lot d'oeufs morts ou mourants sédimentés sur la toile à plancton en faisant le prélèvement destiné à l'aquarium 7.

Par contre, les courbes de survie exprimées en pourcentages du nombre des larves vivantes 24 h après l'éclosion (fig. 2c) se séparent nettement entre le 3ème et le 5ème jour, selon que l'incubation a été achevée en bac ou en panier: la mortalité au cinquième jour dépasse toujours 15% dans le premier cas, alors qu'elle n'atteint jamais 10% dans le second. Dans l'aquarium 6, le manque de nourriture intervient déjà à cette date, tandis que l'abondance de proies dans l'aquarium 4 n'améliore pas la survie.

Le mode d'incubation peut donc avoir une influence sur la survie de la larve durant les premiers jours. Il ne nous est pas possible pour l'instant de préciser si cela est dû au fait que le brassage est réalisé de manière plus douce dans le panier, ou au fait que l'eau y est renouvelée.

Au-delà du cinquième jour, les populations des aquariums 2 et 5, qui reçoivent la même quantité de nourriture alors qu'elles ne diffèrent que par le mode d'incubation, évoluent de façon voisine. L'écart lié à l'incubation ne s'amplifie légèrement que tout à fait à la fin de l'expérience, à un moment où la plus ou moins grande accumulation de déchets non éliminés dans l'eau joue certainement plus que l'origine des larves.

Concentration des *Brachionus* au Début de l'Alimentation. Le huitième jour, alors que la crise du début de l'alimentation est achevée, le pourcentage de survie est nettement plus élevé dans l'aquarium 4 (35%) que dans les aquariums 5 (12%) et 6 (9.5%). Leurs populations, comparables en nombre et de même origine, ont été mises en présence, respectivement, de 5, 3 et 1 *Brachionus* par millilitre.

Les survies peu différentes obtenues pour les concentrations de 3 et 1 *Brachionus* par millilitre indiquent que la gamme choisie est sensiblement trop basse. C'est ce que confirment les observations de Jones (1972a), qui obtient en moyenne environ 55% de survie au 10ème jour, à 16°C, avec une concentration de 10 *Brachionus* par millilitre. Ce chiffre n'est d'ailleurs certainement pas une limite, et il serait intéressant d'expérimenter sur le turbot des concentrations de 10 à 20 Rotifères pas millilitre, comme Theilacker and McMaster (1971) l'ont fait pour l'anchois *Engraulis mordax*.

Quantité de *Brachionus* Offerts à Chaque Larve. La comparaison des aquariums 7, 1 et 2, contenant respectivement à l'origine 509, 974 et 1 700 larves, montre que le nombre de proies disponibles chaque jour par larve est dans notre expérience tout aussi important que la concentration des Rotifères au début de la nutrition: les larves survivantes de ces aquariums ont eu chacune à leur disposition (fig. 2a), du 2ème au 7ème jour, un total de 997, 568 et 450 *Brachionus*; ceci se traduit le huitième jour par 47%, 38% et 15% de survie respectivement, alors que la concentration des proies le 2ème jour était la même partout.

Cela nous conduit à admettre que les capacités d'absorption de la larve sont nettement supérieures à ce que nous avions supposé. La nourriture offerte à partir du 3ème jour ne suffit alors plus à compenser la consommation. Ainsi, les 1 460 larves de l'aquarium 4 avaient à l'origine plus de chances de trouver une proie que les 509 de l'aquarium 7 (5 *Brachionus* par ml contre 3 *Brachionus* par ml). Mais entre le 2ème et le 7ème jour, les survivants n'ont eu au total à leur disposition que 499 Rotifères contre 997, d'où une moins bonne survie le huitième jour (35% dans un cas contre 47% dans l'autre).

Il semble donc que, outre l'usage de concentrations supérieures à 5 proies par ml à la première distribution, l'emploi de régimes de plus de 1 000 *Brachionus* par larve vivante entre le 2ème et le 7ème jour permettrait d'aborder le changement de proie avec des taux de mortalité nettement réduits.

Changement de Proie. Les aquariums 3 et 5, contenant des nombres voisins de larves de même origine, ont reçu les mêmes quantités de *Brachionus* au départ, mais le transfert du régime de Rotifères à celui de nauplii d'*Artemia* a été fait plus tôt et plus rapidement dans le premier (fig. 1a et b): 5ème au 7ème jour contre 6ème au 10ème jour. Cela se traduit le 12ème jour par 4.6% de survie seulement dans le premier cas, et 8.2% dans le second. Les pourcentages sont faibles, mais la différence est quand même sensible: la seconde méthode est nettement préférable.

En fait, nous avions prévu d'interrompre les distributions de *Brachionus* dès le 7ème jour pour l'aquarium 3, dès le 9ème pour tous les autres, et calculé nos élevages de Rotifères en conséquence. Or, manifestement, très peu de larves de 5, et même 6 jours se sont montrées capables d'attraper et d'ingérer des nauplii d'*Artemia* ce qui nous a déterminé à prolonger les distributions de *Brachionus*.

Ainsi, à moins que les améliorations proposées pour le régime de la première semaine n'améliorent la croissance, et par là les possibilités de prédation des larves, il semble bon d'offrir jusqu'au 9ème jour une bonne quantité de *Brachionus*, et de réduire les rations les 10ème et 11ème jour, pour les supprimer le 12ème seulement, tandis que les *Artemia* ne seraient proposés qu'à partir du 7ème jour. Entre 16 et 17°C, Jones effectue le changement de proie entre le 10ème et le 14ème jour.

Quantité d'*Artemia* Offerts à Chaque Larve. En dehors du n° 3, tous les aquariums ont reçu à partir du 5ème jour la même quantité de nauplii d'*Artemia* (fig. 1a et b), puis les doses ont été modulées à partir du 10ème jour en fonction du nombre de larves survivantes. En effet, à ce stade, le manque de proies (fig. 1b) provoquait un début d'augmentation de la mortalité dans les aquariums 4 et 1, qui contenaient les populations les plus importantes (fig. 1c). Les quantités offertes ont été réduites le 12ème jour car une forte proportion des nauplii n'avait pas été consommée. Cette réduction était peut être excessive car très peu de nauplii non consommés restaient dans les aquariums 24 h après. Il n'a malheureusement pas été possible d'en fournir alors suffisamment.

Ce fait a certainement aggravé la mortalité qui a touché tous les aquariums les 12ème et 13ème jour: il s'agissait de larves de petite taille ne disposant plus de *Brachionus*, et n'arrivant vraisemblablement pas à disputer aux larves mieux développées les rares *Artemia* présentes.

A partir du 14ème jour, les *Artemia* ont été fournies largement en excès, et la mortalité a baissé, sauf dans les aquariums les plus chargés (4 et 1). Elle n'a été réduite dans ces derniers que par l'augmentation du taux de renouvellement du milieu, expérimentée d'abord sur le n° 4. Dans le même sens, il faut d'ailleurs noter une absence quasi totale de mortalité les 21ème et 22ème jour, lorsque toutes les larves survivantes auront été transvasées dans des aquariums propres.

La ration quotidienne d'*Artemia* a donc été très irrégulière: excessive le 11ème jour (234 nauplii par larve dans l'aquarium 5), insuffisante les 12ème et surtout 13ème jour (33 et 21 nauplii par larve dans l'aquarium 1), et de nouveau excessive ensuite, au moins dans certains aquariums (plus de 600 nauplii par larve le 19ème jour dans l'aquarium 1). Il nous semble que, pour une offre totale moindre, une progression régulière avec des chiffres de l'ordre de 40 nauplii par larve le 8ème jour, 60 le 10ème, 90 le 12ème, 130 le 14ème, 180 le 16ème et 240 le 18ème, devrait fournir une survie nettement meilleure.

CONCLUSION

Cette expérience, telle qu'elle a été réalisée, ne cherchait pas à apporter des conclusions précises sur les besoins alimentaires de la larve de turbot avant la métamorphose, mais à fournir les données qui éviteraient de lancer une expérimentation fine sur des bases inadéquates. Globalement, le meilleur résultat (19% de survie au vingtième jour) a été obtenu en établissant une concentration de 30 *Brachionus* par litre d'eau le deuxième jour, puis en offrant en moyenne, par larve vivante et par jour, 158 Rotifères du 2ème au 10ème jour, et 175 nauplii d'*Artemia* du 6ème au 19ème jour, avec une concentration au départ de 13 larves par litre d'eau.

Le détail de l'interprétation montre que ce résultat a été obtenu en accumulant dans l'alimentation un nombre non négligeable d'erreurs. De ce fait, avec un régime total mieux équilibré, mais pas nécessairement plus important, la survie pourrait être notablement augmentée. Le respect des chiffres proposés dans l'analyse détaillée de l'expérience devrait, vraisemblablement, fournir une survie au moins égale à 50% le vingtième jour.

Avec des résultats de cet ordre, il deviendra alors possible de disposer des nombres importants de jeunes larves sans lesquels le suivi de la croissance pondérale ne peut avoir que fort peu de valeur, et tester ainsi plus finement la valeur des régimes employés.

A ce moment seulement, avec des animeux qui abordent la métamorphose en bon état physiologique, nous pourrons déterminer si cette phase représente réellement une crise dans la vie du poisson, ou bien si, comme pour la plie ou la sole, il s'agit d'un phénomène que toute larve est normalement apte à supporter.

RESUME

L'expérience décrite porte sur 7 lots de larves maintenues à 21°C et nourries avec le Rotifère *Brachionus plicatilis* O.F. Müller, puis des nauplii du Crustacé Branchiopode *Artemia salina* L., dans des bassins parallélépipédiques transparents ("altuglas"). Elle s'étend de la naissance au vingtième jour, veille du début de la phase critique de métamorphose à cette température. Les larves provenaient d'une ponte provoquée, dont les oeufs avaient été incubés en panier ou en bac à forte aération. Leur concentration variait de 15 à 50 par litre enviro[n] le lendemain de l'éclosion. Les *Brachionus* ont été distribués pendant 7 à 9 jours, et les *Artemia* pendant 13 à 15 jours.

Le meilleur échantillon a fourni près de 20% de survie le 20ème jour. Ce résultat a été obtenu en établissant une concentration de 3 *Brachionus* par millilitre d'eau le deuxième jour, et en offrant en moyenne par larve vivante et par jour, environ 160 Rotifères du 2ème au 9ème jour, et 175 nauplii d'*Artemia* du 6ème au 19ème jour.

Un schéma alimentaire est proposé, qui devrait fournir un pourcentage de survie, à la veille de la métamorphose, au moins égal à 50%.

SUMMARY

Seven batches of turbot larvae were reared in 60 l translucent "Perspex" tanks at 21°C for 20 days and fed first with the rotifer *Brachionus plicatilis* and then with nauplii of the crustacean *Artemia salina*. Eggs were obtained through artificial spawning and incubated in hatching "baskets" or strongly aerated circular tanks. The day after hatching the larval concentration ranged from 15 to 50/l. *Brachionus* was offered over 7 to 9 days, and *Artemia* over 13 to 15 days.

The best result, about 20% survival at the 20th day, was obtained with 3 *Brachionus*/ml offered on the 2nd day, an average daily ration of 160 *Brachionus*/ larva from 2nd to 9th day, and 175 *Artemia* nauplii/larva from 6th to 19th day.

Suggestions are given for better feeding which should ensure at least a 50% survival before metamorphosis.

REFERENCES

Girin, M., 1972. Métamorphose en élevage de deux larves de turbot (*Scophthalmus maximus* (L.)). C.R. Acad. Sc. Paris, 275, 2933-2936.

Jones, A., 1972a. Studies on egg development and larval rearing of turbot, *Scophthalmus maximus* L., and brill, *Scophthalmus rhombus* L., in the laboratory. J. mar. biol. Ass. U.K., 52, 965-986.

Jones, A., 1972b. Rearing larvae of the turbot (*Scophthalmus maximus* L.) to metamorphosis. ICES, Doc C.M. 1972/E, 32.

Theilacker, G.H. and McMaster, M.F., 1971. Mass culture of the rotifer *Brachionus plicatilis* and its evaluation as a food for larval anchovies. Mar. Biol., 10, 183-188.

M. Girin
Centre Océanologique de Bretagne
B.P. 337
29273 Brest / FRANCE

Rearing of Gilt-Head *Sparus aurata*

F. René

ABSTRACT

The levels of external factors (temperature, photoperiod, and feeding) inducing maturation in *Sparus aurata* were ascertained and more than 40 immature fish were captured and taken through to sexual maturity. Hydration of the eggs, then spawning, were induced in females by injection of gonadotropin at a rate of 800 i.u/kg live weight. The eggs released were fertilized by naturally spawning males in the spawning tank.

The eggs, hitherto unknown, show very precise characters during development, which lasts 48 h at 18°C. At hatching, the larvae are 2.5 mm long, mainly motionless, and float near the surface, ventral-side up. The mouth is not open; the eggs are unpigmented. A day later the larvae spread into the body of the water and adopt a vertical, almost immobile position. After two days the eggs become pigmented, the mouth can be opened and the larvae become predatory.

Experiments were conducted to define the type and quantity of prey species required in the rearing medium:

3- 7 days after hatching: *Fabrea salina* or young *Brachionus plicatilis*; 5-10/ml

8-25 days after hatching: *Brachionus plicatilis*; 1/ml

25-70 days after hatching: *Artemia salina* nauplii copepodites and young adult *Eurytemora velox*; 2-5/ml

After about 70 days the larvae show morphological and ecological characteristics similar to the adult and can be fed on dry particles.

F. René
Station de Biologie et Lagunaire
Quai de Bosc prolongé
34-Sète / FRANCE

Mass Rearing of the Bass *Dicentrarchus labrax* L.

G. Barnabé

INTRODUCTION

During a study of sea bass *Dicentrarchus labrax* L. undertaken at the Station Biologique de Sète in 1969 (Barnabé, 1972), an initial artificial-rearing experiment gave a larval survival of 17 days, eggs and sperm being collected from dead trawl-caught fish (Barnabé and Tournamille, 1972). During January 1971 spawning was induced by gonadotropic hormone injections on 3 females maintained in ponds. The larval survival in this instance did not exceed 13 days (Barnabé, 1971).

Following these results new experimental work started in December 1971, at first in the laboratory and then on a broader scale, the object being to induce spawning and to raise the progeny to fingerlings. This experiment was planned to determine the most difficult stages and attempt mass rearing. We partly succeeded during the winter of 1971-1972 for by June 1972, 12 000 sea bass had been reared to a length of 5 cm. In the autumn of 1972 a new 450 m^2 "building" was set up by the University on the basis of this accomplishment.

METHODS

Spawners maintained in ponds became ripe but sometimes had unviable gametes; hormone injections were therefore used on two batches of fishes. The first one contained 27 individuals caught by angling just before or during the spawning season. The second consisted of 47 animals which had been caught one year previously. In the first batch 80-90% of the eggs were viable; in the second only 20-30%. The weight of the smallest ripe spawners was 675 g for the females and 435 g for the males. Batches of 4 to 5 females were injected at the same time with carp pituitary extract at the mean dose of 3mg/kg of fish. They were then put in 2.5 m^3 of water with at least two naturally-ripe males. After 24 to 48 h an abdominal dilation appeared, first spawning occurring between 48 to 60 h after injection. This period depended on the water temperature and on the stage of ripeness of the spawners. The males fertilized the eggs normally.

The eggs floated when the salinity was over 34.5^{o}/oo and the temperature 15°C. Under these conditions dead eggs sank. A 1 kg female spawned up to 200 000 eggs. Ovulation was simultaneous for all oocytes and the gonadosomatic index could reach 30. In fact, not all the eggs were spawned, so the theoretical spawning capacity was higher. The time during which the eggs could be fertilized was very short and half of the females did not spawn, or spawned overripe eggs.

Artificial insemination was also used. The state of oocyte maturation was ascertained by catheter and at the appropriate time the eggs were stripped. Naturally-spawned eggs were collected with a net at the surface and transferred to incubation tanks. These tanks were partly filled with an algal culture, so-called "green water". Such cultures were grown in large tanks of 10-50 m^3 capacity. To start the culture an inoculum (natural-green filtered water from a lagoon) of one tenth of the volume was added. Sea water with 100 g/m^3 nutrients was used to fill the tank. Within 8 days a density of 5×10^5 - 50×10^6 cells/ml was reached when the temperature was 20°C. The tanks were kept under continual daylight illumination during the day with artificial lighting from "grolux" tubes (300 lux at the surface) by night and strong aeration kept the medium mixed.

During incubation of eggs there was constant light with no aeration, so the water was motionless; the parameters of the incubation medium were:

Salinity:	35°/oo
Temperature:	13-16°C
Dissolved O_2:	5-9 ml/l
pH:	7.5 - 8.5
Nitrites:	<1 mg/l

Small amounts of nitrates, phosphates, and sulphates did not have any significant effect. If the mean weight of an egg with ovarian fluid is 1 mg, and the hatching rate is about 95%, the larval density can be predicted from the total weight of eggs when the eggs are put in the hatching tank. With 120 g of eggs in a tank 3 m in diameter and 1.1 m deep, 15 larvae/l should be expected.

After hatching, larvae tend to form dense aggregations at the water surface. This concentration is one of the first causes of larval mortality. The characteristics of the water in the larval rearing tank were as follows:

Temperature:	1st - 10th day	14 - 16°C
	10th - 30th day	16 - 18°C
	30th - 90th day	17 - 20°C
Salinity:	between 25 and 38°/oo	
pH:	7 - 8.5	
Dissolved O_2:	1st - 20th day	5 - 8 ml/l
	20th - 90th day	4 - 9 ml/l
Nitrites:	<2 mg/l	
Continuous lighting:	≃250 lux	
Flow rate:	1 - 30 days :	0% of the volume of the tank per day
	30 - 50 days :	30% of the volume of the tank per day
	50 - 90 days :	100% of the volume of the tank per day

Two days after hatching the rotifer *Brachionus plicatilis* was added to the rearing tank. The method of producing rotifers was similar to that described by Theilacker and McMaster (1971).

During the days following hatching the eyes became pigmented; the mouth was formed and opened on the 4th day, and the yolk was being resorbed. During this time only abnormal larvae died. The rotifer density had to be high, about 5 animals/ml at the beginning of feeding. During later yolk resorption from the 7th to 13th day the mortality rate increased drastically and reached 50 to 70% (Fig. 1).

Some larvae seemed unable to react to the presence of prey. Using various densities of food, or other food species, did not improve the

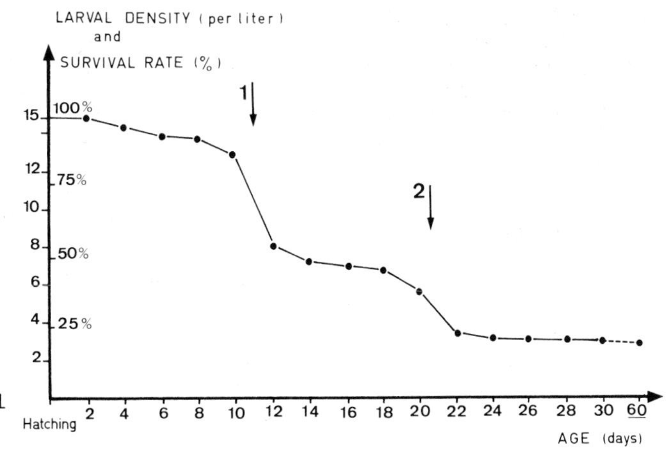

Fig. 1. Mean survival curve of sea bass. 1 and 2: main critical stages

survival rate, though other rotifers such as *Pedalia* or ciliates such as *Fabrea salina* were accepted well. After the 12th day *Artemia* nauplii were added to give a density over 1 nauplius/ml, the rotifer supply being stopped on the 15th day. When the larvae were fed only with *Artemia*, there was an increasing mortality that could perhaps have been due to a low digestibility. From the 15th day wild plankton was added in which the copepod *Eurytemora velox* was dominant. This was offered 5 times a day in excess. Copepods and rotifers tended to remain close to the walls of the tank and often escaped capture. Thus the number of prey available as a food for the larvae was lower than the total number in the tank. From the 30th day *Artemia* and copepods were offered together with minced fish and crab paste. Three days later copepods and *Artemia* were stopped. The paste was given inside net bags of 1 cm mesh so that fish could take pieces of food through the net. The relationship of prey-size to larval length and age is given in Fig. 2.

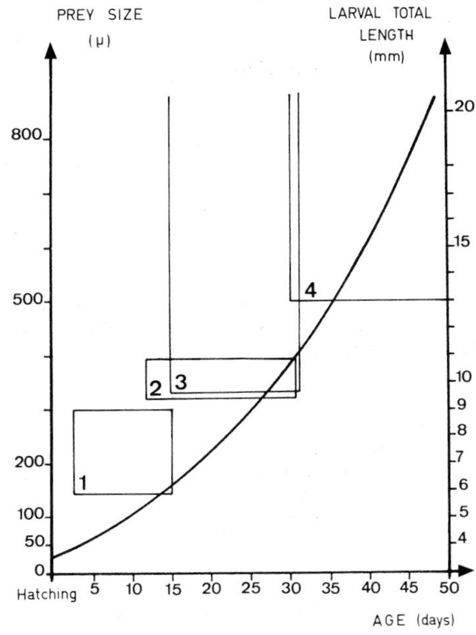

Fig. 2. Age, larval total length and prey-size relationship. (1) *Brachionus plicatilis*. (2) *Artemia salina* nauplii. (3) Copepods. (4) Minced fish and crab

Another system used was an air lift. A screen at the bottom of the air lift and another at the top just below the surface allowed the food to be sieved, and then distributed well in the tank (Fig. 3). This system enables the fish to be trained to have their food distributed from one point only, instead of searching for food throughout the volume of water. A better method is to train the larvae for direct feeding on dry pellets after the 30th day. Due to the lack of feeding hoppers we could not use this method of feeding in all tanks during the experiments. After the 50th day heavy larval mortality ceased.

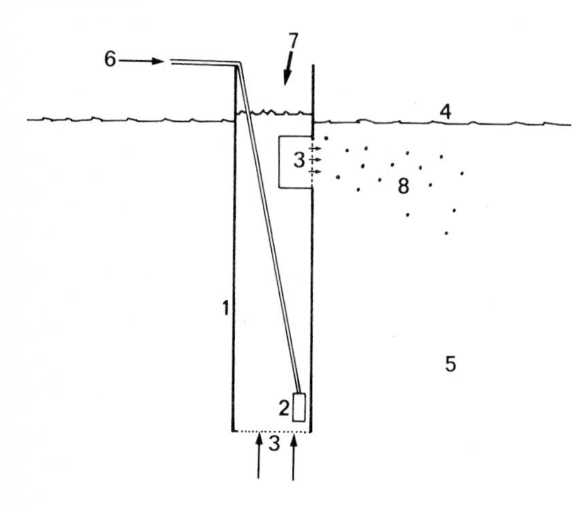

Fig. 3. Feeding hopper. (1) tube. (2) bubbling stone. (3) plankton net. (4) water level. (5) rearing tank. (6) air. (7) food paste. (8) food particles

RESULTS

In 1973, 21 600 sea bass 2 months old were sent out to several other laboratories. Sampling estimates showed the total number of 60-day-old fish to be about 105 000, produced in a total volume of 65 m^3. From this stage, the plywood cylinder tanks used were not suitable for rearing. The number of 100-day-old fish decreased to 45 000.

CONCLUSIONS

There are two main problems in the experiments: first, obtaining food in large quantities - a 10-day-old larva eats 200-500 rotifers/day whereas a 30-day-old fish ingests 50-200 copepods/day; second, there is the problem of making observations in a large volume of water especially in order to assess larval density. Two further points should be emphasized: first, that even in this experiment with a low survival rate more fingerlings were obtained from one female than are usually obtained from one female in trout culture; second, the growth rate of sea bass is higher than that of the trout. These facts together with the hardiness of the fish and their high market value, means that sea bass are especially suitable for marine fish farming.

SUMMARY

Maturing females, after a hormone injection, are placed in a tank of 2.5 m^3 with naturally running males. They then spawn viable eggs which the males fertilize normally. The pelagic eggs above a salinity of 34.5°/oo and a temperature of 15°C are collected and incubated in illuminated still water in which planktonic algae ('green water') has previously been cultured.

Hatching occurs after 70 h. Two days after hatching, the rotifer *Brachionus plicatilis*, separately cultured, is introduced into the medium. The best results are obtained with a rearing density of about 5 rotifers/ml. The rotifers can be replaced by the ciliate *Fabrea salina*. Five days later, the density can be reduced to 1/ml. From the 12th day, nauplii of *Artemia salina* are provided alternately with the rotifers. *Artemia* given for several days alone are inadequate as food. From the 15th to 35th day *Artemia* are given alternately with plankton hauls (mainly the copepod *Eurytemora velox*). From the 30th day to the end of the experiment the food consists of minced fish and crab.

In 5 circular tanks, each of 13 m^3 and consisting of a plastic liner inside a plywood cylinder, 45 000 alevins were produced.

ACKNOWLEDGEMENTS

We received financial support from the CNEXO for this Research.

REFERENCES

Barnabé, G., 1971. Induction de la ponte chez le Loup *Dicentrarchus labrax* L., Rapport CNEXO (unpubl.).

Barnabé, G., 1972. Contribution a l'étude de la biologie du Loup *Dicentrarchus labrax* L. dans la region de Sète, These 3e cycle, Université Sci. et Tech. Languedoc.

Barnabé, G. and Tournamille, J.C., 1972. Expériences de reproduction artificielle du Loup *Dicentrarchus labrax* L., Rev. Trav. Inst. Pêches marit., 36, 185-189.

Theilacker, G.H. and McMaster, M.F., 1971. Mass culture of the rotifer *Brachionus plicatilis* and its evaluation as a food for larval anchovies. Mar. Biol., 10, 183-188.

G. Barnabé
Station de Biologie Marine et Lagunaire
Quai de Bosc prolongé
34200 Sète / FRANCE

Summing-up of the Symposium on the Early Life History of Fish

G. Hempel

The purpose of the Symposium was to review the present state of knowledge in important fields of research in larval fish biology. This summing-up will be made to draw some conclusions for future work. Out of the 62 papers presented only few will be mentioned specifically.

OBJECTIVES OF STUDIES ON FISH EGGS AND LARVAE

Motivations for research on fish eggs and larvae seem to be rather different from author to author and although four major targets are obvious mixed motivations are rather common.

Objective 1: we want to know more about fish eggs and larvae *per se*, for example about embryonic and larval development, about physiology and behaviour, growth rate and mortality, taxonomy, systematics, and zoogeography. The scientific importance of these studies - quite apart from any application, was emphasized by Ahlstrom and Moser who demonstrated the importance of taxonomic research in larval myctophids for clarification of systematics of this group of fish in general. Morphogenesis, and particularly the development of behavioural patterns and of physiological functions, are of general biological interest in view of the changes from a pelagic egg and almost passive yolk-sac larvae to a freely moving and feeding post-larva, and later on to a shoaling or demersal juvenile fish. This is connected with drastic changes in the diet. Amongst the vertebrates, only fish and amphibians pass through several developmental stages with such great biological and ecological differences. Jones pointed to the enormous growth rate in fish larvae which is partly counterbalanced by high mortality, again a feature which is far more pronounced in fish than in higher vertebrates.

Objective 2: we want to know more about marine and freshwater ecosystems. Fish eggs and larvae are sometimes important prey organisms, predators, and grazers. Furthermore, fish eggs and larvae can be used as indicators for the state of pollution and for natural and man-made changes in ecosytems. An evaluation of such a role for ichthyoplankton was not made in detail during the Symposium, but it was stressed several times that larval populations are part of the ecosystem and that they react to any change in the system and its relationship to the abiotic environment.

Objective 3: we want to be able to rear fish eggs and larvae for various purposes for aquaculture, re-seeding and introduction of fish populations in natural and artificial habitats. Furthermore, there is an increasing need for a continuous supply of fish eggs and larvae of well-defined origin for toxicity tests and other physiological experiments, as well as for any selective breeding and for genetical experiments. Rearing of fish larvae is a well established method during identification of undescribed or poorly identifiable larvae. Considerable progress towards this group of objectives was reported both in papers and in verbal discussions during the Symposium.

Objective 4: we want to know more about fish populations and so we use eggs and larvae as indicators of the existence of adult stocks, as measures of parent stock size and as a means to forecast future recruitment. Furthermore, we use egg and larval studies to differentiate between man-made and natural effects on fish stocks, e.g. decreasing recruitment from constant larval production has been taken as an indicator for increasing pre-recruit mortality due to fishing for juveniles. Tanaka has shown how egg and larval surveys can compete successfully with other methods (catch statistics, acoustic and fishing surveys) at least under certain circumstances such as those in the Japanese sardine fisheries. The results of a new method developed by Gjøsaeter and Saetre for the estimate of the Norwegian capelin stock from the number of demersal eggs agreed satisfactorily with other assessments of the same stock. It seems easier and more reliable to describe long-term changes in the abundance and distribution of fish stocks by long time series of ichthyoplankton data than from fisheries which are subject to drastic changes in effort, efficiency and fishing area. Examples of long-term observations were provided for several fish populations of the northeastern Atlantic by Harding, Postuma and Zijlstra, Bainbridge and Hart and by Saville.

It was generally recognized that application of studies of fish eggs and larvae on assessment of fish populations and ecosystems, and on aquaculture and pollution assessments, has to be based on the knowledge of taxonomy, biology, and ecology. On the other hand, it was considered unrealistic to let those applications wait until all basic facts are known. The close links became obvious between pure and applied research in the study of the early life history of fish and in ichthyoplankton populations. Certain ichthyoplankton surveys e.g. herring larval surveys in the North Sea, which were designed to answer specific questions for fisheries management, provide good basic data information on fish biology and biological oceanography. On the other hand experiments on temperature tolerance have practical importance in view of thermal pollution. In studies of fish eggs and larvae ichthyologists, physiologists, ecologists, and fishery biologists are collaborating closely.

Although ecology seems at present the focal point in most studies, many other aspects of the early life-history stages of fish were discussed at the Symposium. Some recent achievements will be discussed with regard to their possible importance for future developments in ichthyoplankton research.

MORPHOLOGY, SYSTEMATICS AND TAXONOMY

Interest in environmental effects on vertebral counts has decreased of late. There is, however, much need for work on the histological and histochemical development of the brain and on the sensory, respiratory, and digestive systems, if we want to understand changes in distribution, behaviour and feeding during the larval stage. Blaxter and his collaborators have produced valuable contributions in this field.

As shown by some of the systematic papers, fish larvae at hatching are at least as different as new-born birds. Later morphological adaptations are connected with feeding and distribution. Those adaptations, however, bring about lesser morphological differences than in adult fish. A comprehensive description of the fish larva in terms of functional morphology is still missing.

It is obvious that some species and families show systematic relationships much more clearly in the larval than in the adult phase. However, the general problem of identification of fish eggs and larvae has not been solved satisfactorily. Series of developmental stages are required to identify eggs and larvae, particularly in warm waters.

PHYSIOLOGICAL ECOLOGY

Physiological studies are still mainly concentrated on salmonids and clupeids and one wonders how far we can generalize those findings. Work on monofactorial tolerance and adaptation to changes in temperature, salinity, low oxygen level, and other chemical factors is still considered important for the understanding of reactions of fish fry to environmental changes. Multifactorial experiments with proper mathematical design and analysis as carried out by Alderdice et al. make the important step from conventional comparative physiology to physiological ecology and should be applied to pollution studies. The work on possible effects of entrainments in power plants and of hydrogen sulfide are examples of the need for incorporating more than just one environmental factor in the analysis. Furthermore, it is now obvious that the various developmental stages show striking differences in their sensitivity to temperature, osmotic, chemical, and mechanical stress. Adverse effects of harmful treatment of the mother fish on the viability of the eggs have been shown and similarly of sublethal harm to the eggs on the subsequent viability of the larvae.

Several field studies have confirmed the high mortality of fish eggs at sea. The causes of the mortality are largely unknown. Experiments show that it is unlikely that wave action and sunlight affect large parts of the egg production.

Starvation experiments are a special kind of physiological ecology study. They ask whether fish larvae in the sea are viable and whether they are approaching or have passed the "point-of-no-return" (or the point of ecological death). Condition factors are a somewhat doubtful measure; whether chemical analysis is a good substitute has still to be shown. Large scale experiments on larval populations in plastic bags should be tried in which changes in size distribution and possibly in meristic characters can be analyzed statistically. The problem of size hierarchies and pecking orders with its relevance to feeding was not dealt with in the Symposium but should receive some attention.

FEEDING AND PREDATION

Starvation of fish larvae at sea has never been demonstrated beyond doubt, although available data on food demand versus food abundance suggest mass mortality by starvation. Additional food which is difficult to trace in the digestive canal, such as naked flagellates, detritus with bacteria, and dissolved organic matter has been suggested but rarely studied in a critical manner. Studies on stomach contents of larvae may not be sufficient to describe the food intake of larvae at sea, while differences in rate and efficiency of digestion with type of food and with size of larvae have to be analyzed. Without doubt match and mismatch of food and larvae in space and time is vital. Any evaluation of food selection requires studies on relative vertical distribution of food organisms and larvae and of the increasing amplitude of their vertical migration with growth. Recent work on the importance of the surface layers as a feeding ground and refuge of certai kinds of fish larvae was mentioned in the discussion. The near-bottom layer might also call for some more attention. Vertical discontinuities in abundance of larvae and of food might be traced by the recent develo ment of closing devices in front of the net and by pumps. On deck control of sampling depth and of changes in filtration due to clogging and sub-surface currents have become technically feasible. Horizontal patchiness, however, has not been satisfactorily studied.

Long term observations show the relationship of egg and larval abundanc with environmental conditions and provide some information on annual differences in mortality rates. Answers on interrelationships and trend in population dynamics and ecosystems will only be reached by a steady continuation of those programmes. Therefore, there is no question of the need for long observational series but there is room for further improvements in scientific and statistical design and in the sampling methods. Comparative tests on filtration rates and sampling capacity of various towed nets were reported and it again became obvious that both net avoidance and extrusion of larvae are largely dependent on towing speed and clogging. So far no net can sample all size categories of fish larvae in an entirely nonselective manner. The choice of gear therefore depends largely on the objectives of the programme. Bongo nets and the various types of Gulf III seem to be preferable in most cases as long as clogging does not reduce filtration, which is rather common for the encased Gulf III sampler during periods of high abundanc of phytoplankton or coelenterates.

Predation is supposed to be one of the causes of mass mortality in fish eggs and larvae. Experimental studies on predation in larval anchovies by pontellids and euphausiids as reported by Lasker are a start in a new field, which was opened up by the development of methods for the continuous production of eggs of some commercially important fish species as well as the controlled rearing of predators such as crustaceans, ctenophores, medusae, and chaetognaths in the laboratory.

AQUACULTURE

Until recently marine aquaculture of fish was largely based on wild fry, the larvae being fed with wild plankton or brine shrimp. During the past few years much work has been done to overcome the shortcomings of these methods and to make the cultivation and selective breeding of marine fish independent of the natural spawning and production cycles. The control of spawning by light and temperature permits egg production at any time of the year. The manufacturing of food particles and the mass culture of food organisms are prerequisites for successful continuous rearing of larvae. Brine shrimp is substituted by other more appropriate, easily digestible food organisms. In the discussion, Lasker reported on controlled spawning in anchovy and other species and several contributors demonstrated recent progress in fertilization and rearing of highly priced fish, e.g. mullets, bass, sole, and turbot. There are still many problems open for further work, particularly in the effect of dissolved organic substances, bacteria, viruses, and parasites on the survival and development of fish larvae in mass culture. Studies of this kind would be also of interest for an interpretation of phenomena at sea.

CONCLUSIONS

The timing of spawning time in the seasonal production cycle of plankton is particularly evident in high latitudes. Synchronization of light, temperature, vertical stability, and nutrient supply is not perfect in nature and spawning is not induced by them in just the same way as the production of planktonic food. Some mismatch should therefore be expected. Why is it that in high latitudes the spawning season of successful species is very short? Why is the amplitude of year-class fluctuations low but still different by one or two orders of magnitude in ecologically and systematically closely related species? Why can races of herring successfully reproduce in different seasons in one and the same area? The review by May on the "critical period" casts some doubt in the validity of any generalization with regard to population dynamics of fish larvae. It is not even certain that in all species regulation takes place at the larval stage.

In summary, a few major points should be listed which seem to require particular attention in the near future:

1. The high mortality of pelagic eggs in the sea.
2. The possible patchiness of food organisms at sea and the reaction of larvae to those patches.
3. Proofs of starvation and predation of larvae at sea.
4. Importance of food which is difficult to trace and to identify in larval guts; digestion rates and digestion efficiency.
5. Further gear development in order to reduce avoidance and extrusion and to improve control of sampling depth and filtration rate.
6. Identification of fish eggs and larvae in tropical and subtropical waters and in the Southern Ocean.
7. Multifactorial studies in physiological ecology, taking into account differences between species and developmental stages.
8. Development of artificial encapsulated food for various kinds of larvae.
9. Continuous production of eggs and larvae of more species under controlled conditions.
10. Theoretical modelling based on actual knowledge of certain species and their ecological relationships.

The Symposium provided a good platform for the discussion of key problems in ichthyoplankton research. The thanks of all participants are due to the Sponsoring Organisations, the Steering Committee, and particularly to the Local Organizers under the chairmanship of Dr. J.H.S. Blaxter.

G. Hempel
Institut für Meereskunde, Universität Kiel
2300 Kiel / FEDERAL REPUBLIC OF GERMANY
Düsternbrooker Weg 20

Subject Index

Some Observations on the Population Dynamics of the Larval Stage in the Common Gadoids

R. Jones and W. B. Hall

INTRODUCTION

The authors have been able to show, in two previous papers (Jones, 1973; Jones and Hall, 1973) that a larval model, based on the idea that food density is critical for survival during the larval stage, can explain much of the early life history of species such as haddock and cod.

In this paper the previous results are summarized. We then used the model to consider theoretically how larval survival might be influenced by the size range of food organisms consumed. This has provided insight into how natural selection, by operating on this variable, might have determined egg size, and hence fecundity, in these species.

ESTIMATES OF EGG AND LARVAL MORTALITY

Mortality during the egg stage varies. According to Runnstrom (1941), the total mortality rate of Norwegian spring-spawned herring eggs ranged from about 10% to 30% depending on egg density. Smith (1970) estimated the mortality rate of the eggs of the Californian sardine at 29% per day. Harding and Talbot (1970) estimated that the mortality rate of plaice eggs in the southern North Sea was about 2-10% per day. Hempel and Hempel (1971) examined herring eggs from haddock stomachs and concluded that about 89-100% of the eggs were alive at the time of ingestion. From this they concluded that, apart from the effects of predation, there was a relatively low rate of egg mortality. Bakke and Bjørke (1971), from underwater observations off the coast of northern Norway, found little or no evidence of capelin egg mortality due to predation.

Estimates of mortality during the larval stage are also variable, and on the whole appear to be quite high. According to Farris (1960), field data on sardine larvae indicated a heavy mortality before the absorption of the yolk sac. Saville (1956), from his studies of Faroe haddock larvae, showed that mortality in this species was about 5-15% per day. In the case of plaice, Harding and Talbot (1970) obtained estimates of larval mortality rates in the southern North Sea of 2-6% per day.

Assuming these estimates are valid, we can then assert that any theory of the population dynamics of the early stages of many species of teleost fishes must be able to account for egg-mortality rates of up to 29% per day and larval mortality rates of 2-10% per day.

CAUSES OF EGG AND LARVAL MORTALITY

Important causes of egg mortality could be predation or dispersal into unfavourable regions. After the hatching and dispersion of the spawning products, however, the situation is likely to be rather different. Once larvae have dispersed throughout the top 50 m or so of water, frequently over an area of many thousands of square miles, a high predation rate is less likely. To account for mortality rates of 5-10% per day by, for example, active predation seems to require the existence of a predator that is capable of searching 5-10% per day of the volume occupied by the larvae. This is conceivable provided the larvae are concentrated into a relatively small total volume, but less likely once they are widely dispersed. Haddock larvae in the northern North Sea, for example, disperse over an area of at least 60,000 km^2. If there are predators that can collectively remove 5-10% per day of the fish larvae over such an area, no one has yet been able to identify them. For example, Cushing and Harris (1973) have shown that chaetognaths, which are predators on fish larvae, are only present in sufficient numbers to account for a larval mortality rate of about 1% per day.

These objections do not constitute proof that larval mortality rates of 5-10% per day cannot be wholly due to deliberate acts of predation. However, they do cast considerable doubt on this hypothesis and prompt us to enquire whether or not there might be another cause of high mortality rates.

One obvious possibility is that mortality is due primarily to starvation; here the evidence seems more convincing. Shelbourne (1957), for example, observed starving plaice larvae in regions of low food density. O'Connell and Raymond (1970) investigated the growth and survival of anchovy larvae in six different levels of zooplankton density. They observed a high mortality during the 6th and 7th days after hatching in containers receiving 1 copepod nauplius/ml/day or less, but not in those containers that received 4 copepod nauplii/ml/day or more. They also noted that those larvae that received 4-8 nauplii/ml/day were 2.5-3 times as long at the end of the experiment as those receiving 1 nauplius/ml/day. Chenoweth (1970) recorded seasonal variations in the condition of larval herring off the coast of Maine, USA and found that larval mortality was unusually high in the winter of 1965, when the larval condition reached its lowest level in four years.

At this point, it is important that we distinguish between mortality primarily due to predation and mortality secondarily due to predation. In the first case the number of larvae dying ought to be a function of the number of predators. Secondarily induced predation could occur, however, if larvae became weakened due to lack of food or disease and were then eaten. In that event, the number of larvae dying would not be a function of the number of predators as much as a function of the amount of food available.

It is also important that we distinguish between mortality in the larval stage and that in the young fish stage. By the time the larvae have grown into young fish, their biomass will have increased and they will have become capable of shoaling and of exhibiting other behaviour patterns. The possibility that mortality at this stage is primarily due to predators cannot be ruled out.

This paper, however, is concerned with an explanation of mortality rates of 5-10% per day during the first month or six weeks of life, when the larvae have become dispersed over a wide area. It is suggested that during this period the quantity of food available to the larvae is the important factor for survival. To quantify this concept it is necessary to estimate the daily food requirements and to compare these with the quantities of food actually available.

LARVAL FEEDING

The principal food of larval cod and haddock are the young stages of copepods such as *Calanus* (Marak, 1960; Sysoeva and Degtereva, 1965; Bainbridge and McKay, 1968). According to Marak (1960), larval gadoids feed on progressively larger individuals of the young stages of copepods until they are about 13 mm in length. On the basis of growth rates of larval haddock it has been calculated (Jones, 1973) that this would require about 45 days. According to the same author and also to Sysoeva and Degtereva (1965), gadoid larvae do not appear to be large enough to eat stage V *Calanus* until they are about 19-25 mm in length. This would require about 56-66 days.

It appears, in fact, that cohorts of larvae grow up at the same time as cohorts of copepods. This would be particularly so if the larvae and copepods grew at the same rate. It is relevant, therefore, to compare the growth rates of the larvae and of the copepods to see if they are indeed the same.

THE RELATIVE GROWTH RATES OF *CALANUS* AND HADDOCK LARVAE

Experimental work on the growth of *Calanus* has been carried out by Mullin and Brooks (1970). These authors give about 43 days for the *Calanus* development time from hatching to Stage VI at 10°C and about 28 days from Stage I to Stage VI. In the northern North Sea in spring, the mean sea temperature in the upper 50 m is about 8°C. Mullin and Brooks give a Q_{10} for developing *Calanus* of 3.67. The development time from hatching to Stage VI at 8°C is therefore estimated to be about $43 \times (3.67)^{0.2} = 56$ days, and from Stage I to Stage VI to be about $28 \times (3.67)^{0.2} = 36$ days.

Mullin and Brooks conducted experiments at temperatures of 10°C and 15°C. By extrapolation from their data Fig. 1 was derived, representing the growth of *Calanus* at a temperature of 8°C. In the calculation, it was assumed that 1 µgC is equivalent to 10 µg wet weight. These results suggest that the body weight increases exponentially with age at a rate of about 15% per day from Nauplius I to Copepodite IV for a period of about 36 days. Subsequent growth also appears to be exponential, but at a lower rate, and the total development time from hatching to the adult should take about 56 days.

In the northern North Sea it therefore takes about 56 days for a *Calanus* to grow from the size at hatching to stage VI. When feeding on *Calanus*, haddock larvae appear to feed on the young stages for about 56-66 days. As a first approximation, therefore, it is suggested that successive cohorts of both species do indeed grow up together.

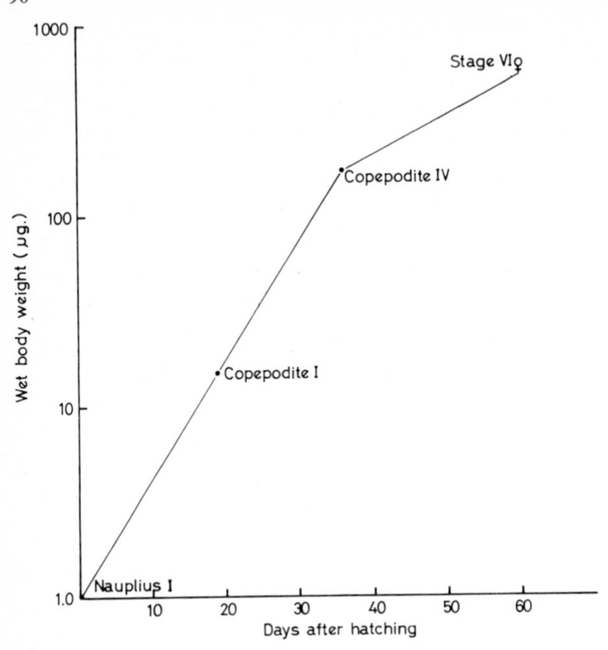

Fig. 1. Growth curve of *Calanus* at 8°C (estimated from data in Mullin and Brooks, 1970)

THE FOOD REQUIREMENTS OF THE GROWING LARVAE

During the early post-yolk sac stage, the larvae increase in weight at a rate of about 12% per day. To account for this at the commencement of feeding, for example, it was estimated (Jones, 1973) that a haddock larva requires about 50 nauplius I *Calanus* or 7 stage I *Calanus* per day. As it grows it eats larger copepod stages and a 56-day-old haddock larva, 19 mm/length would require about 30 adult *Calanus* daily to maintain this growth rate. The daily food requirements of the growing larva would thus appear to be about 7-30 organisms per day, except initially, when it might be higher. When haddock feed on smaller species of copepods, they would require larger numbers per day.

THE BIOLOGICAL BASIS OF THE MODEL

An important consequence of eating young growing stages rather than adult copepods is that groups of fish larvae hatching on successive days will tend to grow up with groups of copepods also hatching on successive days. A situation like this has several features of interest. For example, it appears to be an efficient way of utilizing a growing food species, since it will tend to minimize the actual number of food organisms that have to be eaten to achieve a given amount of growth. For example, a 56-day-old fish larva that consumed stage I *Calanus* rather than adult *Calanus* would require about 1000 organisms per day instead of about 30. Restriction of diet to the largest copepod stage that could be eaten might therefore appear to be an efficient way of utilizing a food resource composed of growing individuals.

Another consequence is that until the larvae are large enough to eat adult copepods, or some other food species, they will tend to be limited to the particular number of copepods in the cohort that they happen to be growing up with. There would, in fact, be a period during which a cohort of larvae would be growing up with a cohort of copepods and grazing on them as on a private food supply. The Stage V *Calanus*, for example, available to the larvae once they were large enough to eat them, would then be the survivors of the actual cohort of copepods with which they had been grazing and growing up.

One consequence of this is that there ought to be an upper limit to the number of larvae that could be supported by a cohort of copepods of any given size. In the absence of any better suggestion, therefore, this could be important as a way of limiting population size in the haddock, and possibly in other species with a similar larval life history.

The simplest possible situation is one in which there is one cohort of larvae and these grow up with one cohort of copepods. A simulation of such a situation has been described by Jones (1973), the basis of the model being effectively described by the following two equations:

$$N_{L/t+1} = N_{L/t} - \psi 1 \ \{N_{L/t} \cdot N_{F/t} \cdot S_t \cdot X_t\} \qquad (1)$$

$$N_{F/t+1} = N_{F/t} - \psi 2 \ \{N_{L/t} \cdot N_{F/t} \cdot S_t \cdot X_t\} \qquad (2)$$

where $N_{L/t}$ = the number of larvae aged t days per unit volume

$N_{F/t}$ = the number of food organisms per unit volume when the larvae are t days old

S_t = the proportion of the unit volume searched in a day by a larva aged t days

X_t = critical daily rate of encounter with food organisms such that larvae failing to encounter food at this rate die.

APPLICATION OF THE MODEL

The input parameters for this model are initial values of $N_{L/t}$, $N_{F/t}$, S_t and X_t. Also required are the functions describing the rate of change of S_t and X_t with time. The number of days the simulation is to be run is also required. The functions ψ_1 and ψ_2 are included as an integral part of the program. The equations have been simulated for a number of functions, ψ_1 and ψ_2, in addition to those specifically used by Jones (1973).

From this investigation the following results were obtained:

1. Extensive simulation has shown that it is probably the effect of grazing during the first few days that is most critical for determining larval survival rate. Basically, for a larva to survive, it must be in a situation where the food density is above some critical value. Simulation shows that as the initial number of larvae is increased, a point is eventually reached at which the effect of grazing reduces the density of food organisms to the critical value. When this happens the subsequent probability of survival is reduced. By starting with too many larvae, however the model is operated, a point is always reached where it is possible for them to reduce the density of the food organisms on which they are feeding to the critical point, after

which they tend to die at a comparatively rapid rate. Within quite wide limits we found this to occur, even when the effective searching capacity of the larva is increased or when the initial density of food organisms is made very large. Starting with too many larvae one reaches at best, the situation in which they are kept alive for a longer than average period initially, until eventually they reduce the density of food organisms to the critical level. In other words, however the model is run it is found that there is effectively an upper limit to the number of larvae that can be supported by a given cohort of food organisms and that the regulatory process occurs mainly during the first 10-15 days of simulation.

2. No difficulty was experienced in generating larval mortality rates of 10% per day. In fact, we found that if the critical rate of encounter with food was made equal to the daily food requirement for growth (i.e. if X_t was made equal to 7 or more) the larval mortality rate was considerably greater than 10% per day. The model is therefore able to account for mortality rates of this order without difficulty.

The tendency to generate high larval mortality rates is particularly apparent if the model is operated on a day-to-day basis. This, however, is equivalent to supposing that the probability of a larva dying on a particular day is due to its rate of encounter with food on that day only and is independent of what happened to it on preceding days. This assumption is obviously unrealistic because death does not result from only one day of starvation. To allow for this, the model was also operated using intervals of 1, 5, and 15 days. The conclusions described below were unaffected by the choice of interval however.

3. As expected, the relationship between initial numbers of larvae and the number of larvae alive after 45 days took the form of a curve that increased to a maximum and then declined. A typical example is shown in Fig. 2.

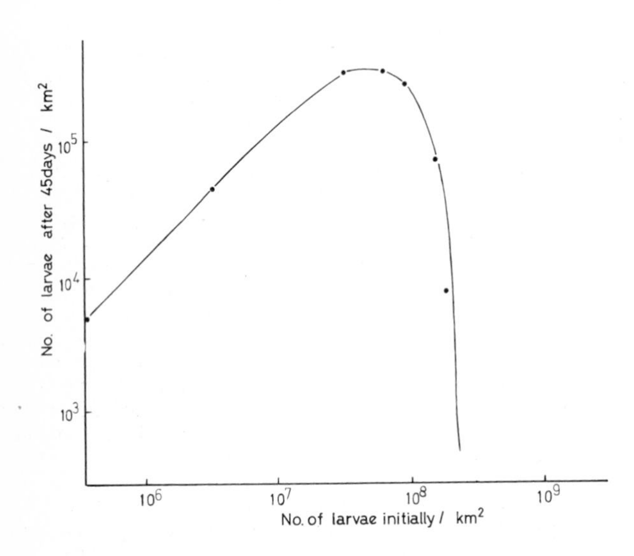

Fig. 2. Typical theoretical relationship between number of larvae initially and number of larvae still alive after 45 days

4. The curve can be moved to left or right, according to the choice of input parameters. Factors that move the curve to the left are, for example: a small number of food organisms initially, a large daily rate of food consumption by the larvae, a long period during which larvae are confined to a particular cohort of food organisms, or a small rate of increase with age in the larval searching capacity. What is important, however, is that for a reasonable range of input parameters, the maximum number of larvae surviving occurs when the number feeding initially is in the range $3 \times 10^7 - 3 \times 10^8/\text{Km}^2$. This in fact tends to agree with observation since the egg densities of a number of fish species fall within this range (Jones, 1973).

5. Although variations in the input parameters moved the curve to left or right through a factor of only 10, they moved it up and down to a much greater extent. In particular the number of larvae surviving is sensitive to variations in the number of food organisms at the commencement of feeding.

In a previous paper (Jones and Hall, 1973) a complete life cycle model was constructed using the larval model to simulate the egg/recruit relationship. In this way it was possible to show that variations in year-class strength of the magnitude observed in North Sea haddock and whiting could be accounted for by assuming quite small annual variations in larval food density. With this model it was also possible to generate year-class fluctuations that were very poorly correlated with stock size, as is observed in practice.

For some other species, such as plaice, a similar investigation suggested that the fluctuations in year-class strength were not as large as might have been expected, allowing for reasonable fluctuations in the larval food supply. For this species, it is quite possible that fluctuations in year-class strength are inhibited by some other factor; one suggestion (Steele and Edwards, 1970) is that the area of suitable bottom when the fish become demersal imposes a limiting factor.

6. Another result of this analysis is that the model tends to generate an asymmetrical frequency distribution of year-class strengths, with poor year classes appearing more commonly than good ones. It is interesting to note that this happens to be just the kind of distribution observed in practice in North Sea haddock and whiting.

7. The larval model is able to show us how density-dependent regulation of population numbers could occur in several species independently in the same area. Because the periods of peak spawning of the major species tend to differ in time, it is possible to see how an overproduction of eggs by one species does not necessarily constitute a threat to the food supply of another species. In the northern North Sea, cod and haddock both spawn in the spring, but peak spawning is about three weeks earlier in the cod than in the haddock. This means that on average haddock larvae begin feeding at a time when cod larvae are 3 weeks old. A 3-week-old larva is estimated to be eating organisms about 20 times larger than a larva at the commencement of feeding, so it seems possible that a 3-week time interval is sufficient to ensure a reasonable degree of independence between these two species.

To investigate the above point further it is necessary to know something about the actual size range of the food particles eaten by larvae of different sizes. Depending on this, there are a number of consequences which are explored in the remainder of this paper.

THE SIZE RANGE OF THE FOOD ORGANISMS CONSUMED

For any fish larva there will be a physical upper limit to the size of food organisms that it can ingest. However, no such physical limitation need apply to the lower limit of the size of food eaten. Consider, for example, the situation when a fish larva is just large enough to eat a Stage I *Calanus* weighing about 15 μg. Suppose further that this fish larva restricted its diet to *Calanus* in the range 3 μg to 15 μg. In Fig. 1 we can see that a 3 μg *Calanus* would be about 8 days old while a 15 μg *Calanus* would be about 19 days old. A diet of 3 μg to 15 μg *Calanus* would therefore be equivalent to a restriction of the food to *Calanus* born within an 11-day period. The size range of the food-particle size can therefore be measured in days; this period will be referred to as the "feeding range" of a larva. At the commencement of feeding cod and haddock larvae feed on copepod nauplii. Since the nauplius stages last about 20 days (Fig. 1) it follows that the feeding range at the commencement of feeding is less than 20 days.

IMPORTANCE OF THE FEEDING RANGE

A number of factors are directly dependent on the feeding range.

1. The feeding range provides a direct measure of the interval of time required between any two cohorts of larvae to make them independent. For example, suppose larvae have a 1-day feeding range. The larvae born on a certain day would then grow up with the copepods born on that day. The food available to them will thus be independent of the food available to larvae born on any other day. Similarly, if larvae have a 10-day feeding range, then batches of larvae born at 10-day intervals will grow up independently of each other.

2. The feeding range will determine the number of food organisms initially available to each cohort of larvae, i.e. larvae with a 5-day feeding range will have 5 days production of copepods to sustain them until they are large enough to eat adult copepods.

3. The minimum size that a larva can afford to be at the commencement of feeding will depend on the feeding range. For example, suppose a larva at the commencement of feeding needs at least 10 days production of copepods to provide a food density in which it can survive. Then such a larva would have to be large enough to eat at least a 10-day-old copepod. If it were smaller than this, it would no longer be able to benefit from a 10-day production of copepods. A larva that only needed a 5-day feeding range, on the other hand, would be able to start its life at a smaller size. Since the size of a larva at the commencement of feeding will presumably depend on egg size, it follows that the feeding range should be one of the factors responsible for determining egg size (and hence, indirectly, fecundity) through the process of natural selection.

THE NUMBER OF COPEPODS INITIALLY AVAILABLE TO EACH COHORT OF FISH LARVA

At the commencement of larval feeding, the number of copepod nauplii or Stage I copepods available to a fish larva will depend partly on its feeding range and partly on the rate of production of copepods.

An estimate of copepod production comes from theoretical work by Steele (1972). He used data from Cushing and Tungate (1963) and estimated that to account for zooplankton production in the northern North Sea would require about 200 000 *Calanus* nauplii/m^2. This value was needed to account for a production cycle of about 70 days. In practice, other species of copepods may also be eaten, so that we may be suppose that copepod cohorts of one species or another would be produced continuously over this 70-day period. As a first approximation, therefore, it will be supposed that copepod production in the northern North Sea in an average year yields the equivalent of about 3 000 *Calanus* nauplii/m^2 per day. This is equivalent to 0.3×10^{10} *Calanus* nauplii/km^2 per day. The number of food organisms initially available to a newly-hatched larva will then depend on its feeding range. If, for example, it started feeding on *Calanus* hatched on a particular day, it would be confined to a cohort of copepods with an initial density of about $0.3 \times 10^{10}/km^2$.

A larva with a feeding range of 10 days, on the other hand, would have at its disposal a cohort of copepods with an initial density of $3 \times 10^{10}/km^2$. Note that if *Calanus* mortality were to be taken into account, the average density available to a newly-hatched larva would be even less than this. Table 1 shows estimates of the initial numbers of copepods available to a fish larva for feeding ranges of 1, 5, 10, 15, and 20 days.

Table 1. Showing the effect of feeding range on various biological parameters likely to affect the survival of fish larvae

Feeding range (days)	1	5	10	15	20
A Initial numbers of food organisms ($\times 10^{-10}$)	0.3	1.5	3.0	4.5	6.0
B Relative weights of larvae at the commencement of feeding	1	2.0	4.0	8.1	16.4
C Relative searching capacities of larvae (by length)	1	1.3	1.6	2.0	2.5
D Time for larva to grow large enough to eat adult *Calanus* (days)	55	51	46	41	36
E Mean *Calanus* body weights (µg)[a]	16.4	12.7	9.5	7.3	5.9
F Relative numbers of *Calanus* juveniles needed per day	1	1.3	1.7	2.2	2.8
G Relative fecundity	16.4	8.2	4.0	2.0	1
H Number of larval cohorts	40	8	4	2.7	2

[a] This row shows the average weight of *Calanus* eaten, once the larvae are large enough to eat 20-day-old *Calanus*.

THE CRITICAL FOOD DENSITY FOR LARVAL SURVIVAL

According to Jones (1973), an individual haddock larva at the commencement of feeding searches per day about 1.5×10^{-10} of the volume under 1 km^2 of surface. Using this estimate, a larva at the commencement of feeding requiring 10 organisms per day, for example, would require an average food density of

$$\frac{10}{1.5} \times 10^{10} = 7 \times 10^{10} \text{ organisms/km}^2\text{, i.e. } 70\ 000/\text{m}^2.$$

As compared with this estimate, observations by Sysoeva and Degtereva (1965) showed that cod larvae that were feeding on *Calanus finmarchicus* did not do so when the density was less than 5 000-18 000 *Calanus*/m^2. However the larvae referred to by these authors were presumably larger, on average, than larvae at the commencement of feeding. The volume that they were able to search in a day was therefore presumably also larger. The critical food density for these larvae therefore ought to have been smaller than the critical level at the commencement of feeding. The theoretical value of 70 000/m^2 thus compares reasonably with this particular observed value of 5 000-18 000/m^2. It should be noted that the value of 1.5×10^{-10} above was based on a searching volume for herring estimated by Rosenthal and Hempel (1969). Subsequent estimates of larval searching volume for herring, pilchard, and plaice, obtained by Blaxter and Staines (1971), were very much lower than the value obtained by Rosenthal and Hempel. If the searching volumes from Blaxter and Staines are used, even larger densities of food organisms are required to ensure larval survival. These observations therefore tend to suggest that, in practice, average food densities in the sea are likely to be inadequate for supporting newly hatched gadoid larvae and that for larval survival to occur, it is necessary for a larva to be in an above average situation of food density.

Comparison of the theoretical value of 7×10^{10} organisms/km^2 with the values in Table 1 in row A suggests that a larva requiring 7×10^{10} organisms/km^2 at the commencement of feeding would need a feeding range of about 20 days or a little more. In general, it is clear from row A, that the larger the feeding range, the more likely it is that a larva at the commencement of feeding would be able to encounter food at the rate necessary to satisfy its needs.

RELATIVE SIZE OF LARVAE AT THE COMMENCEMENT OF FEEDING

Estimates of the relative sizes of larvae at the commencement of feeding can be obtained if it is assumed that the size of a larva is proportional to the size of the largest *Calanus* that it is able to eat. Thus, a larva with a 1-day feeding range could start life so small that it need only eat *Calanus* nauplii weighing about 1 μg. A larva with a 20-day feeding range, on the other hand, would have to be large enough, when it began feeding, to eat a 20-day-old *Calanus* weighing 16.4 μg. Row B of Table 1 shows the relative weights of larvae at the commencement of feeding, based on the weights of the largest *Calanus* that could be eaten.

Because of these size differences, the relative searching abilities of these larvae would presumably be very different. If, for example, searching ability was proportional to body weight, the relative searching capacities would be the same as the relative body weights shown in row B of Table 1. For example, larvae with a 20-day feeding range ought to be able to search 16.4 times the volume/day that could be searched by larvae with a feeding range of only 1 day. If searching capacity were proportional to body length, on the other hand, it would be more appropriate to use the cubed-roots of the body weights; the relative values are shown in row C of the table. These show, for example, that a larva with a feeding range of 20 days ought to be able to search 2.5 times the volume/day that could be searched by a larva with a feeding range of 1 day. In general, it is clear from rows B and C that the larvae with the larger feeding range would have the advantage of a greater searching capacity than the larvae with a smaller feeding range.

PERIOD OF FEEDING ON LARVAL COPEPODS

The smaller a larva at the commencement of feeding, the larger the period required before it is large enough to eat the adult copepods accumulated from all copepod cohorts, regardless of their date of hatching. The length of time can be estimated using the *Calanus* growth curve in Fig. 1, in which 56 days are required for a *Calanus* to grow from the time of hatching to the adult stage.

Thus, a larva with a 1-day feeding range ought to start feeding on 1-day-old *Calanus* and ought to require 55 days to grow large enough to eat adult *Calanus*. A larva with a 20-day feeding range, on the other hand, should be able to grow large enough to eat adult *Calanus* after 56-20 = 36 days.

Values are shown in row D of Table 1 and, in general, it is clear that a larva with a large feeding range has the advantage of being confined to a restricted food supply for a shorter time than a larva with a short feeding range, assuming the latter commences life at the smallest possible size.

THE AVERAGE WEIGHT OF FOOD ORGANISMS AT THE COMMENCEMENT OF FEEDING

Thus far, we have only considered factors that favour a large feeding range. There are other factors, however, that would tend to favour larvae with a short feeding range. For example, by the time a larva with a 1-day feeding range is large enough to eat a 20-day-old *Calanus* it will be eating organisms 16.4 μg in body weight. A larva of the same size with a 20-day feeding range, on the other hand, would be eating *Calanus* aged from 1 to 20 days. If, as a first approximation, we suppose that larvae of different ages within this range are eaten with equal likelihood, it can be determined from Mullin and Brooks' data that the average weight of *Calanus* eaten would be 5.9 μg. Row E of Table 1 gives this and other estimates of the average weight of *Calanus* eaten for larvae with feeding ranges of 1, 5, 10, 15 and 20 days. It shows the average weight of *Calanus* eaten at a time when the larvae are large enough to eat 20-day *Calanus*.

THE NUMBER OF *CALANUS* REQUIRED PER DAY

Given the mean *Calanus* body weights in row E of Table 1, the relative numbers of food organisms required to satisfy the same daily food requirement can be estimated. Estimates are shown in row F of Table 1 relative to the number required for a feeding range of one day.

$$\text{e.g.} \quad 1.3 = \frac{16.4}{12.7}$$

$$1.7 = \frac{16.4}{9.5} \quad \text{etc.}$$

Thus a larva with a 20-day feeding range, for example, would require about 2.8 times as many *Calanus* daily as a larva of the same size with a 1-day feeding range. This is therefore a factor that would presumably operate to the disadvantage of the larvae with the larger feeding ranges.

RELATIVE FECUNDITY

The number of eggs produced should be inversely proportional to the larval weight at the commencement of feeding so that relative values can be estimated from the values in row B of Table 1. Thus, in row G of Table 1 it is shown that 16.4 times as many eggs might be produced from a given ovary weight if the larvae are small enough to have a 1-day feeding range instead of a 20-day feeding range. This is thus a factor that might be expected to favour genotypes with short feeding ranges.

NUMBER OF INDEPENDENT LARVAL COHORTS

Finally, numbers of independent larval cohorts have been estimated, assuming that there is a 40-day spawning period for fish such as the northern North Sea haddock. With a 1-day feeding range, 40 independent cohorts could be produced. With a 20-day feeding range, on the other hand, only two independent cohorts could be expected.

THEORETICAL ESTIMATES OF FEEDING RANGE AND NUMBER OF LARVAL COHORTS

The estimates in Table 1, together with other data relevant to larval survival described in Jones (1973), were used in a modification of the model described by Jones and Hall (1973) to simulate competition between genotypes with different feeding ranges. To simplify the computational procedure, each simulation related only to one pair of genotypes. These were made to compete for a common food supply in the larval stage, but were made independent and with identical growth rates, for the remainder of their lives. Many trials were made, using reasonable ranges for all parameters. Even by deliberately manipulating values of the parameters so as to favour genotypes with small feeding ranges it was not possible to obtain results that favoured genotypes with feeding ranges lower than 5 days. Below these ranges, the relative reduction in searching capacity and in the number of food organisms initially available became so severe that there appeared to be no advantage in having a feeding range less than 5 days. In general, the results suggested that the optimum feeding range for survival at the commencement of feeding ought to be in the range 5-20 days, which is in reasonable agreement with the previous observation that the feeding range ought to be less than 20 days, i.e. reference to Fig. 1 shows that this is in reasonable agreement with observations that gadoid larvae commence feeding on copepod nauplii but soon progress to Stage I copepodites.

It can also be inferred from the above result that North Sea haddock, with a spawning period of about 40 days, ought to produce 2-8 independent cohorts of larvae. It should be noted that these calculations have been made for hypothetical fish larvae theoretically able to grow large enough to eat an adult *Calanus* in something between 36-55 days. These should be treated as relative values only, since in reality haddock larvae probably require longer than this before they are large enough to eat adult *Calanus*. However, the critical number of days in simulation is substantially less than this, so the exact duration of the period of feeding on *Calanus* does not effect the main conclusions in any way.

A food-limiting model seems able, therefore, to account for the evolution of the feeding range in haddock, and hence indirectly provides an explanation of why egg size and fecundity might have evolved to their observed levels.

SOME UNRESOLVED QUESTIONS

It would appear that a food-limiting model is able to explain much of the early life history of species such as haddock and cod. There remain some questions, however, that are unresolved.

1. Even if food limitation is the primary cause of larval mortality, this does not preclude the possiblity that there is a second "critical" stage in the first year of life. Steele and Edwards (1970), for example, have suggested regarding plaice that the area of bottom suitable for settlement when they become demersal may impose a second critical phase in that species. Such a stage might account for the relatively small year-class fluctuations in this and other flatfish species. Whether there is a stage subsequent to the larval stage in the first year of a cod's or haddock's life that also inhibits year-class fluctuations remains to be ascertained.

2. It has already been noted (Jones, 1973) that the egg densities of a number of species fall in the range 3×10^7 - 3×10^8 eggs/km^2, as predicted by the model. Although there are exceptions in the range 3×10^8 - 3×10^9 eggs/km^2, we can allow for some of them by noting that it is not so much egg number, as the number of larvae/cohort at the commencement of feeding with which the model is concerned. For example, North Sea whiting and Peruvian anchovy have relatively long spawning seasons, and if this fact is taken into account the numbers of larvae per cohort for these species might well fall into the range 3×10^7 - 3×10^8/km^2. Also, the estimated number of larvae/cohort at the commencement of feeding for the Californian sardine, if egg mortality is allowed for using data from Smith (1970), appears likely to fall in the range 3×10^7 - 3×10^8/km^2.

Whether or not it will be possible to generalize from these observations by suggesting that many species whose larvae feed on rapidly growing food organisms will be found to have 3×10^7 to 3×10^8 larvae/cohort at the commencement of feeding will require further investigation.

3. Some species spawn in the same area but at different times. Cod and haddock in the North Sea, for example, spawn about 3 weeks apart. There is some overlap of spawning time, however, and presumably there is an intermediate time when competition between the two species takes place. It remains to be discovered what factors might have prevented one species from extending its spawning time at the expense of the other and gradually eliminating it.

It may be significant that cod larvae have larger mouths than haddock larvae of the same size (Hislop, personal communication); this suggests that there are real differences between the larvae of these two species in addition to the fact that they are spawned at different times. In this connection it is also interesting to note an observation by Shirot (1970) that larvae of species with large mouth sizes at the commencement of feeding grew faster than those of species with smaller mouth sizes.

It may be useful, therefore, to investigate further the significance of mouth size in relation to feeding in fish larvae.

4. If the conclusions we reached in this paper are valid, it should be possible to explain differences in egg size within and between species that, like haddock, start feeding on the young growing stages of their food species. For example, within species relatively large eggs might be produced in situations where the rate of production of food organism is relatively small. This might explain, for example, why the eggs of a number of species (including cod, haddock and herring) that spawn in the spring decrease in size during the course of the spawning season, whereas the eggs of those herring that spawn in the autumn increase in size throughout the season (Bagenal, 1971; Blaxter and Hempel, 1963). In comparing egg sizes between stocks or species, however, it might also be necessary to consider various factors such as possible differences in the growth rates of the larvae and their prey species, or differences in temperature and salinity that might affect the buoyancy of the eggs and hence exert an overriding influence on the egg size.

These and other suggestions made in this section are only tentative. They do, however, draw attention to some unresolved features of the early life history and point to possible lines of future research.

SUMMARY

The authors have shown in two previous papers that a larval model based on the idea that food density is critical for survival during the larval stage can explain much that is known about the early life history of fish such as haddock and cod. One characteristic of the feeding of haddock and cod larvae is that they appear to feed on the young growing stages of their food species, giving the impression of cohorts of fish larvae growing up with cohorts of food organisms. In such a situation there would appear to be a choice of feeding strategies. By eating only the largest organisms which it is capable of ingesting, a larva could minimize the number of organisms needed per day, but only at the expense of reducing the availability of suitable food. Alternatively, it could also increase the density of suitable food by eating smaller organisms, but only at the expense of increasing the number required per day. In practice, feeding strategy is likely to be a compromise between these alternatives. Of particular interest is the range of sizes of food organisms when feeding commences.

This will presumably influence larval size at the commencement of feeding and hence could be a factor determining egg size (and hence indirectly in determining fecundity).

Extensive simulation of the effects of competition between genotypes with different feeding strategies has suggested that a species like the haddock ought to be able to maximize its chances of survival by eating juvenile copepods born within a 5-20 day period. This result appears to be reasonably consistent with the size range of copepods such as *Calanus* actually found in gadoid stomachs at the commencement of feeding.

We concluded therefore, that a food limiting model of the kind referred to above, is able to account quantatively, as well as qualitatively, for much that is known about the early life history of species like haddock.

REFERENCES

Bagenal, T.B., 1971. The interrelation of the size of fish eggs, the date of spawning and the production cycle. J. Fish. Biol. 3, 207-219.
Bainbridge, V. and McKay, B.J., 1968. The feeding of cod and redfish larvae. Spec. Publs int. Commn. N.W. Atlant. Fish. (7), 187-218.
Bakke, S. and Bjorke, H., 1971. Diving observations on Barents Sea capelin at its spawning grounds off the coast of northern Norway. ICES C.M. H, 25.
Blaxter, J.H.S. and Hempel, G., 1963. The influence of egg size on herring larvae (*Clupea harengus* L.). J. Cons. perm. int. Explor. Mer 28, 211-240.
Blaxter, J.H.S. and Staines, M., 1971. Food searching potential in marine fish larvae. D.J. Crisp, ed. 4th European Marine Biology Symposium. Cambridge University Press, pp. 467-481.
Chenoweth, S.B., 1970. Seasonal variations in condition of larval herring in Boothbay area of the Maine coast. J. Fish. Res. Bd. Can. 27 (10), 1875-1879.
Cushing, D.H. and Harris, J.G.K., 1973. Stock and recruitment and the problem of density dependence. ICNAF/ICES/FAO Symposium on Stock and Recruitment, Aarhus, Denmark, 7-10 July. Rapp. P.-v. Réun. Cons. perm. int. Explor. Mer 164, 142-155.
Cushing, D.H. and Tungate, D.S., 1963. Studies on a *Calanus* patch. I. The identification of a *Calanus* patch. J. Mar. Biol. Assoc. U.K. 43, 327-337.
Farris, D.A., 1960. The effect of three different types of growth curves on estimates of larval fish survival. J. Cons. perm. int. Explor. Mer 25 (3), 294-306.
Harding, D. and Talbot, J.W., 1970. Studies on the eggs and larvae of the plaice (*Pleuronectes platessa* L.) in the Southern Bight. ICNAF/ICES/FAO Symposium on Stock and Recruitment, Aarhus, Denmark, 7-10 July. Rapp. P.-v. Réun. Cons. perm. int. Explor. Mer 164, 261-269.
Hempel, I. and Hempel, G., 1971. An estimate of mortality in eggs of North Sea herring (*Clupea harengus* L.). Rapp. P.-v. Réun. Cons. perm. int. Explor. Mer 160, 24-26.
Jones, R., 1973. Density dependent regulation of the numbers of cod and haddock. ICNAF/ICES/FAO Symposium on Stock and Recruitment, Aarhus, Denmark, 7-10 July. Rapp. P.-v. Réun. Cons. perm. int. Explor. Mer 164, 156-173.

Jones, R. and Hall, W.B., 1973. A simulation model for studying the population dynamics of some fish species. Institute of Mathematics. In: The Mathematical Theory of the Dynamics of Biological Populations (M.S. Bartlett, R.W. Hiorns, eds.), p. 35-59. London, New York: Academic Press, 1973.

Marak, R.R., 1960. Food habits of larval cod, haddock and coalfish in the Gulf of Maine and Georges Bank area. J. Cons. perm. int. Explor. Mer 25, 147-157.

Mullin, M.M. and Brooks, E.R., 1970. Growth and metabolism of two planktonic, marine organisms influenced by temperature and type of food. In: J.H. Steele (Ed.): Marine Food Chains, p. 74-95. Edinburgh: Oliver and Boyd.

O'Connell, C.P. and Raymond, L.P., 1970. The effect of food densities on survival and growth of early post yolk-sac larvae of the northern anchovy (*Engraulis mordax* Girard) in the laboratory. J. Exp. Mar. Biol. Ecol. 5, 187-197.

Rosenthal, H. and Hempel, G., 1969. Experimental studies in feeding and food requirements of herring larvae. In: J.H. Steele (Ed.): Marine Food Chains, p.344-364. Edinburgh: Oliver and Boyd.

Runnstrom, S., 1941. Quantitative investigations on herring spawning and its years fluctuations at the west coast of Norway. Fisk, Skr. Havundersøk., 6 (8), 71.

Saville, A., 1956. Eggs and larvae of haddock (*Gadus aeglefinus* L.) at Faroe. Mar. Res. (4), 27 pp.

Shelbourne, J.E., 1957. The feeding and condition of plaice larvae in good and bad plankton patches. J. Mar. Biol. Assoc. U.K. 36, 539-552.

Shirota, A., 1970. Studies on the mouth size of fish larvae. Bull. Jap. Soc. Sci. Fish. 36, 353-368.

Smith, P.E., 1970. The mortality and dispersal of sardine eggs and larvae. ICNAF/ICES/FAO Symposium on Stock Recruitment, Aarhus, Denmark, 7-10 July. Rapp. P.-v. Réun. Cons. perm. int. Explor. Mer 164, 282-292.

Steele, J.H., 1972. Factors controlling marine ecosystems. In: D. Dyrssen and D. Jagner (Eds.): The changing chemistry of the oceans. Nobel Symposium, Vol. 20, p. 209-221. Wiley Interscience Division.

Steele, J.H. and Edwards, R.R.C., 1970. The ecology of O-group plaice and common dabs in Loch Ewe. IV. Dynamics of the plaice and dab populations. J. Exp. Mar. Biol. Ecol. 4, 174-187.

Sysoeva, T.K. and Degtereva, S.A., 1965. The relation between the feeding of cod larvae and pelagic fry and the distribution and abundance of their principal good organisms. Spec. Publs int. Commn N.W. Atlant. Fish. (6), 411-416.

R. Jones
Marine Laboratory
Victoria Road
Aberdeen / GREAT BRITAIN

W.B. Hall
Marine Laboratory
Victoria Road
Aberdeen / GREAT BRITAIN

The Possible Density-Dependence of Larval Mortality and Adult Mortality in Fishes

D. H. Cushing

INTRODUCTION

The numbers of fish populations are probably regulated mainly during the larval drift, by a combination of density-dependent growth and density-dependent mortality. During this period, density is maximal; it declines sharply with age. Gulland (1965) distinguished a coarse and a fine control of numbers in a fish population. The coarse control probably occurs during the larval drift and the fine control might occur at a later stage in the life cycle. Cushing (1971) examined the relation between an index of density-dependence and fecundity and found that the index varied with the cube root of the fecundity, that is, with the distance between the larvae in the sea. Harris (in press) examined the critical period during which recruitment was determined by the time taken to grow through it. He looked specifically at the dependence of the critical period on stock and concluded that the stock and recruitment equation of Ricker (1958) could be written:

$$R = AP \exp - BP^{1/3} \qquad (1)$$

where R is recruitment,

P is stock,

A is the coefficient of density-independent mortality,

B is the coefficient of density-dependent mortality.

The cube root of the stock is also a function of fecundity and so of the distance apart of the larvae in the sea. The main part of the regulation of numbers may occur during the larval drift.

The variability of recruitment is low (0.3 to 2.0 orders of magnitude) as compared with the loss in numbers between hatching and recruitment (4 to 7 orders of magnitude); the total mortality is 99.99% or more. The reduction in numbers may be considered as loss in the face of a hostile environment or as a continuous adjustment of numbers in a benign environment. Cushing (in press) has suggested that natural mortality is a density-dependent function of age: the processes that start during the larval drift may continue into the mature age groups.

During their life cycles most fish grow through the trophic levels of a food chain. They start by feeding on a few algae and copepod nauplii, then transfer to copepods, and later to fishes; they ultimately become first- or second-order carnivores. Throughout their lives they are predatory and are themselves subject to predation. Indeed, predation must be the major form of death in the sea; even a sick animal dies by being eaten. A very simplified model of predation states that there are two processes that may be expressed in terms of the time spent in their operation: "handling" and eating food and searching. Handling and eating is probably a small part of the whole and the search time depends on prey density. If predation is density-dependent, the mortality suffered by the prey is also. Fish grow by 4 to 6 orders of magnitude in their lives and they pass through a succession of predatory "fields" in each of which the predator is larger but also much less numerous. The increment in searched volume per predator is much greater relatively than that in weight during a period of time. Growth is a function of age and as the fish grows through the succession of predatory fields it is subject to a density-dependent mortality that is a function of age.

Cushing (in press) developed an equation that describes natural mortality as a density-dependent function of age. The initial assumption is that:

$$N_t = N_{(t-\delta t)} e^{-kN_t \cdot \delta t} \doteq N_{(t-\delta t)} (1 - kN_t \cdot \delta t + \ldots)$$

where N is number, t is time and kN = M; the approximation is valid so long as $M < 0.5$. It can then be shown that

$$N_t = N_o/(1 + M_o t); \; M_o = (N_o - N_t)/N_t \cdot t \qquad (2)$$

where N_o is the initial number,

M_o is the initial mortality rate,

N_t is the number at time, t.

The mortality rate of the plaice larvae is 80%/month (Harding and Talbot (1973); or percentage survival in 30 days, N_{30}/N_o, = 20%. Then, $N_{30}/N_o = 0.2 = 1/1 + 30M_o$, $\therefore M_o = 0.133$. The initial number was established knowing M_o and T, the critical age, at which $N_t = 2$, i.e. $N_o = N_T(M_oT + 1)$, where T is the critical age and $N_T = 2$; the critical age is that at which the specific growth rates and the specific density-dependent mortality rates are equal. In this formulation the specific mortality rates from age to age are the same in each of the adult age groups, so long as N_o is high, i.e. so long as the female fish has a high fecundity. Consequently, the critical age can be established knowing the specific growth rates only, and in general the growth of adult fishes tends not to be density-dependent. In the virgin stock there can be no increment in fecundity with age after the critical age, and replacement within the cohort should be by then complete, which is why $N_T = 2$. Older fish contribute to the population fecundity; if the cohort has not replaced itself by the critical age, the fecundity of the older age groups should do so.

Fig. 1 shows the trend in mortality of the southern North Sea plaice with age, estimated in this way from larval life to the age of sixteen (= T). The average mortality rate during the first month is 80%, as given above. The subsequent mortality rates of 40% per month when they reach the beaches and 10-20% per month during the first winter and 10% per year during adult life are fitted to the curve quite well. From the initial assumption that natural mortality is a density-dependent function of age, the observations of mortality at ages after metamorphosis are quite well fitted.

Jones (in press) and Cushing and Harris (1973) have evolved models that describe the density-dependent mortality of fish larvae. In both models mortality depends upon the availability of food. Cushing and Harris, however, suggest that larvae that feed well swim quickly and avoid predators, but if food is scarce they grow more slowly and mortality increases. Fig. 2 shows the trend of mortality with age under variable food conditions, based on these models. With a sharp increase in food, mortality decreases and after a period of time the quantity of food reverts to the value before the increase. During the time period the rate of change of mortality is less than it would have been,

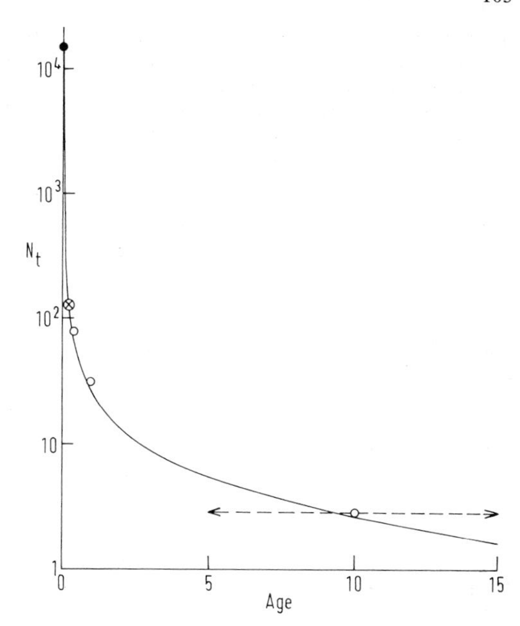

Fig. 1. The trend in natural mortality of North Sea plaice with age (Cushing, in press)

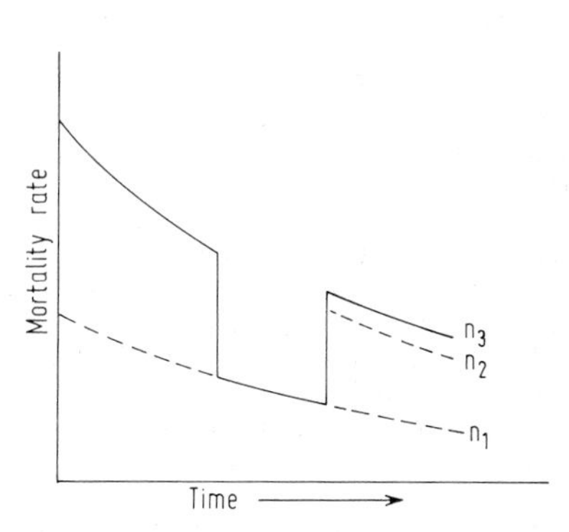

Fig. 2. The trend of mortality with age under variable conditions of food (Cushing, in press); after a period of time food increases sharply, and mortality (full line) is reduced. During the subsequent period the rate of change of mortality is less than it would have been before the food change. At the end of this period food reverts to its earlier density. Because numbers are relatively greater than they would have been, had there been no increment in food, the new mortality rate is higher. In such a way, mortality rate may be modified by patchiness in food. n_2 is the recruitment had there been no food change; n_1 is that had the food change not reverted to its original value; n_3 is that after the increase + decrease of food

had there been no increase in available food. At the end of the time period the numbers of larvae are greater than they would have been with no food increase, and as a consequence mortality is greater. It is well known that the distribution of food is patchy in the sea, so during the life cycle of the growing fish considerable variations may be expected. As the fish grows in its life cycle, the trend of density-dependent mortality is continuously modified in the way suggested in Fig. 2. It could be supposed that a density-independent fall in numbers occurred at a given instant of time; then mortality would decrease instantaneously and be modified subsequently in a density-dependent manner. Thus the variance about an average trend of mortality with age will be that of recruitment at the age of recruitment. Perhaps the density-dependent mortality is greater than the density-independent one, but such a speculation cannot yet be established.

If this mechanism is realistic, the cohort may be said to exploit the carrying capacity of the environment to the maximum. The carrying capacity is determined by the quantity of food available, but it is also sensitive to other environmental conditions. Because the growth rates and death rates of a cohort are established before recruitment, there is a sense in which each cohort has determined the pattern of replacement before recruitment. The cohort may be said to have an exploratory function, by which the food available has been exploited during the immature life cycle. In contrast, the stock has a conservative function in which the exploitations of available food, manifest as successive recruitments, are averaged in each year. In this paper an attempt is made to apply the method to fishes other than plaice.

METHODS

In the study of the plaice, the curve was located by estimating the critical age. In other fishes for which reliable data are available on larval mortality, the required data on growth rate are not available. It can be shown that:

$$N_2/N_1 = (1 + M_o\, t_1)/(1 + M_o\, t_2)$$

$$\therefore \quad M_o = \frac{(N_1 - N_2)}{(N_2\, t_2 - N_1\, t_2)}. \qquad (3)$$

Similarly,

$$N_o/M_o = (t_2 - t_1)\cdot N_1\cdot N_2/(N_1 - N_2). \qquad (4)$$

where M_o is the initial mortality rate, where N_1 is the number at time t_1, where N_2 is the number at time t_2.

It is possible to estimate N_o and M_o from the larval data alone, but it requires that t_1 and t_2 (in equation 3) are properly estimated. Where the data are presented in short time intervals, trial values of t_1 are used until the best fit to the data is obtained.

RESULTS

Graham et al. (1972) have studied the distributions in space and time of larval herring along the western coast of the Gulf of Maine. A Gulf III net and a Boothbay depressor trawl were used at a number of stations every 15 days in the Boothbay harbor area during the autumn between 1962 and 1967. The averaged figures for the 6 years were used.

Fig. 3 shows the original data and the fitted curve which was best fitted to a value of t_1 of 5, using equations 3 and 4. The original numbers were then scaled up by 1.50 to allow for this correction. At 90 days the curve obviously no longer fits the data; after 120 days there is no apparent mortality and the number at 180 days is entered

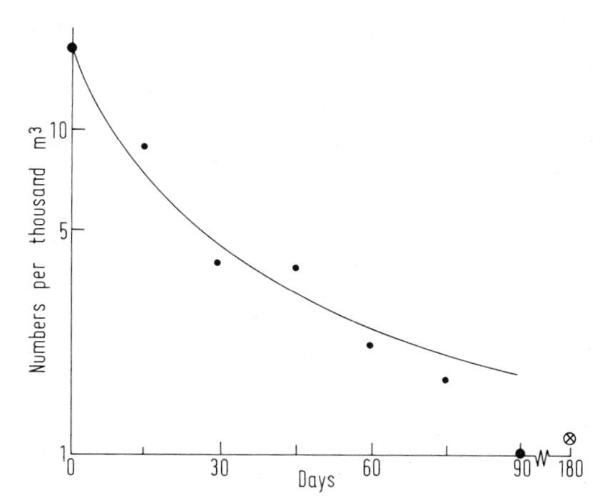

Fig. 3. The mortality of larval and post-larval herring in the area of Boothbay Harbor (from the data of Graham et al., 1972); the full circle shows the observed values at 90 days and the circled cross shows those at 180 days

on the figure. There is an increase in numbers in the interim, probably due to a shoreward migration. The curve in Fig. 3 is fitted from the initial observation. Taking into account a possible availability change at 90 days, the fit to data is reasonable.

Poulsen (1931) published data on the immature life cycle of the cod in the Belt Seas. This population is distinct, in spawning area and season, both from the North Sea cod and from the group that spawns around Bornholm. Eggs were sampled with a Hensen net in April and num-

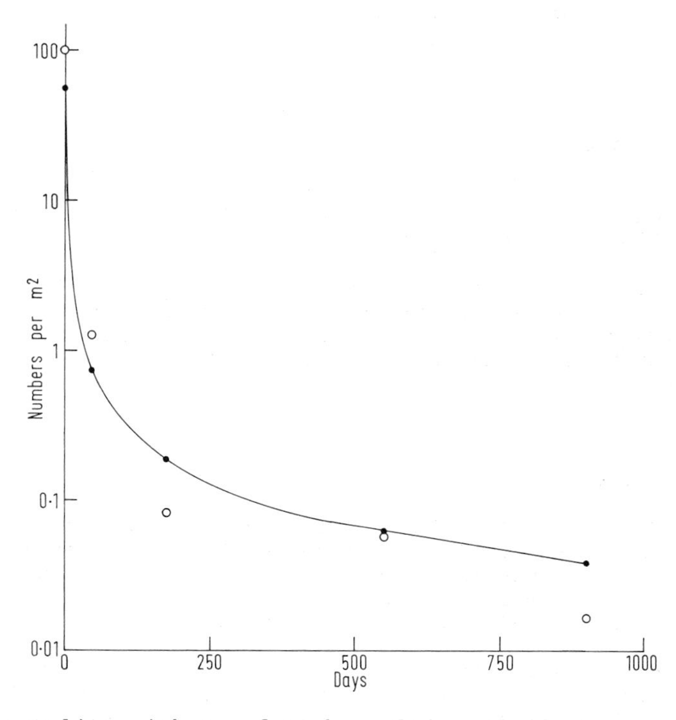

Fig. 4. The trend in mortality with age for the Baltic cod (from the data of Poulsen, 1931); full circle calculated value, circle observed value

bers were expressed as beneath 1 m^2. The larvae were sampled in late April or early May with a Petersen young fish trawl or a ring trawl and were expressed as number per area of 215 m^2. The immature fish were sampled with a small trawl and the numbers were expressed per 18 000 m^2 or per 3 000 m^2, according to whether the trawl caught the fish between the doors or between the wings; as any escape movement of such little fish must be a short distance, the smaller area was used in this paper. Table 1 shows the numbers/m^2 for year classes. The figures for the 5 year classes 1923-1927 were expressed as percentages of the initial numbers, and the percentages were averaged. M_o and N_o were calculated as given above; Fig. 4 shows the observations and the fitted curve.

Table 1. Numbers of cod eggs, larvae and immature cod beneath one square metre for seven year classes (Belt Seas, from Poulsen, 1931)

	1923	1924	1925	1926	1927	1928	1929	
Eggs	15.1	67.2	9.5	2.5	34.1	-	38.2	Apr
Larvae	0.293	0.0093	0.223	0.0326	0.242	0.116	0.195	Late Apr-May
O-group	0.038	0.0013	0.0133	0.00027	0.005	0.0047	0.006	Aug-Oct
I-group	0.018	0.0013	0.015	0.0002	0.004	0.003	-	Aug-Oct
II-group	0.006	0.0018	0.0033	0.00023	0.0013	-	-	Aug-Oct

Fig. 5 shows the mortality curves for the three species under consideration as percentages of initial numbers to the age of 20 years. The percentages for the cod data were based on interpolated values at 45 days, because the series started with eggs and not with larvae. As will be shown in the subsequent figure the initial mortality of eggs and larvae is probably higher in the cod, as might be expected from its higher fecundity. In other words the interpolated value at 45 days excludes this early mortality. The point of the figure is to show that the adult mortality rates of the three species calculated in this way are very similar.

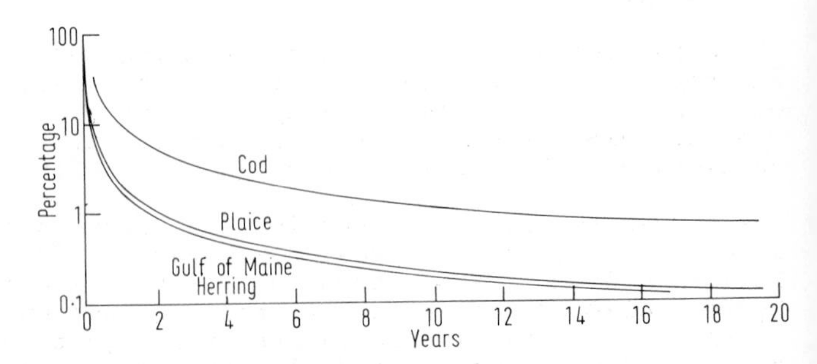

Fig. 5. The percentage numbers for Baltic cod, North Sea plaice and Gulf of Maine herring up to the age of 20 years; the cod series is started at an interpolated value at 45 days

Fig. 6 shows the mortalities calculated during the first year of life at intervals of 45 days. The data for the plaice and herring are probably comparable because larvae were sampled initially at an early stage in both sets of observations. The initial death rate of the cod includes egg mortality. Let us suppose that the cod eggs hatch over a period of 20-30 days and that the samples include eggs of a median age. Then the larvae would appear at 10-15 days; let us suppose that at 15 days there is a representative population of newly-hatched larvae. The arrow on the figure indicates this age of 15 days; it may be considered an overestimate when we recall that the loss of plaice eggs amounts to 80% (Harding and Talbot, 1973). The slope of the cod mortality from the larval stage is greater than that of the plaice

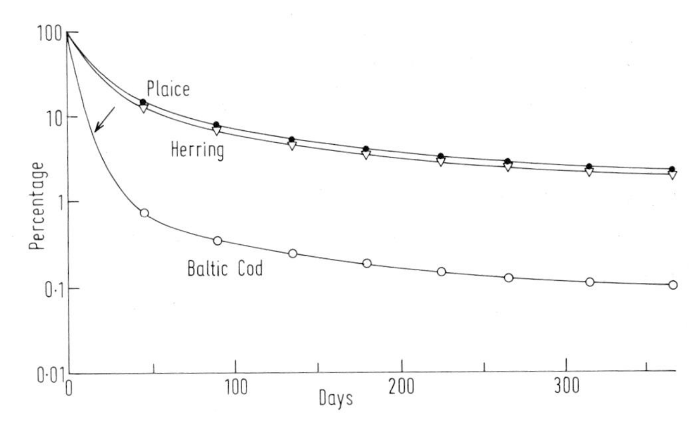

Fig. 6. The percentage mortalities for all three species for the first year at intervals of 45 days

and the herring. From fecundity alone, the mortality of plaice would be expected to be greater than that of herring. However, the death rates of demersal eggs may differ from those of pelagic eggs. The main conclusion from Fig. 5 is that differences between species in natural mortality, as formulated here, are generated at a very early stage.

DISCUSSION

The curves of numbers of plaice, herring and cod decline with age, sharply during the larval drift and less sharply in later ages. The data fit a curve of density-dependent mortality with age quite well. The reasoning leading up to this hypothesis is given in the introduction, and the results given tend to confirm the argument.

The curves for all three species are based on averages and therefore in each species they represent an average condition. Fig. 7 shows Poulsen's data of egg, larval O-, I- and II-group numbers (as n/m^2) by year classes, compared with the averaged curve. The curve is based on the average decline in numbers from egg to middle-staged larvae and it passes through a midpoint in the distribution of year classes at 45 days. However, it overestimates the numbers of II-group fish at 910 days, which suggests that it underestimates the mortality between larvae and O-group fish; this point is shown clearly on the figure.

The mortalities from O- to I-group and I- to II-group are estimated fairly well. A better fit to the later range of data might have been obtained by using the egg to O-group numbers or the larvae to O-group numbers; however, no information would be gained. Because the larvae were samples from 4 to 18+ mm it is difficult to separate the effects of mortality during the larval drift and on the nursery ground, but it would appear that a considerable mortality occurs between the two phases in the life cycle.

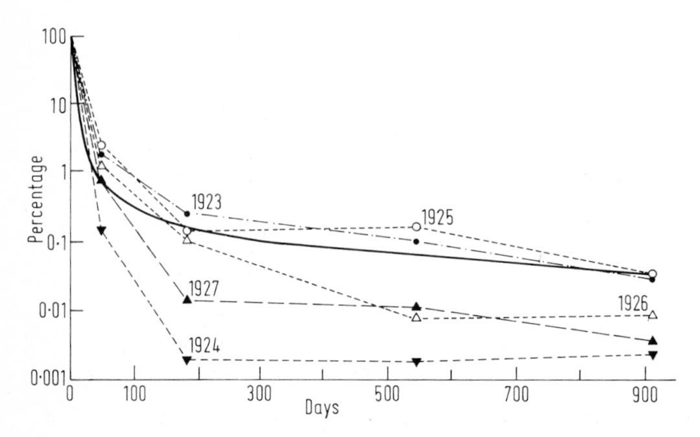

Fig. 7. Poulsen's data for individual year classes compared with the line fitted in Fig. 4

The range of the five year classes is just over an order of magnitude during the larval drift and just over 2 orders of magnitude on the nursery ground, but is reduced to 1 order at the end of the period. It is approximately the range of variability that might be expected in recruitment. The order of year classes is changed during the larval drift, but a change can occur on the nursery ground (Table 2). Thus the major density-dependent or compensatory processes occur during the larval drift, but some adjustment can occur at a later stage on the nursery ground. This is reflected in the reduction in range from 2 to 1 orders of magnitude on the nursery ground.

Table 2. The ranking of year classes at different stages during the life cycle

Stock	Larval drift	O-group	II-group
1924	1923	1923	1925
1927	1925	1925	1923
1923	1926	1926	1926
1925	1927	1927	1927
1926	1924	1924	1924

SUMMARY

The natural mortality of plaice is formulated as a density-dependent function of age; the rationale is that predation is density-dependent because of the time to search and that fish grow with age through a series of increasingly larger predatory fields. In this paper an attempt is made to fit the theory to data on the Boothbay Harbor herring and to the Baltic cod. The analysis of the Baltic cod data showed that the control mechanisms predominated during the larval drift in most years, but in one year class some control occurred on the nursery ground.

REFERENCES

Cushing, D.H., 1971. The dependence of recruitment on parent stock in different groups of fishes. J. Cons. perm. int. Explor. Mer 33, 340-362.

Cushing, D.H. (in press). The natural mortality of plaice.

Cushing, D.H. and Harris, J.G.K. (1973). Stock and recruitment and the problem of density dependence. Rapp. P.-v. Réun. Cons. perm. int. Explor. Mer 164, 142-155.

Graham, J.J., Chenoweth, S.R. and Davis, C.W., 1972. Abundance, distribution movements and lengths of larval herring along the western coast of the Gulf of Maine. Fish. Bull. U.S. Dept. Comm. 70 (2), 307-321.

Gulland, J.A., 1965. Survival of the youngest stages of fish and its relation to year-class strength. Spec. Publs int. Commn N.W. Atlant. Fish. 6, 365-371.

Harding, D.W. and Talbot, J.W. (1973). Recent studies on the eggs and larvae of the plaice (*Pleuronectes platessa* L.) in the Southern Bight. Rapp. P.-v. Réun. Cons. perm. int. Explor. Mer 164, 261-269.

Harris, J.G.K. (in press). The effect of density-dependent mortality on the shape of the stock and recruitment curve. J. Cons. perm. int. Explor. Mer.

Jones, R. (1973). Density dependent regulation of the numbers of cod and haddock. Rapp. P.-v. Réun. Cons. perm. int. Explor. Mer 164, 156-173.

Poulsen, E.M., 1931. Biological investigations upon the cod in Danish waters. Meddr Kommn Danm. Fisk.-og Havunders., Fiskeri IX(I), 149 pp.

Ricker, W.E., 1958. Handbook of computations for biological statistics of fish populations. Bull. Fish. Res. Bd. Can. 119, 300 pp.

D.H. Cushing
Ministry of Agriculture, Fisheries and Food
Fisheries Laboratory
Lowestoft / GREAT BRITAIN

Larval Abundance in Relation to Stock Size, Spawning Potential and Recruitment in North Sea Herring

K. H. Postuma and J. J. Zijlstra

INTRODUCTION

In recent years the idea that larval abundance estimates provide valid (relative) measures for spawning stock size has become more generally accepted in North Sea herring research (Anon, 1971b, 1972). The application of this stock-size parameter was promoted by increasing difficulties encountered in using catch-effort statistics of the rapidly changing herring fisheries. The use of larval data for this purpose presupposes the absence of major variations in egg and larval mortalities between years and over longer periods. Provided these mortalities do not differ greatly between spawning areas, the estimates would also allow a comparison of the size of local spawning stocks (given length composition and fecundity data)(Zijlstra, 1970; Anon, 1971a, b). As larval abundance is supposed to be related to the size of the spawning stock, it provides finally a useful parameter in stock-recruitment studies (Burd and Holford, 1971). The realization of the probable value of larval abundance as a tool in population studies initiated an international collaboration to improve data collection by coordinated surveys with standardized methods (Boëtius and McKay, 1970; Saville, 1970; Wood, 1971; Schnack, 1972; Zijlstra, 1972).

In view of the considerable effort and costs spent annually by research institutes to collect data on larval numbers and distribution, an appraisal of the evidence connecting spawning-stock size and larval abundance seems highly desirable.

Thus far the evidence is mainly derived from the Downs area, where Cushing and Bridger (1966) found that larval abundance followed closely their index for stock size. However, due to the use of different plankton gears in the 14 seasons considered, their case depended largely on the correct application of conversion factors to cope with different net efficiencies. For the Dogger Bank area Zijlstra (1970) suggested a possible relation between a stock-size parameter and the abundance of small larvae, according to 10 years of observations Saville (1971) however, considering 16 years of observations, noted that in the northwestern North Sea changes in larval abundance followed the general trend, but not the annual variations, in his estimates of spawning stock size.

In conclusion, the evidence for a spawning stock size-larval abundance relation, though present, does not seem to be overwhelming, for which difficulties in obtaining complete long-term series of data on larval abundance can partly be blamed. This paper aims at making a reappraisal of the material on larvae and stock size for two areas, the Downs and the Dogger, extending the Downs series with recent data from a period of a slight revival of this stock. Larval abundance estimates will also be compared with data on recruitment and, following Postuma's (1971) suggestion of an effect of temperature on recruitment, with temperature conditions on the spawning grounds and in the coastal nursery.

MATERIAL AND METHODS

Downs Area. Larval abundance estimates were obtained from surveys carried out in the years 1951/52 and 1955/56 - 1971/72 by English, Dutch, and German research vessels. Grateful acknowledgement for permission to include unpublished English data is made to the Director of the Fisheries Laboratory, Lowestoft, England. Table 1 gives an account of the surveys utilized.

English surveys in the winters of 1946/47, 1947/48, and 1950/51 have been omitted from the material. These surveys were carried out with the Hensen net which is inefficient for herring larvae; larval abundance estimates must be raised by a correction factor obtained indirectly by a comparison with a Heligoland Larval Net (HLN)(Bridger, 1961). The surveys in which the HLN was used are included in the present analysis, the correction factors applied being based on a fair number of comparisons with the standard gear, the Gulf III sampler. In Table 1 the gear used in the surveys is indicated.

For each survey larval abundance estimates were computed partly with the density-contour planimeter method and partly with the stations-square method, which should give comparable results (Sette and Ahlstrom, 1948; Schnack, 1972). Recalculation of some of Bridger's original data and the description of his method to compute larval abundance (Bridger, 1961) indicated the possibility of a serious error in his estimates. We are greatly indebted to Mr. J. Wood of the Fisheries Laboratory, Lowestoft, who assisted us in adjusting the English estimates, taking the trouble to recalculate all English survey data.

One problem concerned surveys which were incomplete, covering only part of the spawning area or spawning season. As for the spawning season, the months with the highest larval abundance estimates proved to be December and January (see Bridger) and only surveys in these months have been considered. In most years at least one survey was made in each of these months. Larval abundance estimates of the two months were found to be highly correlated ($r = 0.67$, $n = 15$), with larval abundance in January, being on average 3.4 times higher than in December (Fig. 1). This factor, which is significantly different from unity, was applied to obtain estimates of larval abundance in months where information was lacking (Table 1). Several surveys did not cover the whole spawning area, but only the part north or south of Dover Straits. Comparisons showed the two larval concentrations to be about equally important, which is not a very precise description. As accepting abundance estimates of incomplete surveys meant a serious under-estimation of larval production, the decision that production in both areas should be considered equal was taken as the best approac It was found impossible, because of lack of information, to divide the larval data into small and larger larvae, as was done for the Dogger area.

A final estimate of larval abundance was obtained by taking the means of the surveys in the months December and January separately and avera ing the two monthly means. No attempt was made to obtain an estimate of larval production by integrating larval abundance estimates in time the data available did not allow such a procedure.

Estimates of spawning potential were derived from two sources:

(i) From the catch-per-unit-of-effort estimates of the trawl and drift net fisheries in the area, given in the Report of the North Sea Herrin

Table 1. Outline of Downs larval surveys used, giving season, data, country (E = England, G = Germany, N = Netherlands), gear (HLN = Heligoland Larval Net, G III = Gulf III sampler), area (C = complete area, N = north, and S = south of Dover Straits), and the number or larvae

Season	Date	Country	Number of larvae x 10^{-9}	Gear	Area
1951/52	10–18 Dec.	E	274	HLN	C
	2– 7 Jan.	E	1,584	HLN	N
	23–30 Jan.	E	201	HLN	N
1955/56	7–10 Jan.	E	147	HLN	N
	25–30 Jan.	E	218	HLN	N
1956/57	4– 7 Dec.	E	158	G III	C
	8–17 Jan.	E	229	HLN	N
1957/58	3–10 Dec.	E	34	G III	C
	16–23 Dec.	E	77	G III	C
	2–10 Jan.	E	21	HLN	N
	21–27 Jan.	E	13	HLN	N
1958/59	2–16 Dec.	E	105	G III	C
	15–19 Dec.	E	99	G III	C
	10–29 Dec.	G	214	G III	C
1959/60	30 Nov.– 9 Dec.	N	44	G III	N
	10–18 Dec.	E	1	G III	N
	10–12 Jan.	G	14	G III	C
1960/61	6– 9 Dec.	N	215	G III	C
	13–17 Dec.	N	309	G III	C
	12–27 Jan.	G	64	G III	C
1961/62	29 Nov.–15 Dec.	N	209	G III	C
	13–19 Dec.	E	76	G III	C
	4– 8 Jan.	N	518	G III	S
	23–31 Jan.	E	48	G III	C
	14–18 Jan.	G	184	G III	C
1962/63	14–19 Dec.	E	0	G III	C
	9–15 Jan.	N	87	G III	S
	23–29 Jan.	E	15	G III	N
1963/64	6–14 Jan.	N	18	G III	C
	9–18 Jan.	E	27	G III	C
1964/65	8–17 Dec.	E	15	G III	C
	5– 8 Jan.	N	2	G III	S
1965/66	3–12 Dec.	E	2	G III	C
	7–17 Dec.	N	28	G III	C
	7–14 Jan.	E	7	G III	C
1966/67	7–16 Dec.	N	1,5	G III	N
1967/68	8–18 Dec.	E	37	G III	C
	12–20 Dec.	N	29	G III	C
	2–10 Jan.	E	50	G III	C
1968/69	9–17 Dec.	N	22	G III	N
	18–23 Jan.	E	10	G III	C
1969/70	8–23 Dec.	N	25	G III	C
	3– 6 Jan.	E	39	G III	C
	5–15 Jan.	N	259	G III	C
1970/71	7–18 Dec.	E	61	G III	C
	7–24 Dec.	N	210	G III	C
	5–14 Jan.	E	134	G III	C
	5–23 Jan.	N	149	G III	C
	20–23 Jan.	E	74	G III	C
1971/72	13–22 Dec.	N	10	G III	C
	3–25 Jan.	N	8	G III	C
	18–24 Jan.	E	3	G III	C

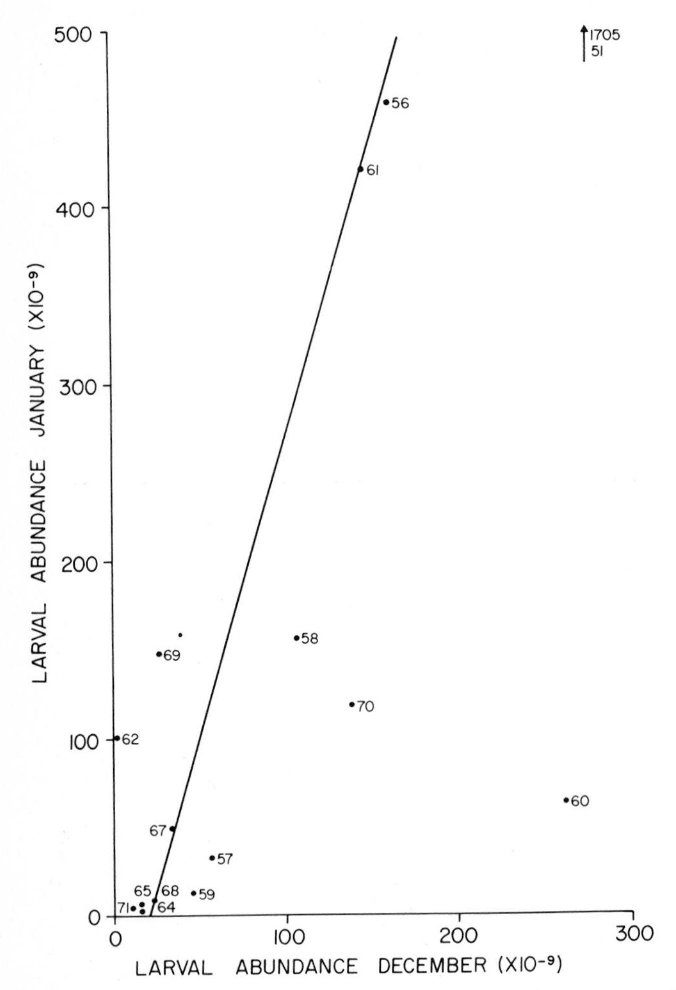

Fig. 1. Relation between December and January larval abundances in the Downs area

Assessment Working Group (Anon, 1972). By converting weights into numbers and using the annual length-distribution data of the drift-net caught herring between October and December, an estimate of egg production was obtained by applying the fecundity-length equation $F = 2.826\ L^3$ for the Downs herring (Zijlstra, 1973).

An index of spawning potential was obtained by combining the two sets of estimates, which were found to be closely related (Fig. 2a, b).

(ii) From the total catch in areas IVc and VIId-e (Southern Bight and English Channel) taken from the North Sea Assessment Working Group Report (Anon, 1972), assuming effort to be constant during the period, one finds some support in a high correlation between spawning potential as defined under (i) and total catch ($r = 0.85$, $n = 19$).

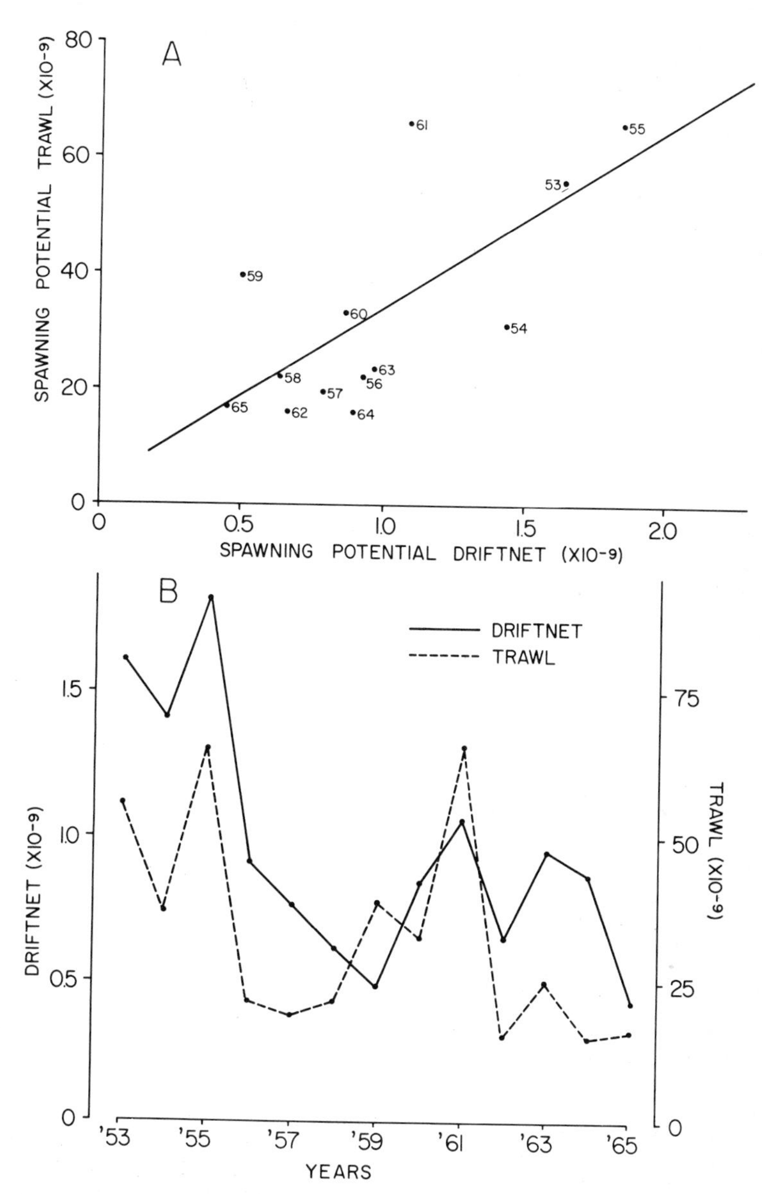

Fig. 2. Relation between spawning potential computed from drift-net and trawl data in the Downs area. (A) shown as a regression (figures refer to years); (B) shown as a time-series

Two sets of estimates of year-class strength were used:

(a) R_3 estimates for the years prior to 1965, obtained from the North Sea Assessment Working Group Report (Anon, 1971b), provided by Burd. This series was completed with unpublished Dutch data for the years 1965-1969.

(b) The total catch in numbers of 3-year-old herring (C_3) derived from the Assessment Working Group Report (Anon, 1972). The use of these actual catch data supposes full recruitment at age 3 years and a constant effort in the years considered.

As with spawning potential and total catch, R_3 and C_3 were found to be highly correlated ($r = 0.83$, $n = 14$).

Doggerbank. The material used for the Dogger area is fully described by one of the authors (Zijlstra, 1970). It includes abundance estimates of small larvae (<11 mm) in a restricted area on the southwestern slope of the Doggerbank and of larger larvae (11-15 mm) in the whole area sampled east of 1° E. The spawning potential of the Dogger stock and an estimate of recruitment at 3 years of age (R_3) were obtained from routine sampling by the Dutch laboratory on herring catches from the area. A new estimate of the abundance of small larvae is introduced here by removing, as far as possible, the yolk-sac larvae from the material, which are thought to be poorly sampled, as they occur in the vicinity of the bottom outside the reach of the sampling instrument.

RESULTS

Downs. Larval abundance estimates have been compared with the two measures for spawning stock size, i.e. spawning potential and total catch. The results are shown in Figs. 3 and 4, both as regressions and as time series. The two stock-size parameters used appear to be related to larval abundance, the correlation coefficients being 0.63 (spawning potential) and 0.67 (total catch) with 18 observations ($p < 0.01$). This demonstrates that a removal of the less reliable larval data of the 1946, 1947 and 1950 seasons, for which reliable estimates of stock size were also hard to obtain (Cushing and Bridger, 1966), does not affect the existence of a relationship. In our case larval abundance has been compared with rather simple and straight forward stock-size parameters.

The time-series, presented in Figs. 3b and 4b, also show a reasonably good correspondence between "stock size" and larval abundance, often in some detail. Thus the increase in spawning potential and total catch, experienced in the years after 1966, is accompanied by a rise in larval abundance. Only in the period 1962-1966 may a discrepancy be noted between the stock-size parameters and larval abundance, the latter being on too low a level. It seems unlikely that poor sampling in the larval surveys was responsible for this discrepancy, as it concerns a period of 5 years (see also Table 1).

Cushing and Bridger (1966) tried to relate spawning-stock size and larval abundance estimates with the strength of the resulting year classes. In the case of the larval abundance-recruitment relation they suggest the presence of a convex curve rising to an asymptote, a conclusion mainly founded on the somewhat doubtful 1946, 1947, and 1950 larval data.

Here the stock-size parameters and larval abundance estimates have been compared with the two recruitment indices R_3 and C_3, as shown

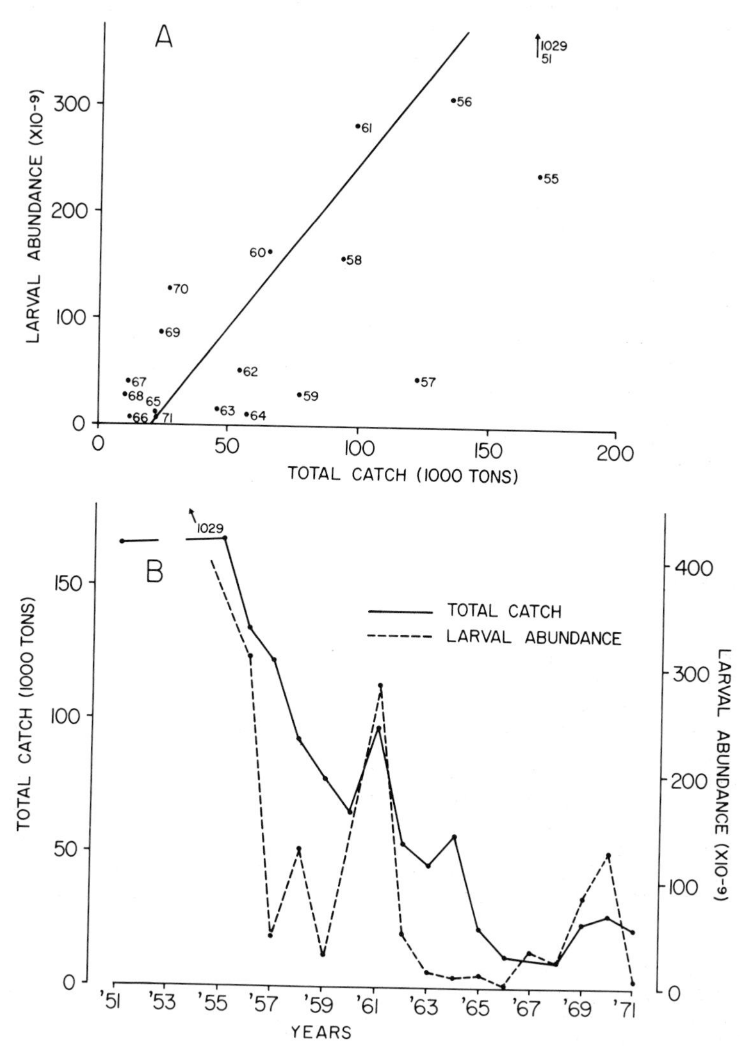

Fig. 3. Relation between spawning potential and larval abundance in the Down area. (A) shown as a regression (figures refer to years); (B) shown as a time-series

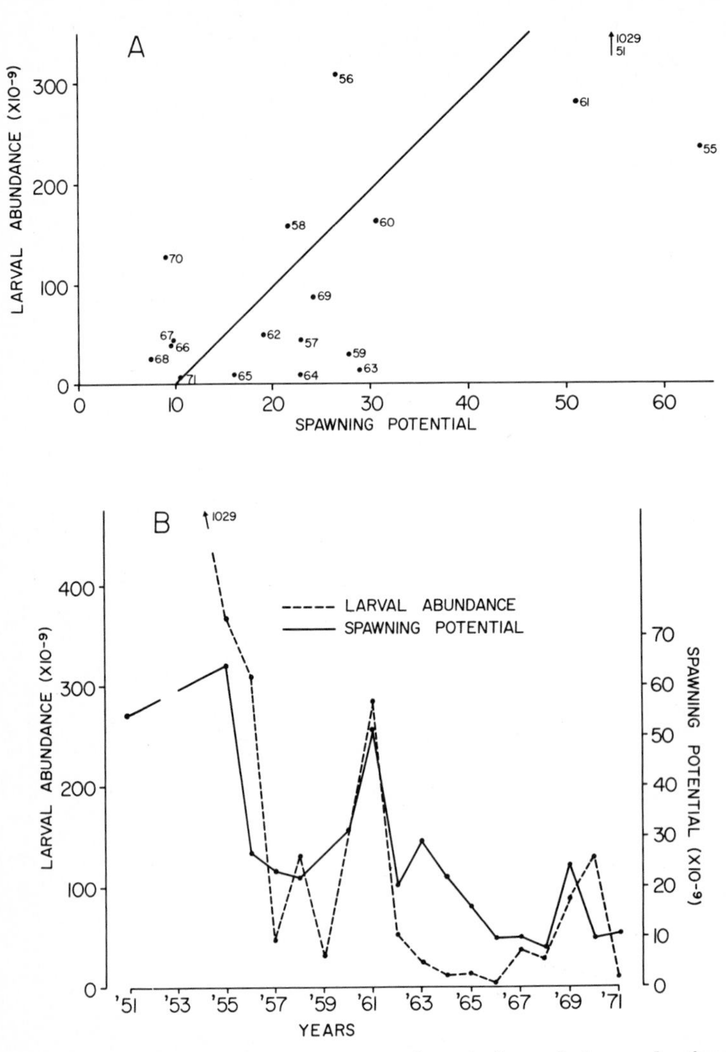

Fig. 4. Relation between total catch and larval abundance in the Downs area. (A) shown as a regression (figures indicate years); (B) shown as a time-series

in Figs. 5 and 6. It will be recalled that C_3 simply represents the total catch of 3-year-old herring in the Downs area. In the absence of a clear suggestion for non-linear relationships, the linear regressions and corresponding correlations have been computed. The correlation coefficients are given in Table 2, indicating a relation of C_3 with both larval abundance and stock-size indices. For the other measure of year-class strength (recruitment), R_3, no association could be established. As the distribution of all parameters used was far from normal, a log-transformation has been applied to the data, which does not alter the correlation coefficients appreciably.

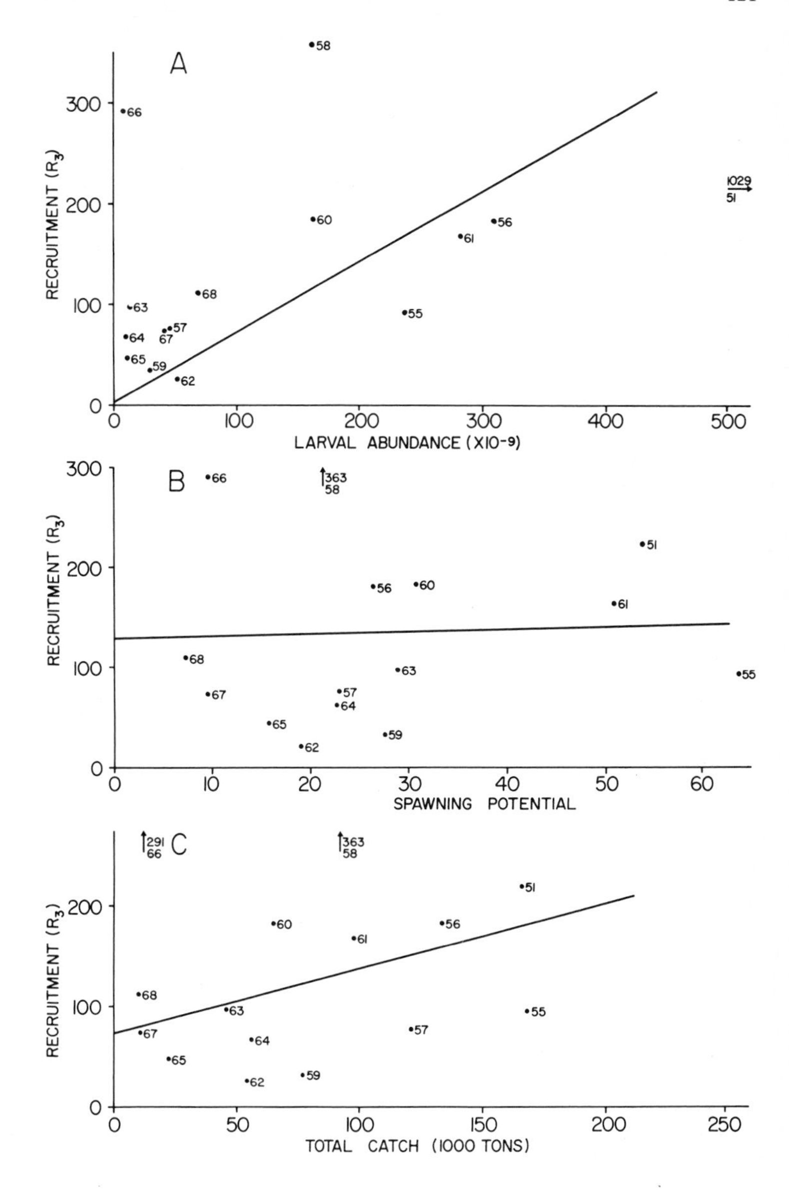

Fig. 5. Regressions of recruitment (R_3) on (A) larval abundance, (B) spawning potential, and (C) total catch, in the Downs area

Finally, following Postuma's suggestion that temperature near the spawning ground and the coastal temperature in the spring following the year of spawning show a relationship with recruitment estimates, these temperatures were related to the recruit indices. In addition, the spawning ground temperature was compared with the larval abundance estimates. The temperatures taken were the same as used by Postuma, being the December surface temperature measured on the Northhinder lightvessel as spawning ground temperature, and the June temperature

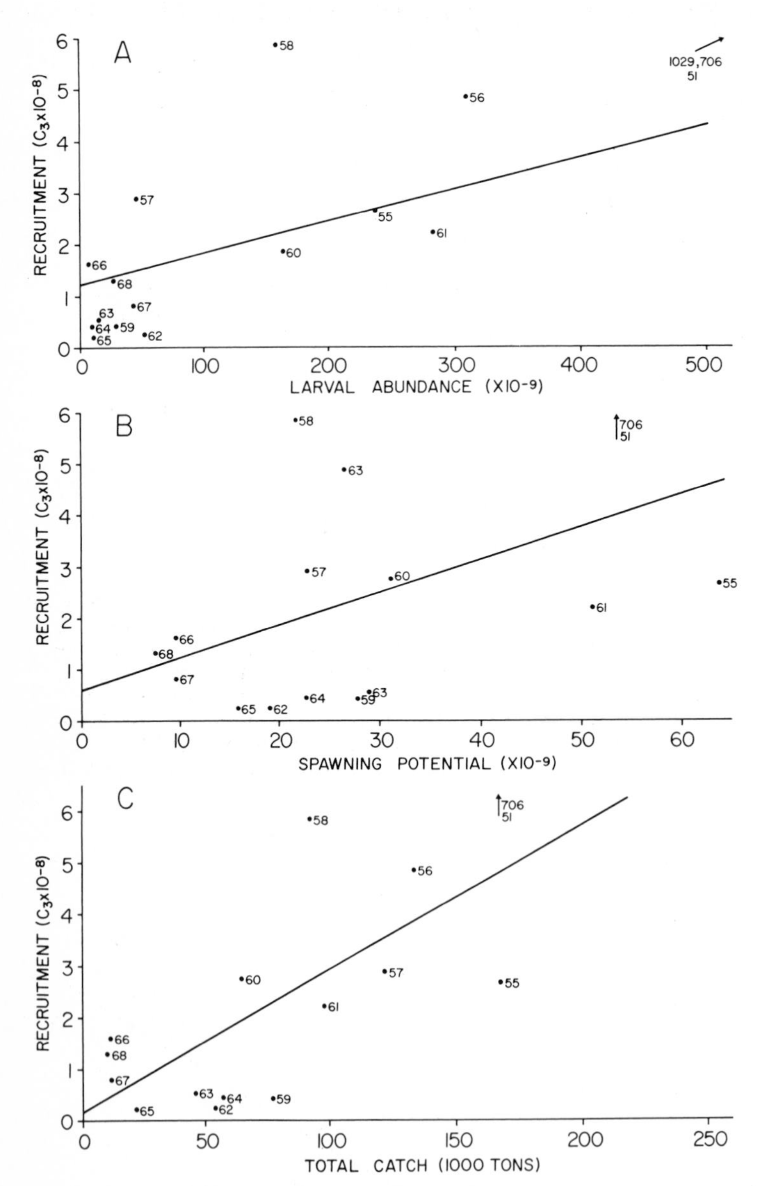

Fig. 6. Regressions on recruitment (C_3) on (A) larval abundance, (B) spawning potential, and (C) total catch, in the Downs area measured on the Texel lightvessel as the coastal spring temperature. Table 3 shows the correlation coefficients found.

It appears that the spawning ground temperature is positively related to both larval abundance and recruitment indices, though the correlation with R_3 is low. This is in general accordance with Postuma's earlier findings, which were based on a largely different set of data. No relationship between coastal temperature and recruitment could be shown.

Table 2. Correlation coefficients of larval abundance and stock-size indices with two measures of recruit-strength, R_3 and C_3 (^{xx} = $p < 0.01$, ^{x} = $p < 0.05$), n = 15

	R_3	C_3
Larval abundance	0.34	0.78^{xx}
Spawning potential	0.00	0.50^{x}
Total catch	0.14	0.74^{xx}

Table 3. Correlation coefficients between temperature near the spawning ground and in the coastal area, following the year of spawning, with larval abundance and recruitment parameters (* = $p < 0.05$), n = 15

	Temperature spawning ground	Coastal temperature
Larval abundance	0.47 *	--
R_3	0.12	0.10
C_3	0.49 *	-0.03

Doggerbank. As already suggested by Zijlstra (1970), spawning potential and abundance of small larvae (<11 mm) were found to be possibly associated ($r = 0.58$, $n = 9$, $p < 0.05$). However, no significant relationship could be established between spawning potential and the abundance of larger larvae or total larval abundance ($r = 0.43$ and 0.50 respectively with 9 observations). Removal of the yolk-sac larvae tended to enhance the association between spawning potential and abundance of small larvae slightly ($r = 0.67$, $n = 9$, $p < 0.025$). This relationship is shown in Fig. 7.

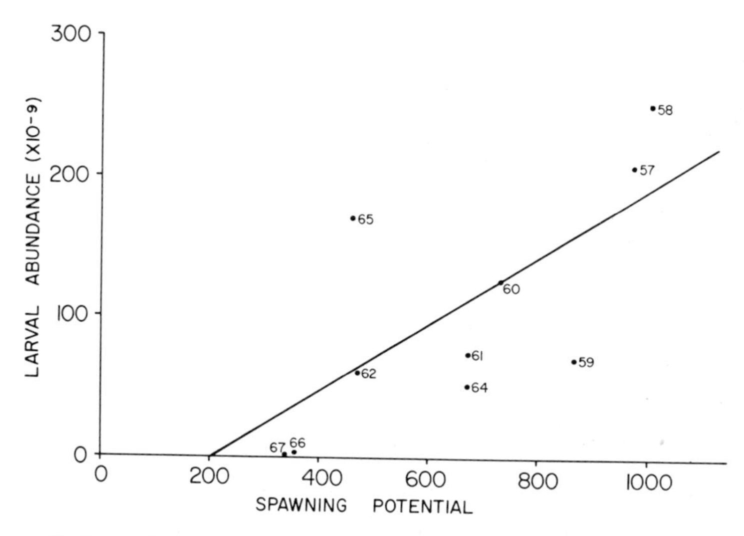

Fig. 7. Regression of larval abundance (larvae <11 mm, yolk-sac larvae omitted) on spawning potential in the Dogger area

Neither spawning potential nor larval abundance indices were significantly related to recruitment, except for a significant correlation with the abundance of larger larvae of 11-15 mm ($r = 0.71$, $n = 9$, $p < 0.025$).

As in the Downs area, estimates of the temperature on the spawning ground and in the coastal area have been compared with recruitment and larval abundance. For the spawning ground, bottom temperatures on the western side of the Doggerbank, at a depth of 20-30 fathoms, collected in the second half of September have been taken, the coastal temperature being measured in April on the Vyl lightvessel (Postuma, 1971). Table 4 shows the correlation coefficients, demonstrating a negative association of larval abundance and recruitment with spawning ground temperatures, which generally agrees with Postuma's observations. Contrary to Postuma's findings, the coastal temperature showed no relation to recruitment in the years considered.

Table 4. Correlation coefficients and their significance between temperatures on the spawning ground (Dogger area) and the coastal area (Danish coast) with larval abundance and recruitment (* $p \approx 0.05$), $n = 9$

	r
Spawning temperature - larval abundance	-0.57 *
Spawning temperature - recruitment (R_3)	-0.50
Coastal temperature - recruitment (R_3)	-0.26

DISCUSSION

From the results presented in the preceding section, three main conclusions can be drawn concerning the relations between spawning stock, larval abundance, recruitment, and the effect of temperature on the spawning ground.

(i) The data from both areas support the idea of a relation between a measure of spawning stock-size and larval abundance. It is true that in both cases studied the relation is largely dependent on a trend in a time-series, but in the long Downs series an apparent revival in the spawning stock in the late sixties was accompanied by a corresponding increase in larval abundance. Moreover, one has to consider that the relation appeared in both areas, was observed before in the Downs area in a partly different set of data, and that a suggestion of a similar relation is present for the Buchan and Shetland regions (Saville, 1971). In the light of these considerations the conclusion, that indeed larval abundance may be taken as an expression of spawning stock size, seems justified.

In this connection it should be noted that the regressions of larval abundance on spawning potential (Figs 3a, 4a and 7) do not pass through the origin, in both regions, a point already discussed by Bridger (196 Cushing and Bridger (1966), and Zijlstra (1970). A lower viability of the spawning products of recruit spawners (Cushing and Bridger, 1966) or an increasing efficiency of the fisheries on a declining stock (Zijlstra, 1970) have been mentioned as possible causes.

(ii) The material from both areas suggests that larval abundance was affected by temperature conditions on the spawning grounds. That this temperature effect was positive in the Downs area, but negative in the Dogger region, is suggestive in the light of former findings of similar relations between spawning temperature and recruitment (Postuma, 1971). Postuma argued, mentioning rearing experiments by Blaxter (1956), that probably both Dogger and Downs eggs would have the highest hatching success with temperatures of approximately 12°C; in the Dogger region temperatures at spawning time range from 12°C upwards and in the Downs area from 12°C downwards. If this suggestion were true, larval abundance would not only be determined by spawning potential but also by spawning temperature. To pursue this point somewhat further by statistical analysis, partial correlations were computed for both areas taking into account spawning potential, total larval abundance, and temperature in the spawning season. The results of this analysis are shown in Table 5. It will be observed that the partial correlations indeed indicate that larval abundance is dependent on spawning potential and spawning temperature. Subsequent analysis of variance showed only 30% of the total variance in Downs and 36% in the Dogger larvae to remain unexplained after elimination of the effect of spawning potential and spawning temperature, accepting linear relations. These findings tend to support Postuma's ideas, based on a study of the relations between spawning potential, spawning temperature, and recruitment.

Table 5. Single and partial correlations for Downs and Dogger herring between spawning potential, total larval abundance and temperature on the spawning ground (1 = spawning potential, 2 = total larval abundance, 3 = temperature at spawning; $^{xxx} = p < 0.01$, $^{xx} = p < 0.025$, $^{x} = p < 0.05$)

	Downs (n = 15)	Dogger (n = 9)
Single Correlations	$r_{12} = 0.75^{xxx}$	$r_{12} = 0.50$
	$r_{13} = 0.14$	$r_{13} = 0.10$
	$r_{23} = 0.47^{x}$	$r_{23} = -0.57^{x}$
Partial Correlations	$r_{12.3} = 0.78^{xxx}$	$r_{12.3} = 0.68^{xx}$
	$r_{13.2} = -0.36$	$r_{13.2} = 0.54$
	$r_{23.1} = 0.56^{xx}$	$r_{23.1} = -0.72^{xx}$

(iii) The results concerning the relationship between spawning potential or larval abundance on one side and measures for recruitment on the other are less easily interpreted than the two preceding items. Cushing and Bridger (1966), Burd and Holford (1971), and Postuma (1971) decided that spawning potential or larval abundance (Burd and Holford) determined recruitment in the Downs stock of herring in the post-war period. This study supports these conclusions only as far as one of the recruitment parameters used (C_3) is concerned, but not for the other (R_3).

For the Dogger area no support is found for a spawning potential-recruitment relationship, in agreement with Postuma's observations for this stock, whereas the evidence for an association between larval production and recruitment seems rather weak, being only based on the relationship of recruitment with the abundance of the larger larvae. A partial correlation analysis, in which spawning potential, temperature, and recruitment (C_3 in the case of the Downs stock) were considered, yielded no results, partial correlation coefficients being similar to those of the single correlations. Of greater interest, however, is the effect of larval abundance, probably a product of eggs spawned and temperature conditions on the spawning ground, and coastal temperature on subsequent recruitment. A partial correlation analysis is shown in Table 6.

Table 6. Single and partial correlations for Downs and Dogger herring between recruitment, larval abundance and coastal temperature (1 = larval abundance, 2 = temperature coast, 3 = recruitment (C_3 in Downs, R_3 in Dogger); $^{xxx} = p < 0.01$, $^{xx} = p < 0.025$, $^{x} = p < 0.05$)

	Downs (n = 15)	Dogger (n = 9)
Single Correlations	$r_{12} = -0.03$	$r_{12} = 0.00$
	$r_{13} = 0.78^{xxx}$	$r_{13} = 0.45$
	$r_{23} = 0.00$	$r_{23} = -0.26$
Partial Correlations	$r_{12.3} = -0.03$	$r_{12.3} = 0.14$
	$r_{13.2} = 0.78^{xxx}$	$r_{13.2} = 0.47$
	$r_{23.1} = 0.00$	$r_{23.1} = -0.29$

It is obvious that the analysis did not support the idea that recruitment is dependent on larval abundance and coastal temperature in the spring, following the year of spawning. It seems unlikely that these largely negative results were caused by the use of a linear model to relate spawning potential or larval abundance to recruitment.

An additional difficulty, when trying to relate recruitment to stock size or temperature for the Downs area, is offered by the situation around the 1966/67 season, when evidently a larger year class was created by this stock. Spawning potential and other parameters of stock size were not particularly high that year, temperature near the spawning ground was even low in December, while in the series of observations on larval abundance the 1966/67 season gave the lowest estimate but was also easily the most poorly sampled (only one survey in December on the northern spawning grounds with the 1966 year class appearing at recruitment on the southern grounds, in the English Channel). It seems therefore impossible from the available data to decide what caused the revival of the Downs stock, an event which evidently started in the 1966/67 season.

In conclusion, evidence to connect recruitment with spawning stock size or with larval abundance is present, but not completely convincing.

SUMMARY

(i) From the association of stock-size parameters with estimates of larval abundance in the Downs and Dogger herring it is concluded that larval abundance offers a reasonable measure for spawning stock size.

(ii) Larval abundance seems not only dependent on the size of the spawning stock, but also on temperature conditions in the spawning area, as could be shown both for the Downs and the Dogger regions in the years considered. The relation was positive in the Downs area and negative in the Dogger herring.

(iii) Though evidence connecting spawning stock-size or larval abundance with subsequent recruitment was present, the data on this point are not in complete conformity.

REFERENCES

Anon, 1971a. Report on the state of the herring stocks around Ireland and northwest of Scotland. ICES, Coop. Res. Rep. Series A, 21, 1-29.

Anon, 1971b. Report of the North Sea Herring Assessment Working Group. ICES, Coop. Res. Rep. Series A, 26, 1-56.

Anon, 1972. Report of the North Sea Herring Assessment Working Group. ICES, Doc. CM 1972/H2.

Blaxter, J.H.S., 1956. Herring rearing. 2. The effect of temperature and other factors on development. Mar. Res., 5, 1-19.

Boëtius, I. and McKay, D.W., 1970. Report on the international surveys of herring larvae in the North Sea in 1968. ICES, Coop. Res. Rep. Series A, 19, 18-30.

Bridger, J.P., 1961. On fecundity and larval abundance of Downs herring. Fish. Invest., Lond., 2 (23), 1-30.

Burd, A.C. and Holford, B.H., 1971. The decline in the abundance of Downs herring larvae. Rapp. P.-v. Réun. Cons. perm. int. Explor. Mer 160, 99-100.

Cushing, D.H. and Bridger, J.P., 1966. The stock of herring in the North Sea and changes due to fishing. Fish. Invest., Lond., 2 (25), 1-123.

Postuma, K.H., 1971. The effect of temperature in the spawning and nursery areas on recruitment of autumn-spawning herring in the North Sea. Rapp. P.-v. Réun. Cons. perm. int. Explor. Mer 160, 173-183.

Saville, A., 1970. Report of the International Surveys of Herring Larvae in the North Sea in 1967. ICES, Coop. Res. Rep. Series A, 19, 2-17.

Saville, A., 1971. The distribution and abundance of herring larvae in the northern North Sea, changes in recent years. Rapp. P.-v. Réun. Cons. perm. int. Explor. Mer 160, 87-93.

Schnack, D., 1972. Report on the International Surveys of Herring Larvae in the North Sea and Adjacent Waters in 1971/72. ICES, Doc. CM 1972/H28.

Sette, O.E. and Ahlstrom, E.H., 1948. Estimation of abundance of the eggs of the Pacific Pilchard (*Sardinops caerulea*) off southern California during 1940 and 1941. J. Mar. Res., 7, 511-542.

Wood, R.J., 1971. Report on the International Surveys of Herring Larvae in the North Sea and Adjacent Waters in 1969/70. ICES, Coop. Res. Rep. Series A, 22, 3-36.

Zijlstra, J.J., 1970. Herring larvae in the central North Sea. Ber. dtsch. wiss. Komm. Meeresforsch., 21, 92-115.

Zijlstra, J.J., 1972. Report on the International Surveys of Herring Larvae in the North Sea and Adjacent Waters in 1970/71. ICES, Coop. Res. Rep. Series A, 28, 1-24.

Zijlstra, J.J., 1973. Egg weight and fecundity in North Sea herring (*Clupea harengus*). Neth. J. Sea Res., 6, 183-204.

K.H. Postuma
Netherlands Institute for Fishery Investigation
Haringkade 1
Postbus 68
Ymuiden / NETHERLANDS

J.J. Zijlstra
Netherlands Institute for Sea Research
P.O. Box 69
Texel / NETHERLANDS

Relations between Egg Production, Larval Production and Spawning Stock Size in Clyde Herring

A. Saville, I. G. Baxter, and D. W. McKay

INTRODUCTION

In 1957 the Marine Laboratory, Aberdeen conducted a preliminary survey of herring egg and larval distributions over Ballantrae Bank, the main spawning ground of the Clyde spring-spawning herring stock. In 1958 more systematic surveys were begun and, with some breaks due to unavailability of research vessels at the appropriate times, these have been continued up to date.

Some results from the 1957 and 1958 surveys were published by Parrish et al. (1959), showing that egg deposition on Ballantrae Bank was confined to limited areas with a gravel and small stone substrate. Subsequent surveys on and over Ballantrae Bank have provided 7 reasonably reliable estimates of egg abundance and 11 of early larval abundance over a period when there have been major changes in the size of the spawning stock. These data, therefore, provide an opportunity to examine the relationship between spawning stock size and egg and early larval production in a herring stock where the duration of spawning, and the extent of the spawning area, are particularly adapted to measurement of these parameters with some precision. The analysis of these data also allows some preliminary conclusions to be drawn on the question of the stock size necessary to safeguard recruitment to the stock.

MATERIAL AND METHODS

Stock Size and Spawning Potential. The herring fishery in the Clyde can be divided into two phases: the fishery in the period January-March that exploits mainly the adult component of the stock as it congregates for, and during, spawning; and the fishery during the remainder of the year mainly in the inner reaches of the Firth of Clyde where a major proportion of the catches are composed of adolescent fish in their second and third years of life. For many years the spawning fishery in the first quarter of the year has been executed by vessels using anchored gill nets on the spawning ground at Ballantrae Bank, and by ring-net vessels fishing principally to the south of Arran, where the fish congregate prior to moving to Ballantrae Bank, and on Ballantrae Bank itself. In 1968 a pair-trawl fishery for herring began in the Clyde and grew fairly rapidly in subsequent years. In 1969, because of concern about a decline in the size of the stock, pair-trawling for herring was prohibited on Ballantrae Bank during February and March. In 1972, after a further decline fishing for herring was prohibited during January to March over the entire Firth of Clyde with exemptions for licensed vessels fishing with anchored gill nets.

Catch and fishing effort statistics are available for each method of herring fishing in the Clyde throughout the period under consideration. Data on the age, size, and racial composition (spring and autumn spawners) of the catches are also available from weekly samples of the catch by each method of fishing since 1960, and from rather less frequent sampling prior to 1960.

These data have been used to estimate the size and spawning potential of the spring-spawning stock in each of the years for which estimates of egg abundance or larval production are available. The size of the spawning stock has been estimated from the equation

$$S = \frac{Z\ C}{F\ (1-e^{-Z})}.$$

S = stock of fish greater than or equal to 3 years old in weight (in crans), Z is the instantaneous rate of total mortality estimated from catch per unit effort data of the anchored gill-net fishery on the spawning ground, F is the instantaneous rate of fishing mortality estimated by assuming a natural mortality rate of 0.2, and C is the catch in weight of fish 3-years-old and older taken in each year in the total Clyde herring fishery.

The spawning potential of the stock in each year was then estimated by multiplying half the weight of the spawning stock in each year by the appropriate mean fecundity for that year. The mean fecundity was obtained by converting the mean length of the spawners in each year to weights and applying the fecundity-weight relationship, F = 283.04 W-9937, given by Baxter (1959) for Clyde spring spawners.

Egg Abundance. Estimates of annual egg production of Clyde herring were obtained from surveys of egg concentrations on Ballantrae Bank during the spawning season. The surveys were conducted mainly by a small spring opening grab with a mouth aperture of 20 cm x 20 cm (Parrish et al., 1959). In 1971 much of the survey work was carried out by underwater television, which gave excellent results.

The position and area occupied by egg concentrations was determined by a Decca marine track plotter from which the total area of spawn was calculated in m^2. Samples of spawn taken by grab were preserved in 10% formalin for detailed analysis at the laboratory. From each grab sample counts were made of the number of eggs in 4 cm^2 subsample blocks and averaged to give density per sample. Since density of eggs between adjacent samples, and sometimes within samples, often varied considerably, contouring of egg densities within a patch was difficult Therefore the total number of eggs within a patch was estimated by raising the average density/m^2 from all samples by the area of the patch.

Larval Production. From the timing of catches of ripe fish by the commercial fisheries, and the estimated time to hatching from temperature data, it appeared that hatching of larvae in the Clyde ought to start, on average, in the first week of March and continue until the end of the first week of April. The initial surveys confirmed these conclusions (Saville, 1965) and in subsequent years surveys were carri out from 1 March to 15 April at 3-4 day intervals. The initial surveys were confined to Ballantrae Bank because this was known, from both fishery evidence and the results of Marshall, Nicholls and Orr (1937), to be a major spawning ground for Clyde spring spawners. However, they mention the possibility that some spawning may also take place to the south of Arran where considerable catches of herring close to spawning condition have been taken by ring-net and pair trawl vessels. To clarify this position intensive surveys were done in three years in this area, alternating with those on Ballantrae Bank. Negligible numbers of recently hatched larvae were caught on these grounds, thereby indicating that Ballantrae Bank is the only spawning area of any importance for Clyde spring spawners.

On each survey this small area, of about 30 km^2 in extent, was sampled by a regular grid of 25 stations, a survey normally being completed within less than 12 h. Sampling was done by oblique hauls from bottom to surface with a 1 m diameter net of 24 mesh/cm nylon.

In the laboratory all herring larvae were removed from the samples and either counted and measured in total, or when the larval numbers were large, the total number and their size distribution estimated by subsampling. For the purpose of estimating annual variations in larval production only larvae less than 9.5 mm in length were considered. Blaxter and Hempel (1963) give the size range of Clyde larvae at hatching, under experimental conditions, as 7.0 - 8.5 mm but state that the non-optimal conditions of laboratory experiments are likely to depress the length at hatching. The numbers of larvae less than 9.5 mm in length in each sample were then converted to the numbers below 1 m^2 of sea surface and these values were then integrated over the whole area of the surface to provide an estimate of the daily "production" of larvae on each day surveyed. To integrate over the entire hatching period these values were plotted against the survey dates and the area under the resulting curve was measured (Saville, 1964).

RESULTS

Stock Size and Spawning Potential. The estimated spawning potential of the Clyde spring spawning stock in each of the years from 1958-1972, given in Table 1, shows wide fluctuations in the period considered, from 16×10^{10} - 1683×10^{10} with high values in isolated years accounting for most of this variation. Indeed, if one omits the two exceptionally high values in 1961 and 1965 the range is quite modest, from 16 - 318×10^{10}. Although there are variations from year to year in average fecundity, and some slight longer term trend in increased fecundity over the period considered due to an increase in mean size per age group, most of this variation in spawning potential is due to variations in the size of the spawning stock. In the history of the Clyde fishery there is evidence of very sharp differences in the strength of recruiting year classes. Thus the very high values of spawning potential in 1961 and 1965 resulted from the recruitment to the spawning stocks in these years of the very strong 1958 and 1962 year classes. Since 1962 no strong year classes have recruited to the stock and this is the main factor responsible for the current depressed state of the fishery and low value of the spawning potential.

The validity of these estimates of spawning potential would appear to be largely dependent on the total mortality rates used in equation 1 and their partition between natural and fishing mortality. Those derived from the catches per unit effort of the anchored drift-net fishery on Ballantrae Bank have been used because, due to the prohibi-

Table 1. Estimates of spawning potential of Clyde spring spawning herring, egg abundance on Ballantrae Bank, and larval production over Ballantrae Bank, 1958-1972

Year	Spawning Potential $\times 10^{-10}$	Egg Abundance $\times 10^{-10}$	Larval Production $\times 10^{-11}$
1958	318	10.63	3.38
1959	16	-	0.32
1960	98	-	5.26
1961	1683	-	8.28
1962	29	-	1.19
1963	21	-	2.11
1964	164	-	4.71
1965	764	75.39	7.67
1966	224	-	-
1967	78	-	5.20
1968	139	48.30	-
1969	273	30.63	-
1970	177	23.28	5.08
1971	42	10.75	-
1972	17	0.46	0.84

tion of other methods of fishing during the spawning fishery in 1972, these are the only ones available in that year. However, for the other years under consideration ring-net mortality rates are also available for the spawning fishery; these are in very close agreement with those derived from the gill-net fishery, which increases one's confidence in their reliability.

However, doubt can be attached to the value of 0.2 used for natural mortality. There is some evidence, from tagging data and from the rather small proportion of post-spawning fish in the catches subsequent to the spawning fishery, that these fish may leave the Clyde after spawning for grounds to the west of Scotland (Wood, 1960). This would not affect the estimated total mortality rates provided such fish returned to the Clyde to spawn in subsequent years. Wood, however, believed that a proportion of these fish did not return to the Clyde but in subsequent years spawned off the northwest coast of Donegal. If this were so then it should be allowed for by increasing the value of the "apparent natural mortality rate" used, above that due to mortality per se. However, the evidence for Clyde herring spawning outside the Clyde is very inconclusive. No herring tagged in the Clyde have been caught outside it in spawning condition and all the evidence suggests that the spring-spawning stocks to the north of Ireland and west of Scotland have been at a very low level in the period considered in this paper. It is felt, therefore, that in view of the fact that in most recent herring stock assessments the natural mortality rate used has been 0.1, by using 0.2 in these estimates, adequate compensation has been made for any emigration to spawning grounds outside the Clyde.

The other way in which emigration could affect the estimates of stock size would be if herring emigrated from the Clyde after spawning, were subsequently caught outside the area, but all of the survivors returned to the Clyde to spawn in the following year. This would result in an underestimate of the stock sizes because although the mortality engendered by the catches taken outside the Clyde would be included in the mortality rates in equation 1, the catches taken would not. However, it would appear that the catches of spring spawners taken in the area designated by Wood (1960), from the north of Ireland to Barra Head, as the one to which the Clyde fish emigrated, have been quite small, at least in the years under consideration. For example, in the total international catch reported from this area in 1971 it is estimated that about 1500 crans were spring-spawned fish. As there are spring-spawning groups in this area other than the Clyde one, it would appear that this would add not more than 500 crans to a total catch of about 14,000 crans taken in the Clyde and thus would have a negligible impact on the estimated size of the spawning stock in that year.

Egg Abundance. Estimates of the annual egg production for herring spawning on Ballantrae Bank are given in Table 1. These estimates are consistently much lower than the spawning potential of the stock and the estimated larval production, which suggests that the egg concentrations sampled by grab represent only a small proportion of the total number of eggs deposited on the Bank during the spawning season.

Ballantrae Bank area is recognized as the main spawning centre for Clyde spring-spawning herring. Spawning in other areas of the Clyde appears to be of negligible importance, at least in recent years. Holliday (1958) reporting on the behaviour of spawning herring showed that the pattern of the bottom substrate was important. Fish would spawn on a plastic sheet with a pattern resembling gravel but would not do so on plain or stone coloured sheeting. Results from egg surveys carried out on Ballantrae Bank showed that egg concentrations were invariably in areas where the bottom substrate consisted of gravel and small stones. Detailed studies of the topography and bottom deposits of the Bank were made by Stubbs and Lawrie (1962) who showed that the ground contains a wide range of different substrates from mud, sand, shingle and gravel to large boulders and rocky outcrops. Gravel and shingle areas suitable for herring spawning are restricted mainly to the central area off Ballantrae extending about 2 miles east to west and 5 miles north to south. The trammel net fishery is confined to this area, and it is only in this locality that concentrations of eggs have been found. The normal procedure adopted during the egg surveys was to start grab sampling in an area where the substrate appeared to be suitable for spawning and where eggs had been located in previous years. After a patch of eggs were located the main effort was directed to determining the limits of the patch and sampling the eggs within it. Subsequently, surveys were conducted over a wider area to include all major deposits of gravel and shingle in an attempt to locate other concentrations of eggs.

The technique of intensive grab sampling in a series of close transects across the gravel and shingle areas makes it unlikely that any major patches of spawn were not sampled. However, it is possible that smaller areas of concentrated eggs (1,000 m^2 and less in area) may have been missed during the surveys, as well as areas where density of eggs was very low.

In addition, there are other banks in the Ballantrae area that have not been surveyed where spawning may take place; but it would seem unlikely that egg concentrations on these banks would be of any major importance since there have been no reports on them of concentrations of spawning fish. Egg deposition may also have been underestimated because of failure to sample the eggs from which the second peak, shown in the larval hatching curves (Figs 1 and 2), are derived. Sampling of eggs normally stopped once appreciable production of larvae commenced. The duration of the hatching period and length of the incubation period are such that appreciable egg deposition could have been unaccounted for in this way.

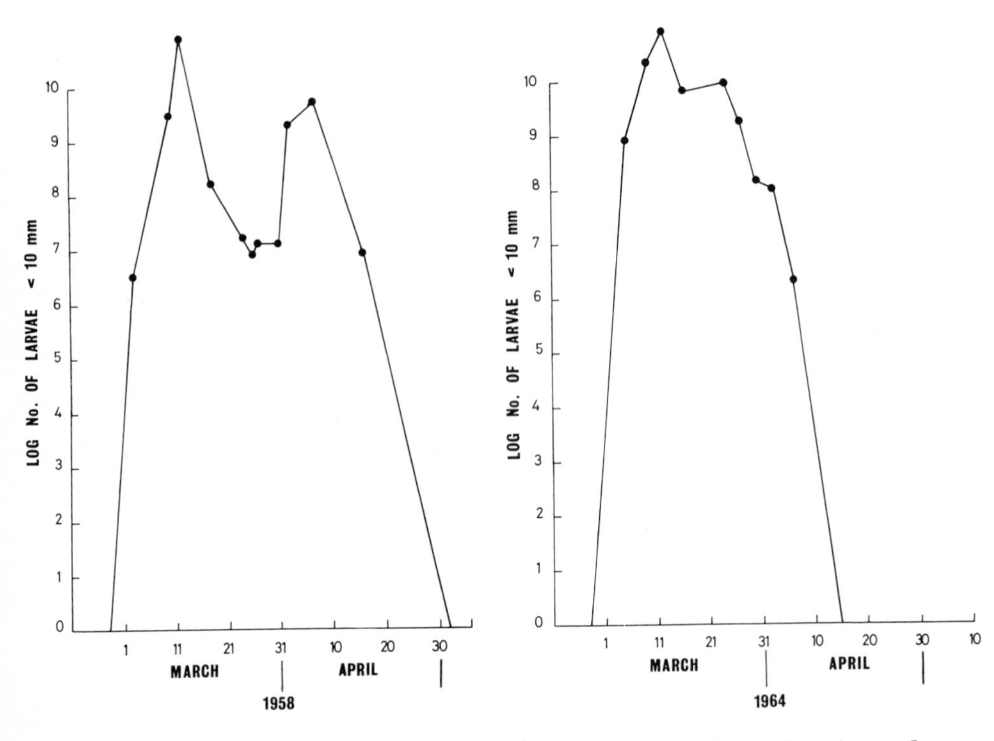

Fig.1. Seasonal production of herring larvae on Ballantrae Bank in 1958

Fig.2. Seasonal production of herring larvae on Ballantrae Bank in 1964

Larval Production. The estimated production of larvae in each of the years, in the period 1958-1972, in which valid surveys have been carried out, are given in Table 1. Due to unavailability of research vessels at the relevant times no surveys at all were done in 1966 and 1969; in 1968 and 1971 sampling was so limited in time that no valid estimate of larval abundance can be made.

Graphs of total early larval abundance against time on Ballantrae Bank are normally very uniform with a major mode some time in the period 10-19 March and in some years a small secondary mode in the period 3-8 April. Two typical examples are given in Figs 1 and 2. As discussed by Saville (1963), the major errors in estimating abundance of larvae from plankton surveys are likely to arise from integrating sample numbers over space from widely separated stations or in integrating survey totals over time from widely spaced surveys. These sources of error are unlikely to have resulted in major inaccuracies in these data where station positions were only about 0.7 miles apart and where the surveys were spaced at about 3-day intervals, at least during the periods of major larval abundance.

As measures of absolute larval production the estimates given in Table 1 may be overestimates due to taking the larvae less than 10 mm in length as representing that day's production. These estimates would still, however, give reliable comparative estimates of production between years as it seems unlikely that there are any major differences between years in size at hatching, or in growth rate during the early stages.

The values of larval abundance given in Table 1 fluctuate appreciably from year to year, and in general agreement with those of spawning potential. It is noteworthy, however, that the range of variation is much less than in spawning potential. This is discussed in greater detail in the next section.

<u>Relations between Spawning Potential, Egg Abundance and Larval Production</u>. The main interest of the parameters discussed in the earlier sections lies in how they are interrelated. This has a bearing not only on the validity of the estimates but could also have important implications in stock management.

It is immediately obvious from the values given in Table 1 that the estimates of egg abundance are in all cases very much lower than the estimated spawning potential of the stock, ranging from 2.7% to 34.7% of the latter. These two parameters are plotted in Fig. 3 and the

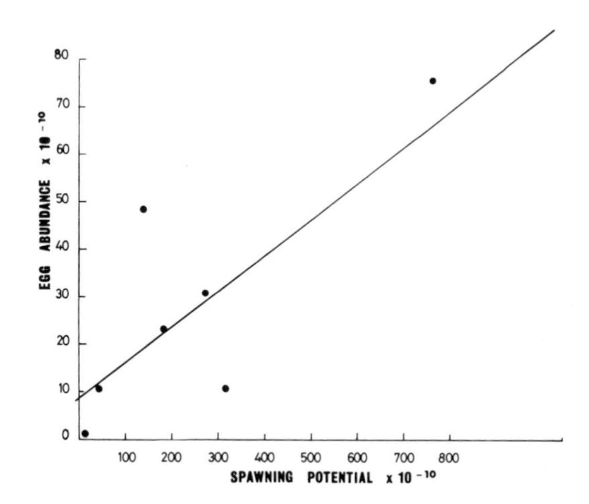

Fig. 3. Annual abundance of herring eggs on Ballantrae Bank plotted against spawning potential of the stock

fitted regression line ($\hat{y} = 8.15\ x + 835$) is drawn in. This regression is significant at the 5% level. It would appear, therefore, that the egg surveys have not sampled all the eggs spawned on Ballantrae Bank, but have sampled them roughly in proportion to their total abundance. This conclusion is also supported by the fact that the egg abundance figures are in all years appreciably lower than those of larval production. The most likely explanation of this discrepancy would seem to be, as discussed in an earlier section, that there are other spawning areas on Ballantrae Bank which have not been sampled and that the proportions of the total spawning on the sampled and unsampled areas is roughly constant from year to year. Indeed the regression line given on Fig. 3 would suggest that only about 10% of the eggs spawned in the Clyde are deposited in the sampled areas.

The values of larval production have been plotted against the values of spawning potential in Fig. 4, and an eye-fitted line drawn through them. This would suggest that larval production is almost linearly related to spawning potential up to a spawning potential value of about 200×10^{10}. There are rather few values above this spawning potential, but they would suggest that the increase in larval production above this point is very small and that one may well have a curve with an asymptote somewhere around a spawning potential of 700×10^{10}.

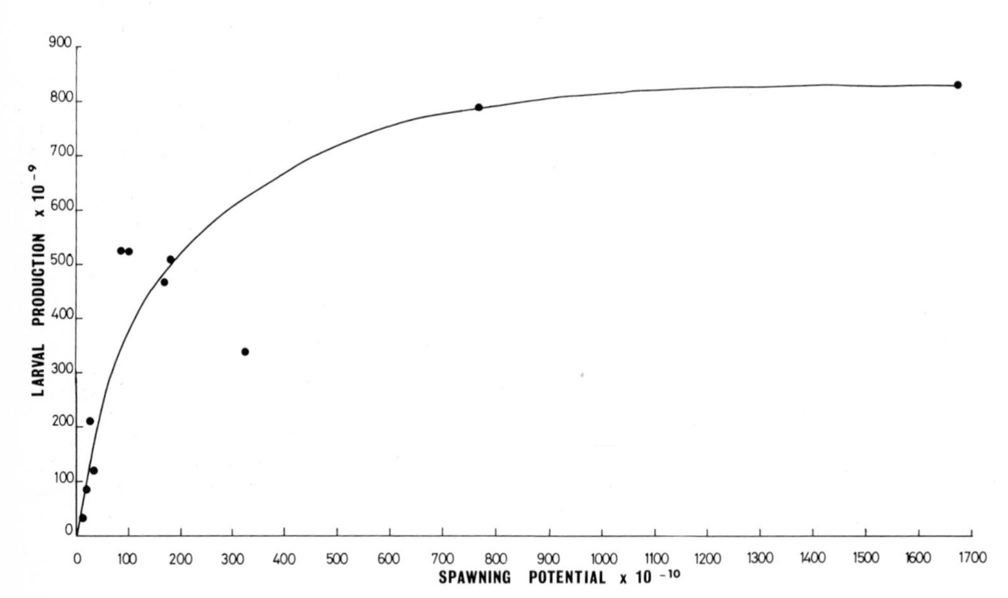

Fig. 4. Annual larval production on Ballantrae Bank plotted against spawning potential of the stock

DISCUSSION

The most interesting outcome of this investigation is the relation shown in Fig. 4 between spawning potential of the stock and the production of larvae. The relation shown in Fig. 3 between spawning potential and egg abundance would suggest that the estimated values of spawning potential have a certain validity but this relationship has, and can have, little further significance. Because of the intensity of sampling of larvae, both over space and time, the estimates of larval production are probably the most reliable ones in existence for any fish stock. It is of considerable interest therefore to examine closely the relationship shown in Fig. 4.

In most high fecundity fish species there is no relationship, at least at high population levels, between spawning potential and the recruitment derived from it. This would, in turn, infer that at high spawning stock levels there is a density-dependent mortality operating somewhere between spawning and recruitment. It has usually been assumed that this density-dependent mortality operates early in the larval stage, and that it operates through food limitation (Gulland, 1965). The relationship shown in Fig. 4 would suggest that in herring there is a density-dependent mortality in the egg stage at high levels of spawning potential, at least in Clyde herring. It is difficult to visualize such a density-dependent mortality in fish species with planktonic eggs, but it has already been shown by previous authors that mortality in herring eggs is high when the eggs are concentrated (Lea, 1930; Runnstrom, 1941; Blaxter, 1956). This may be of particular importance in the Clyde spring-spawning stock where the suitable spawning sites are of very limited area, leading to high concentration of eggs when the spawning stock is large. This could easily result in the asymptotic type of spawning potential-larval production curve shown in Fig. 4.

This may not be the only stage in the reproductive cycle where density-dependent mortality operates; indeed the fact that in these data there is no statistically significant relationship between larval production and year-class strength would suggest that there is a subsequent stage where density-dependent mortality again operates. This would mean that the relationship shown in Fig. 4 cannot be used, by itself, to predict the size of the spawning stock necessary to maintain optimal recruitment. This relationship would, however, suggest that there is little gain in maintaining a spawning stock with a reproductive potential above 6×10^{12} eggs.

A relation between spawning potential and larval production similar to that in Fig. 4 would also mean that larval production would be a poor index of spawning stock size in herring, at high levels of spawning-stock abundance. This type of relation may be peculiar to the Clyde stock because of the limitation of suitable substrate for egg deposition. If, however, it is a more general feature of herring biology it would mean that the decline in the North Sea herring stocks, as measured by the decline in larval production (Anon, 1972) is underestimated.

SUMMARY

From 1958 to date surveys of the abundance of eggs and early larval stages of Clyde spring spawning herring have been carried out over the spawning area of this stock, on Ballantrae Bank. The methods used in obtaining quantitative estimates of annual egg deposition, and of the annual production of early larval stages are described and both of them are related to the corresponding estimates of the spawning potential of the stock.

The estimates of egg abundance are linearly related to the spawning potential of the stock. Estimates of early larval production when plotted against spawning potential follow an asymptotic curve. The significance of the latter relationship in stock management, and its effect on the utility of herring larval surveys as an index of spawning stock size, are discussed.

REFERENCES

Anon, 1972. Report of the North Sea Herring Assessment Working Group. ICES, CM 1972/H13 (mimeo).

Baxter, I.G., 1959. Fecundities of winter-spring and summer-autumn herring spawners. J. Cons. perm. int. Explor. Mer 25, 73-80.

Blaxter, J.H.S., 1956. Herring rearing. II. The effect of temperature and other factors on development. Mar. Res. Scot., 1956, No. 5.

Blaxter, J.H.S. and Hempel, G., 1963. The influence of egg size on herring larvae (*Clupea harengus* L.). J. Cons. perm. int. Explor. Mer 28, 211-240.

Gulland, J.A., 1965. Survival of the youngest stages of fish, and its relation to year-class strength. Spec. Publs int. Comm. N.W. Atlant. Fish. (6), 363-371.

Holliday, F.G.T., 1958. The spawning of the herring. Scot. Fish. Bull. 10, 11-13.

Lea, E., 1930. Mortality in the tribe of Norwegian herring. Rapp. Cons. Explor. Mer 65, 100.

Marshall, S.M., Nicholls, A.G. and Orr, A.P., 1937. Growth and feeding of the larval and post-larval stages of Clyde herring. J. mar. biol. Ass. U.K. 22, 245-267.

Parrish, B.B., Saville, A., Craig, R.E., Baxter, I.G. and Priestley, R 1959. Observations on herring spawning and larval distribution in the Firth of Clyde in 1958. J. mar. biol. Ass. U.K. 38, 445-453.

Runnstrom, S., 1941. Quantitative investigations on herring spawning and its yearly fluctuations at the west coast of Norway. Fiskeridir. Skr. Havundersøk., 6 (8), 1.

Saville, A., 1963. Estimation of the abundance of a fish stock from egg and larval surveys. Rapp. P.-v. Réun. Cons. perm. int. Explor. Mer 155, 164-170.

Saville, A., 1964. Factors controlling dispersal of the pelagic stages of fish and their influence on survival. Spec. Publs int. Comm. N.W. Atlant. Fish. (6), 335-348.

Stubbs, A.R. and Lawrie, R.G.G., 1962. Asdic as an aid to spawning ground investigations. J. Cons. perm. int. Explor. Mer 27, 248-260.

Wood, H., 1960. The herring of the Clyde estuary. Mar. Res. Scot., 1960, No. 1

A. Saville
Marine Laboratory
Victoria Road
P.O. Box 101
Aberdeen / GREAT BRITAIN

D.W. McKay
Marine Laboratory
Victoria Road
Aberdeen / GREAT BRITAIN

I.G. Baxter
Marine Laboratory
Victoria Road
Aberdeen / GREAT BRITAIN

The Use of Data on Eggs and Larvae for Estimating Spawning Stock of Fish Populations with Demersal Eggs

J. Gjøsaeter and R. Saetre

INTRODUCTION

The usual methods of estimating the abundance of fish stocks are by fishery statistics, tagging experiments, acoustic surveys, and counting the spawning products. Hensen and Apstein (1897) first suggested the last method and Hart and Tester (1934) were probably the first to count demersal fish eggs in the British Columbia herring, which spawns between or just below tidal levels. Runnstrøm (1941) estimated the number of eggs on the offshore spawning grounds of Norwegian spring spawning herring, and used his results for estimating spawning stock size. Similar estimates have been reported for Canadian Pacific herring by Stevenson and Outram (1953) and Taylor (1963) and for Clyde herring by Parrish et al. (1959).

Several authors have tried to estimate the spawning stock of herring from the number of larvae. Parrish et al. (1959) assumed that all eggs spawned subsequently hatched and they estimated the daily loss of larvae due to mortality and drift out of the sampling area. Hempel and Schnack (1971) integrated numbers of larvae in space but not in time and did not allow for egg and larval mortality. The numbers of larvae were also increased by a factor of 2 to compensate for the inefficiency of the plankton sampler used. Many authors (e.g. Zijlstra, 1970; Burd and Holford, 1971; Hyronimus, 1971; Saville, 1971) have used the number of larvae sampled to get a relative estimate of 2 or more spawning stocks, or of the same stock in different years. The time of sampling in relation to hatching may have a major influence on the estimates. By using older larvae, this problem will be reduced, but then the danger exists of using larvae which have past the "critical phase" and those larvae will give a very poor estimate of spawning stock size.

This paper discusses methods for calculating the spawning stock size by means of the number of its spawning products. Two new methods for estimating spawning stock size of species with demersal eggs are presented. These methods are applied to the Barents Sea capelin.

PROBLEMS AND METHODS

Basic Equations. For estimating spawning stock size from amount of spawning products, it is necessary to know the relationship between the number of parents and number of eggs and larvae produced.

Thus:

$$P = \frac{E}{F \cdot S \left(1 - \frac{k_1 \cdot d_1}{I}\right)} \qquad (1)$$

where P = spawning stock size

E = total number of living eggs on the spawning ground at the time of sampling

d_1 = days between spawning and sampling

I = incubation time in days

F = mean individual fecundity

S = sex ratio

k_1 = egg mortality until hatching

$$P = \frac{L}{F \cdot S\ (1 - k_1)(1 - k_2)(1 - k_3)} \qquad (2)$$

L = total number of larvae present at the time of sampling

k_2 = hatching mortality

k_3 = larval mortality until sampling

The calculation may be divided into two types:

a) On eggs.

$$P = \frac{1}{F \cdot S\ (1 - \frac{k_1 d_2}{I})} \iint \frac{E_{ij}}{d_2}\, dAdt \qquad (3)$$

E_{ij} = number of eggs at a certain stage per unit area at station i and at time j

d_2 = number of days taken to reach this stage

A = total area of the spawning ground

If sampling is performed after the end of spawning, but before hatching has started, equation (3) may be reduced to

$$P = \frac{1}{F \cdot S\ (1 - \frac{k_1 d_1}{I})} \int E_i dA \qquad (4)$$

E_i = number of eggs per unit area on station i.

b) On larvae.

$$P = \frac{1}{F \cdot S\ (1 - k_1)(1 - k_2)(1 - k_3)} \iint \frac{L_{ij}}{d_3}\, dAdt \qquad (5)$$

L_{ij} = number of larvae at a chosen age per unit area at station i and at time j

d_3 = age of larvae in days.

Sources of Error. One of the main problems in estimating parent stock size by equations (3), (4), and (5) is the difficulty of calculating the confidence limits (English, 1964; Saville, 1964). The factors effecting the errors of the estimates are: determination of fecundity, sex ratio, mortality, the covering of the spawning area in space and time, and the sampling methods. The data available (Pozdnyakov, 1957; Gjøsaeter and Monstad, 1973) suggest that the fecundity of the Barents Sea capelin is fairly stable from year to year. In most cases sex ratio is taken to be 1 : 1. In capelin there is a change in sex composition during the spawning season, but the total number of spawners appear to be almost the same in the two sexes (e.g. Prokhorov, 1965; Monstad, 1971). Prokhorov, however, suggests that there may be a predominance of males in spawning stocks which are dominated by old fish.

In species with demersal eggs the rate of fertilization is high and mortality at the egg stage is usually low. Mortalities under 10% seem to be common (Runnstrøm, 1941; Baxter, 1971; Dragesund et al., 1973; Bjørke et al., 1972). Hatching success seems to be more variable. In herring, the egg and hatching mortality is influenced by egg density (Runnstrøm, 1941; Galkina, 1971). Taylor (1971) has shown that eggs, 8 layers thick, often had a lower production of viable larvae per unit area than eggs 4 layers thick. In capelin no correlation between thickness of egg layer and egg mortality has been observed (Dragesund et al., 1973; Bjørke et al., 1972), but the thickness will probably affect the hatching success. Alderdice et al. (1958) has shown that lack of oxygen is an important cause of mortality in fish eggs. Herring eggs are deposited in "carpets" covering the substratum, while capelin eggs are mixed with gravel on the bottom. The aeration is therefore probably better in dense concentrations of capelin eggs. This may account for the apparent difference in mortality.

Predation is usually a very important cause of mortality at the egg stage (Dragesund and Nakken, 1970). Capelin eggs form an important part of the haddock's diet during the spawning season (Zenkevitzh, 1963; Templeman, 1965; Bjørke et al., 1972). Ducks have also been observed as predators of capelin eggs (Gjøsaeter et al., 1972). Early larval mortality is discussed by May (this symposium). The effect of this on estimates of stock size abundance can be reduced by counting only very young larvae.

The main contribution to the variance of stock estimation is most likely the variability in space and time, of which the latter is probably more important. By introducing tentative mathematical functions for egg abundance, some authors, e.g. English (1964), have tried to reduce this part of the variance. The variance caused by area integration can be reduced by having a sufficiently dense grid of randomly distributed stations but in most cases one has to compromise between a dense grid of stations and a frequent coverage. It must also be supposed that the plankton sampler is giving a good quantitative estimate of abundance, a supposition that is not always fulfilled (see Dragesund et al. - this volume). Sampling by horizontal hauls presupposes that eggs and larvae have a continuous vertical distribution. In fact, vertical "patchiness" can occur (Banse, 1964) and vertical migration is common (see Seliverstov, this symposium).

Calculations. Most species with demersal eggs have a widely distributed spawning area. In that case, calculation using equation (3) is not possible. The method of calculation on larvae according to equation (5) is also problematic because the limited resources available make it difficult to cover the area of larval distribution sufficiently. For these reasons two new methods were applied.

The first method presupposes that the following parameters be determined on both eggs and larvae:

a) Locations of the single spawning beds.
b) The numbers of eggs, e, spawned on one of these spawning beds, B.
c) The number of larvae, l, younger than d days old released from B at the time t_1.
d) The number of larvae, L, younger than d days old released from the total spawning area at the time t_2.
e) Cumulative hatching curves for B and for the total spawning area.

The number of eggs spawned at B and in the total spawning area may be expressed by the following equations:

$$e = \frac{100 \cdot l}{a_1(1-k_1)(1-k_2)(1-k_3)} \qquad (6)$$

$$E = \frac{100 \cdot L}{a_2(1-k_1)(1-k_2)(1-k_3)} \qquad (7)$$

a_1 and a_2 respectively are the ratios between l and L and the total number of larvae hatched at B and in the total spawning area. These ratios are deduced from the cumulative hatching curves and expressed as percentages.

Then $$\frac{E}{e} = \frac{L \cdot a_1}{l \cdot a_2} \qquad (8)$$

and $$P = \frac{L \cdot e \cdot a_1}{F \cdot S \cdot l \cdot a_2} \qquad (9)$$

The second method is based mainly on larval sampling. The total spawning area is divided into sub-areas, and the following parameters are determined for each sub-area.

a) The number of larvae, l_1, $l_2 \dots l_n$, younger than d days old released from the individual sub-areas at time t_1, $t_2 \dots t_n$.
b) Cumulative hatching curves for each sub-area.

The hatching curves are used as in the first method, and the ratios, a_1, $a_2 \dots a_n$ are found for the sub-areas. If mortality factors are disregarded the parent stock size can be estimated from the following equation:

$$P = \frac{100}{F \cdot S}\left(\frac{l_1}{a_1} + \frac{l_2}{a_2} + \frac{l_3}{a_3} \dots \frac{l_n}{a_n}\right) \qquad (10)$$

In the first method the calculations are based on the ratio $\frac{L}{l}$. Hence, the bias introduced by sampling will be strongly reduced. It is also supposed that the mortality of eggs and larvae from spawning bed B is similar to that of the total spawning area. Even if this assumption is not completely fulfilled, the method will obviously reduce the bias caused by the mortality. The hatching curve is usually dome-shaped. Parts of the cumulative hatching curve will then be approximately linear. If we are operating on these quasi-linear parts of the curves and they have equal slopes, then $a_1 = a_2$. The determination of l is probably the most serious source of error. It may be difficult to distinguish between larvae from B and larvae from surrounding spawning beds as the distances between spawning beds are often short.

The second method requires absolute measurements of larval abundance which is open to objection. Mortality of egg and early larval stages is not taken into consideration. Each cruise gives independent estimates of P and the variance, introduced by time integration, is avoided. Apart from the determination of the hatching curve, and the area integration, all sources of errors contribute to an underestimate of the spawning stock size. This method may also be applied to species with pelagic eggs, if eggs at an early stage are sampled.

APPLICATION TO THE BARENTS SEA CAPELIN

General Observations. The spring spawning Barents Sea capelin spawn along the coast of northern Norway and the western Murman coast. The spawning grounds in 1971 and 1972 are shown in Fig. 1. The spawning in Soviet waters was probably of minor importance in these 2 years. Capelin have a short spawning period and long incubation time, the eggs being mixed with bottom substratum forming a layer with a maximum thickness of 15 cm. Egg densities exceeding 5 x 10^6 eggs/m^2 were observed. The highest egg densities were found among gravel and shell-gravel at a depth between 25 and 50 m. The temperature on the spawning ground varied between 2^oC and 6^oC and the salinity between 34.0°/oo and 35.0°/oo. Both direct current measurements and the choice of bottom substratum indicated that the capelin prefer rather strong current on the spawning grounds. The main currents influencing the drift of larvae are shown in Fig. 1.

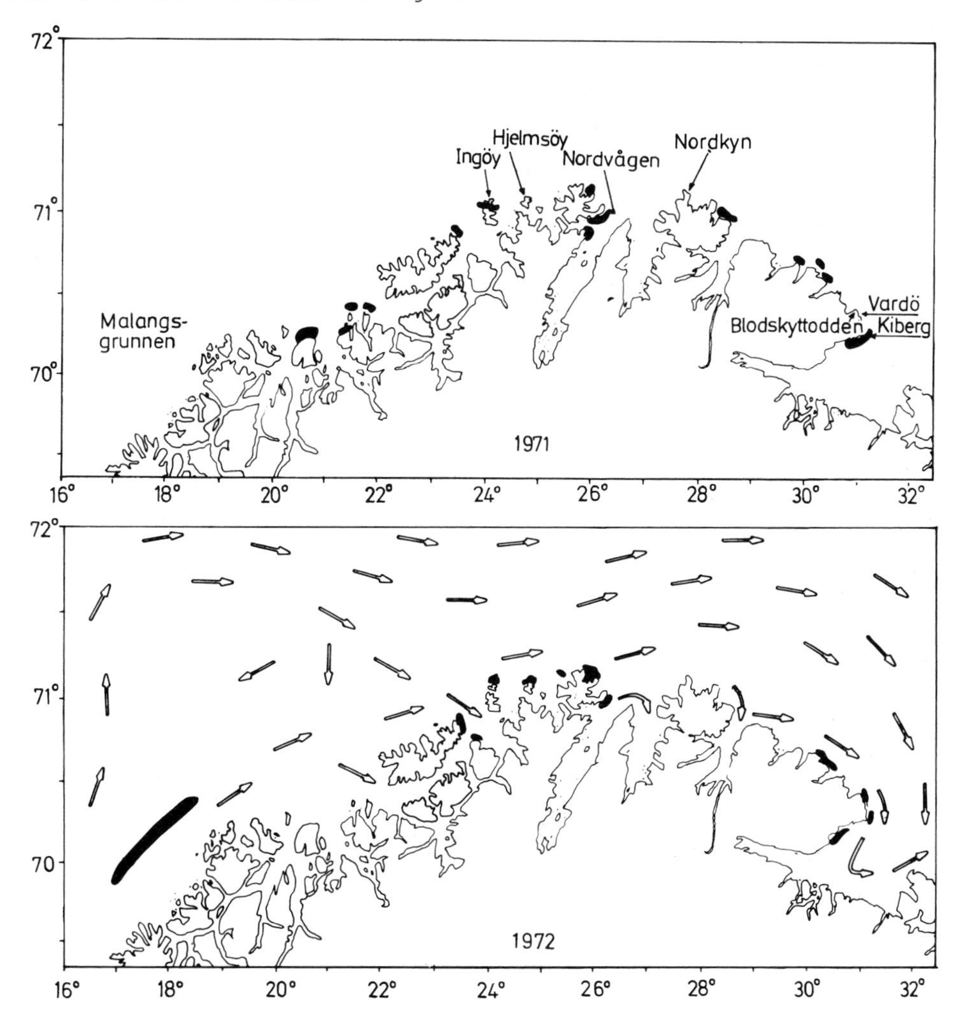

Fig. 1. Capelin spawning beds off northern Norway, 1971 and 1972. The arrows show the dominating currents

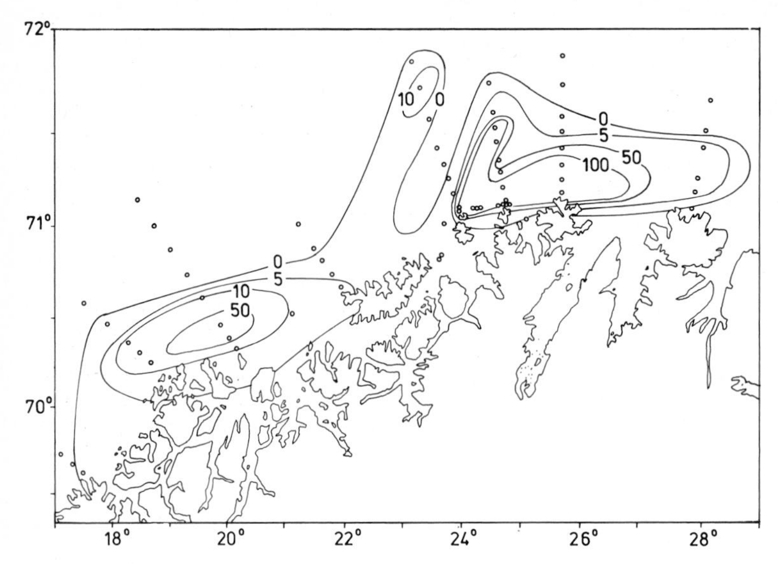

Fig. 2. Grid of stations and distribution of larvae from a general survey, 1972. Figures indicate number of larvae below 1 m^2 surface

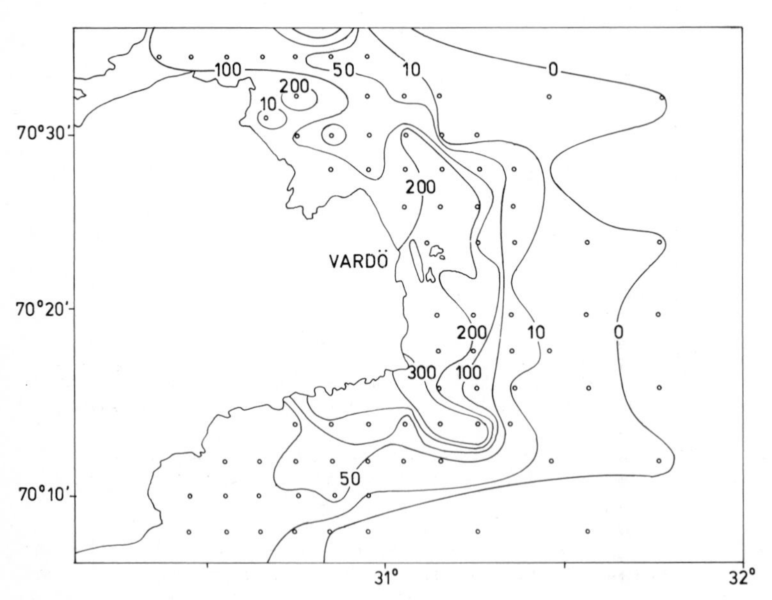

Fig. 3. Grid of stations and distribution of larvae from a detailed survey, 1972. Figures indicate number of larvae below 1 m^2 surface

Methods. The spawning beds were located by an extensive grab survey using Petersen grab. In each egg sample the stage of development was determined in order to calculate the spawning time. From selected sites, eggs were placed in incubators on board ship at prevailing temperatures found on the grounds. In 1971 and 1972 Nordvågen and Blodskyttodden, respectively, were surveyed in detail for both distribution and abundance of eggs and larvae. By using Decca Navigator and maps of scale 1:5000 and 1:10000 accurate surveying was obtained. Divers equipped with sampling tubes were used to determine the egg density as well as the vertical stratification of the bottom substratum (Dragesund et al., 1973; Bjørke et al., 1972).

On the larval surveys the Clark-Bumpus plankton sampler was used in 1971, and in 1972 the Bongo-20 net. Figs 2 and 3 show examples of the larval distribution and grid of stations. As the calculations are based on young larvae only, a relevant age criterion was needed. In 1971 the length was used as a measure of age. Laboratory experiments (unpublished), however, indicate that the yolk-sac size is a better measurement. In 1972 this was used for age determination. The cumulative hatching curves (Fig. 4) were obtained by frequent sampling of young larvae on spawning grounds which were assumed to be representative.

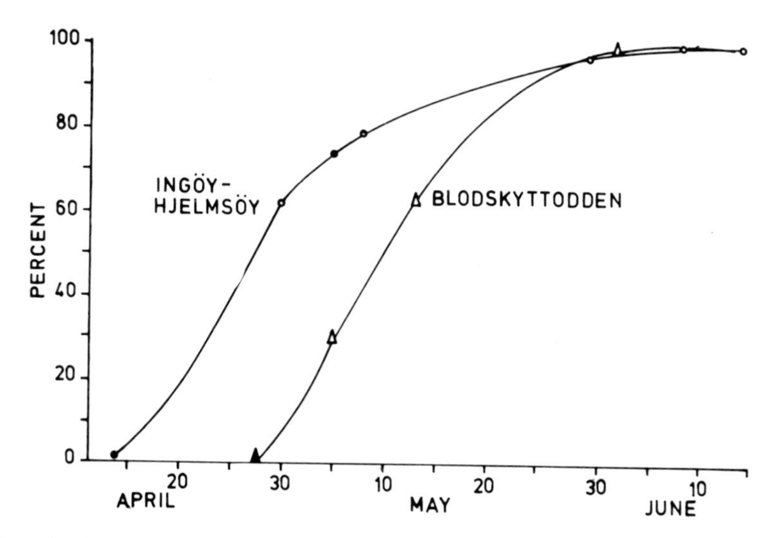

Fig. 4. Cumulative hatching curves from 1972. Filled symbols show the calculated start of hatching. Open symbols represent observed points

Results. In 1971 the estimation of spawning stock was based on larvae ≤7 mm. Detailed investigations were conducted off Nordvågen. The number of eggs were estimated on 26 April, and larvae between 7-9 May. In the area west of Nordkyn the spawning was earlier and the temperature higher than east of Nordkyn. A cruise from 19-25 May was used for estimating number of young larvae west of Nordkyn, and from 7-22 June for the area east of Nordkyn. The following data were obtained:

Eggs at Nordvågen: $e = 7.5 \times 10^{11}$

Larvae ≤ 7 mm off Nordvågen: $l = 8.9 \times 10^{9}$

Larvae ≤ 7 mm at the whole spawning area: $L = 5.2 \times 10^{12}$

The hatching curve is not known in detail, but the data available suggest that the time of larval sampling fell within the quasi-linear part of the curve. The slope of this part of the hatching curve at Nordvågen is supposed to average the slope of the hatching curve in the total spawning area. Then $a_1 = a_2$ and equation (9) will be reduced to

$$P = \frac{L \cdot e}{F \cdot S \cdot l}$$

F is shown to be ca. 10000 (Gjøsaeter and Monstad, 1973) and S = 0.5. 1 ton of capelin corresponds approximately to 30000 specimens. The size of the stock spawning in Norwegian waters in 1971 was therefore:

$$P = \frac{5.2 \times 10^{12} \times 7.5 \times 10^{11}}{8.9 \times 10^{9} \times 10^{4} \times 0.5 \times 3 \times 10^{4}} \text{ tons} = 2.9 \times 10^{6} \text{ tons}$$

The stock estimation in 1972 was based on larvae with yolk-sac ≥ 0.3 mm. This corresponds to an age of approximately 5 days. That year a spawning bed at Blodskyttodden was examined in detail and the number of eggs was estimated in the first half of April and larvae between 13-16 May. During the detailed survey it was impossible to distinguish between larvae 5 days old from Blodskyttodden and from the spawning beds between Vardø and Kiberg. Still younger larvae were therefore considered, and those with yolk-sac larger than 0.5 mm turned out to be fairly well separated. Between the numbers of larvae in the area off Blodskyttodden and in the Vardø-Kiberg area, a ratio of 1.5 was found. As spawning took place simultaneously, and the temperature was equal, we suppose that the ratio between these larvae reflects the ratio between number of eggs deposited in the two areas.

At Blodskyttodden 4.5×10^{12} eggs were spawned, and accordingly 7.5×10^{12} eggs were deposited in the whole area. During the detailed survey 1.3×10^{11} larvae with yolk-sac ≥ 0.3 mm were found in the area. Between Blodskyttodden and Kiberg the spawning population can be estimated directly from number of eggs, and amount to 5.0×10^{4} tons. The other spawning grounds east of Nordkyn were of minor importance.

West of Nordkyn 6.5×10^{11} young larvae were found from 6-12 May. From the hatching curves (Fig. 4), a_1 and a_2 are found to be 19% and 17% respectively for the eastern and western part of the spawning area. Using equation (9) and converting to tons, the quantity which spawned west of Nordkyn is estimated to be 2.8×10^{5} tons.

According to this, 3.3×10^{5} tons of capelin spawned along the Norwegian coast in 1972, the spawning at Malangsgrunnen being disregarded. In this area the spawning took place rather early (1-5 March), and the temperature was exceptionally high (approximately 6°C). The main hatching, therefore, was finished at the time of the larval survey. Accordingly, our data are not sufficient for estimating the stock spawning in this area. Based on data of the development of the fishery in the area it is tentatively concluded that about 1/4 of the stock reaching this area escaped the fishing gear and spawned. This amounts to the order of size of 2×10^{5} tons. Accordingly, between 5×10^{5} and 6×10^{5} tons of capelin spawned in Norwegian waters in 1972.

If the second method (equation 10) is applied to the data from 1972, a number for the stock spawning is arrived at which is approximately 1/10 of that derived from the first method. The main reason for this is that the sampling gear used and the mortality factors cause an underestimate of the number of larvae.

DISCUSSION

The total spawning stock size is derived by adding the amount of capelin caught at the pre-spawning stage and the amount which were able to spawn. The spawning stock of capelin has previously been estimated by acoustic methods and by tagging experiments (Dragesund et al., 1973b). The results are shown in Table 1. The estimates derived by tagging experiments are obviously too high but they indicate that the spawning stock was greater in 1971 than in 1972. In 1972 the spawning stock size calculated from acoustic method was higher than that computed from the egg and larval survey. One reason for this is that the last method only includes capelin spawning in Norwegian waters.

Table 1. Estimated size of spawning stock in million tons

Methods / Year	Eggs and larvae	Acoustic methods	Tagging
1971	4.0	-	5.8
1972	1.7-1.8	1.9-3.7	4.8

SUMMARY

In the first part of this paper the basic equations of the established methods for estimating spawning stock size based on observations of spawning products, are presented.

The sources of error attached to these methods are discussed. Bias and variance are introduced by calculating fecundity, sex ratio, mortality, and by sampling the eggs and larvae. The sampling gears usually applied an underestimation of the spawning stock size. The main contribution to the variance of stock estimation is most likely the variability of larval distribution in space and time of which the time is probably more important.

Two new methods for estimating the size of the spawning stock for fish with demersal eggs are described. Both methods assume that the hatching curve is known, and only newly hatched larvae are used. In the first method the number of eggs deposited at one of the spawning beds is determined. After hatching has started an abundance index for young larvae released from this spawning bed has to be obtained, and referred to a known point at the hatching curve. The ratio between number of eggs and this index, together with the hatching curves and larval abundance index for the whole spawning area are used for estimating the spawning stock size. The main advantage of this method is that the bias introduced by sampling gear and mortality is strongly reduced.

In the second method, the hatching curve is used for calculating the percentage of eggs that are hatched at any time. On the assumption that the number of newly hatched larvae sampled corresponds to this percentage, the size of the spawning stock may be found.

These methods are applied on the Barent Sea capelin in 1971 and 1972. The estimates obtained are fairly similar to those obtained by other methods.

Both methods may be used on marine or freshwater species with demersal eggs and fairly long incubation time. The second method can be modified and used on species with pelagic eggs.

REFERENCES

Alderdice, D.F. and Velsen, J., 1971. Some effects of salinity and temperature on early development of Pacific Herring (*Clupea pallasii*). J. Fish. Res. Bd Can., 28, 1545-1562.
Alderdice, D.F., Wickett, W.P. and Brett, J.R., 1958. Some effects of temporary exposure to low dissolved oxygen levels on Pacific salmon eggs. J. Fish. Res. Bd Can., 15, 229-249.
Banse, K., 1964. On the vertical distribution of zooplankton in the sea. pp. 55-125 in Sears, M. (ed.) Progress in Oceanography 2. Oxford.
Baxter, I.G., 1971. Development rates and mortalities in Clyde herring eggs. Rapp. P.-v. Réun. Cons. perm. int. Explor. Mer, 160, 27-29.
Bjørke, H., Gjøsaeter, J. and Saetre, R., 1972. Undersøkelser pa loddas gytefelt i 1972. Fiskets Gang, 58, 710-716.
Burd, A.C. and Holford, B.H., 1971. The decline in the abundance of Downs herring larvae. Rapp. P.-v. Réun. Cons. perm. int. Explor. Mer, 160, 99-100.
Dragesund, O. and Nakken, O., 1970. Relationship of parent stock size and year-class strength in Norwegian spring spawning herring. ICES/FAO/ICNAF Stock and Recruitment Symposium, Aarhus, 1970, 20, 12 pp. (mimeo).
Dragesund, O., Gjøsaeter, J. and Monstad, T., 1973. Estimates of stock size and reproduction of the Barents Sea Capelin in 1970-1972. Fisk Dir. Skr. Ser. Havundersokelser., 16, 105-139.
English, S.T., 1964. A theoretical model for estimating the abundance of planktonic fish eggs. Rapp. P.-v. Réun. Cons. perm. int. Explor. Mer, 155, 174-184.
Galkina, L.A., 1971. Survival of spawn of the Pacific herring (*Clupea harengus pallasii* Val) related to the abundance of the spawning stock. Rapp. P.-v. Réun. Cons. perm. int. Explor. Mer, 160, 30-33.
Gjøsaeter, J. and Monstad, T., 1973. Fecundity and egg size of spring spawning Barents Sea capelin. Fisk Dir. Skr. Ser. Havundersokelser. 16, 98-104.
Gjøsaeter, J., Saetre, R. and Bjørke, H., 1972. Dykkender beiter pa loddeegg. Sterna, 11, 173-176.
Hart, J.L. and Tester, A.L., 1934. Quantitative studies on herring spawning. Trans. Amer. Fish. Soc., 64, 307-312.
Hempel, G. and Schnack, D., 1971. Larval abundance on spawning grounds of Banks and Downs herring. Rapp. P.-v. Réun. Cons. perm. int. Explor Mer, 160, 94-98.
Hensen, V. and Apstein, C., 1897. Die Nordsee-Expedition 1895 des Deutschen Seefischerei-Vereins über die Eimenge der im Winter laichenden Fische. Wiss. Meeresuntersuch., Abt. Kiel u. Helgol., 2 (2), 1-10

Hyronimus, E., 1971. Abundance and distribution of herring larvae in the western North Sea in 1962-1967. Rapp. P.-v. Réun. Cons. perm. int. Explor. Mer, 160, 83-86.
Monstad, T., 1971. Alder, Vekst og utbredelse av lodde (*Mallotus villosus*) i Barentshavet og ved kysten av Nord-Norge 1968-1970. (Age, grow and distribution of capelin (*Mallotus villosus*) in the Barents Sea and off the coast of North-Norway). Theses (Cand. real), Univ. Bergen. 80 pp.
Parrish, B.B., Saville, A., Craig, R.E., Baxter, I.G. and Priestley, R., 1959. Observations on herring spawning and larval distribution in the Firth of Clyde in 1958. J. mar. biol. Ass. U.K., 38, 445-453.
Pozdnyakov, Yu.F., 1957. The fecundity of capelin in the Barents Sea. Dokl. Akad. Nauk. SSSR, 112, 777-778 (in Russ.).
Prokhorov, V.S., 1965. Ecology of the Barents Sea capelin (*Mallotus villosus* (Müller)) and the prospects for its commercial utilization. Fish. Res. Bd Can. Transl. Ser., 813, 1-131 (mimeo).
Runnstrøm, S., 1941. Quantitative investigations on herring spawning and its yearly fluctuations at the west coast of Norway. Fisk Dir. Skr. Ser. Havundersokelser, 6 (8), 1-71.
Saville, A., 1964. Estimation of the abundance of a fish stock from egg and larval surveys. Rapp. P.-v. Réun. Cons. perm. int. Explor. Mer, 155, 164-170.
Saville, A., 1971. The larval stage. Rapp. P.-v. Réun. Cons. perm. int. Explor. Mer 160, 52-55.
Stevenson, J.C. and Outram, D.N., 1953. Results of investigations of herring populations of the west coast and lower east coast of Vancouver Island in 1952-1953, with analysis of fluctuations in population abundance since 1946-1947. Rep. Br. Columb. Fish. Dept. 1953, 57-84.
Taylor, F.H.C., 1963. The stock-recruitment relationship in British Columbia herring populations. Rapp. P.-v. Réun. Cons. perm. int. Explor. Mer, 154, 279-292.
Taylor, F.H.C., 1971. Variation in hatching success in Pacific herring (*Clupea pallasii*) eggs with water depth, temperature, salinity and egg mass thickness. Rapp. P.-v. Réun. Cons. perm. int. Explor. Mer, 160, 34-39.
Templeman, W., 1965. Some instances of cod and haddock behaviour and concentrations in the Newfoundland and Labrador area in relation to food. ICNAF Spec. Publs, 6, 449-461.
Zenkevitch, L., 1963. Biology of the seas of the USSR. 1-955. London: Allen and Unwin.
Zijlstra, J.J., 1970. Herring larvae in the central North Sea. Ber. dtsch. wiss. Komm. Meeresforsch., 21, 92-115.

J. Gjøsaeter
Institute of Marine Research
5011 Bergen / NORWAY

R. Saetre
Institute of Marine Research
5011 Bergen / NORWAY

Significance of Egg and Larval Surveys in the Studies of Population Dynamics of Fish

S. Tanaka

INTRODUCTION

The Japanese sardine, *Sardinops melanosticta* (Temminck and Schlegel), used to be one of the most important species of fish for the Japanese coastal fisheries, the catch amounting to 1,586,000 tons in 1936. However, the catch has declined since then and reached a minimum of 155,000 tons in 1945 (Kurita and Tanaka, 1956). After the war the nationwide Cooperative Iwashi Resources Investigations[1] was started in 1949 to investigate the decline of the catch and to obtain scientific information for resource management and fishing forecasting (Nakai et al., 1955). The Investigations consisted of three main elements: a survey of the catch and effort statistics, a shore-based statistical survey (length and age data, etc.) at the fishing ports, and a survey at sea for eggs, larvae and hydrographic data, etc. The first aspect was discontinued after 1951 when the Statistics and Survey Division of the Ministry of Agriculture and Forestry commenced a survey of catch and effort statistics under a new system.

The results of statistical analyses were disappointing. The sardine was caught almost everywhere around Japan, with various types of gears and with a large variety of boats. The size and age of fish varied not only with fishing gear but also with areas and season. In 1951, 58 ports were sampled and in 1956, 83. The age composition of the sardine catch was obtained each year, and an estimate of the survival rate of adult sardine (2-age fish and older) found to be about 0.5. However the rate could not be estimated for the immature fish (0- and 1-age fish) because these fish were segregated from the adults and exploited in different seasons or by different types of gear. Standardization of the fishing effort of the various gears was almost impossible. For various fisheries, correlations were made between the seasonal change of the size of available population and fishing effort. In most cases, no correlation was observed between the total decreasing coefficient and the fishing effort, probably due to fluctuation of immigration and emmigration to and from the fishing ground (Yamanaka and Ito, 1957). Tagging of the sardine was not feasible because of the technical difficulties of tagging and a poor rate of recovery. An experiment in which tagged fish were mixed with the catch in the hold of a fishing boat resulted in only several percent of recovery.

A spawning survey at sea was essential both for studies on the ecology of spawning and to provide information on the size of the population and its rate of exploitation by means of an egg census as was done by Sette and Ahlstrom (1948) for the Californian sardine. A spawning survey was feasible because a number of research boats, even though small, of the prefectural fisheries experimental stations were available for the survey. The spawning grounds were concentrated in the coastal areas

---

[1] Iwashi is a general term for sardine, anchovy, and round herring.

and extension of the survey area offshore was not necessary. Since many of the research boats were not equipped with facilities for handling larger nets, a simple method of towing was adopted as the standard: a vertical haul of a small net with a mouth diameter of 45 cm and mesh aperture of 0.3 mm was made from a depth of 150 m to the surface. From 1949 to 1951, 61 boats from prefectural agencies and 4 boats from Fisheries Agency were involved in the survey.

ESTIMATE OF SARDINE EGG ABUNDANCE AND THE ADULT POPULATION IN 1952

The main spawning ground of the sardine was located in the East China Sea west to Kyushu and in the Sea of Japan off central Japan. Only the former is used as an example here (Tanaka, 1955a, 1955b). The sampling stations were distributed as shown in Fig. 1 and each station was occupied once a month during the spawning season from January to May. In order to minimize the effects of the selectivity of net and mortality, only data for the eggs was analysed. The entire spawning ground was divided into 11 sub-areas. The following formula was used as a basis for estimation.

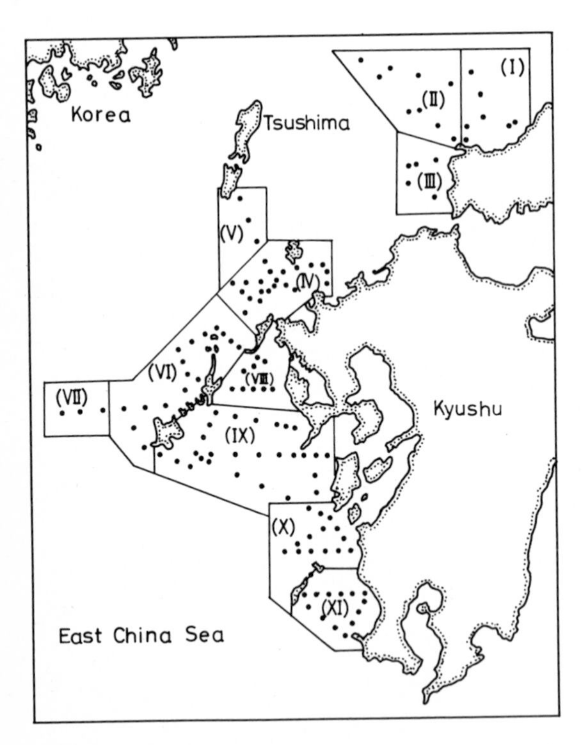

Fig. 1. Distribution of sampling stations and division of areas for estimation of the abundance of sardine eggs. (From Tanaka, 1955b)

$$\hat{E} = \sum_s \sum_t \hat{E}_{st} = \sum_s \sum_t \frac{A_s T_t}{k_{st} b_{st} n_{st}} \sum_i^{n_{st}} \frac{x_{sti}}{3.14 R^2 p_{sti}} \qquad (1)$$

where
- x_{sti}: the number of sardine eggs collected in *i* th sample in *st* th stratum (*s* th sub-area and *t* th month).
- R : radius of the mouth opening, $2R = 45$ cm.
- p_{sti}: filtration efficiency in *i* th sample in *st* th stratum.
- n_{st} : the number of samples in *st* th stratum.
- b_{st} : average incubation time in *st* th stratum (a function of water temperature).
- k_{st} : correction factor of b_{st} for mortality in *st* th stratum.
- A_s : size of *s* th sub-area.
- T_t : length of *t* th month.
- $\hat{E}_{st}$: estimated abundance of eggs in *st* th stratum.

As being unknown, k_{st} and p_{sti} were assumed to be constant regardless of *s*, *t* and *i*, and formula (1) was reduced to

$$\hat{E} = \frac{1}{3.14 R^2 kp} \sum_s \sum_t \frac{A_s T_t}{b_{st} n_{st}} \sum_i x_{sti}. \qquad (2)$$

The variance of the estimate was calculated by

$$V(\hat{E}) = \frac{1}{(3.14 R^2 kp)^2} \sum_s \sum_t \left(\frac{A_s T_t}{b_{st}}\right)^2 \frac{s_{st}^2}{n_{st}}. \qquad (3)$$

where

$$s_{st}^2 = \left[\sum_i x_{sti}^2 - \left(\sum_i x_{sti}\right)^2 / n_{st}\right] / (n_{st} - 1). \qquad (4)$$

The results of calculation under the assumption of $k = p = 1$ are presented in Table 1. The estimated total abundance of eggs spawned in the season was 49×10^{12} with the coefficient of variation of 17%. The number of samples used was 453 and was large enough to give the result with a reasonably small statistical error, in spite of a high degree of heterogeneity of the egg distribution and a small size of the net used. Nakai and Hattori (1962) estimated the survival rate during the egg stage as about 0.5 which corresponds with $k = 0.7$. If the filtration efficiency is assumed to be $p = 0.7$, then the total abundance of eggs spawned would be 100×10^{12}.

Table 1. Estimated abundance of sardine eggs spawned by months in water west of Kyushu in 1952. (From Tanaka, 1955b). Unit: 10^{12} eggs

Jan	Feb	Mar	Apr	May	Total
C.71	29.49	13.07	3.98	1.68	48.93
		Coefficient of variation			16.9%

Effects of mortality of eggs and filtration efficiency of net are omitted.

The abundance of the spawning population of sardine would be given by dividing the total abundance of eggs spawned by the fecundity of the female and the ratio of females among the total. The sex ratio was about 50:50. The fecundity was not known well but Nakai (1962) estimated the possible range as 40 to 120 x 10^3. Here, two levels of fecundity, 60 x 10^3 and 100 x 10^3, were used, and the spawning population size was estimated as 3.3 x 10^9 or 2.0 x 10^9 depending on the assumed value of fecundity.

The total catch in number of large sardine (2- to 7-age fish) along the west coast of Kyushu in 1952 was 0.8 x 10^9, and the rate of exploitation would be 0.25 or 0.4. The survival rate of the large-sized sardine estimated from the age composition in the catch was about 0.5 and so the total mortality coefficient would be Z = 0.7. From these data it was calculated that the fishing coefficient F = 0.35 or 0.56, and the natural mortality coefficient M = 0.35 or 0.14.

The error involved in these estimates may be great. Further, even the total mortality coefficient was not known for the immature fish which formed 90 to 95% or more in number of the total catch. Tanaka (1958) predicted that, in the range of natural mortality coefficient of spawning fish M_M = 0.16 - 0.32 and that of immature fish M_I = 0.16 - 0.8, the highest rate of increase of catch in weight would be obtained by decreasing the fishing intensity for the immature fish and increasing it for the spawning fish. By this modification of the sardine fisheries, the abundance of the spawning population would be kept unchanged. The range of estimates of M given above for the spawning population almost corresponds with the range of M_M and it is suggested that the prediction could be applied to the actual case.

STUDIES ON THE MACKEREL POPULATION

The idea applied to the studies of the sardine population was extended to those of Japanese mackerel, *Scomber japonicus* Houttuyn, in recent years and important contributions to our knowledge were made. Some examples using Watanabe's (1970, 1972) work are cited here. Recently the catch of the mackerel has increased considerably in the Pacific Ocean off central and northern Japan. The abundance of eggs spawned clearly indicates the increase of the spawning population (Fig. 2). The catch in number per boat-day of the pole and line fishery (hanezuri) on the spawning ground shows a very similar trend to the abundance of eggs. It is noteworthy that the abundance of eggs fluctuated in alternate years from 1962 to 1967. The coefficient of variation of these estimates is about 20% and this fluctuation cannot be attributed merely to statistical error. It is interesting that the gonad weight was high in the years of high abundance of eggs and small in the years of low abundance. Watanabe assumed that the high fecundity of one year would affect that of next year.

The abundance of eggs spawned by year classes could be calculated by applying the age composition of spawning population by years and the ratio of the fecundity between age groups. The calculated abundance of eggs spawned by 1 year class through its life span was correlated with the abundance of eggs from which the year class was originated, and a reproduction curve was estimated (Fig. 3). It can be seen from the figure that the reproduction rate has worsened since 1964.

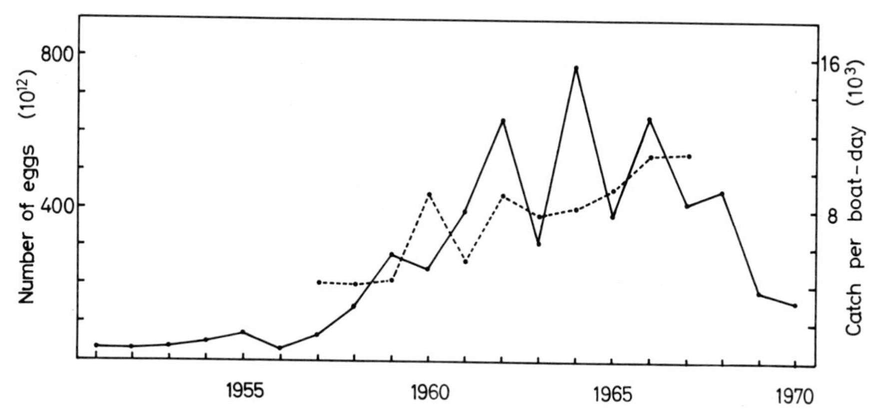

Fig. 2. Abundance indices of the mackerel population in the Pacific off northern Japan. (From Watanabe, 1970, 1972). Continuous line: egg abundance; dotted line: catch in number per boat-day in the spawning season

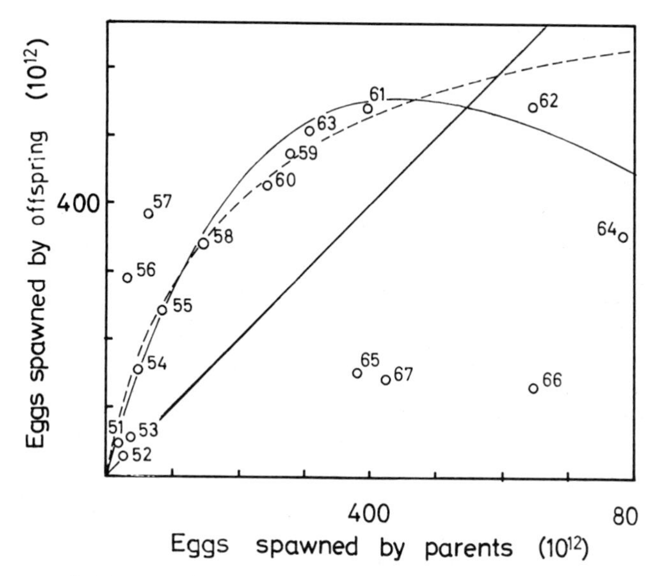

Fig. 3. Reproduction curve of the mackerel population in the Pacific off northern Japan. (From Watanabe, 1970, 1972). Continuous line: Ricker's curve; dotted line: Beverton and Holt's curve. Numerals in the graph indicate year classes

THE COST OF EGG AND LARVAL SURVEYS

It is generally supposed that the effort and cost required for egg and larval surveys is considerable and the possibility of its application is limited, even though the usefulness of such surveys is well known. The cost per sample of surveys at sea is given in Table 2. After certain adjustments it is estimated that the cost is 2200 yen per sample, though a corresponding figure for more recent years would be 4660 yen. Applying this average to the sardine egg samples taken in the water west of Kyushu (cost on average 2200 yen), the total cost for 453 samples would be about one million yen.

Table 2. Budget for the Cooperative Iwashi Resources Investigations entrusted to the prefectural agencies in Tokai and Tohoku Regions. Corresponding data in recent years of the fisheries forecasting program are also shown

Year	Survey on land for biol. statist. (¥1000)	No. of sampling stations planned	Budget per station (¥1000)	Survey at sea (¥1000)	Total no. of egg samples planned	Budget per sample (¥1000)
1951	1 530.0	18	85	400.0	817	0.49
1952	1 332.0	18	74	999.7	802	1.24
1953	1 640.0	20	82	1 497.0	844	1.77
1954	1 540.0	23	67	1 866.0	1 555	1.20
1955	1 378.0	22	63	1 755.2	1 589	1.10
1956	1 200.0	22	55	1 388.0	1 589	0.88
1957	1 138.0	24	47	1 496.3	1 486	1.01
Average x 2 *			135=$375			2.20=$6.1
1971	(whole country)			24 210	11 297	2.14
1972	(whole country)			38 576	15 345	2.51
Average x 2 *						4.66

*To allow for prefectural and national costs

In regard to research boats, it is possible to make sea surveys not by building new research boats but by chartering fishing boats. A simple standard procedure such as that used for the sardine is required. An estimate suggests that the cost of chartering such a boat would be about 6900 yen per sample. Where catch and effort statistics are not available, it would require a tremendous amount of manpower to initiate their collection. In 1951, the Statistics and Survey Division of the Ministry of Agriculture and Forestry had more than 1000 staff working on the catch and effort statistics of the Japanese fisheries using a revised system. Depending on circumstances, fish-egg and larval surveys may thus be the most economical way of approaching the dynamics of fish populations.

REFERENCES

Kurita, S. and Tanaka, C., 1956. Estimation of annual catches of sardine and anchovy in Japan, 1926-1950, using the amount of processed iwashi products. Bull. Jap. Soc. scient. Fish., 22, 338-347.
Nakai, Z., 1962. Studies of influences of environmental factors upon fertilization and development of the Japanese sardine eggs - with some reference to the number of their ova. Bull. Tokai reg. Fish. Res. Lab., 9, 109-150.
Nakai, Z. and Hattori, S., 1962. Quantitative distribution of eggs and larvae of the Japanese sardine by year 1949 through 1951. Bull. Tokai reg. Fish. Res. Lab., 9, 23-60.
Nakai, Z., Usami, S., Hattori, S., Honjo, K. and Hayashi, S., 1955. Progress Report of the Cooperative Iwashi Resources Investigations, April 1949 - December 1951. Tokai Reg. Fish. Res. Lab., Tokyo, 116 pp.
Sette, O.E. and Ahlstrom, E.H., 1948. Estimations of abundance of the eggs of the Pacific pilchard (*Sardinops caerulea*) off southern California during 1940 and 1941. J. Mar. Res., 7, 511-542.
Tanaka, S., 1955a. Estimation of the abundance of the eggs of fish such as sardine. I. Method of estimation by the vertical haul of plankton net. Bull. Jap. Soc. scient. Fish., 21, 386-389.
Tanaka, S., 1955b. Estimation of the abundance of the eggs of fish such as sardine. II. Abundance of the eggs of the Japanese sardine in western area off Kyusyu in 1952. Bull. Jap. Soc. scient. Fish., 21, 390-396.
Tanaka, S., 1958. A consideration on the rational exploitation of the stock of sardine, *Sardinops melanosticta* (T. & S.). Bull. Tokai reg. Fish. Res. Lab., 21, 1-13.
Watanabe, T., 1970. Morphology and ecology of early stages of life in Japanese common mackerel, *Scomber japonicus* Houttuyn, with special reference to fluctuation of population. Bull. Tokai reg. Fish. Res. Lab., 62, 1-283.
Watanabe, T., 1972. The recent trend in the stock size of the Pacific population of the common mackerel off Honshu, Japan, as viewed from egg abundance. Bull. Jap. Soc. scient. Fish., 38, 439-444.
Yamanaka, I. and Ito, S., 1957. Progress Report of the Cooperative Iwashi Resources Investigations, 1954. Japan Sea Reg. Fish. Res. Lab., Niigata, 177 pp.

S. Tanaka
Ocean Research Institute
University of Tokyo
1-15, Minamidai 1
Nakano-ku
Tokyo / JAPAN

Seasonal Fluctuations in the Abundance of the Larvae of Mackerel and Herring in the Northeastern Atlantic and North Sea

V. Bainbridge, G. A. Cooper, and P. J. B. Hart

INTRODUCTION

This paper describes the distribution and seasonal occurrence of larvae of mackerel and herring in the Continuous Plankton Recorder (CPR) Survey during the years 1948 to 1967. Results are considered in relation to seasonal cycles of phytoplankton as a development of the work of Cushing (1967), who showed that the different spawning periods of various herring populations could be linked to the timing of production cycles in the vicinity of their spawning grounds.

Descriptions of the CPR with details of the organisation and logistics of the survey are provided by Glover (1967) and Oceanographic Laboratory, Edinburgh (1973). Fish larvae are identified, counted and measured to the nearest millimetre below the actual length. Data in this paper have been compiled from analyses of all CPR samples from 1948 to 1958 and from alternate 10-mile samples along each route thereafter. The interest of the CPR survey lies in the uniformity of the method, the extensive range over which it is deployed and the regularity of deployment over many years; fish larvae are a minor constituent of the total zooplankton and individual CPR samples are small, representing about 3 m^3 of water filtered, so the numbers of larvae examined are small. To describe patterns of distribution and seasonal variation, therefore, data must be integrated over many years or over wide areas. This paper is based on the examination of about 6,000 mackerel larvae and 22,000 clupeid larvae taken during 20 years.

DISTRIBUTIONS BY MONTHS

The larvae were divided into 2 length groups, the smaller of which was used for the charts showing average monthly distributions (Figs 1 and 2), on the assumption that recently hatched larvae provide the best indication of timing and location of spawning activity.

1. The mackerel, *Scomber scombrus* L. contributed about 14% of the larval fish in the Celtic Sea samples and about 12% in those from the North Sea. At the time of hatching mackerel larvae are about 3.5 mm long; the charts in Fig. 1 present the distributions of larvae under 5 mm in length. In the Celtic Sea they first appeared in March over the continental margin off the mouth of the English Channel and were found progressively further north until they occurred off the southern seaboard of Ireland by June. This apparent shift of the centre of spawning activity is similar to that shown by egg and larval surveys in this area during 1939 described by Corbin (1947). In the North Sea, spawning was more intensive; the highest numbers were in the west during June and became abundant throughout the central North Sea by July. Thus, the populations of larvae in the North Sea and to the west

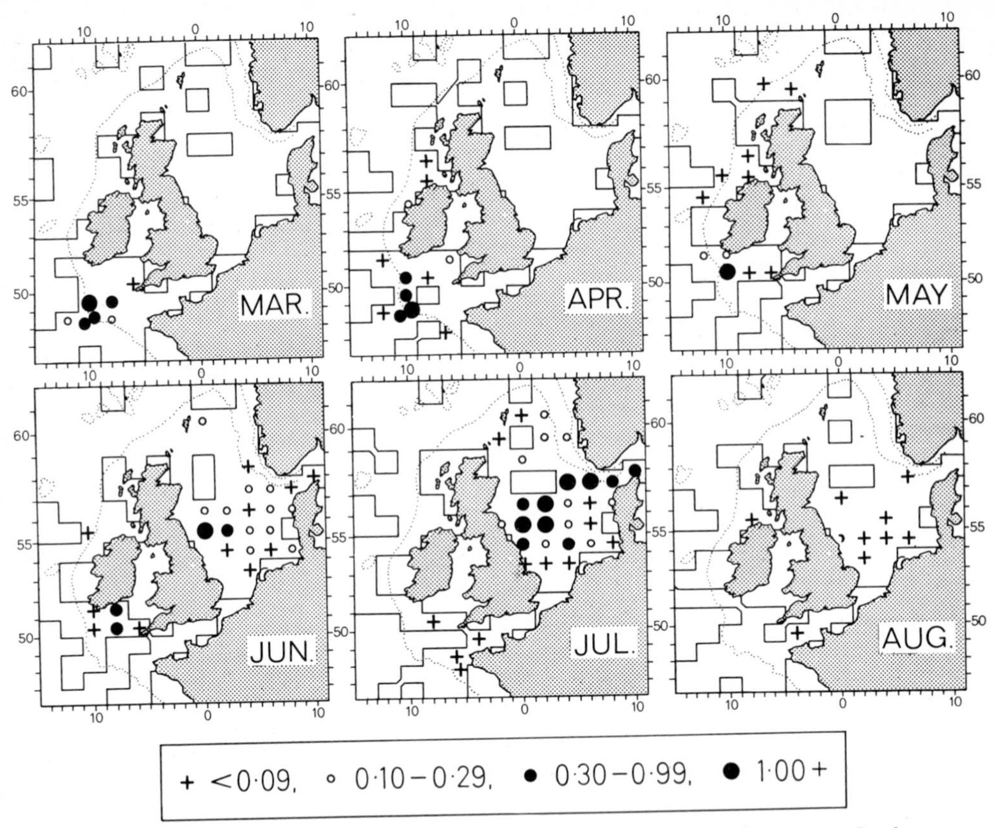

Fig. 1. The distribution and abundance of the early larvae of the mackerel (<5 mm in length), March to August. Symbols represent the average numbers per sample per month in each statistical rectangle over the period 1948 to 1967 (see key). The outlined sampled areas include only rectangles sampled during 3 or more years

of the British Isles were separated spatially, and only partially overlapped in the timing of their spawning periods. These results support the view expressed by Hamre (1970) and Postuma (1972), based mainly on tagging experiments, that the mackerel of the North Sea and Skagerrak form a self-contained stock; distinct from those which spawn south and southwest of the British Isles.

2. Clupeid larvae were mainly found over the continental shelf, forming 20% of fish larvae in the North Sea, 42% to the west of the British Isles and 36% off southern Iceland. Clupeids are considered here as a single group since, as a matter of expediency, they were not normally identified to specific level, although it was generally possible to infer the identity of the various concentrations from the known distribution of adults.

The clupeid larvae include those of herring, *Clupea harengus* (L), which hatch at 6 - 8 mm, pilchards, *Sardina pilchardus* (Walbaum), hatching at 3 - 4 mm and sprats, *Sprattus sprattus* (L), hatching at about 3 mm. Herring larvae predominated, so 10 mm was selected as the upper size limit of early larvae.

Interpretation of the distributions in Fig. 2 would be impossible without independent information on the biology of the various species, particularly the herring. Fortunately, the northeast Atlantic herring has been the subject of intensive research and much that is known of the sub-division into unit stocks as well as the times and places of spawning has been summarized by Parrish and Saville (1965). On the basis of this information, the probable parent stocks of some of the larvae are indicated in Fig. 2 by broken lines. Larvae of the winter-spring spawning Norwegian, Icelandic, Faroese and Scottish (west coast) stocks were recognizable in the CPR samples as distinct patches (Fig. 2). Larvae produced by summer-autumn spawning herring in the central and northern North Sea (the Dogger and Buchan stocks) and off the west coast of Scotland could also be distinguished. As would be expected in this survey area, early larvae of the northern Irish Sea and southern North Sea stocks were absent or rare. Larvae of Icelandic summer spawners were also absent, although this stock is thought to spawn in the same localities as the Icelandic spring spawners.

Most of the early clupeid larvae in the southern North Sea during June and July would appear to be sprats (see Henderson, 1954), while those in the English Channel and Celtic Sea during May and June must include both sprats and pilchards (see Wallace and Pleasants, 1972).

SPAWNING STOCKS AND PRODUCTION CYCLES

Eggs, nauplii, and early copepodite stages of copepods form the principal food organisms of both the mackerel (Lebour, 1918) and the herring (see, for example, Bjørke, 1971; Bainbridge and Forsyth, 1971). Direct estimates of these organisms are not provided by the Recorder survey since most copepods retained are adults or later copepodites (Robertson, 1968). However, an indirect index of the availability of food organisms may be obtained from estimates of the abundance of phytoplankton based on phytoplankton colour on the CPR silks as used by Cushing (1967). The validity of this relationship is based on the dependence of the reproduction of the main herbivorous copepods upon their consumption of food. This has been demonstrated experimentally for *Calanus* by Marshall and Orr (1964) who also showed that seasonal variations in the reproduction of *Calanus* in the sea could be correlated with diatom increases. Using data from the CPR survey, Colebrook and Robinson (1965) have confirmed a close relationship between the timing of the spring outburst of phytoplankton and the spring development of copepods, in contrast to some areas where there was also a distinctive autumn peak of phytoplankton which was not followed by any increase of copepod numbers. However, several classic studies (see Raymont, 1963) have shown that the autumnal phytoplankton peak is accompanied by a brief period of zooplankton development. The absence of any clearly defined autumn increase of copepod numbers in CPR samples has not yet been explained.

In Figs 3 and 4 average numbers of mackerel and clupeid larvae by months in various sub-areas of the CPR survey are compared with the average seasonal fluctuations in phytoplankton abundance (1948 to 1970). The phytoplankton data are taken from Robinson (1973) who also gives the average seasonal cycle of copepod numbers in each sub-area.

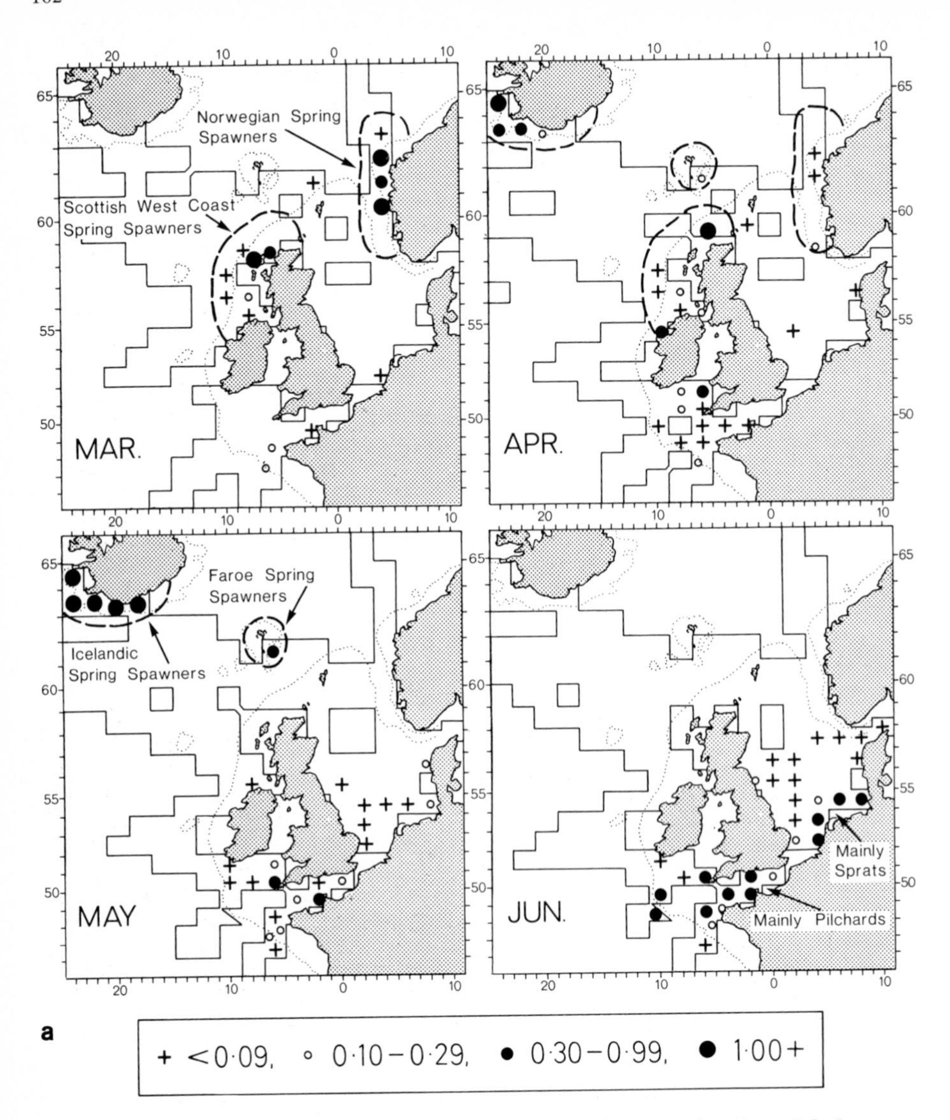

Fig. 2a and b. The distribution and abundance of clupeid larvae (<10 mm in length), March to October. Symbols represent the average numbers per sample per month in each statistical rectangle over the period 1948 to 1967. Larvae considered to mainly represent individual herring stocks have been delimited by dashed lines. The outlined sampled areas include only rectangles sampled during three or more years

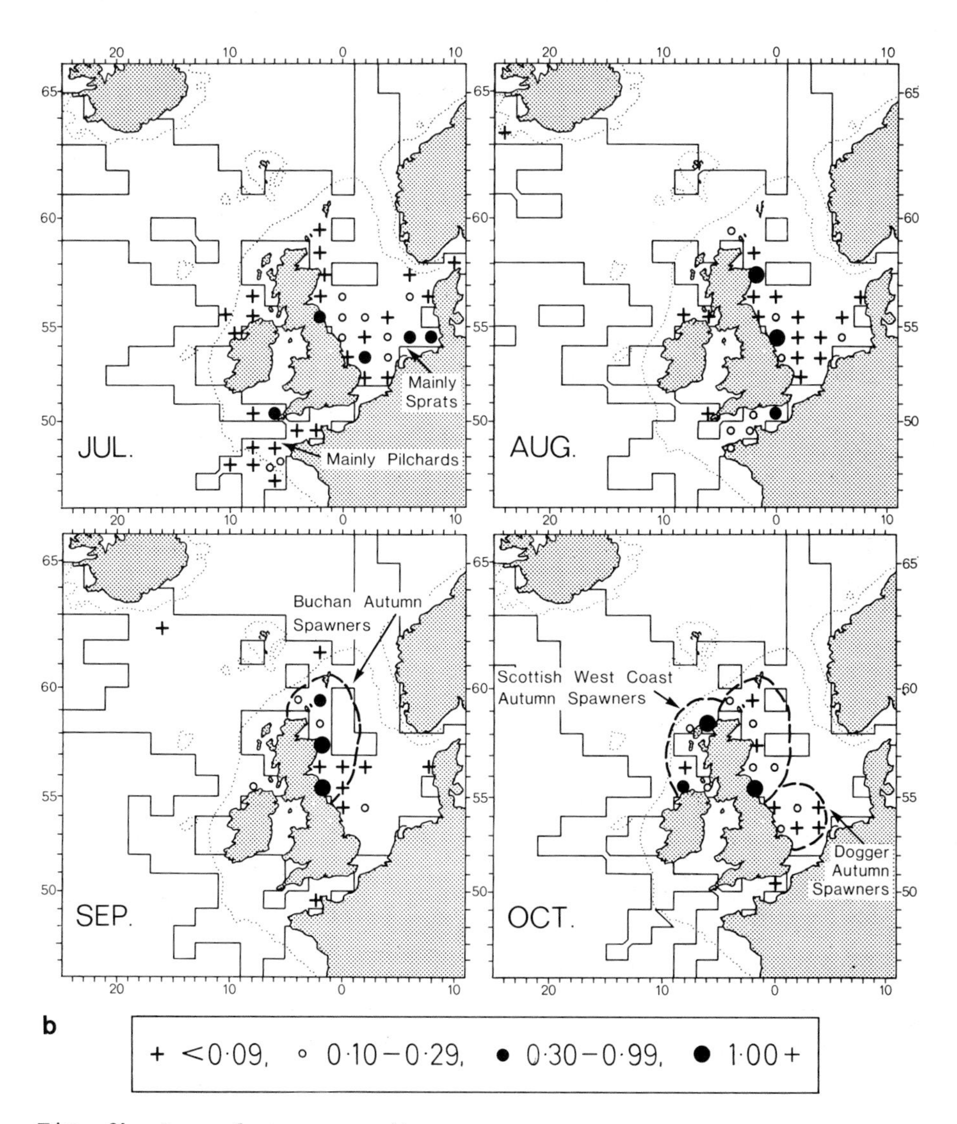

Fig. 2b. Legend see opposite page

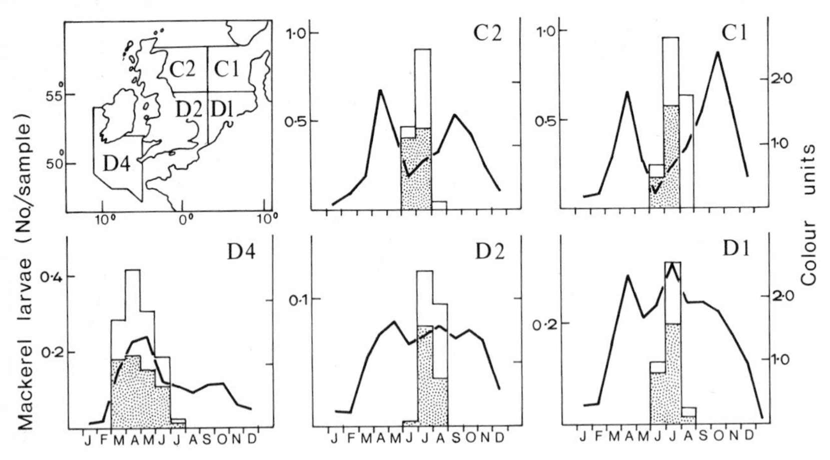

Fig. 3. Average numbers of mackerel larvae in relation to the seasonal cycle of phytoplankton for various sub-areas (see chart) during the period 1948 to 1967. The sets of histograms representing numbers of fish larvae (left-hand scales) are subdivided to show the relative numbers of larvae <5 mm (stippled area) and >5 mm (open area). Phytoplankton colour (right-hand scale) is shown by line graphs

Of the two main populations of mackerel larvae, that centred in the Celtic Sea (sub-area D4) coincides with the spring outbreak of phytoplankton (Fig. 3). The northeasterly spread of larval production in this area as shown in Fig. 1 follows a pattern similar to that of the development of phytoplankton, which also starts near the continental margin (Plate LXII Colebrook and Robinson, 1965). However, in the North Sea, mackerel larvae were most abundant in the central sub-areas (C1 and C2) with the early stages occurring in June and July, clearly after the spring phytoplankton outburst and also after the spring increase in copepod numbers as given by Robinson (1973).

Several spring spawning stocks of herring can be distinguished from the monthly histograms shown in Fig. 4. In March larvae produced by the Norwegian spring spawners were abundant in sub-area B1; in this region the phytoplankton outburst is early and the numbers of copepods increase abruptly between February and April (see Robinson, 1973). The larvae of two other Atlanto-Scandian stocks, Icelandic spring spawners (A6) and Faroese spring spawners (B4) occurred 2 months later with maximum numbers in May, coinciding with the later biological spring in these regions. Larvae of spring-spawning herring were also present to the west of Scotland (C4) and a few were found off the northeast coast also (B2). In both localities the larvae occurred at the same time as the spring development of phytoplankton, this being slightly later in the east than in the west.

Larvae of 4 summer-autumn stocks are discernable from the histograms of Fig. 4. Those of the northern North Sea (Buchan) stock mainly within sub-area C2 can be linked to the autumnal phytoplankton peak in September and larvae of the central North Sea (Dogger) stock, seen as a small peak of early larvae during October in sub-area D2, also coincide with a minor peak of phytoplankton.

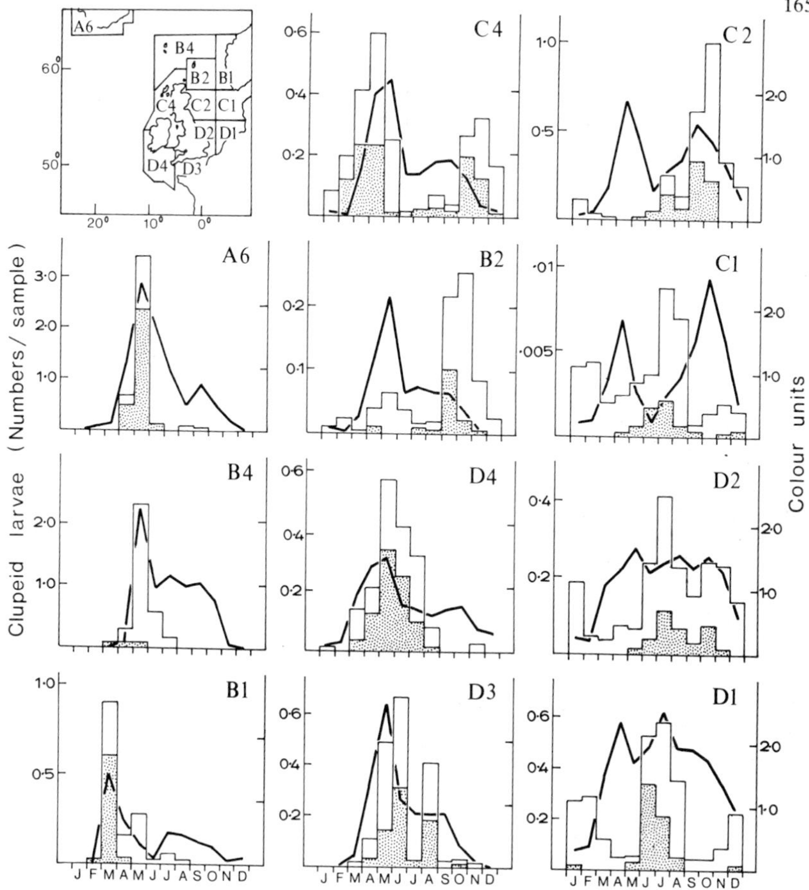

Fig. 4. Average numbers of clupeid larvae in relation to the seasonal cycle of phytoplankton for various sub-areas (see chart) during the period 1948 to 1967. The sets of histograms representing numbers of fish larvae (left-hand scales) are subdivided to show the relative numbers of larvae <10 mm (stippled area) >10 mm (open area). Phytoplankton colour (right-hand scale) is shown by line graphs

In contrast, larvae produced by the Scottish west coast (Minch) stock and the southern North Sea (Downs) stock do not appear from the CPR samples to be related in any simple way to phytoplankton production. The Minch larvae found off the west coast of Scotland in October and November were two months later than the autumn phytoplankton increase while the Downs larvae in the southeastern North Sea (D1) from December to February coincided with the seasonal minima of both phytoplankton and zooplankton (see Robinson, 1973).

The occurrence of larvae of clupeids (pilchards and sprats) in the Celtic Sea (D4) and the English Channel (D3) seems to be linked to the seasonal cycles of phytoplankton, with a time lag of about 1 month in sub-area D3 compared with sub-area D4 for both the spring phytoplankton outburst and the presence of the larvae. Sprat larvae were most numerous in the southeastern North Sea (D1), forming a well defined peak during June and July, after the main phytoplankton increase but at about the period when phytoplankton values were at a maximum in this area.

DISCUSSION

Previous studies of CPR data from the North Atlantic have shown that larval production by two of the commonest oceanic fish in the North Atlantic coincides with what is, on average, the period when food organisms are most plentiful in the areas in which they live. Bainbridge and Cooper (1971) suggested that differences in the timing of the seasonal cycle of plankton in the central North Atlantic and over the North American shelf and slope had been a major influence in the formation of separate spawning stocks of the redfish, *Sebastes mentella* Travin. The short but intensive spawning of the blue whiting, *Micromesistius poutassou* Risso beyond the continental margin west of the British Isles progresses rapidly northwards and is likewise synchronized with the abrupt spring development of copepods in this region (Bainbridge and Cooper, in press). Seasonal fluctuations of mackerel larvae to the west of the British Isles and larvae of the oceanic group of herring stocks described in the preceding pages are consistent with this general pattern of larval production in close association with the average timing of the biological spring. However, both species also include stocks with spawning periods which do not synchronize with any seasonal peaks of phytoplankton and these will be discussed separately.

Mackerel are mainly spring spawners in the Celtic Sea but summer spawners in the North Sea. An examination of CPR larvae of demersal fish, including cod, coalfish, whiting, plaice and dab, showed that only one species, the whiting, *Odontogadus merlangius* (L), resembled the mackerel in having different seasonal patterns of spawning to the west and east of the British Isles. The whiting, together with the mackerel, sprat and pilchard were categorized as south-boreal species by Qasim (1956) and Rass (1959), the other species listed above being assigned to the north-boreal community on the basis of distributions and breeding ranges. Qasim (1956) drew attention to similarities between the spawning seasons of some teleosts and those of marine invertebrate species. Orton (1920) noted that certain marine invertebrates nearer the warmer limits of their range breed during the winter and, conversely, species at their colder limits breed during the summer, a possible adaptive advantage since early stages are less tolerant of extreme conditions than adults. The North Sea, where phytoplankton tends to be more abundant and less restricted seasonally than over the western shelf of the British Isles (Robinson, 1970), may provide suitable feeding conditions for fish larvae during the summer, so allowing south-boreal species to breed at a higher temperature than is attained during the spring in this area. The monthly progression of larval production by mackerel shown in Fig. 1 does, in fact, closely follow the pattern of warming of surface waters as given by charts of monthly mean surface temperatures (ICES, 1962). Throughout the Celtic and the North Sea the first appearance of mackerel larvae on the monthly charts followed the spread of the 10°C and 11°C isotherms.

For the herring, a north-boreal species, Cushing (1967) has suggested that the autumn, winter and spring spawning stocks correspond to the central North Sea, coastal Atlantic and oceanic Atlantic production cycles as described by Colebrook and Robinson (1965). Results reported in this paper confirm the close connection between larval production by spring spawners and the spring phytoplankton outbursts and between larvae of the northern and central North Sea autumn spawners and the autumn phytoplankton outburst. However, larval production by the Scottish west coast (Minch) stock (spawning in the autumn) and the southern North Sea (Downs) stock (spawning in the winter) is less readily related to phytoplankton cycles.

Although the southern North Sea supported a higher standing stock of phytoplankton and copepods over the winter than any other region, herring larvae hatching in December and January seem likely to experience poor feeding conditions. Blaxter and Hempel (1963) noted that the Downs group of herring produced the largest eggs with the greatest yolk reserves and suggested this was an adaptation to a scarcity of food for the larvae. Bainbridge and Forsyth (1971) found that food organisms suitable for herring larvae were scarcer in the English Channel during January than in areas occupied by larvae of the Dogger and Buchan stocks in September or in the Clyde (Scottish west coast) during April. One possible explanation is that the spawning habits of the herring, with the demersal eggs precisely sited on restricted spawning grounds, have permitted it to exploit situations in coastal waters, where slow growth rate of early larvae is not necessarily accompanied by a high level of mortality. The histograms of Fig. 4 show that clupeid larvae greater than 10 mm in length persist throughout the winter in the southern North Sea (sub-areas D1 and D2) and off the west coast of Scotland (C4). These larvae presumably take advantage of the early start of the production cycle in these areas to complete larval development.

Since the great majority of teleost fish larvae die before metamorphosis, forces affecting larval mortality probably play a dominant role in the evolution of spawning stocks. While the maximum availability of suitable food organisms for the larvae would appear to be of critical importance for fish which spawn in North Atlantic oceanic areas and over deep water near the edge of the shelf, this may not be consistently true for those spawning in temperate coastal waters where the average standing crop of phytoplankton is greater and the productive season longer (Robinson, 1970). Both mackerel and herring provide examples of fish stocks for which factors other than larval food supply have played a predominant part in determining the timing of spawning seasons.

SUMMARY

1. The Continuous Plankton Recorder Survey provides information on larval production by mackerel and herring over a wide area of the northeastern Atlantic and North Sea.

2. There were 2 main populations of mackerel larvae, one centred in the Celtic Sea during the spring, the other in the central North Sea during the summer.

3. Larvae produced by most of the main stocks of winter-spring and summer-autumn spawning groups of herring could be delimited from their seasonal and geographical distributions.

4. Larval production by stocks of mackerel and herring spawning in deep water near the continental margin is closely associated with the biological spring when food for the larvae is most plentiful. This is not a general rule for stocks of these fish spawning in coastal waters where the seasonal development of phytoplankton is longer and average standing crops greater.

5. In the North Sea mackerel spawn during the summer, a possible adaptation to temperature conditions at the northern limit of its range. The larvae produced by herring stocks which spawn in the later autumn or winter seem likely to encounter poor feeding conditions until the following spring.

ACKNOWLEDGEMENTS

Thanks are due to the captains and crews of vessels which tow Recorders and to Dr. G.T.D. Henderson who was responsible for the CPR studies on fish larvae until his retirement in 1968. We are also indebted to D.C.T. Forsyth for advice and help with the preparation of this paper and to Miss A. McKenzie for technical assistance. The research was supported by the Natural Environment Research Council as part of the programme of the Institute for Marine Environmental Research.

REFERENCES

Bainbridge, V. and Cooper, G.A., 1971. Populations of *Sebastes* larvae in the North Atlantic. Res. Bull. int. Comm. N.W. Atlant. Fish., No. 8, 27-35.

Bainbridge, V. and Cooper, G.A., in press. The distribution and abundance of the larvae of the blue whiting, *Micromesistius poutassou* (Risso) in the north-east Atlantic, 1948-1970. Bull. Mar. Ecol., 8.

Bainbridge, V. and Forsyth, D.C.T., 1971. The feeding of Clyde herring larvae. Rapp. P.-v. Réun. Cons. perm. int. Explor. Mer, 160, 104-113.

Blaxter, J.H.S. and Hempel, G., 1963. The influence of egg size on herring larvae (*Clupea harengus* L.). J. Cons. perm. int. Explor. Mer, 28, 211-240.

Bjørke, H., 1971. The food of herring larvae of Norwegian spring spawners. Rapp. P.-v. Réun. Cons. perm. int. Explor. Mer, 160, 101-103.

Colebrook, J.M. and Robinson, G.A., 1965. Continuous plankton records: Seasonal cycles of phytoplankton and copepods in the north-eastern Atlantic and the North Sea. Bull. Mar. Ecol., 6, 123-139.

Corbin, P.G., 1947. The spawning of mackerel, *Scomber scombrus* L., and pilchard, *Clupea pilchardus* Walbaum in the Celtic Sea in 1937-1939. J. mar. biol. Assoc. U.K., 27, 65-132.

Cushing, D.H., 1967. The grouping of herring populations. J. mar. biol. Assoc. U.K., 47, 193-203.

Edinburgh, Oceanographic Laboratory, 1973. Continuous Plankton Records: A plankton atlas of the North Atlantic and North Sea. Bull. Mar. Ecol., 7, 1-174.

Glover, R.S., 1967. The continuous plankton recorder survey of the North Atlantic. Symp. zool. Soc. Lond., No. 19, p. 189-210.

Hamre, J., 1970. Internal tagging experiments of mackerel in the Skagerak and the north-eastern North Sea. ICES, Doc. CM 1970/H, 25.

Henderson, G.T.D., 1954. Continuous plankton records: The young fish and fish eggs 1932-1939 and 1946-1949. Hull. Bull. Mar. Ecol., 3, 215-252.

ICES (1962). Mean monthly temperature and salinity of the surface layer of the North Sea and adjacent waters from 1905 to 1954.

Lebour, M.V., 1918. The food of post-larval fish. J. mar. biol. Assoc. U.K., 11, 433-469.

Marshall, S.M. and Orr, A.P., 1964. Grazing by copepods in the sea. In: Grazing in terrestrial and marine environments. D.J. Crisp (Ed.). Oxford: Blackwell Scientific Publications, 227-238.

Orton, J.H., 1920. Sea temperature, breeding and distribution of marine animals. J. mar. biol. Assoc. U.K., 12, 339-366.

Parrish, B.B. and Saville, A., 1965. The biology of the north-east Atlantic herring populations. Oceanogr. Mar. Biol. Ann. Rev., 3, 323-373.

Postuma, K.H., 1972. On the abundance of mackerel (*Scomber scombrus* L.) in the northern and north-eastern North Sea in the period 1959-1969. J. Cons. int. Explor. Mer, 34, 455-465.
Qasim, S.Z., 1956. Time and duration of spawning season in some marine teleosts in relation to their distribution. J. Cons. perm. int. Explor. Mer, 21, 144-155.
Rass, T.S., 1959. Biogeographical fishery complexes of the Atlantic and Pacific Oceans and their comparisons. J. Cons. perm. int. Explor. Mer, 24, 243-254.
Raymont, J.E.G., 1963. Plankton and productivity in the oceans. Pergamon Press, Oxford, 660 pp.
Robertson, A., 1968. The continuous plankton recorder: A method for studying the biomass of calanoid copepods. Bull. Mar. Ecol., 6, 185-223.
Robinson, G.A., 1970. Continuous plankton records: Variation in the seasonal cycle of phytoplankton in the North Atlantic. Bull. Mar. Ecol. 6, 333-345.
Robinson, G.A., 1973. The Continuous Plankton Recorder Survey: plankton around the British Isles during 1971. Annls. biol., Copenh., 28, 59-64.
Wallace, P.D. and Pleasants, C.A., 1972. The distribution of eggs and larvae of some pelagic fish species in the English Channel and adjacent waters in 1967 and 1968. ICES, Doc. CM 1972/J, 8.

V. Bainbridge
Institute for Marine Environmental Research
Citadel Road
Plymouth / GREAT BRITAIN

G.A. Cooper
Institute for Marine Environmental Research
Oceanographic Laboratory
78 Craighall Road
Edinburgh, EH 6 4 RQ / GREAT BRITAIN

P.J.B. Hart
Institute for Marine Environmental Research
Oceanographic Laboratory
78 Craighall Road
Edinburgh, EH 6 4 RQ / GREAT BRITAIN

Present address:

NORDRECO AB
26700 Bjuv / SWEDEN

The Distribution and Long Term Changes in Abundance of Larval *Ammodytes marinus* (Raitt) in the North Sea

P. J. B. Hart

INTRODUCTION

Aspects of distribution, abundance, feeding, and nomenclature of larval sand eels have been described by Ryland (1964), Macer (1965), and Langham (1971). All of these were studies of relatively small areas in the North Sea or west of Scotland, covering short periods of time. Henderson (1961), using the Continuous Plankton Recorder (CPR) survey, described the general distribution around the British Isles but did not analyze variability in time. It is now possible to examine a time series of 21 years (1948 - 1968) with the aim of revealing the composition and variability of the North Sea populations in space and time. This research was supported by the Natural Environment Research Council as part of the programme of the Institute for Marine Environmental Research.

SAMPLING METHODS AND DATA ANALYSIS

Sampling and Data Processing. The Continuous Plankton Recorder survey has been described by Glover (1967) and the methods of analysis by Rae (1952) and Colebrook (1960). Every sample, representing 10 miles of tow, was examined for fish larvae from 1948 - 1958 but only alternate samples from 1959 onwards.

The tracks of all monthly tows in the North Sea, from 1948 to 1968 are outlined in Fig. 1. The distribution of sand-eel larvae along these tracks suggests seven population centres which are shown stippled and numbered I to VII in Fig. 1. The raw data were transformed to $\log_{10}(x + 1)$, where x is the number of fish larvae recorded, and the mean number of larvae per sample calculated for each population centre during January to June of each year.

Species Composition. The five species of sand eel found in the North Sea are characterized by different spawning times depending on latitude (Macer, 1965). The larvae in the CPR samples have never been identified to species but all have been measured. For the present study only the data for the months January to June have been used, this being the period when the majority of the larvae were taken. It is assumed that most of the larvae collected were *Ammodytes marinus* (Raitt), although it is possible that *A. lanceolatus* (Lesauvage) could be caught in May or June. However, as the larvae of this species are smaller than those of *A. marinus* and do not appear until April at the earliest, any records of larvae less than 10 mm in length in May or June were rejected. Macer (1965) showed that larvae of *A. lancea* (Cuvier) also occur in April and May in the southern North Sea but they are too close inshore to be sampled by the CPR. According to Langham (1971), larvae

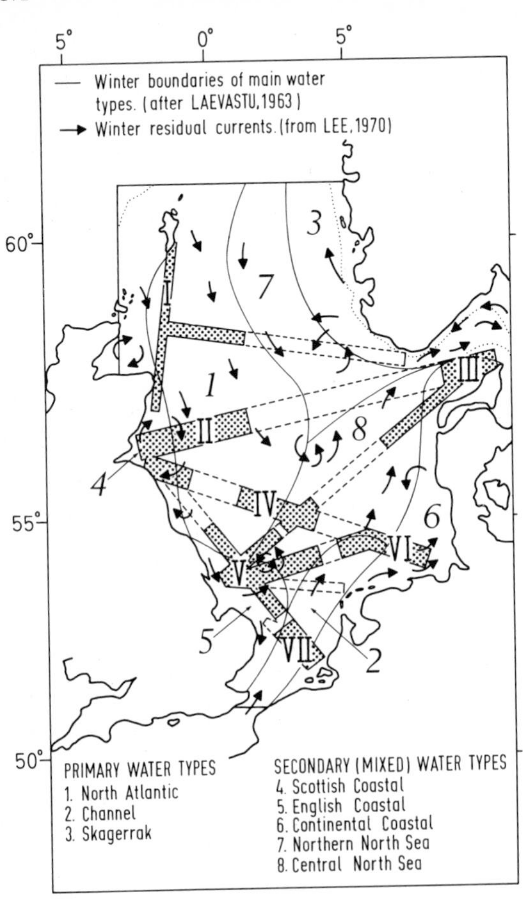

Fig. 1. The North Sea showing sampled areas, main water masses in their winter configuration (arabic numerals 1-8) and winter residual currents. The sampled strips are enclosed by continuous or dotted lines while the main population centres of sand eel larvae within the sampled area are shown stippled and numbered I-VII

of species other than *A. marinus* do occur in the plankton between January and June but only to the west of Scotland, an area not included in the present study.

The Effects of Night and Day on Sample Size. Ryland (1964) showed that during the day larvae of *A. marinus* were concentrated in the upper 20 m of water but at night the numbers became more or less evenly dispersed throughout the water column; thus higher numbers of larvae might be expected in CPR samples taken by day. To test this, the correlation coefficients between numbers of larvae per sample per year and the percentage of night samples were calculated. Six of the 7 centres showed values of $r = 0$ at $P > 0.05$ but in centre IV, $r = 0.8$ with $P \ll 0.01$ (d.f. = 19). This is the opposite of expectation from the findings of Ryland as the positive value of r means that more fish were caught at night. Nevertheless this correlation does mean that part of the variation in abundance in centre IV could result from sampling differences from year to year.

Size Composition of the Samples. As the fish larvae are in the plankton for only a relatively short period, the timing of sampling is critical. Presumably, numbers will be highest soon after hatching and, at this time, will give the best estimate of abundance of larvae for a given year. There are two sources of sampling error that could affect this estimate; the first is variation in the timing of sampling and the second is size selection of larvae. Small larvae (<10 mm) must be more abundant than large ones; accordingly, higher numbers of larvae would be expected in February, March and April, assuming constant sampling efficiency. However, this was not found, indicating that the CPR does not sample the smaller fish as efficiently as the larger ones; therefore, the peak abundance of larvae is underestimated by some unknown amount.

RESULTS

Distributions. The positions of the 7 centres of sand-eel concentration are shown in Fig. 1 together with the main water masses and residual surface currents in their winter configuration (Böhnecke, 1922; Laevastu, 1963; Lee, 1970). The sand-eel larvae tended to be found in secondary water masses (see key to figure). Centres I and VII are almost wholly influenced by high salinity oceanic water; but, although the larger part of centre II is in Atlantic water the majority of the sand eels here came from the small area within the Scottish Coastal water.

Sand eels in centre I are probably at the edge of a main population located to the west, as Langham (1971) found concentrations of *A. marinus* in the southern half of the Moray Firth while Henderson (1961) found high numbers of sand eels immediately to the North of Scotland, areas not included in this study. Macer (1965) found large patches of *A. marinus* larvae in 1960 and 1961 corresponding to centres V and VII. There are two main gaps in the area sampled, one along the Danish coast, where the Danish fishery for adult sand eels first started, and consequently where larvae would be expected, and the other in the northeast corner of the sampled region.

The Timing of Larval Appearance. As the smallest larvae are not caught efficiently by the CPR, any estimate of time of first occurrence will be subject to error. Nevertheless, some estimate can be made if it is assumed that there has not been any systematic trend in the error over

Table 1. The timing of larval appearance in the plankton

CENTRE	% FIRST OCCURRENCE OF LARVAE IN FEB. OR MAR.
I	53
II	40
III	50
IV	34
V	89
VI	52
VII	86

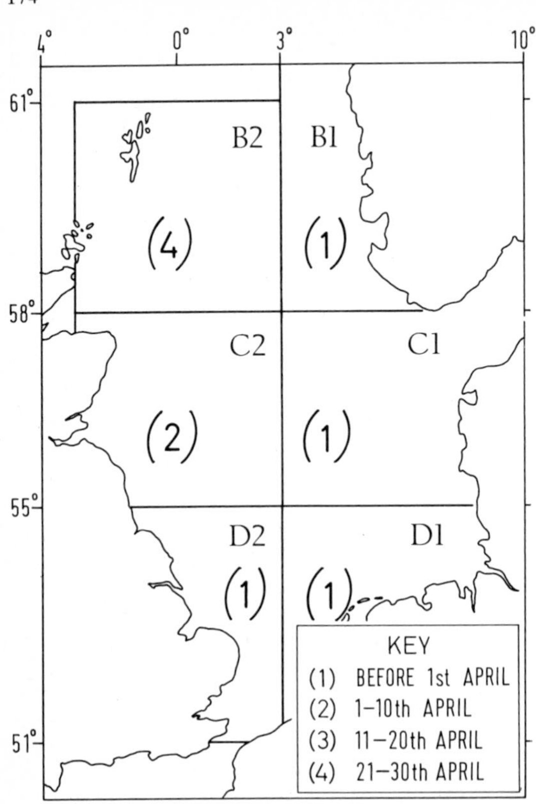

Fig. 2. The timing of the centre of gravity of the spring phytoplankton outbreak in 6 sub-areas of the North Sea (after Robinson, 1970), based on 20 years data

the years. The percentage of years out of 21 when larvae from each centre first occurred in either February or March are shown in Table 1. Robinson (1970) showed that in the North Sea the timing of the peak of the phytoplankton bloom became progressively later from south to north, as shown in Fig. 2, where the lower numbers indicate early peaks and higher numbers the later peaks (see key to figure). Data for the 7 sand-eel population centres can be assigned to the sub-areas of the North Sea used by Robinson (1965). Where a sand-eel population centre (e.g. IV) overlaps several of Robinson's sub-areas the mean timing of phytoplankton in the adjacent sub-areas has been used. From Robinson's data and the results in Table 1, a correlation of $r = 0.8$, ($0.01<P<0.05$ with d.f. = 5), was obtained indicating that the date of appearance of the larvae is correlated with the timing of the phytoplankton cycle.

Larval Abundance 1948 - 1968. For each centre the mean number of larvae per sample per year are shown in Fig. 3. To simplify the analysis of trends and events in different centres, 5-year means were calculated (Fig. 4); these show a partitioning into groups of centres with similar trends in fluctuations of abundance over the 21-year period. In centres I and (to a lesser extent) III, there was an increase over the years but numbers of larvae decreased in all other areas in the 1958-1962 period; subsequently centre VII showed a marked increase. The very low numbers of larvae in centre IV may be the result of inadequate sampling (see section on the effects of night and day sampling). This analysis appears to show that the centres are associated into 3 populations:

Fig. 3. The mean number of sandeel larvae per sample for the months January to June in six of the seven centres of population between 1948 and 1968. Centre IV has been omitted because of poor data available. The horizontal line through each curve shows the long term mean

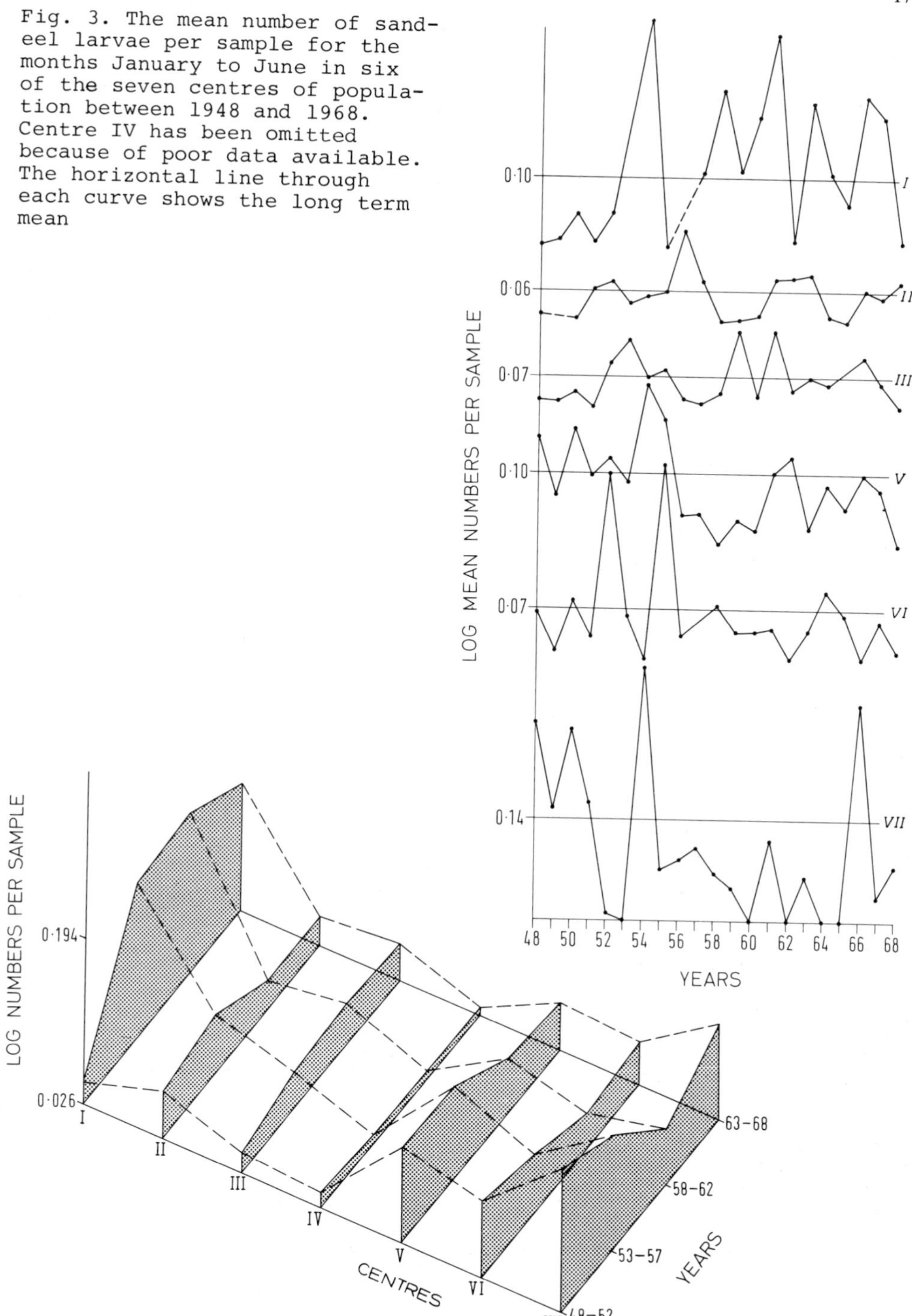

Fig. 4. Mean numbers of larvae per sample per population centre for 5-year periods between 1948 and 1968

1 in the northwest, 1 in the east, and 1 in the central and southern North Sea.

A more rigorous treatment of the data involved principal component analysis (Williamson, 1961 and 1963; Kendal and Stuart, 1966). This analyzed a seven-dimensional space (i.e. the 7 centres) and determined the major axes of variation through the 21 data points; the first two components accounted for 26.8% and 23.2% of the variance. A plot of vector 1 against vector 2, from the transition matrix (Fig. 5) lends support to the association of some centres, as suggested by the five year means. Centre I is again isolated but centre II and V, and VI, and VII are associated in pairs. The analysis also shows a possible association between centres III and IV but this seems doubtful considering the geographical positions of the centres relative to each other.

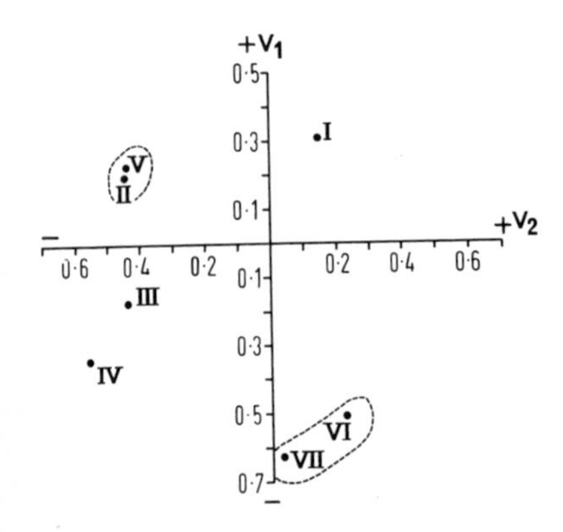

Fig. 5. A plot of the first and second vectors (v_1 against v_2) from the transition matrix showing a grouping of the 7 sand-eel population centres

Principal components were also calculated from the data set treated as a 21 dimensional space (i.e. the 21 years). The first component, which accounted for 43.7% of the variance, is shown in Fig. 6a. Fig. 6b shows a plot of vector 1 against vector 2; it suggests two major groupings of years.

Larvae and the Adult Population. Presumably the abundance of fish larvae is influenced by the adult population but the published data of *A. marinus* in the North Sea are poor. The fishery for sand eels in the North Sea started in 1953 (Macer, 1966; Macer and Burd, 1970; Reay, 1970). At first the fishery was local, but it soon spread to cover most of the southern North Sea (i.e., centres IV, V, VI and VII of this study).

Catch statistics have been very incomplete with little detail of stock abundance but Macer (1966) published year-class abundance for the years 1960 to 1961 from which values of the instantaneous mortality coefficient (z) were calculated. Using these data it has been possible to extrapolate back to calculate the relative abundance of the O-group fish in their year of birth for the period 1955-1960, assuming that the values of z as calculated over the period 1960 and 1961 held for each year class over their whole life. The indices of recruitment which resulted are shown in Fig. 7 which also shows the mean number of larvae per sample for centres IV, V, VI, and VII combined. There appears to be some correspondence between the trend in recruitment and the abundance of larvae, although it should be regarded with caution because it covers only 6 years.

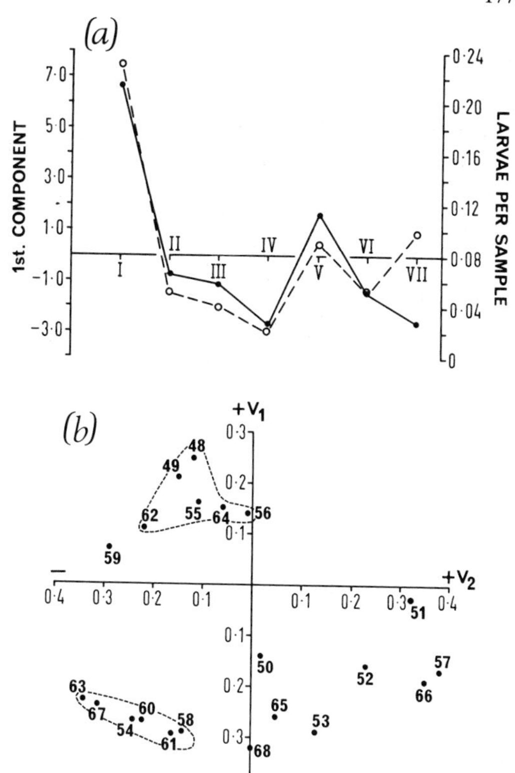

Fig. 6a and b. The first component (continuous line) for all the data combined for the areas I to VII, and the mean number of larvae per sample (dotted line) for the years with negative v_1 and v_2 as shown circled in Fig. 6b. (b) A plot of the first and second vectors (v_1 against v_2) from the transition matrix showing a grouping of some of the years

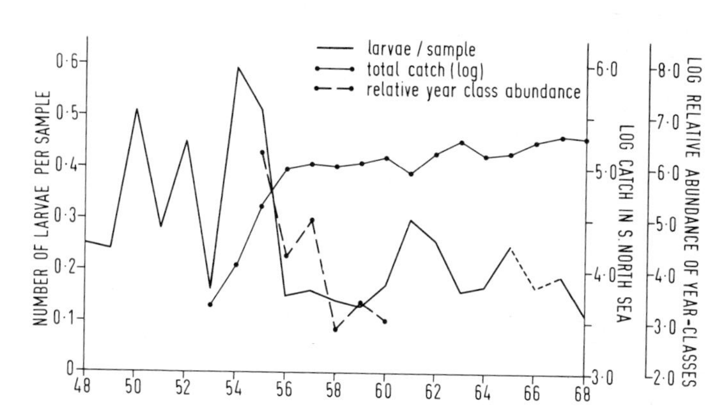

Fig. 7. The mean number of larvae per sample for centres IV, V, VI and VII combined, the log of the total catch (tons) of adult sand eels by all countries involved in the southern North Sea fishery and an index of recruitment, as calculated from the data of Macer (1966), between the years 1955 and 1960. One sample containing an unusually high number of larvae (19) removed

The log of the total catch from the southern North Sea for the years 1953 to 1968 is also shown in Fig. 7. This indicates that, when the fishery increased rapidly between 1953 and 1956, the numbers of larvae caught by the CPR decreased sharply and then stabilized at a relatively low level of abundance after the total catch had settled at a relatively high level. To analyze the situation more objectively the curve for each population centre (except IV) was examined for trend using the Spearman rank correlation coefficient (Kendal and Stuart, 1966) the results of which are shown in Table 2. This analysis confirms the decrease in abundance in the southern centres V, VI and VII (where fishing has been heavy) but, in the northern centres, I and II, where fishing has been light or non-existent, larval abundance has either increased or shown no trend. Centre III does not fit the pattern as no trend is indicated, yet the population in this region has been fished.

Table 2. Detection of trend in time-series for sand eel abundance. r_s = Spearman's rank correlation coefficient

CENTRE	r_s	Standard error of r_s	r_s different from zero?	Trend indicated
I	0.34	0.23	Just	Increase?
II	0.10	0.22	No	None
III	0.20	0.22	No	None
V	-0.45	0.22	Yes	Decrease
VI	-0.25	0.23	Just	Decrease?
VII	-0.53	0.22	Yes	Decrease

There is no relationship between principal components and total catch which is not unexpected considering that the components were calculated from the data for the whole North Sea whilst the fishery only affects the southern populations. In the northern areas, and in the southern area before the fishery started, one would expect natural environmental events to be the main influences on the populations.

Larvae and Water Movements. The population of *A. marinus* in centre I can be classified with the northern intermediate group of plankton as defined by Colebrook (1964) while the remaining centres are in the neritic group. The boundary between these groups fluctuates, partly in response to changes in the extent of influx of Atlantic water throug the Faroe-Shetland Channel and the English Channel. Fraser (1969) has traced the influx of Atlantic water to the east of Scotland by following indicator species whilst Dickson (1971) has studied the long term salinity anomalies in the North Sea between 1905 and 1968.

The principal component analysis on the 21 dimensional space (Fig. 6) showed that there were 2 main groupings of years during the period of study. The grouping of 6 years shown in the lower left hand quadrant of Fig. 6b are those (1954, 1958, 1960, 1961, 1963, 1967), when larval abundance was high in centre I. The mean abundance for each centre

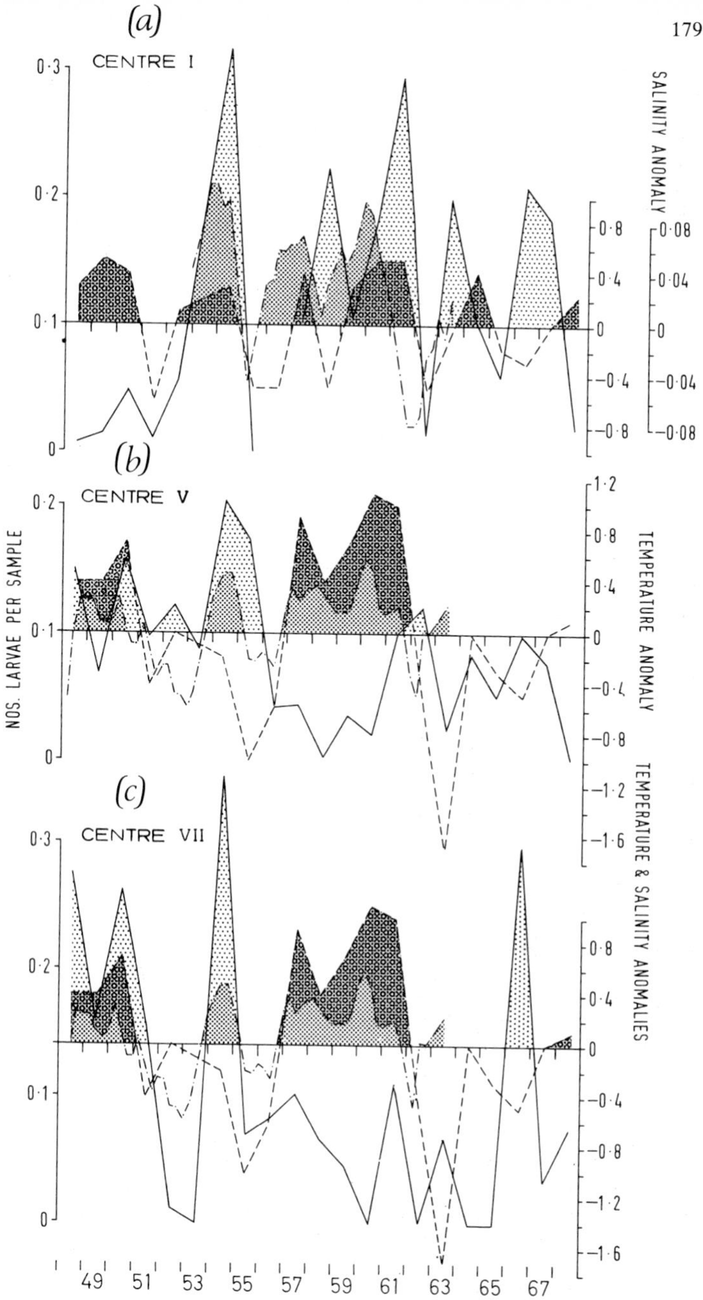

Fig. 8. The relationship between larval abundance and the extent of penetration of the North Sea by Atlantic water between the years 1948 and 1968 (salinity data and their interpretation from Dickson, 1971). Centre I. Light stipple/solid line, larval abundance. Medium stipple/ dot-dash line, salinity anomaly from Dickson's Central North Sea group. Dark stipple/dashed line, temperature anomaly for area B2 (see Fig. 2). Centre V. Larval abundance. Salinity anomaly for Dickson's Southern Bight group. Temperature anomaly for sub-area D2 (see Fig. 2). Centre VII. Larval abundance. Salinity anomaly for Dickson's Southern Bight group. Temperature anomaly for area D2 (see Fig. 2)

belonging to this set of years follows closely the shape of the first component in Fig. 6a which suggests that these years accounted for a major proportion of the variance. Dickson (1971) found a greater than average penetration of Atlantic water in these years. The data for larval abundance in centres I, V and VII are shown in Fig. 8 in conjunctio with the salinity anomalies from Dickson (loc. cit.); a high salinity anomaly indicates increased Atlantic inflow. Also shown are temperature anomalies derived from ICES temperature charts but averaged for the sub areas shown in Fig. 2; these show a correlation with the salinity anomalies. In centre I as shown in Fig. 8, high larval abundance tends to be correlated, in a general way, with the greater penetration of Atlantic water in the North Sea. The same relationship was found in centres V and VII but only in the years before the fishery started in 1956.

DISCUSSION

The main aims of this study were to discover the major patterns of variation in abundance and to account for them. As a generalization, herring populations in different areas around the British Isles produce their larvae in phase with the plankton production cycle (Cushing, 1967). Bainbridge and Cooper (in press) found a similar link in the case of *Micromesistius poutassou*. The correlation between the time of first occurrence of *A. marinus* larvae and the timing of the peak of phytoplankton abundance would seem to be of a similar nature but needs to be interpreted with caution. The estimate of timing used by Robinson (196 is the centre of gravity of the peak of phytoplankton abundance, not the time of the initial outbreak which presumably occurs about a month before. Despite this, some *Ammodytes* larvae occur about a month before phytoplankton begins to bloom; this may indicate that the correlation between the timing of larval appearance and timing of the phytoplankton bloom is not a direct one, but a link by some common factor, for examp temperature.

There is no correlation between the timing of the phytoplankton bloom and the abundance of larvae as might be expected if a link between the two contributed significantly to survival. First among the other facto that might influence variation on larval abundance is the fishery. Between 1952 and 1957 the total catch increased from 0 to about 107,000 tons and it appears that this sufficiently reduced stock size to decrease larval production. Subsequently between 1957 and 1968 the total catch increased much more slowly, by another 100,000 tons, yet larval abundance remained relatively stable. This anomaly could possibly be a result of the faster rate of change of the total catch in the earlier period, but is essentially unexplained by the present study. Where the populations were not exploited, the results suggest that variations in larval abundance were determined by environmental changes related to the extent of penetration of Atlantic water into the North Sea. It is possible that water quality might affect the fish population directly or that changes in the water mass were accompanied by changes in the planktonic community which affected the adults or the larvae (c.f. Russell, 1973). Such changes have been illustrated by Bainbridge and Forsyth (1972), who showed that increased Atlantic inflow in the north western North Sea was accompanied by increased representation of plankton belonging to the northern intermediate group of Colebrook (1964). As the main inflow of Atlantic water through the Faroe-Shetland Channe occurs during the winter half of the year and is probably at its heigh between November and March (Dickson, 1971; Tait, 1957), the period whe *A. marinus* larvae are just appearing, it is probable that the larval stage is the one most affected by the water mass change. Thus, "good" environmental conditions brought about by the presence of Atlantic water and its associated plankton community could produce better larval survival. This hypothesis assumes that the more neritic community, present when Atlantic inflow is slight, contains less food for *A. marinus* larvae or alternatively, more predators.

SUMMARY

1. The Continuous Plankton Recorder survey has provided an opportunity to examine the distribution and long term changes in abundance of *Ammodytes marinus* larvae in the North Sea between 1948 and 1968.

2. Particular attention has been paid to possible sources of error in the data, arising from the sampling method.

3. Initially the North Sea was divided into 7 regional concentrations of sand-eel larvae, but subsequent analysis showed that these 7 centres could be recombined into 3 groups. The time of first appearance at the beginning of the year, within each centre, was found to be correlated with the timing of the spring phytoplankton bloom. It is pointed out that this may not indicate a direct link between larvae and phytoplankton.

4. The hypothesis is proposed that high abundance of the larvae is directly related to greater penetration of Atlantic water into the North Sea, an influence strongly modified by the fishery for adults, which resulted in a decrease in numbers of larvae.

ACKNOWLEDGEMENTS

I should like to acknowledge the work done by Dr. G.T.D. Henderson and Mr. G.A. Cooper who identified, counted and measured the larvae used for this study. Acknowledgements are also due to Mrs. J. Johnson for assistance with data processing.

REFERENCES

Bainbridge, V. and Cooper, G.A. (in press). The distribution and abundance of the larvae of the blue whiting *Micromesistius poutassou* (Risso) in the northeast Atlantic, 1948-1970. Bull. Mar. Ecol.

Bainbridge, V. and Forsyth, D.C.T., 1972. An ecological survey of a Scottish herring fishery Part V: The plankton of the northwestern North Sea in relation to the physical environment and the distribution of herring. Bull. Mar. Ecol., 8, 21-52.

Böhnecke, G., 1922. Salzgehalt und Strömungen der Nordsee. Veröff. Inst. Meeresk. Univ. Berl., 10, 1-34.

Colebrook, J.M., 1960. Continuous Plankton Records: methods of analysis 1950-1959. Bull. Mar. Ecol., 5, 51-64.

Colebrook, J.M., 1964. Continuous Plankton Records: a principal component analysis of the geographical distribution of zooplankton. Bull. Mar. Ecol., 6, 78-100.

Cushing, D.H., 1967. The grouping of herring populations. J. mar. biol. Ass. U.K., 47, 193-208.

Dickson, R.R., 1971. A recurrent and persistent pressure-anomaly pattern as the principal cause of intermediate-scale hydrographic variation in the European shelf seas. Dtsch. hydrogr. Z., 24, 97-119.

Fraser, J.H., 1969. Variability in the oceanic content of plankton in the Scottish area. Progr. Oceanogr., 5, 149-159.

Glover, R.S., 1967. The continuous plankton recorder survey of the North Atlantic. Symp. Zool. Soc. Lond., 19, 189-210.

Henderson, G.T.D., 1961. Continuous plankton records: contributions towards a plankton atlas of the northeastern Atlantic and the North Sea. Part 5. Young Fish. Bull. Mar. Ecol., 5, 105-111.

Kendall, M.G. and Stuart, A., 1966. The advanced Theory of Statistics; Vol. 3 Design and analysis, and time series. London, Griffin.

Laevastu, T., 1963. Surface water types of the North Sea and their characteristics. Serial Atlas of the Marine Environment. Folio 4. Amer. Geogr. Soc.

Langham, N.P.E., 1971. The distribution and abundance of larval sand-eels (Ammodytidae) in Scottish Waters. J. mar. biol. Ass. U.K., 51, 697-707.

Lee, A., 1970. The currents and water masses of the North Sea. Oceanogr. Mar. Biol. Ann. Revs., 8, 33-71.

Macer, C.T., 1965. The distribution of larval sand eels (Ammodytidae) in the southern North Sea. J. mar. biol. Ass. U.K., 45, 187-207.

Macer, C.T., 1966. Sand eels (Ammodytidae) in the southwestern North Sea; their biology and fishery. Fish. Invest., Lond., Ser. II, 24 (6), 1-55.

Macer, C.T. and Burd, A.C., 1970. Fishing for sand eels. Lab. Leaflet (New Ser.), 21, 1-9.

Rae, K.M., 1952. Continuous Plankton Records: explanation and methods, 1946-1949. Hull Bull. Mar. Ecol., 3, 135-155.

Reay, P.J., 1970. Synopsis of biological data on North Atlantic sand eels of the genus Ammodytes: *A. tobianus*, *A. dubius*, *A. americanus* and *A. marinus*. FAO Fish. Synopsis No. 82.

Robinson, G.A., 1970. Continuous Plankton Records: variation in the seasonal cycle of phytoplankton in the North Atlantic. Bull. Mar. Ecol., 6, 333-345.

Russell, F.S., 1973. A summary of the observations on the occurrence of planktonic stages of fish off Plymouth (1924-1972). J. mar. biol. Ass. U.K., 53, 347-355.

Ryland, J.S., 1964. The feeding of plaice and sand eel larvae in the southern North Sea. J. mar. biol. Ass. U.K., 44, 343-365.

Tait, J.B., 1957. Hydrography of the Faroe-Shetland Channel, 1927-1952. Mar. Res., 2, 309 pp.

Williamson, M.H., 1961. An ecological survey of a Scottish herring fishery Part IV: changes in the plankton during the period 1946-1959. Appendix: A method for studying the relation of plankton variations to hydrography. Bull. Mar. Ecol., 5, 207-229.

Williamson, M.H., 1963. The relation of plankton to some parameters of the herring population of the northwestern North Sea. Rapp. P.-v. Réun. Cons. perm. int. Explor. Mer, 154, 179-185.

P.J.B. Hart
Institute for Marine Environmental Research
Oceanographic Laboratory
68 Craighall Road
Edinburgh, EH 6 4 RQ / GREAT BRITAIN

Present address:

NORDRECO AB
26700 Bjuv / SWEDEN

Efficiency Test on Four High-Speed Plankton Samplers

H. Bjørke, O. Dragesund, and Ø. Ulltang

INTRODUCTION

The Gulf III high-speed plankton sampler (Gehringer, 1952, 1961) has been widely used for sampling of fish larvae (Bridger, 1958; Saville, 1966; Zijlstra, 1970). The Bongo sampler (McGowan and Brown, 1966) has more recently been applied (Anon, 1972). In Norwegian coastal waters extensive larval surveys have been carried out in which fish larvae have been collected with Clarke-Bumpus plankton samplers (e.g. Wiborg, 1954, 1960; Dragesund, 1970; Dragesund, Gjøsaeter and Monstad, 1972).

Testing different gears against each other is a difficult task. This is indicated by the rather conflicting results obtained in experiments already carried out. A comparison of the fishing efficiency of a modified Gulf III sampler described as the "Hai" (Hempel, 1960), and a modified Gulf V plankton sampler "Nackthai", i.e. a Gulf III with a longer mouth cone without encasement, was made by Nellen and Hempel (1969). For routine work in sampling fish larvae, the "Nackthai" rather than the "Hai" was recommended. Later, Saville and McKay (1969, 1970) suggested that there was no significant difference in the catching capacities of an encased sampler, a modified Gulf III, as described by Bridger (1958), and the German "Nackthai". A modification of the Bongo sampler (Posgay, Marak and Hennemuth, 1968) has been tested against other high-speed plankton samplers. Comparisons made between catching efficiencies of the Gulf III and BCF Bongo .03 sampler by Sherman and Honey (1968) indicated that the latter provided more accurate estimates.

The objective of the present work was to compare the catching efficiency of four high-speed plankton samplers in order to recommend the most efficient gear in connection with extensive larval investigations recently initiated on capelin in North Norway (Gjøsaeter and Saetre, 1973).

MATERIAL AND METHODS

The material was collected during larval surveys on capelin off the Finnmark coast in 1971 and 1972. Comparisons were made between the following high-speed samplers:

1. BCF Bongo 0.03 (B 20)(see Posgay et al., 1968).
2. Modified Bongo 60 (B 60), Hydrobios Bongo, developed in Kiel, Germany.
3. Clarke-Bumpus plankton sampler (CB)(see Clarke and Bumpus, 1950).
4. Modified version of Gulf III (DG III)(see Zijlstra, 1970).

For the B 20 a TSK flowmeter was placed halfway between the centre and the rim of the sampler's mouth, while for the B 60 a Rigosha flowmeter was placed in the centre. In the CB and DG III samplers the propeller of the flowmeters covered the whole mouth area. Before and after sampling the flowmeters were calibrated, and the results were later confirmed in tank tests.

Some technical data for the nets used are given in Table 1. The depressor used was a Scripps cable depressor (Isaacs, 1953). The hauls were made in a larval concentration marked by a subsurface drogue, released at 25 m depth.

Table 1. Technical data for the nets used

Sampler	Mouth opening of the net in m^2	Porosity[1]	Mesh aperture in μ	Open area ratio[2]
B 60	0.2827	0.51	500	4.92
B 20	0.0323	0.50	500	14.02
CB	0.0125	0.49	500	7.47
DG III	0.0314	0.50	500	20.07

[1]Open area fraction of gauze.

[2]Ratio between open area of gauze and mouth opening.

In 1972 two types of oblique hauls were made alternating with each other 6 times:

1. The B 20, the B 60 and CB were attached in succession to the same wire at intervals of 3 m. Unfortunately this arrangement could not be varied because the CB had to be fastened to a thinner wire (4 mm diameter) attached to the end of the ordinary 6 mm diameter wire. The towing speed of the nets was 3 kn.

2. The DG III sampler was towed separately at 5 kn.

In both types of hauls the samplers were lowered to 30 m and towed obliquely approximately 1200 m along the same tow path through the larval patch. The wire length of the first type of haul was 100 m and pay out and recovery speed 2.5 m/10 sec., while that of the DG III haul was 80 m and pay out and recovery speed 3.3 m/10 sec. A Benthos time-depth recorder was used on every haul. The test was done 16 May 1972 from 11 to 17h GMT.

Some preliminary comparisons of the same nets were made in 1971 using another procedure. This sampling took place during 24h from 9 to 10 May. Oblique hauls were taken near a subsurface drogue released at 25 m depth in a larval concentration of capelin. A total of 36 hauls were made during 3 periods. In each period 12 samples were taken. The B 20, B 60, CB, and DG III were towed separately in random succession. The B 20, B 60, and CB were towed at 3 kn, whereas the towing speed of the DG III was 5 kn. The gears were towed approximately along the same path. The maximum depth of the hauls being about 30 m.

When counting the number of larvae the samples were divided into aliquots ranging from the whole to a hundreth of the sample. At least

2 of the aliquots were counted. When collected with the DG III only half of the sample was counted. When dividing the samples a whirling vessel was used (Wiborg, 1951) or a splitting cylinder (Motoda, 1959). The plankton volume was measured as described by Robertson (1970). For analysis of the data obtained the Wilcoxon matched pairs signed-ranks test (Siegel, 1956) was used for the hauls where 3 nets were attached to the same wire. In order to compare one net with another towed in two separate hauls the Mann-Whitney U test was used (Siegel, 1956). The former tests the null hypothesis that all of the individual samples in two groups of related samples are drawn from the same population. The latter tests the null hypothesis that all of the individual samples in two groups of independently collected samples were drawn from the same population. For all tests a significance level of 0.05 was chosen.

In order to evaluate net avoidance, comparisons were made of the length-frequency distributions of the larvae caught with the 4 nets. The statistical method applied was the Kolmogorov-Smirnov two-sample test (Siegel, 1956). In the 1972 material the length of 50 larvae (when available) from each samples was measured to the nearest 0.5 mm. Usually the length of 100 larvae from each of the 1971 samples was measured to the nearest 0.1 mm. When using the Kolmogorov-Smirnov test the larvae in the 0.1 mm groups were pooled into 0.5 mm groups. Only the larvae from alternate hauls were measured.

RESULTS

In Tables 2 and 3, the catches taken with the different gears are given, and in Table 4 the ratio between larvae and plankton volume is shown. Table 5 shows the result of the tests when comparing the number of capelin larvae per 100 m^3 of filtered water caught with the different gears. It can be seen that the B 20 without a flowmeter catches more larvae than with a flowmeter, and the B 60 with flowmeter more than the B 60 without. On average the B 60 catches more larvae than the B 20. CB catches less larvae than B 60 and DG III, and there is no significant difference between CB and B 20. DG III catches more larvae than any of the other nets, except B 60 with flowmeter, but the difference is not significant.

The logarithmic coefficient of variation was calculated for the 1972 material from the formula:

$$V = 10^s - 1$$

Where s is the standard deviation calculated from the logarithms of the raw data (Cassie, 1968). The following results were obtained:

B 20: V = 160% B 60 : V = 37%

CB : V = 34% DG III: V = 24%

Table 6 shows the result of the tests on the plankton volume per 1000 m^3 of filtered water caught with the different gears. It can be seen that B 60 with flowmeter catches more plankton than B 60 without. DG III catches more than CB and B 20 with flowmeter. CB catches less plankton than any of the other nets.

On the assumption that fish larvae avoided the nets more than plankton organisms, the ratio between number of larvae and plankton volume

Table 2. Catches of larvae taken by the different nets and volume filtered, 1972

Gear	B 20				B 60				CB		DG III	
Station no.	Volume filtered in m^3	Number of larvae per 100 m^3			Volume filtered in m^3	Number of larvae per 100 m^3			Volume filtered in m^3	Number of larvae per 100 m^3	Volume filtered in m^3	Number of larvae per 100 m^3
		With flow-meter	Without flow-meter	Mean		With flow-meter	Without flow-meter	Mean				
203											24.3	2062
204	34.5	841	943	892	246.9	2086	1290	1688	13.2	613		
205											36.7	1920
206	41.3	479	1755	1117	300.6	1622	1114	1368	15.8	329		
207											25.9	2163
208	32.8	159	268	214	242.0	1157	1002	1080	12.8	436		
209											29.8	1341
210	40.7	138	189	163	295.9	870	718	294	16.1	299		
211											35.1	1368
212	39.8	561	980	771	290.1	1000	834	917	15.3	438		
213											34.3	1497
214	39.9	123	132	128	294.1	911	609	766	15.7	592		

Table 3. Plankton volumes ($cm^3/1000\ m^3$) taken by different nets, 1972

Gear	B 20			B 60			CB	DG III
Station no.	Plankton volume			Plankton volume			Plankton volume	Plankton volume
	With flowmeter	Without flowmeter	Mean	With flowmeter	Without flowmeter	Mean		
203								164
204	87	111	99	194	176	185	37	
205								142
206	170	213	192	191	170	181	25	
207								192
208	139	93	116	155	113	134	31	
209								144
210	128	110	119	158	151	155	31	
211								192
212	83	122	103	162	154	158	52	
213								138
214	135	170	153	183	163	173	108	

Table 4. The ratio between number of larvae per 100 m^3 and plankton volume (cm^3/1000 m^3) of the different gears, 1972

Gear	B 20			B 60			CB	DG III
Station no.	larvae/plankton volume			larvae/plankton volume			larvae/ plankton volume	larvae/ plankton volume
	With flow- meter	Without flowmeter	Mean	With flow- meter	Without flowmeter	Mean		
203								126
204	97	85	91	108	73	91	37	
205								135
206	28	83	56	85	66	76	130	
207								113
208	11	29	20	75	89	82	140	
209								93
210	11	17	14	55	48	52	96	
211								71
212	67	80	74	62	54	55	84	
213								109
214	9	8	9	50	37	44	49	

Table 5. The results of the Wilcoxon matched-pairs signed-ranks and the Mann-Whitney U tests, comparing the number of larvae caught per m^3 of filtered water with the different gears in 1972 (two-sided tests). Arrows indicate the gear with highest catch. *: $p < .05$, **: $p < .025$, ***: $p < .01$, n.s.: $p > .05$. w = with flowmeter, w.o. = without flowmeter, M = mean value

Gear		B 20			B 60			CB	DG III
		w.o.	w.	M.	w.o.	w.	M.		
B 20	w.o.								
	w.	↑*							
	M.	↑*	←*						
B 60	w.o.	n.s.	←*	n.s.					
	w.	n.s.	←*	←*	←*				
	M.	n.s.	←*	←*	←*	↑*			
CB		n.s.	n.s.	n.s.	↑*	↑*	↑*		
DG III		←**	←**	←**	←**	n.s.	←*	←**	

Table 6. The results of the Wilcoxon matched-pairs signed-ranks and the Mann-Whitney U tests, comparing the plankton volume per 1000 m^3 of filtered water caught with the different gears in 1972 (two-sided tests). Legend as in Table 5

Gear		B 20			B 60			CB	DG III
		w.o.	w.	M.	w.o.	w.	M.		
B 20	w.o.								
	w.	n.s.							
	M.	n.s.	n.s.						
B 60	w.o.	n.s.	n.s.	n.s.					
	w.	n.s.	←*	n.s.	←*				
	M.	n.s.	n.s.	n.s.	←*	↑*			
CB		↑*	↑*	↑*	↑*	↑*	↑*		
DG III		n.s.	←*	n.s.	n.s.	n.s.	n.s.	←***	

caught was calculated for each net and haul. The tests show that DG III catches more larvae per volume of plankton than the other gears except for the CB, but the difference is not significant (Table 7). The CB showed a higher ratio than the other nets except for B 60 with flowmeter and DG III where the difference was not significant. No significant difference was found between the Bongo nets. When studying Tables 5, 6, and 7 it should be remembered that the number of pairs to compare in the Wilcoxon test were too few to show if the difference between the gears were significant at the 0.025 and 0.01 levels. The ratio between the mouth area of CB, B 20 and B 60 was respectively 1:2.58:22.61, while the ratio between the mean volume filtered was 1:2.58:18.78. This indicates that the waterflow through the B 60 has been reduced.

Table 7. The results of the Wilcoxon matched-pairs signed-ranks and Mann-Whitney U tests, comparing the ratio between the number of larvae and plankton volume, taken per unit volume filtered with the different gears (two-sided tests). Legend as in Table 5

Gear		B 20			B 60			CB	DG III
		w.o.	w.	M.	w.o.	w.	M.		
B 20	w.o.								
	w.	n.s.							
	M.	n.s.	n.s.						
B 60	w.o.	n.s.	n.s.	n.s.					
	w.	n.s.	n.s.	n.s.	n.s.				
	M.	n.s.	n.s.	n.s.	n.s.	n.s.			
CB		←*	←*	←*	←*	n.s.	←*		
DG III		←**	←***	←***	←***	←*	←**	n.s.	

The results of the 1971 experiments are given in Tables 8, 9 and 10. Due to difficulties with the flowmeter of the B 60 sampler at stations 203 and 217, these hauls were omitted when comparing the catching efficiency. A Mann-Whitney U test did not show any significant difference in the catching efficiency between the gears (Table 11). For the Bongo samplers the mean catch of the 2 nets was used. Applying the Wilcoxon test on the catches of the 2 Bongo nets of the same sampler, no significant differences were found.

Table 8. Catches of larvae taken by different gears, first period, 1971

Gear		B 20		B 60		CB		DG III	
Haul	Station no.	Volume filtered in m^3	Number of larvae per 100 m^3	Volume filtered in m^3	Number of larvae per 100 m^3	Volume filtered in m^3	Number of larvae per 100 m^3	Volume filtered in m^3	Number of larvae per 100 m^3
1	203								
	204					16.7	473		
	205							41.5	920
	206	62.4	687						
2	208							47.5	1410
	210	57.2	2874						
	211			405.6	3509				
	212					13.0	792		
3	214					19.2	1249		
	215							46.7	4019
	216	57.5	3212						
	217								

Table 9. Catches of larvae taken by different gears, second period, 1971

Gear		B 20		B 60		CB		DG III	
Haul	Station no.	Volume filtered in m^3	Number of larvae per 100 m^3	Volume filtered in m^3	Number of larvae per 100 m^3	Volume filtered in m^3	Number of larvae per 100 m^3	Volume filtered in m^3	Number of larvae per 100 m^3
1	302	57.3	1616						
	303			439.2	1457				
	305					19.9	1106		
	307							40.4	3760
2	308			421.7	2078				
	309	81.6	1625						
	310							51.1	2983
	311					17.2	1008		
3	313					11.7	283		
	314							52.4	1005
	315	94.2	99						
	316			419.2	80				

Table 10. Catches of larvae taken by different gears, third period, 1971

Gear		B 20		B 60		CB		DG III	
Haul	Station no.	Volume filtered in m^3	Number of larvae per 100 m^3	Volume filtered in m^3	Number of larvae per 100 m^3	Volume filtered in m^3	Number of larvae per 100 m^3	Volume filtered in m^3	Number of larvae per 100 m^3
	401							57.8	364
1	402	92.3	152						
	403			589.7	26				
	404					23.4	542		
	406	88.7	47						
2	408			434.3	1845				
	409					21.2	582		
	410							47.0	715
	412					24.5	465		
3	413	69.8	624						
	414			370.2	201				
	415							51.9	813

Table 11. Results of the Mann-Whitney U test, comparing the number of larvae caught with the different gears, 1971

Gears compared	U-value	Critical values of U for a two-tailed test at = .05	Significance
DG III/B 60	20	12	n.s.
DG III/B 20	27	17	n.s.
DG III/CB	21	17	n.s.
B 60/B 20	30	12	n.s.
B 60/CB	31	12	n.s.
B 20/CB	35	17	n.s.

In order to investigate whether or not the conclusions were the same if using parametric statistical methods, the number of larvae per 100 m^3 for the 1971 material were transformed to log (x + 1). A two-way analysis of variance with gear and period as factors, and hauls as replicates, did not show any significant differences in catches between the gears. Comparing the average catches in the last two periods of the different gears the following ranking was found:

1. DG III (1607 larvae/100 m^3)
2. B 60 (948 larvae/100 m^3)
3. B 20 (694 larvae/100 m^3)
4. CB (664 larvae/100 m^3)

By using the Kolmogorov-Smirnov test to study the length frequency of larvae caught with the different gears, the following ranking of the relative abundance of the largest length groups was obtained for the 1972 material:

1. CB, 2. DG III, 3. B 20, and 4. B 60 (Table 12). All the differences were significant at the 5% level. Also when applying a t-test significant differences in mean lengths between gears were found.

The Kolmogorov-Smirnov test showed the following ranking for the 1971 material:

1. DG III, 2. CB, 3. B 20, and 4. B 60 (Table 13). The differences between DG III and CB were not significant at the 5% level.

Table 12. Length frequency distribution in per cent of larvae caught with different gears, 1972

Gear	B 20			B 60			DG III	CB
Length in mm.	With flow-meter	Without flowmeter	Mean	With flow-meter	Without flowmeter	Mean		
6.0		3.15	1.93	1.13	.76	.95	1.33	1.12
6.5	5.05	17.27	12.51	18.39	19.21	18.76	13.00	5.97
7.0	25.24	28.38	27.05	35.93	40.85	38.25	31.00	26.49
7.5	20.91	17.27	18.58	16.69	14.68	15.70	16.17	19.40
8.0	23.80	23.12	23.28	18.95	17.23	18.10	21.00	23.51
8.5	15.87	6.16	9.84	4.24	3.18	3.72	9.00	14.55
9.0	4.57	2.85	3.50	2.26	1.36	1.83	4.67	5.22
9.5	1.20	.30	.64	.14	.30	.23	.83	.75
10.0	.96	.15	.46	.28	.15	.23	-	.75
10.5	.24	.15	.18	-	-	-	.17	-
11.0	.48	.30	.37	.85	.61	.73	.83	.37
11.5	.24	.30	.28	.14	.45	.29	.50	-
12.0	-	.30	.18	.28	.45	.37	.50	.37
12.5	.24	.15	.18	-	.15	.07	-	-
13.0	.24	-	.09	-	.15	.07	.33	.75
13.5	.24	.15	.18	.14	-	.07	-	-
14.0	.72	-	.28	.42	.15	.29	.17	.75
14.5	-	-	-	-	.15	.07	.33	-
15.0	-	-	-	.14	-	.07	-	-
15.5	-	-	-	-	.15	.07	.17	-
Number of larvae measured	416	666	1082	707	661	1368	600	268
Mean length in mm.	7.838	7.425	7.584	7.395	7.349	7.373	7.619	7.785

Table 13. Length frequency distribution in per cent of larvae caught with different gears, 1971

Gear	B 20			B 60			DG III	CB
Length in mm.	With flow-meter	Without flowmeter	Mean	With flow-meter	Without flowmeter	Mean		
4.5	1.16	-	.56	6.33	-	3.20	-	-
5.0	12.91	-	6.31	20.96	-	10.58	-	-
5.5	20.81	-	10.17	22.27	1.34	11.91	-	-
6.0	20.04	3.13	11.39	19.00	2.67	10.92	.40	.21
6.5	20.23	11.97	16.01	13.97	11.58	12.79	6.21	6.53
7.0	14.84	16.94	15.91	9.61	16.04	12.79	16.23	14.74
7.5	5.78	15.84	10.92	4.15	17.15	10.58	16.83	15.58
8.0	2.31	12.15	7.34	2.62	16.70	9.59	11.42	16.42
8.5	.96	12.71	6.96	.22	13.36	6.73	12.42	14.53
9.0	.77	13.08	7.06	.66	10.24	5.40	17.03	13.05
9.5	.19	7.00	3.67	.22	4.01	2.09	11.22	11.58
10.0	-	4.05	2.07	-	4.45	2.21	4.01	4.21
10.5	-	2.21	1.13	-	2.23	1.16	2.20	1.68
11.0	-	.74	.38	-	-	-	1.80	1.05
11.5	-	.18	.09	-	.22	.11	-	.42
12.0	-	-	-	-	-	-	.20	-
Number of larvae measured	519	543	1062	458	449	907	499	475
Mean length in mm.	6.186	7.973	7.100	5.897	7.375	6.877	8.251	8.222

DISCUSSION

Both the 1972 and 1971 tests have their weak points. Statistically, it is preferable that the gears sample in the same population to permit differences between the gears to be more easily confirmed. The drawback in the 1972 test is mainly of a technical character. The depressor used seemed to be too small when the B 20, the B 60, and CB were attached to the same wire, and the wire angle from the surface was only 18° during the haul. When towing, the gears lay almost in a row with a vertical distance of about 1 m between each gear, and theoretically the gears could affect each other's catching efficiency. B 20, the uppermost gear, might have been affected by the wire. In 1971 the gears were hauled separately in random succession during a 24 h period. However, the drawback in this experiment is that possible changes in larval densities from haul to haul make it necessary to take numerous hauls in order to find significant differences.

In the 1972 material the following ranking between gears was found using the number of larvae per unit filtered water:

1. DG III (1725 larvae/100 m^3)
2. B 60 Mean (1018 larvae/100 m^3)
3. B 20 Mean (548 larvae/100 m^3)
4. CB (451 larvae/100 m^3)

The difference between B 20 Mean and CB was not significant at the 5% level.

In the 1971 material the same ranking was found, but the differences were not significant perhaps because the material was too sparse. If the differences in catching efficiency could be explained primarily by net avoidance, it is expected that the ratio between number of larvae and plankton would be highest with the most efficient gears. The results given in Table 7 indicated that the ratio was highest for the DG III and CB, but the difference between these two gears was not significant at the 5% level.

An examination of the amount of plankton per unit water filtered revealed that CB caught less plankton than the other gears. There was no significant difference between the DG III and the mean catch of each of the Bongo samplers. Since DG III caught more larvae per unit of plankton than the other gears, and the differences in plankton volume per unit water filtered between DG III, B 20, and B 60 Mean did not show any significant difference, it might be concluded that the larvae seemed to avoid B 20 and B 60 more than DG III. Both the larvae and plankton organisms seemed to avoid CB.

Another way to evaluate net avoidance is to study the length frequency distributions of the larvae caught with the different gears. It is assumed that the largest larvae avoid the samplers more easily, and the least efficient gear catches the shortest larvae. Both in 1972 and 1971 the least efficient gear caught the largest larvae. It was impossible to find any reasonable explanation for this.

The test on the 1972 material showed differences in catching efficiency between the 2 nets of the Bongo samplers. The B 20 without flowmeter caught more larvae than B 20 with flowmeter. For the 1971 material no differences in the number of larvae per unit of water filtered was found between the 2 nets. It is difficult to find a reasonable explanation for the differences found. On the one hand, the smaller catch of the B 20 with flowmeter could be explained by regarding the flowmeter as an obstacle in the front of the net, thus making it easier for the larvae to avoid it. On the other hand, this difference was not observed in the 1971 material, although the same statistical method was applied for both materials. The question also arises as to why the B 60 net with flowmeter caught more larvae than the other in the 1972 test. If the number of larvae in the uppermost layers was high, the proximity to the propeller wash might have affected the catch of the 2 nets of the Bongo samplers. However, it is not possible to confirm this with the existing material.

The logarithmic coefficient of variation was by far the highest with the B 20 sampler. An explanation for this could be clogging during the hauls. The high open area ratio and no indication of reduced flow through B 20 compared with the other gears, however, makes this explanation unlikely.

SUMMARY

1. During larval surveys on capelin in 1971 and 1972, comparisons were made of the catching efficiencies of 4 high-speed plankton samplers, viz.:

a) BCF Bongo .03 (Ø 20 cm, B 20),

b) Hydrobios Bongo (Ø 60 cm B 60),

c) Clarke-Bumpus plankton sampler (Ø 12.5 CB), and

d) Dutch version of the Gulf III sampler (Ø 20 cm DG III).

2. In 1972 two types of hauls were made alternating with each other 6 times:

a) The B 20, B 60, and CB were attached to the same wire 3 m apart and towed at 3 kn.

b) The DG III sampler was towed separately at 5 kn.

In 1971 each of the gears were towed separately 9 times in random succession. Both in 1971 and 1972 the tows were double oblique hauls with a maximum depth of 30 m.

3. The test on the 1972 material showed that the number of larvae caught with DG III was significantly larger than those caught with the other gears. The catches of the B 20 and CB were the smallest. In the 1971 material the number of hauls was too few to show significant differences between the gears. However, the results from this test support the results obtained in 1972.

4. When comparing the number of larvae and the volume of plankton caught with the different gears in 1972, it is concluded that the larvae seemed to avoid the B 20 and B 60 more than the DG III. Both larvae and plankton seemed to avoid the CB sampler. Examination of the length frequency distribution of larvae could not explain this avoidance.

REFERENCES

Anon, 1972. Working group on joint survey of larval herring in the Georges Bank - Gulf of Maine areas. ICNAF, Res. Doc., 1972/123.
Bridger, J.P., 1958. On efficiency tests made with a modified Gulf III high-speed tow net. J. Cons. perm. int. Explor. Mer, 23, 357-365.
Cassie, R.M., 1968. Sample design. In: D.J. Tranter and J.H. Fraser (eds): Zooplankton sampling, 105-121. Unesco, Paris.
Clarke, G.L. and Bumpus, D.F., 1950. The plankton sampler - an instrument for quantitative plankton investigations. Spec. Publs Amer. Soc. Limnol. Oceanogr., 5, 1-8.
Dragesund, O., 1970. Factors influencing year-class strength of Norwegian spring spawning herring (*Clupea harengus* Linné). FiskDir. Skr. Ser. HavUnders., 15, 381-450.
Dragesund, O., Gjøsaeter, J. and Monstad, T., 1973. Estimates of stock size and reproduction of the Barents Sea Capelin in 1970-1972. FiskDir. Skr. Ser. HavUnders., 16 (in press).
Gehringer, J.W., 1952. An all-metal plankton sampler (Model Gulf III). Spec. scient. Rep. U.S. Fish Wildl. Serv. Fish., 1952 (88), 7-12.
Gehringer, J.W., 1961. The Gulf III and other modern high-speed plankton samplers. Rapp. P.-v. Réun. Cons. perm. int. Explor. Mer, 153, 19-22.
Gjøsaeter, J. and Saetre, R., 1973. The use of data on eggs and larvae for estimating spawning stock of fish populations with demersal eggs. (This volume).
Hempel, G., 1960. Untersuchungen über die Verbreitung der Heringslarven im Englischen Kanal und der südlichen Nordsee im Januar 1959. Helgoländer wiss. Meeresunters., 7, 72-79.
Isaacs, J.D., 1953. Underwater kite-type cable depressor. U.S. Patent Office Design 168, 999, 3 pp.
McGowan, J.A. and Brown, D.M., 1966. A new opening-closing paired zooplankton net. Univ. Calif. Scripps Inst. Oceanogr. (Ref. 66-23).
Motoda, S., 1959. Devices of simple plankton apparatus. Mem. Fac. Fish. Hokkaido Univ., 7 (1,2), 73-94.
Nellen, W. von and Hempel, G., 1969. Versuche zur Fängigkeit des "Hai" und des modifizierten Gulf-V-Plankton-Samplers "Nackthai". Ber. dtsch. wiss. Komm. Meeresforsch., 20, 141-154.
Posgay, J.A., Marak, R.R. and Hennemuth, R.C., 1968. Development and tests of new zooplankton samplers. ICNAF, Res. Doc., 1968/85.
Robertson, A.A., 1970. An improved apparatus for determining plankton volume. Fish. Bull. Fish Wildl. Serv. U.S., 6, 23-26.
Saville, A., 1966. The distribution and abundance of herring larvae in the northern North Sea, changes in recent years. Rapp. P.-v. Réun. Cons. perm. int. Explor. Mer, 160, 87-93.
Saville, A. and McKay, D.W., 1969. Tests with an encased and an unencased Gulf III sampler. ICES, Doc. CM 1969/H, 27.
Saville, A. and McKay, D.W., 1970. Tests of the efficiency of various high speed samplers for catching herring larvae. ICES, Doc. CM 1970/H, 7.
Sherman, K. and Honey, K.A., 1968. Observations on the catching efficiencies of two zooplankton samplers. ICNAF, Redbook 1968, Part III, 75-80.
Siegel, S., 1956. Non-parametric Statistics for the Behaviour Sciences. New York, Toronto, London: McGraw-Hill, 312 pp.
Wiborg, K.F., 1951. The Whirling Vessel. An apparatus for the fractioning of plankton samplers. FiskDir. Skr. Ser. HavUnders., 9 (13), 1-16.
Wiborg, K.F., 1954. Forekomst av fiskeegg og -yngel i nordnorske farvann våren 1952 og 53. Foreløpig beretning III. Fiskets Gang, 40, 5-9.

Wiborg, K.F., 1960. Forekomst av egg og yngel av fisk i vest- og nord-norske kyst- og bankfarvann våren 1959. Fiskets Gang, 46, 522-528.

Zijlstra, J.J., 1970. Herring larvae in the central North Sea. Ber. dtsch. wiss. Komm. Meeresforsch., 21, 92-115.

H. Bjørke
Institute of Marine Research
5011 Bergen / NORWAY

O. Dragesund
Institute of Marine Research
5011 Bergen / NORWAY

Ø. Ulltang
Institute of Marine Research
5011 Bergen / NORWAY

On the Reliability of Methods for Quantitative Surveys of Fish Larvae

D. Schnack

INTRODUCTION

Many different devices have been developed to increase the efficiency of plankton sampling. High speed samplers have been widely adopted to catch active organisms like fish larvae; such samplers reduce avoidance and act as efficient filters. However, the problem of extrusion becomes particularly important in high speed samplers. The filtration efficiency of the present models is still under discussion but from the UNESCO (1968) review it appears that there are two conflicting requirements for the design of high speed samplers. A relatively small mouth aperture and a large filtering area is needed to obtain a high filtration efficiency depending on a low filtration pressure and low clogging effect. On the other hand, the mouth area should be large enough to prevent net avoidance and to obtain large samples.

This contribution will be restricted to Bongo and Gulf III nets which are those principally used in current surveys.

BONGO NET

The Bongo gear consists of two nets of circular aperture lashed together. The nets are unencased without a mouth-reducing cone and hence a relatively large aperture. They are towed together horizontally or obliquely at speeds up to 6 kn. Models of different size have been built with or without a filtering or non-filtering cylindrical collar ahead of the conical net section (Fig. 1A-C). Initial filtration efficiency of this gear should, according to Tranter and Smith (UNESCO, 1968), closely approach F = 1, at least if fitted with cylinder-cone nets.

The prototype of the Bongo sampler, each net with a 70 cm aperture and a non-filtering collar and fitted with an opening and closing device, was described and tested by McGowan and Brown (1966). The authors found that samples from the left and right nets did not differ significantly. Samplers of different size based on this prototype were subsequently built by the U.S. Bureau of Commercial Fisheries (BCF). All these BCF Bongo models except the smallest, used in one test series, were fitted with cylinder-cone nets. Posgay and Marak (1967) stress the importance of this construction in view of field tests and tank observations made by Smith, Counts and Clutter (1968) which showed the cylinder-cone nets to be partially self-cleaning and the cylindrical section acting as a reserve filtering area. In several tests neither size of mouth area (0.008 to 0.39 m^2) nor towing distance (740 to 5450 m) affected the number of fish larvae and other organisms caught per volume of water (Posgay and Marak, 1967; Posgay et al., 1968a,b). The influence of towing speed was tested, for instance, by Smith (pers. comm.) who found avoidance of larval anchovy at low speeds (1.5 kn) and considerable extrusion of the very young anchovy at 4.5 kn.

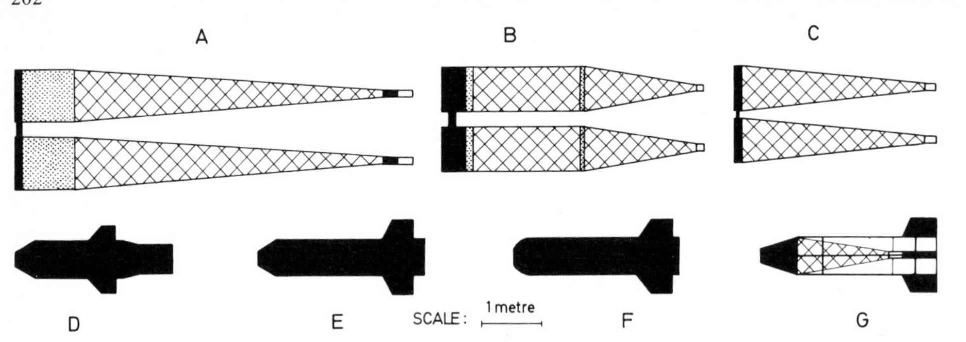

Fig. 1. Schematic drawings of some models of Bongo and Gulf III (view from above). (A) SIO Bongo (McGowan and Brown, 1966): non-filtering collar. (B) BCF Bongo: filtering cylindrical net section. (C) Hydrobios Bongo: no cylindrical net section. (D) Modified Gulf III (Bridger, 1957): aperture 0.2 m diameter, restricted tail unit. (E) Dutch Gulf III and Hai: no restricted tail piece. (F) Hai with hemispherical nose piece. (G) Nackthai: net unencased

Bongo Compared to Other Nets. McGowan and Brown concluded that the Bongo compared favourably with the CalCOFI plankton net while Smith (pers. comm.) found, in extensive tests, no significant difference in the sampling efficiency between the two gears for invertebrate plankton. Comparative tests between different models of the Bongo and other plankton nets carried out in the ICNAF area have been summarized by Posgay (1969). The WP-3 net of the ICES/SCOR/UNESCO Working Group (UNESCO, 1968) was, for example, less efficient in catching fish larvae than the Bongo at a towing speed of 3 kn. Simultaneous tows of Gulf III (Bridger, 1957) and a small BCF Bongo (0.2 m mouth diameter), carried out at 6 kn (Sherman and Honey, 1969) did not indicate a difference in the sampling efficiency for zooplankton organisms larger than 0.4 mm median width. Thus filtration efficiency in the devices seems to be about the same. However, smaller organisms, less than 0.38 mm median width, were caught in greater proportion by the Bongo net. This is explained by extrusion due to a higher filtration pressure in the Gulf III.

Size Selection by Bongo Nets. In order to reduce the effort of sorting fish larvae from plankton samples in survey programmes, it may be desirable to use the largest possible mesh size to retain fish eggs and larvae, thus diminishing the proportion of smaller organisms in the samples. The mesh size will depend on the species sampled and their size spectrum as well as on filtration pressure which varies with sampling gear and towing speed. For ICNAF Larval Herring Surveys in the Georges Bank - Gulf of Maine areas the BCF 61 cm Bongo is used as standard gear, fitted with nets of 0.505 and 0.333 mm mesh size. Data so far available from this programme gave no evidence for differences between the two mesh sizes in sampling efficiency for herring larvae. This has to be checked by further data, particularly for the very young stages of herring, as according to their body height of less than 0.5 mm extrusion seems possible if it is not prevented by the length of larvae.

Sampling with a 60 cm Hydrobios Bongo (Fig. 1C) in the Western Baltic (Müller, in press) enables a comparison between size distributions of larval gobiids caught in nets of 0.5 and 0.3 mm mesh size. From the

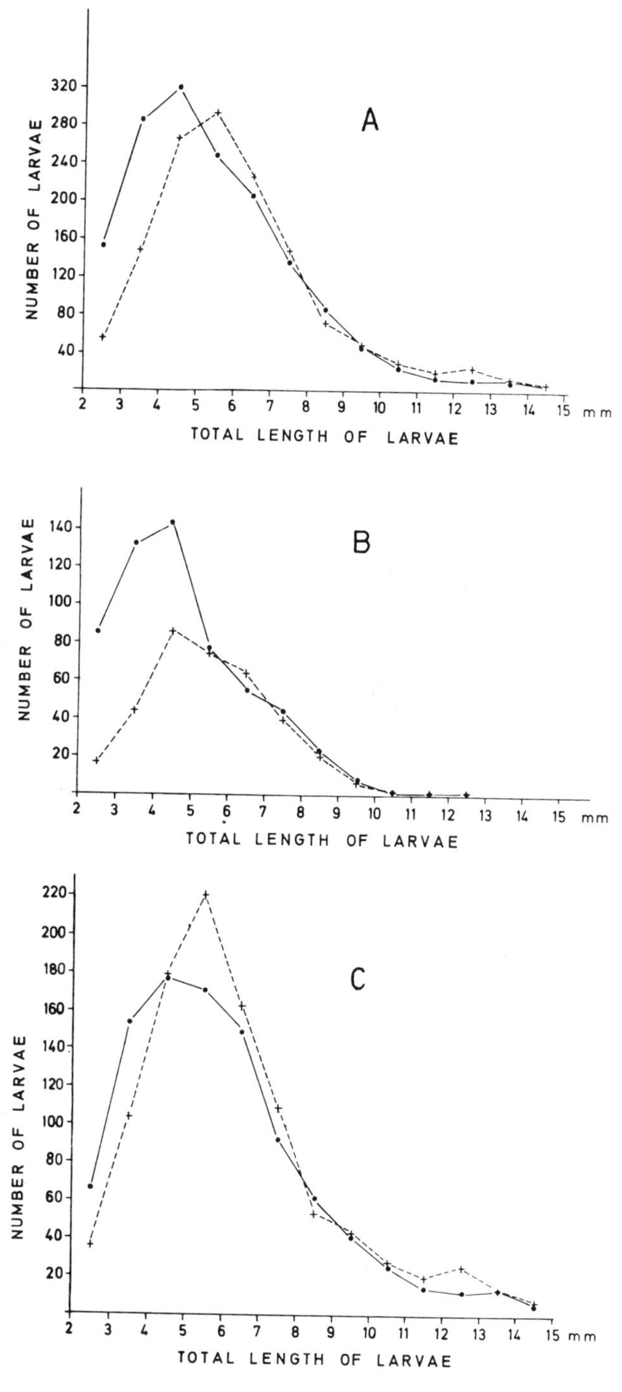

Fig. 2. Length distribution of larval gobiids retained by 0.3 mm meshes (——) and by 0.5 mm meshes (----). (A) Total numbers of larvae from 15 hauls. (B) Numbers of larvae from 8 samples without phytoplankton. (C) Numbers of larvae from 7 samples with phytoplankton

material available 15 stations were selected at which a total of about 100 or more larval gobiids, mainly *Pomatoschistus minutus* (Pal.) and a few *Gobius niger* (L.), were caught. Length distributions of the gobiid larvae are given in Fig. 2 for both mesh sizes. The total number of larvae from all 15 hauls amounted to a slightly lower value (12%) for mesh size 0.5 mm. This difference is apparently due to extrusion of very small larvae through the larger meshes, as the proportion retained by 0.5 mm meshes compared to 0.3 mm was significantly lower for larvae less than 5 mm in length. The sampling efficiency for larger larvae (>5 mm), on the other hand, seemed to be slightly higher with mesh size 0.5 mm (about 10% on average). There is good evidence that the degree of extrusion was somewhat reduced by clogging. This is shown in Fig. 3 where catches of gobiid larvae for stations with high and low phytoplankton are shown. For statistical analysis the numbers of larvae were transformed according to Cassie (UNESCO, 1968) by log (n + 1) for each station and the two size groups. By covariance analysis it can be shown that 0.5 mm meshes retained small larvae (<5 mm) significantly less well than the large larvae, except when phytoplankton occurred in the catches. With phytoplankton present, small larvae were retained significantly better than in catches without phytoplankton and a difference between size groups was reached only when the larvae were separated into groups smaller and larger than 4 mm body length. For larger larvae differences in the relative efficiency of 0.5 mm meshes between the two plankton situations were not significant. The data available in Fig. 3 are not sufficient to fit a proper selection curve. Some evidence is given, however, that extrusion is important in larvae with body heights exceeding the mesh size. This agrees with results reviewed by Vannucci (UNESCO, 1968) and implies that some extrusion might occur for the youngest stages of gobiid larvae even through 0.3 mm meshes.

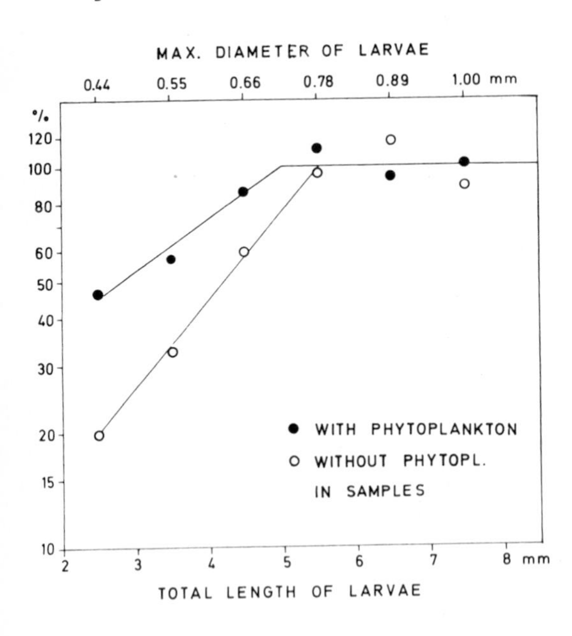

Fig. 3. Percentage of larvae retained by 0.5 mm meshes compared to 0.3 mm meshes. ● in samples with phytoplankton, o in samples without phytoplankton

GULF III NET

The Gulf III is a high speed encased sampler with a mouth-reducing cone. Since the original version, as described by Gehringer (1952), many modifications have been made (Fig. 1D-G). Bridger (1957) reduced the aperture diameter from 40.6 cm to 20.3 cm and showed that herring larvae and *Sagitta* were then caught more efficiently. A Dutch model (Zijlstra, 1970) contains further modifications: no restricted tail unit, a flowmeter mounted in the nose cone, and a filtering cone made of nylon instead of metal gauze. Comparable with this is the Lowestoft model (Harding et al., 1969) but with an aperture diameter of 35.6 cm, and the German model, the "Hai" (Hempel, 1960) with an aperture diameter of 18 cm, later 20 cm. Following tank experiments which showed that the filtration efficiency of plankton samples might be reduced by encasing the net (Tranter and Heron, 1965, 1967), an unencased version of the Hai, called "Nackthai", was built by Nellen and Hempel (1969).

Gulf III Compared to Other Plankton Nets. Several tests have been carried out to compare Gulf III models with other plankton nets, resulting in different conclusions about their relative sampling efficiencies. Comparative fishing with a modified Gulf III (Bridger, 1957) and a small BCF Bongo, mentioned above, resulted in no difference in filtration efficiency. This test did not provide data for comparing sampling efficiency for fish larvae. The Dutch Gulf III was found in some cases to be less efficient in catching fish eggs and larvae than vertically-hauled nets, e.g. the Helgoland larvae net (Oray, 1968; Schnack and Hempel, 1970), whereas in trials with Bongo and Clark-Bumpus samplers a Dutch Gulf III compared favourably as a sampler for fish larvae (Dragesund and Bjørke, this symposium).

Comparative fishing between the Helgoland larva net and the German models of the Gulf III ("Hai" and "Nackthai") have been continued on several cruises of R.V. "Anton Dohrn". Results from the two most instructive series of trials are given in Table 1 and Fig. 4. During February 1968 in the North Sea (Southern Bight) two vertical hauls with the Helgoland Larva Net ("HLN") and one oblique haul with the

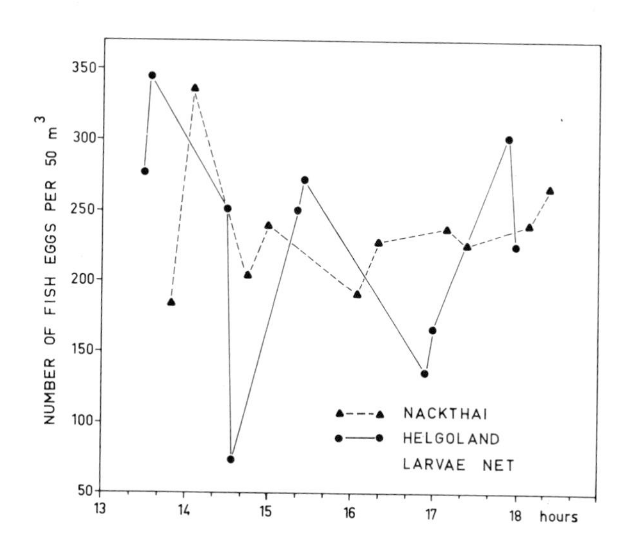

Fig. 4. Numbers of fish eggs caught by Nackthai (-----) and Helgoland Larvae Net (———) per 50 m^3 of water filtered on one station in the North Sea in February 1969 during a 5 h period. Mean number and standard deviation 229 ± 82 for HLN, 235 ± 43 for Nackthai. Time is Central European Time

Table 1. Number of fish eggs and larvae per 50 m^3 of water filtered (100% filtration efficiency assumed) in comparative hauls with Hai and Helgoland Larvae Net (HLN) on a series of stations in the North Sea (Southern Bight) in February 1968

Date	Time	Number of fish eggs/50 m^3			Number of fish larvae/50 m^3		
		Hai	HLN I	HLN II	Hai	HLN I	HLN II
22.2.	16	520	1300	-	2	9	-
23.2.	08	56	72	51	2	0	8
23.2.	10	482	475	436	5	23	35
23.2.	19	111	198	152	14	13	17
23.2.	22	81	45	-	9	4	-
24.2.	03	160	140	159	1	0	2
24.2.	06	334	77	52	0	0	0
24.2.	14	151	170	191	5	0	2
24.2.	16	73	147	274	23	42	47
25.2.	00	157	98	99	3	8	12
25.2.	04	129	83	151	1	3	3
25.2.	10	730	148	109	1	0	0
25.2.	13	560	234	192	1	0	0
25.2.	19	31	37	81	2	1	1
25.2.	21	59	75	111	3	0	0
26.2.	05	218	284	160	1	1	1
26.2.	08	220	467	485	1	2	2
26.2.	12	358	430	225	0	4	3
26.2.	16	138	217	218	9	9	8
27.2.	13	415	128	33	88	95	65
28.2.	07	14	207	650	1	5	8
28.2.	10	422	224	446	4	6	10
28.2.	12	57	87	161	1	4	11
28.2.	15	71	57	287	2	1	3
28.2.	18	52	76	62	115	76	139
28.2.	21	67	58	88	61	63	25
28.2.	23	114	97	240	17	26	29
29.2.	03	105	115	161	3	0	4
29.2.	06	254	150	111	4	8	4
29.2.	18	963	196	313	38	21	22
sum		7102	6092	5698	417	424	461

Correlations:	Eggs	Larvae
HLN I-II	$r = 0.59$, $\alpha < 0.001$	$r = 0.83$, $\alpha < 0.001$
Hai-HLN	$r = 0.31$, $\alpha > 0.5$	$r = 0.95$, $\alpha < 0.001$

Hai were made per station. The total numbers of fish eggs and larvae caught per 50 m^3 of water (assuming 100% filtration efficiency) amounted to almost the same values for both gears, although at individual stations great differences appeared (Table 1). The two successive HLN hauls correlated significantly with one another regarding both eggs and larvae, while the Hai-HLN correlation was only significant for fish larvae. This implies that results for fish eggs were more effected by "patchiness" than by wider trends in egg density within the sampled area. Comparative hauls with an unencased Gulf III (Nackthai) and HLN were also made on one station in the Southern Bight (55°29'N/ 03°30'E) in February 1969. During a period of 5 h 5 pairs of hauls were made with both gears in alternative sequence (Fig. 4). Fish larvae were taken in very low numbers and are therefore not considered here. For both gears the mean number of eggs caught per 50 m^3 of water was almost the same, but the variance between hauls was significantly higher for the HLN. It can be seen from Fig. 4 that differences within pairs of hauls are relatively low for the HLN as well as for the Nackthai, with one exception for each gear. Between pairs of hauls, on the other hand, the difference is low for Nackthai and obviously higher for HLN. The findings from both tests series are of some interest in view of precision of sampling results discussed later, but in no case was any systematic difference in the sampling efficiency of the Hai and HLN found.

Comparisons between Gulf III Models. Bridger (1957), from his tests with 20.3 and 40.6 cm apertures, concluded that herring larvae and *Sagitta* must be able to avoid the larger mouth aperture in spite of the high towing speed (5.8 kn), perhaps alarmed by a "bow wave" due to poor filtration efficiency. A series of different apertures ranging from 2.5 to 12.7 cm and 20.3 to 36.6 cm diameter, mounted to sampler bodies of 15 and 50 cm diameter respectively, were tested at towing speeds of 5-6 knots by Saville and McKay (1970). In agreement with Bridger, they found that the largest aperture for both sizes of sampler caught larval herring (8-22 mm) significantly less efficiently than the smaller apertures. Where aperture size was in proportion to the filtration capacity of the net, no indications of escape were found, regardless of the absolute size of aperture.

Bridger's Gulf III was compared with the Dutch and the Lowestoft model, and unencased samplers ("Nackthai") were tested against encased ones (Saville and McKay, 1969, 1970; Harding et al., 1969). The results did not show significant differences in the sampling efficiencies between these models, although a somewhat lower filtration efficiency, caused by clogging, was obvious in the larger aperture size of the Lowestoft model. Another comparison between encased Gulf III (Hai) and an unencased version (Nackthai) was described by Nellen and Hempel (1969). From flowmeter readings the Hai was found to filter on average 84% of the amount of water filtered by the Nackthai during 10 min tows at 5 kn. Sampling tests were carried out in 3 series, 2 of them resulting in a significantly higher sampling efficiency of Nackthai for both invertebrate plankton and fish larvae, whereas in one series no difference was obvious. The flow pattern through an encased and unencased Gulf III ("Hai" and "Nackthai", respectively) was studied in a wind tunnel (Blendermann, 1969). Total filtration efficiency at 1.5 - 5 kn towing speed proved to be exactly the same in both models (F=1.14) but was reduced (to F=1.06) when a large flowmeter was mounted in the nose cone. Blendermann points out that the very uniform distribution of velocity across the mouth area permits the use of a very small flowmeter, which also reduces damage to the catch. There were some obvious differences in the flow patterns. In the Nackthai the first half of the filtering cone filtered about 45% of the flow, whereas only 27% was filtered by that length in the Hai. Thus filtration pressure increased more rapidly toward the end of the filtering cone in the Hai, reaching values about 10 times higher than in the Nackthai. This leads to more vigorous extrusion of small organisms in the Hai, as actually found by Sherman and Honey (1969) for a modified Gulf III compared with a Bongo net. Moreover, as filtration in the Hai takes place mainly at the end of the net, the small filtering area of that part has to cope with more organisms per unit time, leading to a stronger resistance against the flow and quicker clogging. As this will change the flow pattern within the gear, it seems likely that the differences in filtration efficiency found between Hai and Nackthai in field and wind tunnel trials were caused by this. If so, the relative efficiency of Hai compared to Nackthai should decrease with increasing amount of plankton offered to the Hai in course of its tow.

This may be verified by combining the data of Nellen and Hempel for 3 tests. The relative sampling efficiency of the Hai was obtained by the ratio of catches (Hai : Nackthai) per unit time, the amount of plankton offered to the Hai being calculated from the towing time and the plankton density estimated from the Nackthai sample. In one test series Hai and Nackthai were not towed alternately, but 3 hauls were made successively with each gear and the mean values from the 3 hauls compared. Fig. 5 and 6 show that the sampling efficiency of the Hai relative to the Nackthai for both invertebrate plankton and larval sprat is about the same at low plankton densities but decreases with the increasing amount of plankton offered to the gear during the tow. In both cases a significant negative correlation was found ($r = -0.70$; $p < 1\%$). Moreover, it was established by covariance analysis that the fish larvae were sampled with notably less efficiency than invertebrate plankton in the Hai. In agreement with Bridger (1957) and Saville and McKay (1970), fish larvae seemed able to avoid the Hai, probably due to its reduced filtration efficiency.

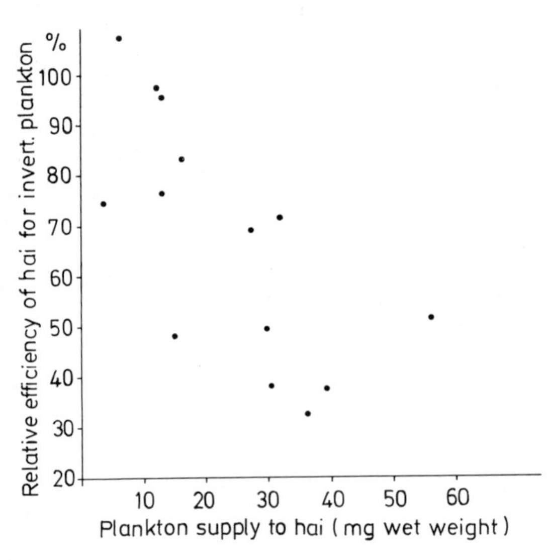

Fig. 5. Relative efficiency of Hai compared to Nackthai for sampling invertebrate plankton plotted against the calculated amount of plankton offered to the Hai during tow

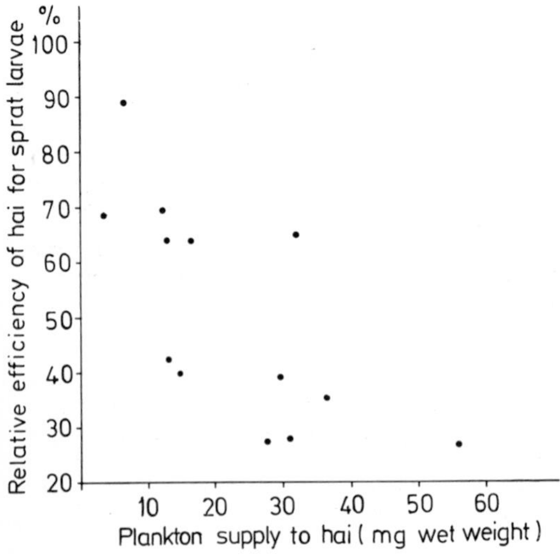

Fig. 6. Relative efficiency of Hai compared to Nackthai for sampling larval sprat plotted against the calculated amount of plankton offered to the Hai during tow

Bias from Handling the Gear. The handling of gear and use of ancillary equipment may greatly influence the accuracy of sampling, depending on variations of current velocity with depth and abundance of larvae. Herring larvae, for instance, may be concentrated in certain depth ranges, depending mainly on light condition (Zijlstra, 1970; Wood, 1971; Schnack, 1972). In that case, estimates of the total number of larvae below 1 m^2 sea surface will be affected very much by differences in the amount of water filtered at different depths. The oblique tow profile used for high-speed samplers helps to equalize the duration of sampling over the whole depth range. Profiles of tows (Bridger, 1957; Schnack and Hempel, 1970) show that the track may, however, be curvilinear. As mentioned by Schnack and Hempel this may particularly affect comparisons between Gulf III, which sample the deeper parts of the water column more intensively and vertically-hauled nets, which easily oversample the upper ranges, depending on the drift of the vessel.

The problem of tow profile in oblique hauls can be solved by an efficient depressor or by careful control of towing (Zijlstra, 1970). But even in the ideal tow the amount of water filtered at different depths may depend on differentials in current speed with depth. Sampling in the North Sea relative to a parachute buoy over a period of 10 h resulted, for instance, in generally higher numbers of herring larvae caught per volume of water filtered in one direction of tow than in the opposite. After a time period of about 6 h a reversal occurred, indicating influence from tidal currents (Fig. 7). The effect of towing direction was only apparent in larvae less than 10 mm in length, suggesting that the mean concentrations of larger larvae (>10 mm) was about the same over depth ranges of different currents. This appears possible because of differences in vertical distribution according to the size of larvae (Zijlstra, 1970; Schnack, 1972). Clearly, towing direction is of great importance when studying sampling efficiency of gears.

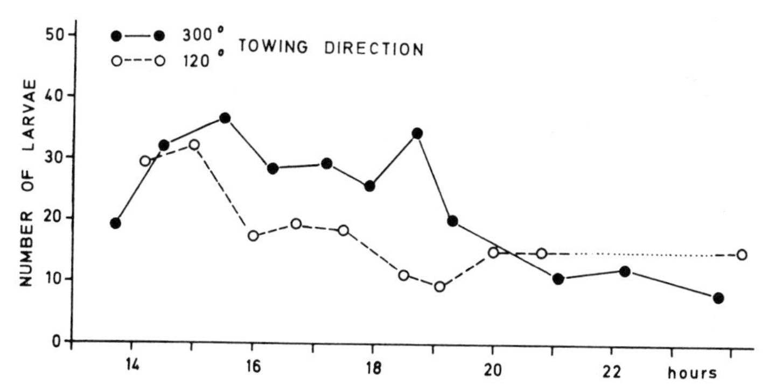

Fig. 7. Differences in numbers of herring larvae (<10 mm) caught per 1000 revolutions of flowmeter between towing direction 300° (———) and 120° (-----). Time is Central European Time

The effect of the vessel on sampling results is not yet known. During sampling with the Bongo net, for instance, when both nets were fitted with a flowmeter, the flowmeter nearest to the propeller of the ship usually showed a slightly higher value. The number of fish larvae caught per haul did not, however, differ significantly between the nets. Simultaneous sampling with Bongo and Neuston nets in the Gulf of Maine indicated either that most larvae were concentrated in the uppermost 30 cm of the water column or that the larvae were undersampled by the Bongo gear in this depth range. It thus appears desirabl to ascertain the influence of the towing vessel.

CONCLUSIONS

The accuracy of quantitative plankton sampling depends on knowledge of the sampling efficiency of the gear, which again depends on filtration efficiency, avoidance and extrusion, and also on handling. The initial filtration efficiency of the Bongo and Gulf III nets in most cases is evidently very close to 100%, being somewhat lower in the Lowestoft model of the Gulf III and somewhat higher in the Dutch type, according to the differences in aperture size. The mean filtration efficiency during the tow is more problematic, as clogging affects each of the samplers to a different degree. Cylinder-cone nets appear to have advantages over simple conical nets for the Bongo net, and an encased modified Gulf III (Hai) was found to be more quickly affected by clogging than the unencased model (Nackthai). Progressive clogging during an oblique haul may lead to unequal sampling at different depths and so affect the results if the vertical distribution of organisms is not homogenous. Conflicting results of comparative tests with the Hai and Helgoland larvae net, in regard to sampling efficiency for fish eggs and larvae, might be due either to clogging or to differences in tow profiles or, if the gears were not used alternately, to sampling in different water.

Reliable evidence is available that indicates fish larvae avoid even high-speed samplers where there is reduced filtration efficiency. On the other hand, experiments with different mouth apertures show that there is no avoidance if there is complete filtration efficiency.

The effect of extrusion on sampling efficiency has to be established separately for each gear and for speed, allowing for clogging; this is one of the main factors determining the accuracy of quantitative sampling of fish larvae and is thus of great importance. The Hai provec at times, to have inadequate filtration capacity during standard sampling operation for herring larvae in the North Sea.

Cassie (UNESCO, 1968) considered the problem of sampling error. The number of hauls within a certain area and also the size of samples influences the precision of sampling if the mean number of organisms considered is less than about 100 per haul. Large versions of high-speed samplers should thus be used for sampling rare fish larvae. From comparisons between the Helgoland larvae net and Hai or Nackthai the tow profile also seems to affect the precision of sampling results. It is suggested that the variance between hauls integrated over the whole water column depends on the distance of horizontal integration during the tow, compared to the horizontal extension of plankton patches. Horizontal integration during a "vertical" tow of the HLN, lasting 5 min for instance, is in the range of metres only, and certain ly not more than some 100 m depending on the drift of the water masses relative to the ship within this time. The actual distance, due to drift, between two successive HLN hauls lies in the same low range. Correspondingly, numbers of eggs caught in two successive hauls at one station did not differ very much (Fig. 4) and were significantly correlated over a series of stations (Table 1).

From Fig. 4 it is apparent that one HLN sample was representative of the station for about 1 h during this test accepting a deviation of 20%. This deviation from mean was exceeded by only one haul of the Nackthai during the whole period of 5 h, indicating either that no obvious density gradients existed or that continuous drift in one direction was small, relative to the towing distance of the Hai (about 2 nautical miles). The significant difference in variability between Hai and HLN hauls may be caused by patches of eggs in a size range between the distance of horizontal integration of Hai and that of HLN. Egg patches of mainly this size may also explain the absence of any correlation between Hai and HLN catches over a series of stations, despite a significant correlation between the two HLN hauls of each station (Table 1). In the case of fish larvae, however, changes in abundance were more obvious on a larger scale between stations, than on a scale below the towing distance of the Hai, as indicated by the very high correlations between Hai and HLN as well as between successive HLN hauls. Thus the precision of results from a certain size and number of samples may be improved by prolonged oblique towing, the degree of improvement depending on the distribution pattern of the organisms.

SUMMARY

A short review is given of results from studies on the efficiency of Bongo high-speed plankton samplers and different models of modified Gulf III in catching fish larvae. Additional results on extrusion and filtration efficiency from comparative fishing are included. The partly contradictory results are discussed in terms of gear handling and the distribution of fish larvae. The dependence of the accuracy of sampling on handling of gear and use of ancillary equipment is shown and the factors influencing precision of quantitative estimates are discussed.

REFERENCES

Blendermann, G., 1969. Windkanalmessungen an einem Planktonnetz. Inst. f. Schiffbau, Univ. Hamb., Schrift Nr. 2152.

Bridger, J.P., 1957. On efficiency tests made with a modified Gulf III high-speed tow net. J. Cons. perm. int. Explor. Mer, 23 (3), 357-365.

Gehringer, J.W., 1952. An all-metal plankton sampler (Model Gulf III). U.S. Fish Wildl. Serv., Spec. Sci. Rep. Fish., No. 88, 7-12.

Harding, D., Nichols, J.M. and Tungate, D.S., 1969. Comparative tests with high speed plankton samplers. ICES CM 1969 L, 19.

Hempel, G., 1960. Untersuchungen über die Verbreitung der Heringslarven im Englischen Kanal und der südlichen Nordsee im Januar 1959. Helgoländer wiss. Meeresunters. 7, 72-79.

McGowan, J.A. and Brown, D.M., 1966. A new opening-closing paired zooplankton net. SIO Reference 66, 23, 56 pp.

Müller, A. (in press). Der Jahresgang des Zooplanktons in der Kieler Bucht. I Das Verdrängungsvolumen. Kieler Meeresforsch., 29.

Nellen, W. and Hempel, G., 1969. Versuche zur Fängigkeit des "Hai" und des modifizierten Gulf-V-Plankton-Samplers "Nackthai". Ber. dtsch. wiss. Komm. Meeresforsch., 20 (2), 141-145.

Oray, I.K., 1968. Untersuchungen über das Laichen der Scholle (*Pleuronectes platessa* L.) in der südlichen Nordsee. Ber. dtsch. wiss. Komm. Meeresforsch., 19 (3), 194-225.

Posgay, J.A., 1969. Recent investigations in the ICNAF area into some of the problems of collecting zooplankton. ICES CM 1969 L, 26, (mimeo).

Posgay, J.A. and Marak, R., 1967. Tests of zooplankton samplers. ICNAF Res. Doc. 67/110, Serial No. 1910.

Posgay, J.A., Marak, R. and Hennemuth, R.C., 1968a. Development and tests of new zooplankton samplers. ICNAF Res. Doc. 68/85, Serial No. 2085.

Posgay, J.A., Marak, R. and Hennemuth, R.C., 1968b. USA - USSR joint work on zooplankton sampling methods. ICNAF Res. Doc. 68/88, Serial No. 2076.

Saville, A. and McKay, D.W., 1969. Tests with an encased and an unencased Gulf III sampler. ICES CM 1969 H, 27, (mimeo).

Saville, A. and McKay, D.W., 1970. Tests of the efficiency of various high speed samplers for catching herring larvae. ICES CM 1970 H, 7, (mimeo).

Schnack, D., 1972. Nahrungsökologische Untersuchungen an Heringslarven. Ber. dtsch. wiss. Komm. Meeresforsch., 22 (3), 273-343.

Schnack, D. and Hempel, G., 1970. Notes on sampling herring larvae by Gulf III samplers. Rapp. P.-v. Réun. Cons. perm. int. Explor. Mer, 160, 56-59.

Sherman, K. and Honey, K.A., 1969. Size selectivity in the catches of the Gulf III and Bongo zooplankton samplers. ICNAF Res. Doc. 68/34, Serial No. 2011.

Smith, P.E., Counts, R.C. and Clutter, R.I., 1968. Changes in filtration efficiency of plankton nets due to clogging under tow. J. Cons. perm. int. Explor. Mer, 32, 232-248.

Tranter, D.J. and Heron, A.C., 1965. Filtration characteristics of Clarke-Bumpus samplers. Aus. J. mar. Freshwat. Res., 16, 281-291.

Tranter, D.J. and Heron, A.C., 1967. Experiments on filtration in plankton nets. Aus. J. mar. Freshwat. Res., 18, 89-111.

UNESCO, 1968. Zooplankton sampling. UNESCO, Monogr. Oceanogr. Methodol., 2, 174 pp.

Wood, R.J., 1971. Some observations on the vertical distribution of herring larvae. Rapp. P.-v. Réun. Cons. perm. int. Explor. Mer, 160, 60-64.

Zijlstra, J.J., 1970. Herring larvae in the central North Sea. Ber. dtsch. wiss. Komm. Meeresforsch., 21, 92-115.

D. Schnack
Institut für Meereskunde
Abt. Fischereibiologie
2300 Kiel / FEDERAL REPUBLIC OF GERMANY
Düsternbrooker Weg 20

Investigations on the Distribution of Fish Larvae and Plankton near and above the Great Meteor Seamount

W. Nellen

The Influence of the Great Meteor Seamount on the Distribution of Ichthyoplankton and Ichthyoneuston

The Great Meteor Seamount considerably influences the distribution of plankton. It forms a barrier to the organisms of the deep scattering layer for horizontal extension; the plankton biomass above the seamount was found to be only one-third as great as that found beside it.

A reduction in number of fish larval species and also in diversity was obvious in the plankton samples which were taken from above the seamount. The oceanic fish larval types were also less abundant above the seamount. The average number of larvae of two neritic forms, *Trachurus* sp. and *Capros aper*, however, was higher above the seamount compared to the samples from the adjacent area. These two species probably inhabit the seamount plateau and due to weak and frequently changing currents their eggs and larvae may develop here without drifting away to deep oceanic zones.

An analysis of the length distribution of fish larvae showed that above the mount mesopelagic oceanic forms in the plankton hauls were represented only by smaller specimens. The length distribution of the larvae of the neritic *Trachurus* sp. and the epipelagic oceanic *Macrorhamphosus* sp., however, was the same in both areas. It is likely that the young of mesopelagic forms decrease above the seamount as soon as they begin to perform vertical day-night migrations which exceed the depth of the seamount plateau.

Young of *Scomberesox saurus*, which are entirely euneustic, were not influenced by the seamount either in terms of frequency or in length distribution. Neuston species, which stay at the surface only during the night hours, were less frequent above the seamount than in the oceanic area. This probably results from vertical migration which exceeds the depth of the plateau, so that the species cannot follow their normal pattern of behaviour.

Probably neither adapted oceanic nor neritic species have been able to "conquer" the relatively small seamount area in a way as to form reproducing stocks over many generations. Thus the seamount must not be considered as a particular biotope. It mainly disturbs the surrounding oceanic biotope, in which it forms a hostile zone. Neritic forms may find a refuge there when they have drifted away from some coastal zones.

---

[1] Published in 1973 in "Meteor" Forsch.-Ergebnisse, (D), (13), 47-69.

Fish Larval Composition of 12 Horizontal Hauls from a Depth of 25 m

Every four hours 12 horizontal ring-trawl hauls were done at a depth of 25 m at a permanent station near the Meteor Seamount. These samples were different from oblique tows in the frequency of several fish larval types. Like the neuston hauls the horizontal ring-trawl hauls also showed that larvae were caught more successfully when only the zone of their main distribution was fished.

Three subsequent hauls of the total of 12 were strikingly different in their fish larval composition. In these samples the numbers of mesopelagic species were much higher and epipelagic species were lower. Because day-night periodicity did not seem to be the reason for this phenomenon and because the hydrographical situation in the area investigated was very stable, it is assumed that internal waves had influenced the vertical distribution of the fish larvae.

W. Nellen
Institut für Meereskunde
Universität Kiel
2300 Kiel / FEDERAL REPUBLIC OF GERMANY
Düsternbrooker Weg 22

Distribution

Nearshore Distribution of Hawaiian Marine Fish Larvae: Effects of Water Quality, Turbidity and Currents[1]

J. M. Miller

INTRODUCTION

Studies of the spatial distribution of fish larvae in relation to variation in their abiotic environment at small spatial scales can yield insight into factors controlling their abundance at larger scales. Such responses appear to be focused at sharp environmental discontinuities, such as islands, facilitating isolation of effects of single factors. The results of the present investigation, while somewhat preliminary, describe the very uneven distributions of fish larvae found near the Hawaiian Islands and the apparent responses of larvae to three factors: water quality, turbidity and currents at the land-water interface. The latter two factors were found to obscure direct relationships between water quality and abundance of larvae. A postulate of active responses of larvae to these factors is suggested to account for the data.

The Hawaiian Islands (Fig. 1) have essentially no enclosed estuaries and have nearshore waters with short residence times. The average speed of nearshore surface currents is about 25-50 cm/sec (Laevastu et al., 1964); especially off leeward (southwest) shores, these currents are subject to reversal in direction, due to the ca. 75 cm semi-diurnal tide. By continental standards, the flow of nearshore water around Hawaii is extremely uniform. The deeper currents, although poorly known, tend to follow the depth contours (Laevastu et al., 1964). The usually diurnal trade winds, though subject to occasional storm reversals and slight seasonal shifts, are consistently 8-10 kn (4-5 m/sec) from 50-60° NE. A variety of relatively uncomplicated effluents are discharged at several points around the periphery of the Islands. Intensive urban developments, except on Oahu, are presently confined to a few isolated segments of shoreline.

The larvae of fewer than 5% of the 448 inshore Hawaiian fishes (Gosline and Brock, 1960) have been described. Owing to their wider distribution, the larvae of the 136 deep-water species are better known. Over the past two years we have learned to identify about 300 different larvae, 100 to genus and 70 to species. Of the larvae taken in inshore samples, 94% can now be identified to family, 72% to genus, and 54% to species.

A year-round bi-weekly survey of fish eggs and larvae in Kaneohe Bay, Oahu (Fig. 1), indicated that many, if not most, families of Hawaiian fishes have spawning seasons of over 6 months (Miller, Watson and Leis, 1973; Watson and Leis, MS in preparation), although peaks of abundance were found within this interval. Many spawned year-round.

Considering that the currents surrounding the Islands have average speeds well in excess of reported sustained swimming speeds of at

---

[1]Contribution No. 436, UNIHI-Seagrant-JC-74-02, Hawaii Institute of Marine Biology.

least early larvae (summarized by Blaxter, 1969), remaining near or returning to the Islands must be a major problem for inshore species. These same considerations lead to predictions of relatively uniform distributions of larvae around the Islands. Such was indicated by Nakamura (1967) and King and Hida (1954) for both fish larvae and zooplankton, mainly outside the 500-m contour, about 10 km offshore.

PART I: METHODS

A survey of inshore environments off Kauai, Oahu and Maui was begun in January 1972 with the working hypothesis that, assuming even distributions offshore with a nearly uniform input of larvae, differences in larval populations should be fundamentally related to water quality. Nineteen locations representing inshore waters adjacent to major types of land developments (or receiving major types of effluent) were selected for study (Fig. 1). Included were 7 harbors, 6 sugar mills, 8 sewers 1 oil refinery, 1 thermal (power) outfall, and 6 sites adjacent to urbanized areas with associated runoff. Thirteen additional locations were selected as reference (control) areas, i.e. adjacent to undeveloped shoreline.

At each of these locations, at least replicate daytime surface tows were made in winter, 20 January to 17 February 1972 and summer, 31 May to 14 June 1972. Each island was sampled within 3 days. A flow-metered 505 μm mesh meter net was towed at the surface for approximately 10 min at about 1.5 kn (75 cm/sec) as close to shore as the boat (draft 2 m) permitted. These shallow near shore stations averaged 5-10 m in depth. A similar number of surface tows over 20-40 m depths were made at most of these locations. None of these deeper stations was visibly affected by shoreline development. Although these stations are referred to as "deep", all were within 5 km of shore. In all, 55 stations were sampled in summer and 56 in winter.

The entire sample was preserved in the field and returned to the laboratory where all larvae were removed and tabulated. Zooplankton volumes were estimated according to the method described by Ahlstrom and Thrailkill (1962). Both larval numbers and zooplankton volumes were standardized to 1000 m^3. Confidence intervals are the width of one standard deviation. At each station O_2, salinity, and temperature were measured. Secchi disc transparency was measured at stations where the bottom was not visible. Water depth was continuously recorded on a fathometer during each tow.

In all, 14,436 larvae of 299 types were taken in the 251 samples. Most larvae were 3-8 mm standard length; the majority of these were less than 5 mm.

PART I: RESULTS

It is necessary to generalize the results of the study here; more detailed systematic treatments will be published later. Many larvae of mesopelagic families were found. Chief among these were myctophids (34 spp.) and gonostomatids (7 spp.). Although many fewer in number, several larvae of offshore pelagic families (e.g. Scombridae, Molidae, and Gempylidae) were also found. The 20 most common families encountered, in order of per cent overall abundance, were Myctophidae (16.3), Blenniidae (13.9), Gonostomatidae (11.8), Gobiidae (11.3), Pomacentridae (8.9), Tripterygiidae (6.7), Molidae (6.4), Mullidae (3.5), Apogonidae (3.3), Carangidae (3.2), Scombridae (3.0), Schindleriidae (2.3), Exocoetidae (2.2), Tetraodontidae (1.4), Gempylidae (0.5), Coryphaenidae (0.5), Dussumieriidae (0.4), Scorpaenidae (0.3), Chloropthalmidae (0.3), and Sphyraenidae (0.3); 92% of all larvae taken belonged to these families. Forty-four other families were represented in the samples.

Four families which dominate the inshore fauna as adults were nearly absent from our samples. Larvae of Scaridae, Acanthuridae, Labridae, and Chaetodontidae comprised less than 0.2% of the larvae in our samples. The previously mentioned year-round survey of Kaneohe Bay, Oahu, where adults of these families dominate the fauna (according to Key, 1973), showed similar larval absences among the 38,000 larvae taken, even in the 116 night surface samples (Watson and Leis, MS in preparation). Recently, Sale (1970) has reported on the occurrence of Acanthuridae larvae in offshore plankton tows. Apparently, the larvae of many of the reef species do not typically inhabit the inshore neuston, especially considering their adult numbers inshore and (occasional) reports of spawning activity there. Certainly the inshore larval fauna does not reflect the abundance of inshore adults, but the extent to which the offshore species depend on these inshore nursery grounds remains to be determined.

In marked contrast to the even distribution of larvae and zooplankton reported offshore (Nakamura, 1967; King and Hida, 1954), values at our more inshore stations ranged as follows:

number of larvae/1000 m^3	1 - 511.5
number of species	1 - 91
Shannon diversity index (Pielou, 1969)	0 - 5.41
zooplankton volume, ml/1000 m^3	1.25- 80.7

The greatest variation (in general, both the highest and lowest values) was found at the shallow (ca. 6 m) stations.

The mean densities of larvae at the shallow stations were: winter - 100/1000 m^3, summer - 140/1000 m^3. Corresponding mean values at the deep stations were 60/1000 m^3 and 75/1000 m^3. The mean numbers of species at the shallow stations were 25 and 28 for winter and summer, respectively; at the deep stations the corresponding values were 17 and 23.

The spatial variations of most species exceeded that of the larvae as a whole. Three examples illustrate this. First, the summer distribution of *Enchelyurus brunneolus*, an inshore blenny with pelagic larvae, is shown in Fig. 2. Numbers ranged from 0-279/1000 m^3. Second (Fig. 3), the summer distribution of *Thunnus albacares*, yellowfin tuna, larvae is shown. The highest numbers, e.g. 48.8/1000 m^3, were at the shallow stations. Finally, the larvae of one mesopelagic species of *Cyclothone* (Gonostomatidae) are shown in Fig. 4. As in the distribution of tuna larvae, the highest numbers were at the shallow stations. The distribution pattern of blenny larvae was somewhat complementary to that of tuna and *Cyclothone* (cf. Figs 2, 3, 4). These examples were not unique. Similar spatial variability was found among the remainder of the species. There were no obvious differences between summer and winter patterns.

Zooplankton volumes were generally higher at the deeper stations. Zooplankton volumes were <u>not</u> significantly correlated with larval fish densities or diversities. This suggests a fundamental difference in

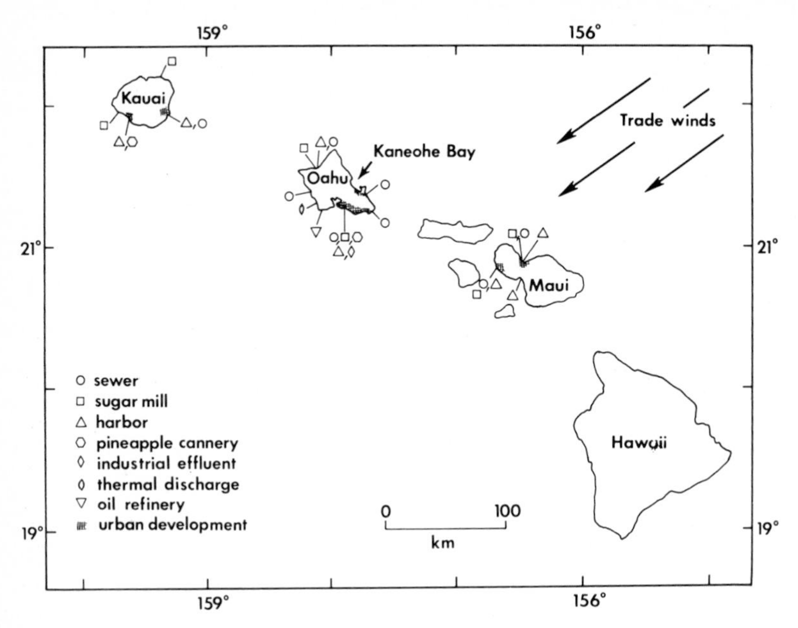

Fig. 1. Hawaiian Islands, USA, showing locations and types of shoreline developments sampled

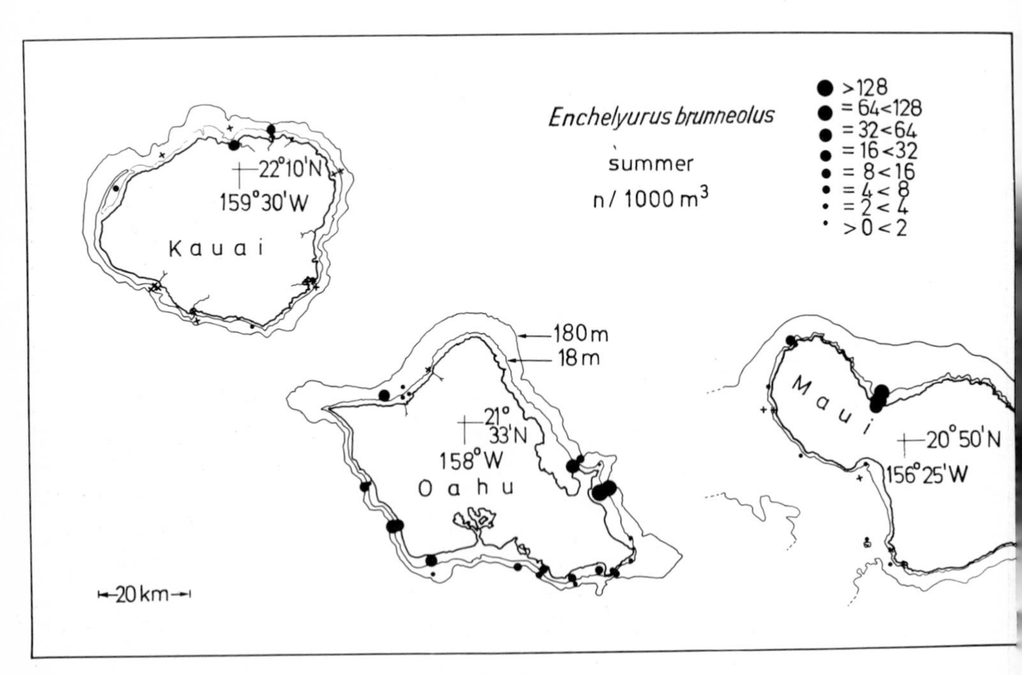

Fig. 2. Summer distribution of *Enchelyurus brunneolus* (Blenniidae) larvae off Kauai, Oahu and Maui, Hawaii. Each island sampled in 3 days between 31 May and 14 June 1972. += station

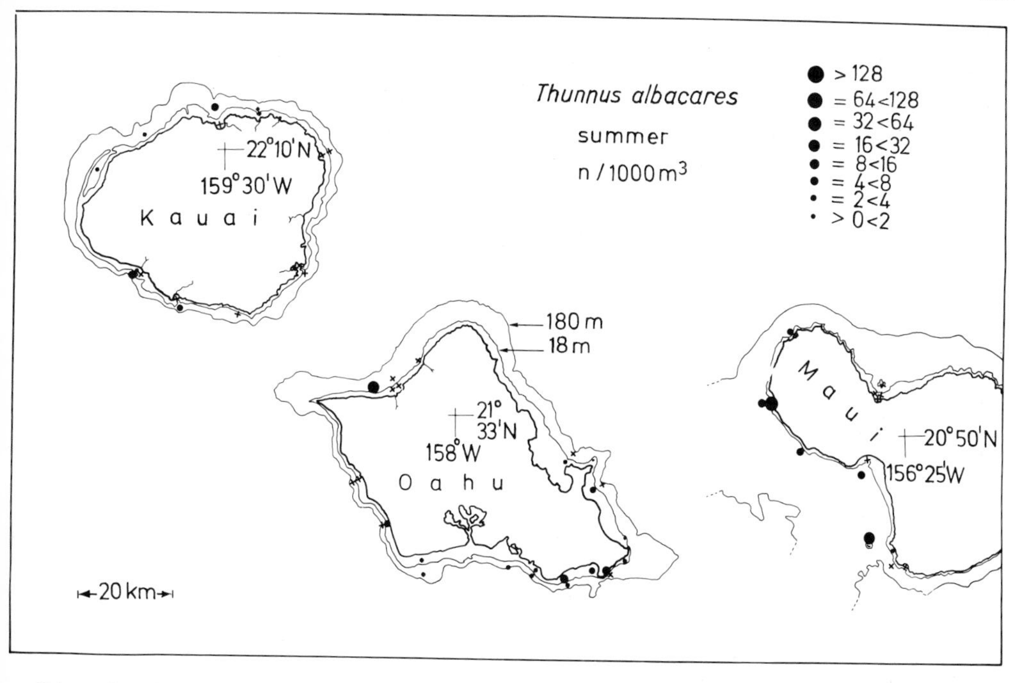

Fig. 3. Summer distribution of *Thunnus albacares* (Scombridae) larvae off Kauai, Oahu and Maui, Hawaii

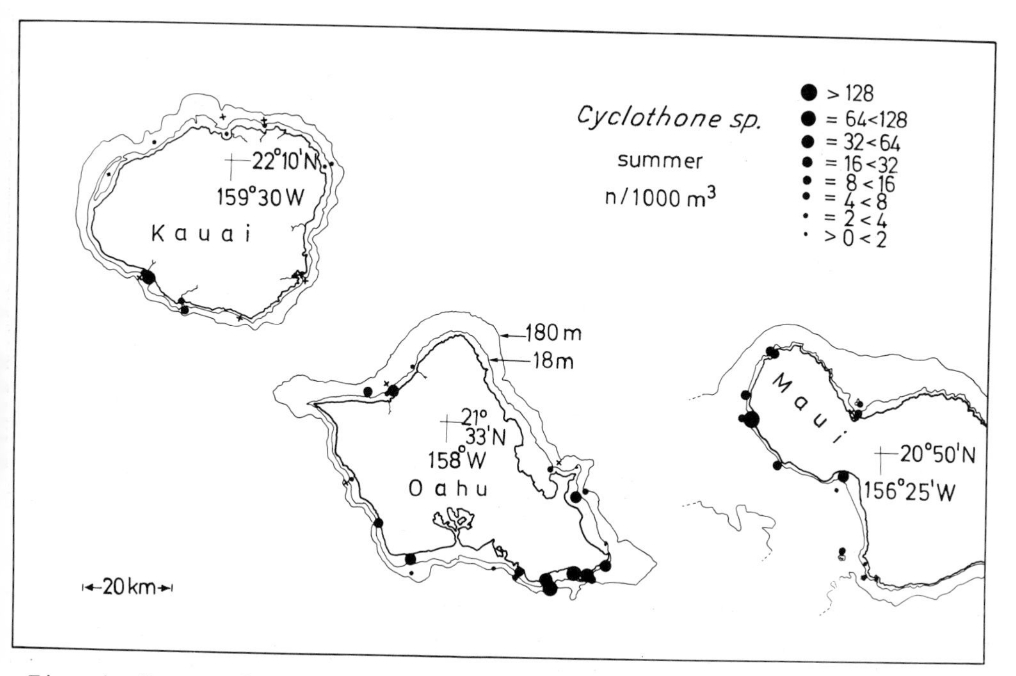

Fig. 4. Summer distribution of *Cyclothone* sp. (Gonostomatidae) larvae off Kauai, Oahu and Maui, Hawaii

the mechanisms underlying the variations in abundance of fish larvae and large zooplankton. I will return to this point.

When the winter concentrations of Blenniidae larvae at the shallow control stations were plotted against those of Myctophidae larvae, a significant negative correlation was found (Fig. 5). The departure from arithmetic linearity must be a result of biological activity of the larvae. A simple physical dilution function would not be expected to yield a straight line on log-log plot. Whatever the cause, this relationship suggests that analyses of species composition of samples, especially as ecotypes, may prove to be useful biological indicators of water mass origin and mixtures. It also follows that an estimate of potential production of inshore larvae could be derived for a location from which an estimate of stress effects could be obtained, once the mixing rate is known.

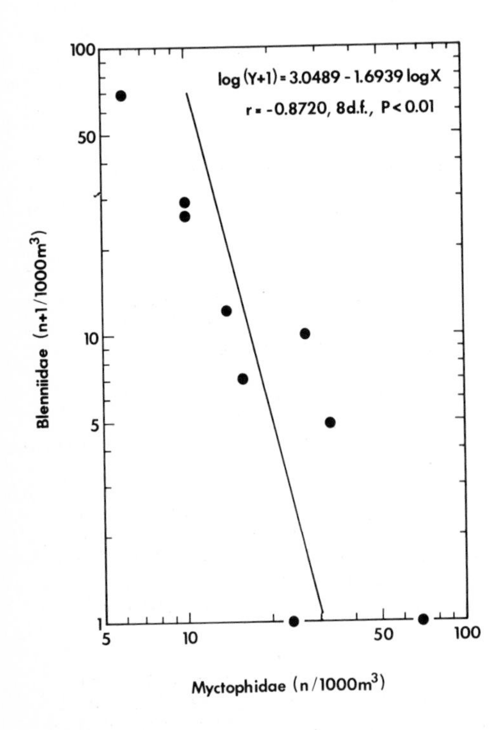

Fig. 5. Relationship between Myctophidae larvae and Blenniidae larvae abundance among shallow (3-10 m) stations not adjacent to shoreline developments, winter 1972. Points are means of replicate samples

When the species numbers, larval densities and diversity indices at the shallow stations were arranged according to season and location (adjacent to shoreline developments, i.e. potential stress)(Fig. 6), the following trends were found. All three numbers were slightly higher in summer than winter. And, with the exception of the summer density, the numbers were lower at the stations adjacent to developments. Although an effect of water quality is indicated, there was much unexplained variation. Certain stations adjacent to developments showed unexpected high values, and certain (presumably) unstressed locations showed low values.

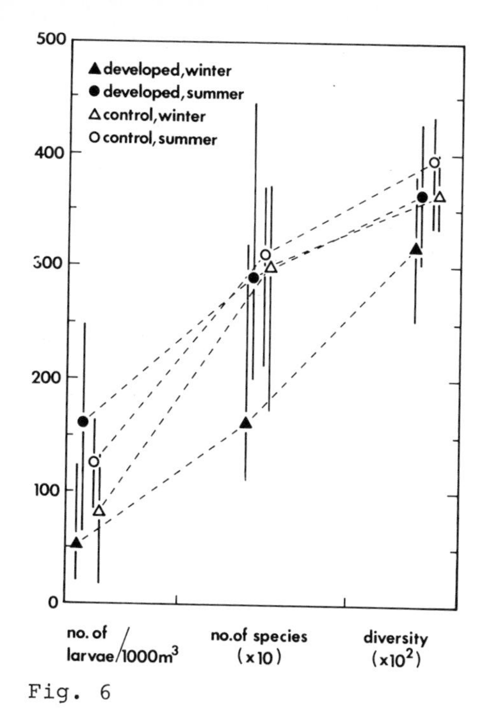

Fig. 6

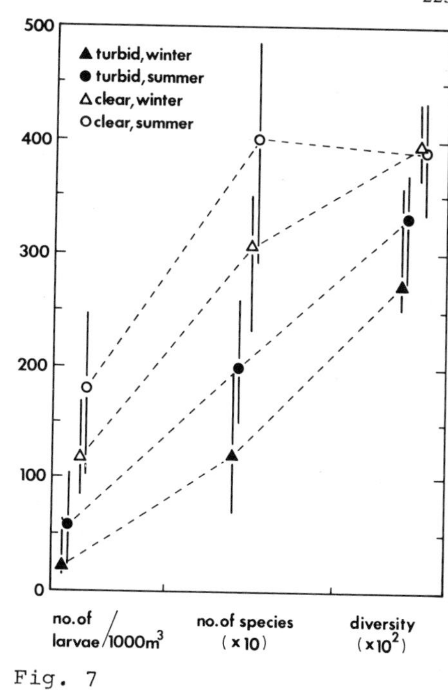

Fig. 7

Fig. 6. Density, number of species and diversity of larval fishes near shoreline developments and at control stations, winter and summer 1972. Points are medians, vertical bars are 95% confidence intervals. 49 stations

Fig. 7. Density, number of species and diversity of larval fishes in turbid and clear water, winter and summer 1972. Points are medians, vertical bars are 95% confidence intervals. 47 stations

It is clear that none of the above numbers was alone sufficient to demonstrate relationships between water quality and larval fish abundance. Certain factors needed to be identified and subtracted before quantified relationships were discernible. Among the factors considered, turbidity and currents (i.e. input of larvae) appeared the most tractable, and the most important.

The data were again arranged as above, except without regard to proximity to developments. Instead, the secchi disc readings were used to divide the data. Owing to the low precision of secchi disc transparency measurements (estimated to be ± 10-20%) and the fact that at most of the stations, the bottom was visible (therefore, no disc reading was made), the data were grouped into 2 categories only: turbid (disc depth less than water depth) and clear (Fig. 7). A significant amount of the variation in the previous analysis was explained. The density and number of species of fish larvae were, respectively, about 75% and 55% lower in turbid water. Turbidity, whether natural or artificial, was negatively correlated with larval fish abundance.

Since turbidity can be an indication of long residence time (therefore, perhaps less input of larvae), and conversely, transparency may indicate dilution (of effluents, for example) it is clear that flow rates and patterns must be determined before cause-and-effect relationships between water quality and fish larvae can be established.

Since there were data from only 3 stations which were "naturally" turbid, and many stations adjacent to developments had extremely clear water, it was not possible to separate respective effects of flow and turbidity. One possible reason why a better correlation between larval fish numbers and shoreline developments was not found is the longshore transport of effluents and their effects away from these developments. However, with the exception of the 3 stations inside enclosed harbors where the lowest densities and diversities of larvae were found, all the stations were exposed to the ca. 25 cm/sec inshore currents reported; therefore, the effect of turbidity is probably real. Assuming it is, the simplest explanation that can be advanced is that in turbid water, larvae cannot see and therefore cannot oppose the currents and thus do not accumulate. The meagre data on larval swimming speeds would seem to indicate that larvae could not accumulate (at least by opposing currents) in waters whose average speed is over 25 cm/sec; however, in this regard, the results of the following investigation are interesting.

PART II: MOLOKINI ISLAND - METHODS

The above observations suggested an active response(s) of larvae to island contact, particularly the higher numbers at the shallower stations; Molokini, an islet off southern Maui was therefore selected for additional study. The island was chosen for its small size (ca. 800 m diameter) to allow more nearly synoptic sampling, the fact that it is surrounded by high quality water of rather uniform depth (ca. 200 m), and the observation that currents in the channel occasionally produce a wake behind the island which could serve as a visual reference for sampling. These currents, though transitory, have been observed to persist through a tidal cycle.

Although Molokini appears semicircular from the air, it is a submerged cone with a reef just beneath (<2 m) the water surface for nearly the remainder of its circumference (Fig. 8). As such, it approximates a cylindrical obstacle to water flow. Water approaching the island diverges at its upstream edge and converges at some distance downstream from the island. On 12 March 1973 such a current pattern developed after noon. The path of surface water, traced by fluorescein dye was

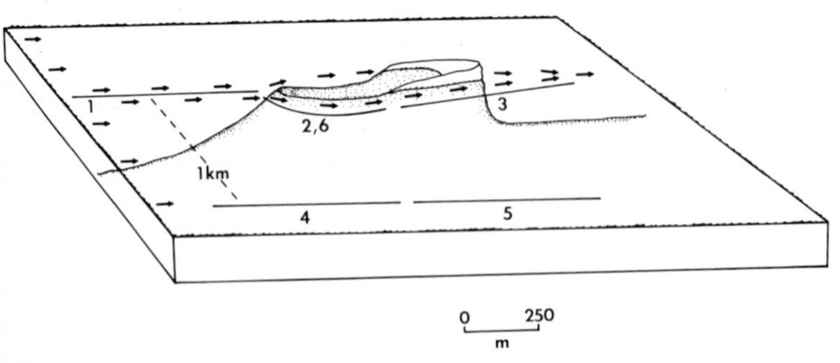

Fig. 8. Molokini, Hawaii, showing size, depth, flow past islet and convergence downstream. 1400 hours, 12 March 1973. Surface sample transects are numbered. Stippled portion under water

as follows: water approached the island from 330° at a speed of 52 cm/sec, was split, and converged again 1 km from the island's downstream edge. This same flow pattern was traced with dye on another occasion. The flow past the island is easily observed as converging surface "slicks" in the absence of strong winds.

The sampling program was begun at 1412 h and completed at 1700 h. The same methods were used as before. All sampling tracks were approximately parallel to the flow (Fig. 8). Station 1 was upstream and centered on the island. Station 2 was just outside of the reef, over about 4 m of turbulent water (typical of the shear zone between water masses) flowing around the island. Station 3 was in one of 2 visible converging slicks, downstream of the island. All 3 station tracks were approximately contiguous. Stations 4 and 5 (controls) were about 1 km from, and parallel to, stations 1 and 3, presumably away from the influence of the island. The following morning the reef station was sampled again (Station 6) when there was no evidence of flow around the island.

PART II: MOLOKINI ISLAND - RESULTS

All samples except those from Station 6 contained relatively large numbers of tunicate eggs. These were removed by sieving and their approximate volume determined before that of the remaining zooplankton. The abundance of pelagic, buoyant eggs such as these, being behaviorally inert, is a conservative biological property of a water mass, at least over short time intervals. As such they can be indicators of water-mass mixing movements, and their variation in abundance may be useful in resolving, by subtractions, active responses of vagile plankters.

The tunicate egg and zooplankton volumes and larval fish density found at each station are shown in Table 1. The densities of fish larvae over the reef (Station 2) and behind the island (Station 3) represent respective increases of 26 and 4 times the density in the surface waters approaching the island (Station 1). Corresponding increases for zooplankton volumes were 3.5 and 1.3 times. The volume of tunicate eggs, however, was the same over the reef and behind the island as in front.

At the outside (control) stations (4 and 5), a much smaller average increase downstream occurred in the fish larvae (0.72) and zooplankton (0.60). The abundance of tunicate eggs decreased.

Changes in the number of species of fish larvae followed the trends in numbers of individuals (Table 2). The individual increases (in larvae/1000 m^3) of the 20 most abundant species in approaching the island (Station 1 to 2) ranged from 9.1 in *Diplophos taenia* (Gonostomatidae) to 464 in *Bolinichthys* sp. (Myctophidae).

Table 1. Abundance of fish larvae, zooplankton and Tunicata eggs at Molokini, Hawaii, 12-13 March 1973

	STATION NUMBER (number of samples)					
	1(3)	2(2)	3(3)	4(2)	5(2)	6(3)
Fish larvae						
$\bar{n}/1000\ m^3$ (S.D.)	69.6(34.1)	1829.6(237.0)	289.3(168.3)	66.2(45.2)	120.0(33.5)	7.6(3.9)
$\bar{n}/1000\ m^3$/Station 1	1.0	26.3	4.2	1.0	1.7	0.1
no. of species	23	55	42	25	24	5
Zooplankton						
ml/1000 m^3 (S.D.)	20.6 (2.9)	72.2 (7.2)	27.0 (6.8)	21.4 (5.7)	32.8 (8.3)	29.0(1.6)
ml/1000 m^3/Station 1	1.0	3.5	1.3	1.0	1.6	1.4
Tunicata eggs						
ml/1000 m^3 (S.D.)	149 (31.1)	142 (5.7)	135 (16.8)	156 (16.9)	83 (2.8)	<1.0
ml/1000 m^3/Station 1	1.0	1.0	0.9	1.0	0.6	

Table 2. Abundance of larval fish species at Molokini, Hawaii, 12 March 1973. Numbers are larvae/1000 m^3

SPECIES	STATION 1	2	3	4	5
Bolinichthys sp.	1.0	464.8	22.0	4.0	0.9
Entomacrodus marmoratus	8.6	315.2	21.6	3.3	4.1
Labridae L-3	2.5	159.1	16.6	6.9	1.7
Cyclothone sp.	0	152.5	21.5	1.1	0
Ceratoscopelus sp.	2.6	108	71.5	2.2	10.4
Epigonus atherinoides	2.6	86.4	9.5	6.2	8.2
Ranzania laevis	1.5	75.9	19.0	1.5	34.6
Cirripectus sp.	5.5	64.7	9.8	3.3	0.9
Gempylus serpens	0	51.9	9.2	5.5	29.7
Abudefduf abdominalis	7.1	46.9	4.2	2.9	0.9
Pomacentridae P-10	0.5	41.1	2.6	2.2	0.8
Pomacentridae P-3	0.5	24.0	3.1	0	0
Synodontidae 1 sp.	0	19.2	7.2	1.1	0
Mullidae 1 sp.	4.5	19.2	2.4	1.8	2.3
Lampadena urophaos	1.2	16.8	3.1	3.6	2.5
Scorpaenidae S-11	0	13.4	0	0	0.9
Pomacentridae P-1	1.2	13.2	2.2	1.1	0
Chloropthalmidae 1 sp.	0	12.0	21.9	0	2.5
Diplophos taenia	0.5	9.6	2.8	0	0.8
Taaningichthys minimus	21.9	8.5	10.1	7.3	5.6

Other species (each <10/1000 m^3)
Myctophidae - *Diaphus* 2 spp., *Lampadena luminosa*; Pomacentridae - 5 spp.;
Scorpaenidae - 2 spp.; Sphyraenidae - *S. barracuda*, *S.* sp.;
Blenniidae - *Istiblennius zebra*, *Exallias brevis*; Coryphaenidae - *Coryphaena hippurus*;
Melanostomiatidae - *Bathophilus* sp.; Engraulidae - 1 sp.; Holocentridae - 1 sp.;
Apogonidae - 1 sp.; Mugilidae - *Mugil cephalus?*; Malacosteidae - 1 sp.;
Carangidae - 3 spp.; Dactylopteridae - *Dactyloptena orientalis*;
Melanocetidae - *Melanocetus johnsoni*; Gigantactinidae - 1 sp.;
Exocoetidae - 3 spp.; Tetraodontidae - 1 sp.;
Callionymidae - *Pogonemus pogognathus*, *Callionymus decoratus*;
Atherinidae - 1 sp.; Berycoid - 1 sp.;
Unidentified - 10 forms

Corresponding individual species increases found at the control stations were much more evenly distributed, the exceptions being *Ranzania laevis* and *Gempylus serpens*, whose respective increases of 33 and 24 larvae/1000 m^3 accounted for nearly all of the total increase between the control stations. Thus, correcting the changes in abundance near the island by corresponding values for other species did not significantly alter the trends. The rank order of abundance of the upper 20 species was the same (within a 1 standard deviation confidence interval) behind the island as over the reef; the rank order of abundance at Station 1 was different.

Thus, unlike the typical island wake, in which downstream increases in abundance are attributed to downstream upwelling, the "action" at Molokini is in front of the island.

PART II: MOLOKINI ISLAND - DISCUSSION

There are two non-mutually exclusive hypothetical origins of the additional larvae over the reef surface and deeper waters, and two mechanisms, active and passive. By "passive" I mean simply, reflecting the abundance of their source, not an actual change in density with respect to their water mass. Concentrations of larvae (or other nearly neutrally buoyant plankters) by purely physical processes probably do not occur outside centrifuges; dilutions can, of course.

It is important in this context to distinguish between concentration as an active process versus a measure of (relatively high) abundance. Too often authors do not make the distinction when referring to "a concentration of plankton in a convergence" (see Boden, 1952, for an example). I suggest accretion or amassing as a substitute for concentration in the active sense. To account for the 26x average increase by larvae in the surface water approaching would require an active response of the larvae; if, on the other hand, their source is deeper water, an active response is not necessarily required, depending on the density of larvae in these waters.

Although the current speed is far greater than reported sustained swimming speeds of larvae of similar sizes and, therefore, beyond their capabilities to oppose, larvae might accumulate over the reef by crossing the shear barrier into slower water over the reef. This sort of mechanism would require that larvae 1. are able to detect the reef from a distance of meters, and 2. possess the appropriate directional swimming response. Neither of these requirements is supported or opposed by published data. Reported perception distances of food particles by larvae of a few cm are probably not applicable to their detection of larger environmental discontinuities. Such mechanisms appear not to have been investigated in any detail. Fish larvae should be even more capable than *Daphnia* (Stavn, 1971) of changing their position relative to water currents. It is, however, important to keep such investigations on a scale meaningful to the organisms. Distribution patterns on a scale of many km will likely never yield insight into individual larval responses to currents, for example.

The similar densities of the buoyant tunicate eggs over the reef and upstream of the island suggest that the water over the reef was of surface origin; however, the 3.5x increase in zooplankton does not. In addition, the zooplankton over the reef consisted of considerably more crustacean larvae than the zooplankton in the surface water approaching the island. These particular crustacean larvae are ordinarily inhabitants of benthic or deep water in Hawaii (D. Ziemann, personal communication). Deeper water with a higher "effective" density of larvae could have brought the additional fish larvae and zooplankton to the surface. The term "effective" density is used because larvae could concentrate from lower densities by swimming upward to avoid contact with the island slope during the water's approach. This process would also require that the larvae oppose the tendency of the water to carry them around the island.

Although accretion of larvae is much more interesting biologically than a steady state, two other considerations argue against it. If the larvae over the reef had to swim to maintain their positions over the reef and the larvae behind the island were those larvae which did not escape the flow around the island, then the larvae over the reef would be larger because the process would be size-selective. By the same reasoning, the larvae behind the island should be of different species composition.

The rank order of abundance was the same over the reef as behind the island. And, the mean lengths of two species, *Entomacrodus marmoratus* and *Ceratoscopelus* sp., were found to be 5.28 ± 1.31 mm (n = 44) over the reef and 5.28 ± 1.20 mm (n = 40) behind the island for *E. marmoratus*, and, 4.09 ± 0.78 mm (n = 55) and 4.08 ± 0.61 mm (n = 79) for *Ceratoscopelus*, i.e. the same. Too few larvae were taken upstream of the island for statistical comparison of lengths of most species. However, the 17 *E. marmoratus* taken in front of the island had a mean length of 6.92 ± 1.35 mm which is significantly larger ($P < 0.01$, t-test) than either of the above. Again this suggests that at least some of the larvae over the reef did not originate in surface water.

The numbers of larvae behind the island are all about equal to the algebraic sum of 0.875 times their density upstream from the island and 0.125 times the density over the reef. The corresponding values for the zooplankton volumes were 0.88 and 0.12, respectively, i.e. in very close agreement. Whether or not an 88% dilution in a distance of about 300 m is consistent with the steady state hypothesis is not known; however, with turbulent flow, these numbers appear reasonable.

The conclusion that there must be at least some vertical transport of water in front of the island is surprising. Surface water approaching the island, would, according to textbooks, be expected to oppose upwelling by producing a local "Anstau" in the vicinity of the reef. However, such theoretical considerations appear to have been overworked. To gain any insight at all into the actual flow pattern around Molokini, or any other real island, it seems biologists must follow the example of even the most theoretically inclined investigators under similar circumstances by building a model (see Uda and Ishino, 1958).

The results of the Molokini investigation were consistent with the observations at other inshore locations: more larvae at the shallower stations. Since the flow of water at most of these other stations is parallel to the shoreline, a different process may be involved; but, whatever the cause, these habitats were found to contain fish larvae of *a priori* unexpected quality and quantity. It was necessary to sample well inshore to see this, however. de Sylva (1972) has recently reported on the occurrence of Stomiatoidea larvae inshore.

If the extremely high nearshore densities of larvae at Molokini and certain other inshore stations around the Hawaiian Islands, are steady state conditions, the implications for the effects of effluents are great. A lethal or debilitating effluent would affect far greater numbers of larvae, its effect being proportional to the volume transport of water, which at least along the shorelines of the Hawaiian Islands is considerable.

The steady-state hypothesis is being tested with time-series investigation. If, on the other hand, it is found that accretion is responsible for these high densities inshore, then it will be interesting to see if larvae lose their orientation and drift at night. If such is the case then "patches" of larvae and plankters must be generated diurnally.

Environmental discontinuities such as islands appear to be sharp focal points of the biological activities of larvae and, as such, afford perhaps unique opportunities to gain ecological insight at a scale between laboratory and oceanic conditions - we hope, with some advantages of each.

SUMMARY

1. Inshore marine fish larvae, unlike those offshore, are unevenly distributed around Maui, Kauai and Oahu, Hawaii. Density, species number and diversity of larvae were higher at shallow (<10 m) stations than deeper (>20 m) ones.

2. The species composition of inshore larvae differed significantly from that of inshore adults. Large numbers of larvae of mesopelagic and offshore pelagic species were found in waters less than 10 m deep. The abundance of inshore larvae from demersal eggs (e.g. Blenniidae was inversely correlated with abundance of mesopelagic larvae (e.g. Myctophidae).

3. Although significant, effects of shoreline developments are obscured by those of turbidity and water currents.

4. Surface zooplankton abundance was higher at the deeper stations, and was not simply correlated to that of fish larvae.

5. Data from Molokini, a small islet off Maui, demonstrate the magnitude and location of the effect of island contact on the abundance of larvae. A 26-fold average increase in density of surface larvae occurred at the upstream edge of the islet. Smaller (4x) increases appeared downstream.

ACKNOWLEDGEMENTS

This work was partly supported by Sea Grants No. 2-35243 and 04-3-158-29 to the University of Hawaii. Dr. W.H. Neill, National Marine Fisheries Service, Honolulu, read and criticized the manuscript.

REFERENCES

Ahlstrom, E.H. and Thrailkill, J.R., 1962. Plankton volume loss with time of preservation. CALCOFI Rep., 9, 57-73.

Blaxter, J.H.S., 1969. Development: eggs and larvae. In: W.S. Hoar and D.J. Randall (Eds.), Fish Physiology, 3, 177-252. New York: Academic Press, 485 pp.

Boden, B.P., 1952. Natural conservation of insular plankton. Nature, 169 (4304), 697-699.

de Sylva, D.P. and Scotton, L.N., 1972. Larvae of deep-sea fishes (Stomiatoidea) from Biscayne Bay, Florida, USA, and their ecological significance. Marine Biol., 12, 122-128.

Gosline, W.A. and Brock, V.E., 1960. Handbook of Hawaiian fishes. University of Hawaii Press, Honolulu, 372 pp.
Key, G.S., 1973. Reef fishes in the Bay. In: S.V. Smith et al. (Eds.), Atlas of Kaneohe Bay: a reef ecosystem under stress, 51-66. University of Hawaii Sea Grant Program, 128 pp.
King, J.E. and Hida, T.S., 1954. Variations in zooplankton abundance in Hawaiian waters, 1950-52. U.S. Fish Wildl. Serv., Spec. Sci. Rep. Fish., 118, v + 66 pp.
Laevastu, T., Avery, D.E. and Cox, D.C., 1964. Coastal currents and sewage disposal in the Hawaiian Islands. Coastal Currents, Haw. Inst. Geophys., Univ. Hawaii, Appen. 2, viii + 101 pp.
Miller, J.M., Watson, W. and Leis, J.M., 1973. Larval fishes. In: S.V. Smith et al. (Eds.), Atlas of Kaneohe Bay: a reef ecosystem under stress, 101-105. University of Hawaii Sea Grant Program, 128 pp.
Nakamura, E.L., 1967. Abundance and distribution of zooplankton in Hawaiian waters, 1955-56. U.S. Fish Wildl. Serv., Spec. Sci. Rep. Fish., 544, vi + 37 pp.
Pielou, E.C., 1969. An introduction to mathematical ecology. New York, London, Sydney and Toronto: John Wiley & Sons, Inc., 286 pp.
Sale, P.F., 1970. Distribution of larval Acanthuridae off Hawaii. Copeia, 4, 765-766.
Stavn, R.H., 1971. The horizontal-vertical distribution hypothesis: Langmuir circulations and *Daphnia* distributions. Limnol. Oceanog., 16 (2), 453-466.
Uda, M. and Ishino, M., 1958. Enrichment pattern resulting from eddy systems in relation to fishing grounds. J. Tokyo Univ. Fish., 44, 105-129.

J.M. Miller
Hawaii Institute of Marine Biology
University of Hawaii
Coconut Island
P.O. Box 1346
Kaneohe, Hawaii 96744 / USA

Vertical and Seasonal Variability of Fish Eggs and Larvae at Ocean Weather Station "India"

R. Williams and P. J. B. Hart

INTRODUCTION

An ecological study of the oceanic plankton at the International Ocean Weather Station 'I' (59°00'N 19°00'W) was started in 1971 by the Institute for Marine Environmental Research. The fish eggs and larvae were identified for 1971 and 1972 and their vertical and seasonal distributions plotted. The extent of the spawning period at this latitude and the vertical distribution of the eggs were examined in relation to the thermal structure of the water. The distribution of the potential predators and the food sources available to the larvae were also investigated. Further work on gut contents of the fish species will be reported at a later date.

METHODS

By kind permission of the U.K. Meteorological Office, scientists were accommodated aboard 6 British Ocean Weather Ships between 26 March and 11 October 1971 and 6 ships between 20 March and 4 October 1972.

Zooplankton sampling was carried out with the Longhurst Hardy Plankton Recorder (Longhurst et al., 1966) attached to a net with a mouth diameter of 0.5 m and a filtering system, within the cod end, of mesh size 0.28 mm. Hauls were taken at approximately 10-day intervals over the periods on station during 1971 and 1972. Each oblique haul provided a series of separate samples from 500 m to the surface from which the zooplankton could be isolated, identified and counted. For each sample in the haul the electronic recorder control unit of the LHPR recorded the temperature or range of temperature of the water through which the Recorder passed, the vertical depth range from which the samples were taken, the time spent within the depth range and the quantity of water passing through the net and the filters per unit time. Other measurements taken during the programme included twice daily temperature profiles (0-300 m, taken with mechanical bathythermographs), the concentration of chlorophyll *a* at standard depths down to 200 m every alternate day and primary production (by the 14C method) at weekly intervals in the upper 70 m.

RESULTS

Environmental Background. Fig. 1 shows the seasonal temperature stratification in 1971 and 1972 based on the bathythermograph casts from 0 to 300 m and the thermistor in the LHPR (calibrated before and after each voyage) from 300 to 500 m. In April 1971, when the sea was becoming warmer, it was evident from the position of the 9°C isotherm that transient thermoclines were being formed. A longer period of stability starting around 28 April (when the 9°C isotherm penetrated down to 150 m) coincided with the beginning of the spring outbreak of phytoplankton. The stabilization of the surface waters in spring is a major factor in determining the timing of the phytoplankton bloom. Greater stability was observed in June when there was a marked increase in the temperature of the surface waters. The stratified surface layers became gradually deeper and were still present when the sampling stopped at the beginning of October although the surface waters had cooled to 11.6°C.

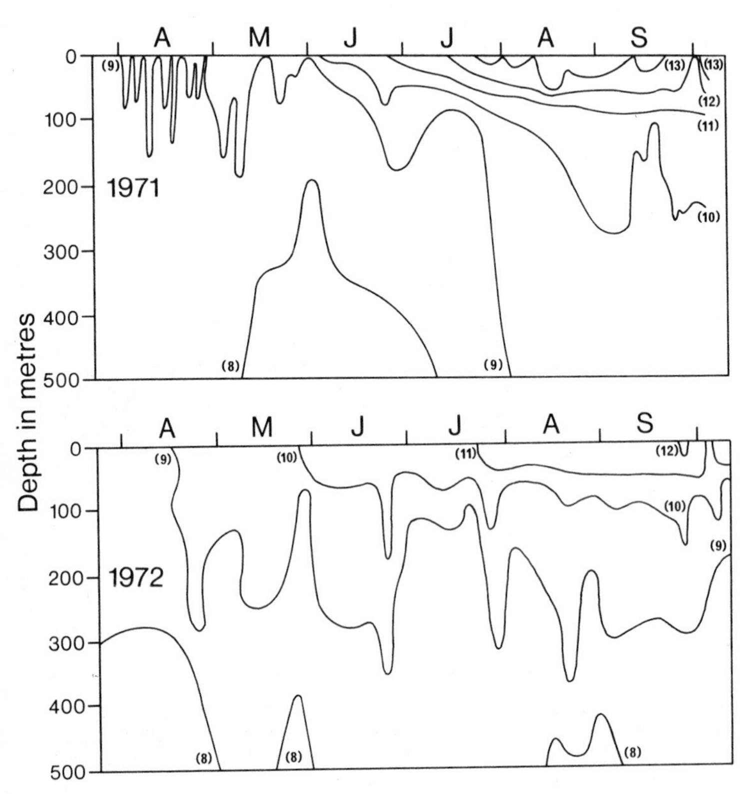

Fig. 1. Isotherms (°C) in the upper 500 metres at Ocean Weather Station INDIA from 31 March to 6 October 1971 and 23 March to 11 October 1972

In 1972 there were no transient thermoclines at the beginning of the year; the 9°C isotherm which first appeared on 16 April had penetrated down to 280 m by 20 April. It remained deep, except for short periods in late May and July.

Fig. 2 shows the seasonal distribution of chlorophyll *a* down to a depth of 200 m. The seasonal cycle followed the expected pattern for temperate waters. The spring peak in 1972 was eight days earlier than in 1971, followed by secondary peaks in summer and autumn that were also earlier in 1972 than in 1971. The 1972 standing crop, integrated over 0-70 m, for the whole of the sampling period was 12.36 g Chl a/m^2 which was higher than that of 1971 (10.02 g Chl a/m^2). These standing crop values are equivalent to 370.65 g C/m^2 or 1.70 g C/m^2/day for 1972 and 300 g C/m^2 or 1.55 g C/m^2/day for 1971; the factor used for converting chlorophyll *a* to total plant carbon was F = 30 (Strickland, 1960).

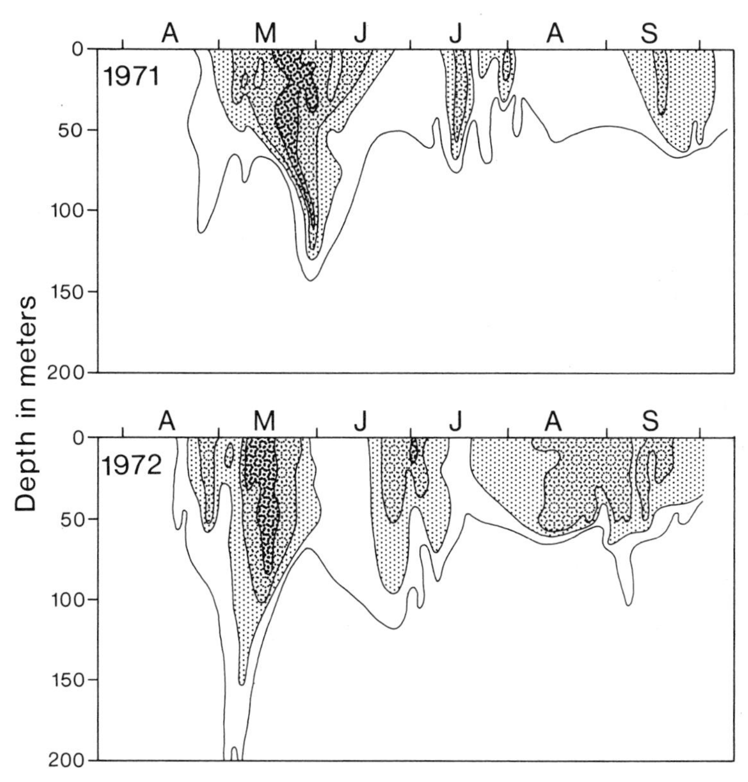

Fig. 2. Chlorophyll *a* concentration in the upper 200 m at Ocean Weather Station INDIA from 22 April to 3 October 1971 and 29 March to the 30 September 1972. The contour levels are drawn at 0.25, 0.5, 1.0, and 2.0 mg Chl a/m^3

The higher standing crop in 1972 was also reflected in higher primary production, 0.109 g C/m^2/day in 1971 compared with 0.169 g C/m^2/day (24 hours) in 1972 and in the higher standing stock of zooplankton over the sampling period.

Fish Eggs. The eggs of *Maurolicus muelleri* (Gmelin) were by far the most abundant of the fish eggs in the LHPR hauls (Fig. 3). The seasonal and vertical distribution of *Maurolicus* eggs are shown in the contour diagrams (Fig. 4). The 1971 values of the contour levels were selected by dividing the sample numbers into four categories of abundance, with equal numbers in each category, and the same contour level values

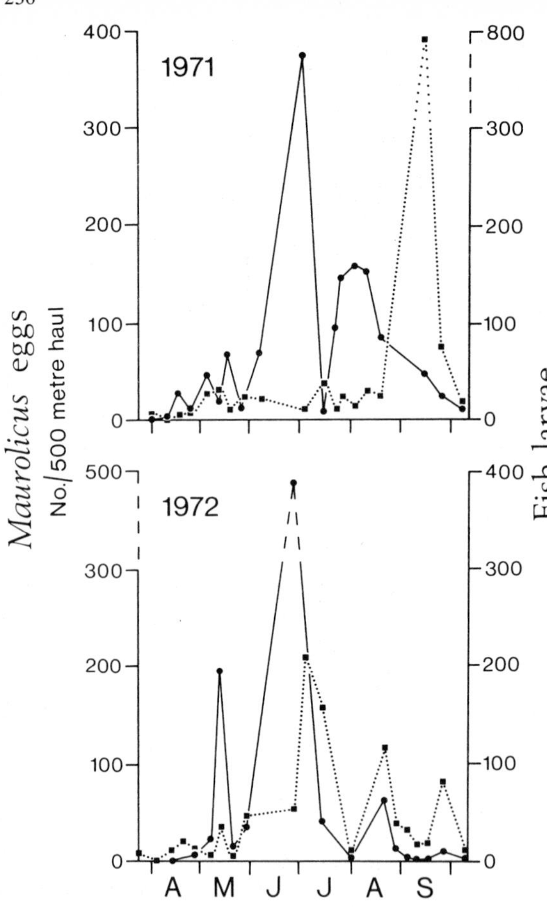

Fig. 3. The number of eggs of *M. muelleri* (solid lines) and total fish larvae (dotted lines) in the LHPR hauls at Ocean Weather Station INDIA during 1971 and 1972. The numbers are those found under a square metre of water from the surface to 500 m

were used to prepare the 1972 diagram. The eggs first appeared in early April at 300 m in 1971 and by the middle of May, in both years, small populations occurred at 100 m reaching densities of 5.4 eggs per m^3 in 1972. In both years the main spawning period occurred in the first half of June, maxima of 5.0 eggs/m^3 occurring at 180 m and 200 m. The eggs were mostly found between 100-300 m although in 1971, when they were more abundant at the end of the season than in 1972, they occurred down to 500 m in August (Fig. 4).

Small numbers of fish eggs of other species also occurred throughout the sampling period in both years, the most abundant being those of *Benthosema glaciale* (Reinhardt). The mean numbers of eggs over the sampling period were 0.14/m^3 in 1971 (31 March to 6 October) and 0.12/m^3 in 1972 (24 March to 11 October)(see Fig. 3).

Fish Larvae. Nine species of fish were identified from the 1971 and 1972 LHPR hauls: *Argyropelecus hemigymnus* (Cocco), *Benthosema glaciale*, *Cyclothone braueri* (Jesperson and Tåning), *Maurolicus muelleri*, *Micromesistius poutassou* (Risso), *Molva byrekelange* (Walbaum), *Myctophum punctatum* (Rafinesque), *Onogadus (Gaidropsarus) argentius* (Reinhardt), and *Stomias boa ferox*

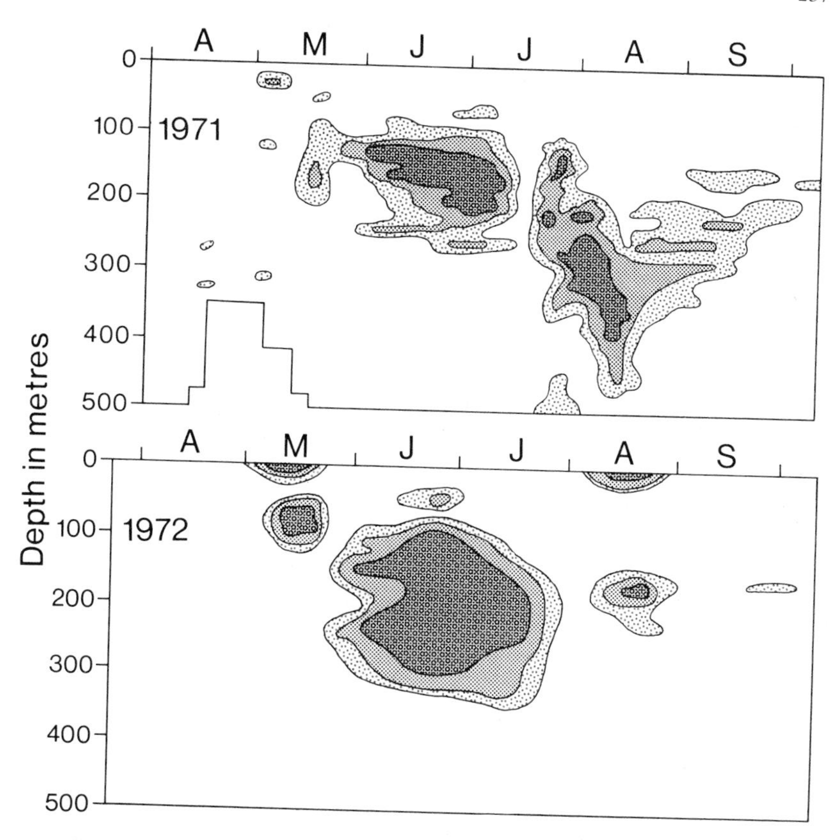

Fig. 4. The vertical and seasonal distribution of eggs of *M. muelleri* at Ocean Weather Station INDIA in 1971 and 1972. Contour levels are drawn at 0.24, 0.36 and 0.68 eggs/m^3 for both years

(Reinhardt). The seasonal and vertical distributions of the total fish larvae for 1971 and 1972 are shown in Figs 3 and 5. The contour diagrams (Fig. 5), based entirely on day hauls, illustrate the pattern of abundance throughout the year; the contour levels have the same values for each year. Virtually all the larvae identified from the 1971 hauls were *M. muelleri* with their peak level of abundance occurring in early September when values of 20 larvae/m^3 were recorded from 0-30 m (Fig. 3). Practically all of the larvae were between 3 and 5 mm long. In 1972 there was a peak in late June - early July of 1.8 larvae/m^3 which were distributed down to 300 m but by the end of August levels of 9.1 individuals/m^3 were recorded in the surface waters (Figs 3 and 5). The total number of larvae occurring throughout 1972 was less than in the previous year.

A small population of *B. glaciale* was present throughout 1971; fish 25 mm long were found in the first haul at 500 m and 30 mm long in surface samples from a night haul taken on 11 May and at 380 m in late May (Fig. 5). The main population of larvae (7-11 mm long) occurred throughout June at 40-70 m and at the surface in July when the size range had increased to 9-14 mm. Small numbers of larvae 4-5 mm long occurred at the surface in August. Most of the larvae of all the

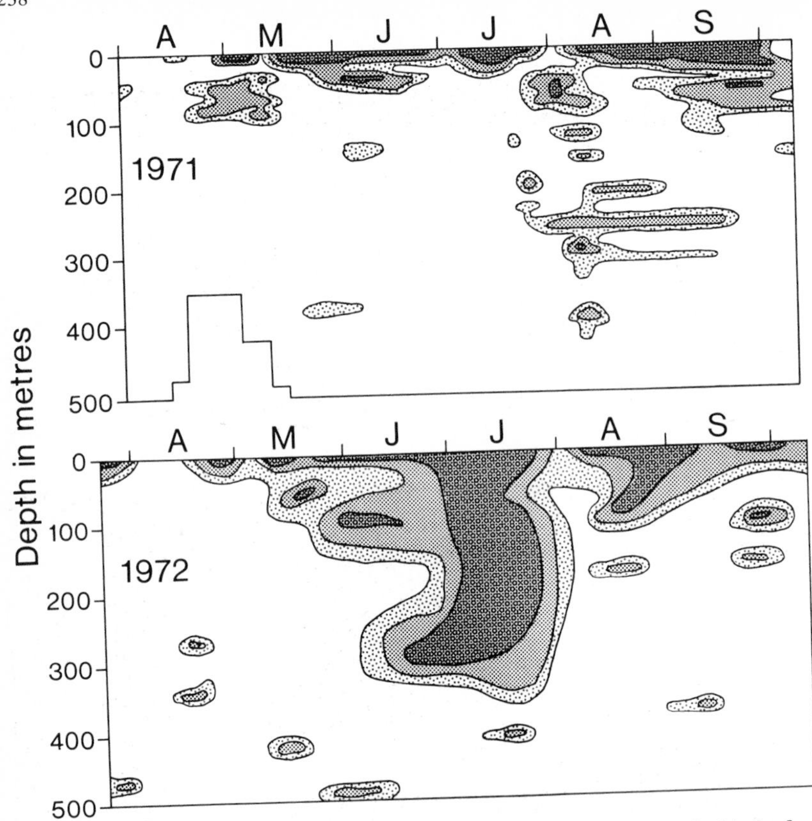

Fig. 5. The vertical and seasonal distribution of fish larvae, based on day hauls, at Ocean Weather Station INDIA in 1971 and 1972. Contour levels are drawn at 0.17, 0.21 and 0.39 larvae/m^3 for both years

other species identified occurred in a patch between 40-100 m (Fig. 5) at the end of April and in early May, although a number were recorded in night hauls throughout the sampling season.

Chaetognatha are possible predators of fish larvae. The vertical and seasonal distributions of the genus *Sagitta*, mostly *Sagitta maxima* (Conant) in 1971 and 1972 are given in Fig. 6. There are many similarities between these distributions and those of the fish larvae (Fig. 5); the differences between the years are reflected by both groups of organisms. In 1971 chaetognaths reached a maximum abundance of 10 individuals/m^3 at the surface on 13 September, but during 1972 the numbers were significantly higher at the surface, reaching a level of 36 individuals/m^3 on 13 May and of 34 individuals/m^3 on 21 August.

One of the main food sources available to the fish larvae are the eggs, nauplii and young copepodites of the copepods. Two of the most abundant copepods are *Calanus finmarchicus* (Gunnerus) and *Metridia lucens* Boeck. The seasonal distributions of the copepodite stages I and II of these

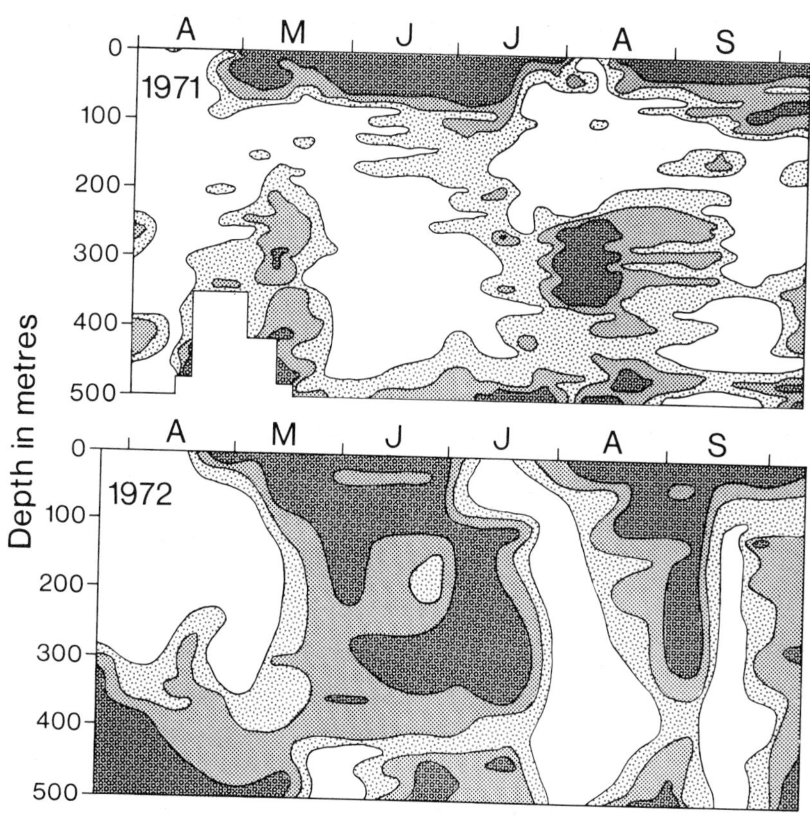

Fig. 6. The vertical and seasonal distribution of *Sagitta* spp. at Ocean Weather Station INDIA in 1971 and 1972. Contour levels are drawn at 0.38, 0.59 and 1.07 individuals/m^3 for both years

copepods in 1971 and 1972 are given in Figs 7 and 8. The main generation peaks of *C. finmarchicus* in both years occurred in May and July-August (Fig. 7). The nauplii and copepodites developed in the surface 60 m, co-occurring with the fish larvae (Fig. 5). The copepodites of *M. lucens*, in 1971, first appeared in large numbers between 0-100 m and 180-280 m in August and September and again in early October between 200 m and the surface (Fig. 8). The development stages of *Metridia* were dispersed over a greater depth range than those of *Calanus* and were also present at the same time as the fish larvae in the latter half of the year (Figs 3 and 8). This was again observed in the 1972 distributions when the main generation of *Metridia* was present between 0-200 m in July and August (Figs 3 and 8).

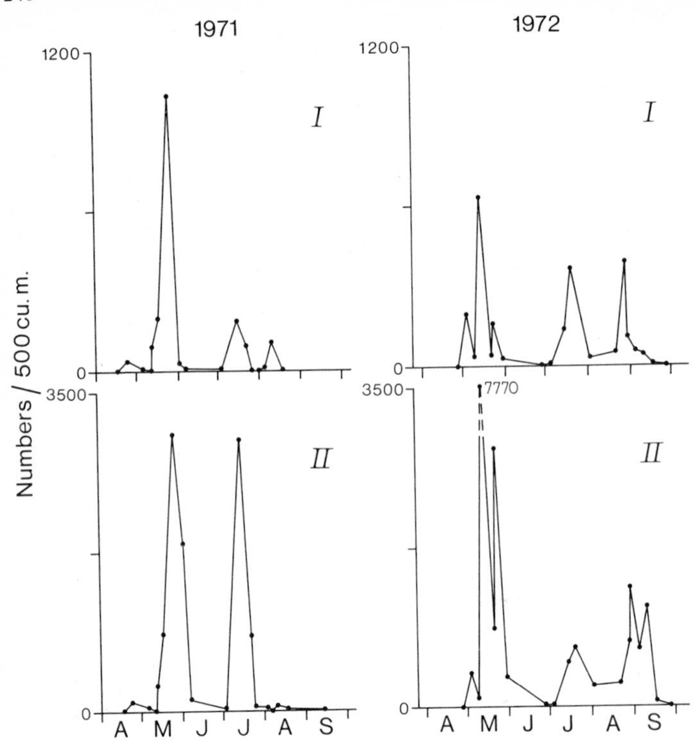

Fig. 7. The seasonal distribution of copepodites I and II of *Calanus finmarchicus* at Ocean Weather Station INDIA in 1971 and 1972. The numbers are those found under a square metre of water from the surface to 500 m

DISCUSSION

The data presented primarily concern the eggs and larvae of *Maurolicus muelleri* with small numbers of other species. The mean number of the eggs and larvae of *M. muelleri* were very similar over the 2-year period. The eggs were concentrated between 100 and 400 m, unlike the distribution found in the Japan Sea by Okiyama (1971), who observed the majority of the eggs of *M. muelleri* in the surface layers. Most of the *M. muelleri* larvae at Station INDIA (Fig. 5) were found in the upper 200 m, except for a period in early June 1972 in which the larvae penetrated in considerable concentrations to 300 m. In the Japan Sea, Okiyama (loc. cit.) did not observe many larvae above 75 m, although Grey (1964) states that the majority of post-larvae are found in depths of 150 m or less. Our data indicate that the eggs of *M. muelleri* are found in water between 8.8 and 10°C. In the southwest Atlantic at about 40°S, Ciechomski (1971) found that spawning of this species occurred between 10° and 16°C. The spawning season seems to be prolonged, as pointed out by Grey (1964) and Ciechomski (1971). Off the coast of Morocco it was found that *M. muelleri* spawned mainly in the winter and spring (Grey, 1964) yet the main distribution of eggs at 59°N occurred between May and early September; this was due perhaps to the later development there of temperature conditions favourable to spawning.

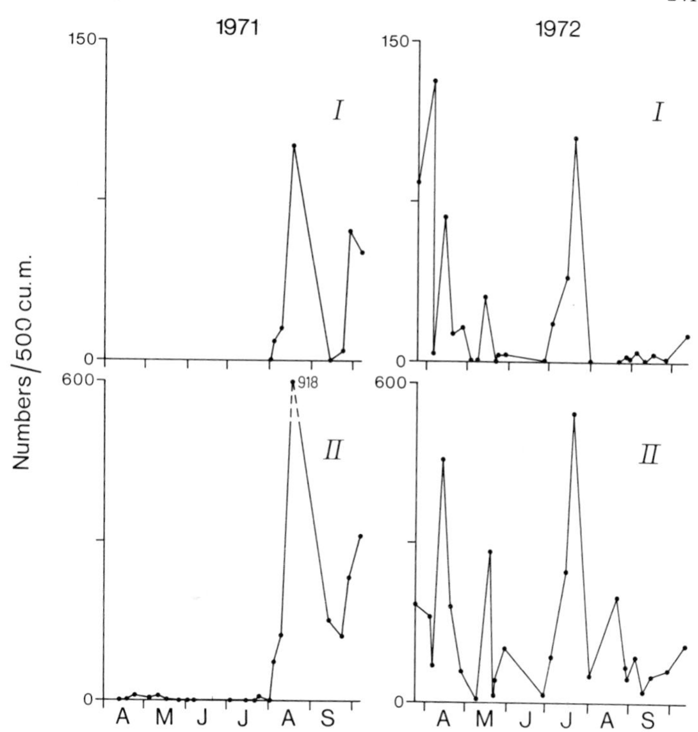

Fig. 8. The seasonal distribution of copepodites I and II of *Metridia lucens* at Ocean Weather Station INDIA in 1971 and 1972. The numbers are those found under a square metre of water from the surface to 500 m

Lebour (1923) showed that *Sagitta* spp. preyed upon herring larvae; Chaetognatha, therefore, might prey upon larvae of *M. muelleri*. The vertical distribution of the Chaetognatha overlapped that of the fish larvae and was similarly matched to the *Calanus* and *Metridia* distributions which are known prey species of the Chaetognatha. Undoubtedly some of the young stages of other fish species such as the Myctophidae will also be predators and indeed large numbers of *Myctophum punctatum* have been caught from Ocean Weather Ships at Station INDIA in the surface waters at night in the latter half of the year.

The diet of *M. muelleri* larvae in the Japan Sea was studied by Okiyama (1971) who found that the smaller larval stages (3-6 mm) fed mainly on *Oncaea* spp. The distribution of *Oncaea* spp. and *M. muelleri* larvae as shown by the present work do not coincide which would suggest that *Oncaea* spp. is not an important item in the diet. The main seasonal abundance of this group occurred in May and June in both 1971 and 1972. However, as shown by Figs 7 and 8, larvae of *M. muelleri* and the young stages of *Calanus finmarchicus* and *Metridia lucens* occur at the same time indicating that these two copepods, at least during their early life history, could provide an important food source for the fish larvae. The coincidence of a food supply with larval distribution is important and probably critical for their survival (Cushing, 1966 and 1967). The relationship of the spawning of *M. muelleri* with the production cycle must be a very general one because of their long spawning season.

Nevertheless it appears that, as shown by Figs 2 and 3, spawning does not start until the spring phytoplankton bloom has begun. Unfortunately we do not have data from a long enough time span to determine whether the timing of the spring outbreak is critical to the survival of *M. muelleri* larvae. This is a point which will be investigated in following years as more data become available.

SUMMARY

The vertical and seasonal distribution of *Maurolicus muelleri* eggs and total fish larvae are given for 1971 and 1972 from the International Ocean Weather Station 'I' (59°00'N 19°00'W). The mean numbers of eggs and larvae over the sampling periods 31 March to 6 October 1971 were 0.14 eggs and 0.12 larvae/m^3 and for 24 March to 11 October 1972 were 0.12 eggs and 0.09 larvae/m^3.

Nine species of fish larvae were identified from the Longhurst Hardy Plankton Recorder hauls. These were: *Argyropelecus hemigymnus*, *Benthosema glaciale*, *Cyclothone braueri*, *Maurolicus muelleri*, *Micromesistius poutassou*, *Molva byrekelange*, *Myctophum punctatum*, *Onogadus (Gaidropsarus) argentius*, and *Stomias boa ferox*. The most abundant of these larvae were those of *M. muelleri* which reached densities of 20 larvae/m^3 in the surface 30 m in September 1971. The majority of the larvae were distributed in the upper 200 m in both years.

Maurolicus eggs occurred between 100 and 500 m and their distribution was correlated with the 9°C isotherm in both years, the eggs remaining in water between 8.8° and 10°C. Maximum densities of 5 eggs/m^3 occurred at 100 m in May, at 200 m in June 1972 and at 180 m in June-July 1971. The seasonal distribution of the larvae is discussed with reference to one of their predators *Sagitta* spp. and the available food sources.

ACKNOWLEDGEMENTS

We wish to thank the officers and men of Ocean Weather Ships "Surveyor" "Reporter", "Adviser", and "Monitor" and many of our colleagues who have helped in the collection of the samples. We would also like to thank J.M. Colebrook whose computer expertise made a number of our diagrams possible.

The work was supported by the Natural Environment Research Council as part of the research programme of the Institute for Marine Environmental Research.

REFERENCES

Ciechomski, J.D. de, 1971. Estudios sobre los huevos y larvas de la sardina fueguina *Sprattus fueguensis* , y de *Maurolicus muelleri*, hallados en aguas adyacentes al sector Patagonico Argentino. Physis. Buenos Aires, 30, 557-567.

Cushing, D.H., 1966. Biological and hydrographic changes in British seas during the last thirty years. Biol. Rev., 41, 211-258.
Cushing, D.H., 1967. The grouping of herring populations. J. mar. biol. Ass. U.K., 47, 193-208.
Grey, M., 1964. Family Gonostomatidae. In: Fishes of the Western North Atlantic. Y.H. Olsen (Ed.). New Haven: Sears Foundation for Marine Research, VI + 599.
Lebour, M.V., 1923. The food of plankton organisms. J. mar. biol. Ass. U.K., 13, 70-92.
Longhurst, A.R., Reith, A.D., Bowers, R.E. and Seibert, D.L.R., 1966. A new system for the collection of multiple serial plankton samples. Deep Sea Res., 13 (2), 213-222.
Okiyama, M., 1971. Early life history of the Gonostomatid fish, *Maurolicus muelleri* (Gmelin) in the Japan Sea. Bull. Jap. Sea Reg. Fish. Res. Lab., 23, 21-53.
Strickland, J.D.H., 1960. Measuring the production of marine phytoplankton. Bull. Fish. Res. Bd Can., 122, 1-146.

R. Williams
Institute for Marine Environmental Research
Oceanographic Laboratory
78 Craighall Road
Edinburgh EH 6 4 RQ / GREAT BRITAIN

P.J.B. Hart
Institute for Marine Environmental Research
Oceanographic Laboratory
78 Craighall Road
Edinburgh, EH 6 4 RQ / GREAT BRITAIN

Present address:

NORDRECO AB
26700 Bjuv / SWEDEN

The Feeding of Plaice and Sand-Eel Larvae in the Southern Bight in Relation to the Distribution of their Food Organisms

T. Wyatt

INTRODUCTION

The quantitative study of planktonic food chains has proceeded rather slowly, and there seem to be two main reasons for this. On the one hand, food remains in the guts of planktonic animals and often appears as a green or brown mush in which few items can be identified with potential prey organisms. This is often the case in herbivores like copepods, but with fish we are more fortunate, since most species are carnivorous during most of their larval life. With patience, we can almost always identify the food that remains in the guts of fish larvae. A second reason may be that food webs are generally regarded as being complex, with a variety of trophic links existing at any one time. According to this view, simple linear food chains do not exist in nature. In the long term, this is a valid generalization, but for brief periods during its life, a larval fish is restricted as to what it can feed on, both by the size of prey it is able to capture, and by the prey organisms available within that size range. By isolating these periods and examining them separately, it should be possible to build up a composite picture of the feeding habits of larval fish which bypasses the theoretical problems of complex food webs.

It is known that in the Southern Bight of the North Sea, the larvae of plaice (*Pleuronectes platessa* L.) and sand eels (*Ammodytes marinus* Raitt) feed on appendicularians during the early part of their life (Ryland, 1964; Shelbourne, 1962). In the same region, in 1968 and 1971, plaice larvae fed almost exclusively on *Oikopleura dioica* for their entire pelagic life, a period of nearly 2 months. The sand eels in 1968 fed mainly on *O. dioica* while growing from 10 mm to 15 mm, again a period of nearly 2 months. Hardy's (1924) classical study of the trophic relations of the North Sea herring provides another example of a simple food chain. Herring between 12 mm and 42 mm in length fed on a diet consisting of 86% *Pseudocalanus*. It is ironic that Hardy's well known work is so often quoted to demonstrate the complexity of food webs.

Gut analyses of the plaice and sand-eel larvae collected in 1968 allowed estimates of the feeding rates of these species to be made, and these in turn were used to account for the mortality of the prey population (Wyatt, 1971). In this paper, the effect of the *Oikopleura* on their predators is dealt with. The distributions of planktonic predators and their prey are not likely to coincide closely except on rare occasions, so that their abundance relative to one another will vary across the region occupied. A similar notion led Hjort (1914) to suggest that larval mortality, and hence subsequent year-class strength, might be controlled by the availability of food during a relatively restricted part of larval life. The "critical period" concept which has emerged from Hjort's idea has been a preoccupation of fish larval research ever since.

It is thought that poorly fed larvae must be identifiable, and various factors have been devised to express their condition. Larvae with the highest condition factors should then be the best fed. The approach certainly works in experimental situations (Hempel and Blaxter, 1963; Wyatt, 1972), but has so far led to negative or equivocal results when applied to field data. The condition of wild herring larvae has been measured by Blaxter (1971) who compared it with the biomass of zooplankton on which the larvae may have been feeding, and by Vilela and Zijlstra (1971) who compared larval condition with subsequent recruitment. Blaxter found an inverse relation between condition and zooplankton biomass, but sought to explain this unexpected result on technical grounds. Vilela and Zijlstra found slight variations in condition, but no relation between these variations and recruitment in different years. Shelbourne (1957) compared the condition of plaice larvae at different concentrations of zooplankton, but his data are open to interpretations other than those he placed on them, so that Blaxter's study is the only one which has a positive bearing on Hjort's original thesis.

MATERIALS AND METHODS

The material on which this study is based was collected by RV CORELLA. These collections are described by Harding (this Symposium). Between January and April 1968, plaice and sand-eel were the most abundant fish larvae in the Southern Bight. In January, plaice larvae comprised about 50% of the larval fish population. By mid-February they had declined in importance, and sand-eel formed about 70%. This species remained dominant until late April, and accounted for 86% of the larval fish population in early April. A similar picture emerges from the 1971 collections, except that the place of *A. marinus* is taken by *A. tobianus*.

The guts of larvae from stations where they were abundant were dissected and all food remains identified, as accurately as possible, and counted. The lengths of all larvae were measured, and in the case of plaice, the height of the body musculature was measured between the bases of the dorsal and anal fins, at right angles to the notochord.

The feeding rates of plaice and sand-eel larvae have been taken as the numbers of *Oikopleura* pellets per gut, without transformation. If the digestion rate is the same in all larvae, the numbers of prey eaten per hour or per day can be calculated from this figure on the basis of information provided by Ryland (1964) and Wyatt (1971).

The relative density of predator and prey has been expressed as encounters per linear metre. This is the product of the cube roots of the densities:

$$\text{Encounters/metre} = (\sqrt[3]{\text{prey/m}^3}) \times (\sqrt[3]{\text{predator/m}^3}).$$

The condition index of plaice has been calculated from the height (H) and length (L) of all larvae from each station where a sufficient number were captured. In laboratory experiments, the slope of a plot of the logarithm of (H/L) on time increases with food density (Wyatt, 1972). Here the logarithm of (H/L) has been plotted on L, and the slope is used as a measure of condition. Each value therefore refers to a group of larvae at different stages of development.

RESULTS

a) Plaice Larvae. Fig. 1 shows a plot of the mean numbers of pellets/ gut on encounters/m for plaice larvae of three different length groups, 5.0 to 5.9 mm (a), 6.0 to 6.99 mm (b), and 7.0 to 7.9 mm (c). Only those stations where more than five larvae were dissected are shown on the graph. It is not easy to fit curves to data of this kind, and any such curve is likely to appear very unsatisfactory in view of the wide scatter of the points. But the general shape such curves must take is clear, and most convincingly is seen in Fig. 1b (6 mm length class) for which the largest amount of information is available. Here the highest numbers of pellets are found in larvae feeding at intermediate encounter rates, while at higher encounter rates, the numbers of pellets fall. In the complete absence of food there can be no feeding, so that the curve must start at the origin. The plankton net used to sample the *Oikopleura* population does not capture the smallest animals, which may account for the fact that some larvae contain food in the apparent absence of their prey. The broken line in Fig. 1b shows these general features, but can only be regarded as semi-quantitative at best. Similar curves might be fitted to Fig. 1a and Fig. 1c, though the data hardly merit it. One gains the impression however, that in all three,

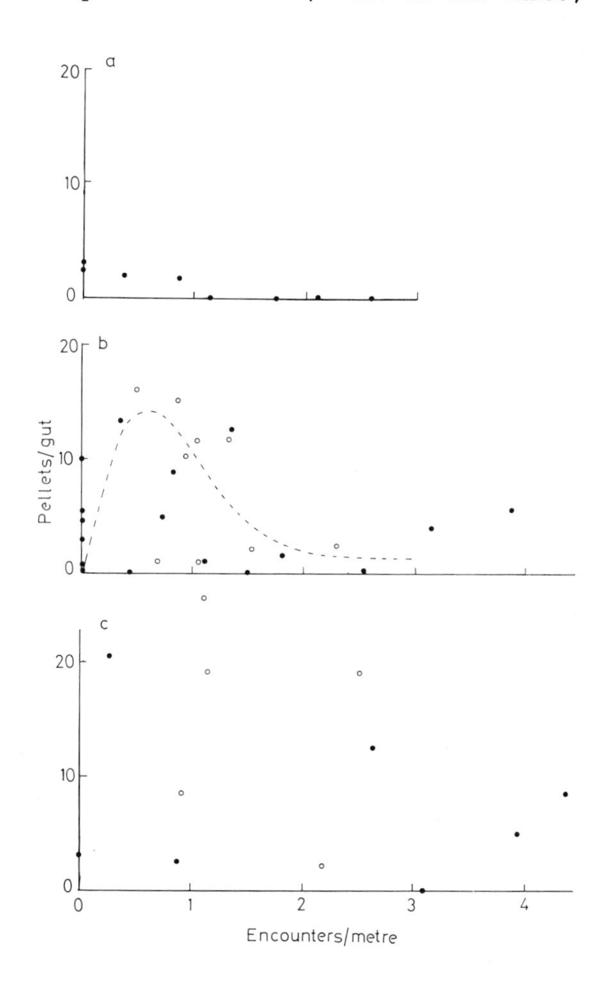

Fig. 1a-c. Relation between feeding rate (in pellets/ gut) and encounters/m in plaice larvae of different lengths: (a) 5.0 to 5.9 mm, (b) 6.0 to 6.9 mm, (c) 7.0 to 7.9 mm. Full circles, 1968 data; open circles, 1971 data

the highest numbers of pellets are found at about 0.5 to 1.0 encounters/m, and it also appears that the height of the peak increases with the length of the larvae.

Some support for the curve shown in Fig. 1b is given by data shown in Fig. 2. Here the condition index has been plotted against encounters/m, and there is a decline in condition as the number of encounters increases. Only those stations have been plotted where the fiducial limits of the regression of $\log_e$ (H/L) on (L) are better than 1%. This has unfortunately resulted in a lack of detail between 0 and 1 encounters/m, so that it is not possible to say whether condition also declines at very low encounter rates. In preparing Fig. 2, the possibility that the food concentrations in which the larvae find themselves has recently changed, has been ignored. As Blaxter (1971) pointed out, it is the availability of food for a period of some days before capture which determines how well the larvae have been feeding, and hence what condition they are in. This may well account for some of the variance seen in Fig. 2.

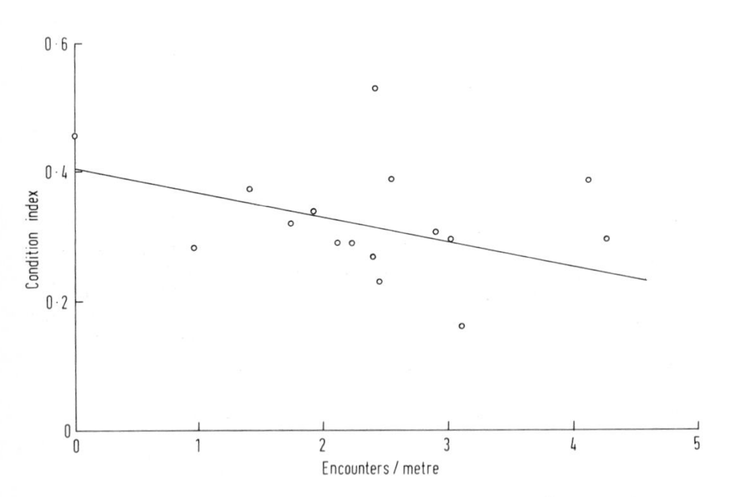

Fig. 2. Relation between condition index and encounters/m in plaice larvae: $y = 0.404 - 0.038\ x$, $r = -0.373$, $f = 51.7$

If the condition index used here is a genuine indication of condition, then from the data presented in Fig. 1 and 2, one may perhaps conclude that higher encounter rates are detrimental to plaice larvae. It is possible that the height of the myotomal musculature responds to low food densities by increased relative growth, and that condition in the sense used here has no implications with respect to survival rates. Nevertheless, a separate explanation is required to account for the small numbers of pellets found in larvae at high food densities.

b) Sand-Eel Larvae. Fig. 3 shows similar data for sand-eel larvae between 7.0 mm and 13.9 mm. The smallest larvae feed to a considerable extent on phytoplankton, and those longer than 12 mm begin to introduce copepods and other items into their diet. Between 8 mm and 11 mm, appendicularian pellets constitute almost the entire gut contents. Fig. 3 therefore shows the relation between gut contents and encounters for the major, but not the only, food items of these larvae. As in the case of Fig. 1, each point on the graph represents a minimum of 5 larvae. The encounter rate at which the numbers of pellets per gut is maximal cannot be determined from this figure, but is certainly higher than 2. In contrast to the situation with plaice larvae, there is no indication of a maximum number of pellets at some intermediate encounter rate. No morphological technique for estimating the condition of sand-eel larvae has been devised.

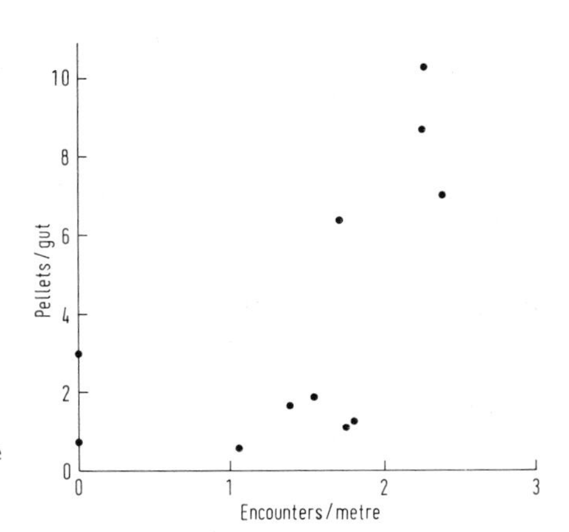

Fig. 3. Relation between feeding rate (in pellets/gut) and encounters/m in sand-eel larvae

c) Plaice and Sand-Eel Larvae in 1963. During 1962, extensive populations of *Oikopleura* were absent from the Southern Bight during the time occurrence of the larval plaice and sand-eel populations. The diet of plaice larvae consisted of 77% *Coscinodiscus*, together with other phytoplankton, small numbers of fish eggs, copepod nauplii and a few *Oikopleura*. The sand-eels ate copepod nauplii and copepodids (44%), fish eggs (21%), and phytoplankton (30%). The condition of the plaice larvae in this material ranged between 0.350 and 0.401. Both these species were able to adapt to the failure of their normal prey.

DISCUSSION

The data presented in Fig. 1 suggest that the optimal prey encounter rate in plaice larvae is about 0.5 - 1.0/m. The numbers of pellets can be converted to numbers of prey by dividing by 2.5, the mean number of pellets found in an individual *Oikopleura*. If we then assume, on the basis of Ryland's (1964) data, that the plaice larvae require 3 h to fill their guts, and that they feed for 16 h a day, the optimal daily ration can be calculated. If we further assume that these larvae can swim at three body lengths a second, then we can calculate the distance they must travel to obtain their daily ration, and the time required to travel this distance. These calculations are given in Table 1. Column (d) of this table shows that very little time is required to obtain their daily ration. On the other hand, if the daily ration given in column (b) is divided into "two square meals", and the time to digest a meal is about 6 h (Ryland, 1964), then with a filling time of 3 h the gut will contain food remains for a total of 18 h. However, it is possible that larvae found to be empty during feeding hours have completed their feeding requirements for the day, and that this argument may account for the large numbers of empty larvae so often reported in work of this kind.

Table 1. Plaice larvae. (a) length class; (b) optimal daily ration; (c) distance larvae must travel at optimal encounter rate (0.5/m) to capture their daily ration; (d) time taken to swim this distance, assuming no stops

(a) mm	(b) *Oikopleura* per day	(c) metres	(d) minutes
5.0 - 5.9	7.5	15	15.3
6.0 - 6.9	30	60	51.7
7.0 - 7.9	47	94	70.1

The most intriguing result of this study is, of course, the decline in feeding at high encounter rates in plaice larvae. This suggests the possibility that fish larvae may be adapted to exploit different relative food concentrations. Plaice and sand-eel larvae occurred together for a period of nearly a month in the Southern Bight in 1968, and since both fed on *Oikopleura* it might be concluded that the two species were competing with one another. In fact, beginning with a high density of prey, the sand-eels could, by their feeding activity, reduce the prey population to a level at which plaice were better able to exploit it. Conservatively, the plaice are able to survive the competition provided by the sand-eels by virtue of their adaptation, and in the absence of that competition, would be at a greater disadvantage. For this process to be effective in numerical terms, sand-eels would need to be more abundant than plaice, as in fact they were. Thus, what seemed at first to make physiological nonsense might make very good ecological sense. To establish whether or not such a phenomenon is real will require a similar study of other systems in which two or more predators feed on the same prey. It is tempting to build a theoretical model of such a system, but this approach is perhaps not warranted until further information is available.

The link, if any, between condition and survival remains unknown. Fig. 2 shows that there can be a significant range in the condidition (as defined in this paper) of larvae captured at different stations. Up to the present no attempt has been made to measure variations in the mortality rates of fish larvae in a single spawning. The best estimates of larval mortality rates so far obtained are for the same larval plaice populations studied here, and give mean values for the whole Southern Bight (Bannister, Harding and Lookwood, this Symposium). The results of Vilela and Zijlstra (1971) have already been referred to.

The data presented on dab, flounder and cod indicate that not all fish larvae are as specialized in their feeding habits as plaice larvae. Each of these three species feeds on several items at any given stage, so that the problems of describing their ecological roles are that much more difficult. They are also, in the Southern Bight at any rate, less abundant than plaice and sand-eels, so that the numbers of larvae available for study are restricted.

SUMMARY

The feeding rates of plaice and sand-eel larvae in the Southern Bight have been determined, and are compared with the distribution of their prey, *Oikopleura dioica*. The effect on the plaice larvae of finding themselves in different concentrations of *Oikopleura* is assessed using a condition index which has been studied experimentally. Some data for 1963 is discussed, a year during which *Oikopleura* was very scarce in this region. Data on the feeding of dab, flounder, and cod larvae are also presented.

REFERENCES

Blaxter, J.H.S., 1971. Feeding and condition of Clyde herring larvae. Rapp. P.-v. Réun. Cons. perm. int. Explor. Mer, 160, 128-136.
Hardy, A.C., 1924. The herring in relation to its animate environment. Part 1. Fish. Invest., Lond., Ser. 2, 7 (3), 1-53.
Hempel, G. and Blaxter, J.H.S., 1963. On the condition of herring larvae. Rapp. P.-v. Réun. Cons. perm. int. Explor. Mer, 154, 35-40.
Hjort, J., 1914. Fluctuations in the great fisheries of northern Europe viewed in the light of biological research. Rapp. P.-v. Réun. Cons. perm. int. Explor. Mer, 20, 1-288.
Ryland, J.S., 1964. The feeding of plaice and sand-eel larvae in the southern North Sea. J. mar. biol. Ass. U.K., 44, 343-364.
Shelbourne, J.E., 1957. The feeding and condition of plaice larvae in good and bad plankton patches. J. mar. biol. Ass. U.K., 36, 539-552.
Shelbourne, J.E., 1962. A predator-prey size relationship for plaice larvae feeding on *Oikopleura*. J. mar. biol. Ass. U.K., 42, 243-252.
Vilela, M.H. and Zijlstra, J.J., 1971. On the condition of herring larvae in the central and southern North Sea. Rapp. P.-v. Réun. Cons. perm. int. Explor. Mer, 160, 137-141.
Wyatt, T., 1971. Production dynamics of *Oikopleura dioica* in the southern North Sea, and the role of fish larvae which prey on them. Thalassia jugosl., 435-444.
Wyatt, T., 1972. Some effects of food density on the growth and behaviour of plaice larvae. Mar. Biol., 14, 210-216.

T. Wyatt
Ministry of Agriculture, Fisheries and Food
Fisheries Laboratory
Lowestoft / GREAT BRITAIN

Vertical Migrations of Larvae of the Atlanto-Scandian Herring (*Clupea harengus* L.)

A.S. Seliverstov

INTRODUCTION

Observations on the vertical distribution of herring larvae given in papers by different authors are rather contradictory. Bridger (1958) found that in the southern North Sea in the daytime when the sky was cloudy a greater number of larvae was collected near the surface and in pelagic layers than near the bottom. In sunny weather the number of larvae was greater in midwater than near the surface and a small number was found near the bottom. At night catches taken at all depths were equal. Investigations by Wood (1971) showed that the vertical distribution of herring larvae to the west of Scotland generally resembled that found by Bridger (1958). In his opinion, the maximum depth of the distribution depended completely on illumination conditions. The distribution of larvae in the daytime agreed with that observed by Woodhead and Woodhead (1955) and at night it resembled the distribution of the other planktonic organisms, as described by Cushing (1951): at dusk they migrated to the surface from the daytime depth and at night they moved to greater depths again.

According to Wood's data, the average length of larvae from catches taken near the surface in the daytime was considerably larger than that at all other depths: according to Bridger (1958), larger larvae were taken in the near-bottom layers. In the area of spawning grounds off the Faroes (Yudanov, 1962) and in the Norwegian shallow water area (Seliverstov and Penin, 1969) average lengths of larvae considerably increase from the bottom to the surface.

Some authors (Bridger, 1956; Tibbo and Legare, 1960; Wiborg, 1961) report that daytime catches of larvae are considerably smaller than those taken at night, but Yudanov (1962) found that the daytime catches of larvae are often large.

All of the above information indicates that the vertical distribution of herring larvae is insufficiently studied. Wood (1971) concluded that it is necessary to carry out investigations on the vertical distribution of herring larvae in all areas under different conditions of illumination; in the areas which are difficult to study it is necessary to sample at stations over a long time period and to collect more representative samples for measurements of the length composition.

The present paper analyzes data from ichthyoplanktonic surveys carried out in 1960-1970 by the research vessels of the Polar Institute along the Norwegian coast, from the Stad Cape to the Vesteralen Islands. In addition, in 1966-1970 an ichthyoplankton station lasting 31 days on the Bogrunden, Frøya, Halten, and Sklinna banks was used to study the distribution of larvae depending on their length-age composition and on time.

METHODS

Annual investigations of spawning grounds were conducted according to standard methods (Yudanov, 1962). On diurnal stations the standard nets were fished every 1.5 - 2.0 h with the ship anchored. The nets were set from the surface to the bottom at 25 m intervals. They were not moved with respect to the ship but collected ichthyoplankton brought by the current. The Juday net was also used and the length composition of the catches it took was compared with that of catches taken by trap nets. We did not allow for larvae from other depths entering the trap net. Only stations where larvae were gathered during 12 series of diurnal observations are included in this paper.

Herring spawning grounds were located by finding larvae less than 24 h old near the bottom as described by Seliverstov (1970, 1971) and Seliverstov and Penin (1969). Temperatures on the spawning grounds were also measured. Larvae in each sample were counted, measured and, when necessary, aged. To determine the age, standard samples and tables compiled from a rearing experiment were used. This experiment was conducted on board the R/V "Akademik Knipovich" in the area of spawning grounds of Norwegian herring in 1969. The conditions of the experiment, temperature, salinity and feeding, were close to natural conditions. Tables were drawn up to compare the development of pigmentation, jaws and gills, and also various length parameters of larvae of known age reared in artificial conditions, with similar characters of larvae collected on the spawning grounds.

Observations on larval behaviour in the aquarium made it possible to clarify their response to light during the first 25 days of the post-embryonic development and to compare these data with those of observations on diurnal stations.

RESULTS

On 1 and 2 April 1967, at a diurnal station on the Sklinna Bank, larvae of 7-10 mm in length (mean 8.25 mm) remained during 24 h in the near-bottom layer at a depth of over 100 m. At depths of 25 and 50 m only 2 larvae were gathered (Fig. 1). All the larvae collected on this station were with yolk sacs; in more than 50 of larvae these were still large.

At the diurnal station on the Bogrunden Bank on 7 April 1967, larvae (mean length 9.49 mm) were distributed from the surface to the bottom, but a greater part of larvae were deeper than 50 m. More than 40% of larvae had small yolk sacs. The diurnal vertical migration was poorly pronounced (Fig. 1).

It seems that smaller larvae remain in the near-bottom layer and do not perform vertical migrations. Under certain conditions, however, they can be observed in mid-water. For instance, on 15 and 16 April 1968, on the Halten Bank in 200 m, the main mass of larvae of 6-10 mm in length (mean 7.28 mm) remained at a depth of 100 m (Fig. 1). This was probably due to the fact that the herring spawn on a slope of the Halten Bank, at a depth of 100-150 m. Small larvae carried away by the current from the spawning grounds thus find themselves in mid-water.

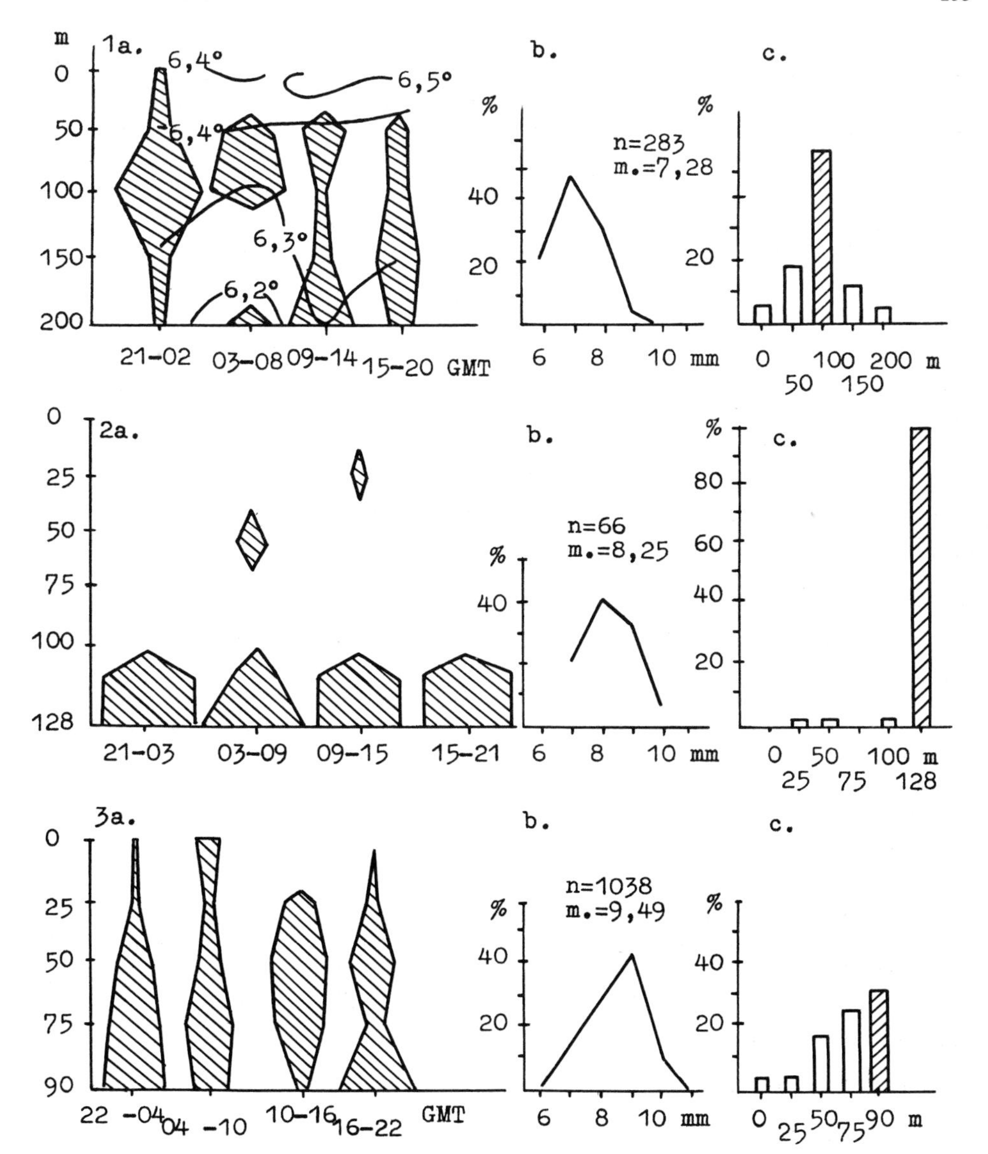

Fig. 1. Vertical distribution of herring larvae of different length. 1. Diurnal station on the Halten Bank on 15 and 16 April 1968. 2. Diurnal station on the Sklinna Bank on 1 and 2 April 1967. 3. Diurnal station on the Bogrunden Bank on 7 April 1967. a) Diurnal vertical migration; depth in metres on ordinates, time in GMT on abscissae. b) Length composition. c) Number of larvae at different depths (in %)

At the diurnal ichthyoplankton station on the Sklinna Bank on 12 and 13 April 1969 larvae were of 6-14 mm in length (Fig. 2). At this station the decrease of larval length with depth can be clearly seen. Near the surface, larvae (mean length 11.47 mm) appeared only at night, from 2100 h to 0200 h, when illumination decreased to 1 lux. The catch

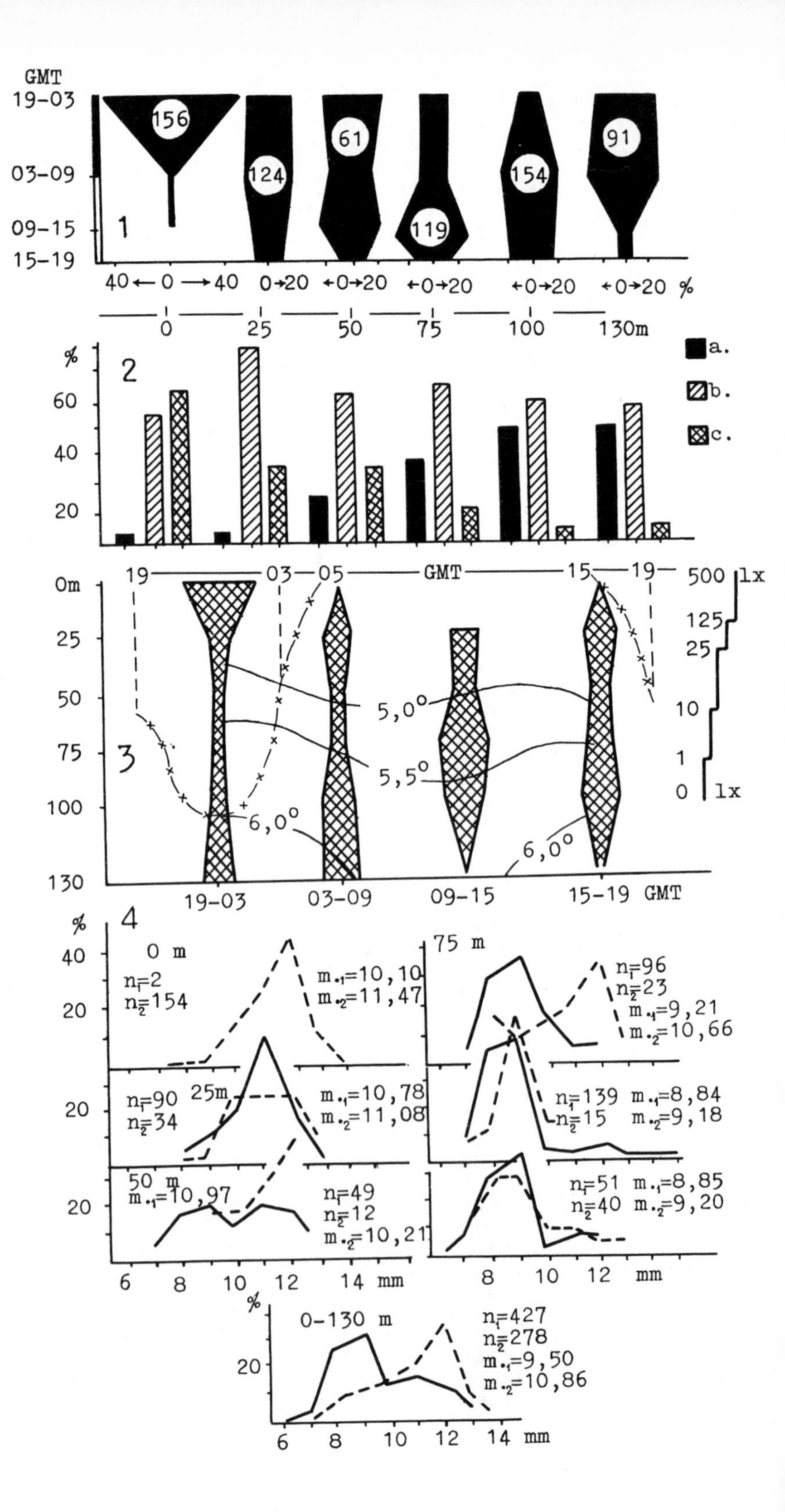
GMT
19-03
03-09
09-15
15-19
1
156
124
61
119
154
91
40 ← 0 → 40
0→20
←0→20
%
0
25
50
75
100
130m
2
%
60
40
20
a.
b.
c.
3
0m
25
50
75
100
130
19
03
05
GMT
15
19
500 lx
125
25
10
1
0 lx
5,0°
5,5°
6,0°
19-03
03-09
09-15
15-19 GMT
4
0 m
n1=2
n2=154
m.1=10,10
m.2=11,47
25m
n1=90
n2=34
m.1=10,78
m.2=11,08
50 m
m.1=10,97
n1=49
n2=12
m.2=10,21
75 m
n1=96
n2=23
m.1=9,21
m.2=10,66
n1=139 m.1=8,84
n2=15 m.2=9,18
n1=51 m.1=8,85
n2=40 m.2=9,20
6
8
10
12
14 mm
0-130 m
n1=427
n2=278
m.1=9,50
m.2=10,86

taken at night was 77 times larger than that taken in the daytime. As the illumination at the sea surface increased, herring larvae started to migrate deeper. Daytime catches taken at depths of 25-50 m were larger than night ones.[1] When the illumination was over 500 lux, no larvae were found in the surface layers. The upper boundary of larval distribution from 0900 h to 1500 h occurred at the depth of 25 m. By 1500 h the first larvae appeared in the surface layers. A great number of large larvae were concentrated from 0300 to 1900 h at a depth of 75 m. At depths of 100-130 m larvae less than 10 mm in length predominated but, even here there was a difference between the length composition of larvae from catches taken by day and by night. A temperature difference of 1.0 deg C observed on the station between surface and bottom did not prevent the vertical migration.

The average length of larvae collected at a diurnal station made on the Frøya Bank on 17 and 18 April 1968 was 11.33 mm. The night catch taken in the surface layer was only twice that taken in the daytime. At depths of 25 m and deeper daytime catches were larger than night catches. This indicates that the diurnal vertical migration covered all the water layers. At the surface the greatest aggregation of larvae was found from 2100 h to 0900 h. By 1500 h larvae almost disappeared from surface layers. From 1500 h to 2100 h a greater quantity of larvae was found at depths over 50 m (Fig. 3). Observations on the behaviour of larvae in the aquarium showed that during the first 12 h they do not respond to light or else have a weak negative phototaxis. 48 h after hatching most larvae are able to remain in mid-water and possess a strong positive phototaxis. In natural conditions herring larvae presumably move into the pelagic layers during the second day. The positive phototaxis weakens after 3-5 days so that not more than 30% of larvae concentrate in the illuminated parts of the aquarium. A positive phototaxis at the age of 1-5 days is an adaptation providing for a movement towards the surface shortly after hatching. Comparison of the size composition of larvae collected on the spawning grounds with that of the larvae reared in artificial conditions shows that in natural environmental conditions larvae at the age of 5 days predominate at depths of 50-100 m, i.e. over 3-3.5 days they migrate 50 m upwards if the spawning depth is 100-150 m.

Larvae at the age of 5-7 days respond only weakly to light. By day 8-day-old larvae show a negative response to artificial or natural light, concentrating in parts of the aquarium where the light is less intense or sinking to the bottom. At night the response of larvae to the light is weakly positive. At this age the average length is 9.40 mm

[1]From here on the total yield taken in the daytime and at night is given and not the catch per unit effort.

Fig. 2. Vertical distribution of herring larvae at a diurnal station made on the Sklinna Bank on 12 and 13 April 1969. 1. Variation in quantity of larvae at depths of 0, 25, 50, 75, 100 and 130 m (near the bottom) during 24 h (in %). Circled figures indicate the number of larvae collected. 2. Length composition of larvae (in %) at depths given above (1.). a) 6-8 mm; b) 9-11 mm; c) 12-14 mm. 3. Diurnal vertical migration, water temperature (continuous line), illumination of sea surface in lux (broken line). 4. Length composition of larvae collected in the daytime and at night at depths of 0, 25, 50, 75, 100 and 130 m. Day hauls - continuous line, n_1 (number) and m_1 (mean length); night hauls - broken line, n_2 (number) and m_2 (mean length)

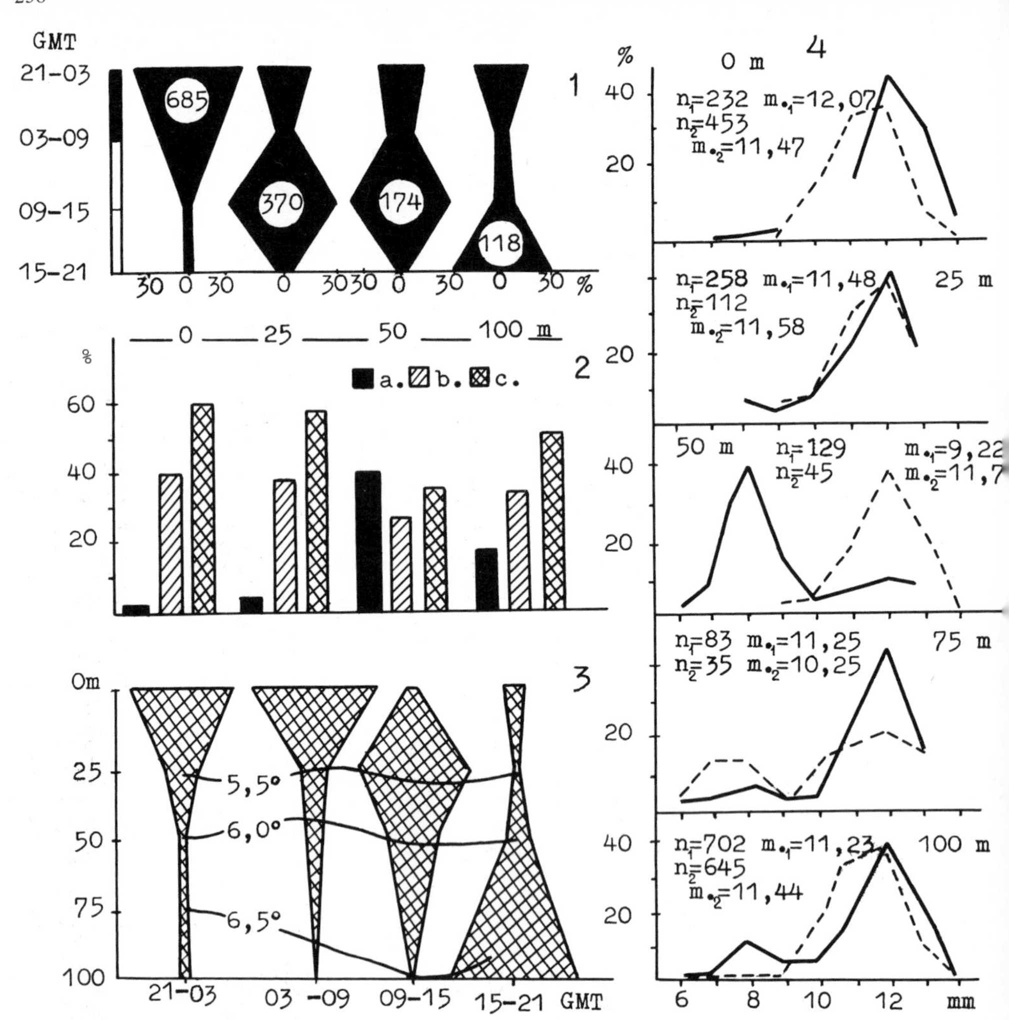

Fig. 3. Vertical distribution of herring larvae made at a diurnal stati on the Frøya Bank on 17 and 18 April 1969. The legend is given in Fig.

(range 8.80 - 10.80 mm). It is at this length that larvae occur in the upper 50 m layer in great quantities.

According to Soleim (1942), the main receptor for catching food is sight. Gusev (pers. comm.) showed that the intensity of feeding of larvae decreases with depth: at 0 m - 36.5% feeding, 50 m - 18.2%, 100 m - 7.3%. Analyzing feeding of larvae in the Lofoten area in 1966, Rudakova (1971) states that an average of 25% of larvae were feeding in the daytime, 11.9% in the early morning and only 3.2% at night. It seems, therefore, that with decreasing illumination the intensity of larval feeding decreases.

To illustrate the vertical distribution of larvae more clearly, data from standard stations in the Norwegian shallow water area were select which are comparable in terms of the geographical position, depth,

Fig. 4. Depth distribution and vertical migrations of herring larvae of different length from standard stations in Norwegian shallow water areas. Ordinate gives depth in metres, abscissae time in GMT. % length compositions also given. n = number sampled; m = mean length in mm

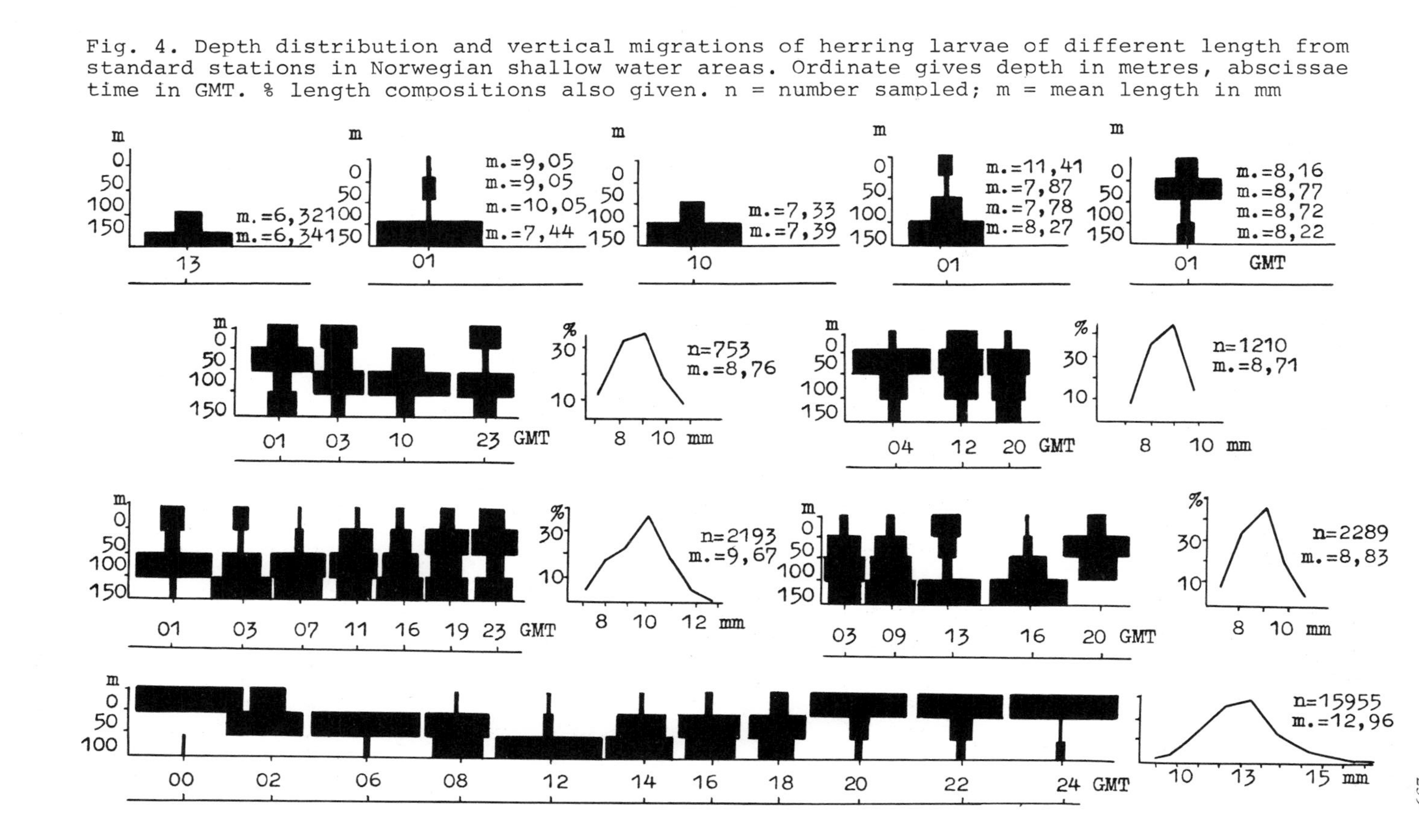

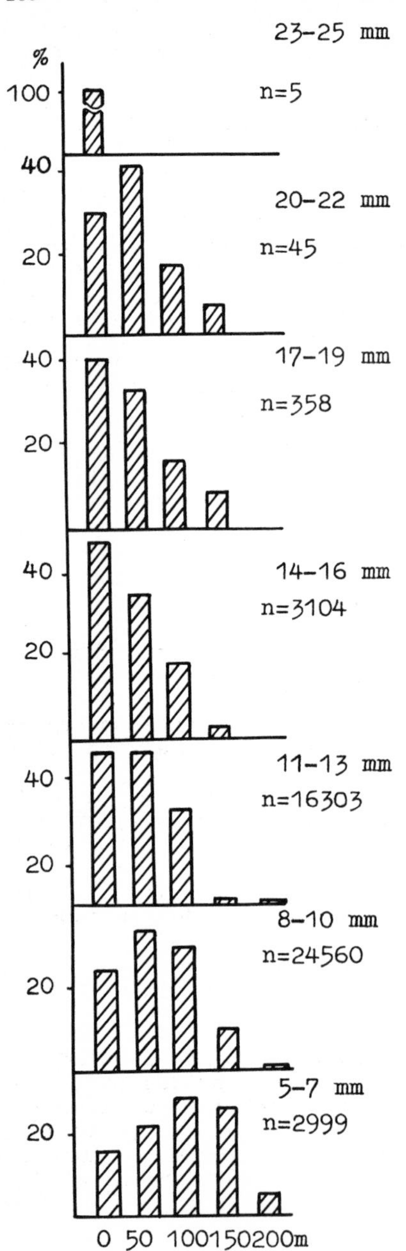

Fig. 5. Distribution of herring larvae by depths depending on their length in the Norwegian shallow water area, 1960-1970. n = number sampled

weather conditions, time, and size-composition of larvae (Fig. 4). This made it possible to determine the depth distribution of larvae of different length. Larvae of 6.63 - 7.44 mm (Fig. 4) stay in the near-bottom layer in the daytime and at night. Evidently, only when larvae reach a length of 7.5 mm do they begin to move to the upper layers of the sea. Stations at which larvae had an average length of 8-10 mm demonstrate an incipient diurnal vertical migration. This migration is not fully formed because by day the larvae still occur in the surface layers. An intensive migration to the upper layers is only typical of larvae at this range of length (Fig. 4).

A vertical migration between surface and bottom is distinctly observed when larvae are 9-17 mm in length (Fig. 4). It was also found out that an increase in numbers of larger larvae between the bottom and the surface is a constant occurrence which is observed in years where hydrometeorological conditions and efficiency of spawning are different (Fig. 5). That is why in March-April, the average length of larvae collected in the spawning area by Norwegian investigators at depths of 0-70 m was larger (Dragesund and Wiborg, 1963; Dragesund and Hognestad, 1966; Dragesund, 1965) than that from surveys conducted by the Polar Institute with the use of ichthyoplanktonic nets (up to the depth of 200 m). The results of observations on diurnal migrations of herring larvae are in agreement to some extent with those of Dragesund (1970) who also found increasing numbers of larvae in the surface layers at night. Our results are, however, greatly in contrast to observations by Bridger (1958) and Wood (1971).

SUMMARY

1. The vertical distribution of herring larvae was studied from data collected at diurnal and standard ichthyoplanktonic stations made over 24 h in the Norwegian shallow water area in 1960-1970.

2. On the second day after hatching, before passing to external feeding, herring larvae make a characteristic migration to the upper 50 m layer.

3. Herring larvae start their regular diurnal migrations only after passing to external feeding, at the age of 6-9 days, the length being 8.8 - 10.8 mm.

4. The amplitude of diurnal vertical migrations of larvae can reach 75-100 m.

5. A temperature difference of 1.0 - 1.5 deg C from bottom to surface does not prevent larvae from vertical migration.

6. A positive phototaxis of larvae at the beginning of feeding has evidently appeared in the process of evolution and provides for migration to the upper, illuminated layers which are rich in zooplankton.

7. An increase in numbers of larger larvae from the bottom to the sea surface occurs in years different both in hydro-meteorological conditions and efficiency of spawning.

REFERENCES

Bridger, J.P., 1956. On day and night variations in catches of fish larvae. J. Cons. perm. int. Explor. Mer, 22, 42-57.
Bridger, J.P., 1958. On efficiency tests made with a modified Gulf III High-Speed Tow Net. J. Cons. perm. int. Explor. Mer, 23, 357-365.
Cushing, D.H., 1951. The vertical migration of planktonic crustacea. Biol. Rev., 26, 158-192.
Dragesund, O., 1965. Forekomst av egg og yngel av fisk i vest og nord norske kyst og bankfarvann våren 1964. Fiskets Gang, Nr. 11
Dragesund, O., 1970. Factors influencing year-class strength of Norwegian spring spawning herring. FiskDir. Skr. Ser. Havundersøkelser, 15, 381-446.
Dragesund, O. and Hognestad, P.T., 1966. Forekomst av egg og yngel av fisk i vest og nordnorske kyst og bankfarvann våren 1965. Fiskets Gang, Nr. 24.
Dragesund, O. and Wiborg, K.F., 1963. Forekomst av egg og yngel av fisk i vest og nordnorske kyst og bankfarvann våren 1963. Fiskets Gang, Nr. 41.
Rudakova, V.A., 1971. On feeding of young larvae of the Atlanto-Scandian herring (*Clupea harengus harengus* L.) in the Norwegian Sea. Rapp. P.-v. Réun. Cons. perm. int. Explor. Mer, 160, 114-120.
Seliverstov, A.S., 1970. Velocity and direction of herring larvae drift in the area of the Norwegian Shallows in March-April 1968. Materialy rybokhozyaistvennych issledovanii Severnogo basseina, 16, 240-253.
Seliverstov, A.S., 1971. Soviet investigations on the Atlanto-Scandian herring in the Norwegian Sea in 1970. ICES, Doc. CM 1971/12.
Seliverstov, A.S. and Penin, V.V., 1969. On the velocity of the herring larvae drift on spawning grounds in the West Skandia shelf area. Trudy PINRO, 25, 64-90.
Soleim, P., 1940. Sildelarvene pa varsildfeltet og havet 1940. FiskDir. Skr. Ser. Havundersøkelser, 6, 39-55.
Tibbo, S.N. and Legare, J.E., 1960. Further study of larval herring (*Clupea harengus* L.) in Bay of Fundy and Gulf of Maine. J. Fish. Res. Bd Can., 17, 933-942.
Wiborg, K.F., 1961. Forekomst av egg og yngel av fisk i vest-og nordnorske kyst-og bank farvann våren 1960. Fiskets Gang, Nr. 9.
Wood, R.J., 1971. Some observations on the vertical distribution of herring larvae. Rapp. P.-v. Réun. Cons. perm. int. Explor. Mer, 160, 60-64.
Woodhead, P.M.J. and Woodhead, A.D., 1955. Reactions of herring larvae to light: a mechanism of vertical migration. Nature, Lond., 176, 349-350.
Yudanov, I.G., 1962. Investigations concerning spawning grounds of the Atlantic-Scandinavian herring. Trudy PINRO, 14, 5-49.

A.S. Seliverstov
Polar Research Institute
6 Knipovitch Street
Murmanks / USSR

Seasonal Variation of Ichthyoplankton in the Arabian Sea in Relation to Monsoons

K. J. Peter

SYNOPSIS

The seasonal variation of ichthyoplankton of the Arabian Sea based on the 694 zooplankton samples collected during the International Indian Ocean Expedition has been discussed in the light of the hydrographical features caused by monsoons. The pattern of distribution of total fish eggs showed that the highest degree of concentration was at the northeastern and southwestern parts of Socotra Island, areas off Kutch and Gujarat, Kerala coast, south of Ceylon and Chagos archipelago. The lowest concentration was noticed in the central part of the Arabian Sea. During the southwest monsoon season there was a slight reduction in these areas of high concentration. During the northeast monsoon, however, the northeastern half of the Arabian coast was changed into an area of lowest density. The areas of high concentration along the west coast of India were retained without much change. In the case of total fish larvae the highest density was noticed in the middle of the Red Sea, also near Aden, around Socotra Island, the Arabian and southern Somali coasts, off Kutch and Bombay, along the Malabar coast, south of southern India and Ceylon, in addition to a few places along the equatorial region. The lowest density was noticed in the central part of the Arabian Sea. During the southwest monsoon season, changes in the pattern of distribution were noticed only along the eastern part of the Arabian Sea. The concentrations along the coastal areas moved to the offshore waters. The central Arabian Sea remained a low production zone during this season without any change.

Though fish larvae were represented in 92% of the collections, their composition showed that only 17% belonged to economically important groups. The percentage of eggs collected during the southwest monsoon season was 59 and in the northeast monsoon season 41, which in the case of fish larvae was 57% and 43%, respectively. The higher salinity, temperature, and oxygen content at the surface layers were also found to be greater in the northeast monsoon season. June, July, and August were found to be the best season for eggs and larvae, the peak being in July. December, January, and February were the poor months for eggs, and April and September for larvae.

Arabian Sea surface water has high salinity, high temperature, and a steep gradient of dissolved oxygen content falling into very low values at 100-150 m/depth. The influx of high-saline water from the Persian Gulf and Red Sea, the intense evaporation, fluctuations in surface temperature and the seasonal shifting of thermocline create very complex conditions. During the southwest monsoon the northerly currents along the western half of the Arabian Sea, and the southerly currents along the eastern half, constitute a clockwise circulation, initiating heavy upwelling at the coastal areas, thus resulting in the higher production of organic matter. But in the central part of the Arabian Sea a cyclonic gyre is developed, and the less mixing of water here accounts for the low production. On the other hand, during the northeast monsoon season a complete reversal of surface currents takes place which in general is rather feeble. Hence this does not effect a thorough mixing of water, because of its lower intensity and shorter duration. The areas of high abundance for eggs and larvae coincide with areas of upwelling or places that are under the influence of divergences.

K.J. Peter
Indian Ocean Biological Centre
National Institute of Oceanography
P.O. Box 1913
Cochin - 682018 / INDIA

Feeding and Metabolism

Food and Feeding of Larval Redfish in the Gulf of Maine

R. R. Marak

INTRODUCTION

The availability of suitable food is one of the important factors influencing the survival of larval fish. Knowledge of their food habits is important if we are to understand their role in the economy of the sea. Recent work by Einarsson (1960) and Bainbridge (1964) on the food of larval redfish (*Sebastes marinus*), Kelly and Barker (1961) on the vertical distribution of redfish larvae (*S. marinus*) and Bainbridge and Cooper (1971) on *Sebastes* populations, has contributed considerably to our knowledge of the early life history of this species. Magnusson et al. (1965), working in the Irminger Sea, found a good correlation between the standing stock of zooplankton and the number of redfish larvae. Little, however, is known of the feeding niche occupied by larval redfish in the plankton community in the Gulf of Maine.

Data from stomach analysis of 402 specimens collected from the Gulf of Maine ranging from 9-48 mm are presented in this paper.

METHODS

Sampling. The larvae were collected in the southwestern part of the Gulf of Maine in late July and early September 1957 on cruises of the

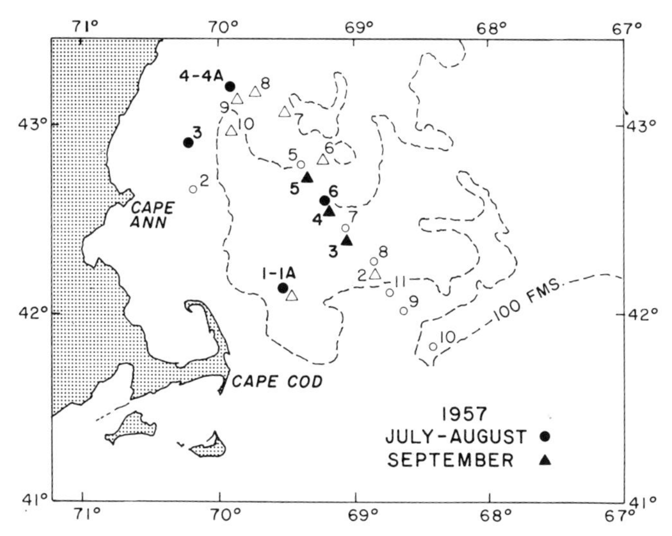

Fig. 1. Mid-water trawl stations occupied in the Gulf of Maine 1957 (solid symbols indicate stations used in analysis)

Table 1. Percent composition of food of larval redfish. Albatross III, Cruise 99, July 1957, Stations 1, 1A, 3, 4, 4A, 6. Albatross III, Cruise 102, Sept. 1957, Stations 3, 4, 5

	Length groups of larvae (mm.)																	
	9-13		14-18		19-23		24-28		29-33		34-38		39-43		44-48		Total	
Cruise No.	99	102	99	102	99	102	99	102	99	102	99	102	99	102	99	102	99	102
#Larvae examined	20	0	69	0	54	1	68	5	53	30	10	56	0	33	0	3	274	128
Invertebrate eggs	6		5		1	10	<1	0	2	5	0	5		6		0	2	4
Larval copepods	32		26		13	30	3	7	1	<1	<1	<1		2		0	12	7
O. similis	20		19		14	50	5	5	1	12	<1	6		4		2	10	13
Fish eggs	6		2		2	0	9	0	3	0	0	0		0		0	4	0
P. minutus	8		5		4	0	11	5	<1	24	<1	36		32		10	5	18
C. typicus	4		8		22	10	40	30	38	38	41	30		30		10	26	25
M. lucens	<1		<1		<1	0	0	0	<1	6	0	2		4		0	<1	2
C. finmarchicus	6		17		20	0	16	0	43	0	55	<1		1		0	26	<1
Larval cestodes	0		0		0	0	0	0	0	20	0	34		58		0	0	34
T. inermis	0		0		0	0	0	0	0	0	0	0		3		75	0	13
Disgested matter	17		18		23	0	14	52	11	14	3	20		15		3	14	17
% empty	5		6		11	0	7	0	11	20	10	14		9		33	8	13
% 1/4 full	25		14		22	0	31	60	26	50	20	27		43		0	23	37
% 1/2 full	35		13		22	0	41	20	36	10	0	27		15		0	25	19
% 3/4 full	25		32		24	0	13	20	21	7	20	21		15		0	23	16
Full	10		35		21	100	8	0	6	13	50	11		18		67	21	15

RV Albatross III. The larvae were caught in a 3 m Isaacs-Kidd mid-water trawl (Isaacs and Kidd, 1953) that was rigged with a Leavitt opening and closing device. The entire net was lined with 9 mm cotton netting. One hour shallow-water tows were made at depths of 10, 20, 30, and 40 m, the series being repeated if large numbers of redfish larvae were taken. Some tows were also made in deeper waters of 60, 80, 100, and 110 m. All larval fish collected were preserved in 10% formalin. Information on gear, station, procedure, and catch are detailed in Kelly and Barker (1961). Station locations are shown in Fig. 1.

Laboratory Methods. Each specimen was measured, the entire digestive tract removed, and an incision was made down the dorso-median section, exposing the food. Initial examination for the degree of fullness was estimated at low magnification (10x) in 5 categories, empty, 1/4, 1/2, and 3/4, full. Sorting and identification of food organisms was done at 45x; the stomach contents of the smallest larvae (9-13 mm) were examined under 100x. Stomach contents were identified, and an estimation of the percentage volume of the different food organisms in each specimen was made.

For clarity of presentation only the major food items (over 98% by volume) are listed in Tables 1 to 3 in order of increasing size. All percentages listed are an unweighted average of individual stomach percent compositions.

Table 2. Percent composition of larval redfish food in an area of high and low larval abundance, July and September 1957

	July		September	
	High	Low	High	Low
	Sta. 6	Sta. 3,4,4A	Sta. 5	Sta. 4
# larvae examined	44	50	25	17
Invertebrate eggs	<1	<1	23	1
Larval copepods	12	27	8	2
O. similis	<1	3	12	18
P. minutus	<1	44	9	16
C. typicus	56	2	42	12
M. lucens	<1	0	0	6
C. finmarchicus	8	10	0	<1
Larval cestodes	0	0	44	0
T. inermis	0	0	0	22
Digested matter	23	11	6	22
% empty	0	2	4	0
% 1/4 full	16	4	0	47
% 1/2 full	36	12	12	35
% 3/4 full	25	52	32	6
Full	23	30	52	12

Table 3. Percent composition of major food items of larval redfish by depth and time of day, July and September 1957

	10 - 20 m				30 - 40 m				60 - 110 m				Total			
	July		September		July		September		July		September		July		September	
	Day	Night	Day	Night	Day	Night	Day	Night	Day	Night	Day	Night	Day	Night	Day	Night
Station No.	1A	1	3,4,	3	1	1	4	3	1	1	3	3				
Tow No.	19	8	2	8,9	3,4	21	3,4	4,11	5	11	12	7				
# Larvae examined	24	91	18	9	23	20	3	37	5	17	16	20	52	128	37	66
Invertebrate eggs	<1	3			11	1			5	0			5	1		
Larval copepods	35	1	5	0	2	8	3	3	7	0	0	0	15	3	2	<1
O. similis	15	0	18	0	13	32	27	0	12	0	0	0	13	10	15	0
Fish Eggs	<1	<1			26	0			0	0			9	<1		
P. minutus			28	29			4	35			55	17			29	27
C. typicus	8	49	15	38	7	8	24	38	2	65	41	22	6	41	27	33
M. lucens			8	0			15	2			0	<1			7	1
C. finmarchicus	38	17			17	40			37	31			30	30		
T. inermis			15	0			0	3			0	0			5	1
Larval cestodes	0	0	0	0	0	0	0	14	0	0	0	55	0	0	0	23
Digested matter	3	27	10	33	24	11	24	15	37	4	4	59	21	14	13	36
Empty	0	7	0	0	17	10	0	16	40	47	69	0	12	13	30	9
1/4% full	25	30	38	89	17	35	67	30	40	53	31	70	23	14	38	50
1/2% full	37	34	39	11	35	20	33	19	20	0	0	25	35	27	22	20
3/4% full	13	13	11	0	22	25	0	24	0	0	0	5	15	13	5	15
Full	25	16	11	0	9	10	0	11	0	0	0	0	15	13	5	6

RESULTS

Food Composition. Larval redfish were feeding mainly on 4 groups of organisms, copepods, euphausiids, fish eggs, and invertebrate eggs. Larval copepods, *Oithona similis*, *Pseudocalanus minutus*, and invertebrate eggs made up over half of the food eaten by larvae in the 9-13 and 14-18 mm size groups in July. *Centropages typicus* and *Calanus finmarchicus* formed the bulk of the food in the larger sizes (Table 1). *C. finmarchicus* was almost completely absent in larval diets in September. *C. typicus* and *P. minutus* were the principal food of larvae up to 44 mm. The largest size group (44-48 mm) fed mainly on the euphausiid, *Thysanoessa inermis* (Table 1). Phytoplankton was not present in any of the digestive tracts examined. Although crab zooea were abundant in the few plankton tows made, they had not been eaten by larvae of any size.

One of the most important factors in the feeding habits of larval fish is the size of available prey. The sizes of the food organisms eaten by the larval redfish are given in Fig. 2. The average length of food organism was plotted for each larval size group for July and September (see Fig. 2). As larval length increases, there is generally an increase in the size of the food eaten.

The percentage of empty stomachs and degree of fullness varied less in July than in September for all size groups of larvae.

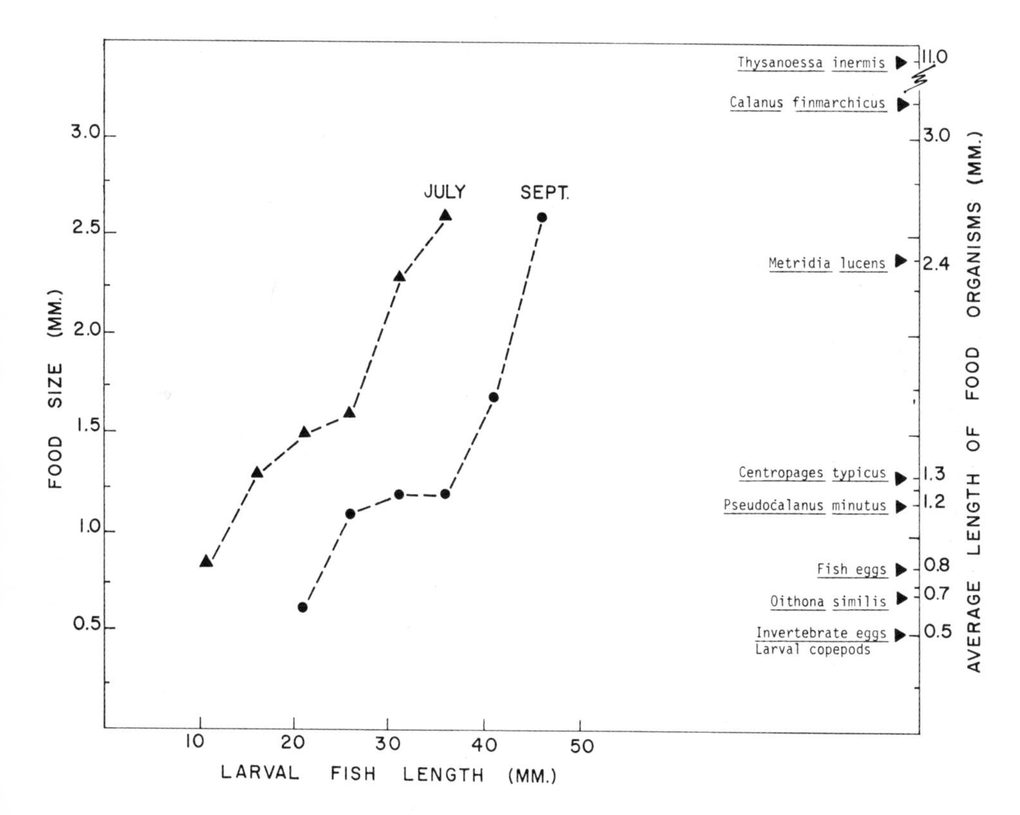

Fig. 2. Relation of food size to larval size and average length of food organisms

Food Differences between Areas. The food of larvae from an area with high larval concentrations and one about 72 km distant with low larval numbers was compared for both months. Although there was no significant difference in the degree of fullness of larvae between these 2 areas, there was a difference in food composition. The dominant food item in July in the area of high larval concentration was *C. typicus* and in the area containing low larval numbers was *P. minutus*. *C. typicus* remained the dominant prey species in the area of high larval concentration in September. In contrast, no single food was predominant in the larvae taken in the area of low larval concentration. The major contributors were: *T. inermis*, *O. similis*, *P. minutus*, and *C. typicus*. Cestode larvae were found in 44% of the digestive tracts in September but none in July in these areas (Table 2).

Food Composition in Relation to Depth and Time. *C. finmarchicus* and *C. typicus* were the major food items of larvae taken at all depths in July with *O. similis* and larval copepods being eaten in moderate amounts (Table 3). *C. typicus* and *P. minutus* formed the bulk of the food (75%) of larvae from all levels in September (Table 3); these organisms are the dominant zooplankters in adjacent Gulf of Maine Coastal waters in late summer and early autumn (Sherman, 1970).

The size range of larvae in July (9-38 mm) and September (19-48 mm) was quite uniform from surface to 110 m. Although the temperature structure was similar for both months, the depths where larvae were most abundant were different (20 m in July, 80-110 m in September).

C. typicus was more prevalent in stomachs of larvae taken at night (1800-0600) than the day at almost all depths in both months. Larvae containing *O. similis* were taken mostly during the day (0600-1800) in each month. The day-night variations at all levels for *C. finmarchicus* in July and *P. minutus* in September were quite similar (Table 3).

The percent of empty stomachs increased with depth in both months. There was very little day-night variation in empty stomachs by depth in July but there were differences in September at the 30-40 and 60-110 m depths. None of the larvae from the 60-110 m level had full stomachs in either month.

DISCUSSION

Larval redfish appear to select food of a certain size range varying with the size of the larvae. Larval copepods, *O. similis* and invertebrate eggs, the principal food of larvae in the smaller size ranges (9-18 mm), become less important as larval size increases. Older copepodite and adult *C. typicus* and *C. finmarchicus*, at least twice the size of the aforementioned food, were the major prey of larger larvae (19-38 mm). Adult *T. inermis* (>8 mm long) were eaten only by the largest larvae (39-48 mm).

There is little evidence of selectivity of larvae among species of suitable size. However, crab zoea, even though they were of a size possible to be eaten by the larger larvae, were not ingested, most likely because of their spiny structure. The few plankton tows taken revealed large numbers of zoea present in the water.

Einarsson (1960) and Bainbridge (1964) studying larval redfish from the Irminger Sea found them to have an opportunistic mode of feeding. Cod, haddock, and pollock larvae in the Gulf of Maine were found to be primarily size-selective feeders (Marak, 1960).

Redfish larvae of similar size ate larger food in the Gulf of Maine than in the Irminger Sea. In the Gulf of Maine, 20 mm larvae fed primarily on adult copepods approximately 0.7 to 3.5 mm in length, whereas 30 mm larvae in the Irminger Sea fed mostly on copepod eggs, gastropod larvae, and some copepodite stages of *C. finmarchicus* which were 0.1 to 1.5 mm.

The feeding habits of a species thus apparently vary locally with the size of food available; fish larvae are forced to eat smaller organisms when larger, presumably more preferred, sizes are not present, thereby expending more energy which may be reflected in their early rate of growth. It would appear that the rate of growth of larval redfish in the Gulf of Maine is greater than in the Irminger Sea, based on the increased availability of larger food organisms in the Gulf of Maine.

There appeared to be an adequate supply of the appropriate quality and quantity of food available to the Gulf of Maine larvae as only 11% had empty stomachs. The same assemblage of zooplankters utilized by larval redfish support commercially important populations of larval cod, haddock, and pollock which showed the same degree of stomach fullness (Marak, 1960). Einarsson's data for the 8-14 mm size group indicated 20% of the stomachs were empty, but of the larger sizes (15-45 mm) he stated only that "...most of the stomachs contain food...".

Seasonal and areal variations in the plankton play an important role in the diets of fish larvae. In this study, the smaller redfish larvae were most abundant where apparently good quantities of larval copepods were available. Sherman and Honey (1971) found considerable variation seasonally in the food of larval herring in the coastal waters of Central Maine. Some changes did occur in the major food eaten by the redfish larvae between July and September and between areas with high and low concentrations. Variations in the dominant food items found in the stomachs were more likely due to the availability of the prey in the particular water mass at the time rather than selectivity.

The curves for the relation of food size to larval size shows an interesting feature for both months which was masked when regression lines were fitted (Fig. 2). The larvae in the 15-25 mm size range in July and the 25-35 mm group in September were feeding mainly on food items in the 1.0 - 1.5 mm size range, suggesting that food of this size was more available to the larvae than food of other size groups.

There is the possibility that the differences in densities of larvae found was related to differences in amount of available food. It appears that redfish larvae have many competitors for these food organisms, based on preliminary stomach analyses of other species in the sample. Ten other species of fish larvae in the same size range were also taken during this study, *Clupea harengus harengus*, *Enchelyopus cimbrius*, *Scomber scombrus*, *Urophycis chuss*, *Merluccius bilinearis*, *Glyptocephalus cynoglossus*, *Tautogolabrus adspersus*, *Cyclopterus lumpus*, *Lumpenus lumpretaeformis*, and *Gasterosteus aculeatus*.

Differences in the vertical distribution of the food organisms are known to exist in the Gulf of Maine (Bigelow, 1924) and are important to the feeding habits of larval redfish. Although no extensive vertical migrations were apparent from Kelly and Barker's data, the presence of *C. typicus* in stomachs at 60-110 m suggests that some of the larvae had eaten in shallower depths than where they were captured. Bigelow (1924) and Sherman (personal communication) have shown *C. typicus* to be a surface (upper 20 m) plankter. Since there is no evidence of food regurgitation due to rapid pressure decrease, the conclusion that the fish were not feeding heavily in the deeper strata is supported by the high percentage of empty stomachs and low percentage of full ones in those samples (Table 3).

The presence of full stomachs day and night indicated that these larvae may not have a diurnal feeding cycle. Although the digestion rate is not known, it seems safe to assume that larvae with full stomachs had eaten recently.

Of all the larvae taken in September (Table 1) 34% contained cestode larvae in their digestive tracts, whereas none were found in the larvae caught in July. The presence of these parasites evidently did not effect the feeding of the larvae because they were present in larvae with both full and empty stomachs. The general condition of the larvae with parasites differed in no way from uninfected ones.

SUMMARY

1. The stomachs of 402 larval redfish were examined. These larvae were taken in an Isaac-Kidd mid-water trawl in the southwestern part of the Gulf of Maine in late July and September 1957.

2. The major portion of the diet of larval redfish consisted of juveniles and adults of 4 species of copepods and larval copepods.

3. The larvae were selective as to the size of food eaten. Larvae of 9-18 mm fed mainly on larval copepods and larvae of 18 mm and larger on adult copepods.

4. The composition of food varied between stations and months, whereas the quantity of the food did not vary significantly.

5. Some larvae appeared to be feeding at depths other than those at which they were caught.

6. Cestode larvae, present in the digestive tracts of larval fish taken in September, had no obvious affect on their condition.

REFERENCES

Bainbridge, V., 1964. A preliminary study of *Sebastes* larvae in relation to the environment of the Irminger Sea. ICNAF Environ. Symp. Contr. B-3.

Bainbridge, V. and Cooper, G.A., 1971. Populations of *Sebastes* larvae in the North Atlantic. Res. Bull. int. Comm. Northw. Atlant. Fish., (8), 27-35.

Bigelow, H.B., 1924. Plankton of the offshore water of the Gulf of Maine. Bull. U.S. Bur. Fish., 40, Part 2, 509 pp.

Einarsson, H., 1960. The fry of *Sebastes* in Icelandic waters and adjacent seas. Rit Fiskideildar, 1 (7).

Kelly, G.F. and Barker, A.M., 1961. Vertical distribution of young redfish (*Sebastes marinus*) in the Gulf of Maine. ICNAF Spec. Publ. (3).

Magnusson, J., Magnusson, J. and Hallgrimsson, I., 1965. The aegir redfish larvae expedition to the Irminger Sea in May 1961. Rit Fiskideildar, 4 (7).

Marak, R.R., 1960. Food habits of larval cod, haddock, and coalfish in the Gulf of Maine and Georges Bank area. J. Cons. perm. int. Explor. Mer, 25, 147-157.

Sherman, K. Copepods and hydrography in coastal waters of the Gulf of Maine. Fish. Bull., U.S. (Ms.).

Sherman, K. and Honey, K.A., 1971. Seasonal variations in the food of larval herring in coastal waters of central Maine. Rapp. P.-v. Réun. Cons. perm. int. Explor. Mer, Vol., 160, 121-124.

R.R. Marak
U.S. Department of Commerce
NOAA - NMFS MARMAP Field Group
RR7A, Box 522A
Narragansett, R.I. 02882 / USA

Food of the Larval Anchoveta *Engraulis ringens* J.

B. R. de Mendiola

INTRODUCTION

The Peruvian fisheries have developed very quickly during the last ten years, resulting in Peru becoming the foremost fishing country in the world. The fishery is largely based on the anchoveta *Engraulis ringens* J., a pelagic fish sufficiently abundant to support catches of up to 10 x 10^6 tons, although 12,000,000 T.M. were taken in 1970.

Because of its economic importance the anchovy is the object of much of the work carried out by the Instituto del Mar del Peru, but no previous work has been done on the food of the larvae. This paper is the first one in this respect, being an examination of the gut contents of anchoveta larvae caught throughout the year.

MATERIALS AND METHODS

The larval material was obtained from plankton samples collected by Hensen nets in hauls from 50 m to the surface.

Anchoveta spawn almost throughout the year but with greater intensity toward the end of winter and end of summer. Most of the material analyzed was taken during a spring cruise in November 1967 which sampled all along the Peruvian coast and gave good catches of larvae of the requisite length. To obtain representative samples 12 stations were chosen at random in each fishing area; where the stations chosen on this cruise did not contain larvae, samples taken at these positions on other cruises were analyzed to bring the number of stations up to 12 in each area (Fig. 1). In all, 83 samples and 12,960 larvae were examined.

Following Pavlovskaia (1961) for *Engraulis encrasicholus* and Ciechomski (1967) for *Engraulis anchoita*, I have defined the larval stage as extending from the end of yolk-sac resorption until the appearance of the stomach diverticula and pyloric cecae, generally at a length of 30 mm, occasionally at 23-25 mm.

The material was classified by area and then by size in groups of 2 mm for larvae smaller than 21 mm and groups of 5 mm for those between 21 and 30 mm.

As in the procedure of Arthur (1956) and Berner (1959), the larvae were clarified and softened in 95% glycerine for 24 h and those that were empty were eliminated. Those with food contents were examined under a compound microscope to identify and count food organisms. The size of the food particles and the width of the mouth of the larvae were measured.

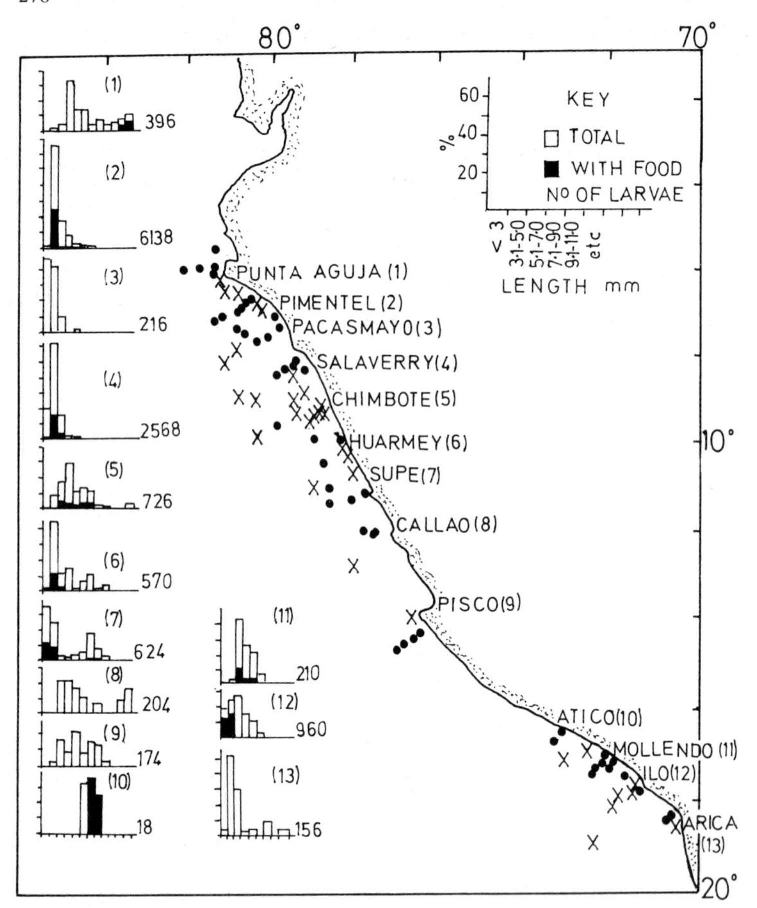

Fig. 1. Map showing stations occupied (x day, • night) and length-frequency histograms at different stations

RESULTS

Alimentary Tract. As in most Clupeidae the intestine of larval *E. ringens* is a long, straight transparent tube attached to the ventral part of the body and visible along its full length. The intestine is undifferentiated until the larva reaches about 6 mm when a light constriction appears about the middle portion, the anterior part being narrow with thin walls, the posterior wider with thicker walls.

Usually the food could be seen through the walls of the intestine but some particles, well-digested and close to the anus, could only be identified after dissection. Round particles seem to be predominant and were always found in the posterior part, indicating a quick passage through the anterior, probably aided by mucous cells (Morris, 1955). This is in conformity with observations by Soleim (1942) and

Bhattacharyya (1957) for *Clupea harengus* larvae and by Berner (1959) for those of *Engraulis mordax*.

Size of Food. To some extent the size of the food of fish larvae depends on the size of its mouth and oesophagus (Lebour, 1919). Fig. 2 shows that for anchoveta the longer the larvae are, the bigger the food particles, we also see that there is a positive correlation between the length of the larva and the size of its mouth. These observations are in accord with Arthur (1956), Berner (1959) for *Engraulis mordax*, and Ciechomski (1967) for *Engraulis anchoita*. Blaxter (1963) suggested that the variations he found in *Clupea harengus* from different areas could be due to the racial differences in jaw gape and the size of the original egg.

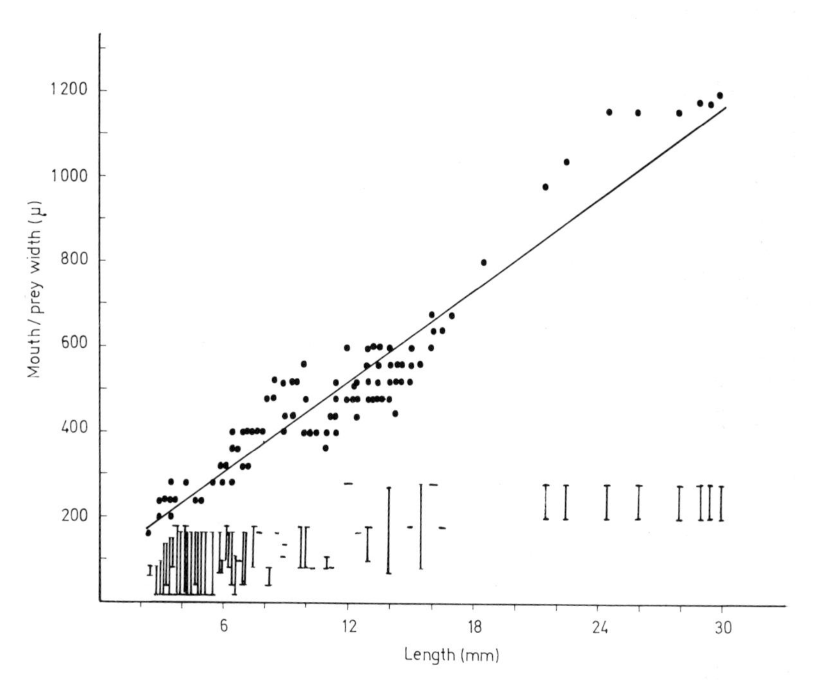

Fig. 2. Width of mouth in μ (●) and width of prey (- I) for larvae of different length. The regression line has a correlation coefficient r = 0.96

Food Composition and its Variation with Size of Larvae. A summary of the food items found in the intestines of the larvae is given in Table 1. The presence of each species stage or group is presented by a "percentage by number", calculated from the total number of organisms counted in all the larvae of a given length, and by "percentage occurrence" calculated from the number of intestines which contained the organism in question out of the total number examined. This table shows that larvae feed mainly on copepods in different stages of development, particularly as eggs and nauplii, and on ciliates and crustacean remains. Some phytoplankton was also found, but only occasionally and in small quantities.

Table 1. Items found in the gut of larval anchoveta of different sizes. %/N = percentage by number; %/occ. = percentage by occurrence; (-1) = less than 1%; (+) = present

Length in mm	<3.0		3.1 - 5.0		5.1 - 7.0		7.1 - 9.0		9.1 - 11.0	
No. total larvae	1080		6576		2280		1230		622	
No. of larvae with food	228		2088		276		84		52	
	%/No.	%occ.	%/No.	%occ.	%/No.	%occ.	%/No.	%occ.	%/No.	%occ.
Phytoplankton										
Diatoms Centricae										
Actinocyclus octonarius	1.4	3	0.8	3						
Actinocyclus ehrenbergi var. *tenella*			0.2	-1						
Actinocyclus tenuissimus			0.5	1						
Asteromphalus heptactis			0.1	-1						
Coscinodiscus decrescens	1.4	3	0.5	-1						
Planktoniella sol			0.5	2						
Roperia tessellata	1.4	3	2.2	6			5.7	7		
Thalassiosira sp.			0.2	-1	10.7	6	7.5	7		
Rest of *Chaetoceros*					+	4	+	7		
Diatoms Pennatas										
Pleurosigma sp.	1.4	3	0.5	-1						
Dinoflagellates										
Dinophysis acuminata	1.4	3								
Diplopsalis lenticula			0.8	3	0.6	2	26.4	21		
Glenodinium sp.			0.2	-1						
Peridinium conicum			0.1	-1						
Peridinium minutum			0.4	1	1.9	7				
Zooplankton										
Cilliates										
Codonellopsis pusilla	1.4	3	6.8	20	7.5	12	15.1	14		
Craterella urceolata	1.4	3								
Helicostomella longa	9.6	18	4.5	14	12.5	11				
Copepods										
Copepodids										
Calanus sp.										
Copepod eggs	53.4	63	50.8	64	23.8	43	28.3	28	88.8	55
Copepod nauplii	27.4	34	11.7	31	23.1	45	7.5	21	11.1	11
Invertebrate eggs			19.2	28	21.9	26	7.5	21		
Rest of copepods										
Rest of crustacea	+	21	+	6	+	19	+	35	+	44

Table 1. (continued)

11.1 - 13.0		13.1 - 15.0		15.1 - 17.0		17.1 - 19.0		19.1 - 21.0		21.1 - 25.0		25.1 - 30.0	
428		390		132		66		12		48		96	
26		35		19		6		0		12		24	
%/No.	%occ.	%/No.	%occ.	%/No.	%occ.	%/No.	%occ.	%/No.	%occ.	%/No.	%occ.	%/No.	%occ.
.2	25	11.1	17	20.0	33							85.7	50
.3	25	11.1	17	10.0	33					100	100	14.3	25
.0	25	55.5	51	60.0	33								
.5	25	22.2	17	10.0	33	100	100						
										+	100	+	50
	75												

From Table 1 it can also be seen that the food composition is related to the length of the larvae. The general features of this relationship are as follows: larvae up to 11.0 mm while feeding mainly on zooplankton (copepods eggs and nauplii) also take ciliates and phytoplankton; larvae larger than 11.0 mm feed exclusively on copepods.

Incidence of Feeding. The percentage of feeding of larvae of different lengths is given in Table 2. This shows that the incidence of feeding changes with the length of the larvae, the greatest being found in larvae of 3.1 - 5.0 mm. The feeding incidence is much lower in larvae from 5-21 mm. Above 21 mm the percentage feeding appears to increase again, although it should be noted that the number of stomachs analyzed in the 2 larger size groups is small.

Table 2. Incidence of feeding related to length

Length of larvae in mm	No. of larvae	Incidence of feeding (%)	Berner (1959)* *E. mordax*
3.0	1080	21.1	23.5
3.1 - 5.0	6576	31.3	50.5
5.1 - 7.0	2280	12.1	14.5
7.1 - 9.0	1230	6.8	4.0
9.1 - 11.0	622	8.4	5.4
11.1 - 13.0	428	6.1	0.0
13.1 - 15.0	390	9.0	
15.1 - 17.0	132	14.4	
17.1 - 19.0	66	9.1	
19.1 - 21.0	12	0.0	
21.1 - 25.0	48	25.0	
25.1 - 30.0	96	25.0	

Number of larvae with food 2,850
*Taken from the author's Fig. 5

Table 2 includes data from Berner (1959) which also show a decrease in the incidence of feeding above a size of 5 mm. Ciechomski (1967), however, found the greatest incidence of feeding in larvae of *E. anchoita* between 3.0 - 4.0 mm body length with a lesser incidence between 5.1 - 9.0 and an increase again at greater length.

The number of larvae with food is only 22% of the total number examined. This is not an isolated observation; in other species of the family Engraulidae, Berner (1959) found that in a total of 13,620 larvae of *E. mordax* only 211 contained food; Ciechomski (1967) found 503 larvae of *E. anchoita* with food from a total of 1,705 larvae. The reasons for this low incidence of feeding are not clear; several hypotheses have been suggested, relating it to: speed of digestion, physical condition of the larvae when collected, influence of light, etc.

Table 3 shows the percentage incidence of food in larvae collected during the day and night in each season. We can see from this table that the percentage of larvae with food is greater by day than by night in all seasons.

Of the total number of larvae with food, 91.6% were caught during the day. This is in close agreement with the results of Berner (1959) for *E. mordax*, who found that of the larvae that had been feeding 95.3

were caught by day. These results suggest that larvae depend on vision for feeding, in agreement with Krogh (1931), Arthur (1959), Schumann (1965) for *Sardinops caerulea*, and Blaxter (1963) for *Clupea harengus*.

Table 3. Seasonal and diurnal incidence of feeding

	Summer	Autumn	Winter	Spring	Total
Total larvae (day)	504	42	5460	341	6347
% feeding	28.6%	0	46.4%	18.6%	43.2%
Total larvae (night)	270	0	4292	2051	6613
% feeding	0	0	1.1%	2.9%	1.6%

Seasonal Variations. From Table 4 it is apparent that eggs and nauplii of copepods were the most abundant food item during the winter (July-September), copepodites during the spring, and *Calanus* sp. in the summer. In the spring, the diet also contains some phytoplankton, as noted in the samples collected at Pimentel and Supe.

Table 4. Seasonal variations in diet

	Summer J-F-M	Autumn A-M-J	Winter J-A-S	Spring O-N-D
Diatoms			x	xx
Dinoflagellate	x			
Ciliates			xx	x
Copepod eggs	xx		xxx	x
Copepod nauplii	xx		xxx	x
Copepodite				xxx
Copepod adults (*Calanus* sp.)	xxx			
Invertebrate eggs	xx		x	
Rest of copepods	x			
Rest of crustaceans	x		x	xx

DISCUSSION

The data given in Tables 1 and 4 show clearly that anchoveta during the larval stage from 3 to 30 mm feed on zooplankton and that the staple food throughout the year is copepods in different stages of development. These results are in agreement with those of Ciechomski (1967) for *E. anchoita*, Arthur (1956) and Berner (1959) for *E. mordax*, and Demir (1965) for *E. encrasicholus*.

Phytoplankton do not contribute significantly to the food of larvae, as they are present in the gut contents only of larvae smaller than 9 mm. The low incidence of phytoplankton in the food of larval anchoveta above this size is perhaps surprising in view of the fact that, when adult, this species is predominantly a phytoplankton feeder. Analysis of the gut contents of juvenile anchoveta has shown that, at lengths above 46 mm the proportion of phytoplankton in the food increases steadily to a length of about 12 cm when the food is almost entirely phytoplankton. This change from a zooplankton to a phytoplankton feeding regime is related to developmental changes in the digestive tract and in the number and length of the gill rakers (Rojas de Mendiola, in press). The results given in Table 3, showing that 92% of the larvae containing food were collected during daylight, would strongly support the hypothesis that feeding in anchoveta larvae, as in the larvae of many other fish species, is a response to visual stimuli (Blaxter, 1963). Even in samples caught during the day, however, the incidence of feeding was low. A low incidence of feeding has been a feature of many investigations of larval food and two possible explanations of this have been put forth. Arthur (1956) suggested that the apparent low level of feeding could be due to the sampling gear catching only larvae in poor condition. Since the present data were obtained from material caught by the slow-moving Hensen net, they might be considered as particularly subject to this bias; however, in as far as a subjective assessment of condition from external appearance is a valid criterion, the material examined did not appear to be in poor condition. The alternative explanation that feeding is intensive over short time intervals and the digestion rate is very fast (Krogh, 1931; Lebour, 1919) therefore seems more feasible in this instance.

From the data given in Table 4 it would appear that there is no marked variation in the composition of the food between seasons, although there is in the quantity of food with the heaviest feeding in summer. With this material taken in different areas along the coast of Peru it was possible to examine the variations between areas in feeding intensity and food composition. The differences found were very small and did not merit detailed description.

ACKNOWLEDGEMENTS

Thanks are expressed to Drs. Aurora Vildoso and A. Landa, who read the manuscript and made valuable suggestions and to biologists Miss Noemi Ochoa and Miss Olga Gómez for their most helpful collaboration.

REFERENCES

Arthur, D.K., 1956. The particulate food and the food resources of the larvae of three pelagic fishes, specially the Pacific sardine, *Sardinops caerulea* G. Unpublished Doctoral dissertation, on file in the Library of Scripps Institution of Oceanography, University of California, La Jolla, 231 pp.

Berner, L.Jr., 1959. The food of the larvae of the northern anchovy *Engraulis mordax*. Inter-Amer. Trop. Tuna Comm. Bull., 4 (1), 1-22.

Bhattacharyya, R.N., 1957. The food and feeding habits of larval and post-larval herring in the northern North Sea. Mar. Res. Scot., No. 3, 14 pp.

Blaxter, J.H.S., 1963. The feeding of herring larvae and their ecology in relation to feeding. CALCOFI Rep., 10, 79-88.

Ciechomski, J.D. de, 1967. Investigations of food and feeding habits of larvae and juveniles of the Argentine anchovy *Engraulis anchoita*. Calif. Coop. Ocean. Fish. Invest. Rep., 40, 72-81.

Demir, N., 1965. Synopsis of biological data on anchovy *Engraulis encrasiocholus* L. (Mediterranean and adjacent seas). FAO Fish. syn. No. 26, Revision 1.

Krogh, A., 1931. Dissolved substances as food of aquatic organisms. Biol. Rev. 6, 4, 412-442.
Lebour, M.V., 1919. The food of young fish. J. Mar. Biol. Ass. U.K., 12 (3), 261-324.
Morris, R.W., 1955. Some considerations regarding the nutrition of marine fish larvae. J. du Cons., 20 (3), 235-265.
Pavlovskaia, R.M., 1961. La survie des larves de l'anchois dela mer noire en foction de leur alimentation. Rapp. comm. int. Mer Medit., 16, 345-350.
Schumann, O.G., 1963. Some aspects of behavior in clupeid larvae. CALCOFI, 10, 71-78.
Soleim, P.A., 1942. Arsaker til rike og fattige argan ger av Sikf Fisjerider-ekto ratets Skrifter, Ser. Havundersøkelssr., 2, 39 pp. (English summary, 2 pp.)

B.R. de Mendiola
Instituto del Mar del Perú
Lima / PERU

Laboratory Studies of Predation by Euphausiid Shrimps on Fish Larvae

G. H. Theilacker and R. Lasker

INTRODUCTION

Despite a large literature on the vulnerability of marine fish larvae to changes in their physical and biotic environment, there is surprisingly little quantitative data available on organisms that eat fish larvae. There is ample evidence that huge mortalities of yolk-sac fish larvae occur (Ahlstrom, 1965); this is not due to lack of food, a factor implicated in the mortality of older larvae (Blaxter, 1969). Predation on yolk-sac larvae may be the most important cause of mortality during the early period in the life history of pelagic fish. There are a number of observations of zooplankters feeding on fish larvae (Garstang, 1900; Lebour, 1925; Wickstead, 1965; Petipa, 1965; Fraser, 1969); copepods, chaetognaths, ctenophores, and a variety of coelenterates have been seen to capture and ingest marine fish larvae. Recently Lillelund and Lasker (1971) quantified the predator-prey relationship between several species of marine copepods and larvae of the northern anchovy, *Engraulis mordax*, and described the behavioural responses involved in this interaction. They found, for example, that a variety of marine copepods (but particularly surface-dwelling pontellids) can capture and ingest or fatally injure young anchovy larvae under laboratory conditions.

Our experiments were designed to determine whether euphausiids could capture fish larvae, and if so, how many and under what conditions. We have combined our laboratory findings with the field data given by Brinton (1967 and unpublished) to estimate the mortality of northern anchovy larvae that may be caused by co-occurring *E. pacifica*.

Of the oceanic zooplankters which frequent the upper mixed layer of the sea where fish larvae are found in greatest abundance, the euphausiid shrimps are often dominant in biomass and number, particularly during the night hours when they migrate to the surface. Although neither Ponomareva (1963) nor Mauchline and Fisher (1969), in their respective monographs, mention fish larvae as a food of euphausiids, most euphausiids are known to be omnivorous and capable of capturing zooplankters, e.g. chaetognaths and copepods; therefore fish larvae may be captured too. The fragile nature of most fish larvae and the lack of hard parts would make recognition of the remains impossible in the euphausiid intestine even if they were an important part of their diet.

Brinton (1962 and unpublished) has estimated the distribution and density of several euphausiid species in the California Current over a number of years, *Euphausia pacifica* Hansen being one of the most abundant species. The general feeding habits and ubiquity of *E. pacifica* in the California Current, and the fact that it can be maintained in the laboratory (Lasker and Theilacker, 1965; Lasker, 1966) prompted us to investigate it as a possible predator of marine fish larvae, particularly of the northern anchovy (*Engraulis mordax* Girard), the dominant clupeid fish in the California Current and to estimate larval mortality from field data.

METHODS

We collected euphausiids by towing a 1-m (mouth-diameter, 0.505 mm mesh) plankton net or a 60 cm diameter Bongo net without bridles with the same mesh size (McGowan and Brown, 1966) in the Pacific Ocean near San Diego and Santa Catalina Island, California, during the spring and summer of 1971 and 1972. On board ship euphausiids were separated from other plankton immediately and after allowing for the initial mortality of damaged animals (10-12 h), the survivors were kept separately in individual containers. Detailed methods for collecting and maintaining euphausiids are given by Lasker and Theilacker (1965).

Juvenile *E. pacifica*, ranging in total length from 6-10 mm and 0.6 - 2.1 mg dry weight, were caught most frequently. Some euphausiids were used for predation experiments on shipboard while others were brought back to the Southwest Fisheries Center aquarium, and used for as long as 25 days. When an experiment was terminated, each animal was measured, rinsed in distilled water, dried at 60°C, and weighed. The anchovy larvae fed to euphausiids were hatched from eggs collected from a spawning school in our aquarium facilities or from hormone-injected anchovies (Leong, 1971). For shipboard experiments the eggs were transferred from the aquarium to the ship, thus eliminating the need for catching anchovy eggs in net tows and sorting the eggs from other plankton.

Temperature was maintained at 17°C, close to the sea-surface temperature at the time euphausiids were captured. *E. pacifica* follows a low intensity light level of about 1 X 10^{-4} μW/cm^2 (Clarke, 1966) during vertical migration; we therefore performed all of our experiments in the dark. Unless otherwise noted, each feeding test lasted for 22 h with a euphausiid in 3500 ml sea water filtered earlier through a Cuno® Aqua Pure Filter (pore size 5 μ). In all experiments the young anchovy larvae appeared to be randomly distributed in the feeding containers.

FEEDING BY ADULT, JUVENILE, AND LARVAL *E. PACIFICA* ON LARVAL ANCHOVIES

Experiments were performed to compare the relative ability of different growth stages of euphausiids to capture anchovy larvae. Larval *E. pacifica* weigh less than 0.6 mg dry weight and are less than 6 mm in total length. Juveniles range from 0.6 - 2.1 mg and from 6 mm to slightly less than 11 mm in length; adults are 11 mm or more long and greater than 2.1 mg in weight (lengths from Brinton, personal communication). In these experiments the smallest *E. pacifica* tested was 5 mm long (0.15 mg) and the largest 20 mm (13.2 mg). The length range of *E. pacifica* in the California Current is from 1-21 mm.

Sixteen larval, 30 juvenile, and 27 adult *E. pacifica* were tested individually to determine their ability to feed on yolk-sac anchovy larvae. Larval and juvenile euphausiids were offered 20 anchovy larvae a day and adults were given 50-80 larvae per day in 3500 ml. Larval euphausiids ate a median number of 2 larvae/day (range 1-5), juveniles ate a median of 7 larvae/day (range 1-19), and adults ate a median of 17 larvae/day (range 5-38)(Fig. 1).

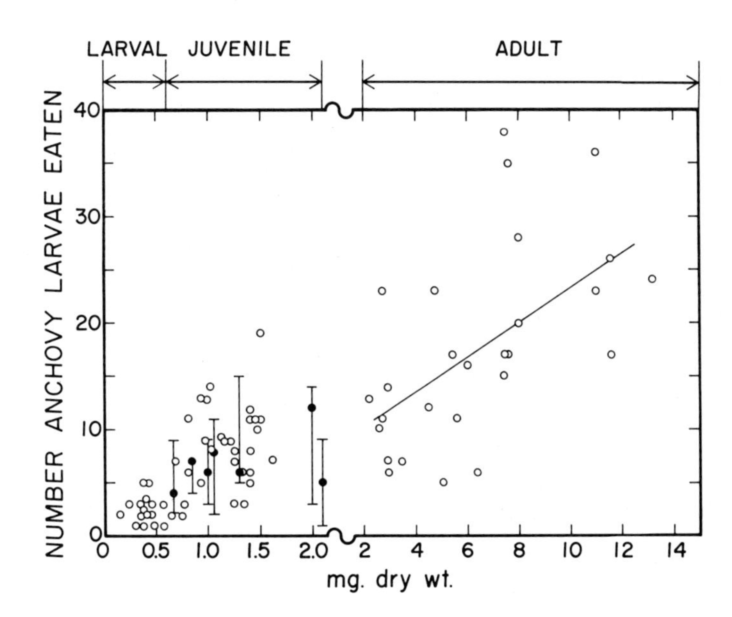

Fig. 1. Number of anchovy eaten per day by individual larval, juvenile, and adult *E. pacifica* (open circles). Closed circles indicate the median number of anchovy larvae eaten (with extremes given by the solid lines) when 7 individuals were fed daily for 5 consecutive days

No weight-specific feeding rate could be demonstrated within the juvenile size class (the slope [b = 4.07] does not differ significantly from zero [$p > 0.1$]). However, within the adult size group, feeding did increase with increasing animal size (Fig. 1). The slope (b = 1.69) of a least squares fit of the data differs significantly from zero ($p < 0.01$).

To test further the hypothesis that animal size, within the juvenile group, does not influence feeding, 7 juveniles were fed 20 anchovy larvae/day on each of 5 successive days. The euphausiids ranged from 7-11 mm (0.67 - 2.1 mg). The median number of anchovy larvae eaten per individual euphausiid/day ranged from 4-12 (Fig. 1). No significant difference in feeding could be demonstrated ($p > 0.20$) when the number of anchovy larvae eaten each day per animal was ranked and compared (Friedman two-way analysis of variance). All succeeding experiments were conducted with juvenile *E. pacifica*, since animal size did not influence feeding.

Lasker (1966) determined the carbon requirement at 10°C needed by various size groups of *E. pacifica* for growth, molting, respiration, and digestive inefficiency. From these data we calculated the necessary number of anchovy larvae which had to be consumed to satisfy the carbon requirement of larval, juvenile, and adult *E. pacifica*. Early anchovy larvae[1] weigh 22 μg and contain 42% carbon; therefore, one

---

[1]Lillelund and Lasker (1971, p. 664) gave 10 μg for the dry weight of a yolk-sac anchovy larva. This figure should have been 22 μg.

larva provides about 9 μg of organic carbon. If no other food is available the carbon requirements of larval euphausiids at 10°C can be satisfied if each eats 1 anchovy larva/day; juveniles require 3-5 larvae/day, and adults from 7-28 larvae/day, depending on the euphausiid's weight.

At 17°C, the temperature of these experiments, the carbon requirements for growth, molting, and respiration would be greater than at 10°C. Respiratory carbon losses at 17°C are increased by 60% ($Q_{10} = 2$; Lasker, 1960) over the losses at 10°C and since respiration alone accounts for most of *E. pacifica's* total carbon requirement (Lasker, 1966) we adjusted the above anchovy larval carbon equivalents by 60%. This increases the larval euphausiids daily requirement, see Table 1.

Table 1. The number of anchovy larvae/day needed by larval, juvenile, and adult *E. pacifica* to satisfy their daily carbon requirements at 17°C (based on Lasker, 1960, 1966) compared to the median and maximum number of anchovy larvae eaten by euphausiids in laboratory experiments

E. pacifica	Anchovy larvae required/day	Anchovy larvae eaten	
		Median	Maximum
Larval	1-2	2	5
Juvenile	5-8	7	19
Adult	11-45	17	38

FEEDING AT DIFFERENT PREY CONCENTRATIONS

The quantity of prey offered to juvenile euphausiids was varied from 1-80 anchovy larvae per 3500 ml/day. When 10 or less larvae were fed to juvenile euphausiids the maximum number eaten (Fig. 2) appears to be limited by the number fed and, probably, the searching capability of the euphausiid. Increasing the density further, offering more than 10 larvae, did not cause any differences in the median feeding rate. Feeding rates were compared statistically at 3 food concentrations (Fig. 2) 18-21, 29-32, and 40-50 larvae fed per container. The median number eaten per day increased from 6 at the lowest density (20 observations), to 8 at the next (9 experiments) and 10 at the highest density (6 experiments); however, no differences could be demonstrated between each paired comparison (Mann-Whitney U test, $p > 0.20$ for each pair tested).

From these experiments it appears that the average number of anchovy larvae juvenile euphausiids can process in a day - catching, eating, digesting, and excreting - is 10 or less. Increasing the prey concentration above this number had no effect on feeding in the volumes tested. This independence between feeding and food concentration has also been described for the crustacean *Daphnia* (Rigler, 1961) and the chaetognath, *Sagitta* (Reeve, 1964). Both authors found that above a critical food concentration the food-intake rate held constant as the food density continued to increase.

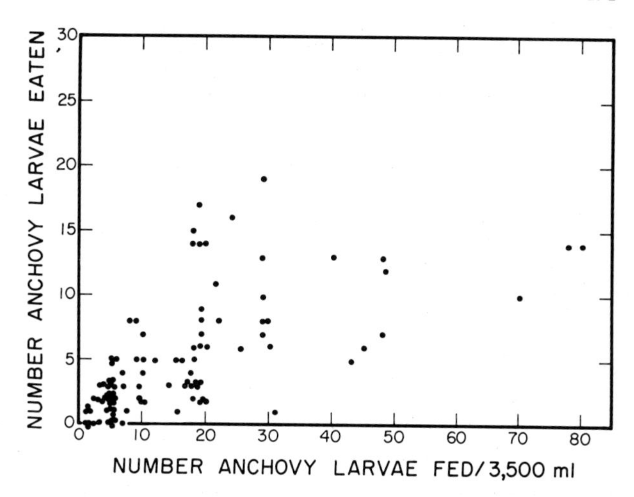

Fig. 2. Feeding of juvenile *E. pacifica* at different concentrations of anchovy larvae

FEEDING RATE

Fisher and Goldie (1959), studying the euphausiid *Meganyctiphanes norvegica*, did not find consistent differences in the amount of feeding between day and night samples taken in the North Sea. However, Ponomareva (1954) found that *E. pacifica* fed at a higher intensity during the evening and night in the Sea of Japan. Roger (1971) also observed a feeding rhythm for most of the 28 euphausiid species he studied in the field. He noted that the intensity of feeding usually increased before the nocturnal ascent, so that the rhythm was not, in most cases, synchronous with the day-night vertical migration.

Feeding intensity may be correlated only with the availability of food. To determine whether a periodicity in *E. pacifica* feeding could be demonstrated when food is not limiting, 11 juvenile euphausiids were fed anchovy larvae for 3 consecutive 12-h periods (midnight and noon were chosen as starting times to avoid any effect of a diel rhythm). Twenty larvae were offered to each euphausiid at the beginning of each 12-h experiment. At the end of the 36-h period, the animals were fed 20 larvae for 24 h. The median number of larvae eaten was 2 (Table 2) during the first 12 h (noon to midnight); 4 were eaten in the midnight-to-noon period and 2 in the second non-to-midnight period. The median for the 24-h period was 5. No periodicity in feeding can be shown when the data were tested statistically. Feeding during the three 12-h periods was homogeneous when compared using Friedman's two-way analysis of variance by ranks ($p > 0.20$). Also, no differences could be demonstrated by comparing the 2 combined noon-midnight periods with the midnight-noon period (Mann-Whitney U, $p = 0.68$). The results suggest a constant feeding rate when food is not a limiting factor.

Table 2. The number of anchovy larvae eaten by 11 juvenile *E. pacifica* for each of 3 consecutive 12-h feeding periods followed by a 24-h perio

Juvenile <u>E. pacifica</u> mg dry weight	Noon-midnight 12 h	Midnight-noon 12 h	Noon-midnight 12 h	24 h
2.00	8	4	8	3
2.10	5	0	2	1
0.52	0	6	6	5
1.28	1	4	3	5
1.23	2	5	3	7
0.67	3	4	6	4
0.56	0	2	0	1
0.82	3	2	2	7
1.00	2	4	1	6
0.84	1	0	0	4
1.02	1	2	2	5
Median number eaten	2	4	2	5

EFFECT OF TIME IN CAPTIVITY ON FEEDING BY JUVENILE *E. PACIFICA*

Twelve animals (0.5 - 2.1 mg dry weight) tested aboard ship on the first day of capture ate a median of 6.5 anchovy larvae when 18-20 were offered in 3500 ml, consuming between 2-13 larvae individually. Eight other juveniles (0.7 - 2.0 mg dry weight) were observed in the laboratory to determine whether or not feeding rate changed during captivity. Five days of feeding (7, 8, 13, 17, and 21 days after capture), when the density of larvae offered was the same as in the above shipboard experiments, were compared (Table 3). The number of larvae eaten per animal on each day was ranked. The sums of the daily ranks were compared and no significant difference in feeding rates was obtained (Friedman two-way analysis of variance by ranks, $0.1 < p < 0.2$). The number of larvae eaten by all animals varied from 1-15 per day; the individual medians for the 5 days of feeding ranged between 2 and 12 and the overall median number of larvae eaten was 6. Therefore, the daily feeding rate of juvenile *E. pacifica* does not appear to change with the time maintained in captivity.

Table 3. Feeding by 8 juvenile *E. pacifica* maintained on anchovy larvae in the laboratory for 3 weeks

Juvenile *E. pacifica*	Number of anchovy larvae eaten					
	Days after capture					Median
	7	8	13	17	21	
no. 1	5	5	8	15	6	6
2	12	3	14	4	14	12
3	7	1	8	3	9	7
4	6	4	2	3	9	4
5	4	7	5	7	7	7
6	1	4	3	1	2	2
7	4	6	3	9	9	6
8	3	5	11	8	8	8

THE EFFECT OF AN ALTERNATIVE PREY

Lillelund and Lasker (1971), in their study on the predation of anchovy larvae by the copepod *Labidocera trispinosa*, found that the addition of *Artemia* nauplii, as an alternate prey, decreased the feeding on larvae. The decrease in feeding rate was proportional to the density of the additional prey; at high nauplii densities fewer larvae were killed. *Euphausia pacifica* will also feed on *Artemia* nauplii, the number eaten determined by Lasker (1966) being 100-350/day for *E. pacifica* ranging in dry weight between 1.8 and 5.6 mg. Nine individual juvenile *E. pacifica* were offered 20 larvae daily for 5 consecutive days with *Artemia* nauplii as an alternative. Four of the experiments were conducted in 3500 ml and 5 in 700 ml containers. On the 2nd, 3rd, and 4th days *Artemia* nauplii were also given at densities of one nauplius per 5, 10, 20, and 40 ml. The median number of the larvae eaten in the 3500 ml containers was 5 when anchovy larvae only were offered, but 3 when *Artemia* nauplii were present with the larvae; in the 700 ml containers the median number eaten was 5 for each treatment. The data were tested statistically (the 2 container sizes tested separately) by comparing the ranked sum of the number of anchovy larvae eaten for the 3 days when *Artemia* were present to the number eaten during the 2 days when anchovy larvae alone were fed. The ratio of the sums of the ranks for the 2 treatments (with and without *Artemia* nauplii) was the same as the ratio of the number of observations (for both container volumes), so the null hypothesis must be accepted ($p > 0.20$, Mann-Whitney U Test).

The same data were rearranged to test the daily feeding on anchovy larvae as a function of *Artemia* density (Fig. 3). The data were homogeneous; feeding on larvae was the same at each *Artemia* density (Friedman's two-way analysis of variance by ranks, $p > 0.20$). Therefore, the presence of *Artemia* nauplii, even at a density of one per 5 ml, does not alter euphausiid feeding on anchovy larvae.

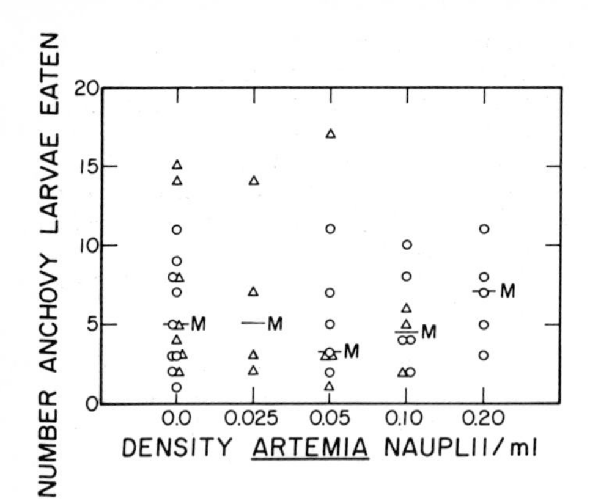

Fig. 3. Predation by *E. pacifica* on anchovy larvae with addition of *Artemia* nauplii as an alternative prey. o = 3500 ml container; △ = 700 ml container; M = median number during control and at each *Artemia* density

In all of the experiments in which *Artemia* nauplii were offered, the number of nauplii eaten increased as the nauplii concentration was increased. Filtering rates (Gauld, 1951) calculated from the number of nauplii eaten at each density remained constant (Friedman's two-way analysis of variance by ranks, $p > 0.20$), indicating that the *Artemia* are filtered as a passive particle. Brooks (1970) concluded that particle-feeding copepods grazed on *Artemia* as a passive particle and that predatory copepods may feed on *Artemia* nauplii in preference to natural nauplii only because they are more easily captured and no active selection of the *Artemia* nauplii over the natural nauplii is involved.

STARVATION

Several juvenile euphausiids, 0.6 - 0.8 mg dry weight, were kept in the laboratory for 24 days without feeding. These animals commenced feeding on anchovy larvae and molted for 2 additional weeks when the experiment was terminated.

To determine whether or not periods of starvation cause a change in feeding rate, 8 individual juvenile *E. pacifica*, (0.70 - 2.00 mg dry weight) were either fed 20 larvae/day on consecutive days or starved from 2-3 days and then fed 20 larvae. Each individual (9-10 observations/animal) was tested separately, comparing the number of larvae eaten during a day of feeding when the animal had been fed the previous day to the number eaten when the animal had not been fed the previous day. One euphausiid ate less than usual after a period of starvation ($p < 0.05$); however, the other 7 animals showed no differences in the daily feeding patterns whether or not the euphausiids had fed the previous day (p values all >0.20, Mann-Whitney U Test). Therefore, under laboratory conditions, when food is not limiting, 2-3 days of previous starvation does not appear to have an effect on feeding.

MOLTING

Lasker (1966) found that molting of *E. pacifica* depressed the feeding on *Artemia* nauplii. To test for this effect, the data from all the anchovy larvae feeding experiments were combined and the number eaten on days the euphausiids did and did not molt were compared. Juvenile euphausiids used in this comparison were fed more than 10 larvae/day and had molted at least once. The median number of larvae eaten for 99 animal-days when the euphausiid did not molt was 7 (range 1-19). There were only 11 molting days that fit the above requirements. The median number of larvae eaten during a molting day was 5 (range 1-14). No significant difference could be demonstrated between the 2 sets of data (Mann-Whitney U Test $p = 0.63$); hence molting does not appear to depress the feeding of *E. pacifica* on anchovy larvae.

The median intermolt period at 17°C for *E. pacifica* fed anchovy larvae (22 observations on 12 animals) was 4 days, the range was between 2 and 6 days; 91% of the observations were between 3 and 5 days. This is the same molting frequency observed by Lasker (1966) for *E. pacifica* fed *Artemia* nauplii at 15°C.

FEEDING ON OLDER ANCHOVY LARVAE

Lillelund and Lasker (1971) noted in their predation study that the ability of the copepod *Labidocera trispinosa* to capture anchovy larvae decreased as the larvae aged and became more active. A 1-day-old anchovy larva rests more than 95% of the time and 2-day-old larvae rest about 70% of the time; at 3 days of age the larva spends almost 50% of its time in intermittent swimming and the next day swims intermittently 85% of the time (Hunter, 1972). Increasing larval activity correlates well with the decreasing ability of *E. pacifica* to capture larval anchovies.

Day-1 (newly hatched) and 2-day-old larvae were fed to 10 individual juvenile euphausiids at a concentration of 1 larva/3500 ml. In this situation the euphausiids were 60% successful in capturing the single larva. Successful capture dropped to 17% when the larvae were 3 days old (12 experiments) and to 11% for 4-day-old larvae (9 experiments), all fed to the euphausiids at the same low density.

FEEDING ON ANCHOVY EGGS

E. pacifica will feed on anchovy eggs although the incidence of ingestion is very low in the laboratory. This is probably because the eggs are unavailable to the euphausiid, floating at the surface of the sea water since there was no water movement in the containers. In the open ocean, anchovy eggs are distributed in the upper 50 m of the water column (Ahlstrom, 1959).

Sixteen juvenile euphausiids were offered 20 anchovy eggs each in 700 ml. Only 3 animals ate eggs; the median number eaten was 5.

CO-OCCURRENCE OF *E. PACIFICA* AND ANCHOVY LARVAE IN THE CALIFORNIA CURRENT

Dr. Edward Brinton of SIO has kindly provided us with unpublished data on the abundance of *E. pacifica* in the California Current. The euphausiids were collected in standard (1-m net towed to 140 m depth) night CalCOFI tows (Kramer et al., 1972) between January 1953 and July 1955. The counts are given as numbers of individuals/1000 m^3 of each mm size group, 1-21 mm, per station per month for 27 stations between Point Conception and San Diego (Fig. 4), an area of 30,000 square miles (8×10^{10} m^2).

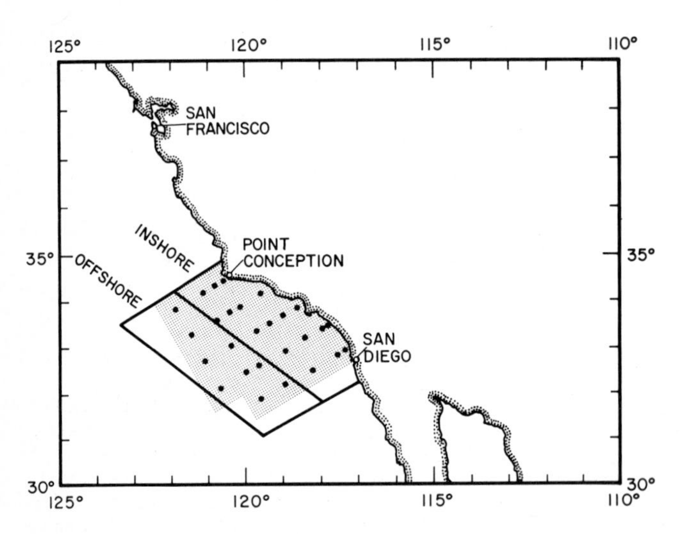

Fig. 4. Map showing anchovy (outlined) and euphausiid (shaded) surveys. ● = euphausiid stations

The abundance of anchovy eggs and larvae in the California Current has been studied intensively since 1949. Data for the anchovy spawning season during 1953, 1954, and 1955 in the same area of the California Current as Brinton's euphausiid study are given by Smith (1972). The anchovy area extends 40 miles further south than the euphausiid area (Fig. 4). Spawning during the 1953-1954 season was between October 1953 and March 1954; in 1955 the major spawning was in the first quarter of the year and is given in numbers of anchovy larvae per m^2 sea surface per positive station in Table 4. There were more positive stations inshore (to 80 miles off the coast) than offshore (between 80 and 160 miles off the coast); the frequencies are given in Table 4. Although there is some spawning throughout the year the number of larval anchovies is only one-tenth to one-fourth of the number obtained during peak spawning. During the spawning period 68% of the larvae sampled were 5.25 mm or less in length (Smith, personal communication), or between hatching and 1-week-old.

About 98% of the anchovy larvae occur above 56 m and 49% are above 8 m (Ahlstrom, 1959). The vertical distribution of each *E. pacifica* size group was calculated from data provided by Brinton (unpublished and 1967). The samples, taken at 25-m increments between 25 and 250 m with a closing net, were collected at 4 night stations. Night catches indicate that an average of 87% of larval, juvenile, and adult *E. pacifica* are above 50 m. However, because of active vertical migration it is conceivable that all of the euphausiids may reach the surface some time during the night.

Table 4. The estimated number of anchovy larvae (< 5.25 mm long) under 1 m^2 of sea surface during the peak spawning season in 1953, 1954, and 1955

Spawning season	Inshore		Offshore	
	No. larvae/m^2	No. positive stations	No. m^2	No. positive stations
1953	36	100	13	86
1954	59	100	11	83
1955	38	95	4	69

The average number of larval, juvenile, and adult *E. pacifica* under 1 m^2 of sea surface in the area of the California Current is tabulated in Table 5 for the months of peak anchovy spawning in 1953, 1954, and 1955 (9 months). At the median feeding rates on anchovy larvae determined in this study and the mean euphausiid densities calculated to include the number of euphausiids co-occurring with anchovy larvae, 831 anchovy larvae could be consumed/day/m^2. For the purpose of this discussion we assume that *E. pacifica* is in the upper 50 m of the water column for 10 h, an hour after sunset to an hour before sunrise. If the total water column *E. pacifica* frequents daily is considered, the *E. pacifica* (average of Brinton's 29-month survey) under 1 m^2 of sea surface could devour a total of 2847 fish larvae/day/m^2 (Table 5) regardless of other zooplankton being present.

Table 5. The average number of *E. pacifica* under 1 m^2 sea surface and the total number of fish larvae that could be consumed daily

	Larval	Juvenile	Adult	Total
No. of *E. pacifica*/m^2 during peak anchovy spawning (9 mo. average)	314	118	38	470
No. of *E. pacifica* above 50 m at night	273	103	33	409
No. of fish larvae *E. pacifica* above 50 m can consume/10 h/m^2	248	328	255	831
E. pacifica /m^2 (29 mo. average)	582	139	43	764
No. of fish larvae *E. pacifica* can consume/2 h/m^2	1,164	952	731	2,847

It is possible that the small number ($<60/m^2$) of anchovy larvae caught in plankton nets (Table 4) is a result of heavy predation. The abundance of euphausiids and other predators may be a factor in determining the patchiness of fish larvae.

SUMMARY

The abundant euphausiid in the California Current, *Euphausia pacifica*, may be an important predator of fish larvae. Laboratory experiments showed that when the number of yolk-sac anchovy larvae (*Engraulis mordax*) offered was not limiting (not less than one larva/350 ml) the median number eaten per day by larval *E. pacifica* was 2; juvenile euphausiids eat 7, and adults 17. The carbon requirements of *E. pacifica* for the 3 growth stages is approximately the amount of carbon provided by the median feeding rates. Within the adult euphausiid size group, feeding on fish larvae significantly increased with increasing animal size, but no size-specific feeding could be demonstrated for juvenile euphausiids. All replicate experiments with individual *E. pacifica* in this study were conducted with juveniles. These showed that the amount of feeding was not affected by the addition of another prey, imposed starvation, molting, or the length of time the euphausiids were in captivity. The amount of feeding increased as the prey density was increased from 1-10 larvae offered per 3.5 l/day; further increases had no effect. Feeding rate was constant regardless of time of day. The percent capture of fish larvae by euphausiids was lower than usual when older, more active anchovy larvae were offered. These studies are discussed in terms of field data collected in the area of the Californi Current where *E. pacifica* and anchovy larvae co-occur.

REFERENCES

Ahlstrom, E.H., 1959. Vertical distribution of pelagic fish eggs and larvae off California and Baja California. U.S. Fish. Wildl. Serv., Fish. Bull., 60, 107-146.

Ahlstrom, E.H., 1965. A review of the effects of the environment of the Pacific sardine. ICNAF Spec. Publ., 6, 53-74.

Blaxter, J.H.S., 1969. Development: eggs and larvae. In: W.S. Hoar and D.J. Randall (Eds.), Fish Physiol., 3. New York: Academic Press, 177-252 pp.

Brinton, E., 1962. The distribution of Pacific euphausiids. Bull. Scripps Inst. Oceanogr., Univ. Calif., 8, 51-270.

Brinton, E., 1967. Vertical migration and avoidance capability of euphausiids in the California Current. Limnol. Oceanogr., 12, 451-483.

Brooks, E.R., 1970. Selective feeding of some adult female copepods on an array of food including *Artemia* and naturally-occurring naupli U. Calif. Inst. Mar. Resour., Res. Mar. Food Chain Prog. Rep., July 1969-June 1970, Part 2, 56-74.

Clarke, W.B., 1966. Bathyphotometric studies of the light regime of organisms of the deep scattering layers. AEC Res. Develop. Rep., SAN-548-1, UC-48 Biol. Med., TD-4500, 47 pp.

Fisher, L.R. and Goldie, E.H., 1959. The food of *Meganyctiphanes norvegica* (M. Sars), with an assessment of the contributions of its components to the vitamin A reserves of the animal. J. Mar. Biol. Ass. U.K., 38, 291-312.

Fraser, J.H., 1969. Experimental feeding of some medusae and chaetognatha. J. Fish. Res. Bd Can., 26, 1743-1762.
Garstang, W.R., 1900. Preliminary experiments on the rearing of sea-fish larvae. J. Mar. Biol. Ass. U.K., 6, 70-93.
Gauld, D.T., 1951. The grazing rate of planktonic copepods. J. Mar. Biol. Ass. U.K., 29, 695-706.
Hunter, J.R., 1972. Swimming and feeding behavior of larval anchovy *Engraulis mordax*. Fish. Bull., U.S., 70, 821-838.
Kramer, D., Kalin, M.J., Stevens, E.G., Thrailkill, J.R. and Zweifel, J.R., 1972. Collecting and processing data on fish eggs and larvae in the California Current region. NOAA, Tech. Rep., NMFS Circ., 370, 38 pp.
Lasker, R., 1960. Utilization of organic carbon by a marine crustacean: analysis with carbon-14. Science, 131, 1098-1100.
Lasker, R., 1966. Feeding, growth, respiration, and carbon utilization of a euphausiid crustacean. J. Fish. Res. Bd Can., 23, 1291-1317.
Lasker, R. and Theilacker, G.H., 1965. Maintenance of euphausiid shrimps in the laboratory. Limnol. Oceanogr., 10, 287-288.
Lebour, M.V., 1925. Young anglers in captivity and some of their enemies. A study in a plunger jar. J. Mar. Biol. Ass. U.K., 13, 721-734.
Leong, R., 1971. Induced spawning of the northern anchovy, *Engraulis mordax* Girard. Fish. Bull., U.S., 69, 357-360.
Lillelund, K. and Lasker, R., 1971. Laboratory studies of predation by marine copepods on fish larvae. Fish. Bull., U.S., 69, 655-667.
Mauchline, J. and Fisher, L.R., 1969. The biology of euphausiids. Advan. Mar. Biol., 7, 1-454.
McGowan, J.A. and Brown, D.M., 1966. A new opening-closing paired zooplankton net. Scripps Inst. Oceanogr. Ref., 66-23.
Petipa, T.S., 1965. The food selectivity of *Calanus helgolandicus*. Invest. Plankton Black Sea, Sea of Azov. Akad. Sci. Ukrainian, SSR, 102-110. Ministry of Agriculture and Fisheries Trans. N.S., (72).
Ponomareva, L.A., 1954. Euphausiids of the Sea of Japan feeding on the copepoda. Dokl. Acad. Sci. USSR (Zool.), 98, 153-154.
Ponomareva, L.A., 1963. The euphausiids of the North Pacific, their distribution and mass species (in Russ., Engl. summary). Moscow, 142 p.
Reeve, M.R., 1964. Feeding of zooplankton, with species reference to some experiments with *Sagitta*. Nature, 201, 211-213.
Rigler, F.H., 1961. The relation between concentration of food and feeding rate of *Daphnia magna* Straus. Can. J. Zool., 39, 857-868.
Roger, C., 1971. Euphausiids of the equatorial and south tropical Pacific Ocean: zoogeography, ecology, biology and trophic relationships. Thesis to be publ. in Mémoires Orstom.
Smith, P.E., 1972. The increase in spawning biomass of the northern anchovy, *Engraulis mordax*. Fish. Bull., U.S., 70, 849-874.
Wickstead, J.H., 1965. An introduction to the study of tropical plankton. Hutchinson and Co., London, 160 p.

G.H. Theilacker
National Marine Fisheries Service
Southwest Fisheries Center
La Jolla, Calif. 92037 / USA

R. Lasker
National Marine Fisheries Service
Southwest Fisheries Center
P.O. Box 271
La Jolla, Calif. 92037 / USA

Chemical Changes during Growth and Starvation of Herring Larvae

K. F. Ehrlich

INTRODUCTION

It is generally accepted that the size of a marine fish year class is determined by the recruitment within the first year of life. The greatest proportion of the mortality occurs before the fish complete their larval stages. Many factors influence survival of young fishes: predation, genetic defects, perturbations of the physical environment, and starvation. The relative importance of each of these forms of mortality is not known, but it is thought that starvation may be a major source.

A "critical period" in the life of larvae is associated with the resorption of yolk, at which time mass mortalities could occur due to insufficient food or feeding difficulties. Whether or not this type of mass mortality does occur in the sea has been debated for some time (see review by May, this symposium). Laboratory studies have demonstrated, however, that larvae of many species, deprived of food for a very short time (often only 1 or 2 days), after resorption of yolk, reach a point of irreversible starvation, resulting in complete mortalities (e.g. Blaxter and Hempel, 1963; Lasker et al., 1970). This suggests that many marine fish larvae may be greatly susceptible to starvation and mass mortalities.

The ability to determine the nutritional condition of sea-caught larvae could be of help to fishery biologists in prediction of larval survival and potential brood strength. Until now estimates of condition have been largely based on morphological measurements such as weight/length^3 or body height, etc. (e.g. Shelbourne, 1957; Blaxter, 1971); but these results have not been explicit, due to concurrent loss of weight and length during starvation. Furthermore, condition factor increases with length in healthy larvae due to allometric growth and ossification of the skeleton, etc. Observation of chemical changes (water, triglyceride, carbohydrate, total nitrogen, total carbon, and ash) was thought to be a better way in which to delineate nutritive condition, at the same time improving the knowledge of the physiology and biochemistry of marine fish larvae.

Considerable chemical work has been done on both young and adult of fresh water species and on starvation of adult marine species (see review by Love, 1970). Chemical work on marine fish larvae is scarce. Apparently only 3 references deal with chemical changes during starvation of marine fish larvae. Marshall, Nicholls, and Orr (1937) determined fat in herring larvae. Lasker (1962) worked on sardine eggs and ovaries but also analyzed changes up to the end of the yolk-sac stage in larvae. May (1971) measured nitrogen, carbon, hydrogen, and ash of grunion during 20 days of growth and starvation.

The research described in this paper is part of a study on chemical, behavioural, and morphometrical changes during growth and starvation of laboratory-reared herring and plaice larvae. This paper describes the chemical changes occurring during ontogenesis and starvation of herring (*Clupea harengus* L.) larvae, with the particular aim of clearly identifying larvae which have reached irreversible starvation. The behavioural aspects are presented separately in this symposium, and the rest of the study will be published elsewhere.

METHODS

Source of Material and Rearing Conditions. Herring gametes obtained in February 1970-1972 from Firth of Clyde spawners were used to rear the larvae. The methods of Blaxter (1968) modified by Ehrlich (1972) were employed to culture substantial numbers of larvae to and beyond metamorphosis. The eggs were kept at about 9.5°C. Larvae were transferred to tanks with water at 7.8°C, which gradually rose to 12-13°C by mid June. The sea water salinity fluctuated somewhat but was usually in the range of 32-33‰. Light was supplied after hatching by 80W fluorescent tubes.

Sampling Procedures. The basic experimental sampling pattern was to rear larvae from hatching through metamorphosis and at various ages, starting at hatching, to allow groups to starve. In this study "starvation" means deprivation of large particulate food such as plankton or *Artemia*. No attempt was made to control dissolved organic material or very fine particles, although all sea water was filtered twice through synthetic fibre wool. However, the extent of feeding on these small organic compounds as well as the benefit derived from them was assumed to be minimal. Feeding on these sorts of substances was described by Pütter (see review by Morris, 1955). Fish were sampled at the start of starvation and usually every three days thereafter until death, or until all larvae in a group being starved had been sampled. The condition of irreversible starvation was termed the "point-of-no-return" by Blaxter and Hempel (1963). Their nomenclature is used in this paper and is frequently referred to as the PNR. The time for different aged fish to reach the PNR is from Blaxter and Ehrlich (this symposium). The end of the yolk-sac stage is referred to as the EYS.

In the first year samples consisted of groups of larvae; in subsequent years individuals were used. Grouped samples were rinsed in distilled water to remove external salt, quickly frozen, freeze-dried, and stored in desiccators at -20°C. After thawing, the group dry weight was measured on a Beckman EMB 1 electrobalance, which could weigh to the nearest 0.5 µg. At hatching about 20 larvae were used per sample, giving a dry weight of at least 2 mg. Absorption of atmospheric water was found to be less than 1% during weighing. The grouped samples were then ground for chemical analyses. In later years when individuals were used wet weights were also determined.

After a larva was rinsed in distilled water, it was briefly dried on a filter paper and transferred to a small aluminium boat for weighing. Wet weight of each sample was recorded at 1, 2, and 3 min. after placing the larvae on the filter paper. Decrease in weight over this time was linear, so it was a simple matter to extrapolate to zero time.

Starvation experiments during the 1st year were started at the EYS and at 16, 24, 41, 55, 74, and 88 days post-hatching. In the 2nd year, whe individuals were used, starvation experiments were done at the EYS and at 30 and 50 days post-hatching; larvae were also sampled 9 times within the 94 days following hatching for examining the effect of growth on body chemistry. In the 3rd year larvae were starved only from the end of the yolk-sac stage. Results from the 3 years were combined for this paper.

Chemical Techniques. Triglyceride determinations were based on the method of van Handel (1961) as modified by Heath and Barnes (1970). Glycerol tristearate was used as the standard. Carbohydrate was determined by the method of Dubois et al. (1956), using a glucose standard with an additional blank without phenol (see Ehrlich, 1972). Total nitrogen and carbon contents were measured on a Perkin Elmer Elemental Analyzer. Samples were combusted for 12 h at 520°C to determine the amount of ash.

Data Handling. Unless otherwise stated, results are reported as percent of dry weight. Percentages are relative values; and so when one component increases, another falls. Percentages thus show the relative size of a particular component or store in an individual but not the actual amount. Although changes in actual amount (% x dry weight in mg x 10 = µg) are useful during growth and especially starvation of a particular size group, they are difficult to use when comparing animals of different sizes. To compare changes during starvation of a particular constituent, independent of the other components, the amount of the constituent has also been shown as a percentage of the initial amount present at the start of the starvation. Thus results have been reported in 3 ways: percent of dry weight, actual amount (µg), and percent of initial amount at start of starvation.

When rearing fishes one finds an increasing size range during growth to metamorphosis. In this period of growth the fish are developing as well as increasing in size. Developmental condition was thought to increase with size more than with age. Accordingly, most of the substances measured in herring showed a better correlation with size than with age of larvae (Figs 1 and 2); for this reason dependent variables have been initially plotted against length of larvae. This was done each day for starved as well as unstarved larvae. Length was chosen instead of weight, because it is an easier reference and did not require a logarithmic scale for plotting the wider weight range. To compare changes during starvation of different sized herring larvae, 6 lengths were chosen as standard sizes. Unstarved lengths of 12, 15, 21, 30, and 35 mm were chosen, because they best illustrated the differences between larvae of increasing size. The use of standard-sized animals is based on the method of Barnes et al. (1963) except that an allowance was made for shrinkage in length of larvae, prior to ossification, during the period of experimental starvation.

Computations and drawings of regression lines and calculation of other statistics were greatly aided by the use of the Hewlett Packard desk computer 9100B. In the present study the lowest level of significance was fixed at 5% ($P \leq 0.05$). Differences distinguishable at this level are referred to as significant; those distinguishable at the 1% level are termed highly significant. In the statistical handling of regression lines the homogeneity of residual variances was tested prior to all covariance analyses. In all cases these variances were homogeneous. When computing changes of some component against length, individual values were used for calculation of regression lines. High individual

Table 1. Best estimates of relationships between various parameters and length (L) for feeding and starved larvae

Days Starved		Equation	n
Weight (W) 10-50 mm			
0		log W = 4.57 log L -5.7052	193
3		-5.7761	36
6		-5.8163	36
9		-5.8569	74
12		-5.8838	25
15		-5.9395	25
Water (Wr) 10-50 mm			
0		% Wr = 15.75 log L +105.98	129
3		+107.15	71
6, 9		+107.96	37
12		+108.67	16
Moribund		+106.10	21
Triglyceride (T) 10-27 mm			
0		% T = $-1.71\ L + 0.04\ L^2 + 22.22$	34
3		$-1.42\ L + 0.04\ L^2 + 18.48$	25
6		$-1.64\ L + 0.04\ L^2 + 20.32$	29
9		$-1.42\ L + 0.03\ L^2 + 17.89$	30
12		$-1.08\ L + 0.03\ L^2 + 13.61$	14
15		$-1.03\ L + 0.02\ L^2 + 12.86$	13
Triglyceride (T) 27-50 mm			
0		% T = 0.08 L +10.10	29
3		+ 9.06	9
6		+ 8.17	9
9		+ 6.54	10
12		+ 6.28	5
15		+ 5.63	15
Carbohydrate (Ch) 10-50 mm			
0	10-21 mm	% Ch = 0.16 L + 1.55	37
	21-29 mm	-0.23 L + 9.48	26
	29-50 mm	0.001L + 3.19	22
3		% Ch = 0 L + 3.26	27
6		+ 3.12	31
9		+ 2.92	44
12		+ 2.77	29
15		+ 2.69	29
Nitrogen (N) 10-21 mm			
0		% N = 0.20 L + 8.02 (days 0–15)	31
3			21
6			28
9			21
12			13
15			14

Table 1. (continued)

Days Starved	Equation	n
Nitrogen (N) 21-50 mm		
0	log % N = -0.14 log L + 1.27	52
3		7
6		7
9		10
12		12
15		16
Ash (A) 10-25 mm		
0	% A = 0 L + 7.50	43
3	+ 8.53	30
6	+ 9.48	30
9	+ 9.68	31
12	+10.17	19
15	+11.43	23
Ash (A) 25-50 mm		
0	% A = 0.15 L + 3.90	34
3	+ 4.71	9
6	+ 5.24	7
9	+ 6.05	9
12	+ 6.76	9
15	+ 8.16	15
Carbon (C) 10-18 mm		
0	% C = 0.22 L +39.57	28
3	+38.96	17
6	+37.83	28
9	% C = 0.57 L +31.85	22
12	+29.27	18
15	+29.22	16
Carbon (C) 18-50 mm		
0	% C = 0 L +44.34	62
3	+43.13	11
6	+41.93	10
9	+40.50	19
12	+40.50	14
15	+40.50	19

variability made composite drawings of successive regression lines during starvation unclear. For this reason the individual points have been only drawn for zero days-starved to show typical variation about lines.

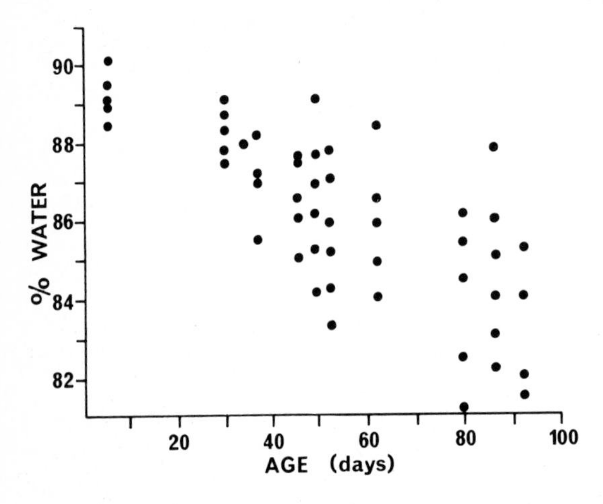

Fig. 1. Relationship between % water and larval age

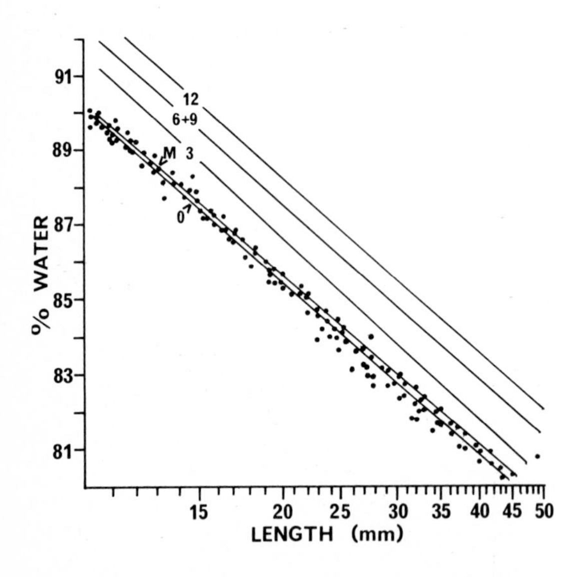

Fig. 2. Relationship between % water and fish length. Lines are drawn for feeding, starved, and moribund (M) fish. The numbers on the lines indicate days of starvation. Typical variation about the lines is shown for unstarved fish only

The regression equations of each component against length were tested for differences between days starved and subsequently for differences between the slopes, if the lines were different. Small slopes, origin or pooled, were tested to see if they were distinguishable from zero. When the lines or slopes were indistinguishable between days starved the appropriate terms were pooled to obtain the best estimates of the regressions, which are given in Table 1. The number of points on each line are also listed, but due to lack of space the original equations have not been given when pooled parameters have been listed. The original regression equations, standard deviations of the slopes and of y for fixed x, and standard errors of the intercepts as well as results of statistical analyses are given by Ehrlich (1972).

RESULTS

Length-Weight

Length-weight relationships were based on live lengths and dry weights. Length-weight lines were determined for unstarved larvae and for starved ones at each day of sampling (Fig. 3 and Table 1). The lines were tested for homogeneity and were found to be different by a highly significant amount, although no significant differences were found between the slopes. The slopes were pooled to get the best estimate of the true slope, and this was then used for determining the equations of the regression lines of weight on length for unstarved and starved larvae. These are the equations, based on the pooled slope, that were used for determining the weight of the standard length larvae at the different days starved. Changes in weight, both actual (mg or μg) and relative (% of initial weight), during starvation of those larvae are shown in Table 2.

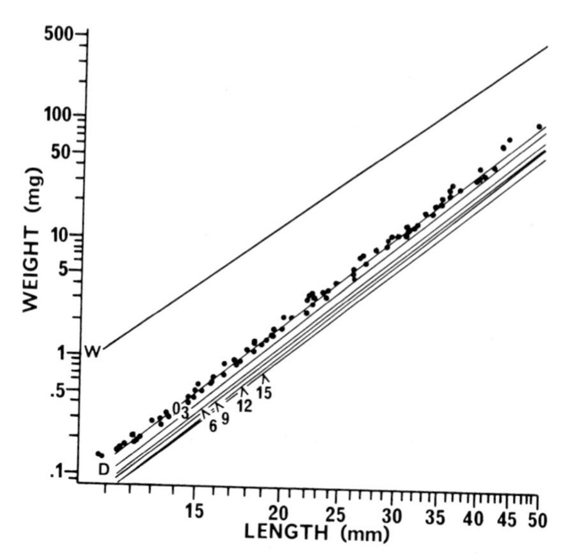

Fig. 3. Length-weight relationships. The line based on wet weights (W) was calculated from the water content of different sizes of larvae: log wet weight = 4.10 log length - 4.2421. The numbers on the dry weight (D) lines indicate days of starvation. Typical variation about the lines is shown for unstarved fish only

Table 2. Larval standard amounts

Length	Days	Wet Wt.	Dry Wt.		Water			Triglyceride			Carbohydrate		
mm	Starved	mg	mg	% init.	%	mg	% init.	%	µg	% init.	%	µg	% init.
12.00	0	1.533	0.169	100	89.0	1.364	100	8.1	13.6	100	3.4	5.8	100
12.06	3	1.484	0.147	87	90.1	1.338	98	6.7	9.9	72	3.2	4.8	83
11.94	6[a]	1.420	0.128	76	91.0	1.292	95	6.7	8.6	63	3.1	4.0	69
11.68	9	1.190	0.105	62	91.2	1.084	80	6.0	6.3	46	2.9	3.1	53
11.60	12	1.185	0.096	57	91.9	1.090	80	4.8	4.6	33	2.8	2.7	46
11.46	15	0.753[b]	0.080	47	89.4[b]	0.673[b]	49[b]	4.4	3.5	26	2.7	2.2	37
15.00	0	3.733	0.468	100	87.4	3.264	100	6.5	30.6	100	3.9	18.2	100
15.42	3	3.902	0.452	96	88.4	3.450	106	5.4	24.2	79	3.2	14.8	81
15.12	6	3.542	0.376	80	89.4	3.166	97	5.1	19.1	62	3.1	11.7	65
	8[a]								14.9			9.5	
14.54	9	2.768	0.286	61	89.6	2.481	76	4.5	12.9	42	2.9	8.4	46
14.47	12	2.716	0.261	56	90.4	2.454	75	3.7	9.6	31	2.8	7.2	40
14.26	15	1.793[b]	0.217	46	87.9[b]	1.576[b]	48[b]	3.3	7.2	23	2.7	5.8	32
21.00	0	14.69	2.181	100	85.2	12.50	100	5.8	127.7	100	4.8	104.2	100
22.13	3	14.98	2.103	96	86.0	12.87	103	5.1	107.6	84	3.2	68.6	66
20.73	6	12.46	1.592	73	87.2	10.86	87	4.3	68.2	53	3.1	49.7	48
20.29	9[a]	10.41	1.314	60	87.4	9.09	73	3.2	42.1	33	2.9	38.4	37
20.16	12	10.10	1.200	55	88.1	8.90	71	2.9	34.8	27	2.8	33.2	32
19.91	15	6.94[b]	0.996	46	85.6[b]	5.94[b]	47[b]	2.4	23.6	18	2.7	26.8	26
25	0	30.17	4.839	100	84.0	25.33	100	7.2	347	100	3.8	185	100
	3	27.64	4.110	85	85.1	23.53	93	6.0	248	72	3.2	134	72
	6	26.67	3.747	77	86.0	29.92	90	5.4	204	59	3.1	117	63
	9	24.29	3.413	71	86.0	20.88	82	3.8	131	38	2.9	100	54
	12[a]	24.03	3.208	66	86.6	20.82	82	3.6	116	34	2.8	89	48
	15	17.71[b]	2.822	58	84.1[b]	14.89[b]	59[b]	2.9	82	24	2.7	76	41
30	0	64.41	11.136	100	82.7	53.27	100	7.6	845	100	3.2	362	100
	3	58.68	9.458	85	83.9	49.22	92	6.6	620	73	3.2	308	85
	6	56.32	8.622	77	84.7	47.70	90	5.7	489	58	3.1	269	74
	9	51.29	7.853	71	84.7	43.44	82	4.0	316	37	2.9	229	63
	12	50.59	7.381	66	85.1	43.21	81	3.8	279	33	2.8	204	56
	15[a]	37.79[b]	6.493	58	82.8[b]	31.30[b]	59[b]	3.1	203	24	2.7	175	48
35	0	122.8	22.53	100	81.7	100.31	100	7.2	1615	100	3.2	732	100
	3	111.4	19.14	85	82.8	92.31	92	6.1	1175	73	3.2	624	85
	6	106.6	17.44	77	83.6	89.18	89	5.2	916	57	3.1	544	74
	9	97.1	15.89	71	83.6	81.22	81	3.6	574	36	2.9	464	63
	12	95.4	14.93	66	84.4	80.48	80	3.4	502	31	2.8	414	56
	15	72.0[b]	13.14	58	81.8[b]	58.92[b]	59[b]	2.7	356	22	2.7	353	48

[a]Start of PNR

[b]Moribund

Chemical Composition

Water. The relationship between percent water and length of herring larvae was a negative logarithmic function of length (Fig. 2 and Table 1). At any length, percent water increased during starvation over the first 12 days; no differences were distinguishable, however, from 6 to 9 days starved. When the larvae became moribund the percentage of water in them decreased. The regression lines of percent water versus length for the various days starved were different by a highly significant amount, but the slopes were not significantly different. The changes in percent water during the starvation of the 6 standard sizes of herring larvae are shown in Fig. 4. The increas-

Table 2 (continued)

Nitrogen			Ash			Carbon			C/N	
%	μg	% init.	%	μg	% init.	%	μg	% init.		% init.
10.4	17.6	100	7.5	12.7	100	42.2	71.3	100	4.05	100
10.4	15.3	87	8.5	12.5	99	41.6	61.1	86	3.99	98
10.4	13.3	76	9.5	12.1	96	40.5	51.7	72	3.89	96
10.4	10.9	62	9.7	10.2	80	38.6	40.6	57	3.72	92
10.3	9.9	56	10.2	9.8	77	35.9	34.4	48	3.47	86
10.3	8.2	47	11.4	9.1	72	35.8	28.6	40	3.47	86
11.0	51.6	100	7.5	35.1	100	42.9	201	100	3.89	100
11.1	50.2	97	8.5	38.5	110	42.4	191	95	3.81	98
11.0	41.6	81	9.5	35.7	102	41.2	155	77	3.73	96
	34.7			30.4			128			
10.9	31.3	61	9.7	27.7	79	40.2	115	57	3.68	95
10.9	28.5	55	10.2	26.5	76	37.6	98	49	3.44	88
10.9	23.6	46	11.4	24.8	71	37.4	82	40	3.44	88
12.2	267	100	7.5	164	100	44.3	967	100	3.63	100
12.4	262	98	8.5	179	110	43.1	907	94	3.46	95
12.2	194	73	9.5	151	92	41.9	667	69	3.44	95
12.1	159	60	9.7	127	78	40.5	532	55	3.35	92
12.1	145	54	10.2	122	75	40.5	486	50	3.36	93
12.1	120	45	11.4	114	70	40.5	404	42	3.37	93
11.8	573	100	7.5	365	100	44.3	2146	100	3.74	100
11.8	487	85	8.5	348	95	43.1	1773	83	3.64	97
11.8	444	77	9.2	345	94	41.9	1571	73	3.54	95
11.8	404	71	9.7	332	91	40.5	1382	64	3.42	91
11.8	380	66	10.3	330	90	40.5	1299	61	3.42	91
11.8	334	58	11.6	328	90	40.5	1143	53	3.42	91
11.5	1285	100	8.3	925	100	44.3	4938	100	3.84	100
11.5	1092	85	9.1	863	93	43.1	4080	83	3.74	97
11.5	995	77	9.6	831	90	41.9	3615	73	3.63	95
11.5	906	71	10.5	821	89	40.5	3180	64	3.51	91
11.5	852	66	11.2	824	89	40.5	2989	61	3.51	91
11.5	749	58	12.6	816	88	40.5	2630	53	3.51	91
11.3	2543	100	9.0	2037	100	44.3	9989	100	3.93	100
11.3	2160	85	9.9	1887	93	43.1	8255	83	3.82	97
11.3	1969	77	10.4	1811	90	41.9	7314	73	3.71	94
11.3	1794	71	11.2	1778	87	40.5	6434	64	3.59	91
11.3	1686	66	11.9	1770	87	40.5	6048	61	3.59	91
11.3	1483	58	13.3	1747	86	40.5	5320	53	3.59	91

ing percentage of water during starvation was only a relative change, for there was an actual loss of water (Table 2). The only exception to this was in 15 and 21 mm larvae where during the first 3 days of starvation there was an actual increase in water. The rate of loss of water during food deprivation to the PNR as a percentage of the initial amount present (Fig. 7) tended to decrease as larvae increased in length, at least up to 21 mm. Above this length the rate of loss stayed about the same for the different sizes of larvae. In moribund larvae the loss of water as a percentage of the initial amount was almost twice that of larvae starved for the same number of days but not yet moribund (Fig. 7).

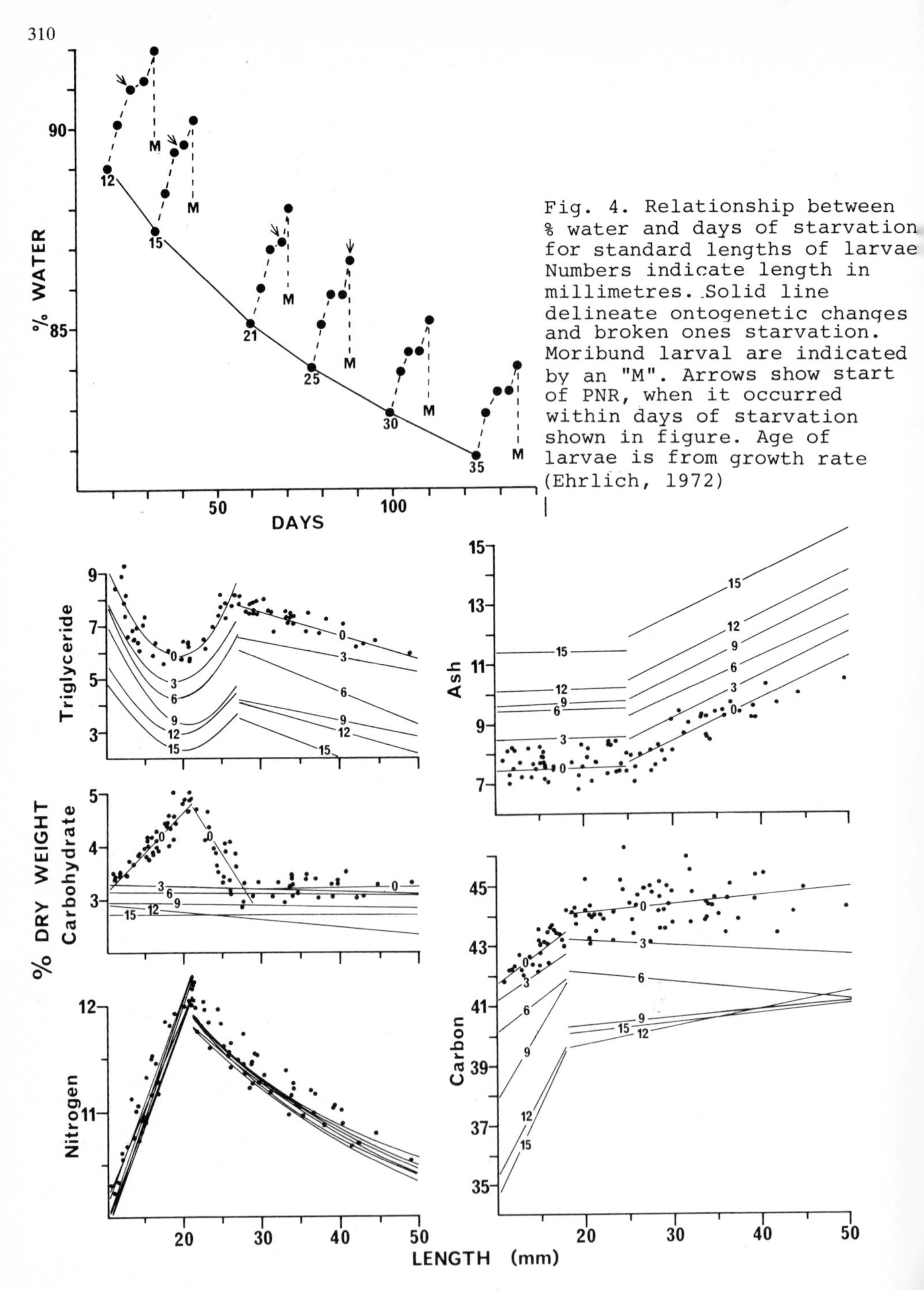

Fig. 4. Relationship between % water and days of starvation for standard lengths of larvae Numbers indicate length in millimetres. Solid line delineate ontogenetic changes and broken ones starvation. Moribund larval are indicated by an "M". Arrows show start of PNR, when it occurred within days of starvation shown in figure. Age of larvae is from growth rate (Ehrlich, 1972)

Fig. 5. Relationship between chemical composition, as % of dry weight and length. Changes are shown for feeding and starved fish; numbers indicate days without food. Typical variation about lines is shown for unstarved fish only

Triglyceride. This constituent showed a concave parabolic relationship with length up to 26 mm, decreasing from 8.1% in larvae at the end of the yolk-sac stage to 5.7% in those just over 19 mm and then increasing to 7.8% by 26 mm. In larger larvae, triglyceride decreased in a linear fashion with increasing length to a value of 7.2% by 35 mm (Fig. 5 and Table 1). The slopes of these curves were maintained throughout starvation.

During 15 days of starvation triglyceride decreased approximately by 3.4 to 4.5% irrespective of the size of the larvae, but the percentage of the dry weight utilized to reach the PNR increased with increasing size from just over 1%, in the smallest herring larvae, to over 4% in those 30 mm long (Fig. 6 and Table 2). The percentage of the unstarved store utilized to reach the PNR increased with the length of the larvae, from 37% in the smallest to 76% in those 30 mm in length (Fig. 7 and Table 2). The decrease in the rate of triglyceride utilization to the PNR with increasing larval size was also shown in terms of the rate of depletion of the actual amount (Table 3).

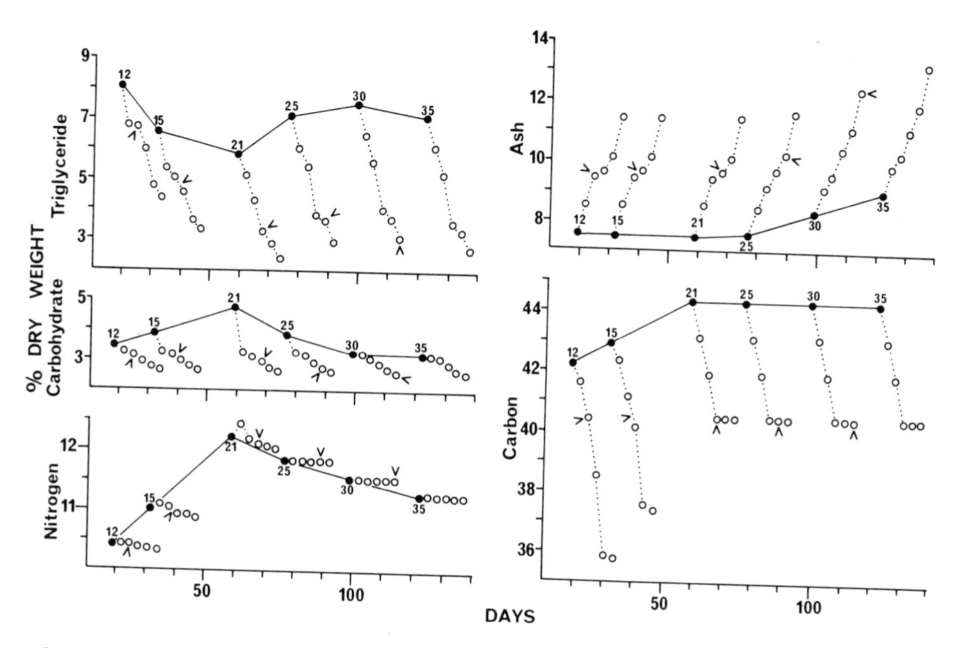

Fig. 6. Relationship between chemical composition (% dry weight) and days of starvation for standard lengths of larvae. Numbers indicate length in millimetres. Solid circles delineate ontogenetic changes and open ones, starvation. Arrows show start of PNR when it occurred within days of starvation shown in figures. Age of larvae is from growth rate (Ehrlich, 1972)

Carbohydrate. Percent carbohydrate during development showed three relationships with length: from the EYS to 21 mm, from 21 mm to 29 mm and from 29 to 50 mm (Fig. 5 and Table 1). In the first phase percent carbohydrate increased with length from 3.4 to 4.8%. In the second, it decreased to 3.2% and then remained constant at this level in the third.

During starvation (days 3 to 15) the regressions of percent carbohydrate against length were highly significantly different, but the slopes were statistically indistinguishable and not different from zero (Fig. 5 and Table 1). The mean values of percent carbohydrate during starvation (days 3 to 15) are also shown in Table 1. In terms of percent changes per day of starvation the only differences between larvae of different lengths occurred during the first three days of food deprivation. In this period the percentage loss of carbohydrate was dependent on the size of the unstarved store (Fig. 6). The percentage of the unstarved store utilized for any given number of days of starvation also increased with the level of carbohydrate in unstarved herring larvae (Fig. 7). There was a trend for the rate of carbohydrate catabolism to the PNR to decrease with increasing size, but this was in part masked by the varying size of the unstarved store (Fig. 7 and Table 3).

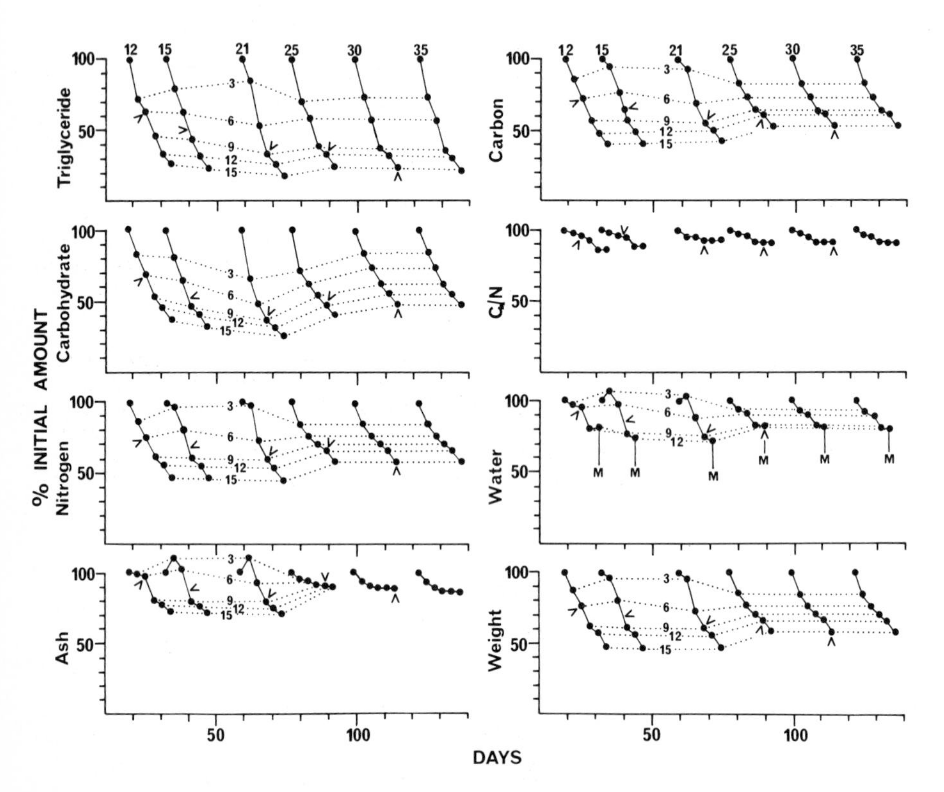

Fig. 7. Relationship between % unstarved amount and days of starvation for standard lengths of larvae. Small numbers show days of starvation, larger ones, at top, length in millimetres. Arrows show start of PNR when it occurred within days of starvation shown in figures. Larval age is from growth rate (Ehrlich, 1972)

Nitrogen. The percentage of nitrogen in herring larvae displayed two relationships with length during growth: from the EYS to 21 mm it increased linearly with length (10.4 to 12.2%); in larger larvae it decreased as a logarithmic function of length to 11.3% at 35 mm (Fig. and Table 1).

In both size groups no significant differences were detected during starvation for the regression lines of percent total nitrogen versus length (Fig. 6). However, when the actual amounts of nitrogen were considered obvious starvation decreases were observed (Fig. 7 and Table 2). The percentage of the unstarved store used to the PNR tended to increase with increasing larval size, although the rate of nitrogen catabolism to the PNR also varied with the size of the unstarved level of nitrogen (Table 3).

Ash. Percent ash showed two relationships with length: from the EYS to 25 mm, feeding herring larvae contained 7.5% ash independent of their length, but in larger larvae the percentage of ash increased with length to 9.0% at 35 mm (Fig. 5 and Table 1).

In all sizes of larvae percent ash significantly increased during starvation, but the rate of change of percent ash per unit length did not significantly vary (Fig. 5 and Table 1). Because of this the differences in the percentage of ash at the PNR were largely due to the differences in the number of days required to reach this condition (Fig. 6 and Table 2). Although the percentage of ash increased up to the PNR the actual amount of ash, in all sizes of herring larvae, slightly decreased (Fig. 7 and Table 2). There was no general pattern of the rate of utilization of ash to the PNR (Table 3).

Carbon. The percentage of carbon in herring larvae increased with the development from the EYS to 18 mm in length (42.2 to 44.3%). In larger fish percent carbon remained constant at the level reached by 18 mm (Fig. 5 and Table 1).

During the starvation of the smaller larvae the rate of increase of percent carbon with length changed between 6 and 9 days of food deprivation. However, within each group of days (0 to 6 and 9 to 15) the regression lines of percent carbon versus length were significantly different, but the slopes were not statistically distinguishable (Table 1). For larvae larger than 18 mm the regression lines of per-

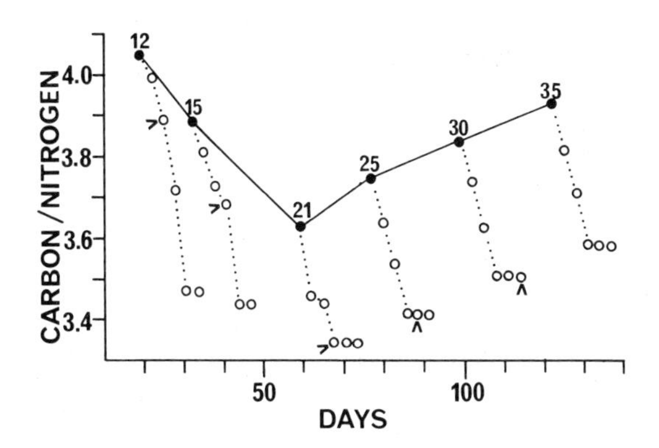

Fig. 8. Relationship between carbon-nitrogen ratio and days of starvation for standard lengths of larvae. Numbers indicate length in millimetres. Solid circles delineate ontogenetic changes and open ones, starvation. Arrows show start of PNR, when it occurred within days of starvation shown in figures. Age of larvae is from growth rate (Ehrlich, 1972)

cent carbon versus length differed up to 9 days of starvation, but after this no further differences were detected. In all larvae from the EYS to 30 mm percent carbon at the PNR was between 40.0 to 40.5% (Fig. 6). However, the percentage of the unstarved store utilized to reach the PNR tended to increase in larger larvae (Fig. 7), although the rate of carbon depletion to the PNR tended to decrease with increasing size (Table 3).

Carbon-Nitrogen Ratio. In feeding herring larvae the carbon-nitrogen ratio decreased with development from 4.05 at the end of the yolk-sac stage to 3.63 at 21 mm in length. It then increased with further growth to 3.93 in 35 mm larvae (Fig. 8).

During starvation the C-N ratio decreased in all sizes of herring larvae, but the size of the ratio at the PNR was dependent on the relative amounts of carbon and nitrogen at the start of starvation. Although the C-N ratio of all sizes of herring larvae decreased during starvation, the percentage of the decrease from the unstarved ratio was small as compared to other chemical components. In all larvae the difference in the C-N ratio between feeding larvae and those starved to the PNR did not exceed 10% (Fig. 7).

DISCUSSION

Length versus Age

The length of herring was found to be a better parameter than age against which to compare changes during growth up to metamorphosis. This has been shown for percent water (Figs 1 and 2) but was also true for the other components. This same situation has also been found for DNA content in the brains of herring larvae (Packard, this symposium). This is quite understandable if one considers the wide range of growth rates (Blaxter, 1968; Ehrlich, 1972) obtained when the larvae were not only growing but also developing. It seems reasonable to assume that chemical changes during growth reflect developmental changes; considering the variation in growth rates, length should be a better parameter than age against which to compare changes.

Relative versus Actual Amounts

Fig. 9, showing starvation changes in actual amounts of components in larvae at the end of the yolk-sac stage, has been included for comparison with Figs 6 and 8 to stress differences between actual and relative amounts of constituents. These figures show that during starvation water increases relatively but decreases in actual amount, the same being true for ash; triglyceride, carbohydrate, and carbon show parallel changes in actual and relative terms; nitrogen, however, barely changes in relative amounts, while the actual amounts drop sharply. Clearly, one must be very explicit when describing chemical changes during starvation as to whether they are relative or actual. Fig. 9 shows actual changes only for end-of-yolk-sac larvae for the sake of emphasis and clarity. The exact amounts for all standard sizes of larvae are given in Table 2.

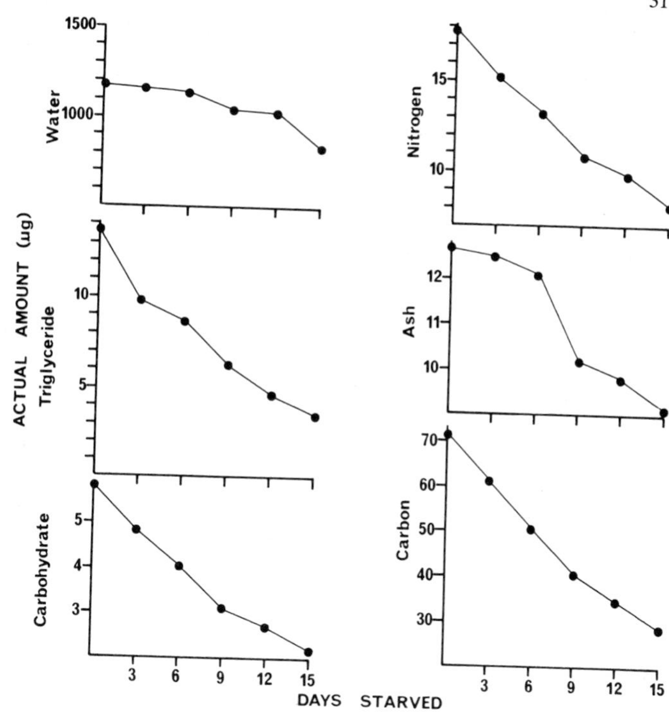

Fig. 9. Changes in actual amounts of chemical components during starvation of larvae from end of yolk-sac stage. PNR was reached after 6 days

Ontogenetic Changes

Percent water decreased with increasing length during growth from the EYS (Fig. 2). This is probably a general phenomenon of growth during development. The probable cause of these changes in young fishes is that the dividing cells lead to high amounts of extracellular space and water content, both of which decrease with growth (Love, 1970). Further evidence for this is shown in that cells increase in size with increasing length of fish (Love, 1958).

It should be noted that all of the changes in Fig. 5 were only relative ones as percent of the dry body weight, and that throughout this period of growth all constituents were increasing in actual amounts but at different rates (Table 2). In the period from the EYS to about 20 mm, nitrogen and carbohydrate were being laid down faster than neutral fat stores. This suggests that it may be to the advantage of herring larvae from hatching to convert food into growth than into purely accumulated energy stores. The total nitrogen deposition representing protein would indicate growth. Their increased size would enable the larvae to feed on a wider size range of food particles (Blaxter and Hempel, 1963). After reaching about 20 mm the larvae started to accumulate neutral fat faster than protein. Larvae of this size probably reached a length that enabled them to capture a sufficiently wide range of food to warrant a change in their metabolism toward increased deposition of fat. The increase in lipid most likely does not reflect a mere increase in consumption of food since food was supplied in excess from the EYS. However, this possibility should not be completely ruled out since it is known that young herring larvae are not very efficient in capturing food (Rosenthal and Hempel, 1970; Blaxter and Staines, 1971). In any case, accumulation of lipid would then increase the energy reserves of the larvae, as fat provides nearly twice the calories per unit weight of protein (Winberg, 1971). Even after the percentage of triglyceride started to decrease (above 25 mm), fat stores were still being laid down faster than protein, as indicated by the ratio of carbon to nitrogen (Fig. 8). In organisms with low amounts of carbohydrate the carbon-nitrogen ratio indicates the relative proportion of lipid to protein (Mullin and Brooks, 1970). Marshall et al. (1937) also found that larval herring increased the relative size of their fat store with growth towards the very high amounts retained by adults. The ash is probably an indicator of the start of ossification. The close correlation between the changing patterns of nitrogen and carbohydrate content may be explained by the requirements of a certain amount of carbohydrate for efficient utilization of protein, as was shown to be the case for carp (Nagai and Ikeda, 1971).

Starvation Changes

Water. During starvation, percent water increased in all sizes of herring larvae (Fig. 4). This change during food deprivation, whether artificially starved in the laboratory or naturally in the sea, from lack of food or during use of reserves for gonad formation, has been reported for many species (see review by Love, 1970). The general range of increase of water during starvation of marine fishes was 2 to 4%. This same range was found for starved herring and also plaice larvae (Ehrlich, 1974) and may be indicative of a general pattern in starvation of marine fishes, while fry of fresh water carp have shown increases of about 10% during starvation (Satomi, 1969). Love et al. (1968) showed that during starvation of cod there was an increase in extracellular space and a decrease in cell size. The same phenomenon was also reported for American plaice (Templeman and Andrews 1956); both of these changes would result in a relative increase in percent water although the actual amount can decrease, as has been found for the larvae (Table 2).

Triglyceride. Throughout starvation of all sizes of herring, triglyceride continually decreased as a percentage of the dry weight (Fig. 6). This component as compared to the others was reduced to the smallest percentage of the original store. Since fat stores offer the larvae the greatest source of stored energy it is quite expected that they should be heavily drawn upon during starvation. That triglyceride was not totally depleted during initial starvation but rather used throughout suggests that a certain amount may be needed to sustain life or possibly for efficient utilization of other components, most likely protein. The percentage of the unstarved triglyceride store utilized to reach the PNR increased with increasing length of the larvae (Fig. 7). This could be interpreted in two ways: first, it could mean that larger larvae were capable of surviving greater depletion, or second, that if triglyceride was used at about a constant rate, then the smaller larvae could have used less of this store due to their shorter survival time. Table 3, however, shows generally for herring that the rate of loss of triglyceride to the PNR decreased with increasing development. This suggests that the efficiency of triglyceride catabolism increased during ontogenesis, or in other words that depletion resistance improved during development.

:arbohydrate. The rate of carbohydrate utilization during starvation lid not vary with larval size when it was about 3.2% of the dry weight or less (Fig. 6). However, when more was present the rate of use was lependent on the store size. The rapid loss of carbohydrate to some ninimal level is quite expected, since it is usually an energy store and expended during a short period of starvation (Love, 1970). Smith (1952) also showed carbohydrate of trout eggs was quickly utilized. The initial loss of carbohydrate probably represented depletion of the carbohydrate stores, while the carbohydrate present after this time nay represent products formed from the conversion of fat or protein, which can mask glucose depletion during starvation (Robertson et al., .961 cited by Love, 1970).

litrogen. Changes in percent nitrogen during starvation of herring arvae were either very small or not at all detectable. However, the actual amount of nitrogen decreased from the beginning of starvation, showing that protein was being catabolized (Table 2). Breakdown of protein from the start of starvation was also found in larval plaice Ehrlich, 1974) and in grunion larvae (May, 1971). Although May eported percent protein increased during starvation, the actual mounts (percent times dry weight) decreased. The breakdown of protein rom the start of starvation may appear to be somewhat surprising in hat lipids are more commonly depleted before protein is mobilized Love, 1970), at least in adult fishes. However, the level of fat in hese fishes, even when unstarved, amounted to only about 2% of their vet weight. This value for lipid was obtained by subtracting the sum of the other components from 100% and is comparable to that found in starved herring (Wilkins, 1967) and trout (Phillips et al., 1960). Considering this it does not seem unreasonable that protein should have been metabolized from the start of starvation.

he rate of nitrogen utilization per unit weight to the point-of-no-eturn decreased with increasing development as it did for triglyceride nd carbohydrate (Table 3). The pattern was the same for nitrogen as or carbohydrate, i.e. an increasing rate of use up to 21 mm length ollowed by subsequent decreases. It seems reasonable that the rate of depletion, during starvation, should reflect the size of the un-tarved store (i.e. when there is more of a component it will be called upon to a larger extent). In terms of percent of the unstarved mount (Fig. 7) it could be seen that the proportion of nitrogen tilized to reach the PNR increased with development. This same general ituation was found for triglyceride, carbohydrate, and nitrogen and ppears to suggest increasing depletion tolerances with growth and evelopment.

:arbon. Changes in carbon were probably in large part reflections of at and carbohydrate, although it was also obviously dependent on pro-ein. This can be seen in rates of loss to the PNR (Table 3). In any ase carbon appeared to be a good index of condition (Fig. 6). In erring larvae, where percent nitrogen did not change much during starvation while triglyceride and carbohydrate decreased, the carbon-itrogen ratio decreased.

sh. Ash content as a percentage of the dry weight markedly increased uring starvation (Fig. 6). However, when the changes were compared o the unstarved amount, it could be seen that these increases were argely relative, reflecting decreasing organic constituents. The de-creasing actual amounts of ash were probably due to losses needed for maintenance of osmotic equilibrium, since actual amounts of water de-reased during starvation (Table 2).

Table 3. Rate of utilization to PNR (μg/mg unstarved dry wt/day)

Length mm	Dry weight mg	Days to PNR	Triglyceride	Carbohydrate	Nitrogen	Carbon	Ash
12	0.1689	6	5.0	1.7	4.2	19.3	0.6
15	0.4685	8	4.2	2.3	4.5	19.3	1.2
21	2.1810	9	4.4	3.4	5.5	22.2	1.8
25	4.8393	12	4.0	1.7	3.3	14.6	0.6
30	11.1358	15	3.8	1.1	3.2	13.8	0.7
35 *	22.5286	15*	3.7	1.1	3.1	13.8	0.9

*Experiment terminated before fish reached PNR. Rate based on larvae at last day of experiment.

Table 4. Calories* used to PNR (calories x 100/mg unstarved dry weight/day)

Length	Dry weight	Days to	ug Protein	Calories x 100			
mm	mg	PNR	N x 6.025	Triglyceride	Carbohydrate	Protein	Total
12	0.1689	6	25.3	4.8	0.7	13.9	19.4
15	0.4685	8	27.1	4.0	0.9	14.9	19.8
21	2.1810	9	33.2	4.2	1.4	18.2	23.8
25	4.8393	12	19.9	3.8	0.7	10.9	15.4
30	11.1258	15	19.3	3.6	0.5	10.6	14.7
35 **	22.5286	15 **	18.7	3.5	0.5	10.3	14.3

*Calorific values for fat, carbohydrate and protein from Winberg (1971).
**Experiment terminated before fish reached PNR. Rate based on larvae at last day of experiment.

Energy Utilization

Energy expenditures in calories used to reach the PNR were estimated by multiplying the quantities of body components used to the PNR by appropriate conversion factors for fat (9.5 cal/mg), carbohydrate (4.1 cal/mg), and protein (5.5 cal/mg) obtained from Winberg (1971). Protein was calculated as total nitrogen times 6.025; this factor is from Love (1970). These computed values were multiplied by 100 for clarity and are shown in Table 4. It can be seen that for any size of herring larvae maximum energy was derived from the catabolism of protein followed by triglyceride and then carbohydrate. There was roughly a three-fold difference between each group. This agrees with reports of energy sources for teleost eggs; Smith (1957) and Lasker (1962) showed the order of importance of food reserves was protein, then lipid, followed by carbohydrate.

An attempt was made to correlate losses of triglyceride, carbohydrate, and protein with oxygen consumption of herring larvae at the EYS. Holliday and Lasker (1964) reported the Q_{O_2} was about 2 for herring larvae during starvation, although there was some tendency for it to decrease in this period. This value was used in conjunction with oxygen combustion values for fat (2 ml/mg), carbohydrate (0.825 ml/mg) and protein (1.22 ml/mg) reported by Winberg (1971). With these values and the amounts of these chemical components utilized to the PNR (Table 2), the time to burn these components and reach the point-of-no-return was calculated (Table 5). The computed time (6 days) agreed with the behavioural observations of the time to the PNR from yolk resorption (Blaxter and Ehrlich, this symposium). These values can only be considered as approximate since total lipid was not known and because protein was not chemically determined. However, it is apparent that the time required to deplete the body components to the extent that the PNR is reached is very much dependent on oxygen consumption.

Table 5. Oxygen and time required to burn proximate compounds used to PNR

Components	Triglyceride	Carbohydrate	Protein
O_2 equivalent (ml/mg) (1)	2000	825	1220
Amount used to PNR (mg) (2)	0.0050	0.0018	0.0250
O_2 used to burn compounds (µl)	10	1.5	30.5
Total O_2 consumed = 42 µl			

Larval dry weight (mg)

EYS	0.1689
PNR	0.1278
Mean weight	0.1483

Rate of larval O_2 consumption Wt x Q_{O_2} (3) $= 0.1483 \times 2 \approx .3$ µl/hr

Time required to use 42 µl O_2 or to reach PNR $= \frac{42}{.3} = 132.5$ hrs ≈ 6 days

If $Q_{O_2} = 3 \Rightarrow$ Rate O_2 consumption $= .45$ µl/hr $\Rightarrow$ days to PNR = 4

(1) O_2 equivalents from Winberg (1971).
(2) Amount used to PNR from Table 1.
(3) $Q_{O_2} = 2$ from Holliday, Blaxter and Lasker (1964).

Holliday et al. (1964) also showed that the QO_2 was dependent upon activity being as much as 9 times the resting value in very active larvae. An increase to even $QO_2 = 3$ would decrease the time to reach the PNR by 2 days or 33%. From this it is obvious that the observed decrease in activity during starvation is of great survival advantage (as long as activity is sufficient for searching), by extending the time over which larvae are capable of feeding. Blaxter and Ehrlich (this symposium) have shown by activity meters that herring and plaice larvae decrease their activity during starvation. This same situation has also been shown for sole (Rosenthal, 1966). This agrees with the general concept of the metabolic rate decreasing during starvation due to a form of conservation metabolism (Phillips, 1969).

Identifying the PNR

One of the main aims of this work was to identify chemical changes during starvation that led to the point-of-no-return. Percent changes are only relative and may not represent real changes in a component, but in terms of identifying the PNR they have been useful, sometimes more so than real changes. Increasing percent water was a very good nutritive indicator more so than the actual amount (Fig. 4 and Table 2). This was even more true for ash where the actual amount changed little, but the percentage rapidly increased. On the other hand, percent changes of nitrogen were of no use in identifying condition; actual amounts of nitrogen, however, gave better indications of the condition (Table 2). Percent changes of both triglyceride and carbohydrate gave good estimates of nutritional condition, but in both actual amounts showed greater differences between healthy and starved larvae. Changes in total carbon in percent or actual amount gave good condition estimates. Ratios of carbon to nitrogen also showed condition variations. There were no abrupt changes peculiar to the start of the PNR. The changes commencing at initial food deprivation just gradually continued. Identification of the nutritive condition of larvae requires a knowledge of their composition as well as size.

In all chemical determinations it was imperative that only larvae of the same size or developmental state be included in one determination. Chemical comparisons between larvae of different sizes, regardless of age, could not yield meaningful results as to their nutritive condition.

It is obvious from this discussion that this work only begins an understanding of the chemistry of herring. Its purpose was to help identify and understand what happens to larvae of different sizes as they starve and reach the point-of-no-return and to show changes during growth and development of the larvae.

SUMMARY

1. Changes in water, triglyceride, carbohydrate, nitrogen, carbon, and ash were followed during ontogenesis from the end of the yolk-sac stage through metamorphosis as well as during starvation of various size groups of herring.

2. Ontogenetic changes were dependent upon larval size rather than age

3. The percentage of water continuously decreased, from about 89 to about 82% throughout development from the end of the yolk-sac stage to metamorphosis.

4. Nitrogen and carbohydrate were laid down faster than triglyceride in the period of initial post-hatching growth (up to 20 mm). The pattern was then altered in larger larvae. It was suggested that the initial preferential deposition of protein to triglyceride was advantageous to the larvae because it enabled them to grow more quickly than if they had been simultaneously accumulating energy stores.

5. During starvation the percentage of water increased about 4% above the unstarved level. The percentage of triglyceride, carbohydrate, and carbon decreased. However, the percentage of nitrogen did not change throughout starvation, although the actual amount decreased. Ash rapidly increased during starvation as a percentage of the dry weight. The carbon-nitrogen ratio decreased in starved animals.

ACKNOWLEDGEMENTS

The author wishes to thank Dr. J.H.S. Blaxter for his suggestions and encouragement throughout this project and for his critique of the manuscript. Thanks are also given to Dr. H. Barnes for his advice on chemical techniques.

REFERENCES

Barnes, H., Barnes, M. and Finlayson, D.M., 1963. The metabolism during starvation of *Balanus balanoides*. J. Mar. Biol. Ass. U.K., 43, 213-223.

Blaxter, J.H.S., 1968. Rearing herring larvae to metamorphosis and beyond. J. Mar. Biol. Ass. U.K., 48, 17-28.

Blaxter, J.H.S., 1971. Feeding and condition of Clyde herring larvae. Rapp. P.-v. Réun. Cons. perm. int. Explor. Mer, 160, 128-136.

Blaxter, J.H.S. and Ehrlich, K.F., in press. Buoyancy and activity changes during growth and starvation of herring and plaice larvae.

Blaxter, J.H.S. and Hempel, G., 1963. The influence of egg size on herring larvae (*Clupea harengus* L.). J. Cons. perm. int. Explor. Mer, 28 (2), 211-240.

Blaxter, J.H.S. and Staines, M., 1971. Food searching potential in marine fish larvae. Four. Europe. Mar. Biol. Symp. (D.J. Crisp, ed.). Cambridge University Press, 467-485 pp.

Dubois, M., Gilles, K.A., Hamilton, J.K., Rebers, P.A. and Smith, F., 1956. Colorimetric method for determination of sugars and related substances. Analyt. Chem., 28 (3), 350-356.

Ehrlich, K.F., 1972. Morphological, behavioural and chemical changes during growth and starvation of herring and plaice larvae. Ph.D. Thesis. Univ. of Stirling, Scotland.

Ehrlich, K.F., 1974. Chemical changes during growth and starvation of larval *Pleuronectes platessa*. Mar. Biol. 24, 39-48.

Heath, J.R. and Barnes, H., 1970. Some changes in biochemical composition with season and during the moulting cycle of the common shore crab, *Carcinus maenas* (L.). J. exp. mar. Biol. Ecol., 5, 199-233.

Holliday, F.G.T., Blaxter, J.H.S. and Lasker, R., 1964. Oxygen uptake of developing eggs and larvae of the herring (*Clupea harengus*). J. Mar. Biol. Ass. U.K., 44, 711-723.

Lasker, R., 1962. Efficiency and rate of yolk utilization by developing embryos and larvae of the pacific sardine *Sardinops caerulea* (Girard). J. Fish. Res. Bd Can., 19 (5), 867-875.

Lasker, R., Feder, H.M., Theilacker, G.H. and May, R.C., 1970. Feeding growth, and survival of *Engraulis mordax* larvae reared in the laboratory. Mar. Biol., 5, 345-353.

Love, R.M., 1958. Studies on North Sea cod. I. Muscle cell dimensions. J. Sci. Food Agr., 9, 195-198.

Love, R.M., 1970. The chemical biology of fishes. London and New York: Academic Press, 547 pp.

Love, R.M., Robertson, I. and Strachan, I., 1968. Studies on the North Sea cod. VI. Effects of starvation 4. Sodium and Potassium. J. Sci. Food Agr., 19, 415-422.

Marshall, S.M., Nicholls, A.G. and Orr, A.P., 1937. On the growth and feeding of the larval and post larval stages of the Clyde herring. J. Mar. Biol. Ass. U.K., 22, 245-267.

May, R.C., 1971. Effects of delayed initial feeding on larvae of the grunion, *Leuresthes tenuis* (Ayres). Fish. Bull. Fish Wildl. Serv. U.S., 69 (2), 411-425.

Morris, R.W., 1955. Some considerations regarding the nutrition of marine fish larvae. J. Cons. perm. int. Explor. Mer, 20 (3), 255-265.

Mullin, M.M. and Brooks, E.R., 1970. Growth and metabolism of two planktonic, marine copepods as influenced by temperature and type of food. In: Marine food chains (J.H. Steele, ed.), 74-95 pp. Edinburgh: Oliver and Boyd.

Nagai, M. and Ikeda, S., 1971. Carbohydrate metabolism in fish. I. Effects of starvation and dietary composition on blood glucose level and the hepatopancreatic glycogen and lipid contents in carp. Bull. Jap. Soc. sci. Fish., 37 (5), 404-409.

Phillips, A.M., 1969. Nutrition, digestion, and energy utilization. In: Fish physiology (W.S. Hoar and D.J. Randall, ed.), Vol. I, 391-432 pp.

Phillips, A.M., Livingston, D.L. and Dumas, R.F., 1960. Effect of starvation and feeding on the chemical composition of brook trout. Progve Fish Cult., 22, 147-154.

Rosenthal, H., 1966. Beobachtungen über das Verhalten der Seezungenbrut. Helgoländer wiss. Meeresunters., 13, 213-228.

Rosenthal, H. and Hempel, G., 1970. Experimental studies in feeding and food requirements of herring larvae (*Clupea harengus* L.). In: Marine Food Chains (J.H. Steele, ed.). Edinburgh: Oliver and Boyd, 344-364.

Satomi, Y., 1969. Veränderung der chemischen Zusammensetzung (Nucleinsäure-, Phosphorlipid-, Kjeldahlstikstoff-, Totalphospor- und Wasser-Gehalt) von Karpfenbrut unter Bedingung der Sättigung, des Hungers und der Wiederauffütterung. Bull. Freshwat. Fish. Res. Lab., Tokyo, 19 (1), 47-72.

Shelbourne, J.E., 1957. The feeding and condition of plaice larvae in good and bad plankton patches. J. Mar. Biol. Ass. U.K., 36, 539-5

Smith, S., 1952. Studies in the development of the rainbow trout (*Salmo irideus*). II. The metabolism of carbohydrates and fats. J. exp. Biol., 29, 650-666.

Smith, S., 1957. Early development and hatching. In: The physiology of fishes (M. Brown, ed.), Vol. I, 323-359 pp. New York: Academic Press.

Templeman, W. and Andrews, G., 1956. Jellied condition in the Americ plaice *Hippoglossoides platessoides*. J. Fish. Res. Bd Can., 13, 147-182

Van Handel, E., 1961. Suggested modifications of the microdetermination of triglyceride. J. clin. Chem., 7 (3), 249-251.

Wilkins, N.P., 1967. Starvation of the herring, *Clupea harengus* L.: survival and some gross biochemical changes. Comp. Biochem. Physiol 23, 506-518.

Winberg, G.G. (ed.), 1971. Symbols, units and conversion factors in studies of fresh water productivity. Int. Biol. Progrm. Sect. PF. Handbook, 23 pp.

K.F. Ehrlich
Scottish Marine Biological Association
Dunstaffnage Marine Research Laboratory
P.O. Box 3
Oban, PA 34 4 AD, Argyll / GREAT BRITAIN

Present address:

Biology Department
Occidental College
Los Angeles, Calif. 90041 / USA

Metabolism of Nitrogenous Wastes in the Eggs and Alevins of Rainbow Trout, *Salmo gairdneri* Richardson

S. D. Rice and R. M. Stokes

INTRODUCTION

Ammonia is a toxic by-product of the normal catabolism of proteins by animals. Although some animals may convert ammonia to a less toxic substance, aquatic vertebrates usually excrete ammonia directly into the environment. Teleosts, both marine and freshwater, excrete ammonia primarily from their gills (Forster and Goldstein, 1970; Smith, 1929; Fromm, 1963; Wood, 1958). Some information on nitrogen excretion and metabolism is available on adults of some teleosts, but practically none is available on embryos except in work by Smith (1947), which however did not deal with the accumulation of waste nitrogen and associated metabolic pathways.

Forster and Goldstein (1970) assumed that because the fertilized eggs of nonviviparous teleosts are free in their environment, the embryos are able to dispose of ammonia by diffusion into the aqueous environment; but these authors do not cite any evidence to support their assumption. Moreover, what little evidence is available suggests that nitrogenous wastes are retained. Potts and Rudy (1969) and Loeffler and Løvtrup (1970) demonstrated very low water and ion exchange between salmonid eggs and the environment. The permeability of ammonia (NH_3) and ammonia ion (NH_4^+) through any of the egg membranes has not been measured. Smith (1947) measured a slight loss of ammonia from trout eggs before hatching. He later (1957) noted that the end products of nitrogen metabolism were unable to escape through the chorion and were accumulated until released at hatching. These studies suggest that ammonia loss from the egg is probably low and may be less than ammonia production. If ammonia is retained within the egg, it may be detoxified to a form such as urea. This possibility is suggested by the fact that during growth, the predominant waste product in fingerling salmon under certain conditions shifts from urea to ammonia (Burrows, 1964).

Except for Burrows' (1964) observations, only small quantities of urea have been observed in the blood or excreta of teleosts. These quantities of urea are so small that they could be accounted for by purine catabolism (Goldstein and Forster, 1965; Cvancara, 1969a) or by dietary degradation of arginine by arginase (Cvancara, 1969b). Goldstein and Forster (1965), however, doubted that an essential amino acid like arginine would be wasted by dietary degradation. In contrast, Cvancara (1969b) concluded that diet may influence liver-arginase activity and that dietary arginine may be the source of much of the urea produced by freshwater teleosts.

Adaptive urea production via the ornithine-urea (O-U) cycle detoxifies ammonia, and occurs in elasmobranchs, coelacanths, and dipnoids (Schooler et al., 1966; Goldstein et al., 1973; Janssens and Cohen, 1966). Missing enzymes of the O-U cycle in teleosts led Brown and Cohen (1960) to conclude that it was through gene deletion that ammoniotelic teleosts lost the ability to synthesize urea via the O-U cycle. Recently, Huggins et al. (1969) assayed for the 5 enzymes of the O-U cycle in several freshwater and marine teleosts; several, but not all, species assayed showed activity for the 5 enzymes. However, as might be expected in ammoniotelic fish, these authors found that levels of activity of these enzymes were quite low, indicating that only insignificant amounts of urea could be synthesized. They hypothesized that traces of a ureotelic phase may be detectable during the embryonic development of teleosts if their ancestors possessed a functional O-U cycle, as proposed by Brown and Cohen (1960).

Our study investigates the excretion and accumulation of urea and ammonia during the embryonic development of rainbow trout (*Salmo gairdneri* Richardson), a freshwater teleost, and attempts to correlate these patterns with activities of O-U cycle enzymes.

MATERIALS AND METHODS

The experiment was conducted on laboratory-reared, fertilized rainbow trout eggs.[1] At selected intervals during development, samples of eggs or alevins were removed for various measurements and assays.

Fertilized rainbow trout eggs were obtained from Bowden National Fish Hatchery, West Virginia (courtesy of the U.S. Bureau of Sport Fisheries and Wildlife). The eggs were transported in insulated containers and reached our laboratory within 6 h of fertilization. In the laboratory they were held in approximately 900 l of water at 10°C continuously recycled through both charcoal and gravel filters. Ammonia levels in the holding water were measured periodically, and never attained 0.1 ppm.

Metabolite Determinations. Ammonia, urea, uric acid, total protein, and free arginine in eggs or alevins were determined on each of at least 3 different homogenates per age group. To supply enough homogenate for all of the determinations, approximately 40 to 50 eggs or alevins were blotted, weighed, and homogenized (Teflon head) with an equal quantity of 0.2 M citrate buffer (pH 6.8). In addition, rate of ammonia and urea excretion into the ambient water was determined on 3 groups of embryos or alevins per age group. Details and verfication of methods for the following analytical procedures have been given by Rice (1971).

Urea was measured by the standard method of Archibald (1945) after pretreatment with urease. Because citrulline and urea were both found in egg homogenates and contributed color about equally in their reaction with diacetyl monoxine, urea was measured by noting the loss of optical density between urease-treated and untreated samples of the homogenate. The loss of optical density was compared to similarly treated Lab Trol (Dade) standards. Alternative methods of urea determinations were found to be incompatible with Trichloroacetic-Acid (TCA), the only effective protein precipitator tested.

---

[1]The term "egg" is used to include the chorion, embryo, yolk, and perivitelline space. An embryo is the developing zygote before hatching; an alevin is the embryo released at hatching.

Total protein was determined by the Biuret method (Layne, 1957).

Free arginine was measured by adding arginase to homogenates and measuring the increase in urea produced by the method of Archibald (1945). It was assumed that arginase is specific for arginine and that a blank compensated for residual urea or citrulline in the sample. Duplicate homogenates had the same optical density after 5 min and 1 h, indicating the amount of arginase was not the limiting factor in the assay. Blanks were prepared by adding the perchloric acid first and then the arginase. Recovery was 100% from aqueous arginine standards and above 95% from known quantities of arginine added to egg homogenates.

Uric acid was analyzed on TCA supernatants by the spectrophotometric procedure of Henry et al. (1957).

The ammonia in TCA supernatants was separated and trapped in disposable center wells by the diffusion method of Conway and Cooke (1939). Ammonia was measured by adding an equal volume of Nessler's Reagent (Kock and McMeekin formula) to the center well fluid and reading the subsequent color at 425 mμ. Ammonium sulphate standards were treated like the samples by adding known quantities to the outer wells of the diffusion dishes. The sensitivity of the method was varied by adjusting the volume of acid in the center well that trapped the ammonia.

The rate of ammonia excretion by embryos and alevins was determined by measuring changes in concentrations of ammonia in static water in which sample organisms were held overnight. The holding water had a pH of 6.8 and was buffered with a 0.02 M concentration of phosphate to trap excreted ammonia and retain it in the relatively nontoxic ionized form. Samples of the holding water were collected and analyzed for ammonia by the modified Conway method. The efficiency of ammonia retention in the phosphate buffer was tested by measuring ammonia in the buffer before and after 18 h, with the result of less than 5% loss of ammonia. Water samples were also tested for urea, but none could be detected.

Enzyme Assays - Chemical Procedures. Each of the five O-U cycle enzymes was assayed by standard chemical procedures. The preparation of the enzyme extract and assay mixture and the subsequent enzyme assay procedures were those described and modified by Huggins et al. (1969) for the assays of carbamyl phosphate synthetase, ornithine transcarbamylase, argininosuocinate synthetase, and argininosuccinate lyase. Some minor changes were made in recommended volumes to increase the concentrations of enzymes in the final assay cocktails. Arginase was assayed by the method of Brown and Cohen (1959).

Each enzyme preparation was assayed at 37°C (to facilitate comparison with published data) and at 10°C (near the temperature for optimal growth in rainbow trout). Because arginase activity varies in the presence of divalent ions and with enzyme concentration (Brown and Cohen, 1959), partial standardization of the method was attempted by pre-incubating a combination of the extract and manganese chloride for exactly 10 min before the addition of the assay mixture. Protein concentration varies with the age of the embryo and thus was not standardized.

The validity of each assay procedure was confirmed by duplicate determinations of enzyme activity for each homogenate; one determination was at half the protein concentration of the other. The observed activities of all duplicate samples were linear with protein concentration (within approximately 10%). The validity of each assay procedure was further confirmed by comparing the enzyme activities of frog liver with those previously documented by Brown and Cohen (1960) Enzyme activities are reported in units of specific activity (1 unit 1 μmole of product formed per hour per milligram of protein). Protein levels of enzyme extracts were determined by the method of Lowry et a (1951).

Enzyme Assays - Isotopic Procedures. The intact ornithine-urea cycle was assayed with the use of radio-labeled substrates in eyed eggs and alevins by the sensitive method described by Schooler et al. (1966). By incubating intact cells with radio-labeled bicarbonate and measuring the labeled urea produced, the rate of urea synthesis of the enti cycle can be determined. Fifteen intact-eyed eggs or alevins were incubated with labeled $HC^{14}O_3$(0.5 μc). The chorions of the eggs in some assays were punctured to facilitate the diffusion and equilibration of labeled HCO_3 and CO_2. Since the embryo may not have access to the labeled substrate because of low membrane permeabilities, the presenc of the label within the punctured eggs was tested on rinsed eggs afte 10 min of incubation. Radioactivity was found. Solution of trout ringers (Stokes and Fromm, 1964) was used as the incubation medium, and was shown to be adequate for survival of punctured eggs for several days. In other assays, eggs were incubated with labeled bicarbonate without puncturing. The assay procedure for separating labeled urea continued as specified by Schooler et al. (1966). After removing excess labeled CO_2, synthesized urea was measured by treating the incubation mixtures with urease and trapping labeled CO_2 in ethanolamine. Labeled CO_2 in ethanolamine was measured by the liquid scintillation procedure of Jeffay and Alvarez (1961).

Argininosuccinate synthetase-lyase was measured by incubating the labeled substrate C^{14}-ureido-citrulline with homogenate in the same assay system described by Huggins et al. (1969) for argininosuccinate synthetase. From previous chemical assays, arginase was known to be present, and added arginase was not needed for the production of labeled urea. Labeled CO_2 was separated from urea as specified by Schooler et al. (1966) and was measured as specified by Jeffay and Alvarez (1961).

Urease was assayed similarly to the above assay, except urea was the labeled substrate that was incubated with homogenate in 0.2 M citrate buffer (pH 6). Any labeled CO_2 produced was without the addition of exogenous urease. Separation and labeled CO_2 measurement proceeded as described above. Although urease activity was clearly indicated, bacteria had not been eliminated as the urease source. Urease activity was found in aquaria water, which would easily contaminate homogenates of eggs and alevins. For this reason, eggs and alevins were incubated in water containing the bacteriacide tetracycline for 2 days before urease was assayed on homogenates of these eggs and alevins.

Fig. 1. Concentrations of ammonia, urea, uric acid, arginine, and protein within tissues of eggs and alevins at the indicated ages. Vertical bars indicate ± 1 standard error

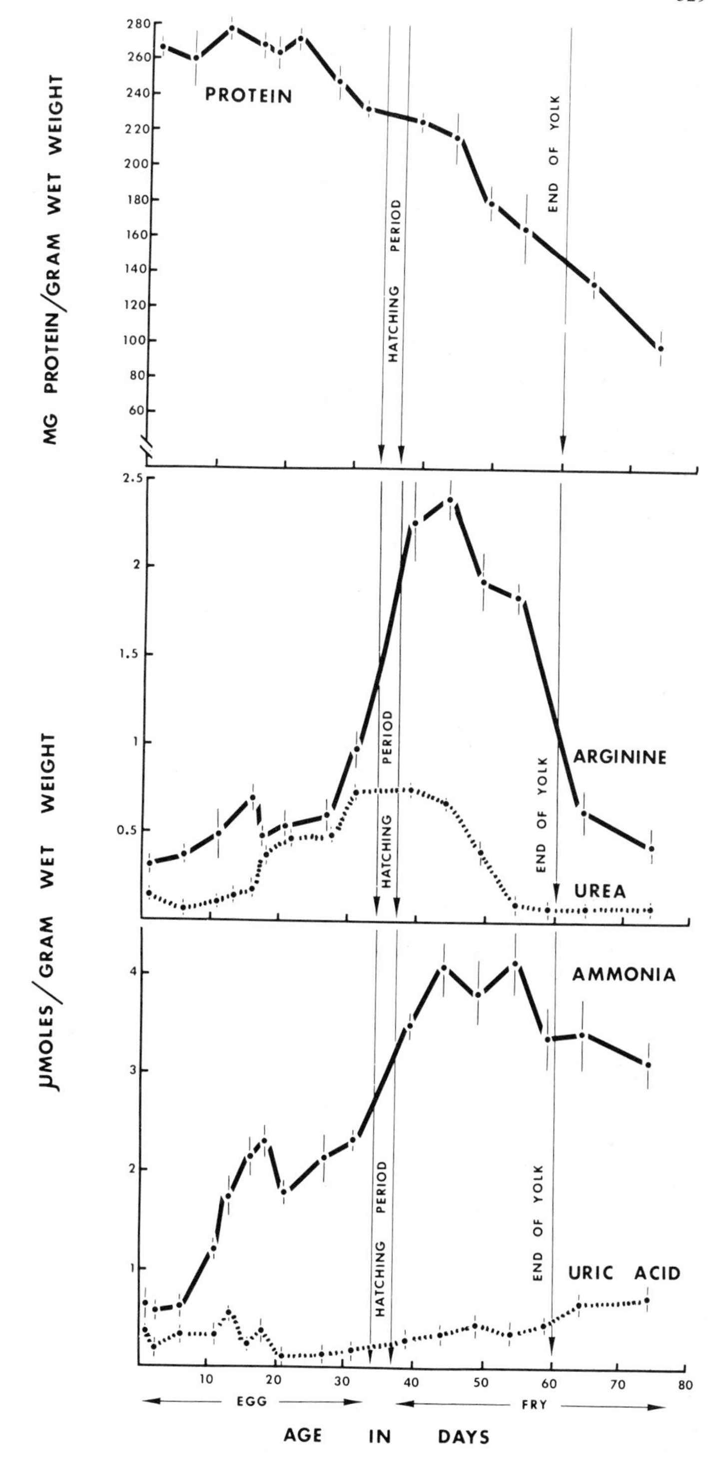

MG PROTEIN/GRAM WET WEIGHT
PROTEIN
HATCHING PERIOD
END OF YOLK
μMOLES/GRAM WET WEIGHT
ARGININE
UREA
AMMONIA
URIC ACID
EGG
FRY
AGE IN DAYS

RESULTS

Changes observed in the endogenous levels of several nitrogenous compounds during the embryonic development of rainbow trout are shown in Fig. 1. Ammonia concentrations increased through hatching and the greatest concentrations were found in alevins. Ammonia levels increased for the first few days after hatching, stabilized somewhat for a few days, then declined as the yolk was absorbed. Uric acid concentrations did not change dramatically during development, but concentrations before and after the hatching period were lower than just shortly after fertilization and when the yolk was nearly absorbed. Urea and free arginine both increased in concentration while the developing embryo was still within the chorion. Peak concentrations of urea and free arginine occurred soon after hatching, but thereafter, concentrations of both compounds decreased. Protein concentrations were relatively constant during the first 25 days of embryonic development and steadily declined thereafter. The average wet weight of the eggs did not change significantly until after hatching, when alevin weight increased.

The rate of ammonia excretion increased continuously throughout development, but the greatest increase was near the end of yolk absorption (Fig. 2). To compare total wastes accumulated with total excreted, we converted ammonia-excretion rates and net ammonia-urea accumulation measurements to total weight of nitrogen accumulated or excreted for a given 10-day period (Table 1). Approximately 10 times more nitrogen was excreted than was accumulated for the first

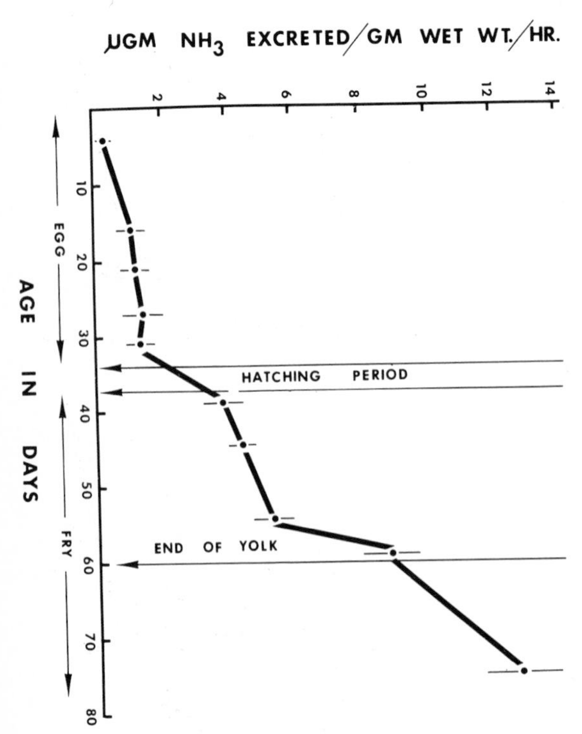

Fig. 2. Rate of ammonia excretion of eggs and alevins at $10^{\circ}C$ at indicated ages. Vertical bars indicate $\pm$ 1 standard error

2 periods (up to 21 days). The ratio of nitrogen excreted to nitrogen accumulated increased dramatically from about 25:1 during the third 10-day period (just before hatching) to about 327:1 during the fourth period (alevin stage).

Table 1. Estimated accumulations of ammonia-urea nitrogen and excreted ammonia nitrogen in 10-day periods for developing embryos and alevins of rainbow trout

Age of embryos (days)	Total µg of ammonia and urea N accumulated in egg per gram wet weight in 10 days	Total µg of ammonia* N excreted per gram wet weight in 10 days
1 - 11	8	72
11 - 21	16.8	192
21 - 31	14.5	360
44 - 54	4.4	1 440

*Urea was never detected as a waste product in the media.

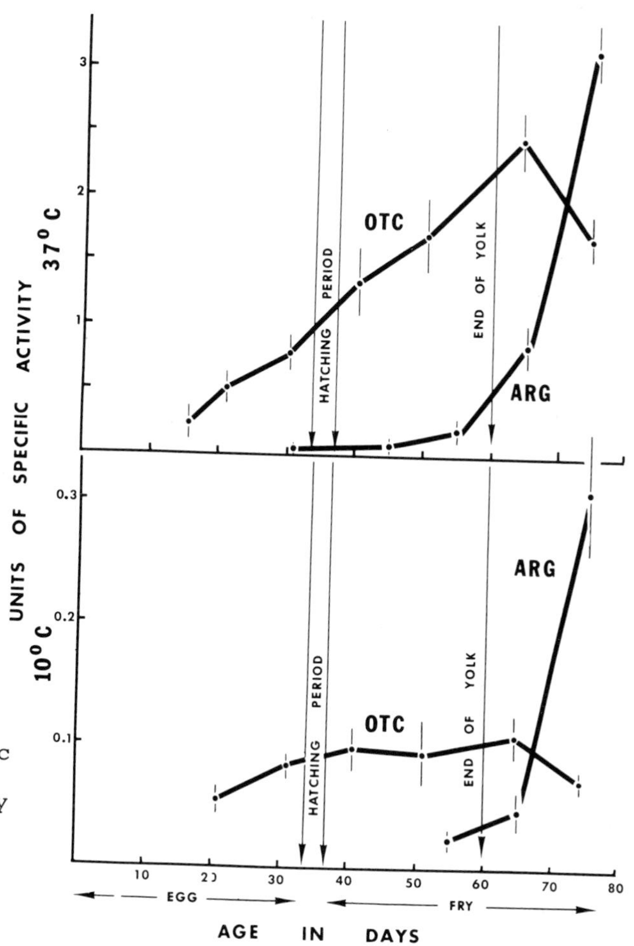

Fig. 3. Units of specific arginase and ornithine-transcarbamylase activity at 10° and 37°C at indicated ages. Vertical bars indicate ± 1 standard error

Chemical assays of enzyme done at 10° and 37°C detected the presence of ornithine transcarbamylase and arginase (Fig. 3). The activities of both enzymes generally increased with development. As expected, enzymes were more active at 37°C than at 10°C, although the increases did not necessarily follow the Q-10 law. The chemical assays were conducted at approximately 5-day intervals from fertilization, but ornithine transcarbamylase activity was not detected until embryos were 16 days old (37°C). Arginase activity was first detected in 31-day-old eggs (37°C). Ornithine transcarbamylase activity increased through development until the yolk was nearly absorbed, after which the activity decreased. In contrast to ornithine transcarbamylase, arginase activity increased rapidly after the yolk was absorbed.

Livers of adult trout were assayed for comparison with the activity of ornithine transcarbamylase and arginase in embryos: ornithine transcarbamylase activity in livers of adult trout (0.0665 units at 37°C and 0.015 units at 10°C) was approximately 1/20 of the activity observed in embryos and alevins. The activity of liver arginase was about 2 times greater in adults (5.06 units at 37°C, 3.04 units at 10°C) than in either embryos or alevins.

Carbamyl phosphate synthetase and argininosuccinate synthetase and lyase were sought but not detected by the standard chemical assays in eggs or alevins. Even the more sensitive isotopic assays did not detect any activity for the complete O-U cycle or for the arginino-succinate synthetase-lyase enzymes in eggs or alevins. Arginino-succinate lyase was detected in adult trout by standard chemical assays, but isotopic assays for the complete O-U cycle and for argininosuccinate synthetase-lyase failed to detect these enzyme complexes. We found urease activity in egg homogenates and suspected it to be due to bacterial activity. Subsequent tests with eggs and alevins held in tetracycline-treated water for 2 days did not show urease activity, but eggs and alevins from normal water did (Table 2).

Table 2. Counts per min of labeled CO_2 released from catabolized labeled urea by urease

Sample	Counts per min
0.5 ml egg homogenate	5 687
0.5 ml aquaria water	1 317
0.5 ml of homogenate from eggs grown in tetracycline-treated water for 2 days before assay	171

DISCUSSION

Because the vitelline membrane of trout eggs has a low permeability, it was suggested that toxic nitrogenous wastes would be converted to a less toxic form and retained within the egg. The observed urea increase before hatching and decrease after hatching indirectly support this hypothesis, but measurements of other parameters in eggs and alevins did not indicate that urea was being synthesized from ammonia

Three enzymes of the O-U cycle could not be detected, thus ruling out urea synthesis via this pathway. The lack of a functional O-U cycle was further indicated when enzyme assays with radioactive substrates also failed to yield labeled urea.

Developing rainbow trout embryos excreted significant amounts of ammonia, which suggests that there is little need to convert ammonia to a less toxic form. The total waste nitrogen accumulated in the egg in 10 days was approximately one-tenth of the amount of nitrogen excreted in 10 days (Table 1). Even though both ammonia and urea accumulated within the egg, the eggs excreted most of the waste nitrogen in the form of ammonia.

Furthermore, studies of the toxicity of ammonia to rainbow trout eggs and alevins also suggest little need for ammonia detoxification (Rice and Stokes, unpublished). Eggs and alevins exposed to concentrations of ammonia up to 100 mg/l showed little or no effect on survival until near the end of the yolk absorption, when a rather sudden susceptibility occurred. When the yolk was nearly absorbed, the 24-h median tolerance limits of the alevins were similar to those of adult trout (·2-4 mg/l). However, growth of pink salmon alevins was poor under chronic exposure to ammonia between 1 and 6 mg NH_3/l over a 10-week period (Rice and Bailey, unpublished data). Ammonia in the ambient water decreases the gradient between source and disposal area, and this could lead to higher internal ammonia levels and distress in salmon alevins. Indeed, Fromm and Gillette (1968) have observed elevated blood ammonia in adult trout in response to increased ambient ammonia. Apparently, detoxification of ammonia by embryos or alevins is not necessary since both can excrete ammonia, are relatively resistant to ammonia toxicity, and are never exposed to high ambient ammonia levels in the natural environment that might hamper excretion. Thus, the original hypothesis of a need to detoxify ammonia is not supported.

Production of ammonia in salmon embryos is expected even though metabolism of fat is the predominant source of energy (Hollett and Hayes, 1946). Because proteins in the yolk are broken down to amino acids before they are resynthesized into proteins in the embryo (Monroy et al., 1961), an amino acid pool is available for catabolism. However, most amino acids are resynthesized into protein rather than catabolized because total egg protein values remain stable until day 21 (Fig. 1), after which catabolism leads to a decrease in total protein. After day 21, arginine levels increase, undoubtedly through protein catabolism since 3 enzymes leading to arginine synthesis were not detected. The levels of other amino acids probably rose similarly, because changes in arginine are more than likely accompanied by changes in most amino acids. Further catabolism of these amino acids, possibly for supplemental energy, would lead to ammonia production. The large increase in endogenous ammonia combined with the increase in ammonia excretion after day 21 correlates with the decrease in total egg protein (Fig. 1) and with the increase in free arginine. The inconsistent rate of accumulation of ammonia and arginine correlates with the nonlinear breakdown of yolk proteins observed by Ando (1965) for rainbow trout eggs.

Because the O-U cycle was incomplete, other explanations for urea synthesis and function are needed. Urea could have been synthesized by the purine catabolism pathway. Although the functional capability of this pathway was not assayed, changes in uric acid, an intermediate precursor, were determined to be less dramatic than the changes exhibited by urea, suggesting the purine catabolism pathway contributed little to urea synthesis. In contrast, changes in free arginine during the development were similar to those of urea, although the concentrations of arginine were higher. Arginase, which degrades arginine to urea and ornithine, was detected at day 31 and rapidly increased in activity after day 45. Low levels of arginase activity together with possible inability of the egg to eliminate urea would explain the rise in urea before hatching. After hatching, arginase activity increased and arginine decreased, but the expected increase in urea failed to appear either as accumulation or excretion. Urea decrease rather than accumulation within the alevins was observed and no urea excretion was detected.

No pathway is known for the utilization of urea in teleost fishes. Brookbank and Whitely (1954) reported urease activity in starfish eggs and suggested a function in recycling nitrogen. Starfish eggs, like trout eggs, have only the nitrogen content within their own yolk to survive on until well after hatching when feeding begins. Brookbank and Whitely (1954) suggested that the ammonia nitrogen released from urea by urease in starfish eggs could be recycled in pyrimidine and amino acid synthesis rather than being excreted and lost. Similar urease activity within trout alevins would explain the drop in urea levels without measurable urea excretion. However, we did not detect urease activity in homogenates of eggs or alevins grown in tetracycline-treated water, whereas urease activity was definitely detected in nontetracycline-treated homogenates and in the water in which they were grown. Bacterial urease could account for some of our observations but Smith (1929) and Wood (1958) reported that bacterial breakdown of urea was negligible in their experiments with adult fish, an assumption that may not have been true in our experiments. Bell et al. (1971) showed live salmon eggs had a characteristic bacterial flora on the outer egg surface which was different from the flora on dead eggs, and described the interaction as a "dynamic ecosystem". Bacterial urease activity was definitely not large in our experiments. Further studies would be needed, however, to prove or disprove urea excretion in trout eggs or alevins.

The complete O-U cycle pathway is present in several marine and freshwater teleosts (Huggins et al., 1969), but the low activities of enzymes involved suggest insignificant synthesis of urea. Since the presence of these enzymes in several teleosts implies some function, Huggins et al. suggest that functional urea synthesis may occur during early developmental stages or certain environmental states. While our data do not support the hypothesis of Huggins et al. (1969) of a functional urea synthesis during early developmental stages, we cannot deny the possibility that fish species from marine or warm water environments may have a functional O-U cycle.

Read (1971) offers an alternative hypothesis for explaining the presence of a complete O-U cycle with little apparent function. Read observed relatively high levels of enzyme activity of the O-U cycle in the marine toadfish (teleost) but a low rate of urea production. The marine toadfish would have no need to convert ammonia to urea for temporary storage of a less toxic nitrogenous waste (like lungfish), nor a need for urea synthesis and retention for osmoregulatory adaptations (like sharks). This observation led Read to conclude that it is quite possible that the enzymes involved in the O-U cycle may also be intermediates in other pathways. Campbell (1965) studied O-U enzymes in several invertebrates and pointed out that the enzyme carbamyl phosphate synthetase is used in pyrimidine biosynthesis and that arginase may be used for dietary arginine control. These biosynthetic functions may have appeared earlier in evolution and were probably modified for excretory functions. Our observations of high activities for ornithine transcarbamylase and arginase and absent carbamyl phosphate synthetase, argininosuccinate synthetase, and argininosuccinate lyase support Read's suggestion that enzymes of the O-U cycle function as intermediates in other pathways.

In conclusion, rainbow trout eggs and alevins excrete ammonia and appear to have little need for detoxification of ammonia. Urea accumulates within the egg shell, apparently from the breakdown of arginine rather than from the O-U cycle as part of ammonia detoxification. The 2 enzymes involved in the O-U cycle that were detected probably function as intermediates in other pathways. The functioning of the O-U cycle as an ammonia detoxification pathway for teleosts in other environments is still possible.

SUMMARY

1. Metabolism of nitrogen waste products was studied in developing rainbow trout eggs and alevins. Ammonia, urea, and arginine accumulated during development of the embryo, and both urea and arginine declined in concentration soon after hatching. Uric acid changed little during development.

2. The hypothesis of urea being stored in the egg as a means of storing ammonia in a detoxified state was not supported. The eggs were shown to be quite capable of excreting nitrogen and only 2 (ornithine transcarbamylase and arginase) of the five O-U cycle enzymes were detected.

3. Urea appeared to be synthesized when arginine from yolk proteins was degraded by arginase. Decrease in total egg protein was accompanied by an increase in free arginine.

4. Data from this study support the hypothesis that the O-U cycle enzymes can function to produce intermediates in other metabolic pathways.

ACKNOWLEDGEMENTS

Dr. Leon Goldstein of Brown University is thanked for teaching us several of the enzyme assay procedures. We are indebted to the Bureau of Sport Fisheries and Wildlife for supplying several thousand rainbow trout eggs from Bowden National Fish Hatchery. We thank the staff at the National Marine Fisheries Service Auke Bay Fisheries Laboratory for reviewing the manuscript and preparing it for publication.

REFERENCES

Ando, K., 1965. Ultracentrifugal analysis of yolk proteins in rainbow trout and their changes during development. Can. J. Biochem., 43, 373-379.

Archibald, R.M., 1945. Colorimetric determination of urea. J. Biol. Chem., 157, 507-518.

Bell, G.R., Hoskins, G.E. and Hodgkiss, W., 1971. Aspects of the characterization, identification, and ecology of the bacterial flora associated with the surface of stream-incubating Pacific salmon (*Oncorhynchus*) eggs. J. Fish. Res. Bd Can., 28, 1511-1525.

Brookbank, J.W. and Whiteley, A.H., 1954. Studies on the urease of the eggs and embryos of the sea urchin, *Strongylocentrotus purpuratus*. Biol. Bull., 107, 57-63.

Brown, G.W. and Cohen, P.P., 1959. Comparative biochemistry of urea synthesis. I. Methods for the quantitative assay of urea cycle enzymes in liver. J. Biol. Chem., 234, 1769-1774.

Brown, G.W. and Cohen, P.P., 1960. Comparative biochemistry of urea synthesis. 3. Activities of urea-cycle enzymes in various higher and lower vertebrates. Biochem. J., 75, 82-91.

Burrows, R.E., 1964. Effects of accumulated excretory products on hatchery-reared Salmonids. U.S. Bur. Sport Fish. Wildl. Res. Rep., 66, 12 pp.

Campbell, J.W., 1965. Arginine and urea biosynthesis in the land planarian: its significance in biochemical evolution. Nature, 208, 1299-1301.

Conway, E.J. and Cooke, R., 1939. LVIII. Blood ammonia. Biochem. J., 33, 457-478.

Cvancara, V.A., 1969a. Distribution of liver allantoinase and allantoicase activity in fresh-water teleosts. Comp. Biochem. Physiol., 29, 631-638.

Cvancara, V.A., 1969b. Studies on tissue arginase and urogenesis in fresh-water teleosts. Comp. Biochem. Physiol., 30, 489-496.

Forster, R.P. and Goldstein, L., 1970. Formation of excretory products. In: Fish Physiology. Ed. by W.S. Hoar and R.J. Randall, Vol. 1, 313-350. New York: Academic Press.

Fromm, P.O., 1963. Studies on renal and extra-renal excretion in a freshwater teleost, *Salmo gairdneri*. Comp. Biochem. Physiol., 10, 121-128.

Fromm, P.O. and Gillette, J.R., 1968. Effect of ambient ammonia on blood ammonia and nitrogen excretion of rainbow trout (*Salmo gairdneri*) Comp. Biochem. Physiol., 26, 887-896.

Goldstein, L. and Forster, R.P., 1965. The role of uricolysis in the production of urea by fishes and other aquatic vertebrates. Comp. Biochem. Physiol., 14, 567-576.

Goldstein, L., Harley-DeWitt, S. and Forster, R.P., 1973. Activities of ornithine-urea cycle enzymes and of trimethylamine oxidase in the Coelacanth, *Latimeria chalumnae*. Comp. Biochem. Physiol., 44B, 357-362.

Henry, R.J., Sobel, C. and Kim, J., 1957. A modified carbonate-phosphotungstate method for the determination of uric acid and comparison with the spectrophotometric uricase method. Amer. J. Clin. Pathol., 28, 152-159.

Hollett, A. and Hayes, F.R., 1946. Protein and fat of the salmon egg as sources of energy for the developing embryo. Can. J. Res., 24, 39-50.

Huggins, A.K., Skutsch, G. and Baldwin, E., 1969. Ornithine-urea cycle enzymes in teleostean fish. Comp. Biochem. Physiol., 28, 587-602.

Janssens, P.A. and Cohen, P.P., 1966. Ornithine-urea cycle enzymes in the African lungfish, *Protopterus aethiopicus*. Science, 152, 358-359

Jeffay, H. and Alvarez, J., 1961. Liquid scintillation counting of carbon-14: Use of enthanolamine-ethylene glycol monomethyl ether-toluene. Anal. Chem., 33, 612-615.

Layne, E., 1957. Spectrophotometric and turbidimetric methods for measuring proteins. In: Methods of Enzymology. Ed. by S.P. Colowick and N.O. Kaplan, Vol. III, 447-454. New York: Academic Press.

Loeffler, C.A. and Løvtrup, S., 1970. Water balance in the salmon egg. J. Exp. Biol., 52, 291-298.
Lowry, O.H., Rosenbrough, N.J., Farr, A.L. and Randall, R.J., 1951. Protein measurement with the Folin phenol reagent. J. Biol. Chem., 193, 265-275.
Monroy, A., Ishida, M. and Nakano, E., 1961. The pattern of transfer of the yolk material to the embryo during the development of the teleostean fish, *Oryzias latipes*. Embryologia, 6, 151-158.
Potts, W.T. and Rudy, P.P., 1969. Water balance in the eggs of the Atlantic salmon *Salmo salar*. J. Exp. Biol., 50, 223-237.
Read, L.J., 1971. The presence of high ornithine-urea cycle enzyme activity in the teleost *Opsanus tau*. Comp. Biochem. Physiol., 39B, 409-413.
Rice, S.D., 1971. A study of nitrogen waste product metabolism in the eggs and fry of rainbow trout, *Salmo gairdneri*. Thesis, Kent State University, 77 pp.
Schooler, J.M., Goldstein, L., Hartman, S.G. and Forster, R.P., 1966. Pathways of urea synthesis in the elasmobranch, *Squalus acanthias*. Comp. Biochem. Physiol., 18, 271-281.
Smith, H.W., 1929. The excretion of ammonia and urea by the gills of fish. J. Biol. Chem., 81, 727-742.
Smith, S., 1947. Studies in the development of the rainbow trout (*Salmo irideus*). I. The heat production and nitrogenous excretion. J. Exp. Biol., 23, 357-373.
Smith, S., 1957. Early development and hatching. In: Physiology of Fishes. Ed. by M.E. Brown, Vol. I. New York: Academic Press, 323-356.
Stokes, R.M. and Fromm, P.O., 1964. Glucose absorption and metabolism by the gut of rainbow trout. Comp. Biochem. Physiol., 13, 53-69.
Wood, J.D., 1958. Nitrogen excretion in some marine teleosts. Can. J. Biochem. Physiol., 36, 1237-1242.

S.D. Rice and R.M. Stokes
Department of Biological Science
Kent State University
Kent, Ohio / USA

Present address:

S.D. Rice
National Marine Fisheries Service
Auke Bay Fisheries Laboratory
P.O. Box 155
Auke Bay, Ak. 99821 / USA

Feeding, Starvation and Weight Changes of Early Fish Larvae

N. Ishibashi

INTRODUCTION

The present study concerns weight changes, first feeding, and early starvation in the larvae of *Tilapia sparmanii* A. Smith, a tropical freshwater fish with attached eggs, and in *Paralichthys olivaceus* (Temmik et Schlegel), a common marine flatfish in Japan, with discrete pelagic eggs. Such studies are of importance when investigating survival and growth of fish larvae.

MATERIAL AND METHODS

Tilapia larvae, hatched under constant environmental conditions, were obtained from adults reared in the aquaria of the Tokyo University of Fisheries. *Paralichthys* larvae were obtained by artificial fertilization of eggs and milt collected from mature adults landed at the Amatsu fish market in Chiba Prefecture. The larva were hatched in the aquarium at the Kominato laboratory of Tokyo University of Fisheries. Only 1 male and 1 female were used for each species in the artificial fertilizations. In the study of *Tilapia* larvae, a square aquarium of 10-litre capacity was used. A bag of netting was placed inside the aquarium and the water was circulated into the bag by an air-lift apparatus. The water was filtered through active charcoal and glass fibre, the temperature being held at 27 ± 1°C by heater and thermostat. Soon after hatching, the larvae were separated to 2 groups; 1 group was fed, the other was not. A 2-litre plastic aquarium was used for *Paralichthys* larvae and the water temperature was kept at 20 ± 1°C. Two similar groups were separated off after hatching. The effect of temperature on *Tilapia* larvae was assessed by incubating at 27°C and separating the larvae into 3 temperatures, 30°C, 27°C, and 24°C, again with a fed and unfed group.

Morphological observations and body measurements were made by microscope. Samples of 20 *Paralichthys* and 3 *Tilapia* were used for weighing. The larvae were put in a net, and superficial water was absorbed by filter paper. To measure dry weight of *Tilapia* 6 larval samples were dried at 110°C for 30 min.

Bivalve larvae (*Mactra veneriformis*), rotifers (*Brachionus plicatilis*), and *Artemia* nauplii were used as food for *Paralichthys*, and *Artemia* larvae only for *Tilapia*.

Oxygen consumption measurements were made on *Tilapia* larvae by placing 4 larvae in a 100 ml closed bottle for 5 h and measuring the change in oxygen content of the water by Winkler's method.

RESULTS

The incubation time for *Tilapia* was 48 h at 27°C. The total length was 4.2 mm at hatching, the larvae being inactive and without a functional mouth. After 2 days the mouth opened, the caudal fin started to move actively, and after 3 days the larva began to swim. The wet weight changes are shown in Fig. 1. Weight increased rapidly after hatching; the weight on day 3 was 0.65 mg heavier than at hatching. At this time the larva began to take food and a weight of 8.80 mg was reached on day 5. As the yolk of unfed larva was resorbed, the weight decreased and on day 9, the unfed larvae were inactive and the weight was 1.24 mg, 25% less than that on day 3. Only 1% of the unfed larvae survived to day 12.

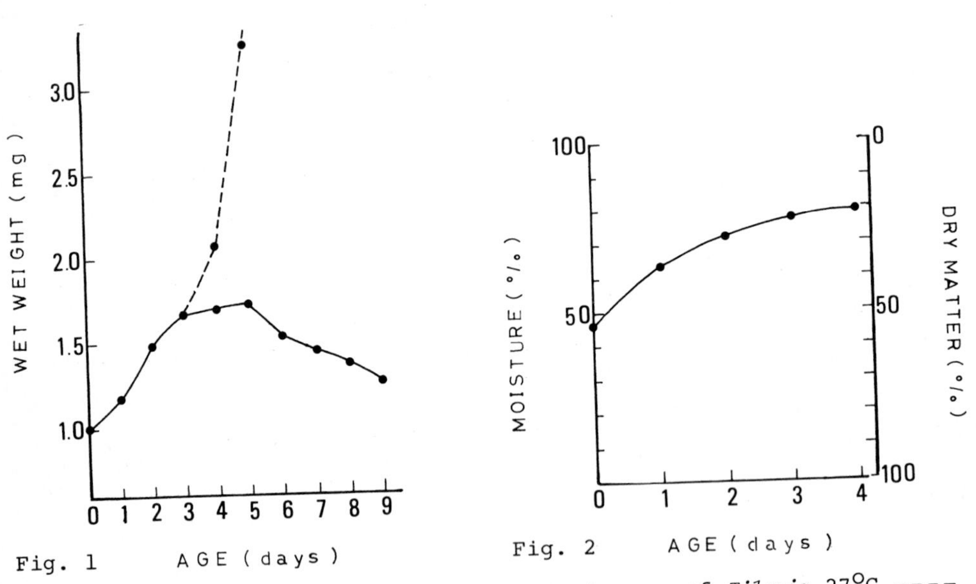

Fig. 1. Weight changes after hatching in larvae of *Tilapia* 27°C.---- indicate feeding individual

Fig. 2. Dry weight and water content of *Tilapia* after hatching

The dry weight of *Tilapia* larvae after hatching is shown in Fig. 2. Soon after hatching, the dry weight was 53% of the wet body weight. Afterward, it decreased steadily, and on day 4 was 22%. In this period the larval body weight increased 2.2 times. The yolk-sac volume was reduced to 25% by first-feeding. The yolk of unfed larva was absorbed faster than that of fed larva.

The development of the gut of *Tilapia* is shown in Fig. 3. The gut only developed on day 3; in unfed larvae it had atrophied by day 10. Fig. 4 shows the effect of temperature on body weight changes in *Tilapia*. This began to increase on day 2 at 30°C, on day 4 at 27°C, and on day 6 at 24°C. This was related to first-feeding, the larva beginning to take food on day 2 at 30°C, day 3 at 27°C, and day 4 at 24°C. The weight of unfed larva decreasing steadily from day 3, day 4, and day 12 respectively, 50% survival being reached at day 7, day 10, and day 12 At first-feeding the larval weight was about 1.3 mg in all 3 temperatures.

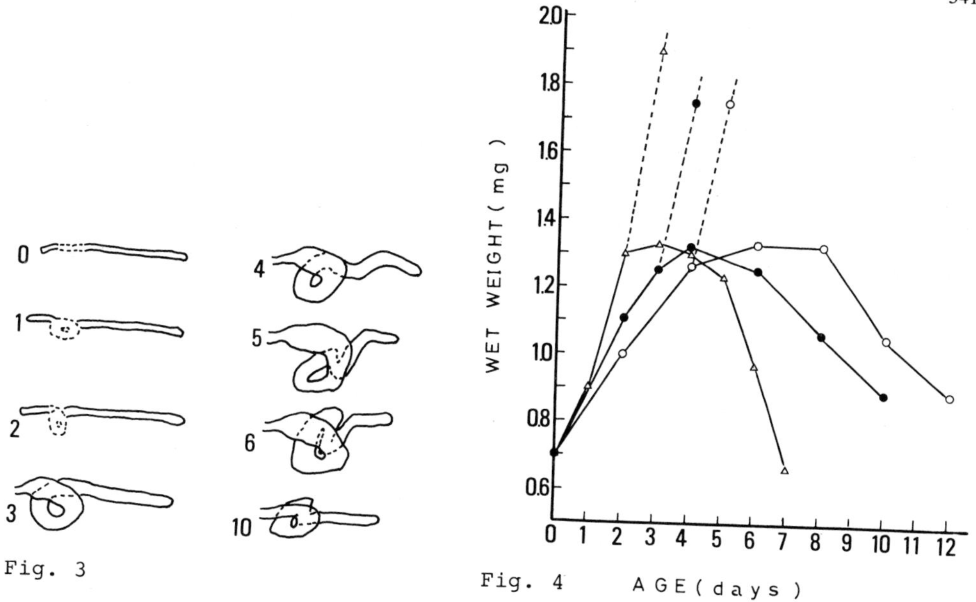

Fig. 3. Development of the gut of *Tilapia*. Numbers give days post-hatching; 10 indicates gut atrophied

Fig. 4. Effect on temperature of weight changes after hatching in *Tilapia* larva. (△: 30°C, ●: 27°C, o: 24°C)

The oxygen consumption of *Tilapia* larva after hatching is shown in Fig. 5. Soon after hatching, it was 0.010 ml/h/larva reaching 0.023 ml/h/larva on day 4. From day 6, it decreased steadily in unfed larva, being 0.017 ml/h/larva on day 8.

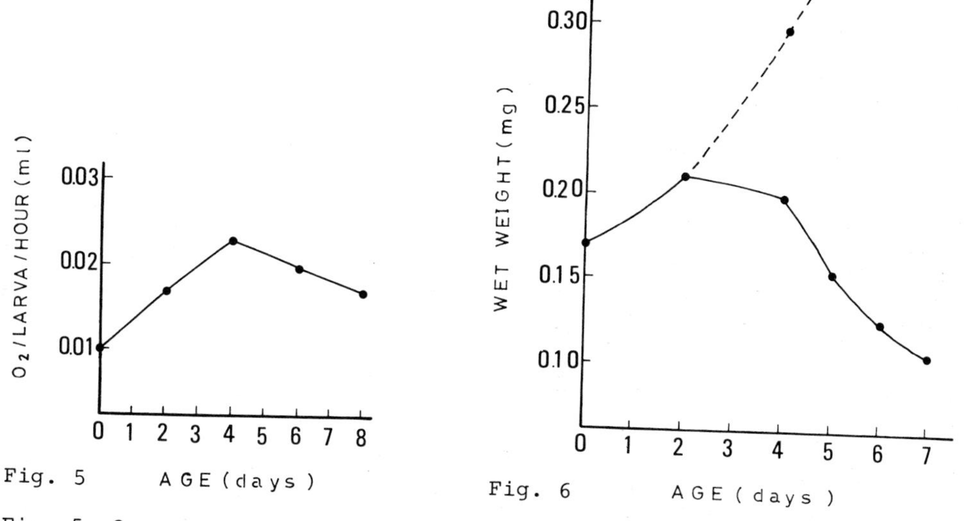

Fig. 5. Oxygen consumption of *Tilapia* larvae at 27°C

Fig. 6. Weight changes of *Paralichthys* after hatching at 20°C. ---- indicate feeding individual

The incubation period of *Paralichthys* was 48 h at 20°C, the total length of the larvae being 3.0 mm at hatching. Soon after hatching, the larvae floated motionless at the water surface but after 12 h they began to sink head-downward. On day 2 the mouth opened and the larvae began to swim. The weight changes of *Paralichthys* after hatching are shown in Fig. 6. The larval body weight increased, until at day 2 it was 0.04 mg heavier than that of hatching. The larvae began to take food 2 days after hatching, before disappearance of the yolk. After feeding the weight of fed larva increased immediately and on day 4, the weight was 0.30 mg. On the other hand, the body weight of unfed larvae decreased from day 4 and on day 7 was 52% lower than day 2. The development of the lower jaw of *Paralichthys* larva is shown in Fig. 7. It started to form about 48 h after hatching, before complete yolk resorption. The larvae began to take bivalve larvae before full development of the jaw but rotifers were first taken later.

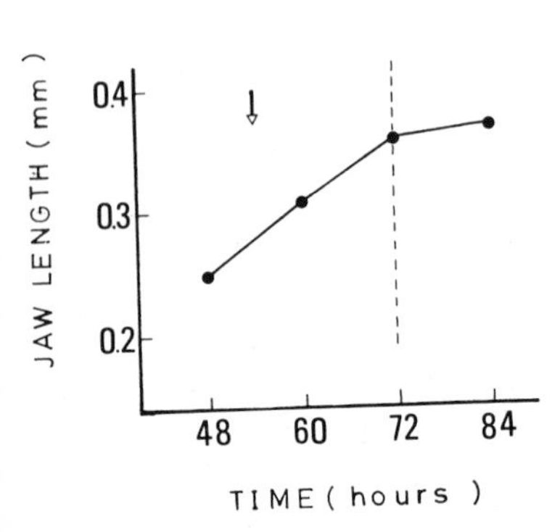

Fig. 7. Growth of lower jaw of *Paralichthys*, ⟶ first-feeding, --- yolk resorbed

DISCUSSION

In both species body weight increased after hatching and first-feeding occurred at maximum weight. If the larvae did not feed at this time, the weight decreased. Gray (1928) and Suyama and Ogino (1958) showed a similar weight increase after hatching in brown and rainbow trout larva, respectively. In marine fish, weight changes after hatching have been shown for herring (Blaxter and Hempel, 1963), grunion (May, 1971), and sardine (Lasker, 1962). In brown and rainbow trout and in *Tilapia sparmanii*, the increase of weight after hatching was due to water imbibed from the exterior. Osmoregulation is of importance in this respect. Lasker and Theilacker (1962) investigated the osmotic pressure of sardine larva and more recently (Lasker and Theilacker, 1968) studying the role of the chloride cells.

At first-feeding when body weight gained on the yolk alone is maximal, the yolk is not fully resorbed nor is the gut fully developed. Iwai (1972) reported that the yolk and oil globules had not disappeared at first-feeding in the larvae of many species. In sardine larvae, Lasker (1962) showed that the jaw is still developing up to the disappearance of the yolk. Using 18 species of fish larvae, Tanaka (1969) found that the development of the alimentary canal was almost the same at first-feeding, the tissue being similar to the digestive tissue of "stomach-less" fish. May (1971) reported the yolk of unfed larvae disappeared faster than that of fed larvae in the grunion. In this experiment, *Tilapia* larvae showed the same tendency.

In *Tilapia* larvae, higher temperatures led to earlier feeding but there were no distinct effects on larval weight at first-feeding. In brown trout, Gray (1928) reported that the final size of the embryo was smaller at the higher temperature while Braum (1967) showed the length of *Coregonus* larva was smaller at higher temperature at first-feeding.

In *Tilapia* and *Paralichthys* the loss in weight during starvation followed a sigmoid pattern as Kleiber (1961) found. Blaxter (1963, 1969) described the problem of recovery from starvation in the larval stage as the "point-of-no-return", a term used for the point at which the larvae were alive but too weak to feed. May (1971) could not find a point-of-no-return for grunion larvae.

SUMMARY

After hatching, the body weight of the larvae of *Tilapia sparmanii*, a tropical freshwater fish with attached eggs, and *Paralichthys olivaceus*, a marine flatfish with discrete pelagic eggs, increases during yolk resorption. At first-feeding the weight gained on the yolk reserves is maximal. When the larvae first feed, some yolk remains and the gut and jaw are not fully developed. Increasing weight after hatching is due to uptake of water into the body of *Tilapia*. In *Tilapia*, oxygen consumption is highest at maximum weight without food. The higher the water temperature, the faster the larvae reach first-feeding, but there is no distinct effect of temperature on the weight gained on the yolk reserves.

ACKNOWLEDGEMENTS

I wish to express my sincere thanks to Prof. Ogasawara, Prof. Uno, and Assist. Ohno in Tokyo University of Fisheries for advice and encouragement during this study and to Mr. Kwon for helping with the English in this paper.

REFERENCES

Blaxter, J.H.S., 1969. Development: eggs and larvae. In: W.S. Hoar and D.J. Randall (eds.), Fish Physiol., 3, 177-252 pp. New York: Academic Press.

Blaxter, J.H.S. and Hempel, G., 1963. The influence of egg size on herring larvae (*Clupea harengus* L.). J. Cons. perm. int. Explor. Mer, 28, 211-240.

Braum, E., 1967. The survival of fish larvae with reference to their feeding behaviour and the food supply. In: The Biological Basis of Freshwater Fish Production (ed. S.D. Gerking). Oxford: Blackwell Scientific Publications, 113-131 pp.

Gray, J., 1928. The growth of fish. II. The growth-rate of the embryo of *Salmo fario*. J. exp. Biol., 6, 110-124.

Iwai, T., 1972. Feeding of teleost larvae: a review. La Mer (Bull. Soc. franc.-jap. d'Oceanogr.), 10 (2), 71 pp.
Kleiber, M., 1961. The fire of life. New York, London.
Lasker, R., 1962. Efficiency and rate of yolk utilization by developing embryos and larvae of Pacific sardine, *Sardinops caerulea* (Girard). J. Fish. Res. Bd Can., 19, 867-875.
Lasker, R. and Theilacker, G.H., 1962. Oxygen consumption and osmoregulation by single Pacific sardine eggs and larvae (*Sardinops caerulea* Girard). J. Cons. perm. int. Explor. Mer, 27, 25-33.
Lasker, R. and Theilacker, G.H., 1968. "Chloride cells" in the skin of larval sardine. Exp. Cell. Res., 52, 582-590.
May, R.C., 1971. Effects of delayed initial feeding on larvae of the grunion, *Leuresthes tenuis* (AYRES). Fish Bull., U.S., 69 (2), 411-425.
Suyama, M. and Ogino, C., 1958. Changes in chemical composition during development of rainbow trout eggs. Bull. Jap. Soc. Sci. Fish., 23, 785-788.
Tanaka, M., 1969. Studies on the structure and function of the digestive system in teleost larvae. II. Characteristics of the digestive system in larvae at the stage of first-feeding. Jap. J. Ichthyol., 16 (2), 41-49.

N. Ishibashi
Fisheries Resources Research Laboratory
Tokyo University of Fisheries
4 - 5 - 7 Konan
Minatoku
Tokyo / JAPAN

Quantitative Nucleic Acid Histochemistry of the Yolk Sac Syncytium of Oviparous Teleosts: Implications for Hypotheses of Yolk Utilization

W. E. Bachop and F. J. Schwartz

INTRODUCTION

In the yolk-sac syncytium of the muskellunge embryo, *Esox masquinongy ohioensis* Kirtland, there are nuclei whose size and staining properties with basic dyes suggest that their chromatin content is greater than that of other embryonic nuclei (Bachop, 1965). This study is intended to test that supposition, using a histochemical test specific for one of the constituents of chromatin and using a microspectrophotometer to quantify the histochemical reaction in different nuclei relative to each other. The rationale of microphotometry is that insertion of the microscope in the optical train, and substitution of a dyed tissue on a slide for a dye solution in a cuvette, does not militate against using Beer's law to deduce relative concentration of dye molecules from relative absorption of light (Pollister and Ornstein, 1959).

Two main requirements for microphotometry in the visible spectrum are 1. a specific quantifiable chromogenic chemical reaction and 2. preservation of colored chemical reactants *in situ*. Among various histochemical tests purported to be specific for DNA, a principal constituent of chromatin, the Feulgen reaction seems to be the test preferred by the majority of investigators for meeting both requirements (Pearse, 1960).

According to Beer's law, per cent transmission of a nucleus would be inversely proportional to the number of absorbing molecules in that nucleus, and the relationship would be logarithmic rather than linear. If one were to measure the per cent transmission (T) of nucleus A and the per cent transmission of nucleus B, the dye content of nucleus A would be to the dye content of nucleus B as the product, log (1/T) for nucleus A times volume of nucleus A, is to the product, log (1/T) for nucleus B times volume of nucleus B (Moses, 1952). By itself, log (1/T) is a measure of nuclear dye concentration, not nuclear dye content.

In practice, attempts to measure directly the light transmitted by a single whole nucleus, while excluding all other light from the photometer, fail to prevent stray light from passing around the nuclear border and introduce a source of error called the Schwarzschild - Villiger effect. In nuclei whose shapes approximate these of geometrical figures whose volume formulae are known, the above source of error can be circumvented by measuring the light transmitted by the central portion of the nucleus (known as the "plug") and using this value, in conjunction with the appropriate volume formula, to compute the light that supposedly would have been transmitted by that single whole nucleus.

Because of their asymmetrical shapes, yolk-sac syncytium nuclei (known also as "periblast nuclei") are rarely measurable by the aforementioned one-wave length, "plug" method, since the ratio of the volume of the "plug" to the volume of the whole nucleus is an unknown in the case of such irregularly-shaped nuclei. However, if one measures the *Is/Io* of a single whole nucleus at two different wave lengths, one can use these readings to calculate a factor that will correct for light that passes around rather than through the nucleus (Swift and Rasch, 1956, p. 376; Ornstein, 1952).

Theoretical justification for the claim that light passing around rather than through the nucleus does not interfere with accurate measurement in the two-wave length method was given by Patau (1952). He also gave the theoretical basis and procedure for selecting the two wave lengths to be used. Direct calculation of correction factors is no longer required since tables have been prepared (Mendelsohn, 1958).

Therefore, a measure of the amount of dye in a single whole asymmetrical nucleus can be calculated by multiplying the area of the photometric field (whose limits encompass but are not coterminal with the nuclear boundary) by the product of the correction factor and the per cent absorption, i.e. 100 per cent minus the per cent transmission (Swift and Rasch, 1956). As with the one-wave length "plug" method, dye contents of different whole nuclei are compared with each other rather than to an absolute standard.

An objection occasionally put forth against Feulgen microphotometry contends that if a certain amount of DNA were to combine with a given amount of dye in the Feulgen reaction, there would be no assurance that half the amount of DNA would combine with half the amount of dye. In short, that the Feulgen reaction might not conform to the law of definite proportions - it might be a non-stoichiometric reaction. Pollister and Ornstein (1955) said, in response to such theoretical objections, that ".... the empirical approach has shown that the Feulgen reaction actually can be used for relative quantitative estimation of DNA throughout the tissues of an animal and, quite possibly, in the same tissue in a variety of animals". Kasten (1960) concurred.

Still another objection sometimes put forth against Feulgen microphotometry is that chromatophore distribution in Feulgen-processed nuclei is not homogeneous, a requirement for compliance with Beer's law upon which photometry is based (Swift and Rasch, 1956, pp. 394-395; Pollister and Ornstein, 1955, pp. 7, 40-42, 45). The force of this objection is lost when the "two-wave length" method is used, since this modified photometric method does not require homogeneity of chromatophores (Patau and Swift, 1953; Garcia and Iorio, 1966).

MATERIALS AND METHODS

Living muskellunge embryos belonging to the subspecies *Esox masquinongy ohioensis* Kirtland (Salmoniformes: Esocidae) were taken directly from a fish hatchery jar into which fertilized eggs had been placed 95 h earlier. These embryos, along with water from the hatchery jar, were quickly placed into a petri dish and inspected with a dissecting microscope. Within 4 min, those embryos judged alive and normal by hatchery criteria were put into a glass-stoppered bottle containing a very large volume of fixative (100% ethanol, 3 volumes: glacial acetic acid, 1 volume) for 4 h at about 25°C. Similar withdrawals were made from hatchery jars whose fertilized eggs had been incubating for 163 and 238 h (just hatching), respectively, at water temperatures rising from 13.3°C to 18.3°C. The incidence of dead or abnormal embryos in the hatchery jars was not noticeably greater than usual. All chemicals used were analytical grade.

Embryos attached to their yolks, and still within their chorions, were Feulgen processed following the procedures of Leuchtenberger (1958), with the optimal hydrolysis time being 12 min in 1 normal hydrochloric acid at 60°C. After dechorionation by use of jeweler's tweezers in the ascending 25% ethanol, embryos were dehydrated through ethanol to toluene, and then placed in a Cargille index-of-refraction liquid whose refractive index, 1.554, matched that of the embryonic cytoplasm (Deitch, 1966). With the aid of iridectomy scissors, jeweler's tweezers, pin vises, and insect pins, the cellular layers overlying the yolk sac syncytium (periblast) were stripped away as much as possible. Fragments of the underlying periblast were than pulled off the yolk and placed in index-of-refraction liquid (1.554) on a scrupulously clean glass slide below a comparably clean cover glass for microphotometric measurement by the two-wave length method.

The microphotometer system components included a Bausch and Lomb monochromator, Type 33-86-44; Leitz Aristophot (Binocular-monocular photographic tube) with Leitz Aplanat condenser, Leitz 2 mm objective, Leitz Periplan 10 x ocular and Leitz 6 x ocular; and Photovolt Corporation Model 520 (photomultiplier tube, power supply, and galvanometer). The electric current for the monochromator was regulated by a Sola Constant Voltage Transformer and a Jefferson Transformer (969-001-059). The installation, housed in the Department of Pathology, The Ohio State University, had been aligned and maintained by Dr. Ruth Kleinfeld of that department. The optical components and monochromator components were essentially the same as those pictured in Figure 1A and Figure 1B, respectively, by Swift and Rasch (1956).

The galvanometer reading for light transmitted through a nucleus, Is, divided by the galvanometer reading for light transmitted through an equal area of nuclei-free protoplasm, Io, is a quotient, Is/Io, referred to as the per cent transmission, T (Swift and Rasch, 1956, p. 367). The logarithm of the quotient, 1/T, is referred to as the Extinction, E, i.e. log (1/T) is E. Extinction varies as a function of wave length. Such extinctions were plotted on one periblast nucleus, and it was found that at 498 mμ the extinction was one-half the extinction at 538 mμ. These two wave lengths were thereafter used in the two-wave length measurement of periblast nuclei, having fulfilled the theoretical requirements for two-wave length microphotometry (Swift and Rasch, 1956).

In periblast shreds, the spatial relations of periblast nuclei to each other, to the torn margin of the shred, and to the small non-periblast nuclei were such that there were rarely any nuclei-free areas of protoplasm of the same size as the periblast nuclei. To circumvent this difficulty, a small field or "plug" of nuclei-free protoplasm was measured microphotometrically and called Rp, and the light transmitted by a field of equal area containing no periblast at all was measured microphotometrically and called Rg. Next, a field of the same area as the one containing the periblast nucleus but containing no periblast at all was measured microphotometrically, and called R. Finally, the light transmission one would like to measure but cannot - the reading one would expect for a field of nuclei-free protoplasm of the same area as the one containing the periblast nucleus - was called Io. The proportion, Rp/Rg equals Io/R was used to compute Io.

The similarity of the *Rp* values to the *Rg* values incline the authors to the belief that this refinement was perhaps unnecessary. Korson (1951), after using a comparable procedure, expressed a similar opinion.

On the day of hatching, yolk-sac larvae of the zebra fish, *Brachydanio rerio* (Hamilton - Buchanan), were prepared for Feulgen microdensitometry following the methods previously described for the muskellunge, except for the following modifications. Optimal Feulgen hydrolysis time was ten minutes in 1 normal hydrochloric acid at 60°C. Correct matching of refractive index of embryonic cytoplasm required mounting periblast shreds in Cargille index-of-refraction liquid 1.568. Nuclei from four yolk sac larvae were measured with the 40 x objective of a Vickers M85 Scanning Integrating Microdensitometer operating at a wave length of 560 mμ (Mendelsohn and Richards, 1958; Hale, 1966). The instrument was housed in the Molecular Cytogenetics Laboratory of the Cold Spring Harbor Laboratory for Quantitative Biology, Long Island, New York.

RESULTS

Two sample measurements, for a periblast nucleus and a non-periblast nucleus, are shown for the muskellunge in Table 1. The calculations corresponding to these two measurements are shown in Tables 2 and 3. Two series of replicate determinations on two nuclei are included to show how much variation is attributable to the instrument and/or its operator (Table 4).

Table 1. Sample raw data for microphotometric calculations

Nucleus	Replicate Measurement	*Is* *		*Io* *		C
		1	2	1	2	
Non-periblast #4	1	56.2	56.0	66.7	72.5	10.25
	2	56.8	56.4	67.3	73.0	
Non-periblast #6	1	37.5	40.2	43.2	50.9	10.5
	2	38.0	40.8	43.8	51.1	
Periblast #4	1	19.9	18.1	25.4	27.3	19.0
	2	19.7	17.6	25.0	27.0	
Periblast #20	1	20.7	19.6	25.8	28.1	19.0
	2	20.3	19.2	25.2	27.9	

C Radius of field containing nucleus measured; 1 498 mμ; 2 538 mμ; * see text for explanation.

In 1/3 developed muskellunge embryos (95 h of incubation), the majority of periblast nuclei measured had approximately four times as much dye content, i.e. presumably four times as much DNA content, as the supposedly diploid non-periblast nuclei.

Table 2. Calculation of relative DNA content of periblast nuclei in arbitrary units*

Periblast Nucleus Number	Variables Used in Calculation						M
	T_1	T_2	$1-T_1$	$1-T_2$	$(1-T_2)/(1-T_1)$	D	$= (1-T_1)DC^2$
4	0.786	0.657	0.214	0.343	1.60	1.107	$= (0.214)(1.107)(19)^2$ $= 85.5$
20	0.804	0.693	0.196	0.307	1.565	1.138	$= (0.196)(1.138)(19)^2$ $= 80.5$

T_1 *Is/Io*, fraction of light transmitted by nucleus at λ_1. T_2 *Is/Io*, fraction of light transmitted by nucleus at λ_2. $1-T_1$, fraction of light absorbed by nucleus at λ_1. D, correction factor corresponding to $(1-T_2)/(1-T_1)$. See Swift and Rasch, 1956, Table V. M, amount of DNA-dye complex in nucleus, in arbitrary units, relative to amounts in other nuclei, i.e. "relative DNA content". *For raw data, whose averages were used in Table 2, and for definitions of C, *Is*, *Io*, see Table 1.

Table 3. Calculation of relative DNA content of non-periblast nuclei in arbitrary units*

Non-Periblast Nucleus Number	Variables Used in Calculation						M
	T_1	T_2	$1-T_1$	$1-T_2$	$(1-T_2)/(1-T_1)$	D	$= (1-T_1)DC^2$
4	0.844	0.772	0.156	0.228	1.461	1.247	$= (0.156)(1.247)(10.25)^2$ $= 20.4$
6	0.868	0.794	0.132	0.206	1.545	1.157	$= (0.132)(1.157)(10.5)^2$ $= 17.8$

*For raw data, whose averages were used in Table 3, see Table 1. For definitions of symbols, see Tables 1 and 2.

Table 4. Replicate determinations on 2 Feulgen-stained nuclei*

Replication	Periblast Nucleus	Non-Periblast Nucleus
1	82.4	20.7
2	84.6	19.8
3	85.0	20.8
4	80.6	19.2
5	78.4	19.4
6	81.8	18.9
7	83.4	20.4
8	85.4	20.2
9	79.2	19.6
10	79.8	19.0
Mean	82.0	19.8
Standard deviation	2.4	0.6
Standard error	0.8	0.2

* Table illustrates that variation within a column, attributable to experimental error on part of microphotometer and operator, is of different order of magnitude from variation between columns, attributable to difference of periblast and non-periblast nuclei in relative DNA content, expressed in arbitrary units.

In 2/3 developed muskellunge embryos (163 h of incubation), the majority of periblast nuclei measured still had approximately four times as much DNA content as the supposedly diploid non-periblast nuclei. However, the number of periblast nuclei whose shape and size precluded their being measured had increased.

A histogram is used to show the frequency distribution and relative DNA contents in arbitrary units for 25 periblast nuclei and 20 non-periblast nuclei from the 1/3 developed muskellunge embryos, and for 40 periblast nuclei and 20 non-periblast nuclei from the 2/3 developed muskellunge embryos (Fig. 1).

With the scanning integrating microdensitometer, the total optical density of a very large periblast nucleus from a zebra fish (Cypriniformes: Cyprinidae) was measured and then compared with the total optical density of each of 5 small, non-periblast nuclei that had been dissected from the same yolk-sac larva. Such a comparison yielded a ratio of total optical density of a periblast nucleus to total optical density of a non-periblast nucleus. Twenty giant nuclei were thus measured and compared. Their ratios ranged from 4:1 to as high as 20:1, the latter ratio being found in mammoth trilobed and bilobed periblast nuclei. Such total optical density ratios implied that the relative DNA contents of those giant nuclei measured were 4 to 20 times the diploid complement of DNA.

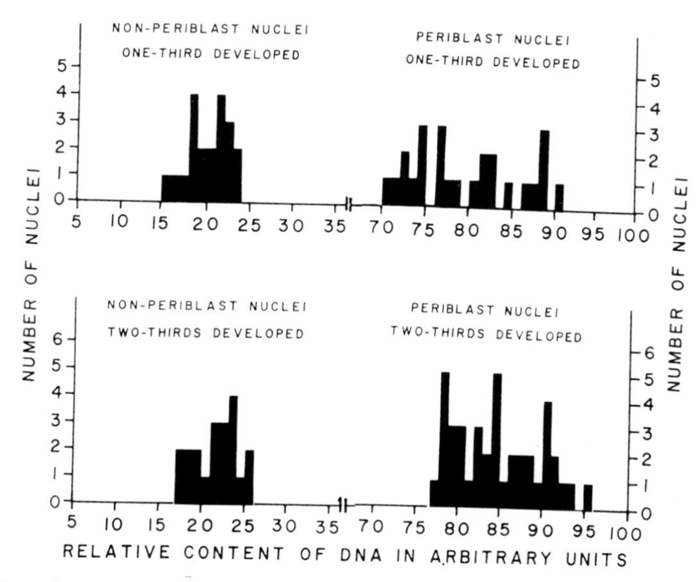

Fig. 1. Histograms: relative DNA content of nuclei

DISCUSSION

The yolk of the embryo of an oviparous teleost is completely enclosed within a syncytium. This yolk-sac syncytium (periblast) differs from syncytia in other vertebrates in that many, if not all, of its nuclei become giants. These giant nuclei have been found by Feulgen microphotometry and scanning integrating microdensitometry to contain amounts of DNA far in excess of the amount found in nuclei of diploid cells of the same embryo. Study of the mode of formation of these giant nuclei shows that the excess DNA is compartmentalized in the form of extra sets of chromosomes per nucleus (polyploidy) without apparent increase in amount of DNA per chromosome (Bachop and Price, 1971).

The yolk-sac syncytium is interfaced with the yolk on one side and with the yolk sac blood capillaries and blood sinusoids on the other, and is the embryonic structure through which substances derived from yolk must pass to enter the embryonic blood stream. These substances derived from yolk bypass the embryonic gut entirely. That is, 1. the yolk is not contained within the lumen of the embryonic gut; 2. the yolk is not contained within a cavity lined by endodermal cells; 3. the lining of the embryonic gut is not continuous with the lining of the yolk sac; 4. the lumen of the embryonic gut is not confluent with the lumen of the yolk-filled yolk sac. The yolk-sac syncytium is finally resorbed along with the last of the yolk, but the embryonic gut develops into the adult gut.

The presence of giant polyploid nuclei in the syncytium subserving yolk resorption in oviparous teleosts suggests that biochemical hypotheses of teleostean yolk resorption predicated on the mistaken assumption that yolk is resorbed from the lumen of the embryonic gut through a layer of endodermal cells containing small diploid nuclei should be supplanted by hypotheses that not only accord with the topographic and microscopic anatomy of the yolk sac of oviparous teleosts, but also ascribe a functional role to somatic polyploidy (polysomaty) and syncytialization in these embryos.

SUMMARY

1. Normal-appearing muskellunge embryos (Salmoniformes: Esocidae: *Esox masquinongy*) and zebra fish yolk-sac larvae (Cypriniformes: Cyprinidae: *Brachydanio rerio*) were placed alive into the fixative considered by histochemists to be optimal for quantitative *in situ* preservation of deoxyribonucleic acid - absolute ethanol (3 volumes), glacial acetic acid (1 volume). Nuclear deoxyribonucleic acid was visualized by the histochemical reaction considered by histochemists to be most specific and stoichiometric for DNA - the Feulgen nucleal reaction.

2. DNA contents of individual nuclei in an individual muskellunge embryo were compared with each other by two-wave length Feulgen microphotometry. In each muskellunge embryo measured, the yolk-sac syncytium contained many giant nuclei whose DNA contents were 4 or more times the DNA contents of nuclei within presumably diploid cells of the same embryo.

3. DNA contents of individual nuclei in an individual zebra fish yolk-sac larva were compared with each other by scanning integrating Feulgen microdensitometry. In each yolk-sac larva measured, the yolk-sac syncytium contained many giant nuclei whose DNA contents were 4 or more times the DNA contents of nuclei within presumably diploid cells of the same larva. In mammoth bilobed and trilobed nuclei of the larval yolk-sac syncytium, DNA contents as great as 20 times the diploid DNA content were found.

REFERENCES

Bachop, W.E., 1965. Size, shape, number, and distribution of periblast nuclei in the muskellunge embryo (Clupeiformes: Esocidae). Trans. Am. microsc. Soc., 84, 80-86.

Bachop, W.E. and Price, J.W., 1971. Giant nuclei formation in the yolk-sac syncytium of the muskellunge, a bony fish (Salmoniformes: Esocidae: *Esox masquinongy*). J. Morph., 135, 239-246.

Deitch, A.D., 1966. Cytophotometry of nucleic acids, 327-354 pp. In: Introduction to Quantitative Cytochemistry. G.L. Wied (ed.). New York - London: Academic Press, 623 pp.

Garcia, A.M. and Iorio, R., 1966. Potential sources of error in two-wave length cytophotometry, 215-237 pp. In: Introduction to Quantitative Cytochemistry. G.L. Wied (ed.). New York - London: Academic Press, 623 pp.

Hale, A.J., 1966. Feulgen microspectrophotometry and its correlation with other cytochemical methods, 183-199 pp. In: Introduction to Quantitative Cytochemistry. G.L. Wied (ed.). New York - London: Academic Press, 623 pp.

Kasten, F.H., 1960. The chemistry of Schiff's reagent. Intern. Rev. Cytol., X, 1-100.

Korson, R., 1951. A microspectrophotometric study of red cell nuclei during pyknosis. J. exp. Med., 93, 121.

Leuchtenberger, C., 1958. Quantitative determination of DNA in cells by Feulgen microspectrophotometry, 219-278 pp. In: General Cytochemical Methods, Vol. I. J.F. Danielli (ed.). New York - London: Academic Press, 471 pp.
Mendelsohn, M.L., 1958. The two-wave length method of microspectrophotometry. J. biophys. biochem. Cytol., 4, 407-424.
Mendelsohn, M.L. and Richards, B.M., 1958. A comparison of scanning and two-wave length microspectrophotometry. J. biophys. biochem. Cytol., 4, 707-709.
Moses, M., 1952. Quantitative optical techniques in the study of nuclear chemistry. Exp. Cell Res., Suppl., 2, 75-102.
Ornstein, L., 1952. Distributional error in microspectrophotometry. Lab. Invest., 1, 250-265.
Patau, K., 1952. Absorption microphotometry of irregular shaped objects. Chromosoma, 5, 341-362.
Patau, K. and Swift, H., 1953. The DNA-content (Feulgen) of nuclei during mitosis in a root tip of onion. Chromosoma, 6, 149-169.
Pearse, A.G.E., 1960. Histochemistry Theoretical and Applied, 2nd ed. Boston: Little Brown and Co., 998 pp.
Pollister, A.W. and Ornstein, L., 1955. Cytophotometric analysis in the visible spectrum, 3-71 pp. In: Analytical Cytology, R.G. Mellors (ed.). New York: McGraw-Hill Book Co.
Pollister, A.W. and Ornstein, L., 1959. The photometric chemical analysis of cells, 431-518 pp. In: Analytical Cytology, 2nd ed., R.G. Mellors (ed.). New York: Mc-Graw-Hill Book Co., 543 pp.
Swift, H. and Rasch, E., 1956. Microphotometry with visible light, 353-400 pp. In: Physical Techniques in Biological Research, Vol. III, G. Oster and A.W. Pollister (eds.). New York - London: Academic Press, 728 pp.

W.E. Bachop
Department of Zoology
Clemson University
Clemson, S.C. 29631 / USA

F.J. Schwartz
Institute of Marine Sciences
The University of North Carolina
Morehead City, N.C. 28557 / USA

Physiological Ecology

Effects of Thermal Shock on Larval Estuarine Fish – Ecological Implications with Respect to Entrainment in Power Plant Cooling Systems[1]

D. E. Hoss, W. F. Hettler, Jr., and L. C. Coston

INTRODUCTION

In the United States most existing and proposed power plant cooling systems are of the "once-through" type in which water is passed through the condensers and released back to the cooling source (lake, river, estuary or ocean) at an elevated temperature. Organisms small enough to pass through intake screens (usually 9.5 mm) such as larval fish may be entrained in the cooling system and subjected to an acute thermal shock which may be detrimental to individuals and the population.

In a recent review of information on once-through cooling systems Edsall and Yocom (1972) cite reports of heavy kills of entrained fish larvae at various power plants. Marcy (1971) found that no young fish of 9 species survived passage through the Connecticut Yankee Atomic Power Plant when discharge temperatures were above 30°C. He estimated an average yearly larval fish kill of 179 million at this plant. Other studies have estimated larval fish kills as large as 164.5 million/day (Edsall and Yocom, 1972).

The above data do not identify the specific cause of death during entrainment but show that plants with once-through cooling systems may act as "a large artificial predator" (Coutant, 1971). This predatory potential is particularly important when power plants are located in estuarine areas that are used extensively as nursery grounds by many species of fish.

The stresses on entrained organisms may be mechanical, chemical, and thermal (Edsall and Yocom, 1972). Our paper examines one of these, thermal stress, in detail using laboratory results on both acute (percent survival) and chronic (changes in metabolism and growth) effects.

MATERIALS AND METHODS

We collected larval fish in a tide net (Lewis et al., 1970) and by dip net in the estuary near Beaufort, North Carolina. Following capture, but before experimentation, fish were acclimated in the laboratory for 3-7 days at experimental temperatures and salinities. Usually the acclimation temperatures corresponded to the environmental temperatures at which the fish were caught. All fish in these experiments were fed to satiation once a day using brine shrimp nauplii (*Artemia salina*) or dry commercial food. Artificial lights regulated to natural photoperiods were used throughout the experiment.

---

[1]This research was supported by U.S. Atomic Energy Contract AT(49-7)-5.

The larval fish used were: Atlantic menhaden, *Brevoortia tyrannus* (Latrobe); spot, *Leiostomus xanthurus*, Lacépède; pinfish, *Lagodon rhomboides* (Linnaeus); and three species of flounder, *Paralicthys dentatus* (Linnaeus); *P. lethostigma*, Jordan and Gilbert; and *P. albigutta*, Jordan and Gilbert. They are important components of the estuarine ecosystem along the southeastern Atlantic coast of the United States. All of these fish use the estuary as a nursery and are therefore subject to man-imposed stresses such as entrainment in power plant cooling systems located in estuarine areas.

To determine the critical thermal maximum (CTM) we followed the basic procedure described by Hutchinson (1961). Individual fish were placed in 500 ml distillation flasks containing 300 ml of sea water at acclimation temperature. The distillation flask was heated at approximately 1°C per minute by a hemispherical mantle heater connected to a variable transformer. The water in the distillation flask was aerated and stirred by compressed air. Flaring of the opercula was used to indicate the CTM.

We measured routine metabolism of larval fish as individuals and as groups. A differential respirometer (Umbreit et al., 1964) was used for individual measurements. Larvae were transferred into 15 ml respiration flasks containing 3 or 5 ml of Millipore-filtered sea water. Oxygen consumption measurements were made at hourly intervals on each fish for a period of 4 h and no fish was used in more than one experiment. Wet weights were obtained after the last oxygen measurement. Routine metabolism of groups of larval fish was measured in flowing water respirometers. The respiration chambers are air-sealed doughnut-shaped Plexiglas swimming channels which provide a constant supply of air-saturated sea water at controlled temperature and salinity. The difference in oxygen concentration between the incoming and outgoing water times the flow rate of water through each chamber gave the total amount of oxygen used by each group of larvae. The number and total wet and dry weights of each group was determined at the end of the experiment.

The relationship between oxygen consumption and wet weight was calculated so metabolism for fish of different weights could be estimated. In the equation, $\log Q = \log a + k (\log W)$, Q is the oxygen consumption, W is the weight of the fish (mg) and a and k are constants for the species obtained from least squares regression lines of oxygen consumption on weight (Winberg, 1956).

We measured the percent survival after thermal shock at various temperature-salinity-time combinations. The fish used were acclimated to temperatures at which they normally occurred and were subjected to thermal shocks within the predicted range of condenser temperature elevations (5.6° to 18°C)(Auerbach et al., 1971). Our basic experimental design consisted of one salinity (29°/oo) three acclimation temperatures (5°, 10° and 15°C) three shock increments (12°, 15° and 18°C) and five exposure times (0, 10, 20, 30, and 40 min).

Because of the sensitivity of larval fish, proper handling is of great importance to ensure that experimental results are due to the test conditions and not to the handling procedure. Our method of handling the fish for shock experiment is shown in Fig. 1. The transfer container is a plastic bell jar with holes drilled in the sides to permit the exchange of water. A residual amount of water is always maintained in the container so that the fish are not exposed to air when being transferred between tanks. A rubber stopper fitted in the bottom of the container is removed to release the fish after exposure to heated water. Using this method, handling of the fish is reduced to a minimum and percent survival is high.

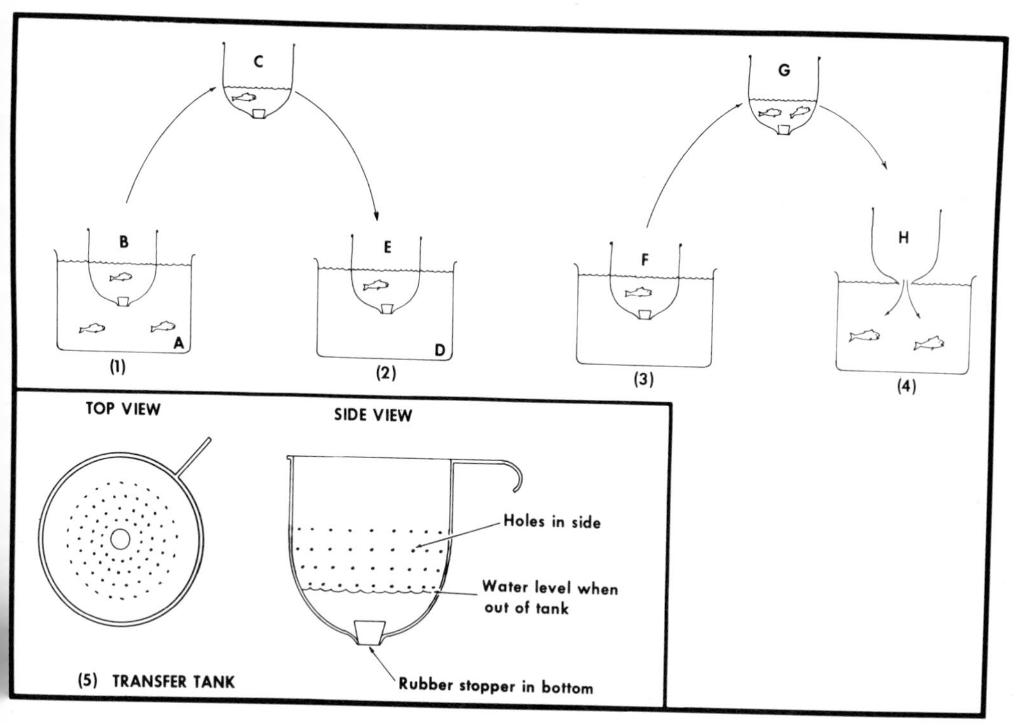

Fig. 1. Method for transferring fish from acclimation aquaria to heated water. (1) Fish moved from acclimation tank (A) to transfer container (B). (2) Transfer container removed from acclimation tank, partially drained (C) and placed in heated water tank (D). Heated water enters transfer container (E). (3) Fish remain in heated water for specified period of time (F). (4) Transfer container removed from heated water (G) and placed in acclimation tank where plug is pulled (H) and fish are released back into water at acclimation temperature and salinity. (5) Detail of transfer tank

Selected median resistance times to high temperatures were determined for comparison with the results obtained in our thermal shock experiments. In general, the method described by Fry et al. (1946) was used to determine resistance time. In this method, a sample of fish was transferred directly from acclimation temperature to the increased temperature. Fish were held at the elevated temperature until they died or for an arbitrary period of 7 days. Time of death of individual fish after transfer was noted.

RESULTS

Critical Thermal Maximum

We can make the following general observations concerning CTM measurements for menhaden, spot, pinfish, and flounder (Table 1). 1. A 5 deg C increase in acclimation temperature on the average increased the CTM value about 1 deg C. 2. We could not acclimate menhaden to 5°C. At 10° and 15°C acclimation, they had lower CTM values than the other species tested. 3. Flounders acclimated to 15°C had a higher CTM (32.0°C) than other species tested. 4. The difference between an acclimation temperature and a corresponding CTM value (last column Table 1) is an estimation of the increase in temperature that a fish can tolerate. Thus, at 15°C flounder would be the most resistant to thermal change and menhaden the least.

Table 1. CTM values for larval estuarine fish acclimated at 5°, 10° and 15°C and an average salinity of 29°/oo (15°C data in part from Hoss et al., 1971)

Larvae	Acc. temp. (°C)	No. of fish	Av. wet wt. (mg)	CTM (°C)	Std. error	Diff. between CTM and accl. temp. (°C)
Menhaden	10	11	56.6	28.9	.099	18.9
	15	23	51.0	29.7	.128	14.7
Spot	5	14	41.8	28.4	.127	23.4
	10	10	43.0	30.4	.180	20.4
	15	45	22.9	31.1	.411	16.1
Pinfish	5	11	38.5	30.0	.250	25.0
	10	11	34.8	30.3	.286	20.3
	15	24	29.4	31.0	.070	16.0
Flounder	5	12	22.3	29.0	.271	24.0
	10	20	31.6	30.5	.137	20.5
	15	12	30.3	32.0	.147	17.0

Thermal Shock

A summary of the results on thermal shock for fish tested is presented in a series of three dimensional figures showing percent survival for various combinations of acclimation temperature, exposure time and increases in temperature (Figs 2-5). Zero exposure time shown on these figures represents control fish, i.e. fish that were transferred between containers but were not subjected to a thermal shock. Less than 100% survival in control groups would indicate that factors other than increased temperature were causing death. In those cases where some control fish died (89% survival in Atlantic menhaden acclimated at 10°C, and 93% survival in pinfish acclimated at 5°C) percentages were adjusted by dividing the control fractional survival into the experimental fractional survival. Percent survival one day after the thermal shock is shown in the figures although we followed the survival for 7 days. In most cases we found survival after 7 days essentially the same as after one day. Usually, when shocked fish die between days 1 and 7 we also found a reduced survival in the control group. This indicates that mortality after day one was not related to thermal shock.

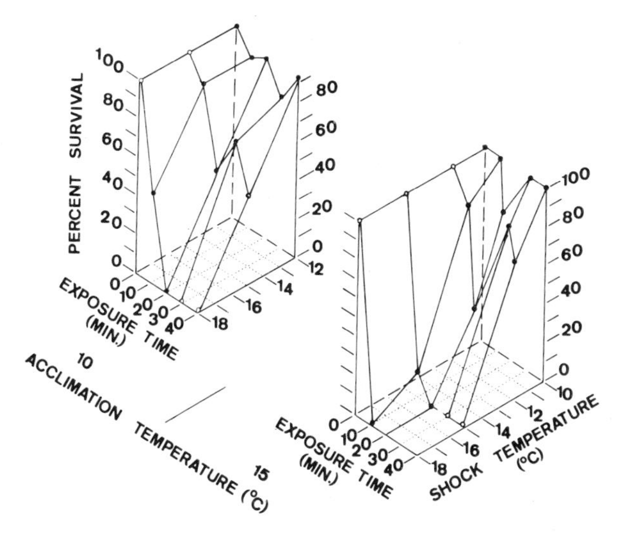

Fig. 2. Percent survival of larval Atlantic menhaden after exposure to thermal shock for various periods of time. Each point represents an average of 11 fish. The average wet weight of the fish was 80 mg. Closed circle, actual data point; open circle, estimated point

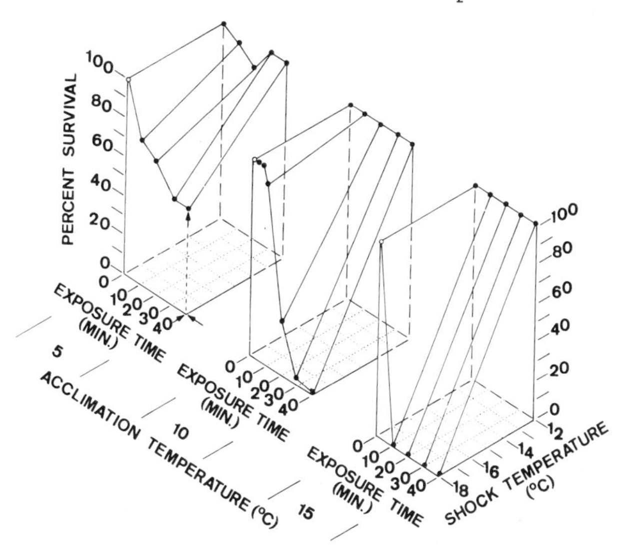

Fig. 3. Percent survival of larval spot after exposure to thermal shock for various periods of time. Each point represents an average of 28 fish. The average wet weight of the fish was 36 mg. Closed circle, actual data point; open circle, estimated point. Arrows indicate 18°C shock and 40 min exposure

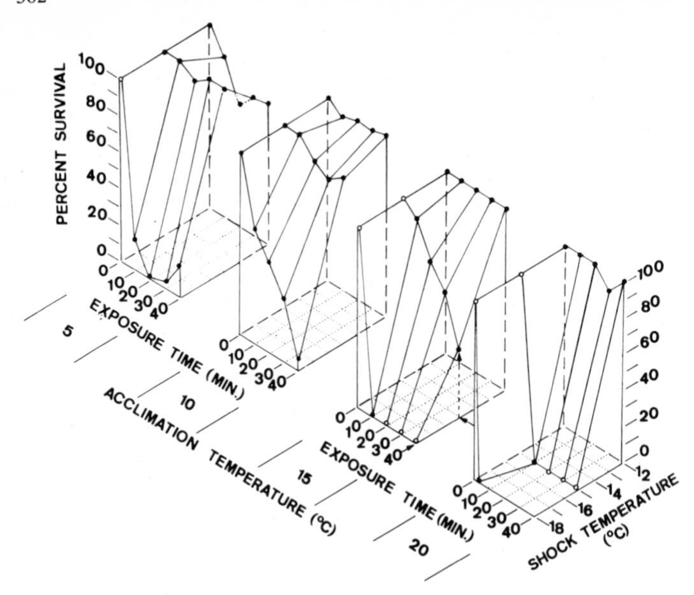

Fig. 4. Percent survival of larval pinfish after exposure to thermal shock for various periods of time. Each point represents an average of 27 fish. The average wet weight of the fish was 35 mg. Closed circle, actual data point; open circle, estimated point. Arrows indicate 15°C shock and 40 min exposure

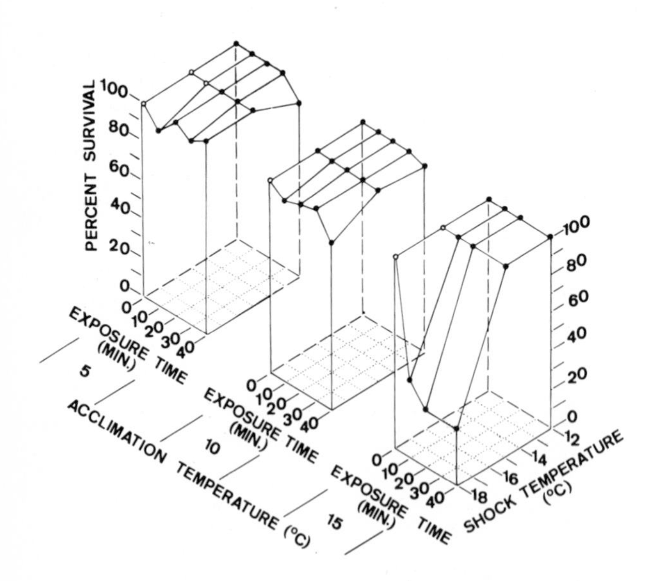

Fig. 5. Percent survival of larval flounder after exposure to thermal shock for various periods of time. Each point represents an average of 13 fish. The average wet weight of the fish was 26 mg. Closed circle, actual data point; open circle estimated point

Atlantic Menhaden. We were not able to hold menhaden at 5°C. This temperature appears to be at or below the lower incipient lethal level for the species (Lewis, 1965). Menhaden survived well at 10°C and 15°C and results for these acclimation temperatures are shown in Fig. 2. Menhaden acclimated to 10°C and given a 12°C shock were not appreciably affected but both 15° and 18°C increases in temperature caused significant mortality. When acclimation was increased from 10° to 15°C, percent survival was reduced after 12°, 15°, and 18°C increases but not after a 10°C increase.

Spot. Spot were held at acclimation temperatures of 5°, 10°, and 15°C (Fig. 3). A 12°C shock did not cause appreciable mortality at any of the acclimation temperatures. An 18°C shock reduced survival at all acclimation temperatures. Over 50% of the larvae could survive an 18°C shock at 5°C regardless of exposure time. However, fish acclimated to 10°C showed greatly reduced survival to the same shock when exposure was greater than 10 min. Larvae acclimated to 15°C could not survive an 18°C shock for any of the exposure periods tested.

Pinfish. Larval pinfish are available over a considerable temperature range, thus we were able to conduct experiments on pinfish acclimated to 5°, 10°, 15°, and 20°C. Most of the larvae could withstand a 12°C temperature shock regardless of acclimation temperature (Fig. 4). A 15°C shock was detrimental only at the higher acclimations (15° and 20°C), while 18°C shocks were generally fatal. At the higher shocks, duration of exposure became important, particularly when acclimation temperatures were also high.

The effect of a rapid change in salinity (20°/oo) on survival of thermally shocked pinfish was also measured (Table 2). Larval pinfish acclimated to 20°C and 15°/oo were given a 12°C thermal shock for 0, 10, 20, 30, and 40 min. One half of the fish were then returned to 15°/oo sea water, the other half to 35°/oo sea water and percent survival at the two salinities was compared. The high percentage survival conformed with previous +12°C shock experiments at 20°C acclimation. The additional stress of a 20°/oo salinity increase did not affect survival up to 10 min later. However, there was a significant reduction in survival after 20 min (as measured by a "t" test on the pooled samples) although there was no consistent trend with exposure time. This indicates that there was some salinity stress.

Table 2. Survival of larval pinfish which experienced rapid changes in salinity after thermal shock. Fish were acclimated to 20°C and 15°/oo. Parenthesis indicate the numbers of fish per exposure

Shock temp. (°C)	Exposure time (min)	% survival after one day at	
		15°/oo	35°/oo
Controls	0	100 (10)	100 (10)
32°	10	100 (10)	100 (10)
32°	20	100 (9)	88 (8)
32°	30	100 (10)	100 (11)
32°	40	100 (9)	80 (10)

Flounder. Three species of paralichthid flounder were collected. Separation of these species at the larval stage requires examination with a microscope which in most cases is harmful to the fish. Therefore we identified the flounder at the end of the experiment or at the time of death. Of the 395 flounder examined, the three species were found to be present in the following percentages: 1. *Paralichthys lethostigma* 55.7%, 2. *P. dentatus* 39.5%, 3. *P. albigutta* 4.8%. We calculated percent survival by species, noticed no difference between them, and pooled the three for final calculation of percent survival.

Of the larval fish examined, flounder had the highest survival following thermal shock (Fig. 5). The flounder showed high survival at 12° and 15°C increases in temperature. Even under the most harsh conditions tested (18 deg C increase above 15°C acclimation for 40 min) 30% of the flounder survived.

The effect of a sudden salinity change following thermal shock was measured in flounder. The fish were acclimated to 13°C and 25°/oo. After a thermal shock of either 12° or 18°C for 0, 10, 20, 30, and 40 min the fish were returned to either 25°/oo or 35°/oo sea water. An 18°C shock plus a 10°/oo increase in salinity did cause a significant (Student's t = 5.38) reduction in survival after a 10 min exposure (Table 3). As in the pinfish experiment, salinity had no effect in conjunction with a 12°C shock. Indications are that the salinity effect is slight but may become important at maximum shock temperatures

Table 3. Survival of larval flounder acclimated to 13°C and 25°/oo. After thermal shock fish were returned to either 25°/oo or 35°/oo water. Numbers in parenthesis are numbers of fish in each group

Shock temp. (°C)	Exposure time (min)	% survival after one day at	
		25°/oo	35°/oo
Controls	0	100 (20)	100 (21)
25°	10	100 (20)	100 (21)
	20	100 (19)	100 (20)
	30	100 (24)	100 (23)
	40	100 (25)	100 (23)
31°	10	88 (26)	18 (22)
	20	4 (23)	5 (21)
	30	5 (22)	0 (19)
	40	0 (21)	- -

Median Resistance Time

In order to compare the predictive capabilities of our percent survival figures with that of Coutant's (1971) survival nomogram for high temperature shock we measured the median resistance times of spot acclimated to 5°C and pinfish acclimated to 15°C. The time in min to 50% mortality was plotted on a log scale against temperature (Fig. 6). Least square regression lines were calculated and a line fit to the median survival times for both species. Values for a, the intercept and k, the slope of the lines were 22.759 and -0.708 for pinfish and 14.506 and -0.535 for spot. Although this is only a small part of the data needed for a complete nomogram for these species it can be used within limits. For example, a thermal shock of 30°C for 40 min would be expected to kill half or more of the larval pinfish exposed to it (line A-B Fig. 6). A shock of 23°C for 40 min (line C-B Fig. 6) would kill less than 50% of the spot.

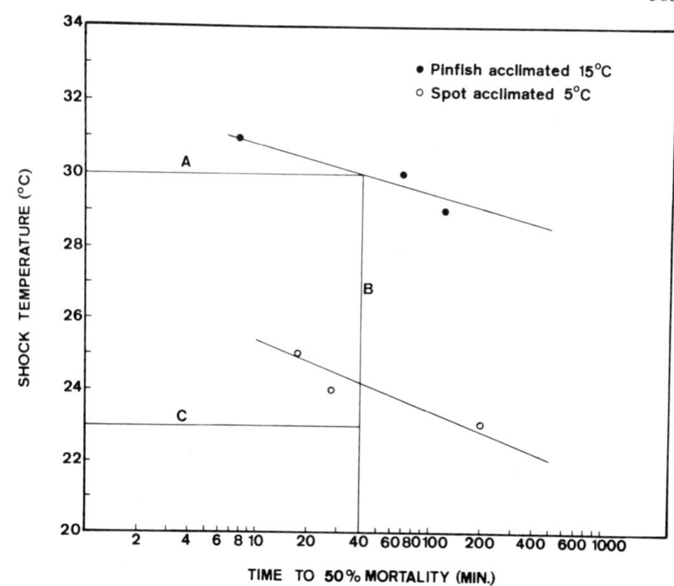

Fig. 6. Time-temperature relations for 50% survival of pinfish and spot after thermal shock. Closed circles represent pinfish acclimated to 15°C. Open circles represent spot acclimated to 5°C. Line A-B indicates 30°C shock temperature and 40 min exposure. Line C-B indicates 23°C shock temperature and 40 min exposure

Metabolism, Growth and Behavior

Our shock experiments with spot and pinfish showed that a 12°C increase in temperature for 40 min would not appreciably reduce survival (Figs 3 and 4). Additional experiments were designed to determine if this sublethal thermal 12°C shock had affected growth or oxygen consumption in the fish.

Growth. A group of 173 larval pinfish acclimated to 15°C was divided into a control and an experimental group. A subsample of 15 fish was taken from each group and the dry weight obtained at the start of the experiment. The experimental group was then given a 12°C thermal shock for 40 min and returned to 15°C sea water. The control group was handled in the same manner, but were not given the 12°C thermal shock.

After 9 days the experiment was terminated and the dry weight of the remaining fish obtained. The results are presented in Table 4 and show that growth between the two groups was not significantly different.

In a second experiment larval spot acclimated to 15°C were divided into experimental and control groups of 50 fish each. A subsample was taken from each group and the mean dry weight obtained. As in the previous experiment with pinfish, experimental fish were given a 12°C shock for 40 min and returned to 15°C sea water. Control fish were not given the thermal shock.

Table 4. Comparison of growth of control and experimental pinfish and spot larvae. Experimental fish were subjected to a 12°C thermal shock for 40 min

Species	Days	Treatment	No. of fish	Mean dry wt. (mg)	t
Pinfish	0	Control	15	5.15	0.03 n.s.[1]
	0	Experimental	15	5.14	
	9	Control	71	8.86	1.50 n.s.
	9	Experimental	59	7.89	
Spot	0	Control	15	12.19	1.33 n.s.
	0	Experimental	13	11.23	
	14	Control	29	21.38	1.21 n.s.
	14	Experimental	9	23.70	

[1]Not significant at .05 level

Two to four days after the start of the experiment we noted fungal infection in both groups of fish. After 14 days, only 9 fish survived in the experimental group and 29 in the control group. Dry weights were obtained on the survivors and, like pinfish (Table 4) the growth was not significantly different after 14 days.

Metabolism. Metabolism measured as oxygen consumption, can be a useful means of determining sublethal stress such as increased temperature (Steed and Copeland, 1967). In a previous paper (Hoss et al., 1971) we measured respiration in larval fish subjected to rapid temperature changes. In those experiments respiration was measured while the fish were at the elevated temperature for approximately 6 h and showed that respiration did increase with increased temperature. The 6-h exposure period is, however, longer than that normally experienced in an entrainment situation. In addition, a more important ecological question is whether or not respiration in surviving fish is appreciably changed after the fish are returned to ambient conditions. We therefore measured oxygen consumption of larval pinfish and spot before and after exposure to a 12 deg C increase in temperature.

In individual respiration experiments a group of pinfish acclimated to 15°C were given a 12°C shock for 40 min and then all but 18 of the fish were returned to a 15°C holding tank. The 18 larvae were placed in individual respiration flasks at 15°C and oxygen consumption was measured. Subsamples were taken from the holding tanks after 1 and 7 days and additional respiration measurements were made. These respiration measurements were used to calculate values for metabolism-weight coefficients and the results were then compared to data previously collected on larval pinfish held at 15°C without thermal shock. The results (Table 5) indicate a decrease in oxygen consumption per mg of fish with time. This is to be expected since the larvae grew during the experiment (Table 5) and it has been previously established that oxygen consumption per unit weight is greater for smaller than for larger fish (Winberg, 1956; Hickman, 1959; Hoss, 1967).

Table 5. Oxygen consumption of pinfish following a 12°C increase in temperature for 40 min. Fish acclimated at and returned to 15°C

Time after shock	Wet wt. (mg)	No. of fish	Observed oxygen consumption ($\mu g\ O_2/h/mg$)	Oxygen consumption estimated from base line equation[1] ($\mu g\ O_2/h/mg$)
2-4 h	24	18	0.45	0.44
1 day	25	18	0.44	0.44
7 day	31	18	0.42	0.43

[1] $\log Q = -0.2320 + (0.91)(\log W)$ (Hoss et al., 1971)

To determine if the changes in oxygen consumption with weight were within the expected range we estimated the oxygen consumption of non-shocked fish of the same size acclimated to 15°C. We did this by using the metabolism-weight coefficients for larval pinfish (Hoss et al., 1971). It can be seen in Table 5 that observed and predicted oxygen consumption values are approximately the same. It appears that oxygen consumption of fish exposed to a 12°C shock returns to normal levels within a few hours and remains there.

The effect of a 12 deg C increase in temperature on the routine metabolism of groups of larval spot was measured in flowing water respirometers. The fish were not disturbed or handled during the 5-day experiment. The temperature regime to which fish were exposed was as follows: a 15°C acclimation period for 3 days; metabolic rate measurements for 48 h at 15°C; an increase in temperature to 27°C in less than 6 min and maintained for 40 min; a decrease in temperature to 15°C in 30 min followed by respiration measurements at 15°C for 2 additional days. This regime was applied to 2 groups of larvae simultaneously. Two other groups of larvae, in similar chambers, were maintained at 15°C throughout the 5 days as controls. The fish were fed daily following respiration measurements. Control groups totaled 905 larvae (mean wet weight 58.3 mg). Experimental groups totaled 876 larvae (mean wet weight 59.7 mg).

Oxygen consumption of control and experimental fish was compared before and after exposure to the 12°C shock. Before the 12°C shock oxygen consumption of control fish averaged 0.58 mg O_2/g wet wt/h and experimental fish averaged 0.56 mg O_2/g wet wt/h, a difference of 3%. After the shock oxygen consumption was approximately the same, 0.55 mg O_2/g wet wt/h control and 0.53 mg O_2/g wet wt/h experimental, a 4% difference.

Pinfish respiration was also measured in flowing water respirometers after a 12°C temperature shock superimposed on a 23°/oo salinity shock. The fish were acclimated at 15°C and 5°/oo salinity. The temperature was elevated to 27°C for 10 min and returned to 15°C after an additional 30 min. Two of the groups were maintained at 5°/oo; the other two groups were increased to 28°/oo salinity during the cooling period to simulate a cooling water effluent discharge into the ocean environment. Fish maintained at 5°/oo totaled 3290 (mean wet wt 30.8 mg). Fish changed to 28°/oo salinity totaled 2826 (mean wet wt 31.3 mg). Respiration following temperature shock and a combination of temperature-salinity shock was compared to control respiration as in the previous experiment. Oxygen consumption of pinfish given a temperature shock alone (0.57 mg O_2/ g wet wt/h) was 1% less than control fish

(0.58 mg O_2/g wet wt/h). Oxygen consumption of those fish given both a temperature and salinity shock was (0.54 mg O_2/g wet wt/h) 3% less than control (0.56 mg O_2/g wet wt/h). These changes in respiration were not significant (t = 1.1). These results agree with our previous oxygen consumption measurements on individual pinfish following thermal shock. In those experiments the percentage change in oxygen consumption after thermal shock was 2%.

Behavior. During the course of our thermal shock experiments we were able to make some general observations on the behavior of the larvae during and after thermal shock. Fish subjected to a 12°C increase in temperature generally showed little or no distress. Increases of 15° or 18°C, however, usually caused an immediate reaction. Typical reactions were loss of equilibrium, erratic swimming, violent jumping, and convulsions. If the larvae survived, this initial period of reaction was often followed by a return to relatively normal behavior for the remainder of the exposure period. Often the only sign of stress would be an increased opercular movement indicating increased respiration.

Those fish that survived the initial increase in temperature also received a thermal shock when they were returned to the acclimation temperature. The reaction to this second shock was often more violent than to the first and seemed to be more pronounced at the colder acclimation temperatures. Pinfish acclimated at 5°C, for example, showed no stress when given a 12 deg C increase, but on their return to colder water appeared to be stunned.

DISCUSSION

We have looked at ways to evaluate lethal and sublethal effects of thermal shock on larval fish hoping to find methods that will facilitate prediction of detrimental effects due to entrainment in a power plant. The sublethal effects of heat shock on those fish that survive entrainment could be important ecologically. We did not find any long term effect of thermal shock on oxygen consumption or post-shock growth rates of the estuarine fish tested. Previous work (Hoss et al., 1971) showed that sudden increases in temperature increased respiration rates and that there was a difference in response between species. Our present investigations, however, indicate that the metabolic increases are temporary and return to normal in surviving fish after a few hours.

We did observe one sublethal effect, loss of equilibrium, that has important ecological implications. Coutant (1969, 1971) has noted that equilibrium loss due to high temperature may lead to death through immobilization and inability to avoid predators. We observed equilibrium loss not only after increases in temperature but also after decreases in temperature to ambient. Thus, in many cases fish that were not affected by the initial increase in temperature were completely immobilized by the sudden decrease in temperature that could occur in a rapid dilution type of outfall. This second shock could be important ecologically in two ways. First, the second shock alone or in conjunction with the initial shock may be a direct cause of mortality.

Second, even if the fish is not killed directly by the shock it is certainly more vulnerable to predation. Predation studies of the type suggested by Coutant should take this second shock into account.

Of methods tested measuring lethal effects, the percent survival most nearly simulated the temperature conditions that entrained fish would encounter in a cooling water system. The CTM would be the least useful for predictive purposes. The application of the CTM to entrainment situations has been discussed extensively by Coutant (1970). He pointed out that the CTM is the result of two variables, time and temperature, but only provides temperature as an end point. The inadequacy of this can be shown by using as an example fish acclimated to 15°C and CTM values from Table 1. The CTM indicates that a 15 deg C increase in temperature to 30°C would probably be fatal to menhaden but not to pinfish and flounder. Our shock survival data show, however, that some menhaden will survive a 15 deg C increase if exposure time is 10 min or less and that some pinfish will not survive a 15 deg C increase if exposure time is 10 min or longer. The usefulness of CTM measurements would appear to be in rating species as to their overall sensitivity to thermal stress.

Estimates of median resistance time provides better predictions than the CTM but are still lacking in several respects. Although the resistance concept could be enlarged, by current definition it cannot be used to estimate relative mortality above or below 50%. If a power plant "predation" rate of 30% or perhaps 70% were arbitrarily selected as an ecologically acceptable criteria, published median resistance times would not be useful. A more serious fault of the resistance concept is that it ignores two important factors: 1. the effect of exposure time to increased temperature and 2. the shock of returning from high to ambient environmental temperature. We think that our approach to estimating percent survival from the overall thermal shock overcomes the disadvantages inherent in resistance time measurement.

Data on shock survival for each species can be used in conjunction with plant operating characteristics to predict the destructive potential of the power plant on that species. As an example, assume a power plant elevates cooling water 15° to 30°C, and larval fish drawn into the intake are exposed to this temperature for 40 min. Using the figures for each species at the proper acclimation temperature we can predict survival of the entrained larvae. Entering the pinfish 15°C acclimation graph (Fig. 4), at 40 min exposure time and 15°C shock temperature (indicated by arrows), we find that 35% of the pinfish survived. Under the same conditions survival for the other species are estimated to be; menhaden 0%, spot 50%, and flounder 100%. In comparison, using the median resistance time and similar input data for pinfish all we could say is that 50% of the pinfish would die (Fig. 6, line A-B).

The difference in predictive ability between the two methods can also be shown with spot. Assuming an 18 deg C increase in temperature for 40 min and an acclimation temperature of 5°C our graph (Fig. 3) indicates 53% survival. Using the same conditions, the intersection of lines C-B on the median resistance time graph (Fig. 6) falls well below the 5°C acclimation line giving the impression that no mortalities would occur.

The shock survival method we used for predicting entrainment effects is better than resistance time for several reasons. We could modify exposure time and temperature to correspond with actual plant operating characteristics. In addition to the initial thermal shock the fish receive during intake, we can simulate any potential second shock on return to ambient conditions. Our results indicate that this second shock may be an important factor, especially at low acclimation temperatures. The second shock is not accounted for by the thermal resistance patterns because in this method the fish are kept at the elevated temperature for an indefinite period of time. We are also able to estimate the actual percent survival as opposed to the greater or less than 50% survival predicted by the thermal resistance patterns.

These studies and others already cited indicate that entrainment in power plant cooling systems could have serious ecological consequences, especially when power plants are located on estuaries. The effect of the power plant on the ecosystem will, of course, depend on the design and operating characteristics of the plant and the types of organisms subjected to entrainment.

Because plant design, species, and physical characteristics of the water will vary from site to site there is a need to examine each individual estuarine power plant site, as suggested by Adams (1970). Water temperature criteria should be flexible and based on the most sensitive important species present at the site. This could change with season. For example, in our experiments menhaden were the most sensitive species, but the larvae are only found during the winter months. Standards for summer months should not be based on larval menhaden, but on a larval species that is present during the summer.

SUMMARY

Critical thermal maximum (CTM), oxygen consumption, survival, and behavior of larval fish were measured as a function of the magnitude of temperature change, exposure time, and salinity for larvae of Atlantic menhaden (*Brevoortia tyrannus*), spot (*Leiostomus xanthurus*), pinfish (*Lagodon rhomboides*), and 3 species of flounder (*Paralichthys* spp.).

Measurements of CTM were inadequate for evaluating the effects of entrainment on larval fish, although they were useful in establishing ranges of thermal tolerance. Oxygen consumption of larval fish responded quickly to thermal shock. The response was dependent on the magnitude and direction of the temperature change. Survival was dependent on magnitude of temperature change, length of exposure, salinity, and species of fish. Observations on behavior of the experimental fish indicate that at least two shocks are discernible during entrainment: the first when the fish passes from environmental temperature to the entrainment temperature and a second when the fish passes back to the environmental temperature.

The results of these experiments are discussed in relation to the possible effects of entrainment on larval fish populations in estuarine waters.

REFERENCES

Adams, J.R., 1970. Thermal effects of electric power plants, 18 pp. In: Proceedings of the Joint Committee on Atomic Energy Hearings on Environmental Effects of Producing Electric Power, 24-26 February 1970, Washington, D.C.

Auerbach, S.I., Nelson, D.J., Kaye, S.V., Reichler, D.E. and Coutant, C.C., 1971. Ecological considerations in reactor power plant siting In: Environmental aspects of nuclear power stations. Vienna: International Atomic Energy Agency, 803-820 pp.

Coutant, C.C., 1969. Temperature, reproduction and behavior. Chesapeake Sci., 10, 261-274.
Coutant, C.C., 1970. Biological aspects of thermal pollution. I. Entrainment and discharge canal effects. Chem. Rubber Co. Crit. Revs in Environ. Control, 1, 341-381.
Coutant, C.C., 1971. Effects on organisms of entrainment in cooling water: steps toward predictability. Nucl. Saf., 12, 600-607.
Edsall, T.A. and Yocom, T.G., 1972. Review of recent technical information concerning the adverse effects of once-through cooling on Lake Michigan. U.S. Fish and Wildl. Serv., Bur. Sport Fish. and Wildl., Great Lakes Fish. Lab., Ann Arbor, Mich., 86 pp.
Fry, F.E.J., Hart, J.S. and Walker, K.F., 1946. Lethal temperature relations for a sample of young speckled trout, *Salvelinus fontinalis*. Univ. Toronto Stud. biol. Ser. 54; Publs Ont. Fish. Res. Lab., 66, 5, 35 pp.
Hoss, D.E., 1967. Rates of respiration of estuarine fish. Proc. SE Ass. Game Fish Commnrs, 21, 416-423.
Hoss, D.E., Coston, L.C. and Hettler, W.F., Jr., 1971. Effects of increased temperature on postlarval and juvenile estuarine fish. Proc. SE Ass. Game Fish Commnrs, 25, 635-642.
Hickman, C.P., 1959. The osmoregulatory role of the thyroid gland in the starry flounder, *Platichthys stellatus*. Can. J. Zool., 37, 997-1060.
Hutchinson, V.H., 1961. Critical thermal maxima in salamanders. Physiol. Zoöl., 34, 92-125.
Lewis, R.M., 1965. The effect of minimum temperature on the survival of larval Atlantic menhaden, *Brevoortia tyrannus*. Trans. Amer. Fish. Soc., 94, 409-412.
Lewis, R.M., Hettler, W.F., Jr., Wilkens, E.P.H. and Nelson, G.N., 1970. A channel net for catching larval fishes. Chesapeake Sci., 11, 191-198.
Marcy, C.B., Jr., 1971. Survival of young fish in the discharge canal of a nuclear power plant. J. Fish. Res. Bd Can., 28, 1057-1060.
Steed, D.L. and Copeland, B.J., 1967. Metabolic responses of some estuarine organisms to an industrial effluent. Contri. Inst. mar. Sci. Univ. Tex., 12, 143-159.
Umbreit, W.W., Burris, R.H. and Stauffer, J.F., 1964. Manometric techniques. Minneapolis, Minn.: Burgess Publishing Co., 305 pp.
Winberg, G.G., 1956. Rates of metabolism and food requirement of fish. (Transl. from Russian). Fish. Res. Bd Can., Transln Ser. 194, 239 pp.

D.E. Hoss
National Marine Fisheries Service
Atlantic Estuarine Fisheries Center
Beaufort, N.C. 28516 / USA

W.F. Hettler, Jr.
National Marine Fisheries Service
Atlantic Estuarine Fisheries Center
Beaufort, N.C. 28516 / USA

L.C. Coston
National Marine Fisheries Service
Atlantic Estuarine Fisheries Center
Beaufort, N.C. 28516 / USA

Survival of Australian Anchovy *(Engraulis australis)* Eggs and Larvae in a Heat Trap

P. M. Powles

INTRODUCTION

While the total amount of biological literature on thermal effects on organisms is immense, practically no information could be found pertaining to effects on Australian fish species (Powles, 1972), and thus almost none on their eggs and larvae. Yet after the 1974 completion of its "Snowy Mountains hydro-scheme", Australia will become completely dependent on Steam Electric Stations to produce electrical energy (Howard, 1972). Most of the plants are on estuaries which are important nursery areas. Some entrainment work has been done on invertebrate eggs and larvae (see Barnett, 1972 and Powles, 1972) but only one author, Marcy (1971), has specifically studied survival of fish larvae which had passed through a power plant. The present work is the first attempt aimed at estimating survival of fish eggs and prolarvae carried through the cooling condenser pipes of a power station.

The academic rationale for attempting this project stems from the very extant but somewhat controversial view that most tropical and subtropical aquatic forms live close to their thermal death point, and are thus very sensitive to upward thermal stress (Krenkel and Parker, 1969). With this in mind, Electricity Commission Engineers of New South Wales requested data for planning, namely: an ultimate upper incipient lethal, approximate survival percentage, and an $L.D._{50}$ level for summer temperatures. This paper provides quantitative estimates of the above for the eggs of anchovy, *Engraulis australis*, (White) and "leatherjacket", *Meuschenia trachylepis* (Munroe).

METHODS AND MATERIALS

Most so-called lakes in Australia are in fact estuaries, or drowned valleys, with various degrees of openness to the sea. Lakes Macquarie and Tuggerah are no exception, and their fauna are basically marine. Salinities range from 15 to 32°/oo. In 1971 and 1972 a plankton survey (Fig. 1) revealed that during September to February, surface waters at many stations contained anchovy eggs, at times in great abundance. These particular stations are close to Vales Point and Munmorah, two important "once-through" Australian power stations operated on coal. Commencing in October, 1972, 10-min plankton sets for eggs were made at the mouth of the Munmorah inlet canal. Insufficient eggs were taken, possibly because of 1. particularly low salinity levels (17°/oo in October), 2. a virtual heat trap is formed between the two lakes as the plant displaces total lake volume every 4 days, 3. lack of anchovy spawning. Operations were therefore transferred to Vales Point, 4 miles away.

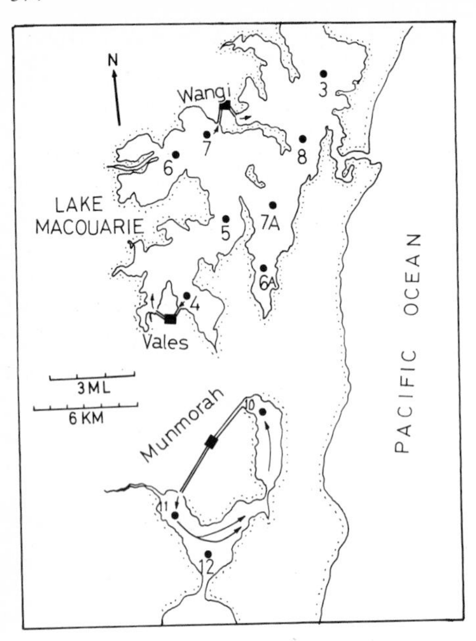

Fig. 1

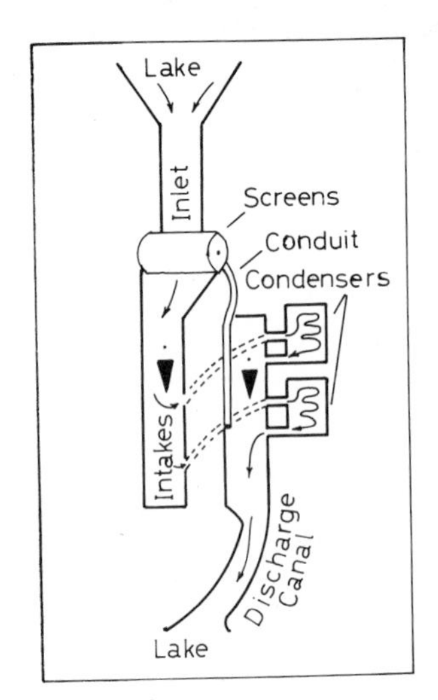

Fig. 2

Fig. 1. Map of the study area showing plankton stations (●) and electric power stations (■)

Fig. 2. Diagram of power plant showing the part through which a pelagic fish egg passes. Sampling sites are denoted by the two black triangles

At peak loads this station pumps about 4.8 x 10^5 gals/min (2.5 x 10^9 ml/min) of cooling water through its condensers. The path an egg might take is illustrated diagrammatically in Fig. 2. Inlet water is taken from the lake and passed through 0.93 cm-meshed screens which remove macro-organisms. (These are sprayed from the screens, diverted around the condenser by means of a large pipe, and released into the discharge canal). The cold inlet water passing through the screens, probably containing 99% of the pelagic eggs, flows down the inlet canal into the "intakes" and thence to water boxes which contain 30 m lengths of 2.3 cm I.D. tubing with two 90° turns. Velocity in the pipes is 2 m/sec and heat exchange takes place in less than 20 sec. The water is released under considerable pressure and flows out the discharge canal (9 x 3.6 m) for 1 km, reaching the lake in about 20 min, and areas of normal lake surface temperature in about 30 min.

Collection of Eggs. 10-min surface plankton hauls were obtained by suspension of a 0.39 mm nylon plankton net (1 m opening) with jar attached to codend. The two sampling sites are shown in Fig. 2. Current was sufficient in both inlet and discharge canal to keep the net taut, but turbulence was much greater in the outlet canal. Net contents were washed into a 1-litre jar, aerated, placed in a cooler, and transported to Munmorah Power Station's biological laboratory. Inlet samples were maintained at the prevailing inlet temperature. Discharge samples were allowed to cool to inlet temperature over a 30-min period, and maintained there.

Treatment of Eggs. Eggs were removed using a 50 ml aliquot Australian C.S.I.R.O. plankton sorting ring and illuminator (Herron, 1969) within a 2-h period after capture, and the rest of the sample discarded. Individual eggs were placed in 15 x 60 mm vials with 3 ml of filtered sea (lake) water containing 100 ppm soluble antibiotic dusting powder, Neotracin. From each site one sample was kept at outside, uncontrolled temperatures, for periods from 36 to 72 h, and another sample maintained at controlled inlet temperature. The former series was used to assess hatching success and hardiness, the other was used for hatching success and $L.D._{50}$ estimates.

The majority of anchovy eggs arriving at the plant contained developed embryos (stage 19 of Blackburn, 1941). At 25°C such eggs usually hatched within 12 h. This stage was chosen for $L.D._{50}$ tests, since the embryos could be seen to wriggle, when alive. Acclimation temperatures are denoted as inlet temperature at the time of capture.

Data Analysis. An approximate $L.D._{50}$ was first estimated by Fry's method (1971), in which temperature is raised 1°C every 4 min. The number dying at each temperature was recorded, and the sample observed over 24 h. Inevitably, some eggs hatched during the experiment or within the 24-h period. However, no difference in sensitivity could be determined between eggs and prolarvae.

For the accurate 30 min $L.D._{50}$ tests, different sets of eggs (10 where possible) were immersed directly into the following temperatures: 30, 32, 34, 35, 36, and 38°C and observed over a 24-h period. Eggs which turned opaque and sank were judged dead, as were larvae which did not move when gently touched. Data were analyzed by the "Litchfield-Wilcoxin" (1949) method as recommended by Sprague (1972) and by the simpler "Reed-Muench" (Woolf, 1968) method.

RESULTS AND CONCLUSIONS

Inlet versus Outlet. Survival of eggs which had passed through cooling condensers was compared to survival of eggs taken at the inlet (Table 1). Of 106 individual late-stage eggs reared from Inlet samples, 11, or 10% either did not hatch or survive over a 24-h period (Nov.-Dec.). Discharge-site samples showed much higher mortality, with 20 of 69 eggs (40%) not surviving over the comparable period. The last series sampled at a time when outlet temperature was 32.1°C, showed a survival value of only 30%.

Numbers of leatherjacket samples were low until December-January, but the data suggest that mortality was about the same or a little higher than for anchovy. It is concluded that pelagic eggs of anchovy and leatherjacket passing through the condensers of Vales Point Power Station suffered a 30 to 40% mortality above a 10% non-hatching rate, which may or may not be the level of non-hatching in nature.

Holding Characteristics. Some series were kept simply at outside ambient temperatures, to observe larval behavior and hardiness (Table 2). Series 3, taken on 18 December and exposed to a heat wave at 54 h after capture, is of interest. The temperature rose to 40°C. All larvae had previously completed yolk-sac absorption and 8 of 8 survived the next 12 h. Hatching success in the other samples was as good as the inlet controlled-temperature series (about 10%).

Table 1. Relative hatching success of eggs taken from Inlet and Discharge canals of Vales Point Power Station, Australia, and maintained at controlled "Inlet" temperatures

	Inlet Samples (22.1-26.5 deg C)				Outlet Samples (26.5-32.1 deg C)			
	Date	No. Alive	No. Dead	% Dead	No. Alive	No. Dead	% Dead	Differential °C
Anchovy	11 November 1972	13	0	0	12	6	50	24.7/26.8
	18 November 1972	16	2	12.5	15	3	20	22.4/28.8
	28 November 1972	17	3	17.7	15	7	46.6	23.3/28.4
	29 November 1972	9	0	0	8	5	62.5	22.4/27.4
	10 December 1972	20	1	5	9	1	11.1	22.1/28.0
	18 December 1972	14	3	21.4	No Sample - Chlorinating			—
	9 January 1973	17	2	11.7	10	7	70	26.5/32.1
	Total	106	11	10.4	69	29	42.0	
Leatherjacket	18 November 1972	1	0	0	—	—	—	22.9/28.8
	28 November 1972	4	1	25	4	1	25	23.3/28.4
	29 November 1972	1	0	0	2	1	50	22.4/27.4
	10 December 1972	4	3	75	3	3	100	22.1/28.0
	18 December 1972	17	1	5.9	No Sample - Chlorinating			
	9 January 1973	12	0	0	13	6	46.1	26.5/32.1
	Total	39	5	12.8	22	11	50	

Table 2. Survival of late-egg/prolarval stages held at outside temperatures for various time periods up to 72 h

	Date	°C Temp. Range	No.	0-6	12	18	24	30	36	42	48	54	60	66	72	No. not hatched	% Survival Per 36 h.	% Survival Per 72 h.
Anchovy	28 November 1972	22-27	13	13	13	9	9	9	9	5	4	4	4	-----	0	4	70	31
	8 December 1972*	22-25	20	20	20	----	20	----	19	— preserved —						0	95	—
	18 December 1972	25-30[1]	15	15	15	14	13	13	13	13	13	8	-----	8	3	1	87	20
	9 January 1973*	26-28	18	18	16	----	16	16	16	16	— preserved					2	89	—
		Total	66													7/66	85.2	25.5
Leather-Jackets	28 November 1972	22-27	4	4	3	3	3	2	1	0						1	75	0
	8 December 1972*	22-25	4	3	3	-----	3	1	1	— preserved						1	75	—
	18 December 1972	25-30[1)]	17	17	16	15	14	14	14	14	11	9	-----	5	3	1	82	18
	9 January 1973*	26-28	10	10	10	----	10	10	10	— preserved						0	100	—
			35													3/35	85.5	18.0

* Observed over only 36 h period.

[1] At 54 h, temperature rose to 40°C ambient, for 10 h.

Most eggs hatched in the first 12 h at 22-25°C. It was estimated that yolk-sac absorption of late-stage anchovy eggs took place about 21 h after capture at 22-25°C (Series 2, 3, 4), or 10-15 h after hatching. It is suggested (see Table 2) that marked mortality usually did not occur until 42 to 48 h after capture, that is, about 24 h after yolk-sac absorption.

It is concluded that 1. Australian anchovy eggs from Lake Macquarie will remain viable within diurnally fluctuating water temperatures in excess of the 2-3 deg C range found in nature. Mortality over the first 2 to 3 days of life in midsummer, at temperatures allowed to fluctuate with outside ambient temperatures (2-5 deg C), was about 10%. 2. Mortality 48 h after yolk absorption, however, was about 80%. 3. The upper ultimate incipient lethal for larvae is probably 40°C, and for prolarvae, possibly higher.

Thermal Tolerance. The CTM or "critical thermal maximum" was approximated (Fry, 1971) for 10 anchovy and 10 leatherjacket eggs (Fig. 3). This test usually produces a level at least 3°C above the 48 or 96-h TLM, because the animals can acclimate as the temperature is raised. The TLM by this method was 37.6°C for anchovy and 39.6°C for leatherjacket.

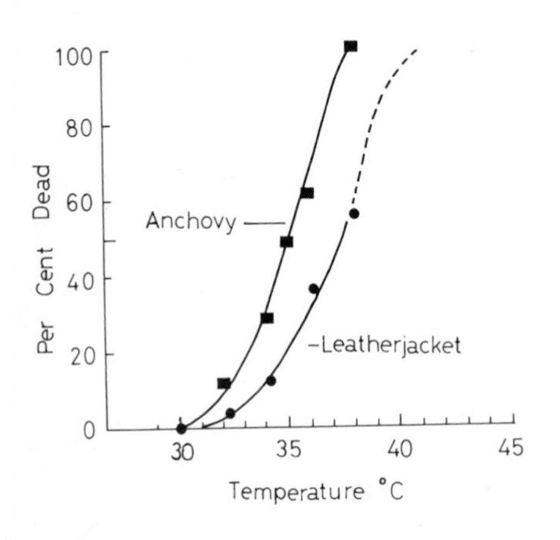

Fig. 3. Approximate L.D.$_{50}$'s (temperature) estimated by heating eggs 1°C per 4 min. The curves are fitted by eye

L.D.$_{50}$ Estimates. Computation of the 30 min L.D.$_{50}$ value for anchovy was made from 4 series. In Fig. 4 the anchovy egg/prolarvae L.D.$_{50}$ for an acclimation range of 25.5 - 27°C was 35.1 ± 0.9°C, using the Reed-Muench method. In Fig. 5, L.D.$_{50}$ was computed by the Litchfield-Wilcoxin method as 34.3 ± 0.9°C. The two methods thus gave virtually the same value, but the lower temperature is probably the safer.

DISCUSSION

Marcy (1971) showed that larval entrainment of several species through a nuclear power plant on the Connecticut River produced 66% mortality with a temperature rise from 22 to 28°C, 83% mortality from 22 to 33°C and 100% mortality when the differential was 24 to 35°C. Moss (1970) states that American shad (4-9 cm) experienced rapid mortality when temperatures were quickly raised from 22-25°C to 32.5°C. Coutant (1972)

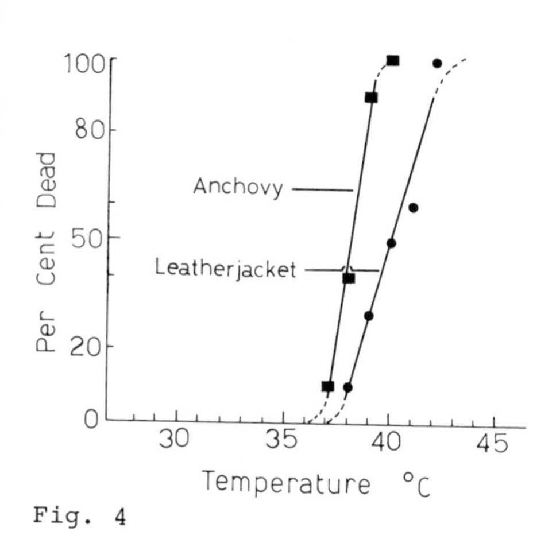

Fig. 4

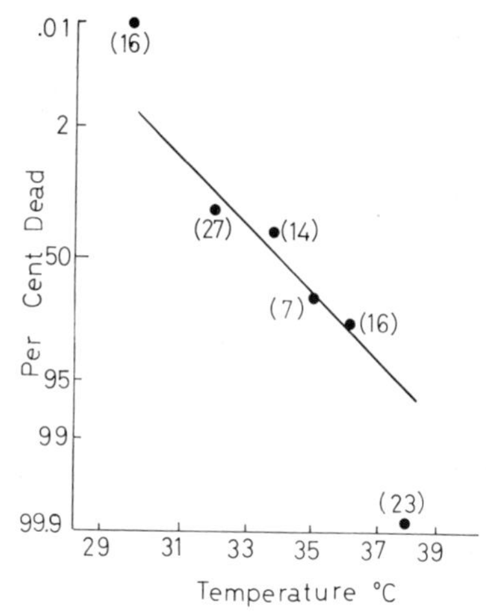

Fig. 5

Fig. 4. 30 min L.D.$_{50}$'s of eggs computed by the Reed-Muench method and fitted by eye (L.D.$_{50}$ = 35.1 $\pm$ 0.9°C)

Fig. 5. Thirty-min L.D.$_{50}$ for anchovy (*Engraulis australis*) computed by the Litchfield-Wilcoxin (1949) method. (L.D.$_{50}$ = 34.3 $\pm$ 0.9°C). The number of eggs tested at each temperature is bracketed

gives, for young-of-the-year Atlantic menhaden (1-4 cm) acclimated at 27°C, an L.D.$_{50}$ of 32.5°C. All the above are boreal or cool-temperate species.

A more precise comparison would perhaps be with fish of warmer regions. No entrainment data are known, but Timet (1959), working with Adriatic or warm-water littoral species, estimated temperature tolerance of *Mullus barbatus*, *Gobius paganellus* and *Scorpaena porcus* kept at 22 to 25°C, and obtained a 3-4 L.D.$_{50}$ of 32°C. These, however, were adult fish with lower acclimation temperatures than used here and no data were given for larvae or eggs. However, Timet obtained a very sharp death rate; nearly all his fish died between 30 and 32°C, a very narrow range of temperature. Australian anchovy eggs and larvae taken at 25 to 27.5°C survived a temperature rise of 30 to 38°C, a considerably wider range. Blaxter (1956) and Alderdice and Forrester (1968) have data on temperature tolerances of pelagic eggs from cool-water species, *Clupea harengus* and *Parophrys vetulus*. The former had an upper lethal 12°C over its acclimation (4°C) temperature, and the latter only 5-7 deg C over its acclimation temperature (5-7 deg C). The clupeoid eggs seem less vulnerable, and in this respect, anchovy eggs follow the same pattern. Fry (1971) states that temperature tolerance in adult tropical fish is not much greater than temperature species for those so far investigated. This does not appear to apply to Australian anchovy and leatherjacket eggs and larvae, which have substantial tolerances above their normal seasonal temperature range (15-25 deg C).

One of the weaknesses of this preliminary study is that no samples were taken at the far end of the discharge canal. It is possible that velocity and turbulence through the canal may increase mortality. Marcy (1971) found this to be the case, although the discharge canal he was sampling was almost twice as long as Vales Point's. Barnett (1971) suggests that the interaction between turbulence, pressure, and chlorine should be investigated as a mortality factor in power plant entrainment.

The effects of entrainment on anchovy and leatherjacket eggs does not appear to be catastrophic, but this depends, of course, on the number or proportion of the total population affected. This would have to be determined by means of a special survey.

A number of courses of action are suggested by this study. Serious mortality probably occurs mainly in late summer with the higher surface water temperatures. This time period corresponds to the onset of leatherjacket spawning and the last half of anchovy spawning. Studies on leatherjacket eggs and spawning should therefore be encouraged, especially since this is a commercially exploited species in the lake.

Since organisms generally recover more quickly the less the time of exposure, piping in of cold water to the source of the discharge canal should be seriously considered by engineers, as should means of lessening the turbulence as water comes out of the condensers.

Quantitative estimates of the number of eggs passing through should be made and related to the adult spawning potential.

Long-term effects, such as abnormalities caused by high temperature, should be considered.

Devices such as "skimmers", which remove grease and detritus from the upper layers of water in discharge canals, should be sampled to determine what proportion of eggs they take up.

SUMMARY

The Australian anchovy spawns in sea bays and estuaries from September to February. Vast egg numbers must pass through cooling condensers of Australian steam electric stations, but this is the first attempt to document mortality caused by entrainment. At Vales Point, on Lake Macquarie, N.S.W., survival of late eggs from inlet and discharge canals was 90% and 60%, respectively (November to December), but only 30% when discharge canal temperatures rose above 32°C (January to February). Part of this mortality could be due to turbulence and/or abrasion in the collecting net, since $L.D._{50}$ values for anchovy eggs acclimated at 25.5 to 27°C averaged 34.3 ± 0.9°C, a temperature rarely reached in condenser pipes. Preliminary $L.D._{50}$ estimates for "leatherjacket" (Triggerfish, *Meuschenia trachlyepis*) eggs were somewhat higher, at 37°C. The "critical thermal maximum" temperature for both species's eggs lies between 37 and 40°C.

ACKNOWLEDGEMENTS

I would like to thank the School of Zoology, University of New South Wales, for their kind hospitality and use of facilities. Dr. John MacIntyre gave continual assistance and support, for which I am grateful, as did Mr. and Mrs. Kenneth Pulley. The New South Wales Electricity Commission also supported the project and provided laboratory facilities, for which I thank in particular, Mr. Lewis Drummond and Mr. Garth Coulter. Trent University generously provided financial support, thus enabling me to travel to the electricity plants. The librarians at the C.S.I.R.O. Fisheries and Oceanography Station, Cronulla, N.S.W., were most helpful in finding reference material.

REFERENCES

Alderdice, D.F. and Forrester, C.R., 1968. Some effects of salinity, and temperature on early development and survival of the English sole (*Parophrys vetulus*). J. Fish. Res. Bd Can., 20, 525-550.
Barnett, P.R.O., 1972. Effects of warm water effluents from power stations on marine life. Proc. R. Soc. Lond. B., 180, 497-509.
Blackburn, M., 1941. Economic biology of some Australian clupeoid fish. Bull. C.S.I.R.O. No. 138: VIII. The anchovy, 72-89.
Blaxter, J.H.S., 1956. Herring rearing. II. The effect of temperature and other factors on development. Mar. Res., 5, 1-19.
Coutant, C.C., 1972. Biological aspects of thermal pollution. II. Scientific basis for water temperature standards at power plants, MS for CRC Critical Reviews in Environmental Control.
Fry, F.E.J., 1971. The effect of environmental factors on the physiology of fish. In: Fish Physiology, Vol. VI. W.S. Hoar and D.T. Randall (Eds.). New York - London: Academic Press, 1-87 pp.
Herron, A.C., 1969. A dark-field condenser for viewing transparent plankton animals under a low-power stereomicroscope. Mar. Biol., 2, 321-324.
Howard, K.A., 1972. Environmental considerations of waste dispersal from power generation. Thermo-fluids Conference, Sydney, Thermal discharge, engineering and ecology. The Instit. Engineers, Australia, 7-14.
Krenkel, P.A. and Parker, F.L., 1969. Biological aspects of thermal pollution. Vanderbilt University Press, Nashville, Tenn., 407 pp.
Litchfield, J.T. and Wilcoxin, F., 1949. A simplified method of evaluating dose-effect experiments. J. Pharmac. Exp. Therapeutics, 96, 99-113.
Marcy, B.C., Jr., 1971. Survival of young fish in the discharge canal of a nuclear power plant. J. Fish. Res. Bd Can., 28, 1057-1060.
Moss, S.A., 1970. Responses of young American Shad to rapid temperature changes. Trans. Amer. Fish. Soc. 99, 2, 381-384.
Powles, P.M., 1972. Lethal effects of hot water. Thermo-fluids Conference, Sydney, Thermal discharge, engineering and ecology. The Instit. Engineers, Australia, 52-59.
Sprague, J.B., 1972. The ABC's of pollutant bioassay using fish. Symp. Environ. Monitoring, Los Angeles, Calif., Amer. Soc. Test. Materials.
Timet, D., 1962. Studies on heat resistance in marine fishes. I: Upper lethal limits in different species of Adriatic littoral. Thalassia Jugoslavia 1, 3, 1956-1967.
Woolf, C.M., 1968. Principles of Biometry. D. van Nostrand, 359 pp.

P.M. Powles
Trent University
Trent / CANADA

Present address:

School of Zoology
University of N.S.W.
Kensington 2033 / AUSTRALIA

Effects of Cadmium on Development and Survival of Herring Eggs

H. Rosenthal and K.-R. Sperling

INTRODUCTION

Cadmium is a biological nonessential heavy metal found only in minute traces in natural waters. In sea water, the normal concentration is considered to be between 0.01 and 0.10 ppb (Goldberg, 1965; Riley and Taylor, 1968; Bryan, 1971; Preston, 1973). In aerated sea water, the concentration of cadmium which can remain in solution is several orders of magnitude higher than that usually found in the sea (Krauskopf, 1956). Preston et al. (1971) and Abdullah et al. (1972) reported that in coastal areas of the Irish Sea the cadmium content was higher than offshore. Similar observations have been made by Windom and Smith (1972) in continental shelf waters of the Atlantic coast of the United States. High concentrations of cadmium were also found in bottom samples from sludge dumping areas and in sediments of estuaries (Butterworth et al., 1972; Nauke, personal communication).

In fresh water, the tolerance limits for cadmium are influenced by the hardness of the water. In soft water, Ball (1967) recorded 0.01 mg Cd/l for the 7-day TLm (rainbow trout), whereas Pickering and Henderson (1966) measured 4-day TLm (fathead minnow) values of 73.5 mg Cd/l for hard water; 0.11 mg Cd/l was lethal in long-term experiments (Pickering and Gast, 1972). Eisler (1971) found in an acute toxicity bioassay with $CdCl_2$ at $20^{o}C$ and $20^{o}/oo$ salinity that the 4-day TLm ranged between 0.32 mg Cd/l and 55.0 mg Cd/l in organisms of various marine species. Concentrations greater than 0.1 mg Cd/l in sea water are considered to be toxic to marine organisms.

Little work has been done on the influence of cadmium on the embryonic development of fish eggs. Pickering and Gast (1972) reported a decrease in survival of embryos of the fathead minnow at experimental concentrations of 0.057 ppm Cd. Sangalang and O'Halloran (1972) found testicular injuries in brook trout exposed to nonlethal doses of cadmium down to a concentration of 0.01 ppm. The present investigation deals with the effects of various concentrations of cadmium on the embryonic development of herring eggs. In addition the effect of mixtures of cadmium-EDTA, cadmium-zinc and cadmium-ascorbic acid were investigated.

Interest in EDTA arises from its ability to combine with metal ions to form nonionic water-soluble complexes or chelates. In sea water, in the presence of abundant calcium ion, EDTA will still chelate cadmium because of the much greater stability constant of the cadmium-EDTA-complex. Cadmium therefore will displace calcium from its combination with EDTA.

Some symptoms of cadmium poisoning are similar to those of zinc deficiency and it seems as if cadmium acts as an antimetabolite of the chemically related and metabolically essential element zinc, and that this effect is due to the zinc-cadmium ratio. Experiments in zinc-cadmium mixtures were therefore undertaken to prove whether or not an increased zinc concentration would diminish the effects of cadmium. Cadmium-ascorbic acid solutions are of interest because ascorbic acid is also a good complexing agent, although it is rapidly metabolized by living tissues.

MATERIAL AND METHODS

After transport of parent fish (length 25-31 cm, weight 85-175 g) to the laboratory, herring eggs from Baltic spring (Lindaunis, Schlei) and autumn (Spodsbjerg, Denmark) spawners were fertilized artificially and incubated in 1.5 l containers at 10°C in continuously aerated sea water of 16°/oo salinity. Test solutions, containing different concentrations of cadmium and mixtures of cadmium-EDTA, cadmium-zinc, and cadmium-ascorbic acid, were derived from standard solutions and were renewed every two days. The molar concentration of EDTA was kept at 8.5×10^{-6} in the first and 42.5×10^{-6} in the second experimental series. Zinc was presented at a concentration of 0.02 ppm, which was additional to the normal concentration in sea water. Ascorbic acid was maintained in experimental sea water at a molar concentration of 8.53×10^{-6}. Temperature, salinity, and pH values were measured daily. Mean temperature was held between 9.9 and 10.1°C. Salinity between 15.9 and 16.3°/oo and pH between 7.96 and 8.13, all with very small fluctuation. Table 1 shows the cadmium levels maintained throughout the experiments.

Table 1. Test conditions maintained during incubation experiments (intended values for salinity: 16.0°/oo, temperature: 10.0°C, pH: 8.0) with eggs from *Clupea harengus*. n = number of measurements, $\overline{x}$ = mean, s = standard deviation

Experimental design		Working level Cd concentration (ppm)		
		n	$\overline{x}$	s
Cd concentration (ppm)				
Control		-	-	-
0.1		28	0.086	± 0.035
1.0		23	0.398	± 0.128
5.0		21	3.660	± 0.168
10.0		25	8.484	± 0.429
Control	+ EDTA	24	0.064	± 0.034
0.1	+ EDTA	22	0.063	± 0.026
1.0	+ EDTA	28	0.395	± 0.085
5.0	+ EDTA	18	3.810	± 0.514
10.0	+ EDTA	19	8.594	± 1.002
Control	+ zinc	27	0.101	± 0.040
5.0	+ zinc	15	4.519	± 0.413
Control	+ ascorbic acid	-	-	-
1.0	+ ascorbic acid	-	-	-
10.0	+ ascorbic acid	-	-	-

The cadmium content of test solutions was determined daily. From each test medium, two samples of 5 herring eggs were collected at intervals of 24 h and preserved for determination of cadmium uptake. All determinations were accomplished by atomic absorption spectrophotometry after wet combustion of biological samples, using a Perkin-Elmer AAS 300 single beam apparatus equipped with a HGA 70 graphite furnace and a D_2-underground compensator. Technical details of the modified AAS method will be described in a separate paper (Sperling, 1973).

The following criteria were used to asses the effects of cadmium on herring eggs: embryonic mortality, incubation time (time from fertilization to 50% hatching), hatching rate, percentage viable hatch, size of larvae at hatching, yolk-sac length, eye diameter of newly-hatched larvae and diameter of the otic capsule at hatching.

RESULTS

Embryonic Survival and Incubation Time. Embryonic survival remained high in all cadmium concentrations at early developmental stages. In concentrations of 5 and 10 ppm cadmium a remarkable mortality occurred from the 5th day of incubation (Fig. 1), when the heart normally starts beating. At the same time an increasing number of malformations could be observed. None of them could be interpreted as typically cadmium-induced. They were comparable to those found under various stress situations (Rosenthal and Stelzer, 1970; Alderdice and Velsen, 1971). A softening of the egg membrane was detected in eggs incubated in cadmium concentrations of 5 and 10 ppm several days after beginning of exposure.

Fig. 2 shows the hatching distribution at 24 h intervals after commencement of hatching. Even the lowest cadmium concentration (0.1 ppm) led to a shortened incubation period. Time from fertilization to 50% hatching is 258, 254, 293, 332 and 361 h in 10.0, 5.0, 1.0, 0.1 ppm cadmium and the controls, respectively. In the presence of EDTA or zinc the effect of cadmium on the incubation period was diminished slightly.

Hatching Rate and Percentage Viable Hatch. Hatching rates are relatively high in all experiments up to a cadmium concentration of 5 ppm (Table 2). However, percentage of viable hatch is substantially reduced in concentrations of cadmium higher than 0.1 ppm. If cadmium is complexed by EDTA, or displaced by zinc, the percentage of viable hatch increases considerably, even in cadmium concentrations in which all hatched larvae would otherwise have been malformed. On the other hand, the presence of ascorbic acid in a test solution containing 1 ppm cadmium did not lead to any viable hatch. The poisonous effect of cadmium seems to be increased. It may be assumed, that the ascorbic acid acts as a vehicle for cadmium, transporting it through the protective egg membrane into the embryonic cells, where it would be liberated quickly to toxic ionic levels as the ascorbic acid carrier is metabolized.

Total Length and Yolk-Sac Length of Larvae at Hatching. Measurements of total length of larvae at hatching were made only on what was considered to be viable hatch. No data are available concerning the higher

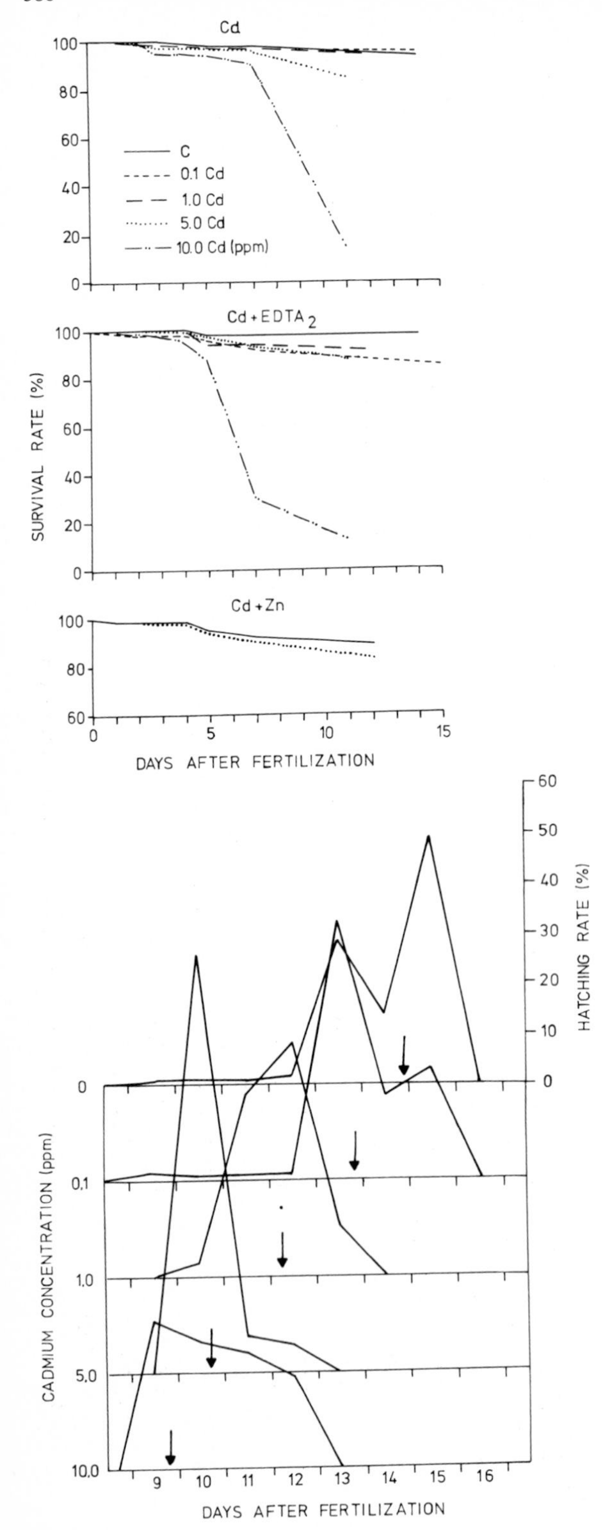

Fig. 1. Percentage survival of herring eggs incubated in different test solutions containing cadmium (0 to 10 ppm), cadmium-EDTA (42.5×10^{-6} molar), and cadmium-zinc (20 µg/l) mixtures

Fig. 2. *Clupea harengus* larvae. Hatching distribution (%/day) and 50% hatching time (arrows). Eggs incubated in different cadmium concentrations

Table 2. Hatching rate (%) and percentages of viable hatch of herring larvae in relation to cadmium concentration and cadmium + EDTA, cadmium + zinc and cadmium + ascorbic acid mictures. n = total number of hatched larvae, a = hatching rate as %, b = viable hatch as %

Experimental Cd-concentration (ppm)	Cd^{2+}			Cd^{2+}			Cd^{2+} + $EDTA_1$			Cd^{2+} + $EDTA_1$			Cd^{2+} + ascorbic acid			Cd^{2+} + Zn^{2+}		
	n	a	b	n	a	b	n	a	b	n	a	b	n	a	b	n	a	b
Controls	749	88.9	87.0	200	96.2	93	109	82.6	81.0	575	93.3	93.0	89	83.3	82.1	675	85.6	84.5
0.1	722	94.6	82.7	159	97.0	-	-	-	-	641	83.8	83.6	-	-	-	-	-	-
1.0	590	84.0	16.3	53	25.7	0	164	81.6	36.5	847	73.2	54.3	48	26.1	0.0	-	-	-
5.0	742	74.8	0.0	41	26.8	0	-	-	-	908	83.3	0.6	-	-	-	1047	80.2	11.4
10.0	151	14.4	0.0	53	27.0	0	171	81.4	0.0	86	13.4	0.0	53	34.6	0.0	-	-	-

$EDTA_1$ = 8.5×10^{-6} molar

$EDTA_2$ = 42.5×10^{-6} molar

ascorbic acid = 8.53×10^{-6} molar

Zn = 20 µg/l above natural concentration

cadmium concentrations (5 and 10 ppm). Generally, hatched herring larvae tended to be smaller with increasing cadmium concentration. This tendency existed also in the experiments with the cadmium-EDTA-complex (Fig. 3 and Table 3). A comparison of the total length of newly-hatched larvae incubated at the same cadmium concentration, but in different test solutions containing EDTA and zinc, shows that in the controls larvae tended to hatch at a smaller size if EDTA and zinc were added (Table 4). At the 0.1 ppm cadmium level, larvae were of the same size in both experiments, either with cadmium alone or with the cadmium-EDTA-complex. At the 1.0 ppm cadmium level, larvae from eggs incubated in the cadmium-EDTA-complex were larger than in the test medium containing cadmium in the ionic form. At the 5.0 ppm cadmium level viable hatch from eggs incubated in the cadmium-EDTA-complex tended to be smaller than in the cadmium-zinc interaction system, whereas in eggs exposed to 5.0 ppm ionic cadmium no viable hatch occurred.

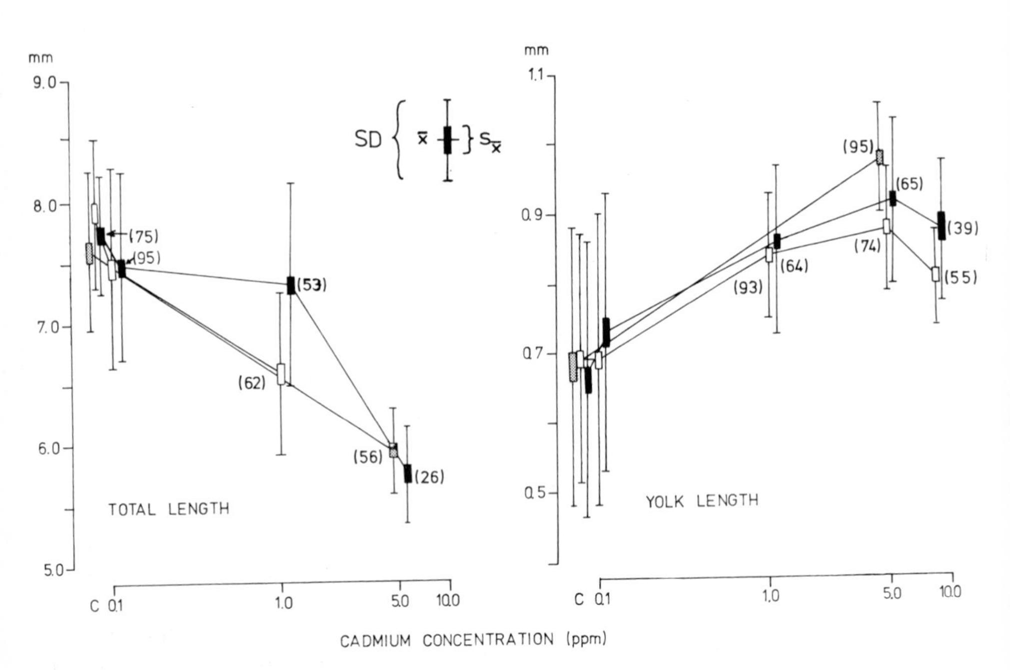

Fig. 3. *Clupea harengus* larvae. Total length and yolk-sac length at hatching. Eggs were incubated in different cadmium concentrations (white column), cadmium-EDTA solutions (black column), and cadmium-zinc mixtures (spotted column). $\bar{x}$ = mean; SD = standard deviation; $s_{\bar{x}}$ = s. error of the mean; C = control. Number of measurements other than 100 are given in brackets

Considering larval size and yolk sac size at hatching, the differences in total length of larvae hatched in different test media seem to be related more to the shortened incubation period than to the toxic effect of cadmium. Generally, early-hatched larvae were smaller than those hatching later, but had bigger yolk sacs. In concentrations up to 5.0 ppm cadmium it was not possible to distinguish between effects caused directly by the metal ion, and such effects caused indirectly by the shortening of the incubation period. In the experiments at the 10.0 ppm cadmium level, the yolk-sac length was shortened in comparison to the values obtained at the 5.0 ppm cadmium level. Furthermore, in the cadmium concentrations of 5.0 and 10.0 ppm yolk sacs of hatched larvae were significantly longer when the eggs were incubated in cadmium-zinc and cadmium-EDTA mixtures than in cadmium alone (Fig. 4).

Table 3. Total length (mm) of hatched larvae (viable hatch) of *Clupea harengus*. Eggs were incubated in different test media containing cadmium, cadmium + EDTA and cadmium + zinc solutions. n = number of measurements, $\bar{x}$ = mean, V = coefficient of variation (%), P = probability, c = controls

	Cadmium concentration (ppm)				
Experiment	c	0.1	1.0	5.0	10.0
Cadmium					
n	81	93	62	-	-
$\bar{x}$	7.91	7.46	6.58	-	-
V	7.7	10.9	10.3	-	-
P	<0.002				
		<0.0001			
		<0.0001			
Cadmium + EDTA					
n	75	95	53	26	-
$\bar{x}$	7.74	7.47	7.32	5.74	-
V	6.3	10.3	6.0	6.9	-
P	<0.007				
		<0.001			
			<0.0001		

Diameter of Eye and Otic Capsule. Eye diameter as well as the diameter of the otic capsule was reduced with increasing cadmium concentration in all test media. In higher cadmium concentrations this decrease was greater than could be explained by the shortening of the incubation period alone. Exceptionally large eyes were obtained from eggs incubated in the cadmium-zinc mixture (Fig. 4). Reduction of diameter of the otic capsule was diminished in experiments where chelating agents were added. In 10.0 ppm cadmium some malformed larvae hatched without otic capsules. It is well known that cadmium interferes with calcium metabolism (Kobayashi, 1971) and calcium is required in the development of the otic capsule of fish larvae.

Uptake of Cadmium During Embryonic Development. The uptake of cadmium by herring eggs exposed to different cadmium concentrations shows rather a different picture than might be expected. Typical Langmuir curves (an increase of the cadmium content to an asymptotic maximum with time) did not occur. Uptake was rather rapid, reaching a more-or-

Table 4. Comparison of total length in mm of newly-hatched herring larvae (viable hatch), incubated in test media containing corresponding concentrations of cadmium and mixtures of cadmium + EDTA and cadmium + zinc. n = number of measured larvae; $\overline{x}$ = mean; V = coefficient of variation; P = probability

Cadmium concentration (ppm)		Test combinations: Cadmium	Cadmium + EDTA	Cadmium + zinc
Controls	n	81	75	89
	$\overline{x}$	7.91	7.74	7.60
	V	7.7	6.3	8.7
	P	Cadmium vs. Cadmium + EDTA: <0.04	Cadmium + EDTA vs. Cadmium + zinc: <0.07	Cadmium vs. Cadmium + zinc: <0.003
0.1	n	93	95	-
	$\overline{x}$	7.46	7.47	-
	V	10.9	10.3	-
	P	Cadmium vs. Cadmium + EDTA: >0.25		
1.0	n	62	53	-
	$\overline{x}$	6.58	7.32	-
	V	10.3	6.0	-
	P	Cadmium vs. Cadmium + EDTA: <0.0001		
5.0	n	-	26	56
	$\overline{x}$	-	5.74	5.95
	V	-	6.9	6.1
	P	-	Cadmium + EDTA vs. Cadmium + zinc: <0.02	

less stable level some hours after beginning of exposure. The maximum levels attained depended on the initial cadmium concentration of the test media (Fig. 5). Immediately before hatching, the amount of cadmium determined in herring eggs dropped considerably, and hatched larvae, even in the highest cadmium concentration, contained not much more cadmium than the controls.

Eggs incubated in cadmium-EDTA mixtures generally took up virtually no cadmium, when incubated in concentrations up to 5.0 ppm (Fig. 6). Higher cadmium levels could not be sequestered by EDTA, because the chelating capacity was fully utilized at that level. The uptake in the 10.0 ppm cadmium-EDTA solution therefore appears to give a comparable picture but with lower values than obtained in eggs incubated at the 10.0 ppm cadmium level without EDTA.

In the cadmium-zinc interaction system, there was a rapid uptake at the beginning of the exposure and a rather quick loss during the incubation period (Fig. 7).

In order to consider the actual cadmium concentration in the embryo, the content of cadmium in yolk, embryo and chorion was determined immediately before hatching at the 5.0 ppm cadmium level. Lowest cadmium values were obtained from yolk sacs in all test media. The highest content of the metal was found in the egg capsule (Table 5).

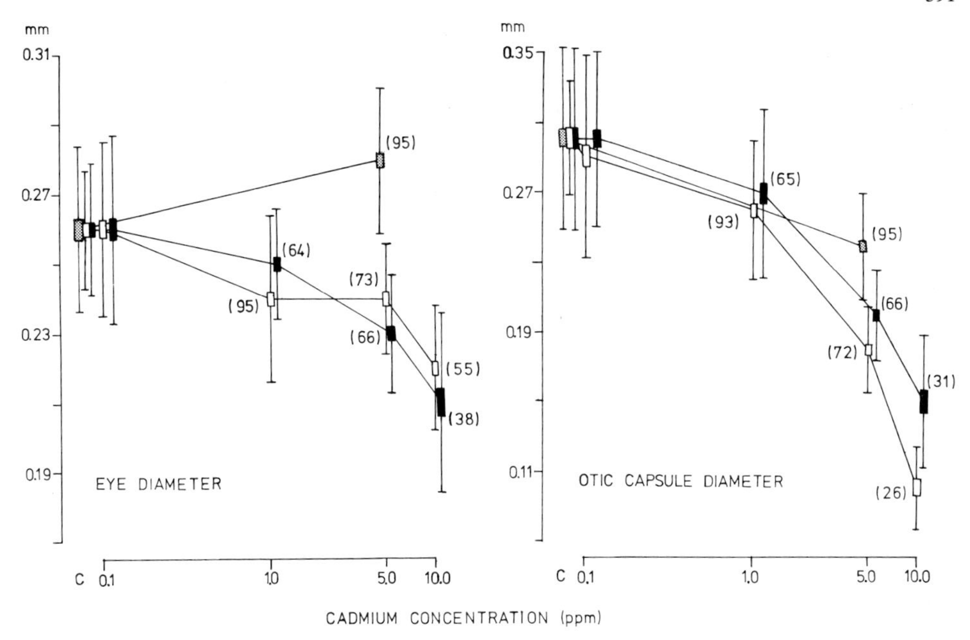

Fig. 4. *Clupea harengus* larvae. Diameter of eye and otic capsule at hatching. Eggs were incubated in different cadmium concentrations, cadmium-EDTA solutions and cadmium-zinc mixtures. Symbols as in Fig. 3. Number of measurements other than 100 are given in brackets

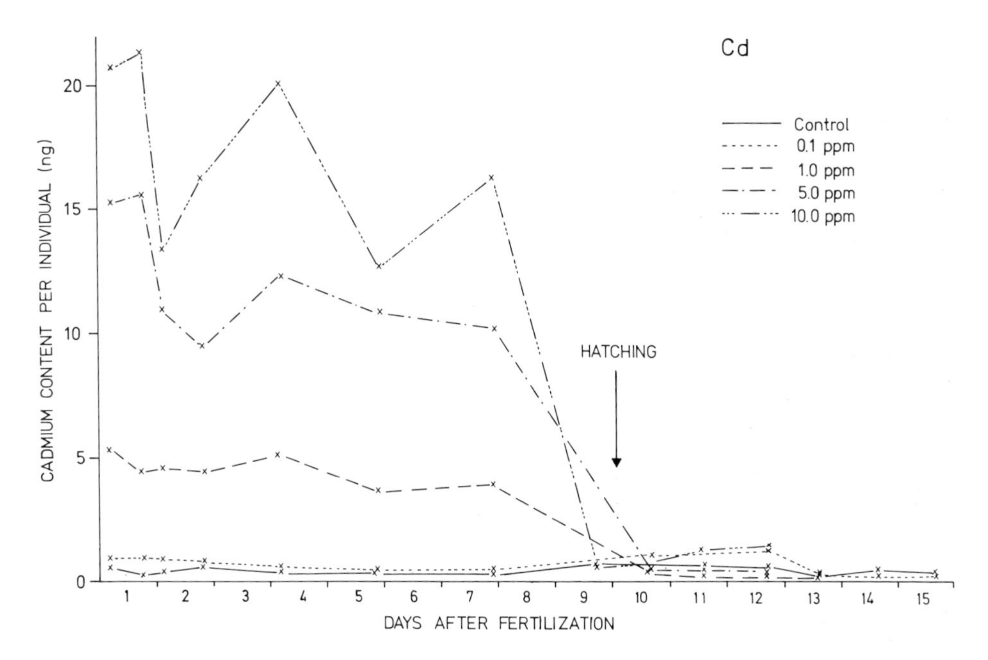

Fig. 5. Uptake of cadmium by *Clupea harengus* eggs, incubated in different cadmium concentrations, and the cadmium content of hatched larvae

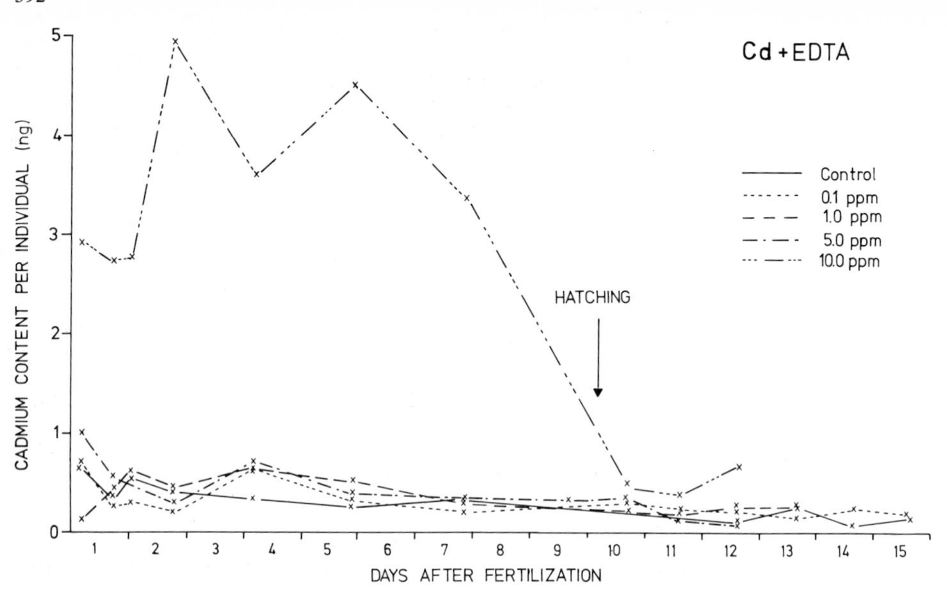

Fig. 6. Cadmium uptake by herring eggs incubated in different cadmium-EDTA solutions and the cadmium content of hatched larvae. Molar EDTA-concentration = 42.5×10^{-6}

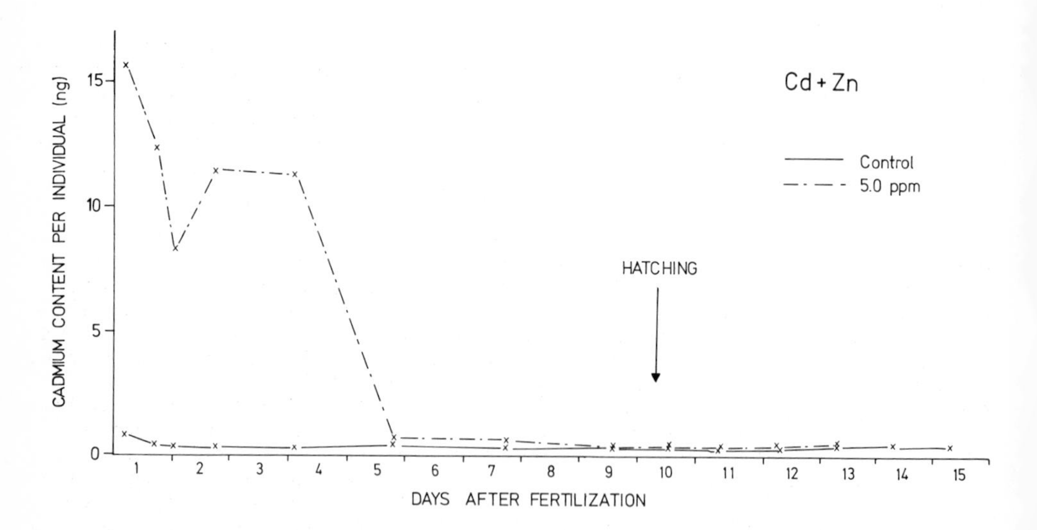

Fig. 7. Cadmium uptake by herring eggs incubated in a cadmium-zinc mixture. Concentrations: Cd = 5.0 ppm; Zn = 20 μg/l

Table 5. Cadmium content (ng) of yolk, embryo and chorion of *Clupea harengus* eggs, immediately before hatching. The eggs were incubated in different solutions containing a = 5 ppm Cd; b = 5 ppm Cd + EDTA; and c = 5 ppm Cd + zinc

	a	b	c
Yolk	0.360	0.158	0.450
Embryo	1.331	0.675	0.832
Chorion	3.820	0.940	1.961

Again the cadmium content (ng/individual) was considerably lower in embryo and chorion from eggs incubated in test solutions which contained chelating agents. From the data of Table 5 it is possible to estimate the cadmium content of herring larvae and chorion on a dry weight basis. Immediately before hatching the dry weight was about 0.095 mg for the herring embryo and about 0.046 mg for the chorion. If herring eggs are exposed to 5.0 ppm cadmium during the whole incubation period they take up about 1.69 ng cadmium per individual, which is equal to a cadmium concentration of about 18 ppm. On the other hand, the chorion contains 85 ppm.

DISCUSSION

The concentrations of cadmium determined in the test media were lower than those initially added to experimental sea water. Losses of cadmium in solution may be related to absorption on the surface of the containers, to precipitation, or to uptake by the herring eggs. In contrast to these findings Jackim et al. (1970) found no loss of cadmium in experimental water and Eisler (1971) reported less than 5% loss of cadmium in sea water after 264 h of aeration.

Heavy metals in sea water are not totally free in the ionic form. They are bound to hitherto unknown compounds with various strength. It can therefore be expected that not all of the total amount of cadmium added to experimental sea water is available to the exposed organism in an ionic form. Zirino and Yamamoto (1972) have recently calculated the degree of complexity between major cations and anions in sea water, including higher-order complexes, in order to estimate the extent of complexation of the divalent ions of Cu^{2+}, Zn^{2+}, Cd^{2+} and Pb with the anions OH^-, Cl^-, SO_4^{2-}, HCO_3^- and CO_3^{2-}. They found that slight variations of pH values, normally encountered in sea water, influence the distribution of the chemical species of all these metals, except cadmium, in which the effect is only negligible. Uncomplexed Cd^{2+} constitutes about 2.5% of the total. Mostly it is strongly associated with chloride ions. If this is true, the changes of cadmium content of herring eggs during incubation cannot be explained by changes in the distribution of the chemical species of the metal with changing pH. On the other hand, it is unknown in which form cadmium is bound to the egg surface. In mammals it is exclusively fixed to the liver-produced protein metallothionein, which contains many sulfhydryl-groups (Livingstone, 1972; Kägi and Vallee, 1960). The sulfhydryl-boundage of cadmium is known to be in concurrence with other forms of fixation. The occurrence of different saturation levels of cadmium in the membrane of herring eggs suggests the presence of a labile equilibrium of metal-complexes with a high dissociation rate. Therefore, we may assume that there is no connection between cadmium and sulfhydryl-groups of proteins, but mucopolysaccharides may act as ion exchangers.

The observed loss of cadmium in herring eggs immediately before hatchi leads to the conclusion that the metal is bound mainly on the egg capsule and that it will be removed partly when the hatching enzymes digest the chorion. Similar findings with zinc were reported from Coho salmon eggs by Wedemeyer (1968).

The results of decreased toxic effects of cadmium in the presence of chelating agents are not in contrast to other results known from the literature. White and Munns (1951) found that the inhibitory effect of cadmium on yeast growth was antagonized by the addition of zinc to the medium. Parizek (1957) concluded from experiments with male rats that the effects on spermatogenesis caused by cadmium may be counteracted with large amounts of zinc salts. Eisler et al. (1972) have shown that cadmium uptake is influenced by zinc and copper. Johnson and Turner (1972) found similar effects in rats.

From the data available we tentatively conclude that the effects of cadmium on embryonic development of herring eggs may occur in the following ways:

1. Cadmium uptake takes place mainly at the surface of the egg capsule, forming protein or mucopolysaccharide complexes, and causing a change in the physico-chemical properties of the chorion, more or less depending on concentration. This change in the physico-chemical properties may stress developing embryos, resulting in a shortened incubation period, lower percentage of viable hatch, and probably smaller larvae.

2. Cadmium may enter through the egg membrane and act directly on the embryo. The actual toxic concentration of cadmium in the embryo is much lower than that taken up by the egg capsule. From that point of view, the embryonic stages must be considered as very sensitive.

3. The ionic form of cadmium in the water is important for the action of toxicity mechanism.

Probably both indirect and direct effects of the cadmium ion influences development and survival of herring eggs.

SUMMARY

1. Herring eggs from Baltic spring and autumn spawners were incubated at different concentrations of cadmium and in different test solutions containing mixtures of cadmium-EDTA, cadmium-zinc, and cadmium-ascorbi acid.

2. Incubation time was reduced considerably with increasing cadmium concentration.

3. Percentage of viable hatch was 16.3%, 82.7%, and 93.0% in 1.0, 0.1 ppm Cd and the controls, respectively. In a concentration of 5.0 ppm Cd no viable hatch occurred.

4. With increasing cadmium concentration hatched larvae had a smaller size, but longer yolk sacs.

5. The effect of cadmium was diminished in the presence of zinc and EDTA, resulting in larger larvae with larger otic capsules and higher percentages of viable hatch.

6. Uptake of cadmium by herring eggs was rather rapid, reaching a stable level some hours after the beginning of exposure.

7. Immediately before hatching, the amount of cadmium determined in herring eggs dropped considerably. Hatched larvae contained virtually no cadmium.

8. Eggs incubated in cadmium-EDTA-mixtures generally took up virtually no cadmium until the chelating capacity of EDTA was fully utilized.

9. Cadmium uptake took place mainly at the surface of the egg capsule, causing a change in the physico-chemical properties of the chorion.

10. The ionic form of cadmium in sea water is important for the action of the toxicity mechanism.

REFERENCES

Abdullah, M.J., Royle, L.G. and Morris, A.W., 1972. Heavy metal concentration in coastal waters. Nature, 235 (5334), 158-160.

Alderdice, D.F. and Velsen, F.P.J., 1971. Some effects of salinity and temperature on early development of the Pacific herring (*Clupea pallasii*). J. Fish. Res. Bd Can., 28, 1545-1562.

Ball, J.R., 1967. The toxicity of cadmium to rainbow trout (*Salmo gairdnerii*, Richardson). Water Res., 1, 805-806.

Bryan, G.W., 1971. The effects of heavy metals (other than mercury) on marine and estuarine organisms. Proc. Roy. Soc. Lond. B 177, 389-410.

Butterworth, J., Lester, P. and Nickles, G., 1972. Distribution of heavy metals in the Severn estuary. Marine Poll. Bull., 3 (5), 72-74.

Eisler, R., 1971. Cadmium poisoning in *Fundulus heteroclitus* (Pisces: Cyprinodontidae) and other marine organisms. J. Fish. Res. Bd Can., 28 (9), 1225-1234.

Eisler, R., Zaroogian, G.E. and Hennekey, R.J., 1972. Cadmium uptake by marine organisms. J. Fish. Res. Bd Can., 29, 1367-1369.

Goldberg, E.D., 1965. Minor elements in sea water. Chemical Oceanography 1. London - New York: Academic Press, 163-196 pp.

Jackim, E., Hamlin, J.M. and Sonis, S., 1970. Effects of metal poisoning on five liver enzymes in the killifish (*Fundulus heteroclitus*). J. Fish. Res. Bd Can., 27, 383-390.

Johnson, A.D. and Turner, P.C., 1972. Early action of cadmium in the rat and domestic fowl. VI. Testicular and muscle blood flow changes. Comp. Biochem. Physiol., 41 A, 451-456.

Kägi, J.H.R. and Vallee, B.L., 1960. Metallothionein: A Cd- and Zn-containing protein from equine renal cortex. J. Biol. Chem., 235, 3460.

Kobayashi, J., 1971. Relation between the "Itai-Itai" disease and the pollution of river water by cadmium from a mine. Adv. Water Poll. Res., I-25, 1-7.

Krauskopf, K.B., 1956. Factors controlling the concentrations of thirteen rare metals in sea-water. Geochim. Cosmochim. Acta 9, 1-32.
Livingstone, H., 1972. Measurement and distribution of zinc, cadmium and mercury in human kidney tissue. Clin. Chem., 18, 67-72.
Parizek, J., 1957. The destructive effect of cadmium ion on testicular tissue and its prevention by zinc. J. Endocrin., 15, 56-63.
Pickering, Q. H. and Gast, M.H., 1972. Acute and chronic toxicity of cadmium to the Fathead minnow (Pimephales promelas). J. Fish. Res. Bd Can., 29 (8), 1099-1106.
Pickering, Q.H. and Henderson, C., 1966. The acute toxicity of some heavy metals to different species of warm water fishes. Air Water Poll. Int. J., 10, 453-463.
Preston, A., 1973. Heavy metals in British waters. Nature, 242, 95-97.
Preston, A., Jefferies, D.F., Dutton, J.W.R., Harwey, B.R. and Steele, E.K., 1971. British Isles coastal waters: The concentrations of selected heavy metals in sea water, suspended matter and biological indicators - a pilot survey. ICES, Doc. CM 1971/E:7.
Riley, J.P. and Taylor, D., 1968. Chelating resins for the concentrations of trace elements from sea water and their use in conjunction with atomic absorption spectrophotometry. Anal. chim. Acta 40, 479-485.
Rosenthal, H. and Stelzer, R., 1970. Wirkungen von 2,4- und 2,5-Dinitrophenol auf die Embryonalentwicklung des Herings *Clupea harengus*. Mar. Biol., 5 (4), 325-336.
Sangalang, G.B. and O'Halloran, M.J., 1972. Cadmium-induced testicular injury and alterations of androgen synthesis in brook trout. Nature, 240, 470-471.
Sperling, K.-R., 1973. Eine modifizierte Methode zur atomabsorptionsspektralphotometrischen Bestimmung von Cd und anderen Schwermetallen in Seewasser und biologischem Material (in preparation).
Wedemeyer, G., 1968. Uptake and distribution of ^{65}Zn in the Coho salmon egg (Oncorhynchus kisutch). Comp. Biochem. Physiol., 26, 271-279.
White, J. and Munns, D.J., 1951. Inhibitory effect of cadmium on yeast growth. J. Inst. Brew., 57, 175.
Windom, H.L. and Smith, R.G., 1972. Distribution of cadmium, cobald, nickel and zinc in southeastern United States continental shelf waters. Deep Sea Res. 19, 727-730.
Zirino, A. and Yamamoto, S., 1972. A pH-dependent model for the chemical speciation of copper, zinc, cadmium, and lead in seawater. Limnol. Oceanogr., 17 (5), 661-671.

H. Rosenthal
Biologische Anstalt Helgoland
Zentrale Hamburg
2000 Hamburg - 50 / FEDERAL REPUBLIC OF GERMANY
Palmaille 9

K.-R. Sperling
Biologische Anstalt Helgoland
Zentrale Hamburg
2000 Hamburg - 50 / FEDERAL REPUBLIC OF GERMANY
Palmaille 9

Resistance of Plaice Eggs to Mechanical Stress and Light

T. Pommeranz

INTRODUCTION

The mortality of the eggs of plaice (*Pleuronectes platessa* L.) in the southern North Sea is estimated at 50-90% (Buchanan-Wollaston, 1923, 1926; Bückmann, 1961; Bannister, Harding and Lockwood, this symposium). Neuston investigations show that fish eggs concentrate near the surface in both calm and rough weather (Zaitsev, 1961, 1971; Tveite and Danielsen, 1969; Hartmann, 1970; Nellen and Hempel, 1970; Solemdal, 1970; Pommeranz, 1973). They are thus exposed to the sunlight and to mechanical stress from wave action (Hempel and Weikert, 1972). Wave action has for some time been considered a cause of severe mortality of pelagic fish eggs (Rollefsen, 1930, 1932; Devold, 1935; Zaitsev, 1968). Light, on the other hand (see review by Blaxter, 1969) seems more detrimental to demersal eggs which are usually incubating in dim illumination. Near the surface ultraviolet light may be detrimental to pelagic eggs (Marinaro and Bernard, 1966).

The present study was designed to quantify the effect of both physical stresses and light on plaice eggs under experimental conditions.

METHODS

Plaice eggs were obtained by artificial fertilization at the Biological Station Trondheim in Norway and incubated at temperatures which slowly increased from 5 to 7.5°C. For the measurements of mechanical properties an experimental device (Fig. 1), by which the eggs could be deformed by a lever, was constructed (see Zotin, 1961; Shelbourne, 1963; Ciechomski, 1967). Counterweights compensated the weight of the lever and the spring balance; thus the reading of the force needed no correcting. A needle hinge kept the friction low and a combination of worm gear and cog prevented the expanded spring balance from contracting. Illumination entered through the plexiglass lever. The degree of deformation was defined as the reduction of the height of the eggs in percent of the initial height.

Apstein's (1909, 1911) system of 25 stages was adopted so that the data of the different egg batches could be related to each other, independent of their incubation temperature. As an example the deformation of a stage 10 egg is plotted against the applied force in Fig. 2. 25% deformation was reached at 3.7 g, 50% at 42 g, and 75% at 195 g. At 180 g, which caused 73% deformation, a thin outer membrane of the chorion broke. At this point an extrusion of a small amount of fluid was observed, and the deformation was no longer reversible. The extreme deformation connected with this phenomenon is illustrated schematically in Fig. 2. 410 g were required to crush the egg.

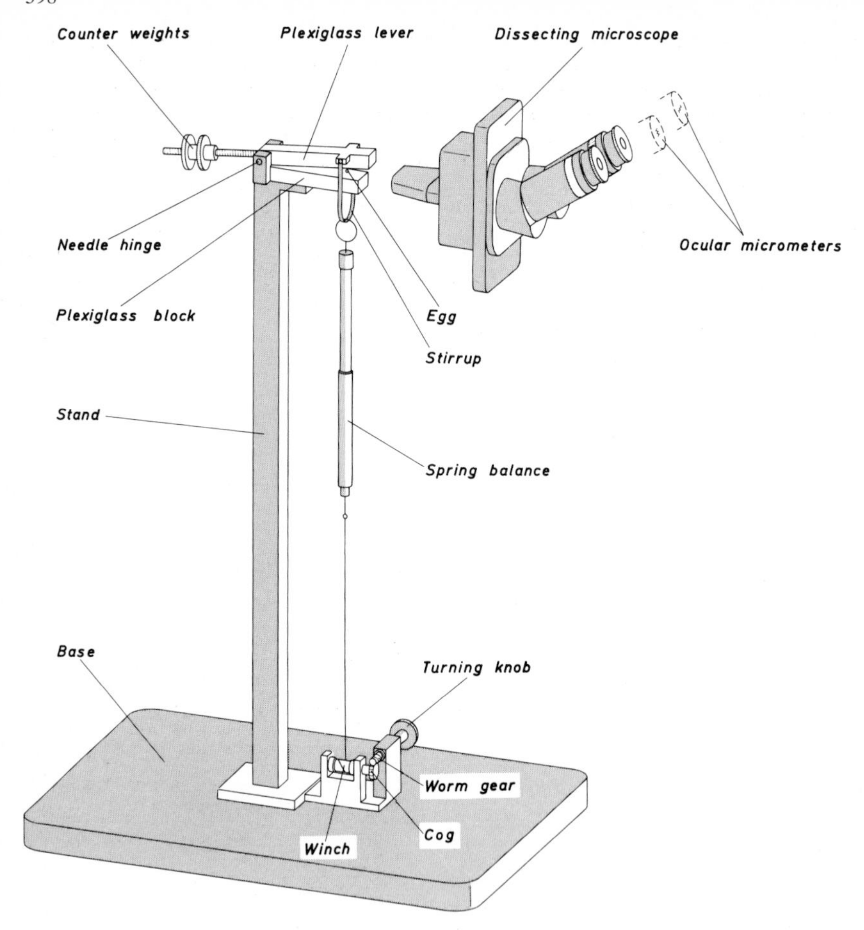

Fig. 1. The experimental device for measurements of the mechanical properties of plaice eggs

The pressure applied during deformation was determined by measuring the contact area between the egg and the lever. To investigate the subsequent effects of deformation the eggs were reexamined after 24 h. Controls were handled but not deformed.

Foam and bubbles are found near the sea surface in rough weather. These conditions were simulated by exposure of the eggs (and later, hatched larvae) to spray from a sprinkler giving about 40 ml/sec 20 cm above the surface of 10 l tanks (Fig. 3). During the experiment the temperature slowly rose from 6.1°C to 6.8°C. Eggs were also exposed to the permanent bubbling of an aerator with an air output of 10.2 ml/sec (Fig. 4). The temperature was between 6.4°C and 9.9°C. Each experiment was done on 500 stage-3 eggs. Control experiments were performed in similar tanks but with the eggs kept outside the area of spraying or bubbling by a gauze cylinder running from the bottom to above the surface.

The usual form of breaking waves in the open sea is the "spilling breaker" (Carstens, 1968). These are initiated when gusts of wind accelerate the water particles in the wave crest so that their speed becomes higher than the spreading velocity of the wave. The water mass at the wave crest then detaches from its orbital movement and surfs on the front slope of the wave. Such conditions were simulated by accelerating water by gravity from a 5-litre header tank down a chute (Fig. 5) 2.1 m long at an angle of 60° to the horizontal. At the beginning of the test, plaice eggs were placed at the bottom of the chute where they were suspended at the surface in a circular aggregation, with a 5-10 cm diameter and a depth of 1 cm. The centre of the aggregation was about 10 cm away from the edge of the chute. When a trigger was released a watertight flap on the header tank opened. At the end of the chute the accelerating water was deflected by a curvature, thereby surfing horizontally onto the water surface.

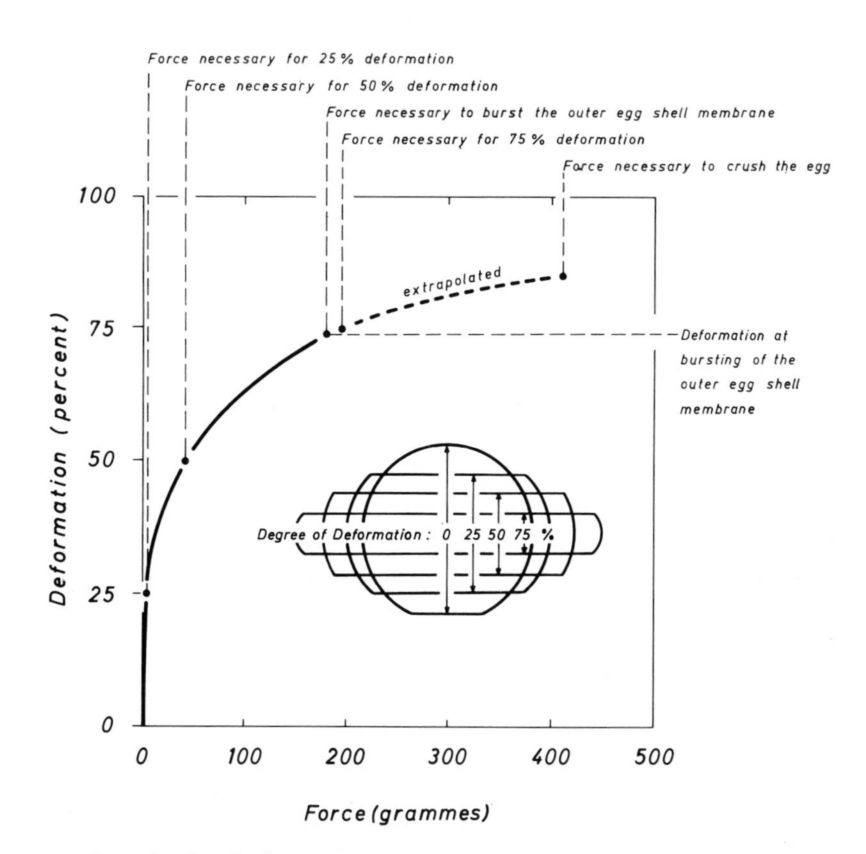

Fig. 2. An example of the deformation of a plaice egg plotted against the applied force

The impact of the water caused the eggs to drift away to about 75 cm. They were retained by a gauze box and representative samples were collected and examined for mortality after 24 h. Suitable control experiments were run.

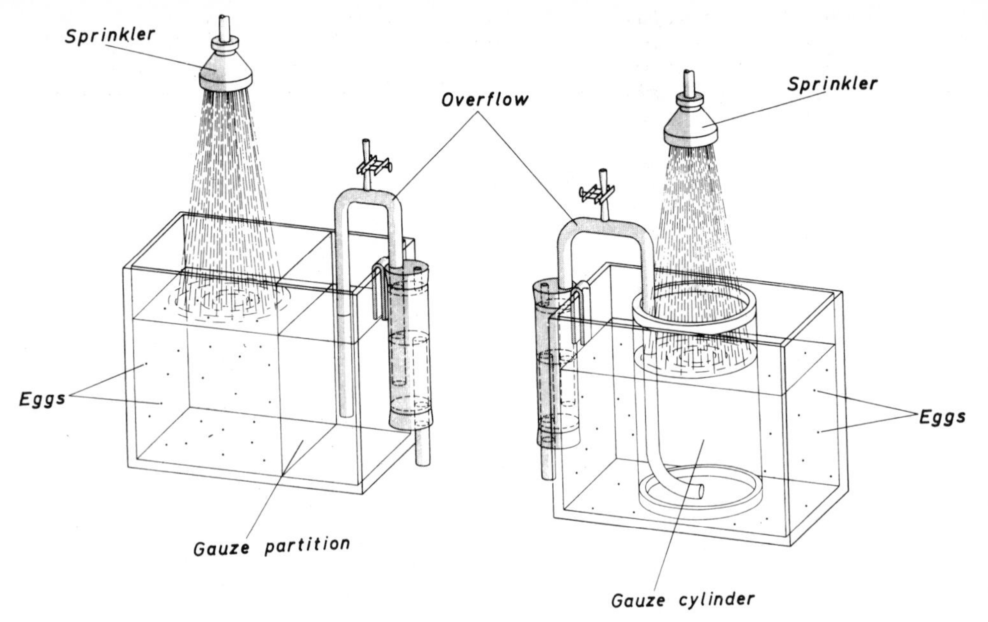

Fig. 3. The spray experiment; left: test, right: control

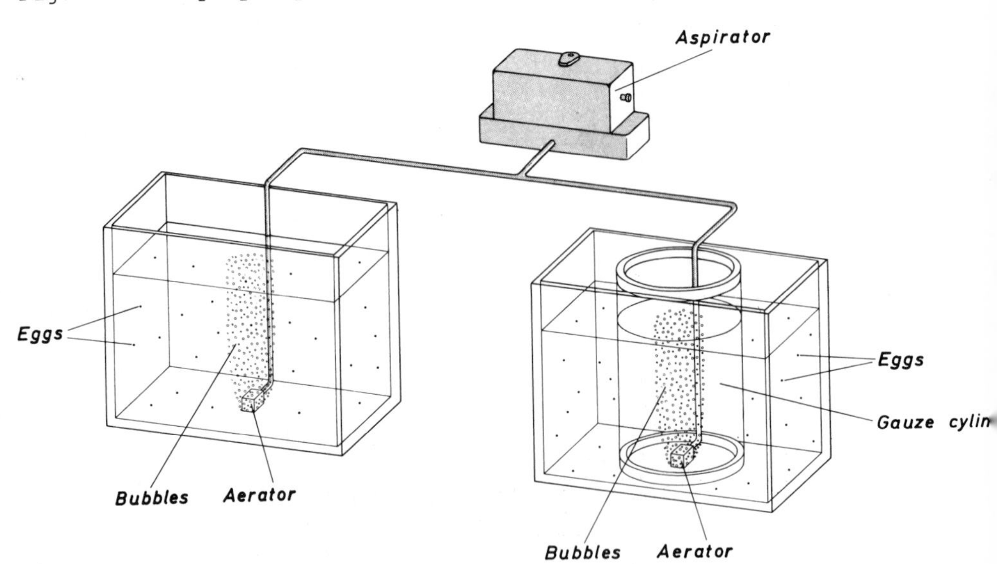

Fig. 4. The experiment with bubbles; left: test, right: control

For exposure to light, funnel-shaped containers (Fig. 6) of about 350 ml capacity were used with a renewal rate of 200 ml/min. The eggs were retained by gauze liners and exposed to artificial ultraviolet and infrared light in a darkroom with 3 containers. In the first container the eggs were continuously exposed to the light, in the second they were exposed over 12 h periods and in the third they were kept in continual darkness. The ultraviolet light was produced by mercury lamp (HQV, 125 W; Osram) installed between two containers 20 cm above the water surface. The light intensity was 0.05 ly/min (ly = 1 Langley = 1 cal/cm^2), the exposure causing a slight warming of the water of 0.15 deg C. The infrared light was produced by a tungsten lamp (Sl/r, 250 W; Osram) with red filter 1 m above the water surface. The light intensity was 0.40 ly/min and the warming due to the lamp was 0.3 deg C.

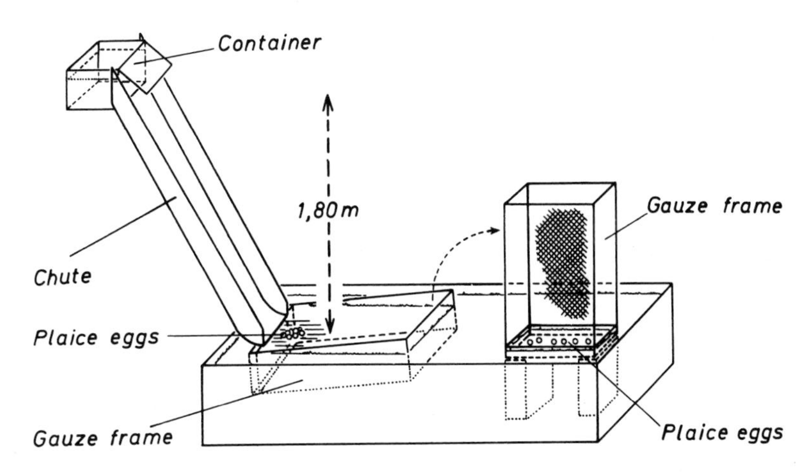

Fig. 5. The experiment with surf water

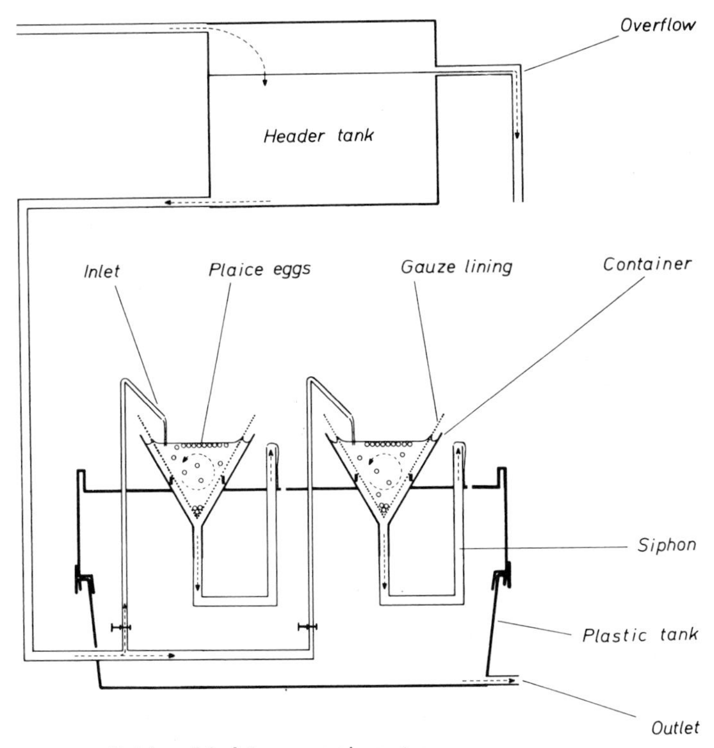

Fig. 6. The water system of the light experiments

For exposure of eggs to natural daylight four different experimental conditions were used. The first container was open to the light. Above the second was placed a 3 mm thick sheet of glass, which filtered out the ultraviolet component of the spectrum. The third container was covered by two such sheets with a 1 cm layer of sea water between them. This also filtered out the long-wave infrared above 1400 nm. The last container was kept dark. The solar radiation was measured by an actinograph (Robitzsch-Fuess; Fuess, Berlin-Steglitz). The sky was uninterrrupted from N through E to WSW but the containers with light filters were partly shaded by the frames during the morning and the afternoon.

RESULTS

The force to crush the unfertilized eggs was about 1.5 g and there was little change during the first 30 min after fertilization (Fig. 7). Afterwards the average resistance increased to about 20 g at 90 min post-fertilization. The continuous curve of the bursting point of one egg batch indicates a rapid increase in subsequent hardening until the eggs have reached the 4-cell stage (Fig. 8). The curve of the force necessary to break the outer chorion is not always reflected in the curve of the force necessary to crush the egg. Confirmatory tests were done on other egg batches. The decrease in resistance to bursting which begins about 11 h post-fertilization, cannot be fully explained and may not be significant. In controls it was found that unfertilized eggs in water only showed an increase of the crushing force to 5.7 g during the first 6 h.

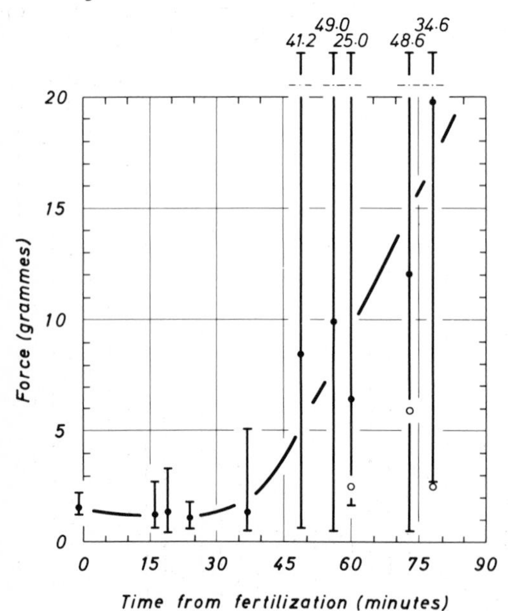

Fig. 7. The forces necessary to burst the outer chorion (open circles: single values) and to crush the egg (solid circles: mean values of 10 or 20 measurements; vertical lines: ranges of values) during the first 1.5 h after fertilization

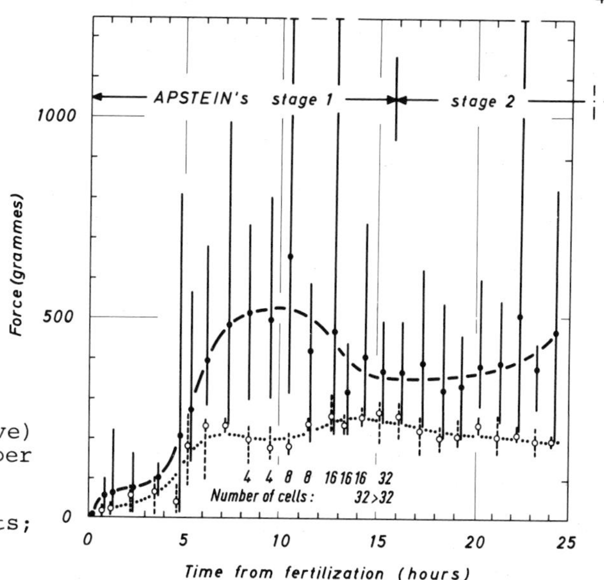

Fig. 8. The forces necessary to burst the outer chorion (lower curve) and to crush the egg (upper curve) during the first 24 h. Each mean value represents 10 measurements; the vertical lines give the range of value

The further curve during development (Fig. 9) based on stages shows an increase until the end of gastrulation (Stage 6), a decrease until closing of the blastopore (Stage 10). At Stages 20-23 some eggs resisted more than 3500 g, thus reaching the upper end of the measuring range of the device. Prior to hatching (Stage 25) the curve reached a minimum. The curve of the force necessary to break the outer chorion is rather constant. At nearly every stage eggs of particularly low resistance could be observed.

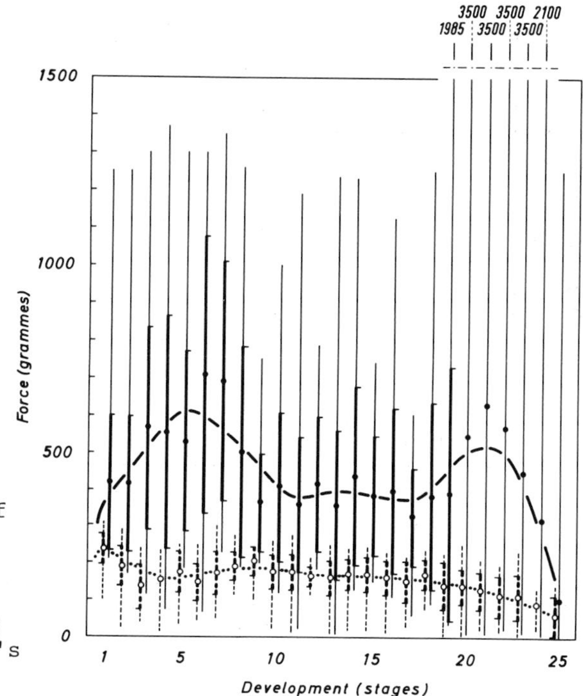

Fig. 9. The force necessary to burst the outer chorion (lower curve) and crush the egg (upper curve). The mean values (solid and open circles) during the whole of development, as far as possible, the standard deviations (thick vertical lines), and the range of values (thin vertical lines) are plotted against Apstein's stages

In Fig. 10 the mean values of the force resulting in 25%, 50%, and 75% deformation are given together with curves from Fig. 9. The curves of resistance to 25% and 50% deformation remain fairly level with a maximum at the early stages and a decrease later. The curve of resistance to 75% deformation runs similarly in its middle part and shows that bursting of the outer chorion takes place near the 75% deformation point. The average percentage of deformation resulting in bursting of the outer chorion decreases slowly from about 70% at Stage 5 to 63% at Stage 25 (Fig. 11).

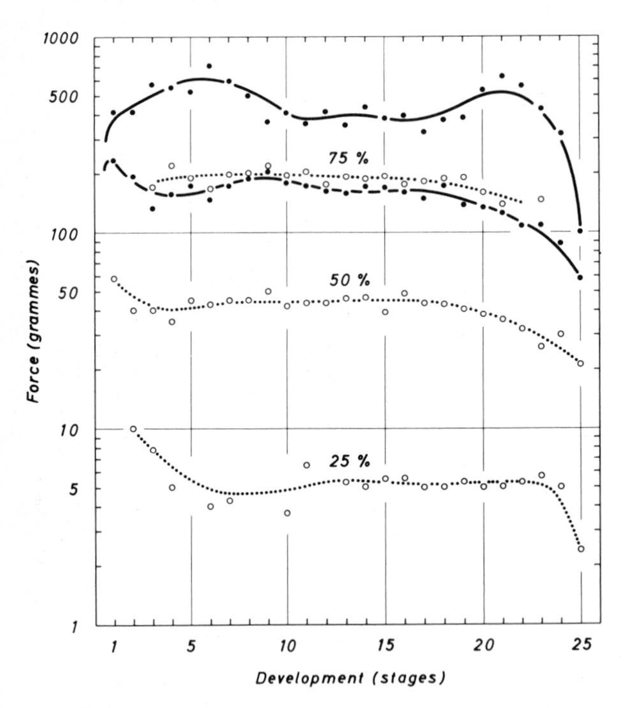

Fig. 10. The resistance to deformation (dotted curves), to the bursting of the outer chorion (lower solid curve), and to the bursting of the egg (upper solid curve)

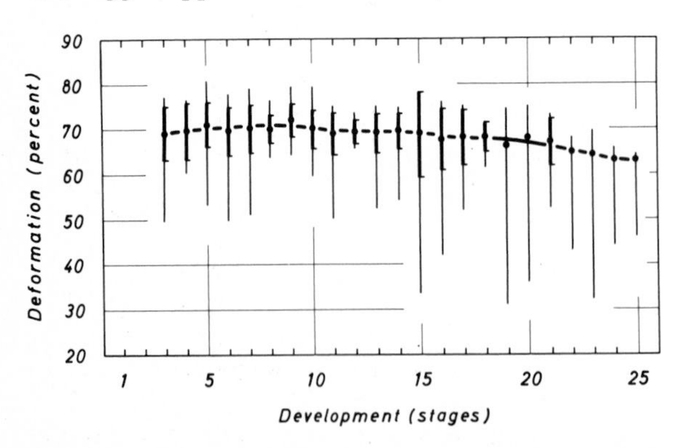

Fig. 11. The deformation at bursting of the outer chorion. Given are the mean values (circles), the standard deviations (thick vertical lines), and the range of values (thin vertical lines)

The average pressure causing 25% deformation decreased from 0.7 kg/cm^2 at Stage 2 to 0.2 kg/cm^2 at Stage 25 (Fig. 12). The pressure at 50% deformation 1 h post-fertilization amounted only to 0.12 kg/cm^2; it showed a maximum at Stage 17. Only a few measurements of the pressure resulting in the bursting of the outer chorion could be made for technical reasons during the second half of development. The mean values decreased from 2.7 kg/cm^2 at Stage 16 to about 1.5 kg/cm^2 at Stage 24.

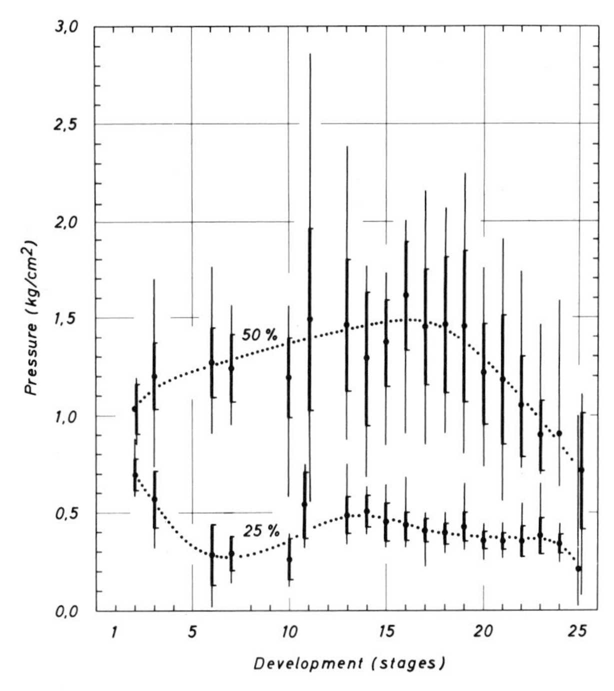

Fig. 12. The resulting pressures at 25% and 50% deformation (mean values: circles; standard deviations: thick vertical lines; ranges of values: thin vertical lines)

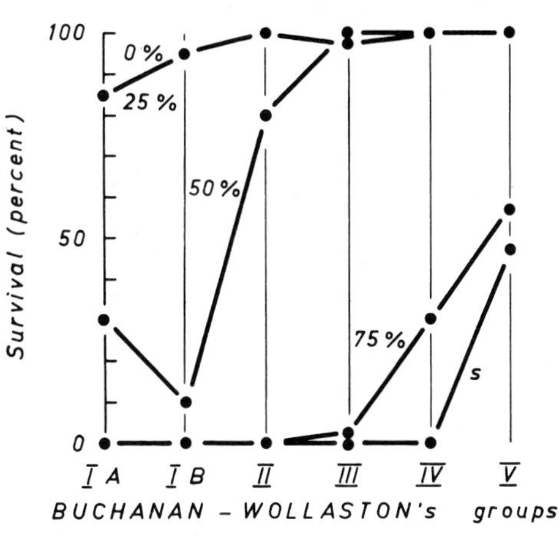

Fig. 13. The survival after 25%, 50%, 75% deformation and bursting of the outer chorion (s). 0% = control

The effect of deformation is shown in Fig. 13. The survival curves of the control and of the 25% deformed eggs are identical. Deformation of 50% produced low survival until the end of gastrulation. Deformation of 75% caused very low survival until the embryo had surrounded 3/4 of the yolk's circumference. Bursting of the outer chorion was always lethal until the embryo has surrounded the yolk completely. Shrinking of the embryo (see Rollefsen, 1930, 1932) could often be observed 10-30 min after 75% deformation, but some of the extremely deformed embryos at Stage IV and V survived and managed to hatch.

In addition to pressure, which acts vertically on the egg surface, shearing forces were applied tangentially to the egg surface in the later experiments. Exposure to spray produced slightly higher mortalities in the test sample than in the control, 4 and 5 days post-fertilization (Fig. 14). This caused only little difference of survival with a maximum of 5%. During the hatching period many dead larvae were collected in the experimental tank but not in the control. Obviously the newly hatched larvae were not resistant to the spray. The survival curve (Fig. 15) shows that an exposure of 1 h caused no distinct difference in mortality but after 24 h of exposure to the spray less than 10% had survived.

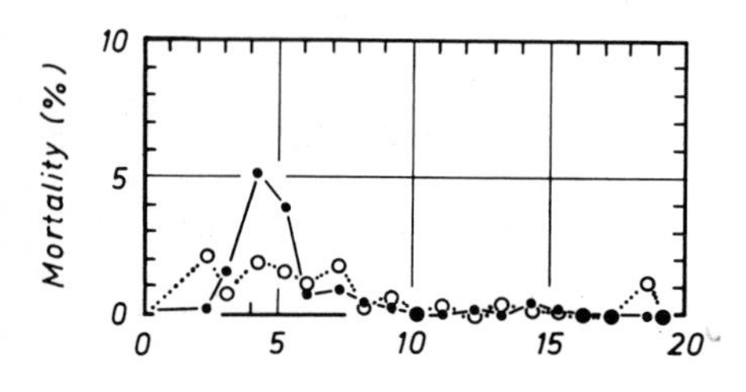

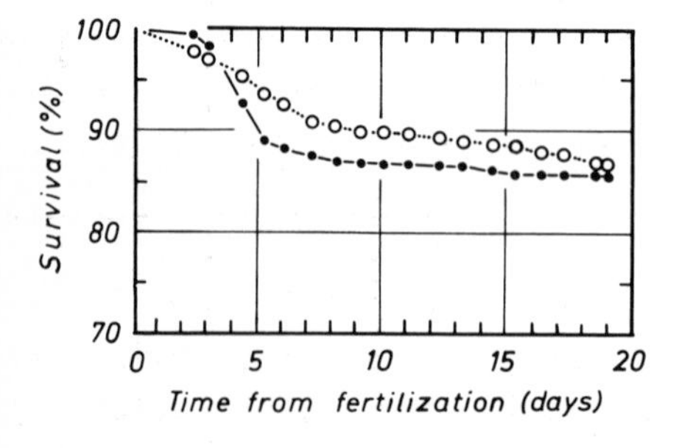

Fig. 14. The mortality and the resulting survival produced by the spray (test sample: solid circles; control: open circles)

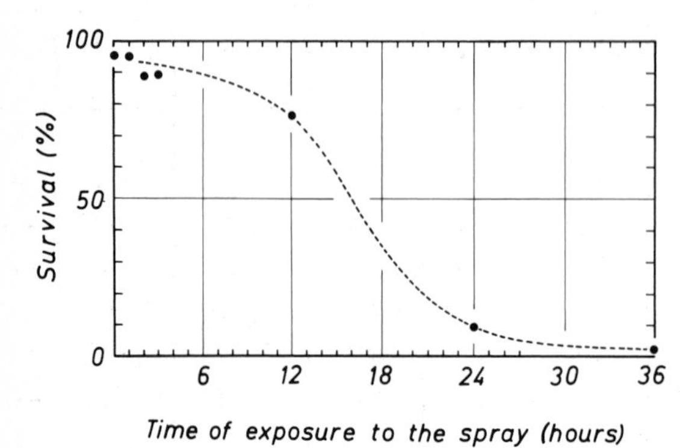

Fig. 15. The resistance of yolk sac larvae to the spray of a sprinkler

In the test batch exposed to bubbling the mortality was slightly higher than in the control during the period from the 3rd until the 6th day (Fig. 16). The final survival was 8.5% lower than in the control.

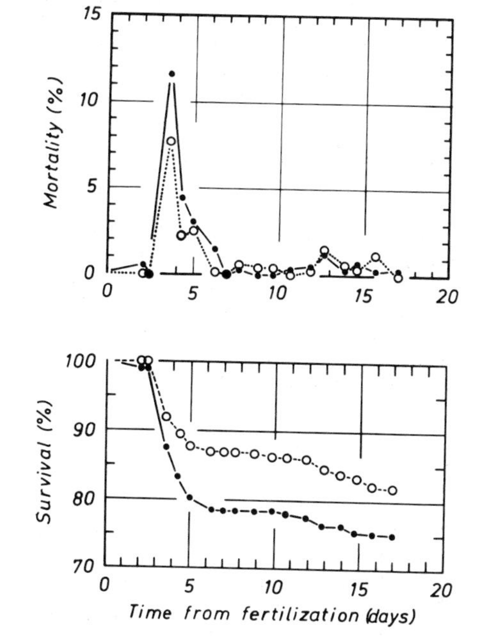

Fig. 16. The mortality and the resulting survival produced by permanent bubbling (test sample: solid circles; control: open circles)

Table 1. Results of the exposure of plaice eggs to surf water

No.	Stage		Time from fert.	Test		Control		Test kill./surv.	Contr. kill./surv.	α (%)
	B.-W.	AP.		kill.	surv.	kill.	surv.			
			hrs.							
1	I A	1	0,92	68	70	48	76	0,971	0,632	---
2	I A	1	1,23	90	222	97	276	0,405	0,352	---
3	I A	1	1,85	113	488	62	514	0,232	0,121	0,1
4	I A	1	2,53	39	185	55	369	0,211	0,149	---
5	I A	1	2,67	100	621	112	586	0,161	0,191	---
			days							
6	I B	4	2,22	4	52	3	82	0,077	0,037	---
7	II	9	4,51	0	59	0	24	0,000	0,000	---
8	III	16	9,41	1	77	1	106	0,013	0,009	---
9	IV	20	10,10	0	98	8	146	0,000	0,055	2,5
10	V	24	14,18	12	101	3	170	0,119	0,018	0,1
11	V	25	12,17	38	154	27	160	0,247	0,169	---

The tests with accelerated water were made at short time intervals immediately after fertilization (Table 1) since it could be assumed that the resistance of the eggs was lowest at this time. In most cases there were slightly more dead eggs in the test samples than in the control but χ^2-tests showed that the differences were not usually significant.

The results of exposure to artificial ultraviolet light are somewhat inconclusive (Fig. 17). χ^2-tests showed that the survival at the end of the second experiment was significantly lower, both in the containers which were continuously exposed, and in those exposed for 12-h period compared with the dark control. In the first experiment, however, the survival was nearly the same in the continuously exposed container as in the control. In the periodically exposed container it was significantly lower.

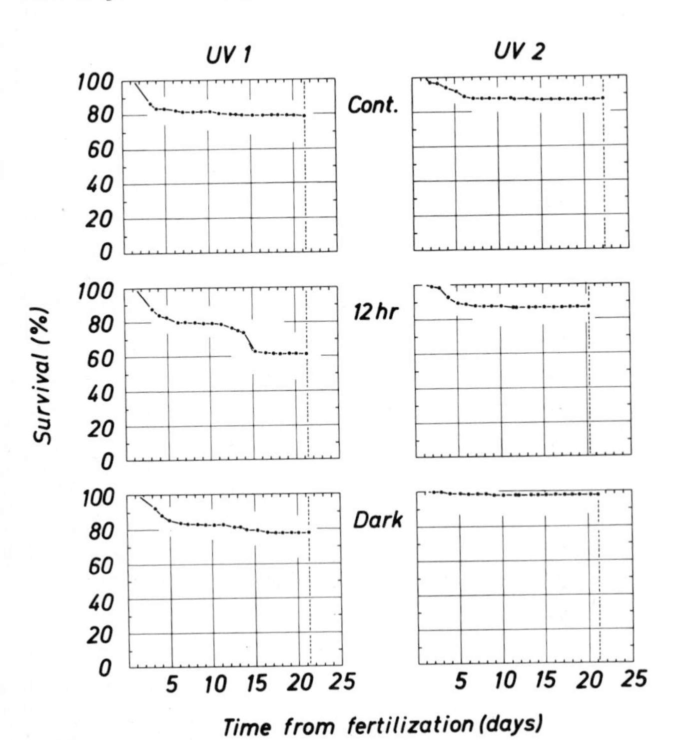

Fig. 17. The survival during the experiments with artificial ultraviolet light (Cont. = continuously exposed; 12 h = exposed in 12-h periods; Dark = control). The dotted vertical lines in this and the following figures indicate the end of the hatch period

The artificial infrared light could be more deleterious. The final survival in the dark containers was nearly the same (Fig. 18). In the continually exposed containers, however, it was much higher during the first experiment than during the second one. In the periodically exposed containers the results were just the opposite.

At the beginning of the first test with natural daylight there was a low mortality in the open container (Fig. 19). During this period an average of 257 ly/d was radiated. The short-term maxima were already high (up to 1 ly/min) but this would not have caused marked damage. From the 13th day on, however, the mortality increased, perhaps due to the high radiation during this day and the day before. The mortality in the other containers of this test series was always low except for an initial mortality also observed in the open container. During the first week of the second test there was a very high average radiation of 462 ly/d. This caused extremely high mortality.

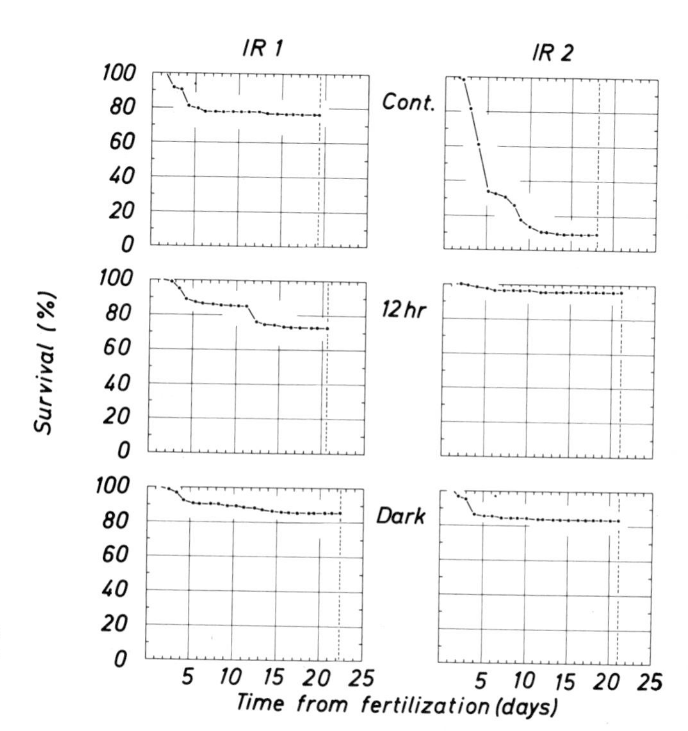

Fig. 18. The survival during the experiments with artificial infrared light (symbols as in Fig. 17)

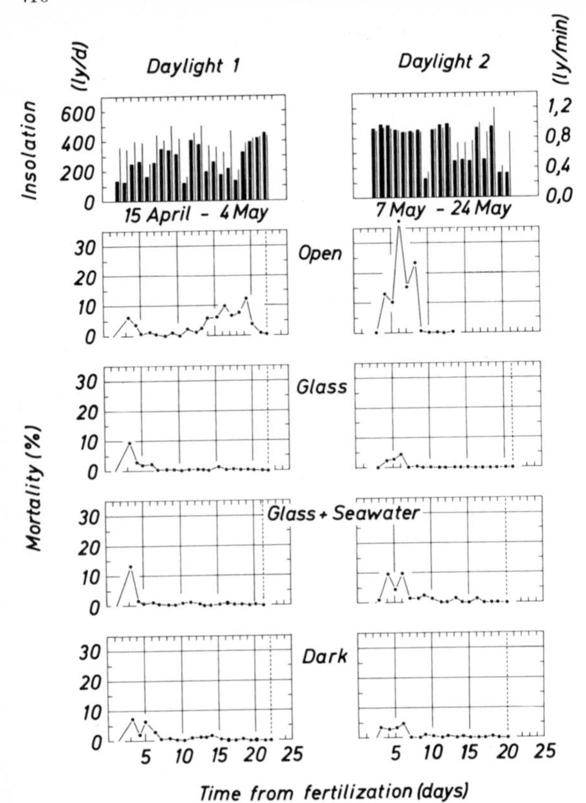

Fig. 19. The daily insolation in ly/d (thick vertical lines), the corresponding short-termed radiation maxima in ly/min (thin vertical lines), and the mortality (circles) during the experiments with natural daylight (Open = no light filters; Glass = containers exposed under a single glass-pane; Glass + Seawater = containers exposed under two glass-panes with 1 cm seawater between them; Dark = control)

The combination of visible and total infrared light under the single sheet of glass proved to be optimal (Fig. 20). There was lower final survival, especially in the second test, in the containers under the double sheet with sea water. In the dark containers, also, a slightly lower total egg mortality was observed.

DISCUSSION

During the first 1-2 h after fertilization Zotin (1958) observed an initial decrease of the shell toughness of the eggs of salmon, trout, and whitefish. Fig. 7 shows that there is a similar period of about 30 min in plaice eggs. Applying the results of the measurements during the whole activation indicates that eggs in hatcheries should not be manipulated until 6 h or better, 9 h post-fertilization. According to Yamamoto (1961), the unfertilized eggs of most fish species are activated only by contact with water but this had only little effect

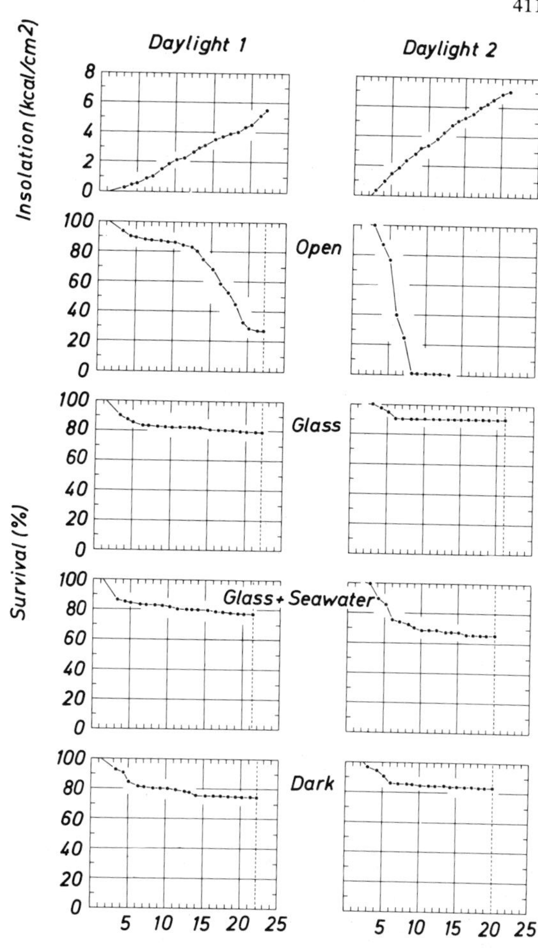

Fig. 20. The cumulative insolation (kcal/cm^2) and the survival during the experiments with natural daylight (symbols as in Fig. 19)

on the unfertilized plaice eggs. The thin outer chorion, whose breakage preceded the bursting of the egg, is supposed to be the zona radiata externa. Two chorionic membranes were also observed in marine eggs (Götting, 1966, 1967; Hagström and Lönning, 1968). It is known that the hardening of salmon chorion is mainly due to changes in the inner membrane (Yamamoto, 1961).

The curves of the force necessary for 25%, 50%, and 75% deformation of the eggs and the breaking of the outer chorion did not always correlate with the force necessary to break the egg (Fig. 10). Therefore it is questionable whether this procedure of measuring the mechanical resistance, prevalent up until now, is really appropriate. Except for a much lower crushing force of eggs from a pond-wintered stock during the gastrula stage the data measured by Shelbourne (1963) fit approximately to the results presented here. Measurements during 2 voyages of R.V. "Anton Dohrn" to the area Southern Bight-Channel of the North Sea in January 1973 and 1974 showed that the average bursting point was usually much higher in the natural than in the test material. The causes of differences in the shell strength are not known. In freshwater species it is influenced by the calcium content of the water in which the eggs are activated (Zotin, 1958).

From Fig. 10 it could be seen that nearly the same forces resulted in the same degree of deformation after hardening of the egg. Fig. 13, however, demonstrates that the same deformation produced completely different survivals in the different stages. Mechanical resistance to deformation is thus not synonymous with viability. Due to combinations of these two factors, the plaice eggs pass through 4 different phases:

1. In the first hours after fertilization the mechanical resistance is extremely low.

2. From about 10 h post-fertilization until the embryo has surrounded half of the yolk's circumference the mechanical resistance is high but viability is relatively low.

3. From this stage until the end of the development both mechanical resistance and viability are relatively high.

4. Prior to hatching the mechanical resistance becomes low (Suga, 1963; Poy, 1970) but the viability is high.

It can thus be concluded that the first hours after fertilization represent the most sensitive period and that phase 3 is optimal.

Mechanical stress at the sea surface can occur in two principal ways: shearing force and pressure. The orbital movement of the water particles causes no, or only small mechanical stress and the deformation of the water surface is far above the dimensions of the plaice eggs. It is difficult to extrapolate the results of the pressure measurements to natural conditions because data on mechanical stress at the sea surface are scarce. Only one rough estimate could be found that spilling breakers develop pressures in the order of 0.1 kg/cm^2 (Führböter, pers. comm.). According to Fig. 13 the pressure connected with 25% deformation is not deleterious for the plaice eggs after activation. The lower end of the standard deviation of pressures for more than 25% deformation was lowest during Stage 6 with a value of 0.13 kg/cm^2 (Fig. 12). Hence the resistance of the plaice eggs to pressure connected with deformation between plates is in the same order of magnitude as the estimated pressure in spilling breakers. Since an average pressure of only 0.12 kg/cm^2 at 50% deformation was measured this is true especially during the activation. Slow deformation between plates, however, represents completely unnatural conditions since eggs in the sea are surrounded by water from all sides. Under such circumstances an increase of pressure transmits into the water and causes no deformation. Potential change of pressure will also be applied more rapidly at the sea surface. Plunging breakers are very rare in the open sea. They develop shock pressures (Führböter, 1966) which are much higher than the measured resistance of the plaice eggs between plates. Since these pressures only exist for 0.1 sec in small areas, a damage of plaice eggs is only occasionally possible.

The effect of the shearing forces could not be estimated directly because the inertia of the plaice eggs was not known. If it is small, and this can be assumed because of their spherical shape, the shearing forces would result in a rotation. If their inertia is high, damage could be caused by strain on the chorion or the embryo. The results of the experiments with a spray, with bubbles, and with surfing water indicate that only very few plaice eggs cannot tolerate these conditions. The relatively high susceptibility of the larvae to the spray show that the egg shell is an effective shelter for the embryos while the perivitelline space may serve as a shock absorber (Blaxter, 1969).

Thus it seems that mechanical forces at the surface are only slightly deleterious to plaice eggs. Bearing in mind also that plaice spawn near the bottom of the sea (Saemundson, 1926), so that the soft early eggs do not get directly into the uppermost water layers, and that the turbulence occurring with higher wind forces disperses egg concentrations near the surface (Pommeranz, 1973), it seems even less likely that high mortalities should be caused by wave action.

The high mortality in the open containers of the experiment with natural daylight is most probably caused by the ultraviolet component. The fact that there was not always a marked influence of the artificial ultraviolet light was possibly due to the discontinuous spectrum of the lamp. A deleterious effect of artificial infrared light has already been established for the embryos of perch and smelt (Eisler, 1961). The high survival in natural daylight under glass is contradictory to the observations of Dannevig and Hansen (1952) and Dannevig and Sivertsen (1933). They established higher mortalities in containers near the window than in containers which were kept dark. It is suggested that plaice eggs should be incubated in the dark, since this would also give a high survival without temperature increases. Not all species need to be sheltered, since exposure to light did not affect the eggs of angler fish (Klugh, 1929), sole (Flüchter, 1965), and pike (Lillelund, 1967).

From the experiments with natural daylight it can be concluded that the tolerance limit of plaice eggs in a surface layer is situated below the deleterious intensity of 462 ly/d during the second test and above the non-dangerous radiation of 257 ly/d during the first test. The mortality was not directly dependent upon the applied dose since 50% survival was achieved at 4 kcal/cm^2 in the first test and at 2 kcal/cm^2 in the second test.

These results can be discussed in terms of their significance for the southern North Sea. There, the spawning season of the plaice begins in December and lasts until March. In December and January the mean total solar radiation amounts to about 50 ly/d, in February to 100 ly/d, and in March to 200 ly/d (Black, 1956). Since that is always below the supposed tolerance limit it is unlikely that there can be damage to plaice eggs by daylight. In addition the light intensity is reduced near the sea surface by reflection, absorption, and scattering; ultraviolet components are absorbed by "Gelbstoff" (see Jerlov, 1968) and plaice eggs will be continually moved by turbulent conditions. It seems that the search for the main causes of plaice eggs mortality must be continued.

SUMMARY

1. Extreme deformation of plaice eggs by a lever resulted in bursting of the thin outer chorionic membrane before final crushing.

2. The average crushing force for the unfertilized eggs was 1.5 g. After fertilization it increased to about 500 g during the first 10 h and to 700 g at the end of gastrulation. There was then a decrease until hatching.

3. The mortality caused by deformation decreased considerably during the development. Due to combinations of different viability with varying mechanical resistance to deformation, the eggs pass through an extremely vulnerable phase during the first hours after fertilization; a second phase with low viability but with high mechanical resistance after the activation until the embryo has surrounded half of the yolk' circumference; a third optimal phase with high viability and high mechanical resistance; and a final phase with a decrease of the mechanical resistance immediately before hatching.

4. The pressure at lethal deformation between plates can be similar to the pressure in breakers. This is probably not deleterious for the egg since the pressure transmits into the surrounding water under natural conditions.

5. By means of a spray, bubbling, and surfing water the eggs were also exposed to shearing forces. Only few eggs could not tolerate these, but yolk sac larvae were relatively susceptible to the spray.

6. Artificial ultraviolet light was not very harmful, though artificial infrared light occasionally caused a high mortality.

7. The lethal effect of natural daylight was mainly due to the ultraviolet component. The tolerance limit of eggs in 9.5 cm deep containers lay between 257 ly/d and 462 ly/d above the natural insolation during the spawning season.

ACKNOWLEDGEMENT

Grateful thanks are due to Prof. Dr. G. Hempel for advice and for the extensive discussion of the results. Konservator T. Strømgren kindly provided facilities to work at Trondheim Biologiske Stasjon Norway.

REFERENCES

Apstein, C., 1909. Bestimmung des Alters pelagischer Fischeier. Mitt. dtsch. Seefischerver., 25, 364-373.
Apstein, C., 1911. Verbreitung der pelagischen Fischeier und Larven in der Beltsee und den angrenzenden Meeresteilen 1908/09. Wiss. Meeresuntersuch., 13, 225-284.
Black, J.N., 1956. The distribution of solar radiation over the earth's surface. Arch. Met., Wien (B), 7, 165-189.
Blaxter, J.H.S., 1969. Development: eggs and larvae. In: Fish Physiology. W.S. Hoar and D.J. Randall (eds.). New York - London: Academic Press, 177-252 pp.
Buchanan-Wollaston, H.J., 1923. The spawning of plaice in the southern part of the North Sea. Fish. Invest. Lond. (2), 5, No. 2, 1-36.
Buchanan-Wollaston, H.J., 1926. Plaice-egg production in 1920-21 treated as a statistical problem with comparison between the data from 1911, 1914, 1921. Fish. Invest. Lond. (2), 9, No. 2, 1-36.
Bückmann, A., 1961. Über die Bedeutung des Schollenlaichens in der südöstlichen Nordsee. Kurz. Mitt. Inst. Fisch. Biol. Univ. Hamb., 11, 1-40.

Carstens, T., 1968. Wave forces on boundaries and submerged bodies. Sarsia, 34, 37-60.

Ciechomski, J.Dz. de, 1967. Influence of some environmental factors upon the embryonic development of the argentine anchovy *Engraulis anchoita* (Hubbs, Marini). Rep. Calif. ocean. Fish. Invest., 11, 67-71.

Dannevig, A. and Hansen, S., 1952. Faktorer av betydning for fiskeeggenes og fiskeyngelens oppvekst. Fiskeridir. Skr. Havundersøk., 10, 1-36.

Dannevig, A. and Sivertsen, E., 1933. On the influence of various physical factors on cod larvae; experiments at Flødevig sea-fish hatchery. J. Cons. perm. int. Explor. Mer, 8, 90-99.

Devold, F., 1935. The susceptibility of plaice eggs to shock. K. norske vidensk. Selsk. Forh., 8, 71-74.

Eisler, R., 1961. Effects of visible radiation on salmonoid embryos and larvae. Growth, 25, 281-346.

Flüchter, J., 1965. Versuche zur Brutaufzucht der Seezunge (*Solea solea* L.) in kleinen Aquarien. Helgoländer wiss. Meeresunters., 12, 395-403.

Flügel, H., 1967. Licht- und elektronenmikroskopische Untersuchungen an Oozyten und Eiern einiger Knochenfische. Z. Zellforsch., 83, 82-116.

Führböter, A., 1966. Der Druckschlag durch Brecher auf Deichböschungen. Mitt. Franzius-Inst. Grund- und Wasserbau Hannover, 28, 1-206.

Götting, K.-J., 1966. Zur Feinstruktur der Oocyten mariner Teleosteer. Helgoländer wiss. Meeresunters., 13, 118-170.

Götting, K.-J., 1967. Der Follikel und die peripheren Strukturen der Oocyten der Teleosteer und Amphibien. Z. Zellforsch., 79, 481-491.

Hagström, B.E. and Lönning, S., 1968. Electron microscopic studies of unfertilized and fertilized eggs from marine teleosts. Sarsia, 33, 73-80.

Hartmann, J., 1970. Verteilung und Nahrung des Ichthyoneuston im subtropischen Nordostatlantik. "Meteor" ForschErgebn. dt. ForschGem. (D), 8, 1-60.

Hempel, G. and Weikert, H., 1972. The neuston of the subtropical and boreal north-eastern Atlantic Ocean. A review. Mar. Biol., 13, 70-88.

Jerlov, N.G., 1968. Optical oceanography. Amsterdam-London-New York: Elsevier Publishing Co.

Klugh, A.B., 1929. The effect of the ultra-violet component of sunlight on certain marine organisms. Can. J. Res., 1, 100-109.

Lillelund, K., 1967. Versuche zur Erbrütung der Eier vom Hecht, *Esox lucius* L., in Abhängigkeit von Temperatur und Licht. Arch. FischWiss., 17, 95-113.

Marinaro, J.Y. and Bernard, M., 1966. Contribution à l'étude des oeufs et larves pélagiques de poissons méditerranéens. 1. Note préliminaire sur l'influence léthale du rayonnement solaire sur les oeufs. Pelagos, 6, 49-55.

Nellen, W. and Hempel, G., 1970. Beobachtungen am Ichthyoneuston der Nordsee. Ber. dt. wiss. Kommn Meeresforsch., 21, 311-348.

Pommeranz, T., 1973. Das Vorkommen von Fischeiern, insbesondere von Eiern der Scholle (*Pleuronectes platessa* L.), in den oberflächennahen Wasserschichten der südlichen Nordsee. Ber. dt. wiss. Kommn Meeresforsch., 22, 427-444.

Poy, A., 1970. Über das Verhalten der Larven von Knochenfischen beim Ausschlüpfen aus dem Ei. Ber. dt. wiss. Kommn Meeresforsch., 21, 377-392.

Rollefsen, G., 1930. Observations on cod eggs. Rapp. P.-v. Réun. Cons. perm. int. Explor. Mer, 65, 31-34.

Rollefsen, G., 1932. The susceptibility of cod eggs to external influences. J. Cons. perm. int. Explor. Mer, 7, 367-373.
Saemundsson, B., 1926. Fiskarnir (Pisces islandiae). Bókaverzlun Sigfúsar Eymundssonar, Reykjavik.
Shelbourne, J.E., 1963. Marine fish culture in Britain. II. A plaice rearing experiment at Port Erin, Isle of Man, during 1960, in open sea water circulation. J. Cons. perm. int. Explor. Mer, 28, 70-79.
Solemdal, P., 1970. Intraspecific variations in size, buoyancy, and growth of eggs and early larvae of Arcto-Norwegian cod, *Gadus morhua* L., due to parental and environmental effects. ICES, Doc CM 1970/F: 28.
Suga, N., 1963. Changes of the toughness of the chorion of fish eggs. Embryologia, Nagoya, 8, 63-74.
Tveite, S. and Danielsen, D.S., 1969. Funn av neustonorganismer i norske farvann. Fiskets Gang, 18, 292-296.
Yamamoto, T., 1961. Physiology of fertilization in fish eggs. Int. Rev. Cytol., 12, 361-405.
Zaitsev, Yu.P., 1961. The pelagic surface biocoensis of the Black Sea. Zool. Zh., 40, 818-825 (in Russian).
Zaitsev, Yu.P., 1968. La neustonologie marine: objet, méthodes, réalisations principales et problèmes. Pelagos, 8, 1-48.
Zaitsev, Yu.P., 1971. Marine neustology. Israel Program for Scientific Translations, 207 pp.
Zotin, A.J., 1958. The mechanism of hardening of the salmonid egg membrane after fertilization or spontaneous activation. J. Embryol. exp. Morph., 6, 546-568.
Zotin, A.J., 1961. The physiology of aqueous exchanges in the eggs of fishes and cyclostomes. Isdat. Acad. Nauk. SSSR, Moscow, 318 pp. (in Russian).

T. Pommeranz
Institut für Meereskunde
Abt. Fischereibiologie
2300 Kiel 1
Düsternbrooker Weg 20-22
FEDERAL REPUBLIC OF GERMANY

Effect of Hydrogen Sulfide on Development and Survival of Eight Freshwater Fish Species[1]

L. L. Smith, Jr. and D. M. Oseid

INTRODUCTION

The properties of hydrogen sulfide toxic to aquatic organisms have been substantially documented in the literature, but not at concentrations commonly observed in natural freshwater systems. The rapid oxidation of H_2S and its usual occurrence in areas of low dissolved oxygen have led to the erroneous assumption that it is of little significance in the life history of fish and most fish-food organisms except in areas of heavy pollution. Constant evolution from organic bottom detritus and break-down products from some pollutants keeps a low concentration in many areas where early life history stages of fish may be found. Sampling near the mud-water interface, especially in flowing systems, has not been done consistently, and very low concentrations have been considered unimportant.

Undissociated H_2S is the toxic form; its concentration is dependent on pH. At pH 9.0 about 1% is undissociated; at pH 6.7, 50%; and at pH 5, approximately 99% is present as H_2S. As we have shown in a number of experiments, especially with juvenile fish stages, the toxicity is strongly influenced by temperature.

The physiological effects of H_2S on fish, at the very low levels found to be lethal or detrimental, are not well understood. The toxic action on the various life history stages is presumed due to interference with enzyme action or to mitochondrial changes, but no satisfactory documentation is available.

The significance of our experimental work in the context of this symposium is that levels of H_2S that may commonly occur in freshwater systems near the bottom, in lakes at the time of vernal turnover, in water from underground sources, or in contaminated rivers have been shown to be highly toxic, or at least inhibitory to vital functions of fish. Several field studies have shown H_2S concentrations sufficiently high to kill eggs and fry. Colby and Smith (1967) demonstrated levels within 20 mm of the mud-water interface varying from 0.02 to 0.20 mg/l in a stream noted as a walleye fishery, but where the population was dependent on inward migration of adults. In natural spawning areas of northern pike Adelman (1969) found H_2S concentrations near the bottom as high as 0.22 mg/l and commonly in the range of 0.03 - 0.08 mg/l during the spawning period. Other authors (Ellis, 1937; Ziebell et al., 1970; Bell et al., 1972) have shown higher levels in a variety of fish habitats. Many studies report high concentrations in polluted areas. The Minnesota Department of Natural Resources abandoned 2 hatchery operations because H_2S concentrations (0.20 mg/l) were encountered in ground-water supplies (personal communication). In another operation at an egg-stripping station adult

---

[1]Paper No. 8340 Scientific Journal Series, Minnesota Agricultural Experiment Station, St. Paul, Minnesota, U.S.A.

fish and fertilized eggs were killed in water taken from a lake during early spring overturn. Hydrogen sulfide was shown to be high. Scidmore (1956), working on a Minnesota lake during winter, found H_2S concentrations of 0.3 and 0.4 mg/l with 6.0 and 3.6 mg/l O_2, respectively.

Adelman and Smith (1970) showed that eggs and fry of northern pike were affected by very low levels of H_2S. These authors (1972) demonstrated the sensitivity of goldfish eggs and larvae to low concentrations of H_2S and the influence of temperature and O_2 on its toxicity. Smith and Oseid (1972) reported short-term effects of H_2S on the early stages of 3 fish species. The experimental studies summarized here were conducted on 8 species of freshwater fish: brook trout (*Salvelinus fontinalis* (Mitchill)), rainbow trout (*Salmo gairdneri* Gibbons), white sucker (*Catostomus commersoni* (Lacépède)), northern pike (*Esox lucius* L.), fathead minnow (*Pimephales promelas* Rafinesque), goldfish (*Carassius auratus* L.), walleye (*Stizostedion vitreum vitreum* (Mitchill)), and bluegill (*Lepomis macrochirus* Rafinesque).

The studies reported here were sponsored in part by the University of Minnesota Agricultural Experiment Station, the U.S. National Institute of Health, and the U.S. Environmental Protection Agency.

MATERIALS AND METHODS

The work described here was carried on in the University of Minnesota laboratories with well water delivered through polyvinyl chloride pipes. In a few cases chlorinated well water from the same aquifer was delivered through black iron pipe and carbon filters. Nitrogen stripping columns were used to adjust dissolved oxygen concentrations below saturation. Flow-through apparatus was used for all tests. It was of the type described by Colby and Smith (1967), or an apparatus modified from that of Mount and Brungs (1967). The first test equipment used H_2S gas, and the second used sodium sulfide and adjusted pH to produce the desired level of H_2S in the test chamber. The pH was adjusted by heavy aeration or by addition of H_2SO_4. Test chambers were 7.6 x 7.6 x 5 cm acrylic boxes or cylinders in which the eggs and fry were confined by nylon screens.

Hydrogen-ion concentration ranged from 7.6 to 8.0 and temperature from 9° to 25^oC in the various tests. Test water was alkaline (Table 1). Undissociated H_2S ranged from 0.005 to 0.087 mg/l and was calculated from total sulfides, pH, and temperature. Tests of H_2S concentrations were made on samples taken from the test chambers near the outlet 3 times daily. Volume of the test chamber was replaced in 1.5 - 2.0 min.

Walleye and sucker eggs were secured from State of Minnesota egg-stripping stations and were put under treatment within 48 h of fertilization. Trout eggs came from State or commercial hatcheries and from adults stripped in the laboratory. Bluegill and northern pike eggs were stripped from adult fish and fertilized in the laboratory. Fathead and goldfish eggs were from natural spawning in the laboratory. All acute tests were run for 96 h, until eggs or fry died, or until eggs hatched. Bluegill, fathead, and goldfish were treated from fry stage through maturation, or adults were held in H_2S for a minimum of 46 days prior to start of spawning and thereafter until all spawning ceased. Adult brook trout were secured from a State hatchery and held in treatment for 37 days prior to start of spawning and thereafter for 39 days to the end of spawning.

Table 1. Analysis of laboratory water in which bioassays were conducted

Item	mg/l
Total hardness	210.0
Total alkalinity	230.0
Iron	<0.1
Manganese	0.02
Calcium (as $CaCO_3$)	140.0
Magnesium (as $CaCO_3$)	70.0
Sodium	5.0
Fluoride	0.23
Total phosphorus	0.04
Chloride	<1.0
Sulfate	<5.0
Potassium	1.0

A specified number of live eggs was placed in each chamber (50-1000) for the tests. Survival of eggs and fry in H_2S treatments was reported as a percentage of the controls. Values for LC_{50} were calculated by standard procedures.

EFFECT OF ADULT EXPOSURE

The effect of exposure of fish to H_2S on spawning behavior, egg deposition, and egg viability was determined for fathead minnows, bluegills, and brook trout. These fish were exposed prior to and through spawning for various periods (Table 2).

Fathead minnows were exposed for two reproductive cycles at levels of H_2S between 0.0004 and 0.0069 mg/l and at 23°C. In the first cycle the number of eggs laid per female was not significantly different at the various levels. In the second cycle a slight reduction in eggs per female occurred with 912 in a control and 791 at 0.0069 mg/l H_2S. Survival of spawning adults was lower in both cycles at the highest concentration, being 33% in the first cycle and 41% in the second at 0.0061 and 0.0069 mg/l, respectively. Except at these levels survival of spawning adults varied from 90% to 93%. Bluegills held for 826 days had no reproduction at 0.0018 mg/l and significantly reduced fingerling-to-adult survival at 0.0073 mg/l.

Bluegills held in H_2S for 46 days before spawning (total exposure 97 days) had reduced egg deposition. The controls deposited 17,562 eggs per female; fish held at 0.0014 mg/l, 6,157 eggs per female; and those at higher levels deposited none. Failure at the higher levels was related to reduction or absence of male spawning activities. At

Table 2. Effect of adult fish exposure to H_2S on egg viability and number of spawners - exposure time in parenthesis

Species	H_2S (mg/l)	Adult Survival to Spawning (%)	Number of Eggs per Female	Activity[1/]
Brook Trout (76 days)	Control	100	949	Normal
	0.0055	80	444	"
	0.0079	100	433	"
		100	96	Greatly reduced
	0.0121	100	275	" "
	0.0128	100	286	" "
Fathead #1 (297 days)	Control	93	1 179	Normal
	0.0004	90	1 332	"
	0.0012	93	444	"
	0.0031	93	1 296	"
	0.0061	33	1 460	"
Fathead #2[2/] (274 days)	Control	97	912	Normal
	0.0007	91	1 130	"
	0.0013	90	412	"
	0.0037	90	535	"
	0.0069	41	791	"
Bluegill #1 (826 days)	Control	100	9 928	Normal
	0.0018	100	0	Reduced
	0.0037	90	0	"
	0.0073	70	0	"
	0.0081	0	-	-
Bluegill #2 (97 days)	Control	100	17 562	Normal
	0.0007	100	12 795	"
	0.0014	100	6 157	Slightly reduced
	0.0027	100	0	Greatly reduced
	0.0078	100	0	No activity

[1/] Spawning behavior as judged by nest attendance, nest building, courting activity.

[2/] Second cycle started from eggs deposited by fish held in H_2S from egg to adult.

0.0014 mg/l H_2S there was slight alteration in spawning behavior, at 0.0027 mg/l it was greatly reduced, and at 0.0078 mg/l none occurred. Brook trout held for 37 days prior to spawning (total exposure 76 days at levels of 0.0055 - 0.0128 mg/l deposited fewer eggs per female than controls. At 0.0128 mg/l H_2S the number of eggs laid was 30% of controls and at 0.0055 mg/l was 46%. No apparent effect on behavior was noted at 0.0079 mg/l, but at higher concentrations spawning activity was greatly reduced.

Goldfish were exposed for 234 days prior to spawning. Because their eggs were attached to artificial vegetation and egg predation by adults was significant, exact comparative counts from various levels of H_2S were not possible. However, at concentrations of 0.010 and 0.028 mg/l H_2S the number of spawnings per female was 0.5 and 0.75, respectively, in contrast to 1.0 in the control (Table 3). At 0.005 mg/l there appeared to be a stimulation with 3.0 spawnings per female.

Table 3. Spawning behavior and success of goldfish held in H_2S treatment 234 days prior to spawning

H_2S (mg/l)	Number of spawnings	% survival	Number of females	Number of spawnings per female
Control	2	100	2	1.0
0.005	12	100	4	3.0
0.010	2	100	4	0.5
0.028	3	100	4	0.75

EFFECT OF HYDROGEN SULFIDE ON EGG HATCHING

Walleye, fathead minnow, bluegill, rainbow trout, brook trout, northern pike, and goldfish eggs were subjected to various levels of H_2S from 0.005 to 0.059 mg/l in various experiments to determine its effect on incubation (Table 4). The response was measured in terms of days to hatch, percentage hatch, and LC_{50}. In egg tests on walleye percentage hatch dropped from 86% of control at 0.012 mg/l to 9% at 0.159 mg/l, and hatching time was extended from 22 days in the control to 26-27 days at all levels of H_2S treatment. Survival of fathead minnow eggs incubated at 0.016 - 0.058 mg/l dropped from 87% at 0.016 mg/l to 17% at 0.058 mg/l. Hatching was extended from 6 days in the control to 8-9 days at the higher H_2S concentrations. Bluegill eggs tested at 0.011 - 0.035 mg/l survived at 63% of controls in the lowest concentration and 13% in the highest. Hatching time was short and did not vary significantly at the various levels. The hatch of brook trout eggs incubated at 0.005 - 0.011 mg/l H_2S varied from 67% at 0.005 mg/l to 50% at 0.011 mg/l. Brook trout eggs laid by adults held in the same test conditions during maturation and egg deposition had much poorer survival with 45% at 0.007 mg/l and no survival at 0.011 mg/l. Rainbow trout eggs incubated at 0.006 - 0.047 mg/l survived at the rate of 76% in the lowest to 4% in the highest concentration. Northern pike eggs incubated at 0.018 - 0.058 mg/l varied from 88% of controls at the lowest level of H_2S to 6% at the highest. Goldfish eggs held at 0.010 - 0.029 mg/l hatched at the rate of 95% in the lowest level and 12% in the highest.

Median tolerance limits (LC_{50}) were calculated for various time periods varying from 3 to 29 days in different species (Table 5). Among the species of fish eggs tested LC_{50} values varied from 0.016 mg/l at 3 days (72 h) in bluegills to greater than 0.054 mg/l for brook trout in 4 days (96 h). Rainbow trout eggs at 7 and 29 days had LC_{50} values of 0.022 and 0.015 mg/l H_2S, respectively. Of the 8 species tested goldfish and bluegill eggs were the most sensitive with 0.020 and 0.016 mg/l LC_{50} in 72 h. The most resistant were brook trout with greater than 0.054 mg/l and walleyes with 0.052 mg/l in 96 h.

Table 4. Percentage hatch and days to total hatch of fish eggs incubated at various concentrations of H_2S

Species	°C	H_2S (mg/1)	% of Control hatched	Days to hatch
Brook Trout	9	0.005	67	-
		0.007	50	-
		0.011	50	-
Rainbow Trout	12	0.006	76	29
		0.012	78	29
		0.024	55	29
		0.028	7	29
		0.047	4	29
Northern Pike	13	0.018	88	-
		0.028	64	-
		0.045	52	-
		0.058	6	-
Fathead Minnow	23	0.016	87	6
		0.026	83	8
		0.045	60	9
		0.058	17	8
Goldfish	22	0.010	95	3
		0.017	85	3
		0.029	12	3
Walleye	12	Control	100	22
		0.012	86	26
		0.037	77	27
		0.048	20	27
		0.059	9	26
Bluegill	23	0.011	63	-
		0.016	65	-
		0.024	10	-
		0.035	13	-

Table 5. Median tolerance limits (LC_{50}) as mg/l H_2S of eggs, fry, and juveniles of various species - duration days in parenthesis

Species	°C	Eggs	Sac fry	Swim-up fry	Juvenile
Brook Trout	9	0.054(4)	0.031(4)	--	--
	12	--	--	0.022(4)	--
	13	--	--	--	0.017(4)
Rainbow Trout	12	0.022(7)	--	0.013(5)	0.013(4)[1/]
	12	0.015(29)	--	0.006(20)	0.009(17)
Sucker	15	0.028(4)[3/]	0.020(4)[3/]	--	0.019(4)
Northern Pike[2/]	13	0.037(4)	0.026(4)	--	--
Fathead	24	0.035(4)	0.007(4)		0.016(4)
	24	0.035(6)	0.006(6)		0.013(7)
Goldfish	22	0.020(3)	0.025(3)	--	0.090(4)[4/]
Walleye	12	0.052(4)	0.013(2)[3/]	--	0.020(4)
Bluegill	22	0.016(3)	0.030(3)	0.009(3)	0.032(3)
	22	--	0.029(7)	0.008(7)	0.032(7)

1/ At 20 °C.
2/ Adelman and Smith (1970).
3/ Smith and Oseid (1972).
4/ Adelman and Smith (1972).

LENGTH OF FRY AT HATCHING

The length of normal fry of 5 species at time of hatch showed the retarding influence of incubation in H_2S (Table 6). Rainbow trout, sucker, northern pike, fathead minnow, and walleye were examined. Eggs were incubated at levels of H_2S varying from 0.012 to 0.065 mg/l in the various species. The greatest percentage reduction in fry length occurred in eggs with the longest incubation period. Rainbow trout were reduced from 13.7 mm in the control to 11.5 mm at 0.047 mg/l, or 16%. Walleye fry varied from 7.3 mm in the control to 6.8 mm at 0.048 mg/l, a reduction of 7%. At 0.065 mg/l sucker length was reduced 9.6%. The least reduction in length was 6% in northern pike and fathead minnows.

Table 6. Length of fry at hatch after incubation at various concentrations of H_2S

Species	°C	H_2S (mg/1)	Length (mm)	Days to hatch
Rainbow Trout	12	Control	13.7	29
		0.012	13.5	29
		0.028	13.1	29
		0.047	11.5	29
Sucker	15	Control	10.4	11
		0.015	10.3	14
		0.037	9.6	13
		0.065	9.4	13
Northern Pike	13	Control	8.2	-
		0.018	8.1	-
		0.028	8.1	-
		0.045	8.2	-
		0.058	8.0	-
Fathead	24	Control	4.7	4
		0.016	4.7	6
		0.026	4.8	8
		0.045	4.3	9
		0.058	4.4	8
Walleye	12	Control	7.3	22
		0.024	7.2	22
		0.037	7.0	27
		0.048	6.8	27

DEFORMITY OF FRY

A significant effect of H_2S on incubating eggs in addition to lowered survival was indicated by the percentage of deformed fry which occurred at higher concentrations. The sucker, northern pike, fathead minnow, goldfish, and walleye were observed (Table 7). Fathead minnows were the least deformed with 9% of control affected at 0.026 mg/l H_2S. Deformity was greatest in the sucker with 63.6% affected at 0.023 mg/l and 66.7% at 0.041 mg/l. Goldfish were also sensitive with 62% deformity at 0.029 mg/l. It is believed that all deformities noted were lethal and that the deformity rate in conjunction with the survival rate would eliminate essentially all fry survival at the higher test concentrations. Sucker egg survival was between 20% and 60% better at the lowest levels of testing than the controls although deformity rate was slightly greater. This test in conjunction with other observations suggested that in this species very low levels of H_2S appeared to improve survival. Walleye deformity varied from 0.5% at 0.012 mg/l to 54.2% at 0.059 mg/l H_2S.

Table 7. Percentage of fry deformed at hatch after egg incubation at various levels of H_2S

Species	°C	H_2S (mg/l)	Deformed (%)	Survival through hatch (%)
Sucker	13	Control	0.9	100
		0.006	1.9	158
		0.011	6.2	140
		0.023	63.6	12
		0.041	66.7	7
Northern Pike	13	Control	0.0	100
		0.018	4.0	88
		0.028	5.0	64
		0.045	14.0	52
		0.060	18.0	6
Fathead	24	Control	0.0	100
		0.016	0.0	87
		0.026	9.0	83
		0.058	0.0	17
Goldfish	22	Control	0.0	100
		0.010	0.0	100
		0.017	16.0	85
		0.029	62.0	12
Walleye	12	0.012	0.5	86
		0.037	3.5	77
		0.048	34.8	20
		0.059	54.2	9

EFFECT OF OXYGEN CONCENTRATION ON EGGS

The effect of dissolved oxygen concentration on egg response to H_2S was evaluated on the basis of LC_{50}, survival, and time to hatch in several species (Tables 8 and 9). Survival of walleye eggs hatched at 3 and 6 mg/l O_2, 12°C, and H_2S concentrations of 0.012 - 0.060 mg/l was increasingly poor as H_2S concentration increased. Hatching after incubation at 3 mg/l O_2 took 30 to 33 days in all concentrations of H_2S in contrast to 23 days for the control. At 6 mg/l O_2 hatch took 26 to 27 days in all H_2S levels except 0.024 mg/l with 22 days for the control. Survival to hatch was greater at the lower H_2S concentrations in 6 mg/l O_2 than in 3 mg/l O_2, but at the higher H_2S concentrations survival was not greatly different. Suckers exposed for 12 days prior to hatch at 3 and 6 mg/l O_2, 13°C, and H_2S concentrations of 0.006 - 0.049 mg/l took significantly longer to hatch at low oxygen in H_2S treatments, but survival rates at lowest levels of H_2S in both concentrations of oxygen were higher than controls. At the highest H_2S concentration survival was somewhat greater with low oxygen.

Table 8. Effect of various levels of oxygen and H_2S on survival of eggs and time to hatch of sucker and walleye[1/]

°C	O_2 (mg/l)	H_2S (mg/l)	Percentage hatch	Days to hatch
		SUCKER		
13	3	Control	100	13
		0.009	113	17
		0.016	93	13
		0.031	55	17
		0.049	9	17
13	6	Control	100	11
		0.006	154	12
		0.012	135	14
		0.028	8	15
		0.049	1	13
		WALLEYE		
12	3	Control	100	23
		0.013	62	33
		0.025	59	30
		0.039	21	32
		0.049	14	32
		0.060	7	-
12	6	Control	100	22
		0.012	86	26
		0.024	73	22
		0.037	77	27
		0.048	20	27
		0.059	9	26

1/ Sucker treated 12 days prior to hatch; walleye exposed 19-20 days and removed to fresh water. (Data from Smith and Oseid (1972).)

Values of LC_{50} with time exposures from 96 h to 19 days at various temperatures and dissolved oxygen levels were determined for rainbow trout, sucker, northern pike, and walleye (Table 9). There was a tendency for sucker eggs to be more resistant at low oxygen concentrations but little effect in the other species.

Table 9. Effect of dissolved oxygen and temperature on resistance of fish eggs to H_2S expressed as LC_{50} at various exposure times

Species	°C	O_2 (mg/l)	LC_{50}	
			96 h	(days (exposed)
Rainbow Trout	12	6	--	0.015 (29)
	15	6	0.049	--
Sucker[1/]	15	6	0.028	--
	13	6	--	0.019 (12)
	13	3	--	0.033 (12)
Northern Pike[2/]	13	6	0.037	--
	13	2	0.034	--
	13	2	--	0.024 (10)
	13	6	--	0.028 (10)
Walleye[1/]	15	6	0.080	--
	12	6	0.052	0.022 (10)
	15	4	0.066	0.066 (7)
	12	3	0.064	0.036 (19)

1/ Smith and Oseid (1972).
2/ Adelman and Smith (1970).

EFFECT ON FRY

The effect of H_2S on fry was more severe than on eggs at the higher concentrations tested and varied greatly among species (Table 10). Of the five species subjected to H_2S for 48 h walleye fry at 15°C were particularly sensitive. Northern pike at 13°C were much more resistant. Goldfish, suckers, and fathead minnows showed intermediate resistance. Brook trout sac fry subjected to 0.026, 0.029, 0.032, and 0.036 mg/l H_2S at 9°C after 96 h had a survival of 100, 50, 0.0, and 0.0%, respectively. In 0.026 mg/l H_2S 20% survived 192 h. At 0.029 mg/l H_2S 40% remained at 108 h and none at 116 h.

The LC_{50} of sac fry varied with time and species (Table 5). Walleye LC_{50} was 0.013 mg/l in 48 h, northern pike 0.026 mg/l in 96 h, brook trout 0.031 mg/l in 96 h, bluegill 0.030 mg/l at 72 h, fathead 0.007 mg/l at 96 h, sucker 0.020 mg/l at 96 h, and goldfish 0.025 mg/l at 72 h.

Table 10. Survival of fry in various concentrations of H_2S expressed as percentage of control

Species	°C	H_2S (mg/1)	Survival (%)
Brook Trout	9	0.026	100
(96 h)		0.029	50[1/]
		0.032	0
		0.036	0
Sucker	15	0.006	100
(48 h)		0.013	96
		0.028	16
		0.035	10
Northern Pike	13	0.008	100
(48 h)		0.024	98
		0.035	98
		0.046	34
		0.086	0
Fathead	24	0.004	95
(48 h)		0.010	20
		0.020	30
		0.044	30
		0.072	0
Goldfish	21	0.014	96
(48 h)		0.021	98
		0.027	38
		0.049	2
Walleye	15	0.006	97
(48 h)		0.009	6
		0.022	0

[1/] 0.0% survival after 116 h.

RESISTANCE OF VARIOUS STAGES OF LIFE HISTORY

To the extent that acute toxicity tests demonstrate relative vulnerability to H_2S it is apparent that a wide variation exists between life history stages of fish (Table 5). In all species studied, except bluegill and goldfish, eggs were less vulnerable than fry or juveniles. In the fathead minnow the fry were five times more vulnerable than the egg, and half as resistant as juveniles. Juveniles were more resistant than sac fry in walleye, goldfish, and fathead minnow, less resistant in brook trout, and similar in sucker and bluegill. The most vulnerable stage in bluegill was the swim-up fry which was half as resistant as the egg and one-third as resistant as the juvenile. In rainbow trout with long exposure the swim-up fry was half as resistant as the egg, but only slightly less resistant than the juvenile. When various stages are compared on the basis of H_2S concentrations with no demonstrable effect (Table 11), it becomes apparent that LC_{50} values may be misleading and that stages other than egg and fry may determine the well-being of the population with reference to toxicants. Spawning success or juvenile growth rate may be better indicators than survival of eggs or fry.

Table 11. No-effect levels of H_2S (mg/l) at any stage compared to egg and fry resistance

Species	No-effect levels				LC_{50} [2/] (days) juvenile
	Spawning	Growth [1/] juvenile	Fry survival	% hatch	
Brook Trout	< 0.005	< 0.0015	< 0.026	< 0.005	0.017 (4)
Rainbow Trout	--	--	--	< 0.006	0.009 (17)
Northern Pike	--	--	0.004	--	--
Fathead	0.0013	0.003	--	0.005	0.013 (7)
Goldfish	0.005	< 0.007	0.014	< 0.007	0.090 (4)
Walleye	--	0.004	< 0.009	< 0.012	0.020 (4)
Bluegill	0.007	0.002	< 0.008	< 0.011	0.037 (7)

1/ Smith, Oseid, and Adelman, unpublished data.
2/ Smith and Oseid (1971).

DISCUSSION

The low concentrations of hydrogen sulfide which are toxic to fish eggs and fry frequently occur in natural and polluted situations. The present studies indicate much lower tolerance levels than have been reported in earlier literature. It is suggested that the stress on early life history stages may limit fish production in areas or water strata suitable for older fish. Fish populations which deposit eggs on bottoms where there is organic decomposition or where water containing low concentrations of H_2S flows over eggs and fry resting on non-organic bottoms, may have their reproductive ability limited. Periodic conditions which bathe eggs or fry with water containing H_2S during vernal overturn may destroy or limit year classes in some lakes. The experimental work reported here indicates that very low-level treatment of adults with H_2S can substantially reduce fecundity, inhibit normal spawning activity, and increase mortality of spawners. The significance of our work is the demonstration that hydrogen sulfide in heretofore ignored concentrations, especially in the zone near the substrate-water interface where eggs are deposited, may be a controlling factor in maintenance of some fish populations.

SUMMARY

The eggs and fry of 8 species of North American freshwater fish were subjected to concentrations of hydrogen sulfide varying from 0.006 - 0.086 mg/l. In general survival was reduced, fry length at hatch shortened, and fry deformity increased with increasing H_2S concentration. Treatment of fish of some species prior to spawning reduced fecundity and egg survival. Low O_2 concentration increased resistance of fry but decreased resistance of eggs. The no-effect levels of H_2S were substantially lower than acutely toxic levels as described by median lethal concentrations. Levels of H_2S commonly found in natural systems may inhibit or prevent reproduction of some fish species. It is concluded that many habitats with unexplained absence of satisfactory fish reproduction should be examined for possible natural contamination with H_2S in potential spawning and nursery areas.

ACKNOWLEDGEMENTS

The authors wish to thank various members of our laboratory staff who have assisted with the work and especially Melvin Matson, Ira Adelman, and Gary Kimball for unpublished data contributed to this summary.

REFERENCES

Adelman, I.R., 1969. Survival and growth of northern pike (*Esox lucius* L.) in relation to water quality. Thesis, Univ. of Minnesota, St. Paul, Minn., 195 pp.

Adelman, I.R. and Smith, L.L., Jr., 1970. Effect of hydrogen sulfide on northern pike eggs and sac fry. Trans. Amer. Fish. Soc., 99, 501-509.

Adelman, I.R. and Smith, L.L., Jr., 1972. Toxicity of hydrogen sulfide to goldfish (*Carassius auratus*) as influenced by temperature, oxygen, and bioassay techniques. J. Fish. Res. Bd. Can., 29, 1309-1317.

Bell, D.A., Ramm, A.E. and Peterson, P.E., 1972. Effects of tidal flats on estuarine water quality. J. Water Pollut. Control Fed., 44, 541-556.

Colby, P.J. and Smith, L.L., Jr., 1967. Survival of walleye eggs and fry on paper fiber sludge deposits in Rainy River, Minnesota. Trans. Amer. Fish. Soc., 96, 278-296.

Ellis, M.M., 1937. Detection and measurement of stream pollution. U.S. Bur. Fish. Bull., 48, 365-437.

Mount, D.I. and Brungs, W.A., 1967. A simplified dosing apparatus for toxicological studies. Water Res., 1, 21-29.

Scidmore, W.J., 1956. An investigation of carbon dioxide, ammonia and hydrogen sulfide as factors contributing to fish kills in ice covered lakes. Minn. Dep. Conserv., Bur. Res. Plann., Invest. Rep., 177, 9 pp.

Smith, L.L., Jr. and Oseid, D.M., 1971. Toxic effects of hydrogen sulfide to juvenile fish and eggs. Proc. 25th Purdue Ind. Waste Conf., Eng. Ext. Ser. No. 137, 739-744.

Smith, L.L., Jr. and Oseid, D.M., 1972. Effect of hydrogen sulfide on fish eggs and fry. Water Res., 6, 711-720.

Ziebell, C.D., Pine, E., Mills, D. and Cunningham, R., 1970. Field toxicity studies and juvenile salmon distribution in Port Angeles Harbor, Washington. J. Water Pollut. Control Fed., 42, 229-236.

L.L. Smith, Jr., Department of Entomology, Fisheries and Wildlife, University of Minnesota, St. Paul, Minn. 55101 / USA

D.M. Oseid, Department of Entomology, Fisheries and Wildlife, University of Minnesota, St. Paul, Minn. 55101 / USA

Effect of Parental Temperature Experience on Thermal Tolerance of Eggs of *Menidia audens*

C. Hubbs and C. Bryan

INTRODUCTION

The geometric increases in electrical energy consumption may produce significant quantities of heated waste water that may cause major ecosystem alterations. This poses an especially critical problem for estuaries that are major nurseries. Although all life history stages are temperature-sensitive the majority of fishes are most stenothermal during early developmental stages (Hubbs, 1966). Because most fish incubation tests use a set of fertilized eggs derived from parents maintained at a single temperature regime, this experiment was designed to determine the impact of parental acclimatization on F_1 thermal tolerance.

MATERIAL AND METHODS

We selected *Menidia audens* as the experimental animal because: 1. the eggs are easily incubated, 2. ripe adults are plentiful in the vicinity of the University of Oklahoma Biological Station and 3. preliminary studies suggest that parental maintenance temperatures affected survival of F_1's (Hubbs, Sharp and Schneider, 1971). The major handicap is the fragility and consequent high mortality of adults. *Menidia audens* is one of two freshwater representatives of a brackish water-salt water genus. It survives salinities of 25°/oo+ at all life history stages (Hubbs et al., 1971).

The stocks were obtained from a small embayment (Mayfield Flat) of Lake Texoma adjacent to the University of Oklahoma Biological Station. After capture, the females were checked for ripe ova and then transferred to stock aquaria (190 l) and maintained for 1 to 3 days at 15 (cold), 24 (warm), and 31°C (hot). Except for maintenance of adults the standard experimental conditions followed the methods of Hubbs et al. (1971) that is: isolation of individual eggs in screw-capped culture tubes within 1 h of mixing of gametes, placing the culture tubes in an aluminum temperature gradient block, and daily examination of developmental stages. Developmental rates paralleled those previously reported and are therefore not repeated here. The sublethal incubation temperatures used here, 32-34°C and 12-18°C, cover temperatures between the developmental optima of 17-33.5°C reported by Hubbs et al. (1971) and approach or exceed lethal conditions (above 34.2°C and below 13.2°C). In order to determine the impact of parenthal acclimatization temperature, pregastrulation mortality must be included in the data. The number of eggs isolated was therefore used as the sample size, not the number of gastrulated eggs as has been done previously. Two sets of eggs did not develop at any temperature and are excluded from the calculations.

When it was discovered that the primary impact of parental incubation temperature was on gastrulation success, a series of two sets of tests was run to ascertain whether the impact was merely temperature shock. The first was to place eggs from warm-maintained parents in sublethal cold for 24 h and completion of incubation at 18-19°C; the second was to maintain incubated eggs from warm- and cold-maintained parents for 2 to 3 days at 16°C, then place them at 12-13°C for 3 to 6 days and complete the incubation at 18-19°C.

RESULTS

Constant Incubation Temperature. Females isolated for 24 or more h at hot temperatures (31°C) produced only inviable eggs; the males had free flowing milt that could successfully fertilize eggs from females in other stock tanks. Apparently the mid-July termination of spawning results from warm lake temperatures. The low frequency of ripe adults during an extremely warm week in late June and high frequency of ripe adults in mid-July 1972 is in accord with these data.

Females isolated for more than 48 h at 24°C failed to produce viable ova, probably reflecting nutritional deficiencies in our stock tanks. Females isolated for more than 48 h at 15°C produced ova that gastrula- ed at rates approximating those from females held 24 h at 15°C when in- cubating at 15-18°C, but less than 1% gastrulated when incubated at 32-34°C (and both gastrulated eggs were at 32-32.5°C).

The one-day parental acclimatization temperature has contradictory impact dependent upon the developmental stage studied (Table 1). Althoug the percentages to gastrulate are positively correlated with acclimati- zation temperature, each subsequent survival percentage is inversely correlated. Both of the individual gastrulation comparisons have probability values of marginal significance (0.04 and 0.06) but the combined value is 0.0016. Each of the hot temperature survival comparisons are highly significant, as is the eye pigmentation contrast at cold temperature.

Table 1. Average survival percentages of eggs from warm and cold acclimatized parents. The survival percentage was obtained for each of 21 incubation temperatures in each experimental condition and these were averaged

Incubation temperature	32-34°C			12-18°C		
Parental acclimatization temperature	24°C	15°C		24°C	15°C	
n	696	409	p =	844	734	p =
fert. to gastr.	79	74	0.06	17	21	0.04
gastr. to eye pigm.	79	87	0.003	66	49	0.003
eye pigm. to hatch	64	74	0.007	69	57	0.13
hatch to yolk depletion	62	78	0.0002	82	69	0.65

The values seem to be maintained over much of the temperature range studied (Fig. 1). The only potential exceptions are at the hottest temperatures at which cold-acclimatized eggs tend to gastrulate more frequently than warm-acclimatized eggs, and the low survival of larvae from warm-acclimatized eggs at temperatures below 15°C. The latter may reflect the infrequent gastrulation, i.e. if an egg is unable to gastrulate it is unlikely to produce a successful larva. At most experimental temperatures, the eggs from warm-acclimatized parents were relatively more successful at cold temperatures and those from cold-acclimatized parents produced more larvae at warm temperatures.

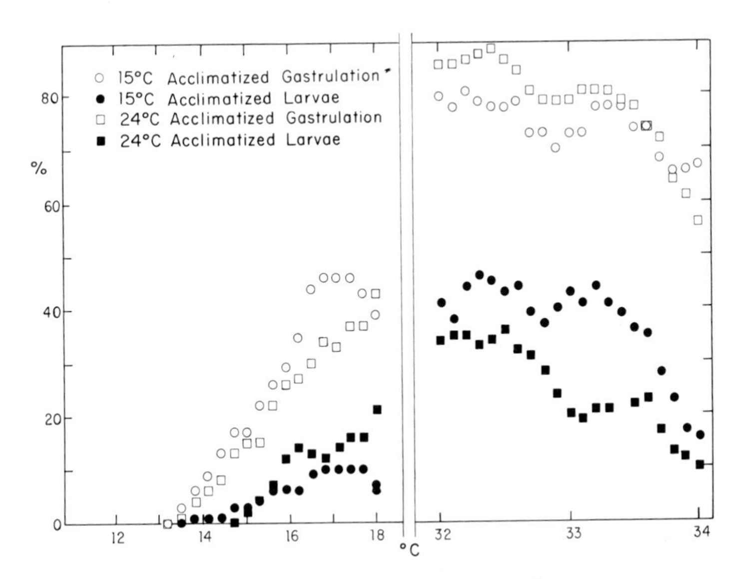

Fig. 1. Percentages of innoculated eggs that gastrulated or successfully depleted yolk reserves at sublethal warm and cold temperatures. Parental stocks were acclimatized 24 h at 15 and 24°C before the gametes were removed

Changing Incubation Temperatures. A correlation of acclimatization temperature and incubation success may result exclusively from thermal shock. An organism transferred from 24 to 13°C may be more adversely affected by shock than one transferred from 15 to 13°C. If shock were involved, a second thermal shock would likely exacerbate mortality; yet if simple thermal stress were involved, a return to standard survival temperatures would hinder mortality. Those warm-acclimatized eggs, stressed 1 day and incubated at optimal temperatures until yolk depletion, consistently had higher survival than comparable eggs that completed their incubation at the stressed temperature (Fig. 2).

The reversal of the acclimatization effect after gastrulation may result from mortality of the less genetically adapted eggs during previous stages. If so, a change in the degree of stress should enhance the survival of cold-acclimatized eggs contrasted to warm-acclimatized eggs. To test for this hypothesis, eggs from 15 and 24°C acclimatized stocks (1 day) were cultured at 16°C until neural groove formation and then stressed for 3 days at 12-14°C; then incubation was completed at 19-20°C. As all experimental eggs were in neural groove, gastrulation mortality was excluded from consideration. At each stage the percentage of innoculated cold-acclimatized eggs to achieve any developmental stage was greater than that of comparable warm-acclimatized eggs: pigmented eye 43% vs 11%, $p = 0.005$; hatch 30% vs 7%, $p = 0.001$; and normal larvae 17% vs 3%, $p = 0.001$. Clearly the primary problem is in the initial stress as the pigmentation-hatch (62% vs 70% and hatching-normal larval (40% vs 27%) fractions are more nearly equivalent and do not approach statistical significance. Additional exposure to cold stress (6 vs 3 days) prior to return to 19-20°C cause the survival of cold-acclimatized eggs equivalent to those of warm-acclimatized eggs stressed 3 days: 10% vs 11%, 4% vs 7%, and 2% vs 3% survival to the same 3 stages.

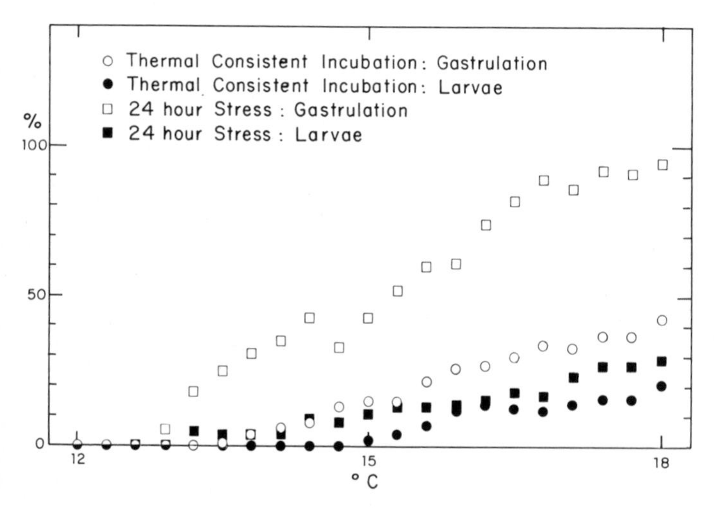

Fig. 2. Percentages of innoculated eggs that gastrulated or successfully depleted yolk reserves at sublethal cold temperatures. All parental stocks were incubated 24 h at 24°C. The 24 h stress experiments were incubated at 19°C after 24 h exposure

DISCUSSION

Parental acclimatization temperature has a significant impact on the thermal limits of successful incubation. This impact is proportional to the time of parental exposure, with 3 days parental maintenance having much more impact than 1 day of exposure. Short term acclimatization has an effect during initial stress, but the impact lessens or is reversed during later ontogeny. It was not possible to make similar observations on zygotes from parents with a prolonged acclimatization as the initial exposure killed so many of the stressed embryos. Thermal shock seems not to be the cause for those observations because subsequent equivalent shocks increased survival and Hagen (1964) showed that shock susceptibility was inversely correlated with age.

It is not surprising that parental maintenance temperatures have an impact on survival of offspring, as eggs are likely to be exposed to environmental temperatures approximating those of their parents.

Furthermore, the longer the parental exposure, the more likely temperature will occur in that region throughout incubation. If the sublethal exposure is short, those thermal conditions are more likely to parallel parental exposure in the next time interval (gastrulation) than in one after considerable delays (post-gastrulation). Each is, therefore, in accord with "good" reproductive strategy.

Future studies on incubation temperatures should record parental acclimatization and if possible use eggs from parents maintained at more than one temperature to ensure that the entire developmental range is reported.

SUMMARY

Thermal tolerance of eggs of the silverside, *Menidia audens*, is shown to be influenced by prior temperature experience of the parent. A 9°C difference in parental acclimation temperature resulted in an upward shift in thermal tolerance of eggs from the higher of two parental acclimation groups. This shift, measured in terms of the thermal range at which gastrulation would occur, was of the order of 1 deg C. However, a reversal occurred when the thermal stress, used to define the range, was maintained after gastrulation. That is, the thermal tolerance range of eggs from the higher parental temperature acclimation was displaced downward by about 2 deg C, in comparison with the range obtained for eggs from the lower parental acclimation group. This phenomenon is not likely to involve temperature shock; eggs exposed to sublethal thermal stresses and returned to optimal temperatures showed high survival, despite being subjected to two thermal shocks in such transfers.

ACKNOWLEDGEMENTS

This study was done at the University of Oklahoma Biological Station with support of the Oklahoma Biological Survey. We are grateful for the support of the directors, Loren P. Hill and Paul S. Risser, respectively.

REFERENCES

Hagen, D.W., 1964. Evidence of adaptation to environmental temperatures in three species of *Gambusia* (Poeciliidae). Southwestern Naturalist, 9, 6-19.

Hubbs, Clark, 1966. Fertilization, initiation of cleavage and developmental temperature tolerance of the cottid fish, *Clinocottus analis*. Copeia, 1966, 29-42.

Hubbs, Clark, Sharp, H.B. and Schneider, J.F., 1971. Developmental rates of *Menidia audens* with notes on salt tolerance. Trans. Amer. Fish. Soc., 100, 603-610.

C. Hubbs
The University of Texas at Austin
Department of Zoology
Austin, Tex. 78712 / USA

C. Bryan
Baylor University
Waco, Tex. / USA

Effects of Constant and Rising Temperatures on Survival and Developmental Rates of Embryonic and Larval Yellow Perch, *Perca flavescens* (Mitchill)

K. E. F. Hokanson and Ch. F. Kleiner

INTRODUCTION

The yellow perch, *Perca flavescens* (Mitchill), is an important sport, commercial, and forage fish species indigenous to the eastern part of the United States and Canada. Available data on its thermal requirements consist of thermal tolerance and preference of juveniles. Hart (1947) reported that the ultimate upper incipient lethal temperature of yellow perch was 29.7°C when acclimated to 25°C. The final preferendum of young yellow perch was 24.2°C; older perch preferred 21°C (Ferguson, 1958). Mansueti (1964) described the egg strands of the yellow perch and the morphology and anatomy of embryonic, larval, and juvenile stages and noted the coincidence during development with the Eurasian yellow perch, *Perca fluviatilis* (Linnaeus). Swift (1965) published a table of percentage mortality and average time to hatch for *P. fluviatilis* at various temperatures. Kokurewicz (1969) also described the effect of constant temperature on developmental rate of *P. fluviatilis* up to the time of hatch.

The study reported here was conducted in the spring of 1971. It is part of the continuing program at the National Water Quality Laboratory to determine thermal requirements of aquatic organisms and to develop laboratory culture methods for toxicant bioassays. This study defines the thermal tolerance limits and optimal culture conditions of embryonic and larval stages, describes the effect of temperature on development and hatchability, and compares the physiological requirements of the North American and Eurasian yellow perch.

METHODS

Apparatus

The embryo-survival and developmental rate experiments were conducted in the apparatus described in detail by McCormick and Syrett (1970). One modification was the arrangement of three embryo- and larvae-rearing units, each 15 x 10 x 10 cm deep, in series (Fig. 1). A manifold provided triplicate series for a total of 9 incubation units at each temperature. Water flowed continuously through each unit at an average rate of 216 ml/min (SD = 21, range 114-354).

Several types of experimental chambers were required in this study; these allowed flow-through conditions, minimized handling of delicate life stages, and permitted observing development and vitality of embryonic and larval stages under a dissecting microscope. Eggs were incubated in chamber 1A which was placed inside 1B (Figs 1 and 2). Newly-hatched larvae fell through the coarse screen on the bottom

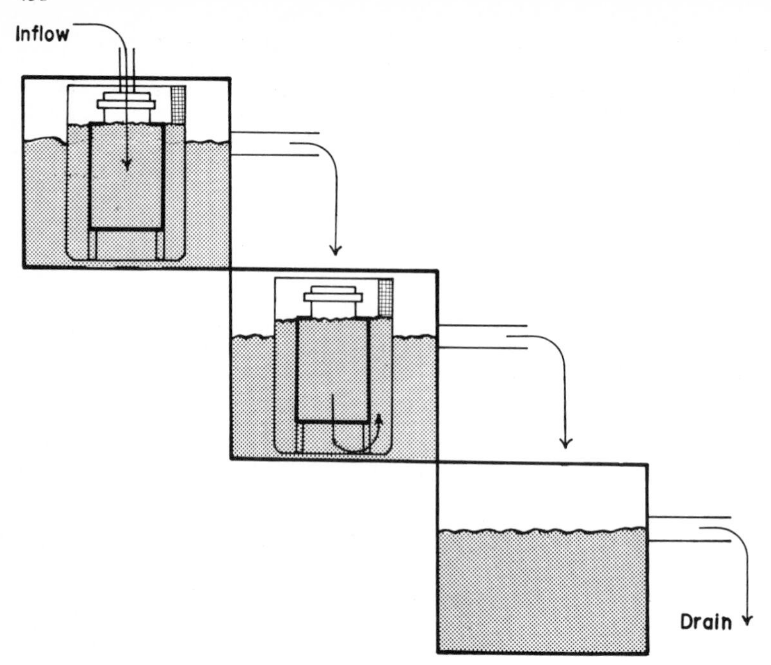

Fig. 1. Side view of incubation units showing their stepwise arrangement and flow pattern through egg incubation chambers

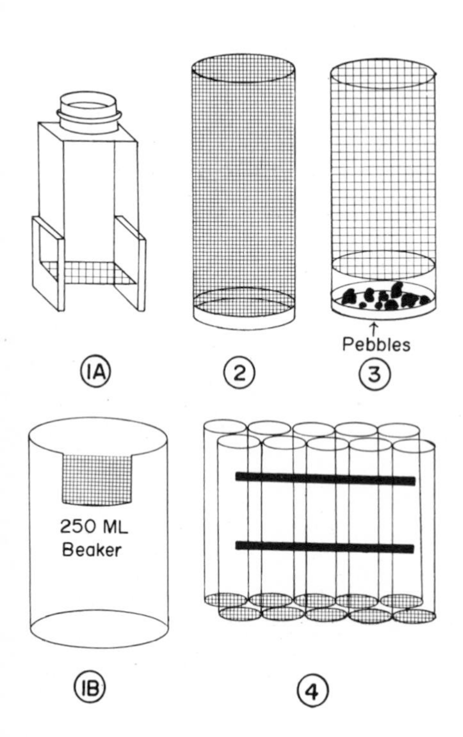

Fig. 2. Experimental chambers for retaining yellow perch embryos (coarse screen) and larvae (fine screen). 1A embryo incubation, 1B outer chamber for 1A to retain larvae after hatching, 2 for newly hatched larvae, 3 for early embryo incubation, 4 for individual larval experiments (horizontal bar holds "cells" together)

(8 meshes/cm) of 1A and were retained by chamber 1B (Figs 1 and 2); they were transferred to chamber 2 until swim-up or 50% mortality of unfed larvae. Embryos incubated at 12°C to formation of the neural keel were retained in chamber 3 until they were transferred to test temperatures in chambers 1A and 1B. Pebbles in chamber 3 prevented embryos from settling in stagnant areas on the bottom of the container. Chamber 4 permitted observations on survival and activity of individual larvae. Coarse screens (8-12 meshes/cm) were used to increase water circulation past embryos (chambers 1A and 3, Fig. 2). Fine screens (16-20 meshes/cm) were used to retain free-swimming larvae (chambers 1B, 2, 4).

The temperature-controlling apparatus usually maintained the standard error of the mean (SE) temperature within ± 0.06°C and the range for each temperature within 0.9°C. The maximum SE was ± 0.12°C at 25.4°C and the maximum range was 1.5°C at 3.1°C. Temperature increased by 0.1 - 0.2°C from the upper to lower incubation units below 13°C. Daily temperatures of each unit were read to the nearest 0.1°C with a mercury thermometer and ranges were observed from recording thermographs.

Temperature-controlled untreated water from Lake Superior was vigorously aerated in the headbox to maintain dissolved gases near air saturation. The mean dissolved oxygen concentration was maintained at 97% (range 94-101%) air saturation at each temperature determined from biweekly samples. The range of pH values was 7.7 - 7.9. Other water quality characteristics of this supply have been described by Hokanson et al. (1973).

Experimental Procedure

Adult yellow perch were collected from Little Cut Foot Sioux Lake, Itasca County, Minnesota. Two lots of fish were collected and transported to the laboratory, one on 29 April and the other on 5 May 1971. The temperature of the water at the time of collection was 6° and 11°C, respectively. Fish were held at 12.0° - 12.6°C in the laboratory for 16 and 5 h, respectively, to increase egg fertility. Eggs were fertilized at 12°C in a constant temperature room. Females were killed, and egg strands were excised and fertilized by the dry method. Each egg strand was cut into subsamples of 93-293 eggs and distributed to 24 test chambers, each randomly assigned a treatment. One pair of fish from the first lot was used in the study of developmental rate, and 2 pairs from the second lot were used in the study of embryo survival through early larval stages.

Embryo-Survival Experiment. 24 treatments in 3 groups (Table 1) were used to determine the effects of constant or rising temperatures on survival of yellow perch embryos of known age to the larval swim-up stage. In the first group, eggs were exposed to temperature extremes soon after fertilization to include the most sensitive developmental stage (see Hokanson et al., 1973). Eggs were exposed to 9 constant temperatures from 3° to 22°C (2-3 deg C intervals). Egg subsamples were adjusted to test temperatures within 1 h of fertilization before they were placed under continuous-flow conditions. In the second group the survival of older embryos was observed. Egg subsamples were incubated at 12.0°C until the neural keel formed (63.5 h) before they were adjusted to 11 constant temperature levels from 3° to 26°C over a 1-h period. In the third group egg subsamples were exposed to 4 separate rising temperature regimes, since the thermal tolerance of

Table 1. Survival and hatchability of yellow perch embryos exposed to constant or rising temperatures - replicates combined. Groups 1 and 3 - fertilization to larval swim-up stage; group 2 - fertilization to neural keel at 12.0°C, neural keel to larval swim-up stage at test temperatures

Treatment group	Mean temperature (°C)	Total hatch (%)	Normal hatch (%)	Dead and abnormal hatch (%)	Swim-up larvae (%)	Duration of hatching (Days)
1	3.1	0	0	0	0	30.0[a]
	5.0	52	26	26	0	20.0
	7.1	75	44	31	24.5	13.5
	10.1	83.5	68	15.5	50	12.5
	13.1	84.5	81	3.5	75	8.5
	16.1	87	83.5	3.5	72.5	6.5
	18.2	75.5	67	8.5	53.5	6.0
	20.0	58	48	10	30.5	6.5
	22.1	7.5	1.0	6.5	0	2.5
2	3.3	61	2	59	0	30.5
	5.2	67	20	47	0.4	20.5
	7.2	74.5	50	24.5	28	23.5
	10.2	80.5	56.5	24	40	11.0
	13.1	81.5	77.5	4	70.5	8.0
	16.1	81.5	79.5	2	73.5	4.0
	18.2	68	65	3	61.5	2.0
	19.9	93.5	90.5	3	79.5	5.0
	22.1	81	69	12	58.5	5.0
	23.8	67.5	23.5	44	14.5	4.5
	25.4	24.5	0	24.5	0	2.5
3	4.9° + 0.5°/Day	80.5	76	4.5	66	2.0
	4.9° + 1.0°/Day	80.5	77.5	3	72	3.5
	6.9° + 1.0°/Day	84.5	82.5	2	71.5	3.0
	10.0° + 1.0°/Day	88	85.5	2.5	85	2.0

[a] Developmental rate exposure at 3.1 °C.

successive developmental stages is known to increase for some species (Hokanson et al., 1973). Eggs were incubated at 4.9°, 6.9°, and 10.0°C for 24 h after fertilization; then they were increased by 2°-3°C at an average rate of rise of 0.5° or 1.0°C/day. Two egg strands provided common replicates for all groups to facilitate comparison of treatment effects.

Three measured parameters used for this study were total hatch, normal hatch, and swim-up larvae (attached below surface film) expressed as percentages of the total number of eggs in each subsample. The numbers of normal, dead, and abnormal yolk-sac larvae were recorded daily at hatching. Total hatch is an expression of embryo survival to hatch regardless of the condition of the newly hatched yolk-sac larvae. Normal hatch excludes dead and abnormal yolk-sac larvae that hatch. The percentage of swim-up larvae gives a measure of the vitality of newly-hatched larvae. Larvae were retained until the swim-up stage or until death. Three endpoints are used in this paper to describe yellow perch responses to temperature - optimum, optimum range, and median tolerance limit (TL50); the distinction between egg and embryo has been defined by Hokanson et al. (1973). The maximum total hatch was adjusted to 100% for each egg strand, and the percentages of normal hatch and swim-up larvae were normalized by the formula of Abbott (1925) before derivation of the TL50 limits. The arcsin $\sqrt{\text{percentage}}$ transformation was used for the 2-way analysis of variance tests of treatment differences (Steel and Torrie, 1960).

Embryonic and Larval Developmental Rates. This experiment was designed to describe the median time to attain various morphological stages and physiological events in the early life stages of yellow perch. Subsamples from one egg strand were adjusted to 8 constant temperatures from 3.1° to 19.7°C within 1 h of fertilization. The following 7 stages and events, which could be readily recognized in living embryos and larvae, were observed:

1. Neural keel; yolk plug large.

2. Heart beat.

3. Retinal pigmentation; eye appears totally black, including the lens of the eye.

4. Branchial respiration; mouth movements synchronized with opercula.

5. Mass hatch; 50% of total hatch.

6. Swim-up larvae; 50% of normal hatch attached to surface film or free-swimming.

7. Mortality of unfed larvae; 50% of normal hatch.

Observations were made twice a day and recorded to the nearest 0.5 day. Total lengths of normal larvae at mass hatch were measured with an ocular micrometer to the nearest 0.1 mm.

The survival of newly-hatched larvae was observed at 12.0°C after mass hatch at 5.4°, 7.3°, and 10.4°C. Twenty normal larvae were adjusted to 12.0°C at 2-3 deg C/day, and individual survival and activity were observed.

RESULTS

Embryo Survival

Different proportions of yellow perch embryos developed in the range of 3.3° to 25.4°C, depending on their age and thermal history (Fig. 3). The highest mean total hatch was 93.5% at 19.9°C in the second treatment group; therefore, an average 6.5% of the eggs in all subsamples were assumed to be infertile. The remaining proportion that failed to hatch were assumed to be dead embryos. The proportions of dead and abnormal larvae at hatch represent the differences between the percentage total hatch and normal hatch values. Common abnormalities included spinal curvatures and edema as reported by Kokurewicz (1969). Percentage dead larvae represents the differences between the normal hatch and swim-up larvae values. Differences in temperature response between egg strands were not significant ($P > 0.10$, F-test); therefore replicates were combined. The effects of temperature on the response of each treatment group are described below.

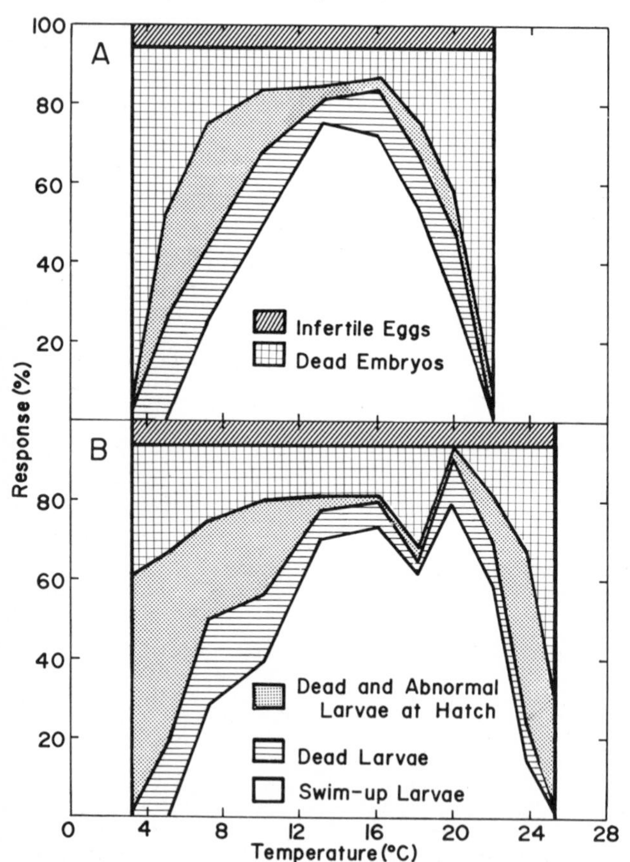

Fig. 3. Effect of constant incubation temperatures on percentage total hatch, normal hatch, and swim-up larvae of known aged yellow perch embryos-replicates combined. (A) Embryos incubated at test temperatures from fertilization to larval swim-up stage; (B) embryos incubated at 12.0°C until neural keel formed, then at test temperatures to larval swim-up stage. Fate of eggs shown as percentage differences

Group 1. Hatching occurred over a narrower range of temperatures when embryos were exposed to constant temperatures soon after fertilization (Table 1, Fig. 3A). The maximum mean total hatch was 87% at 16.1°C. Only 3.5% of these were dead or abnormal at hatch. The incidence of dead or abnormal larvae increased at temperature extremes. Greater than 67% normal hatch occurred in the optimum range of 10.1° to 18.2°C. Variations in normal hatch within the optimum range are considered similar to the optimum response (Table 2). Hatching occurred in the range of 5.0° to 22.1°C, but did not occur at 3.1°C. Only 1% normal hatch occurred at 22.1°C. The duration of the hatching period was about 6 days at 16°-20°C and 20 days at 5°C. The lower and upper TL50 define the range of temperatures for which survival is equal to or greater than 50% of the optimum response, and these limits were 6.8° and 19.9°C for normal hatch (Table 2).

A maximum of 75% of the eggs survived to the larval swim-up stage at 13.1°C (Table 1). More than 53% survived to the swim-up stage in the optimum range of 13.1° to 18.2°C. A relatively greater proportion of larvae that hatched at low temperatures failed to reach the swim-up stage than at high temperatures. Larvae were inactive at 5.0°C and all died. Survival of swim-up larvae at 22.1°C was limited by embryo survival. The lower and upper TL50 for embryo survival to the swim-up stage were 9.8° and 18.8°C (Table 2).

Table 2. Median tolerance limits (TL50) and optimum range of yellow perch embryos incubated at a series of constant temperatures

Treatment group	Response	Endpoint TL50 ($x \pm SE$) Lower	Upper	Optimum range (°C)[a]
1[b]	Total hatch	4.8 ± 0.20	20.5 ± 0.20	-
	Normal hatch	6.8 ± 0	19.9 ± 0.05	10.1 - 18.2
	Swim-up larvae	9.8 ± 0.80	18.8 ± 0.25	13.1 - 18.2
2[c]	Total hatch	<3.3	24.6 ± 0.05	-
	Normal hatch	7.0 ± 0.05	22.9 ± 0.20	13.1 - 18.2
	Swim-up larvae	9.3 ± 0.05	22.5 ± 0.10	13.1 - 22.1

[a]Range of temperatures over which the response was not significantly different from the highest value (P = 0.05, Tuckey's multiple range test).

[b]Embryos reared from fertilization to larval swim-up stage.

[c]Embryos incubated at 12.0°C from fertilization to formation of neural keel, then exposed to test temperatures to swim-up stage.

Group 2. Hatching occurred over a wider range of temperatures when embryos were exposed to temperature extremes at an older stage of development (Table 1, Fig. 3B). The maximum mean total hatch for this group was 93.5% at 19.9°C. Relatively greater proportions of embryos survived to hatch at the temperature extremes from 3.3° to 25.4°C, but many were dead or abnormal at hatch. Only 2% normal hatch occurred at 3.3°C and none at 25.4°C. A total of 65-79.5% normal hatch occurred in the optimum range of 13.1° to 18.2°C. The duration of the hatching period was only 2.0 - 5.0 days above 16°C and required up to 30.5 days at 3.3°C.

Similar proportions of embryos survived to the swim-up stage below 18°C as in the first treatment group, but more survived at higher temperatures. A maximum of 79.5% of the eggs survived to the swim-up stage at 19.9°C. Greater than 58.5% survived in the optimum range from 13.1° to 22.1°C. The lower and upper TL50 for survival of older embryos to the larval swim-up stage were 9.3° and 22.5°C (Table 2).

Group 3. The responses to all rising temperature regimes were within the optimum range reported for constant temperature groups. A total of 76 - 85.5% normal hatch and 66-85% swim-up larvae were produced (Table 1). Incidence of dead and abnormal larvae were also low (2-4.5% for this treatment group. Best results were obtained with an initial temperature of 10°C increasing by 1.0°C/day.

Hatching was generally completed in 2 - 3.5 days for rising temperatur regimes (Table 1). Mass hatch was completed in 21 days at 16.1°C for the 4.9° + 0.5°C/day treatment and in 10 days at 20.0°C for the 10.0° 1.0°C/day treatment. Larval swim-up was completed in 25 days at 16.2°C and in 14 days at 24.3°C for these respective treatments.

Embryonic and Larval Developmental Rates

The median time required for morphological differentiation of the embryo (stages 1-4) decreased exponentially with increasing temperatur (Fig. 4). More than 80% of the eggs developed to the neural keel stage at 3.2°C, but most perished by the time the heart beat and circulation began. Median time to hatch ranged from 6 days at 19.7°C to 51 days at 5.4°C. Mass hatch occurred within a day after the start of branchia respiration above 7.3°C. At lower temperature levels hatching was premature and occurred before the heart beat started at 3.2°C. Larval swi up occurred on the day of hatch above 13.1°C and 1-2 days later at low temperatures. None reached the swim-up stage at 5.3°C. Feeding trials at higher temperatures suggested that feeding commenced once the larva became free-swimming. Activity of larvae declined rapidly in the absen of food even though 50% survived for 15 days at 19.8°C and 65 days at 5.3°C. The median period between swim-up and mortality of unfed larvae

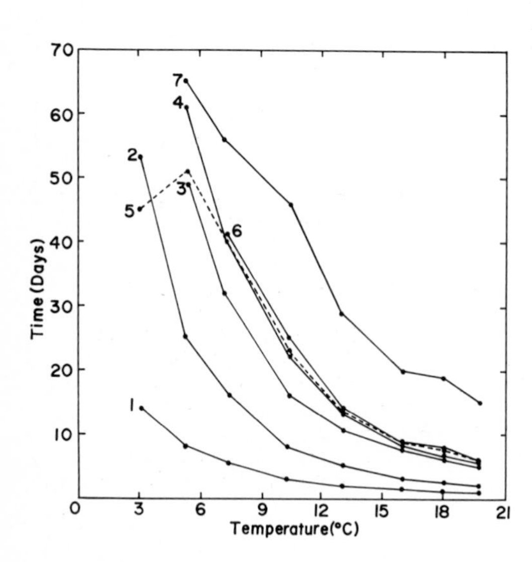

Fig. 4. Relationship between temperature and median time of development of yellow perch embryos and larvae. (1) Neural keel; (2) heart beat; (3) retinal pigmentation; (4) branchial respiration; (5) mass hatch; (6) swim-up larvae; (7) unfed larvae mortality

was 9 days at 19.8°C and 21 days at 10.5°C. At lower temperatures this period was shortened, and death occurred only 4 days after branchial respiration began at 5.3°C.

Larvae that hatch at low incubation temperatures have an increased chance for survival when transferred to a higher temperature. A total of 25%, 85%, and 90% reached the swim-up stage at 12°C after hatching at 5.4°, 7.3°, and 10.4°C, respectively.

P. flavescens vs *P. fluviatilis*

Thermal requirements for total hatch and time to mass hatch of *P. flavescens* and *P. fluviatilis* are shown in Fig. 5. The lower and upper TL50 for total hatch of *P. flavescens* were 4.8° and 20.5°C (Table 2). The calculated lower and upper TL50 for *P. fluviatilis*, from the data of Swift (1965), were 6.1° and 20.2°C. The median time to hatch was nearly identical for both species above 13°C. At lower temperatures *P. fluviatilis* tended to hatch about 3-6 days sooner.

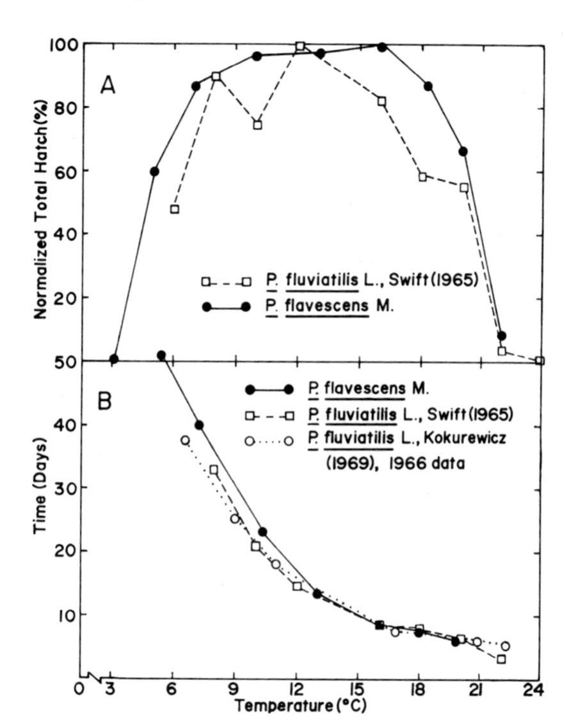

Fig. 5. Effects of constant incubation temperatures on *Perca flavescens* and *P. fluviatilis*. (A) Relative effects on total hatch; (B) time to mass hatch

The total length of *P. flavescens* larvae at mass hatch ranged from 4.7 to 6.6 mm. The mean lengths were maximal (6.1 - 6.3 mm) at 7.1° - 13.1°C, intermediate (5.6 - 5.8 mm) at 16.1° - 20.0°C, and minimal (5.1 - 5.2 mm) at 5.0° and 22.1°C. Newly-hatched larvae at 10°C appeared elongated with smaller yolk sacs and well-developed jaws, whereas those at higher temperatures were shorter with larger yolk sacs. Most larvae were premature when incubated below 7°C. This description of size and condition of newly-hatched yellow perch at various temperatures is identical to that of *P. fluviatilis* (Kokurewicz, 1969).

DISCUSSION

Thermal tolerance of successive embryonic and larval stages of yellow perch increased with morphological differentiation. Survival of early embryonic stages (before neural keel) was generally favored at 3.1° to 19.9°C. The lower and upper TL50 for normal hatch were 7.0° and 22.9°C for older embryonic stages (after neural keel). Newly-hatched larvae required temperature in excess of 9.8°C for normal activity and survival. Larval survival at high temperatures was limited only by embryo survival in this study. The optimum range for sustained growth and survival of feeding larval perch needs to be established. The optimum range (13.1° - 18.2°C) for embryo survival from fertilization to the larval swim-up stage was quite narrow at constant incubation temperatures which affected both embryo and larval requirements. In contrast, embryos exposed to rising incubation temperatures (range 4.9° - 24.3°C) had optimum survival to the larval swim-up stage.

The optimal thermal regime for culture of yellow perch was an initial exposure of fertilized eggs to 10°C and exposure to 20°C before hatching. This procedure was used successfully by Hale and Carlson (1972) in laboratory culture of larval perch. Larvae must be fed shortly after the larval swim-up stage. Larval survival thereafter will depend on the amount of food especially at elevated temperatures (Kudrinskaya, 1970). The period between larval swim-up and mortality of unfed larvae is progressively shortened at higher temperatures. The "critical period for endogenous feeding of larval perch would be even shorter since starvation of larval fishes is evident before death occurs (Toetz, 1966).

Morphological differentiation of the embryo and hatching are independently influenced by incubation temperatures. Hatching is generally caused by the joint action of temperature on secretion of hatching enzymes and on enzyme and embryonic activity (Hayes, 1942). Hatching of yellow perch follows shortly after commencement of branchial respiration above 7°C, and it is morphologically premature at lower temperatures. Ishida (1944) reports in *Oryzias* that the initiation of respiratory movements appears to rupture the hatching glands which line the pharynx. At temperatures above 13°C hatching of yellow perch is probably favored by increased enzyme and embryonic activity. Hatching was especially favored by rising temperature regimes that produce more viable larvae, shorter hatching periods, and fewer abnormal larvae than constant temperatures. Lower temperatures often produce longer larvae with somewhat smaller yolk sacs at hatching (Blaxter, 1969). This phenomenon may be related to lower efficiencies of conversion from yolk to embryonic tissue in salmon at low temperatures (Hayes and Pelluet, 1945).

Premature hatch at low incubation temperatures has been reported in other species and is discussed by Kokurewicz (1969). The cause of premature hatch below 7°C is not clear since embryonic and enzyme activity would be minimal. It is doubtful that hatching glands were active by the time of hatch at 3°C. Mansueti (1964) observed that the yellow perch egg case tended to become thin as the embryo grew. A few days before hatching the egg case began to soften and take on a ragged appearance. The variability of the hatching period also increased at lower temperatures. In preliminary experiments we observed mass hatch of yellow perch 23 days sooner at 6°C and 16 days sooner at 8°C when egg subsamples were handled (pipette) daily for observations of development.

The apparent earlier hatching of *P. fluviatilis* at lower incubation temperatures (Fig. 5) may be caused by different degrees of handling of the egg strands. We designed incubation chambers (Figs 1 and 2) to minimize handling of *P. flavescens* eggs. Kokurewicz (1969) noted considerable variation among his experiments on *P. fluviatilis* in size and time to mass hatch at low incubation temperatures, but he did not mention the amount of handling his eggs received. The relative effects of temperature on total hatch and size at hatch of these two species are remarkably similar and tend to support the theory that *P. flavescens* and *P. fluviatilis* may be conspecific.

SUMMARY

1. The effects of 24 treatments of constant or rising temperatures (3.1° - 25.4°C) on survival of yellow perch embryos of known age were determined in continuous-flow incubators. The median times to attain 7 morphological stages and physiological events in living embryos and larvae were also observed at 8 constant temperatures (3.1° - 19.7°C).

2. Early embryonic stages (before neural keel) tolerated constant temperature exposures from 3.1° to 19.9°C. The median tolerance limits (TL50) for normal hatch of older embryonic stages (after neural keel) were 7.0° and 22.9°C. The lower TL50 of swim-up larvae was 9.8°C. The optimum range for swim-up larvae was only 13.1° to 18.2°C when embryos were reared at constant incubation temperatures after fertilization. Survival of swim-up larvae at higher temperatures was limited only by embryonic survival.

3. Optimum yields of swim-up larvae (66-85%) were produced when initial incubation temperatures (4.9° - 10.0°C) were increased 0.5° - 1.0°C/day (range 4.9° - 24.3°C). Rising temperature regimes also favored shorter hatching periods and lower incidence of abnormalities at hatch.

4. Mass hatch occurred in 6 days at 19.7°C and in 51 days at 5.4°C. Hatching followed commencement of branchial respiration above 7.3°C, but was morphologically premature at lower temperatures.

5. Larval swim-up occurred on the day of hatch above 13°C, and within 2 days at lower temperatures. Unfed larvae survived 15 days at 19.8°C and 65 days at 5.3°C. The median period between swim-up and mortality of unfed larvae was 9 days at 19.8°C and 21 days at 10.5°C.

6. Total length of normal larvae at mass hatch varied from 4.7 to 6.6 mm. Mean lengths were 6.1 - 6.3 mm at 7.1° - 13.1°C, 5.6 - 5.8 mm at 16.1° - 20.0°C, and 5.1 - 5.2 mm at 5.0° and 22.1°C.

7. Effects of incubation temperatures on total hatch, median time to hatch, and size at mass hatch of *Perca flavescens* were compared to *P. fluviatilis*. The data tend to support the theory that *P. flavescens* and *P. fluviatilis* may be conspecific.

ACKNOWLEDGEMENTS

The authors thank the Minnesota Department of Natural Resources, Division of Game and Fish, for its assistance in collecting adult yellow perch, Joyce Hokanson for preparation of the illustrations, and those who made helpful criticisms of the manuscript, especially Dr. Peter J. Colby, Ontario Ministry of Natural Resources.

REFERENCES

Abbott, W.S., 1925. A method of computing the effectiveness of an insecticide. J. Econ. Ent., 18, 265-267.
Blaxter, J.H.S., 1969. Development: Eggs and larvae. In: W.S. Hoar and D.J. Randall (eds), Fish Physiology, Vol. III. London - New York Academic Press, 178-241 pp.
Ferguson, R.G., 1958. The preferred temperature of fish and their midsummer distribution in temperate lakes and streams. J. Fish. Res. Bd Can., 15, 607-624.
Hale, J.G. and Carlson, A.R., 1972. Culture of the yellow perch in the laboratory. Prog. Fish-Cult., 34, 195-198.
Hart, J.S., 1947. Lethal temperature relations of certain fish of the Toronto region. Trans. Roy. Soc. Can., Sec. V, Biol. Sci., 41, 57-71.
Hayes, F.R., 1942. The hatching mechanism of salmon eggs. J. Exp. Zool., 89, 357-373.
Hayes, F.R. and Pelluet, D., 1945. The effect of temperature on the growth and efficiency of yolk conversion in the salmon embryo. Can. J. Res., 23, 7-15.
Hokanson, K.E.F., McCormick, J.H. and Jones, B.R., 1973. Temperature requirements for embryos and larvae of the northern pike, *Esox lucius* (Linnaeus). Trans. Amer. Fish. Soc., 102, 89-100.
Ishida, J., 1944. Hatching enzyme in the freshwater fish *Oryzias latipes*. Annot. Zool. Jap., 22, 155-164.
Kokurewicz, B., 1969. The influence of temperature on the embryonic development of the perches: *Perca fluviatilis* L. and *Lucioperca lucioperca* (L.). Zool. Poloniae., 19, 47-67.
Kudrinskaya, D.M., 1970. (Food and temperature as factors affecting the growth, development and survival of pike-perch and perch larvae). Vop. Ikhtiol., 10, 779-788. (Trans. Amer. Fish. Soc.)
Mansueti, A.J., 1964. Early development of the yellow perch, *Perca flavescens*. Chesapeake Sci., 5, 46-66.
McCormick, J.H. and Syrett, R.F., 1970. A controlled temperature apparatus for fish egg incubation and fry rearing. MS. Nat. Water Quality Lab., Duluth, Minn., 9 p.
Steel, R.G.D. and Torrie, J.H., 1960. Principles and procedures of statistics. New York: McGraw-Hill Book Co., 481 p.
Swift, D.R., 1965. Effect of temperature on mortality and rate of development of the eggs of the pike (*Esox lucius* L.) and the perch (*Perca fluviatilis* L.). Nature, (Lond.), 206, 528.
Toetz, D.W., 1966. The change from endogenous to exogenous sources of energy in bluegill sunfish larvae. Invest. Indiana Lakes and Streams, 7, 115-146.

K.E.F.Hokanson, U.S.Environmental Protection Agency, National Water Quality Laboratory, Box 5oo, Monticollo, Minn. 55362/USA

Ch.F. Kleiner, U.S.Environmental Protection Agency, National Water Quality Laboratory, 6201 Congdon Boulevard, Duluth, Minn. 55804/USA

Temperature Tolerance of Early Developmental Stages of Dover Sole, *Solea solea* (L.)

D. N. Irvin

INTRODUCTION

The successful controlled rearing of marine fish, though sometimes achieved by the development and practice of empirical techniques, requires a more specific knowledge of ranges of tolerance to limiting factors for successive developmental stages, to ensure that optimal rearing conditions are maintained. Eventual unforeseen or unavoidable variations in such factors must also be considered. Temperature is a ubiquitous parameter that has been extensively investigated, although the majority of work to date concerns either incubation ranges of eggs or post-larval stages - the latter mainly for freshwater species.

Brett (1970), in his review of temperature tolerance relationships of marine fish, includes data for embryonic and post-embryonic stages. Much of the additional work on early developmental stages has considered temperature tolerance as a factor in isolation. Results are given for Alaska pollack (Hamai et al., 1971), sea trout (Grodzinski, 1971), and striped bass (Turner et al., 1971), although the comprehensive series of papers by Alderdice and Forrester (1968, 1971a, b) and Alderdice and Velsen (1971) on a variety of marine species extended the scope of such investigations by the use of multifactorial experiments incorporating the additional variable of salinity.

The ecological significance of possible temperature tolerance ranges of egg and larval stages is of considerable interest, and mortalities of larval yellowtail flounder in the sea have been attributed by Colton (1959) to lethal temperature effects. Marcy (1971) has considered the effect of entrainment and temperature shock within a power station cooling system and attributes subsequent larval mortalities in excess of 90% - mainly of alewives and blueback herring - largely to the heat shock. The strength of year classes of *Parophrys vetulus* has been inversely correlated with water temperature during larval development by Ketchen (1956) and for cod and herring by Hela and Laevastu (1962). In general, tolerance ranges are likely to vary with development stage, as has been demonstrated by Lewis (1965) for Atlantic menhaden and generally reviewed by De Sylva (1969).

With the exception of the work on clupeids by Blaxter (1960) most of the available data concerns North American and Pacific species, comparatively little work having been published on European species. The work of Flüchter (1970), however, and the preliminary work of Riley (pers. comm.) on sole development, conflicted with previous estimated tolerance ranges for early developmental stages of captive stocks of Irish Sea sole and prompted this further, more detailed, investigation of certain aspects of the problem. In order to determine temperature tolerance ranges for successful development the requirements of three important stages were considered, namely: full-term incubation of eggs, successful first feeding of larvae following yolk-sac absorption, and complete development through to metamorphosis. The temperature requirements for the initiation of first feeding in sole larvae will be reported separately.

METHODS

A spawning stock of *S. solea* has been maintained at Port Erin for 6 years. Natural spawning occurs in late April and early May, the pelagic eggs being collected by sweep net. Because the ponds were either swept or flushed daily and because of the low temperature of spawning (11°C), eggs were always obtained at Stage 1A (Simpson, 1959). The quality of these eggs varied markedly, though as yet merely subjective criteria based on morphological appearance are used for assessing potential viability.

Effect of Temperature on Survival of Sole Eggs. A collection of such eggs was made during May 1972 and the material transferred to a 40 l bath containing 700 ppm streptomycin sulphate and 300 ppm sodium penicillin E (potency 745 i.u./mg). Duplicates, each of approximately 600 eggs, were transferred by dip net directly into beakers containing 800 ml of equivalent antibiotic-treated sea water stabilized at each of 6 constant temperatures (± 0.05°C). A second series of eggs was transferred to similar beakers of treated water but at the tempera ture of spawning and then transferred to the constant temperature bath This established a degree of acclimation for the second series in the order of 10 deg C/h. Non-viable eggs became opaque and sank to the bottom whence they were removed daily. Subsequent mortality was recorded and the percentage fallout of the initial number stocked was calculated at the termination of the experiment.

Effect of Temperature on Survival of Sole Larvae. The first of a series of lethal-limit experiments was set up with Stage 1A yolk-sac larvae - approximating to the 1A Stage of Ryland (1966). About 100 larvae, hatched and reared at 12.5°C, were transferred by wide-bore pipette to each of 4 replicate glass jars containing 500 ml of sea water at each of 5 constant temperatures, the range for which had been previously determined by rough tests to establish a meaningful median LC_{50} value. Replicates were aerated at a sufficient level to maintain 90% oxygen saturation as measured, but excessive disturbance of the water was avoided because of the delicate nature of the larvae at this stage. *Artemia* nauplii were added to replicates once functional mouthparts developed. Further tests were carried out on active first-feeding larvae, approximating to Ryland Stage 2A, and metamorphosing Stage 5 larvae in order to determine corresponding median 4 day LC_{50}[1] values. Numbers of larvae were proportionally reduced in these latter tests, i.e. 50-60 Stage 2 and 10-20 Stage 5

[1]Sprague (1969) has recommended that standard terminology be adopted in toxicity studies and his suggestions are followed in this paper. I have avoided the use of the term 'L.T.50' here because of confusion concerning the term 'L.T.' referring either to "lethal time" or "letha temperature". The term L.T.50 should be restricted to "median resistan time" or "median lethal time". The term L.C.50 used in this paper is synonymous with "median lethal concentration" or "median lethal temper ture" and may be prefixed by a specific time reference, namely "4 day L.C.50" or "incipient L.C.50", the latter implying 50% survival with infinite time at the stated temperature.

larvae. Similar determinations of lower lethal limits were made in aerated 1 l beakers containing 500 ml of water, the fish having been acclimated to a range of constant temperatures (25°C, 17°C, and 10°C) for 2 weeks prior to experimentation. Controls were run in all lethal limit experiments, maintained in similar containers at acclimation temperatures.

Time/temperature-tolerance relationships were analyzed by transforming the generally sigmoid percentage mortality functions to linear probabilities, as conventially employed in bioassay methods for toxicity testing (Sprague, 1969). Linear regressions were fitted from probit values and confidence limits about the abscissa were calculated following the method of Litchfield and Wilcoxon (1949) and Litchfield (1949). Median survival times plotted against test temperatures permitted interpolation to a theoretical 4 day median lethal-limit value for successive developmental stages.

RESULTS

Effect of Temperature on the Survival of Sole Eggs. The temperature range for successful full-term incubation determined for Stage 1A sole eggs during the peak of the 1972 spawning season was found to be 8-16°C. A typical mortality curve is shown in Fig. 1 for eggs incubated at 16°C. Successful hatching of eggs occurred during the 96-h test period at this temperature and had partially commenced in the 14°C series. Initial mortality curves for all temperatures were fitted by eye and replicates pooled to fit linear regressions of the type "$y = a + b \log_e x$" where y = probit value (Table 1). Fitting linear functions permitted the comparison of mortality over the range of incubation temperatures (Fig. 2). The intercepts of the fitted regressions in the region of 75% mortality show the equivalent mortality of sole eggs at various incubation temperatures. Though mean egg mortalities were slightly lower in the middle of the range than at the extremes, the differences were not significant. The absolute percentage mortality of eggs within the incubation range of 10-16°C was therefore essentially equivalent. There was no significant difference between mortality of non-acclimated and acclimated 1A eggs, though the acclimated 1A eggs showed considerably advanced mortalities during the 96-h test period. This anomaly is attributable to the experimental procedure. The initial random stocking of eggs in the acclimated series of replicates contained a certain percentage of poor quality eggs which, though still floating, were slowly but consistently losing buoyancy and sinking. The non-acclimated series stocked approximately 1 h later contained proportionally fewer of such eggs and therefore subsequent mortalities were slightly affected.

The 100% mortality of eggs acclimated at 8°C which was consistent for both replicates, suggested that 8°C might be close to the lower lethal limit for the normal incubation of this source of eggs. However, subsequent experiments, not reported here, demonstrated that the majority of normal eggs from natural pond spawnings would complete full term incubation at 8°C, so the lower limit for this phase of development might be better estimated at 7°C.

There is a clear time/temperature relationship for mortality during incubation as can be seen by the slope functions included in Fig. 2. The time to 50% mortality of sole eggs is shown in Fig. 3 as a function of incubation temperature.

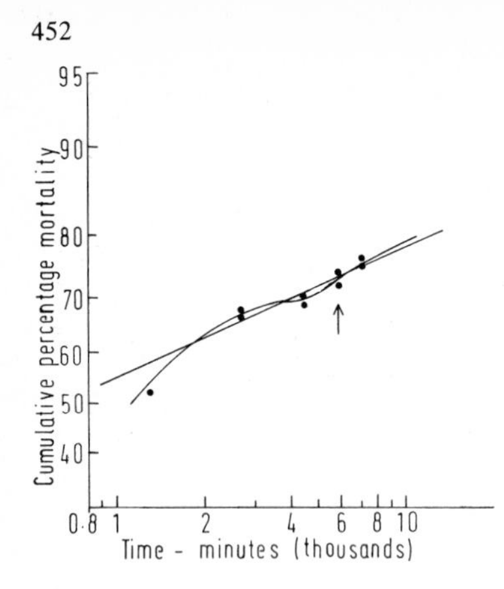

Fig. 1. Typical cumulative mortality of sole eggs, for acclimated and non-acclimated treatments, showing initial mortality curves fitted by eye and fitted linear regressions as shown in Fig. 2 - incubation temperature of 16°C. Time of commencement of hatching indicated

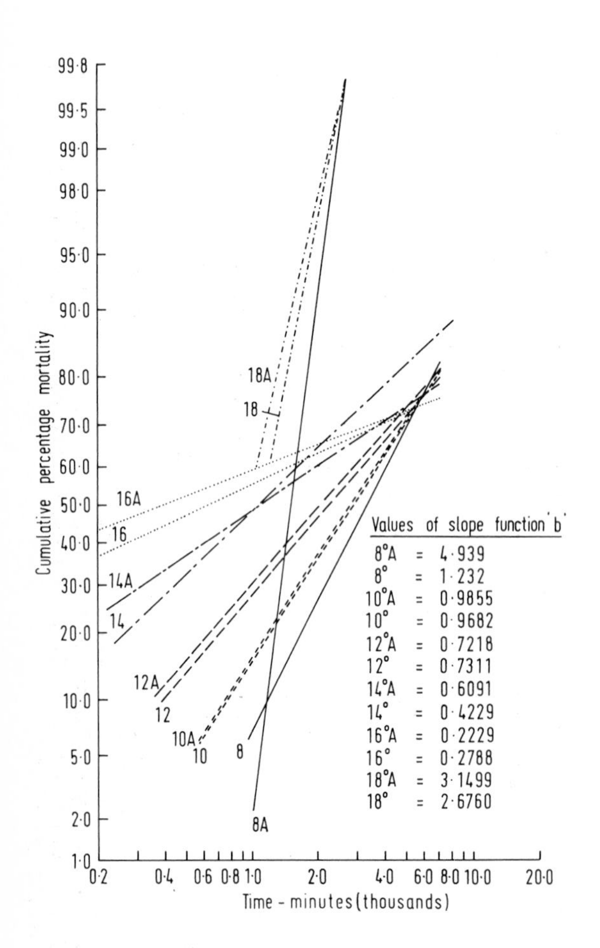

Fig. 2. Fitted regressions for cumulative mortality of sole eggs in relation to incubation temperature, for acclimated and non-acclimated treatments

Table 1. Regression data for acclimated (A) and non-acclimated Stage 1A sole eggs at different temperatures (test salinity 34.3 ^{o}/oo). For all tables linear regression of the type: $y = a + b \log_e x$ (where y = probit value). N.B. The results of interest are when the values of P are high (i.e. there is no mortality trend with time)

Test temp. and treatment, °C	Mean terminal % mortality (96 h)	Equation	Correlation coefficient 'r'	Regression value for 't' and 'n'
8 (A)	100	$y = -31.27 + 4.94\ x$	1.0000	1 474.58 (n = 2)
8	75.5	$y = -5.02 + 1.23\ x$	0.9657	10.52 (n = 8)
10 (A)	75.5	$y = -2.85 + 0.99\ x$	0.9286	7.08 (n = 8)
10	74.1	$y = -2.72 + 0.97\ x$	0.9228	6.77 (n = 8)
12 (A)	76.1	$y = -0.66 + 0.73\ x$	0.8617	4.80 (n = 8)*
12	73.7	$y = -0.52 + 0.72\ x$	0.8915	5.57 (n = 8)
14 (A)	85.5	$y = 0.73 + 0.61\ x$	0.9818	8.95 (n = 3)*
14	75.6	$y = 2.04 + 0.42\ x$	0.8637	4.85 (n = 8)*
16 (A)	76.6	$y = 3.70 + 0.22\ x$	0.8768	5.16 (n = 8)
16	77.5	$y = 3.22 + 0.28\ x$	0.9717	11.63 (n = 8)
18 (A)	100	$y = -17.14 + 3.15\ x$	0.9914	10.71 (n = 2)*
18	100	$y = -13.40 + 2.68\ x$	0.9984	25.30 (n = 2)*

Probabilities related to values of 't', on null hypothesis of 'b' = 0, 'r' = 0: (1) Probabilities quoted: $P > 0.05$, (2) Asterisk notation: ***, **, * equivalent to $P = 0.05$, $P = 0.02$, $P = 0.01$ respectively, (3) No value: $P < 0.001$

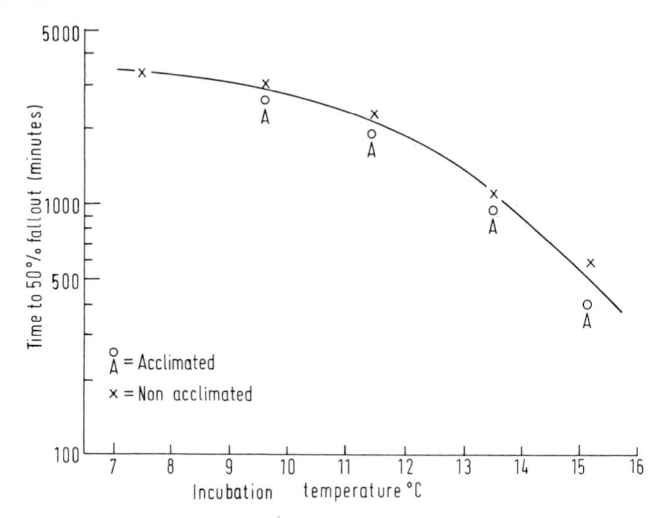

Fig. 3. Time/temperature relationship for 50% mortality of sole eggs during incubation, for acclimated and non-acclimated treatments

Effect of Temperature on the Survival of Sole Larvae. Typical fitted regressions for time/temperature mortality relationships for Stage 2A larvae are shown in Fig. 4 and regression data for all larval stages are given in Table 2. Median survival times for the three larval stages considered are shown in Fig. 5, with fitted 95% confidence limits. Lines fitted to these points by eye interpolate to a theoretical 96-h median upper lethal limit for the successive stages, with corresponding 100% survival treatments within 1-2 deg C of these points. There were no mortalities in controls maintained at rearing temperatures for any of the stages of development tested. This treatment provides estimates of lethal limit values of 23°C for yolk-sac larvae, 24°C for first-feeding larvae and 28.1°C for metamorphosing larvae.

The similar treatment of lower lethal limit results for Stage 5 sole are given in Figs 6 and 7 and regression data in Table 3. The acclimation period of two weeks prior to testing had a significant effect on the lower LC_{50}. Values of 5°C, 7.3°C, and 8.7°C were estimated for the three acclimation temperatures of 10°C, 17°C and 25°C, respectively, a rise of 1°C of the LC_{50} accompanying every 4°C rise in acclimation temperature.

It was not possible to estimate actual 'incipient' lethal limits for these stages, since the fitted line to median survival times, over the range of test temperatures, did not approach an infinite value, the incipient level, within the 96-h test period, that could be equated to the cessation of acute effects as conventially employed in toxicity assays (Sprague, 1969). The apparent absence of the typical sigmoid response relationship and the high values for the slope function 'b' of the probit/time regressions were characteristic features of many of the mortality responses to test temperatures (Tables 1, 2, and 3). It is likely that these factors reflect to a considerable extent the genetic uniformity of the material being tested, which had in all cases been taken from a single day's spawning, the likely products of only a very limited number of crosses.

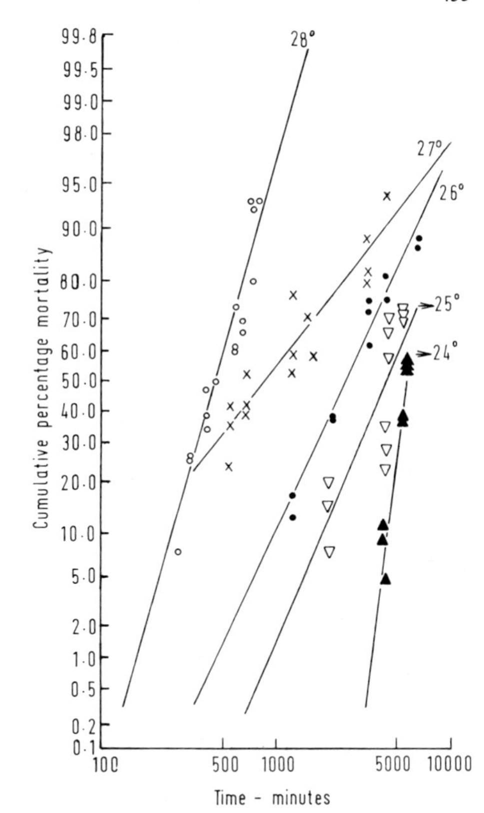

Fig. 4. Typical mortality of sole larvae as a function of upper lethal test temperatures, with fitted linear regressions: Stage 2A, first-feeding larvae

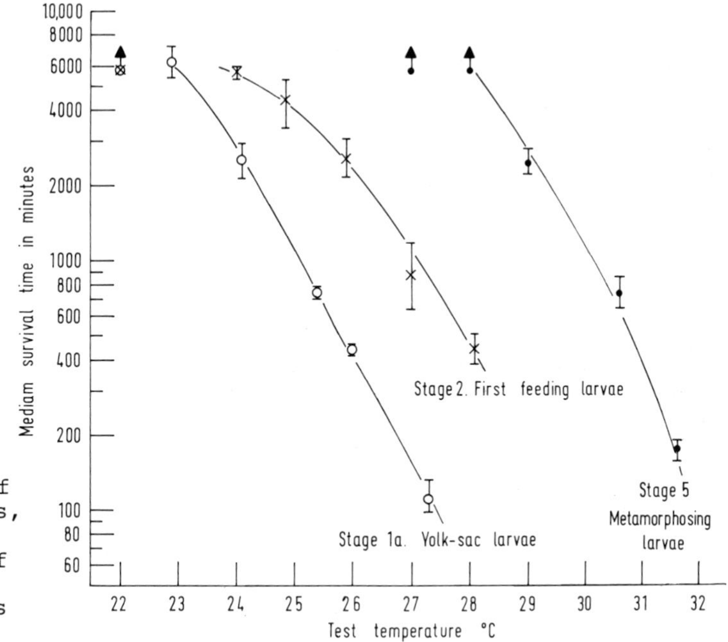

Fig. 5. Median survival times for successive larval stages over the range of test temperatures, estimated from regressions as of Fig. 4; 95% confidence limits are fitted

Table 2. Regression data for yolk-sac larvae, first-feeding larvae and metamorphosing larvae (test salinities stated). Probabilities related to values of 't' and regression type as for Table 1

Test temp. ^{o}C	Mean terminal %mortality (96 h)	Equation	Correlation coefficient 'r'	Regression value for 't' and 'n'
	Stage 1A: yolk-sac larvae (34.2^{o}/oo)			
27	100	$y = -5.41 + 2.22x$	0.9854	14.17 (n = 6)
26	100	$y = -8.65 + 2.24x$	0.9780	12.40 (n = 7)
25	100	$y = -9.85 + 2.25x$	0.8700	6.36 (n = 13)
24	82	$y = -3.07 + 1.03x$	0.8607	8.45 (= 25)
23	51	$y = -24.45 + 3.37x$	0.5342	1.79 (n = 8, P = 0.2)
	Stage 2A: first-feeding larvae (34.5^{o}/oo)			
28	100	$y = -8.76 + 2.26x$	0.9520	11.64 (n = 14)
27	100	$y = -0.42 + 0.80x$	0.9366	9.64 (n = 13)
26	100	$y = -5.99 + 1.40x$	0.9861	20.59 (n = 12)
25	71.3	$y = -7.99 + 1.55x$	0.8593	5.31 (n = 10)
24	56.5	$y = -39.23 + 5.13x$	0.9833	13.22 (n = 6)
	Stage 5: metamorphosing larvae (34.1^{o}/oo)			
31	100	$y = -13.61 + 3.61x$	0.9079	5.30 (n = 6)*
30	100	$y = -12.74 + 2.69x$	0.9302	8.78 (n = 12)
29	100	$y = -17.95 + 2.93x$	0.9194	7.75 (n = 11)

Table 3. Regression data for lower lethal limits of Stage 5 sole, over a range of acclimation temperatures (salinity 34.1°/oo). Probabilities related to values of 't' and regression type as for Table 1

Test temp. °C	Mean terminal % mortality (96 h)	Equation	Correlation coefficient 'r'	Regression value for 't' and 'n'
	25°C acclimated stock			
4	100	$y = -6.99 + 4.58x$	0.9278	8.612 (n = 12)
5	100	$y = -2.17 + 2.06x$	0.9200	7.79 (n = 11)
6	100	$y = 0.30 + 1.01x$	0.9064	6.79 (n = 10)
7	100	$y = 0.06 + 0.86x$	0.9049	6.01 (n = 8)
8	100	$y = -7.77 + 1.76x$	0.9110	5.41 (n = 4)*
	17°C acclimated stock			
3	100	$y = -1.03 + 1.55x$	0.8272	4.42 (n = 9)*
4	100	$y = -0.19 + 0.94x$	0.9206	7.85 (n = 14)
5	100	$y = -1.66 + 1.03x$	0.9216	7.51 (n = 10)
6	100	$y = -12.26 + 2.24x$	0.9471	10.64 (n = 13)
7	62.5	$y = -18.80 + 2.81x$	0.8514	4.59 (n = 8)*
	10°C acclimated stock			
2	100	$y = -20.71 + 8.73x$	0.9682	9.47 (n = 6)
3	100	$y = 0.60 + 0.81x$	0.9167	9.46 (n = 17)
4	100	$y = 0.60 + 0.63x$	0.8638	6.18 (n = 13)
5	24.3	$y = 1.78 + 0.30x$	0.7957	3.72 (n = 8)*

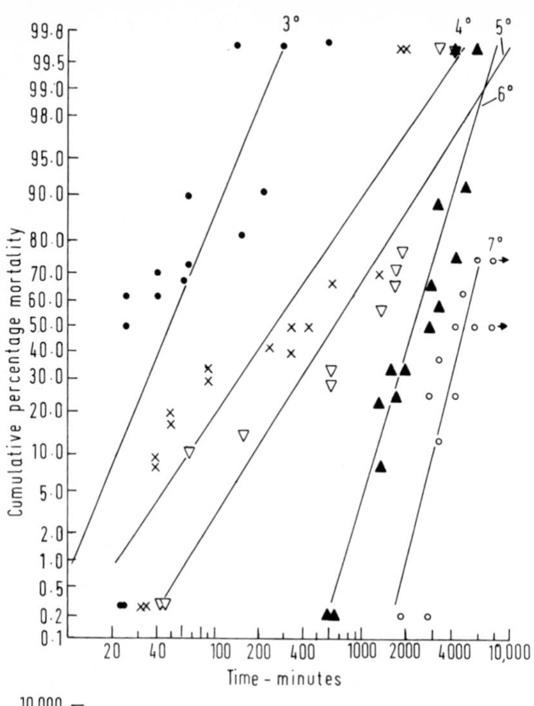

Fig. 6. Typical mortality of Stage 5 metamorphosing sole larvae as a function of lower lethal temperatures, with fitted linear regressions: 17°C acclimated

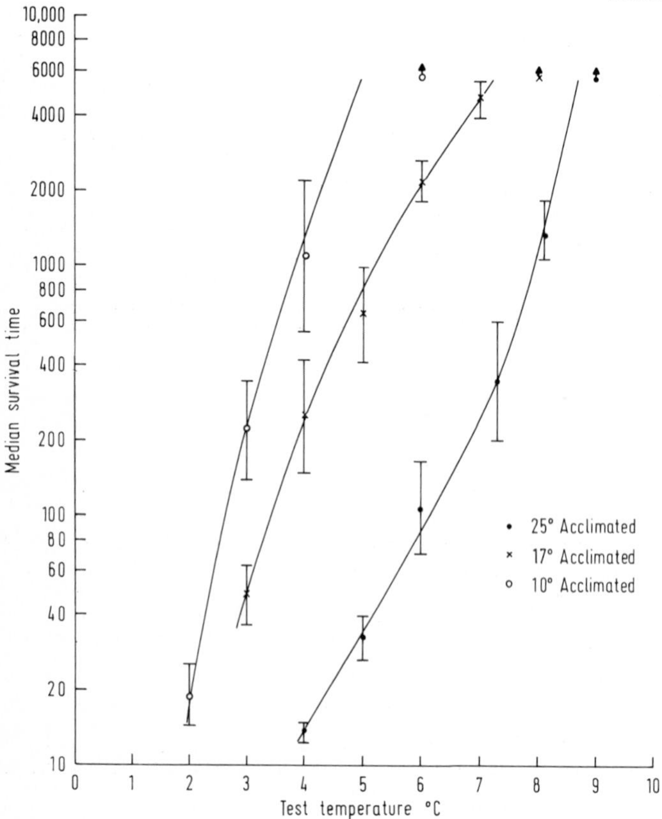

Fig. 7. Legend see opposite page

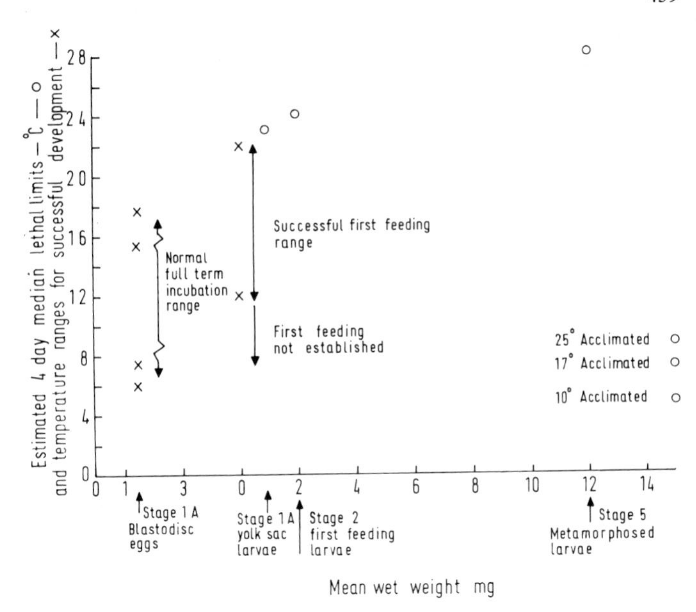

Fig. 8. Summarized temperature tolerance ranges for successive early developmental stages of sole

The progressive development of temperature tolerance with successive developmental stages is shown in Fig. 8. In order to demonstrate this trend graphically the mean wet weight is related to LC_{50} and temperature-tolerance ranges. This should not be construed as implying any causal relationship between the two parameters. The temperature range for normal full-term incubation of local Stage 1A sole eggs is determined to be above 7°C and below 17°C; at temperatures of 6°C and 18°C normal development does not occur. The embryonic stages may therefore be considered to be markedly stenothermal. During the course of larval development there is an increase of 5 deg C in the upper median lethal limit from yolk-sac larvae to metamorphosing Stage 5 larvae. At this latter stage in development the absolute tolerance range between upper and lower median lethal limits is approximately 21 deg C, double that of Stage 1A eggs.

DISCUSSION

With the exception of lower lethal limits for yolk-sac and first-feeding larvae temperature ranges have been determined. The main object of the work was to establish tolerance ranges that could be applied to controlled rearing techniques for the species, as a prerequisite for further assessing the possible suitability of *S. solea* for intensive cultivation. Emphasis was therefore placed on defining temperature

---

◀ Fig. 7. Median survival times of larvae from three acclimation temperatures over a range of test temperatures, estimated from regressions as of Fig. 6; 95% confidence limits are fitted

tolerance ranges for survival as limiting boundaries within which detailed work could be undertaken on optimal temperatures for the consistent rearing of large numbers of juvenile sole. This latter work has been completed and will be published elsewhere. Temperature tolerance was therefore considered in terms of percentage survival responses to ranges of constant temperatures, to estimate sustainable thermal limits for finite periods approaching theoretical incipient levels, as opposed to short-duration temperature shocks and subsequent effects on survival.

It is of considerable interest that conventional incipient levels could not be exactly determined within the 96-h test period. A similar effect has been demonstrated for upper lethal limits of the guppy with mortalities continuing even after one week (Arai et al., 1963). In the majority of previous investigations, incipient upper lethal levels have been established within 24-h test periods, though 5-6 days were usually required to define incipient lower levels (Doudoroff, 1945; Blaxter, 1960). In many other determinations of upper temperature tolerance limits, LC_{50} values estimated at the end of various test periods have been assumed to be incipient levels. It is apparent, however, that for sole larvae acute responses to high temperatures continue to occur after 24 h.

This 'acute' response could in part be due to the gradual selective mortality that occurs during the early stages of development. Such mortality is likely to occur sooner at higher temperatures, whilst not occurring at lower control temperatures within the test period. Therefore no accurate correction can be made from controls to account for this factor. However, such an effect is not likely in Stage 5 larvae, where routine mortalities are very low. As a consequence of not being able to define exactly an incipient level, the temperature within the test period corresponding to recorded 100% survival, 1 deg C below the median survival time, has been found in practice to define the tolerance range.

It is interesting to compare tolerance ranges presented here for sole with some of the work previously published for marine fish. Flüchter (1970) gives ranges of 6.75 - 19.00°C for development to the point of hatching of Stage 1A sole eggs in good agreement with the present results. His data suggest that 'food acceptance' - 'Nahrungsaufnahme' - is only successful between 8.75°C and 16°C, implying that outside this range of temperature first-feeding larvae are not viable. This range is considerably wider than that of 12-16°C found by Irvin (unpubl.).

The range for successful full-term incubation of similar eggs produced from artificial fertilizations is suggested by Riley from unpublished data (pers. comm.) to be 6-15°C. However, viable first-feeding larvae only developed in his experiment at 15°C; larvae hatching below this temperature died after utilizing their yolk reserves. This phenomenon was thought to suggest either that a rise in temperature was required to initiate first-feeding in sole larvae or that there were different temperature ranges for successful incubation and first-feeding. It has been shown that a rise in temperature *per se* is not required to initiate first-feeding. Stage 1A eggs incubated, hatched, and maintained at 12-16°C yielded larvae which successfully commenced first-feeding at constant temperatures, whereas at 8°C and 10°C hatching was successful but first-feeding did not commence. There are indeed therefore two discrete thermal requirement ranges for these stages (Irvin, unpubl.)

Successful full-term incubation of good quality Stage 1A eggs can be predicted between 7°C and 16°C; there is, however, a more restricted range for initiation of first-feeding between 12°C and 16°C for larvae hatched from eggs incubated at constant temperatures within this range. It is of course possible that larvae hatched from eggs incubated at temperatures below 12°C would establish first-feeding if transferred to warmer water. Such conditions might be expected to occur in the sea, with developing eggs being subjected to a gradual rise in temperature from the time of spawning at depth to hatching and subsequent development in the surface waters. However, this has not yet been demonstrated experimentally.

Although there are obviously certain anomalies between available data for sole tolerance ranges, it is felt that these may not entirely be due to differences in adult stocks or methods of obtaining eggs. It is possible that, with regard to egg stages at least, differences may be attributable to the temperature at which the eggs were fertilized. Hubbs et al. (1971) have demonstrated that in *Menidia audens* the temperature regime in which the spawning stock is maintained has a marked effect on temperature tolerance of eggs at gastrulation. Stock acclimated for 2-3 days at 15°C produced eggs that gastrulated successfully at 14°C but not at 32°C, while eggs from stocks acclimated for a similar duration at 28°C gastrulated at 34°C but not at 18°C. Parental thermal history may therefore have an important effect on tolerance ranges for subsequent egg and larval stages.

Extrapolation of this work to ecological situations is uncertain, considering the origin of egg and larval material and the artificial experimental conditions employed; nevertheless, it suggests that temperature is of considerable importance as a potential limiting factor of normal egg and larval development. In practice, however, local pond spawning usually occurs within the tolerance range for normal egg incubation. It commences regularly during late April and early May, coinciding with the rapid annual rise in sea temperature of about 4 deg C per month from the normal spring temperatures in April of 7°C. The tolerance of larvae after hatching is well within the range of sea temperatures likely to occur at these latitudes.

SUMMARY

1. Temperature tolerance of *S. solea* increases during ontogenesis. Tolerance ranges for acclimatized Irish Sea stocks are in good agreement with similar stocks from the North Sea. Embryonic stages are notably stenothermal in relation to the generally eurythermal larval stages, though tolerance continues to increase during larval development up to the period of metamorphosis.

2. Successful full-term incubation of Stage 1A eggs is predictable between 7°C and 16°C. Absolute mortality is not significantly related to temperature within this range, but time to 50% mortality is related to temperature.

3. The successful first-feeding of larvae hatched from Stage 1A eggs incubated and maintained within this range was only obtained over the narrower range of 12-16°C. However, larvae which were hatched within the tolerance range at 13°C and transferred to temperatures of up to 22°C commenced first-feeding.

4. 96-h LC_{50} values of 23°C, 24°C, and 28.1°C were derived for upper lethal limits of yolk-sac larvae, first-feeding larvae, and metamorphosing larvae respectively. Lower 96-h LC_{50} values for Stage 5 larvae were significantly affected by acclimation temperature. At this stage of development the absolute tolerance range is 20-23°C.

ACKNOWLEDGEMENTS

This work was undertaken in part requirement for the degree of Ph.D. of the University of Liverpool and was supported by a Fisheries Research Training Grant awarded by the Natural Environmental Research Council. I wish to thank Dr. R. Alderson for helpful criticism.

REFERENCES

Alderdice, D.F. and Forrester, C.R., 1968. Some effects of salinity and temperature on early development and survival of the English sole (*Parophrys vetulus*). J. Fish. Res. Bd Can., 25, 495-521.

Alderdice, D.F. and Forrester, C.R., 1971a. Effects of salinity and temperature on embryonic development of the Petrale sole (*Eopsetta jordani*). J. Fish. Res. Bd Can., 28, 727-744.

Alderdice, D.F. and Forrester, C.R., 1971b. Effects of salinity, temperature, and dissolved oxygen on early development of the Pacific cod (*Gadus macrocephalus*). J. Fish. Res. Bd Can., 28, 883-902.

Alderdice, D.F. and Velsen, F.P.J., 1971. Some effects of salinity and temperature on early development of Pacific herring (*Clupea pallasi*). J. Fish. Res. Bd Can., 28, 1545-1562.

Arai, M.B., Cox, E.T. and Fry, F.E.J., 1963. An effect of dilutions of sea water on the lethal temperature of the guppy. Can. J. Zool., 41, 1011-1015.

Blaxter, J.H.S., 1960. The effect of extremes of temperature on herring larvae. J. mar. biol. Ass. U.K., 39, 605-608.

Brett, J.R., 1970. Temperature; Animals; Fishes; Functional Responses. In: Marine Ecology, Vol. 1, O. Kinne (ed.), 516-560. London - New York: Wiley-Interscience.

Colton, J.B., Jr., 1959. A field observation of mortality of marine fish larvae due to warming. Limnol. Oceanogr., 4, 219-222.

De Sylva, D.P., 1969. Theoretical considerations of the effects of heated effluents on marine fishes. In: Biological aspects of thermal pollution. P.A. Krenkel and F.L. Parker (eds.), 229-293. Nashville, Tenn.: Vanderbilt University Press.

Doudoroff, P., 1945. The resistance and acclimatization of marine fishes to temperature changes. 2. Experiments with *Fundulus* and *Atherinops*. Biol. Bull. mar. biol. Lab., Woods Hole, 88, 194-206.

Flüchter, J., 1970. Zur Embryonal- und Larvalentwicklung der Seezunge *Solea solea* (L.). Ber. dtsch. wiss. Komm. Meeresforsch., 21, 369-376.

Grodzinski, Z., 1971. Thermal tolerance of the larvae of three selected Teleost fishes. Acta biol. cracov. (Sêrie zoologique), XIV, 289-298.

Hamai, I., Kyûshin, K. and Kinoshita, T., 1971. Effect of temperature on the body form and mortality in the development and early larval stages of the Alaska pollack, *Theragra chalcogramma* (Pallas). Bull. Fac. Fish. Hokkaido Univ., 22, 11-29.

Hela, I. and Laevastu, T., 1962. Fisheries hydrography. Fishing News Ltd., London, 137 pp.
Hubbs, C., Sharp, H.B. and Schneider, J.F., 1971. Developmental rates of *Menidia audens* with notes on salt tolerance. Trans. Amer. Fish. Soc., 100, 603-610.
Ketchen, K.S., 1956. Factors influencing the survival of the lemon sole (*Paraphrys vetulus*) in Hecate Strait, British Columbia. J. Fish. Res. Bd Can., 13, 647-694.
Lewis, R.M., 1965. The effect of minimum temperature on the survival of larval Atlantic menhaden, *Brevoortia tyrannus*. Trans. Amer. Fish. Soc., 94, 409-412.
Litchfield, J.T., Jr., 1949. A method for rapid graphic solution of time per cent effect curves. J. Pharmac. exp. Ther., 97, 399-408.
Litchfield, J.T., Jr. and Wilcoxon, F., 1949. A simplified method of evaluating dose-effect experiments. J. Pharmac. exp. Ther., 96, 99-113.
Marcy, B.C., Jr., 1971. Survival of young fish in the discharge canal of a nuclear power plant. J. Fish. Res. Bd Can., 28, 1057-1060.
Ryland, J.S., 1966. Observations on the development of larvae of the plaice, *Pleuronectes platessa* L., in aquaria. J. Cons. perm. int. Explor. Mer, 30, 177-195.
Simpson, A.C., 1959. The spawning of the plaice (*Pleuronectes platessa*) in the North Sea. Fish. Invest., Lond., Series 2, 22 (7), 111 pp.
Sprague, J.B., 1969. Measurement of pollutant toxicity to fish. I. Bioassay methods for acute toxicity. Wat. Res., 3, 793-821.
Turner, L.J. and Farley, T.C., 1971. Effects of temperature, salinity, and dissolved oxygen on the survival of striped bass eggs and larvae. Calif. Fish and Game, 57, 268-273.

D.N. Irvin
Ministry of Agriculture, Fisheries and Food
Fisheries Laboratory
Port Erin, Isle of Man / GREAT BRITAIN

Present address:

Ministry of Development and Engineering Services
Fisheries Bureau
P.O. Box 235
State of Bahrain

Development of the Respiratory System in Herring and Plaice Larvae

C. de Silva

INTRODUCTION

This study was initiated to gain some insight into the development of the respiratory system in 2 teleost species, herring (*Clupea harengus* L.) and plaice (*Pleuronectes platessa* L.) well separated taxonomically. Plaice hatch with gill arches (Holliday and Jones, 1967) as do herring, but the arches are better developed in the plaice. The gill filaments develop later in both species. The larvae are completely transparent at hatching without respiratory pigment, the blood becoming pink, weeks or months later at metamorphosis.

The gill- and body-surface areas were measured at different stages as a means of determining the surface area available for cutaneous respiration. The appearance of the respiratory pigment was also determined and quantified.

Although many workers have measured the surface area of the body in an attempt to relate it to metabolism (Gray, 1953) as well as the gill areas (Muir, 1969; Muir and Hughes, 1969) the only developmental study of gill area has been made by Price (1931) who worked on bass in a range of 2 to 40 cm. Harder (1954) studied the development of branchial elements in herring and confined his measurements to that of the gill filaments alone. Although Hawkins and Maudesley-Thomas (1972) have described the haematology of adult fish in the younger stages only, Radinskaya (1960) attempted to determine the first appearance of haemoglobin in embryonic sturgeon, and Ostroumova (1962) determined its appearance in embryos of rainbow trout. Later Radzinskaya (1966, 1968) measured its concentration in embryonic and juvenile salmon.

METHODS

Herring larvae used for the experiments were reared from eggs according to the method of Blaxter (1968). Eggs were obtained from spawning females of Ballantrae Bank in the Firth of Clyde in February 1971 and 1972. Plaice eggs were obtained from the White Fish Authority hatchery at Ardtoe, Scotland in March 1971 and 1972. They were reared using Shelbourne's (1964) method. Wild O-group plaice were also caught locally by push-nets.

For the external surface area measurements larvae were anaesthetized with MS 222 (1/15,000) and the outline of the body and fins was immediately drawn on graph paper using a WILD M5 binocular microscope with a camera lucida attachment. The wet weight was then determined on a Beckmann EMB 1 microbalance, and the larvae frozen for 30 min, freeze-dried and stored till the dry weight was determined.

In herring the body was observed to be cylinder-shaped and the area was calculated as πdl where d is the average depth of 5 measurements taken at various points along the body and l the total length (de Silv 1973). The data were grouped into length classes to facilitate calculation. In the early stages of plaice the method used to calculate the surface area was similar to that of herring. From Stage IV (Ryland, 1966) onward the value obtained for the surface area of one side was doubled to give the total surface area. No attempt was made to measure the surface area of larvae with yolk sacs or to determine the extent of buccal and intestinal respiration, if any, with development.

The material for the gill measurements was fixed in 10% sea water-formalin, after the body length was measured. The wet weight was estimated for these lengths from the surface area and weight relationships. For the larger plaice and herring fixed long after metamorphosis both the wet weight and body length were determined before fixation. Prior to measurement by binocular microscope, the gills and pseudobranch of one side were dissected out and the arches separated. No allowance was made for shrinkage but Gray (1954) and Hughes (1966), measuring the gills of large fish, observed that shrinkage was slight.

The length of every fifth or sixth inner and outer filament was measured on the epibranchial, ceratobranchial, and hypobranchial, the averag of all these measurements being taken as the filament length for that particular arch. This was done for all 4 arches of one side. In addition, the longest filaments on the epibranchial and ceratobranchial, the shortest filament at the junction of the epibranchial and ceratobranchial, and the filament at the junction of the ceratobranchial and hypobranchial, were used to measure the height and width of the secondary lamella, as well as the spacing of the lamellae. The gill area was calculated in a manner similar to that of Muir and Hughes (1969) assuming the area of the secondary lamella to be equal to that of a triangle. Some material was also embedded in paraffin wax and sections cut at 4-6 μ to confirm some of the gill measurements and to examine the circulatory fluid.

To identify haemoglobin histochemically the benzidine method of Slonimski and Lapinski (1932) as cited by Radzinskaya (1960) was used with slight modification. This method indicates haemoglobin because of its peroxidase activity. Larval tissue was also homogenized and treated chemically, according to the method of Korzhuev and Radzinskaya (1957), prior to measuring haemoglobin levels spectrophotometrically. Details of the standards used and the analytical method are given by De Silva (1973).

RESULTS

Surface Area

The relationships between surface area and length are shown in Figs 1 and 2. The results were grouped into length classes to facilitate handling of the data and curves corresponding to a line of best fit obtained by log-log regression analysis. In herring the total surface area (i.e. including fins) varied by a slope value of 2.13, while body surface area varied by a value of 2.27. In plaice, on the other hand, the body area to length relationship has a slope value of 2.09

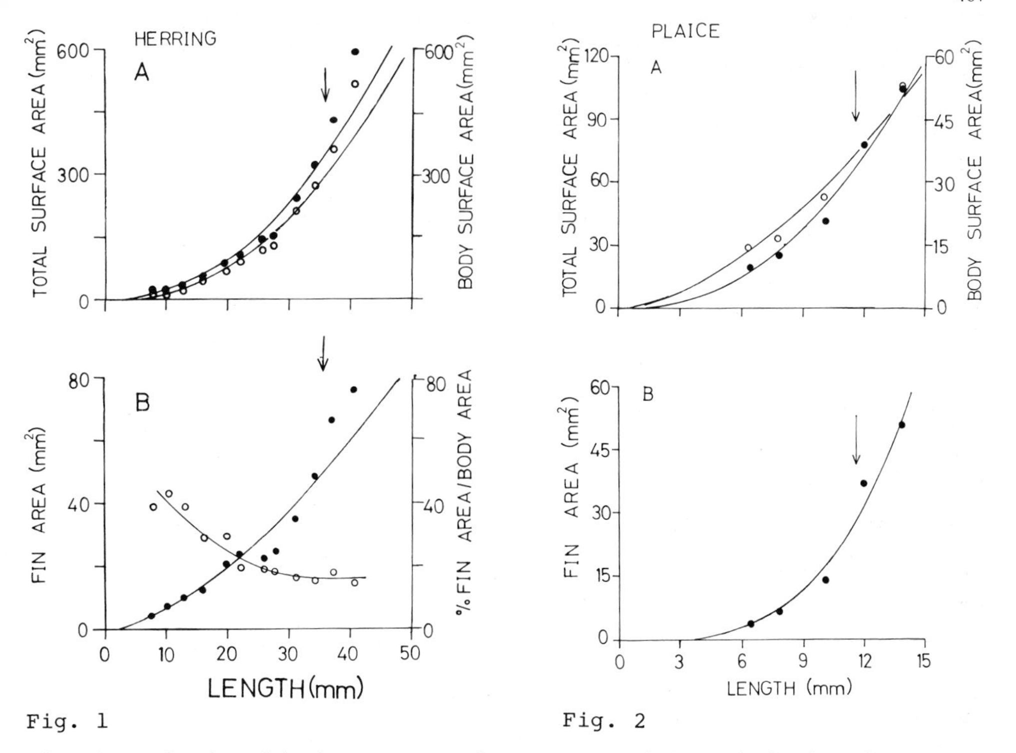

Fig. 1   Fig. 2

Fig. 1. Relationship between surface area and length in herring. (A) Total surface area •, and body surface area o. (B) Fin surface area •, and length and percentage fin area/body area ratio with length. Arrows indicate metamorphosis

Fig. 2. Relationship between surface area and length in plaice. (A) Total surface area •, and body surface area o, with length. (B) Fin surface area and length. Arrows indicate metamorphosis

while the total surface area increases at a rate of 2.57. This is to be expected because the plaice is changing shape throughout development, becoming increasingly flatter and broader as metamorphosis approaches. In herring the fin area, as expected, showed a slower rate of increase with length (slope value of 1.59) when compared to the body and total surface area. The fin area - body area ratio reaches a minimum at a body length of about 25 mm. Significantly this is the time of appearance of the gills (Harder, 1954 and present study). Plaice did not show this pattern, probably due to its different body proportions.

In Table 1 the relationship between surface area and weight are given. The total surface area increases at a higher rate compared to the body surface area in herring, while the reverse is observed in plaice. This is probably a reflection of the changing shape of the plaice as it approaches metamorphosis. Similarly the fin area increases at a higher rate.

Table 1. Equations for the relationship between weight (W) and surface area (Y) in herring and plaice

Herring		S.D. of slope
Total Y	Log Y = 22.93 + 0.576 log W	.015
Body Y	Log Y = 16.59 + 0.608 log W	.015
Fin Y	Log Y = 7.079 + 0.409 log W	.026
Plaice		
Total Y	Log Y = 23.42 + 0.497 log W	.067
Body Y	Log Y = 16.030 + 0.402 log W	.050
Fin Y	Log Y = 6.728 + 0.679 log W	.011

Gill Area

A plot of gill area against weight and log-log coordinates (Fig. 3) shows a curve at the upper ends of the graph in both herring and plaice with the point of inflexion corresponding to metamorphosis. The data are therefore plotted with separate lines fitted up to metamorphosis and beyond it. The regression analyses are summarized in Table 2. The relationship between body weight and the different components of the gill area were determined using the method of linear logarithmic transformation as described by Muir and Hughes (1969). From the equations it appears that there is a sharp decrease in the value of the slope at metamorphosis in herring from 3.36 to 0.79, and from 1.59 in plaice, but examination of Fig. 3 shows that this is not so. There is a gradual change in the slope at metamorphosis but for convenience the two stages are treated separately.

Gill Dimensions

a) Total Filament Length. This is plotted for the two spaces in Figs 4A and 5A and further details are given in Table 2.

In herring filaments with secondary lamellae first appear on arches III and IV, at a body length of 20 mm, followed soon after by the appearance of lamellae on the filaments of arch II and lastly arch I. Although the filaments as well as secondary lamellae appear last on arch I the number of filaments increases rapidly until, after the 35 mm stage, it has the greatest number of filaments. In plaice the pattern of development is slightly different; filaments are present on arches I, II, and III at 8 mm body length. Later (about 9 mm) the filaments appear on arch IV. Lamellae appear first on arch II at a body length of 8 mm followed by arch I. From the start, arch II has the largest number of filaments, in contrast to herring.

b) Area of an Average Secondary Lamella. The results are given in Figs 4A and 5C and in Table 2. Each point represents the total area of a single average lamella, i.e. for both sides of the lamella. From the intercept values 'a' in Table 2 it is evident that a 1 mg plaice has a higher average secondary lamellar area than a herring of the same weight. In contrast to this the rate of development of the secondary lamellar area is higher in herring than in plaice, for the stages both up to and beyond metamorphosis.

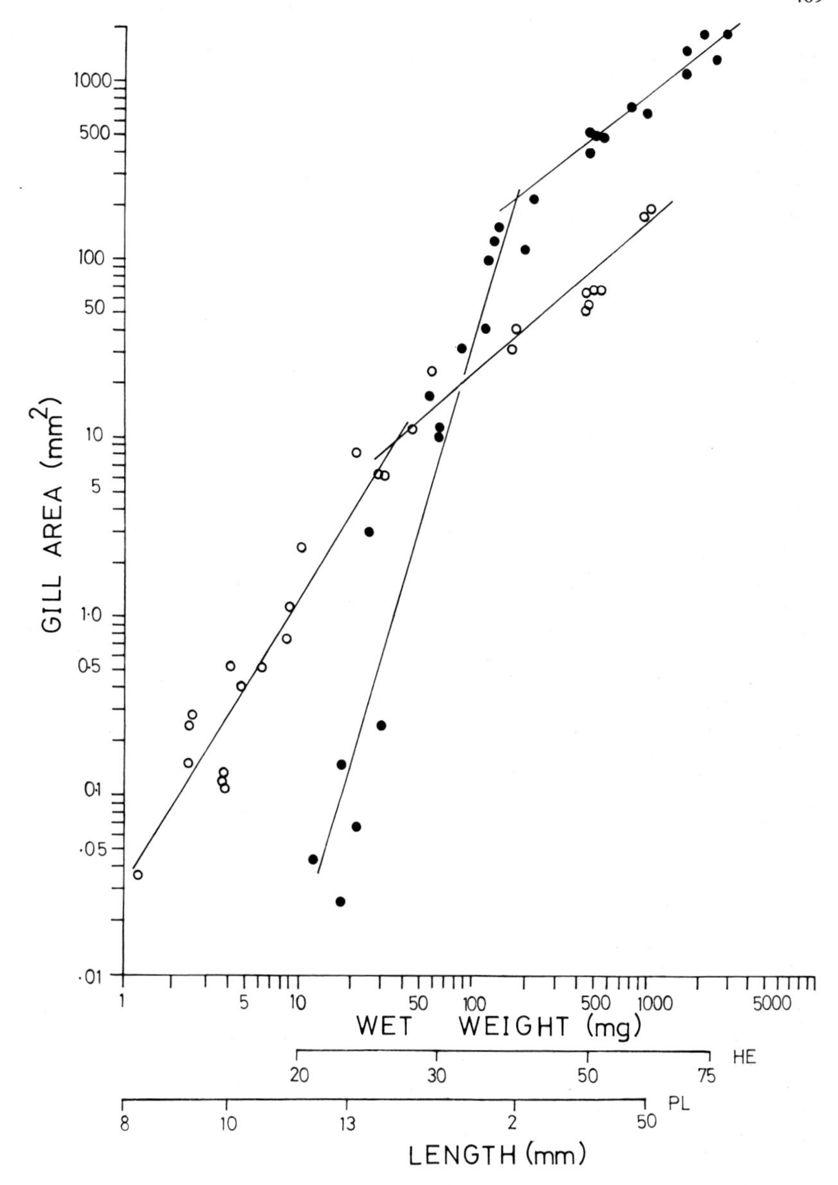

Fig. 3. Relationship between gill area and weight in herring (●), and plaice (o). Corresponding lengths given as a guide; HE - herring, PL - plaice

Table 2. Relationships between gill area and gill dimensions with weight in herring and plaice

	Species	Stage	a	b	.95 C.L. of B	u
Gill area	Herring	up to 40 mm	0.000007	3.36	$2.8 < b < 3.82$	16
		beyond "	3.764	0.78	$0.62 < b < 0.97$	11
	Plaice	up to 12 mm	0.0312	1.59	$1.0 < b < 2.18$	13
		beyond "	0.465	0.85	$0.79 < b < 0.90$	16
Total filament length	Herring	up to 40 mm	.0342	1.84	$1.56 < b < 2.12$	16
		beyond "	20.89	0.57	$0.44 < b < 0.70$	11
	Plaice	up to 12 mm	.3586	0.88	$0.64 < b < 1.12$	13
		beyond "	6.609	0.59	$0.55 < b < 0.61$	16
Area of an average secondary lamella	Herring	up to 40 mm	.00001	1.30	$1.01 < b < 1.59$	16
		beyond "	.00088	0.36	$0.04 < b < 0.68$	11
	Plaice	up to 12 mm	0.00024	0.71	$0.43 < b < 0.85$	13
		beyond "	0.0016	0.21	$0.17 < b < 0.26$	16
Spacing/mm of filament	Herring	up to 40 mm	10.65	0.201	$0.13 < b < 0.27$	16
		beyond "	24.89	0.035	$0.02 < b < 0.04$	11
	Plaice	up to 12 mm	21.21	0.045	$0.07 < b < 0.11$	13
		beyond "	21.45	0.040	$0.05 < b < 0.03$	16

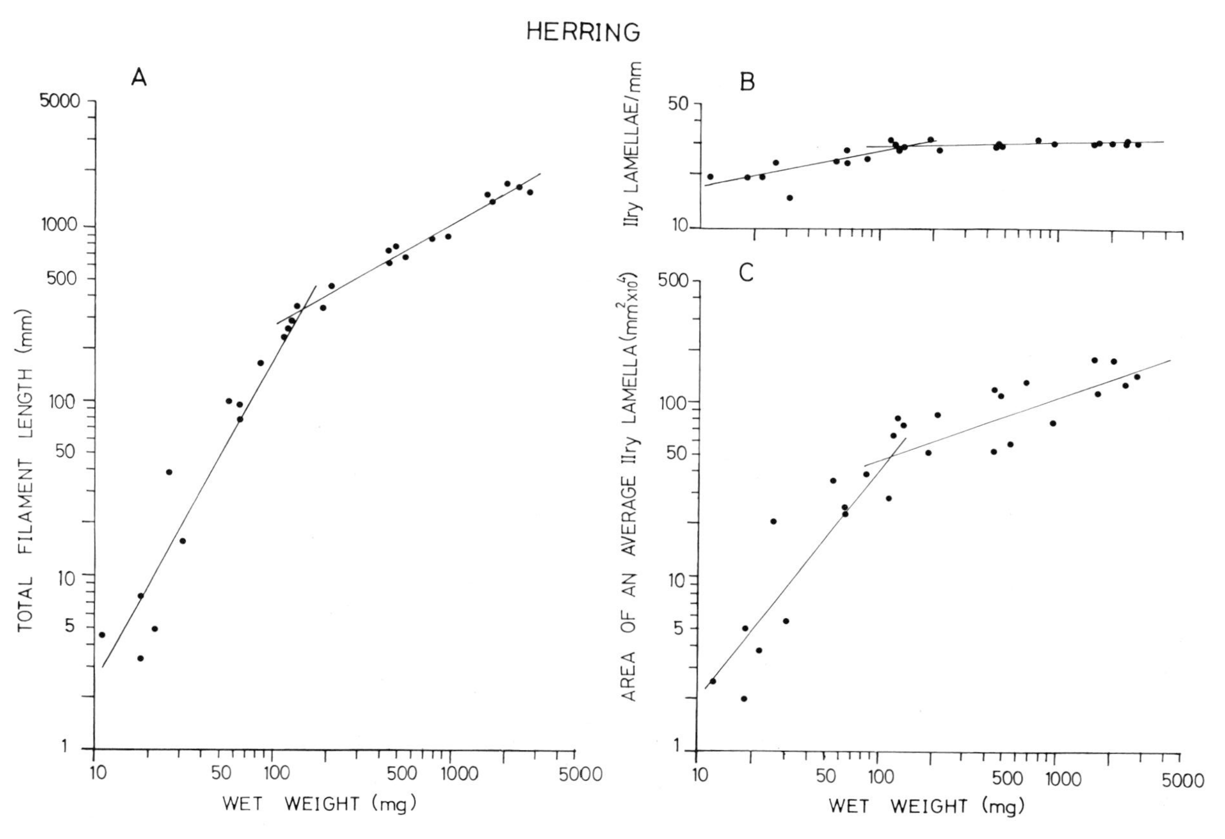

Fig. 4. Relationship between gill dimensions and weight in herring. (A) Total filament length. (B) Secondary lamellae/mm of filament. (C) Area of an average secondary lamella

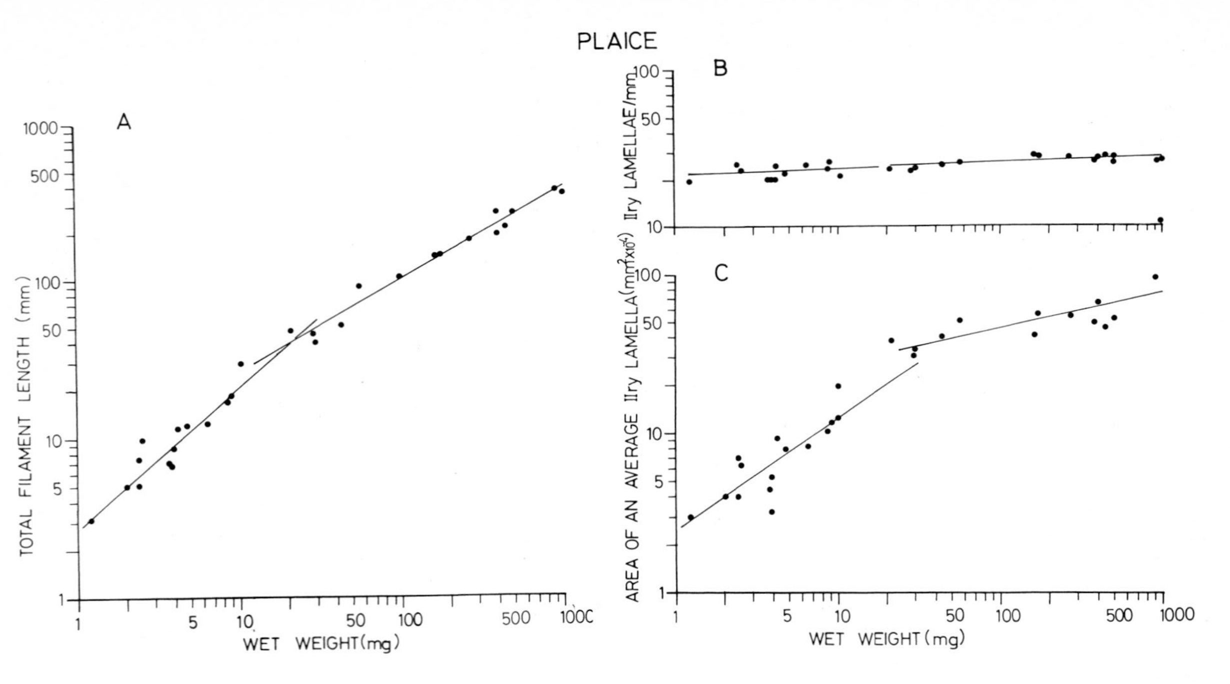

Fig. 5. Relationship between gill dimensions and weight in plaice. (A) Total filament length. (B) Secondary lamella/mm of filament. (C) Area of an average secondary lamella

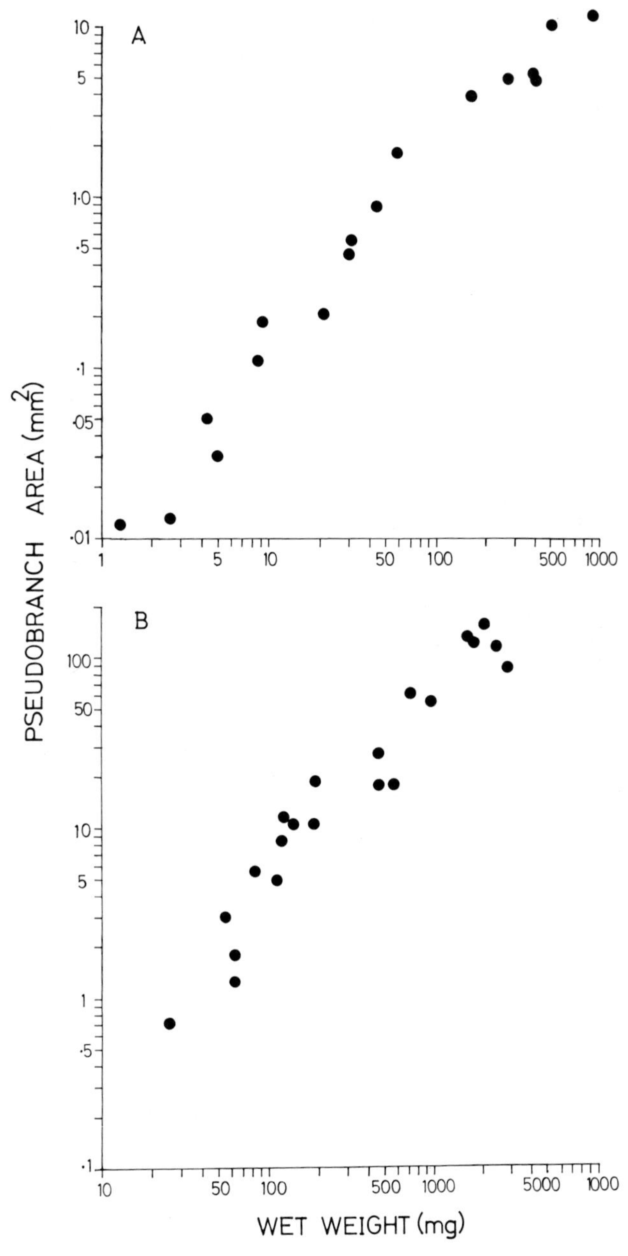

Fig. 6. Relationship between pseudobranch areas and weight in (A) plaice, (B) herring

c) Spacing of the Secondary Lamellae. The results are given in Figs 4B and 5B and in Table 2. It is evident that in the early stages the spacing for a 1 mg herring is about half that of plaice. Beyond metamorphosis, however, the spacing per unit length of filament, as denoted by the intercept "a" values, is approximately equal. Comparison of the slopes gives a different interpretation: in herring there is an initial high rate of development in the spacing per unit length of filament compared to plaice ("b" values of 0.20 and 0.05 respectively); after metamorphosis the rate slows down in herring approximately to that of plaice.

The rate of development of the pseudobranch in terms of its area is shown in Fig. 6. The small surface area in the early stages shows that its contribution toward uptake of oxygen, if any, is negligible at this time.

Blood

Newly-hatched herring larvae show no evidence of corpuscles in the blood even though the heart starts to beat prior to hatching and a circulatory system is present in the larvae. From histological sections it was evident that at about 14-mm body length there was still no evidence of corpuscles, although a matrix was present, in the cavity of the ventricle and around the gut region; this stained light green with Mallorys trichrome stain. Later (16-mm body length) definite corpuscles were present but it was not possible to distinguish different types. By 29-mm body length both erythrocytes and leucocytes could be distinguished. In metamorphosed fish the number of corpuscles greatly increased and the different types of leucocytes were also distinguished easily. In plaice larvae there were no corpuscles in the blood at hatching nor at the start of feeding. Corpuscles of both types were observed at about 10-mm body length as the larvae metamorphosed.

Haemoglobin

The results are summarized in Table 3 and are illustrated by Fig. 7A and B. According to Radzinskaya (1960), the slight bluish colour observed in the early stages is not a true peroxidase reaction, the bluish-green colour being the true one. Peroxidase synthesis appears to take place very early in development in both species.

Plaice at stage 4b (Ryland, 1966 - body length 10 mm) show no visible pink colouration of the blood vessels due to haemoglobin, and no peroxidase activity is observed. There is however a light pink colouration of the cardiac region particularly above (dorsal to) the ventricle which is thought to arise from haemoglobin or a precursor in the pericardial fluid. The fluid when released gives a deep blue colour with the benzidine reagent. In stage 5 plaice (body length 12 mm - near metamorphosis the blood corpuscles in the vessels to the gills turn a visible pink colour. In addition the paired-dorsal aortae are also pink at this stage. The fluid from these vessels gives a deep blue colour when released. It is probable that both corpuscles and fluid give the reaction although no attempt was made to distinguish between them.

In herring the following changes are observed around metamorphosis.

a) At about 32 mm there is no visible pink colour in the heart or gills and the larva is still transparent.

Table 3. Summary of the staining characteristics of the two species using the benzidine test

Stage of development	Species	Staining characteristics
Eggs	Herring	48-68 h eggs show a slight bluish reaction at the region of the blastopore, and at the attachment discs. 4-day-old eggs show a more intensive staining of membrane but not of larvae. Later (10 days) membranes stained very rapidly (2-3 min) and after about 5 min the larva took on the stain. Separation of the larva from the egg membrane showed that the region of the body just behind the head and until about the last quarter of the body gave a slightly bluish stain. No stain was observed in the heart region or connection to the yolk sac, indicating the absence of peroxidase in these areas.
	Plaice	Membranes do not give a reaction in 3-day-old eggs, but the larva gives a slight reaction. Later stage eggs (10-day-old) give a more intense stain in the larva.
Yolk sac	Herring	Gill region and heart give a bluish-green colour, within 1-2 min the rest of the body took on a light green stain.
	Plaice	Gill region and heart give a bluish-green colour, within 1-2 min the rest of the body took on a light green stain.
1 week of feeding	Herring	Similar to the yolk-sac stage. In addition branches of the mandibular arterial arch also gave a stain, and at the base of pectoral fins an intense stain appeared in the form of granules (Fig. 7A (i)).
2 weeks feeding	Plaice	Similar to yolk-sac stage except that gill and heart region stain more intensely, and blood vessels to gill and ventral head region take on the stain as well (Fig. 7B (i)).
3 weeks feeding	Plaice	The ventricle does not stain probably due to a thickening of its wall, but sinur venosus stains intensely, as well as the blood vessels of the gills and ventral part of the head. In addition the base of the caudal fins also give an intense stain (Fig. 7B (ii)).
4 weeks feeding	Herring	Similar to 1 week of feeding. Gills staining more intensely than heart, probably due to increased muscle development in or around the heart.
5 weeks feeding	Herring	Similar to 4 weeks feeding. In addition an intense staining was observed at the base of the caudal fin (Fig. 7B (ii)).

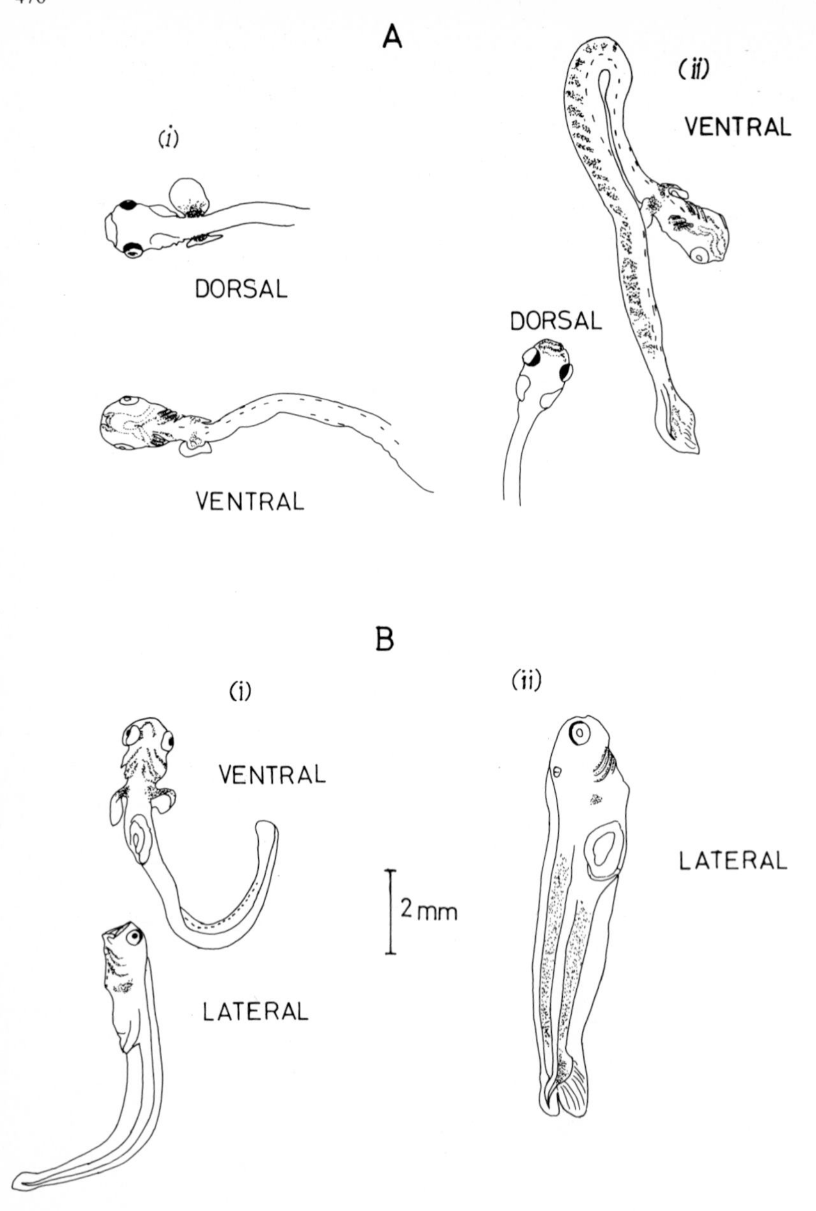

Fig. 7. Sites of peroxidase synthesis in larvae as determined by the benzidine test shown by fine dots. (A) Herring at (i) 1 week of feeding, (ii) 5 weeks of feeding. (B) Plaice at (i) 2 weeks of feeding, (ii) 3 weeks of feeding

b) Later (35-36 mm) the heart is light pink in colour, the gills being lighter. Slight pigmentation was observed dorsally and there was no silvery colour on the body (i.e. no scales are present).

c) At 37-38 mm the heart and gills are bright pink in colour. The belly region is silvery at this stage. The pericardial fluid when released gives a dark blue colour with the benzidine reagent.

d) About 44-mm stage the heart and gills are bright red and the vessels along the length of the body are a deep reddish-pink. The dorsal part of the body is deeply pigmented at this stage and the branchiostegal apparatus is silvery in colour.

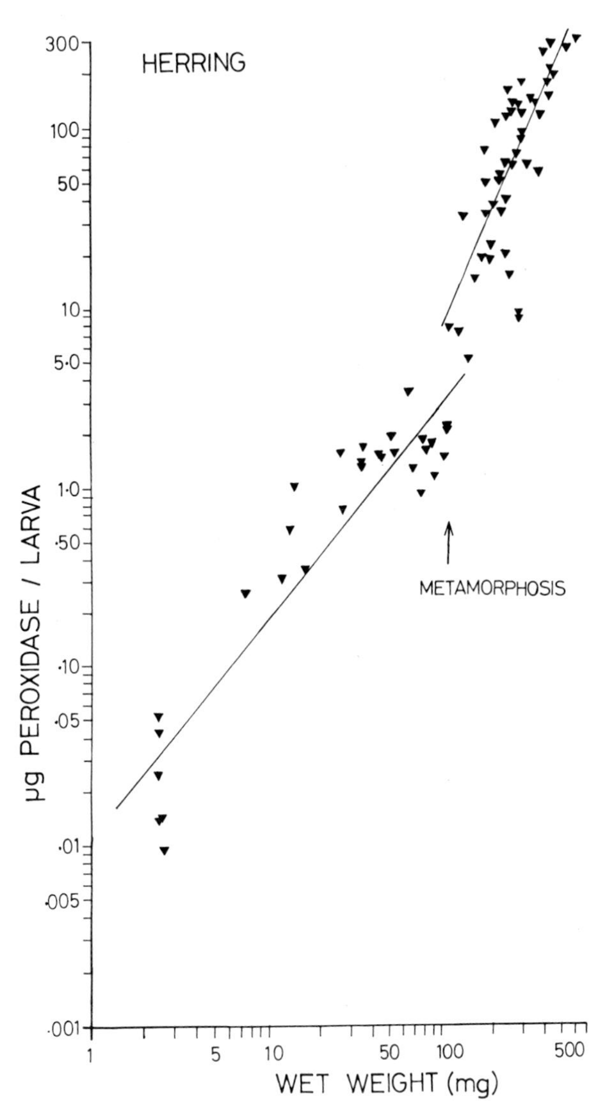

Fig. 8. Relationship between peroxidase concentration per larva and weight in herring

The spectrophotometric results are expressed graphically as total peroxidase concentration per larva and peroxidase concentration per unit weight against weight on a log-log scale (Figs 8-10 and Table 4). In herring the concentration per larva increased rapidly from its time of appearance up to metamorphosis, followed by an apparent further increase in the slope in newly-metamorphosed fish. This difference is significant ($p < 0.001$). In plaice the slope decreased after metamorphosis as compared with "wild" plaice caught locally. The difference is highly significant ($p < 0.001$).

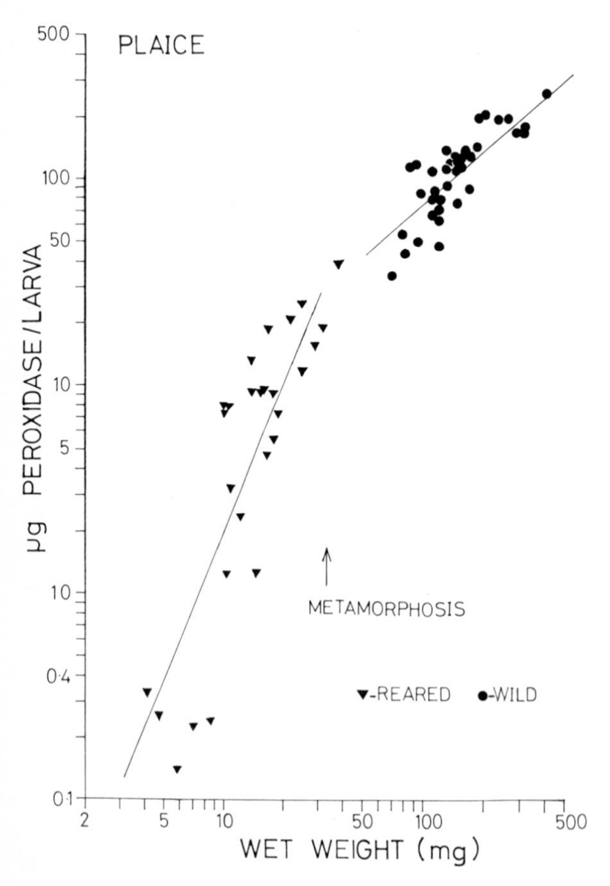

Fig. 9. Relationship between peroxidase concentration per larva and weight in plaice

Peroxidase expressed per unit weight (Fig. 11) showed an increase in the slope at metamorphosis in herring which is highly significant ($p < 0.001$). Plaice on the other hand, showed a decrease in the slope after metamorphosis ($p < 0.001$). Comparison of the relative amounts of peroxidase shows these values to be higher in plaice than in herring.

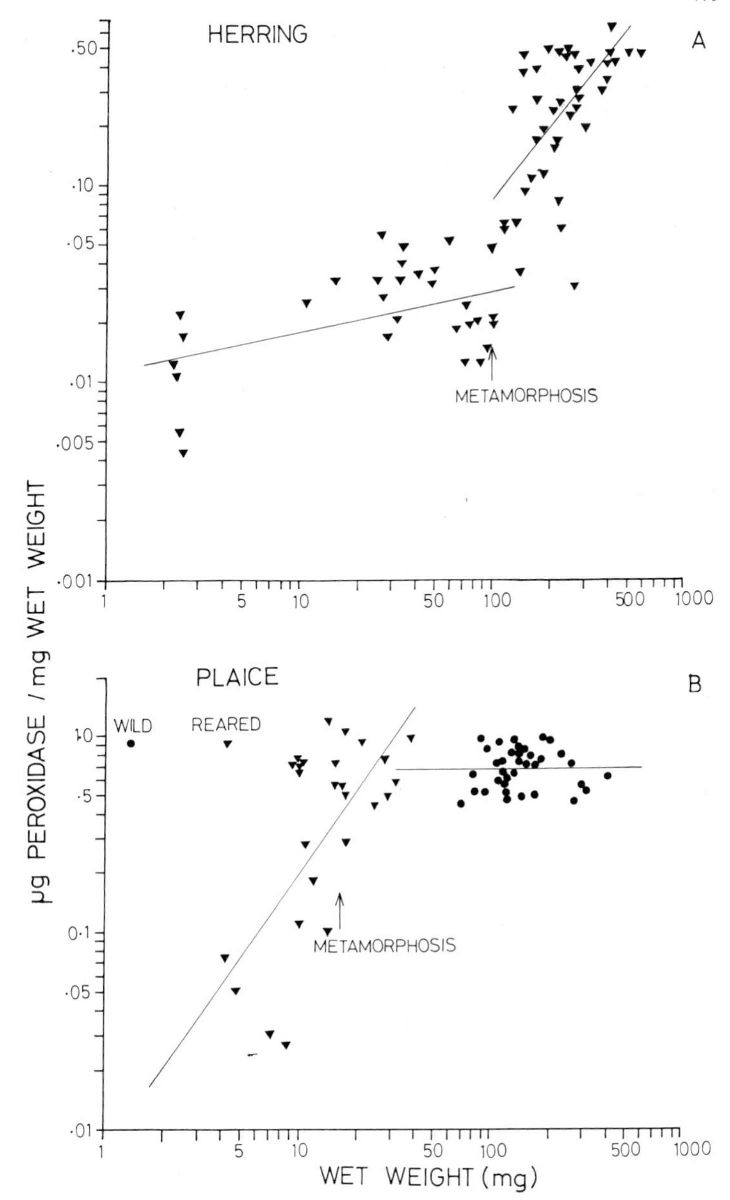

Fig. 10. Relationship of peroxidase concentration per unit weight, to weight in (A) herring, (B) plaice

Table 4. Equations for the relationship between concentration of peroxidase and wet weight in herring and plaice

Herring		Equation	S.D. of Slope	N
µg Peroxidase/Larva	Reared - up to met.	$\text{Log } Y = 0.110 + 1.222 \text{ Log } X$	0.08	31
	Metamorphosed	$\text{Log } Y = 0.002 + 2.267 \text{ Log } X$	0.22	49
µg/mg Wet Wt.	Reared - up to met.	$\text{Log } Y = 0.110 + 0.214 \text{ Log } X$	0.08	31
	Metamorphosed	$\text{Log } Y = 0.002 + 1.262 \text{ Log } X$	0.22	49
Plaice				
µg Peroxidase/Larva	Reared	$\text{Log } Y = 0.089 + 2.353 \text{ Log } X$	0.28	27
µg/mg Wet Wt.	Reared	$\text{Log } Y = 0.090 + 1.356 \text{ Log } X$	0.28	27
µg Peroxidase/Larva	Wild	$\text{Log } Y = 1.521 + 0.848 \text{ Log } X$	0.10	37
µg/mg Wet Wt.	Wild	$\text{Log } Y = 0.662 + 0.009 \text{ Log } X$	0.10	37

DISCUSSION

The fins appear to play an important role in cutaneous respiration after hatching, but an inverse relationship is observed between fin area and gill area in the pre-metamorphic stages in both herring and plaice. Presumably as the gills develop, the role of the fins becomes reduced. Holeton (1971a) found that in young trout larvae the rhythmic movement of the pectoral fins tended to create considerable currents of water around them, the pectoral fins probably acting as accessory ventilatory organs in the early stages. Such currents may well be of importance in fish larvae generally.

The slope of the gill area: body-weight relationship is greater before than after metamorphosis in both herring and plaice. In herring it decreases from values of 3.36 to 0.79 while in plaice it decreases from 1.59 to 0.85. The values for the slopes after metamorphosis are similar to those of the oxygen consumption: weight relationships 0.82 and 0.65 for newly-metamorphosed herring and plaice (see De Silva and Tytler, 1972). Winberg (1960) and Paloheimo and Dickie (1966) considered 0.8 to be a value representative of a large number of species. Nevertheless a great deal of variation from this value has been observed; Muir and Hughes (1969) suggest that species variation in the gill area to weight relationship could be expected as well.

When the <u>total</u> respiratory surface area, including that of the gills when they appear, is plotted against body weight on a log-log basis as in Fig. 11, slope values of 0.59 and 0.51 are obtained for herring and plaice. Since oxygen uptake is proportional to the surface available for gas exchange the total respiratory surface area to body weight relationship should be similar to that of the oxygen consumption to body weight relationship. The low values obtained may be explained only after establishing the importance of diffusion and ventilation rates as well as the thickness of the water-blood pathway in the secondary lamellae of the gills in oxygen uptake. Furthermore, the anatomical and functional areas of the gill need not coincide. The present measurements represent only the maximum possible functional area. Hughes (1966) estimated that the respiratory surface was between 60-70% of the total lamellar surface, based on the area of blood channels in the secondary lamellae. Shunt mechanisms, too, may come into operation and therefore the functional area may be less still.

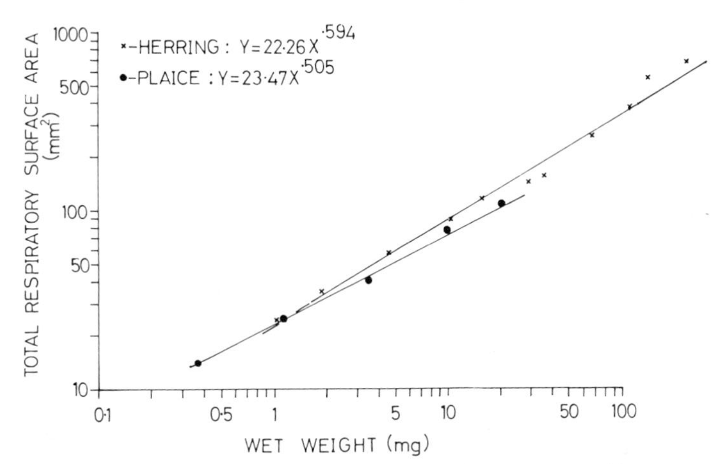

Fig. 11. Relationship between total respiratory surface area (including gills) with weight up to metamorphosis. (x) herring, (•) plaice

Price (1931) obtained a value of 2.38 for the slope in the gill area-length relationships in bass of a size range of 2-40 cm. This is similar to the values of 2.76 and 2.61 for the gill area-length relationship in herring and plaice in the present study. Ursin (1967) analyzing Gray's (1954) data reported a value of 0.82 for the slope of the gill area-weight relationship. Muir and Hughes (1969) obtained values of 0.85 for 3 species of tuna weighing 1-40 kg and 0.90 for the blue fin tuna alone.

Table 2 shows that larval herring have a higher gill area than plaice of the same weight. This may be correlated with the much greater activity of the herring. In the 'O' group stage, however, plaice show a more rapid increase in gill area than herring. It is concluded that this higher rate of gill development may be of some adaptive value to young plaice migrating to the shallower parts of bays.

Table 5 gives a comparison of the present data with that of other authors for fish of the same weight. It is evident that a 1-g herring and bass have approximately the same gill area while that of a plaice

Table 5. Total gill area of herring and plaice (postmetamorphosis) compared with values of other species based on a log-log relationship of gill area to body weight

Species	1g	100g	Author
Herring	882.36 mm^2	33,546.00 mm^2	Present data
Plaice	165.02 mm^2	8,269.00 mm^2	Present data
Bass	865.00 mm^2	33,000.00 mm^2	Price, 1931 - based on Muir and Hughes calculations of Price's data
Skipjack tuna	5,218.00 mm^2	262,000.00 mm^2	Muir and Hughes, 1969
Yellowfin and blue fin tuna	4,025.00 mm^2	200,000.00 mm^2	Muir and Hughes, 1969
Roach	398.00 mm^2	19,000.00 mm^2	Muir and Hughes, 1969

is 5 times less. Tuna, on the other hand, have gill areas about 5 times that of a herring or bass. The relationship to activity of the species is apparent.

Price's (1931) observation that in the early stages an increase in gill area is brought about mainly by an increase in the total filament length, is confirmed by the present data. From Figs 4 and 5 the total filament length for a 1-g herring and plaice was found to be 1071 mm and 389 mm, respectively, giving a ratio of 2.75. One interesting observation is that the numbers of lamellae per unit length of the filament increases with size, in the size range examined, for both species. In other species of fish the spacing may decrease (Muir and Hughes, 1969; Muir, 1969). The spacing of the secondary lamellae varies greatly in different species of fish. Hughes (1966), and Hughes and Shelton (1962) have considered its significance from an ecological viewpoint. They state that in general, the more active fish possess a larger number of filaments of greater length, and therefore a greater number of secondary lamellae. Further, the secondary lamellae are smaller and more closely-packed than those of a more sluggish fish. The smaller pore size in the active species is believed to be advantageous because the diffusion distance is decreased although the gill resistance is increased. More sluggish fish have widely-spaced lamellae and a reduced total filament length. From the histological sections it was evident that herring and plaice showed a similarity to this "active" and "sluggish" category, the pore size of metamorphosed herring and plaice being 0.058 and 0.077 mm, respectively.

The sites of peroxidase activity in herring and plaice larvae are similar to those of Radzinskaya (1960, 1962) on embryonic sturgeon and salmon, and Ostroumova (1962) on embryonic rainbow trout. The slight reaction they observed in the egg membranes was probably due to other respiratory enzymes such as cytochrome, cytochrome oxidases, and catalases, since they are the sites of gas exchange at these stages (Radzinskaya, 1960). The staining of the body musculature as well as the intensive staining at the base of the pectoral and caudal fins is probably indicative of myoglobin or its precursors in the feeding stages. This probably plays a respiratory role in the life of the larva.

Quantitative measurements show a higher rate of haemoglobin synthesis per individual at the pre-metamorphic stage in plaice when compared to herring, but not at metamorphosis. It is evident that even in the early stages of its appearance both absolute and relative values of peroxidase are higher in plaice than in herring.

The change in the slope at the post-metamorphosic stage from 3.00 to 0.95 in plaice (Fig. 9 and Table 4) is indicative of a slowing down of its rate. The high rate of haemoglobin synthesis around metamorphosis in herring as compared to the pre-metamorphic stages (Fig. 8 and Table 4) is probably a reflection of the increasing activity of the fish at these stages. Although the value of peroxidase per larva are low in Radzinskaya's (1960) data in comparison to the present study, expressed per unit of wet weight they are similar to that of "wild" plaice (range of 0.6 - 0.7) while the values for herring are higher.

Haemoglobin increases the affinity and oxygen carrying capacity of the blood even at low pO_2's. Nevertheless there is a great deal of evidence to show that some adult fish are fairly independent of the presence of haemoglobin. One striking observation is the discovery by Ruud (1954) of the haemoglobin-free fish of the Antarctic *Chaenocephalus*, whose blood has an oxygen capacity similar to that of sea water. Further most species of fish appear to have the ability to lose most of their haemoglobin without suffering any great damage. Steen and Berg (1966) and Ryback (1960) give reports of fish with severe anaemia but otherwise appearing normal. Many authors have used carbon monoxide to block more than 90% of the oxygen-carrying capacity of the haemoglobin in a variety of adult fish (see Holeton, 1971b) and larval fish (Holeton, 1971a) without any harmful effects. Thus in fish larvae, the lack of respiratory pigment in the early stages of development in the plankton, where no oxygen depletion is likely to occur, should not limit the metabolism of the larva, which has a high body surface area-weight ratio. Around metamorphosis, when activity increases as the surface area per unit weight diminishes the appearance of respiratory pigment offsets any disadvantages that might otherwise accrue, especially under hypoxic conditions (see De Silva and Tytler, 1972).

ACKNOWLEDGEMENTS

I wish to thank Dr. J.H.S. Blaxter for his interest and constructive criticism of the manuscript and also Dr. P. Tytler and Miss Lucia Solarzano for their interest and encouragement in the project. The supply of plaice eggs from the White Fish Authority (Mr. J. Dye and Mr. M. Bell) and the advice of Mr. D.A. Conroy of the Zoological Society of London on the haemoglobin work is gratefully acknowledged.

REFERENCES

Blaxter, J.H.S., 1968. Rearing herring larvae to metamorphosis and beyond. J. mar. biol. Ass. U.K., 48, 17-28.
De Silva, C.D., 1973. The ontogeny of respiration in herring and plaice larvae. Thesis, University of Stirling, Scotland.
De Silva, C.D. and Tytler, P., 1972. The influence of reduced environmental oxygen on the metabolism and survival of herring and plaice larvae. Neth. J. Sea Res., 7, 345-362.
Gray, I.E., 1953. The relation of body weight to body surface in marine fish. Biol. Bull. mar. biol. Lab., Woods Hole, 105, 285-288.
Gray, I.E., 1954. Comparative study of the gill area of marine fishes. Biol. Bull. mar. biol. Lab., Woods Hole, 107, 219-225.
Harder, W., 1954. Die Entwicklung der Respirationsorgane beim Hering (*Clupea harengus* L.). Z. Anat. Entwgesch., 118, 102-123.
Hawkins, R.I. and Maudesley-Thomas, L.E., 1972. Fish haematology - a bibliography. J. Fish. Biol., 4, 193-232.
Holeton, G.F., 1971a. Respiratory and circulatory responses of rainbow trout larvae to carbon monoxide and to hypoxia. J. exp. Biol., 55, 683-694.
Holeton, G.F., 1971b. Oxygen uptake and transport by rainbow trout during exposure to carbon monoxide. J. exp. Biol., 54, 239-254.
Holliday, F.G.T. and Jones, M.P., 1967. Some effects of salinity on the developing eggs and larvae of the plaice (*Pleuronectes platessa*). J. mar. biol. Ass. U.K., 47, 39-48.
Hughes, G.M., 1966. The dimensions of fish gills in relation to their function. J. exp. Biol., 45, 177-195.
Hughes, G.M. and Shelton, G., 1962. Respiratory mechanisms and their nervous control in fish. Adv. comp. Physiol. Biochem., 1, 275-346.
Korzhuev, P.A. and Radzinskaya, 1957. A micromethod of haemoglobin determination. Vop. Ikhtiol, 9, 192-196.
Muir, B.S., 1969. Gill dimensions as a function of fish size. J. Fish. Res. Bd Can., 26, 166-170.
Muir, B.S. and Hughes, G.M., 1969. Gill dimensions for three species of tunny. J. exp. Biol., 51, 271-285.
Ostroumova, I.N., 1962. The first appearance of haemoglobin in the embryos of the rainbow trout. Dokl. Akad. Nauk SSSR, 147, 263-264.
Paloheimo, J.E. and Dickie, L.M., 1966. Food and growth of fishes. II. Effect of food and temperature on relation between metabolism and body weight. J. Fish. Res. Bd Can., 23, 869-908.
Price, J.W., 1931. Growth and gill development in the small-mouthed black bass, *Micropterus dolomieu* Lacepede. Ohio State Univ. Stud., 4, 46 pp.
Radzinskaya, L.I., 1960. The peroxidase reaction and the formation of haemoglobin during the embryonic development of sturgeon. Dokl. Akad. Nauk SSSR, 130, 1173-1176.
Radzinskaya, L.I., 1966. Changes in the blood indices of young and spawning Neva Salmon (*Salmo salar* L.). Vop. Ikhtiol., 6, 568-573.

Radzinskaya, L.I., 1968. Localization and amount of peroxidase in the embryonic development of Neva salmon (*Salmo salar* L.). Vop. Ikhtiol, 8, 304-306.

Ruud, J.T., 1954. Vertebrates without erythrocytes and blood pigment. Nature (Lond.) 173, 848-853.

Ryback, B., 1960. A pale hag-fish. Nature (Lond.) 185, 777.

Ryland, J.S., 1966. Observations on the development of larvae of the plaice, *Pleuronectes platessa* in aquaria. J. Cons. perm. int. Explor. Mer, 30, 177-195.

Shelbourne, J.E., 1964. The artificial propagation of marine fish. Adv. mar. Biol., 2, 1-83.

Steen, J.B. and Berg, T., 1966. The gills of two species of haemoglobin free fishes compared with those of teleosts with a note on severe anaemia in an eel. Comp. Biochem. Physiol., 18, 517-526.

Ursin, E., 1967. A mathematical model of some aspects of fish growth, respiration and mortality. J. Fish. Res. Bd Can., 24, 2355-2453.

Winberg, G.G., 1960. Rate of metabolism and food requirements of fishes. Fish. Res. Bd Can., Transl. No. 194.

C. de Silva
Dunstaffnage Marine Research Laboratory
Oban, PA 34 4 AD / GREAT BRITAIN

Present address:

Department of Zoology
University of Sri Lanka
Colombo Campus, Thurstan Road
Colombo 3 / SRI LANKA

Effects of Reduced Oxygen on Embryos and Larvae of the White Sucker, Coho Salmon, Brook Trout, and Walleye

R. E. Siefert and W. A. Spoor

INTRODUCTION

This study was made to determine the effects of continuous-reduced dissolved oxygen concentrations on the development and survival of the white sucker, (*Catostomus commersoni* (Lacepede)), coho salmon, (*Oncorhynchus kisutch* (Walbaum)), brook trout, (*Salvelinus fontinalis* (Mitchell)), and walleye, (*Stizostedion vitreum vitreum* (Mitchell)), from fertilization until the larvae were feeding. The effects of reduced oxygen concentrations on survival and hatching of embryos have been described for white suckers (Oseid and Smith, 1971b), coho salmon (Shumway et al., 1964; Phillips et al., 1966), brook trout (Garside, 1966), and walleyes (Oseid and Smith, 1971a; Van Horn and Balch, 1956). However, these studies extended only through hatching and give no information on larval development and survival. Mason (1969) subjected coho salmon embryos and larvae to 2 reduced oxygen concentrations, but developmental delay at the reduced oxygen levels was compensated for by adjusted water temperatures.

MATERIALS AND METHODS

The experimental chambers and the water delivery systems were identical to those described by Siefert et al. (1973). The dissolved oxygen concentrations during the white sucker, brook trout, and walleye experiments were lowered with a vacuum-degassing system similar to that described by Mount (1964). For the coho salmon test the dissolved oxygen concentration was lowered by a counter-current of nitrogen gas in a stripping column. Water was delivered to the chambers in all tests at 60 ml/min, with an approximate velocity of 3.3 cm/min past the embryos. Brook trout were also tested at 15 ml/min (velocity 0.8 cm/min). Oxygen concentrations were measured each day by the unmodified Winkler method. Nominal oxygen concentrations are used for simplicity in subsequent discussion.

The untreated water taken from Lake Superior was analyzed during the experiments according to methods described by the American Public Health Association (1965). Chemical characterictics of the water were as follows (in mg/l): total alkalinity, 38-43; total hardness, 44-47; sodium, 1.10 - 1.36; potassium, 0.44 - 0.60; calcium, 13.1 - 14.2; and magnesium, 2.98 - 3.61. Ranges of pH in the experimental chambers were: white sucker, 7.5 - 7.9; coho salmon, 7.0 - 7.5; brook trout, 7.6 - 7.9; and walleye, 7.5 - 7.6. Experimental temperatures were chosen from the optimum range for each species tested (Allbaugh and Manz, 1964; McAfee, 1966; Anonymous, 1967; and National Water Quality Laboratory, unpublished data). Walleyes were also tested at a temperature above optimum.

White suckers were captured in Greenwood Lake, Cook County, Minnesota on 3 June 1970, and were immediately transported to the laboratory and held overnight at 16°C. The eggs were stripped and fertilized by the dry method (Sorenson et al., 1966). Embryos were selected at random and placed in experimental chambers, 100 to a chamber, immediately after fertilization. (The dry method of fertilization and randomized placing of embryos was used in all experiments.) A 15-h photoperiod, regulated as described by Drummond and Dawson (1970), was used. All chambers were maintained at 18°C (range ± 0.2°C). Testing was done at dissolved oxygen concentrations of 12.5, 25, 50, and 100% saturation (1.2, 2.3, 4.6, and 9.3 mg/l).

Coho salmon were netted in Lake Superior near the mouth of the French River on 21 December 1970. The fish were stripped and the eggs were fertilized immediately. Water temperature was 2°C at stripping. The embryos were held in a covered container, and the water temperature was gradually raised to 7°C over a period of 4 1/2 h. Each experimental chamber was allotted 54 embryos 5 h after stripping. The chambers were darkened until hatching began; thereafter a photoperiod of 10 1/2 h was used. An average temperature of 7.4°C (range ± 0.4°C) was used for the first 93 days of the test. It was then raised gradually to 10°C over a period of 5 days, where it remained (range ± 0.3°C) for the last 21 days of testing. Testing was done at dissolved oxygen concentrations of 12.5, 25, 50, and 100% saturation (1.5, 3.0, 6.0, and 12.0 mg/l at 7°C, and 1.4, 2.8, 5.6, and 11.2 mg/l at 10°C).

Brook trout were seined from Mud Creek, Cook County, Minnesota, on 5 November 1971, and were immediately stripped. After the water temperature had been raised gradually from 2°C to 8°C, 90 embryos were placed into each experimental chamber 5 h after fertilization. The chambers were darkened during incubation; after hatching a photoperiod of 10 1/2 h was used. Test temperature was maintained at 8°C (range ± 0.2°C). At 60 ml/min the fish were tested at dissolved oxygen concentrations of 12.5, 20, 35, 50, and 100% saturation (1.5, 2.3, 2.9, 4.1, 5.8, and 11.7 mg/l); at 15 ml/min flow fish were tested at 25 and 50% saturation.

Walleyes from Cut Foot Sioux Lake, Itasca County, Minnesota, were held in the laboratory 4 days at 10°C prior to stripping. Eggs were fertilized on 4 May 1972. The water temperature was raised gradually to 17°C in 2 h and 20°C in 3 h before 100 embryos were placed into each experimental chamber. These temperatures were maintained with ranges of ± 0.2°C. A 16-h photoperiod was used. At 17°C dissolved oxygen concentrations were adjusted to 12.5, 20, 25, 35, 50, and 100% saturation (1.2, 1.9, 2.4, 3.4, 4.8, and 9.6 mg/l). At 20°C oxygen concentrations were 25, 50, and 100% saturation (2.2, 4.5, and 8.9 mg/l).

All larvae were fed an excess ration at least three times each day. White suckers were given live brine shrimp (*Artemia salina*) and natural plankton. Coho salmon and brook trout were given Oregon Moist[1,2] and Glencoe Mills[3] trout foods. Walleyes were fed live brine shrimp nauplii and natural plankton strained through No. 253 Nitex screen to remove

---

[1]The United States Environmental Protection Agency neither recommends nor endorses any commercial product; trade names are used only for identification.

[2]Product of Moore and Clark Company, La Conner, Washington.

[3]Product of Glencoe Mills, Glencoe, Minnesota.

adult predaceous copepods. Fecal material and excess food were removed daily. The experiments were ended when all larvae were feeding, with the exception of the coho salmon test, in which fish at 25% saturation of dissolved oxygen had not started to feed.

RESULTS

White sucker survival was similar at all oxygen concentrations except at 12.5% saturation, where total mortality occurred before hatching began (Table 1). Hatching and start of feeding were identical in the controls and at 50% saturation. Fish at 25% saturation started and completed hatching about 6 h earlier than those at the higher oxygen concentrations, and they began to feed 4 days later. They also had a distinctly different behavioral pattern, remaining near or at the surface after first feeding. Fish in the higher concentrations dispersed throughout the chamber within a day after start of feeding. Fish length at the end of the experiment was similar at 50% saturation and in the controls, but was markedly less at 25%.

Coho salmon survival was moderately reduced and inhibitory effects on development increased at each successive reduction in oxygen concentration to and including 25% saturation; at 12.5% saturation all died before or during hatch (Table 2). Hatching was delayed 3 - 5 1/2 days at each lower oxygen concentration from the controls to 25% saturation. Start of feeding was delayed 12 days at 50% saturation and more than 22 days at 25%. Length was slightly reduced at 50% saturation and noticeably reduced at 25%.

Survival of brook trout was high at oxygen concentrations of 25% saturation and above at a flow rate of 60 ml/min (Table 3). At 20% saturation survival was markedly reduced, however, and at 12.5% saturation all fish died before they started to feed. At 20% saturation most mortality (85%) occurred during the pre-feeding larval stage; at 12.5% saturation most mortality (59%) occurred during the embryo stage.

Although start of hatching of brook trout was not affected by reduction of oxygen to 20% saturation, duration of hatch was prolonged several days at this tension. First feeding was delayed 2 to 8 days with each successive reduction including 25% saturation; at 20% saturation feeding started 36 days after the control. Mean fish length decreased with decrease in oxygen concentration. Total decrease from the control was as follows: at 50% saturation, decrease in length was 0.7 mm; at 35%, 1.3 mm; at 25%, 3.0 mm; at 20%, 5.9 mm. Survival and development were similar at the two water flow rates at 50% saturation, but at 25% saturation and the low flow rate the hatching period was prolonged and feeding was delayed by a few days.

Walleyes at 17°C survived nearly as well at 50% saturation as the controls; at 35% saturation survival was reduced markedly, and all larvae at the lower tensions died before they started to feed (Table 4). No hatching occurred at 12.5% saturation. Development was not affected at 50% saturation; at saturations of 35% and lower larval size at hatching was reduced. At 25% and 20% the larvae were noticeably weak swimmers. At 35% saturation and above most mortality took place during the larval stage; at saturations below 35% most of the deaths were during the embryonic stage. Development at all comparable oxygen concentrations was faster at 20°C than 17°C (Table 4), but few fish survived to the end of the test.

Table 1. Times to hatching and feeding, and total lengths and survival at 22 days after fertilization of white suckers at the oxygen concentrations indicated, a flow rate of 60 ml/min, and 18°C. Values are for duplicate experimental chambers with 100 embryos initially in each chamber

Oxygen Concentration				Time in days to:			Total length at 22 days (mm)	Survival at 22 days (%)
% saturation[a]		mg/l						
Nominal	Actual (mean)	Mean ± SD	Range	Start of hatching	At least 90% hatched	Start of feeding		
100 (Control)	98	9.1 ± 0.1	9.0-9.3	7 ¼	7 ¼	12 ¼	16.7 (15.9-17.2)[b]	92.0 (92)[b]
50	53	4.9 ± 0.1	4.6-5.2	7 ¼	7 ¼	12 ¼	16.4 (13.2-17.0)	95.5 (94-97)
25	27	2.5 ± 0.1	2.3-2.8	7	7	16 ¼	13.4 (12.0-14.1)	89.5 (89-90)
12.5	13	1.2 ± 0.1	1.0-1.5	---	---	---	--- ---	0

[a]In relation to atmospheric air.

[b]Mean and range.

Table 2. Times to hatching and feeding, and total lengths and survival at 119 days after fertilization of coho salmon at the oxygen concentrations and temperatures indicated, and a flow rate of 60 ml/min. Values are for duplicate experimental chambers with 54 embryos initially in each chamber

Temperature[b]	Oxygen Concentration				Time in days to:			Total length at 119 days (mm)	Survival at 119 days (%)
	% saturation[a]		mg/l						
	Nominal	Actual (mean)	Mean ± SD	Range	Start of hatching	At least 90% hatched	Start of feeding		
7 °C	100 (Control)	97	11.6 ± 0.3	11.0-12.5	60	63 1/4	93	33.8 (30.0-40.0)[c]	79.6 (76-83)[c]
10 °C		92	10.4 ± 0.2	9.7-11.1					
7 °C	50	50	6.0 ± 0.3	5.1-6.9	65	66 1/4	105	31.6 (30.0-34.0)	73.1 (72-74)
10 °C		49	5.5 ± 0.3	4.8-6.4					
7 °C	25	24	2.9 ± 0.4	2.3-4.3	70	71 3/4	119[d]	26.3 (22.5-28.0)	70.4 (68-72)
10 °C		24	2.7 ± 0.2	2.3-3.0					
7 °C	12.5	12	1.4 ± 0.1	0.9-1.9	85	89 1/4	---	--- ---	0
10 °C		12	1.4 ± 0.02	1.37-1.44					

[a]In relation to atmospheric air.

[b]First 93 days the test temperature was 7 C. Days 94 through 99 it was gradually raised to 10 C where it remained to end of test.

[c]Mean and range.

[d]No feeding activity during test period.

Table 3. Times to hatching and feeding, and total lengths and survival at 133 days after fertilization of brook trout at 8°C and the oxygen concentrations and flow rates indicated. Values are for duplicate experimental chambers with 90 embryos initially in each chamber

Oxygen concentration				Time in days to:			Total length at 133 days (mm)	Survival at 133 days (%)
% saturation[a]		mg/l						
Nominal	Actual (mean)	Mean ± SD	Range	Start of hatching	At least 90% hatched	Start of feeding		
				60 ml/min				
100 (Control)	90	10.5 ± 0.4	9.4-11.1	62	67	93	24.9 (20-30)[b]	90.5 (90-91)[b]
50	50	5.8 ± 0.2	4.8-6.2	51[c]	66	95	24.2 (18-30)	85.0 (84-86)
35	36	4.2 ± 0.2	3.6-4.7	57[c]	66	100	23.6 (20-28)	85.5 (81-90)
25	25	2.9 ± 0.2	2.4-3.6	63	68	108	21.9 (18-26)	91.5 (91-92)
20	20	2.3 ± 0.1	1.9-2.7	62	74	129	19.0 (18-21)	14.5 (4-25)
12.5	14	1.6 ± 0.2	1.2-2.6	71	85	--	-- --	0
				15 ml/min				
50	49	5.7 ± 0.3	4.8-6.6	62	66	96	24.6 (18-30)	89.0 (86-92)
25	25	2.9 ± 0.3	2.5-3.7	62	73	110	21.2 (19-24)	85.0 (84-86)

[a] In relation to atmospheric air.

[b] Mean and range.

[c] One hatch until day 62.

Table 4. Times to hatching and feeding, and total lengths and survival at 20 days after fertilization of walleye at the oxygen concentrations and temperatures indicated, and a flow rate of 60 ml/min. Except as noted, values are for duplicate experimental chambers with 100 embryos initially in each chamber

Oxygen Concentration: % saturation[a] Nominal	% saturation[a] Actual (mean)	mg/l Mean ± SD	mg/l Range	Time in days to: Start of hatching	Time in days to: At least 90% hatched	Time in days to: Start of feeding	Total length at 20 days (mm) Mean	Total length at 20 days (mm) Range	Survival at 20 days (%)
17 °C									
100 (Control)	93	8.9 ± 0.3	8.1-9.3	7 3/4	11	14	9.5	(8.4-10.4)[b]	41.5 (37-46)[b]
50	50	4.8 ± 0.2	4.0-5.0	7 3/4	11	14	9.5	(8.9-10.2)	38.5 (25-52)
35	35	3.4 ± 0.3	2.7-3.9	7 3/4	11	14 3/4	9.2	(8.9-9.5)	14.5 (2-27)
25	25	2.4 ± 0.2	2.1-2.9	9 3/4	13	--	--	--	0
20	21	2.0 ± 0.1	1.8-2.3	8 3/4	12	--	--	--	0
12.5	13	1.3 ± 0.1	1.0-1.7	--	--	--	--	--	0
20 °C									
100[c]	92	8.4 ± 0.5	7.2-9.5	5	8	9	9.2	9.2	2
50	49	4.5 ± 0.3	4.0-5.6	5	8 1/4	10	9.0	9.0	0.5 (0-1)
25	25	2.3 ± 0.2	2.0-2.6	8	11	--	--	--	0

[a] In relation to atmospheric air.

[b] Mean and range.

[c] Single experimental chamber with 100 embryos initially.

DISCUSSION

Rate of development, along with survival, should be used as criteria for judging the effects of reduced dissolved oxygen concentrations on the early life stages of fish. The ecological significance of a delay in development must be considered for each species and its particular habitat.

Since a reduction in size at emergence is most likely detrimental to coho salmon survival (Mason, 1969), even the highest reduced oxygen concentration that we tested (50% saturation) could have an effect on the success of a year class. Some minor retarding effects on the development of brook trout occurred at all reduced oxygen concentrations, but marked delays in development and poor survival did not appear until the tension was reduced to 20% saturation.

A reduction of the oxygen tension to about 50% saturation does not appear to be harmful to the development or survival of early-stage white suckers or walleyes raised at temperatures near their optima. The delaying effects of 25% saturation on first feeding and growth of white suckers, as well as the behavioral change, might be a disadvantage for later survival. The sharp decrease in survival of walleyes at 35% saturation clearly makes this oxygen tension inadequate for this species. Even at high oxygen concentrations, 20°C is near the lethal temperature for early-stage walleyes.

SUMMARY

1. The effects of continuous-reduced dissolved oxygen concentrations were determined on the embryos and larvae of the white sucker, coho salmon, brook trout, and walleye from egg fertilization until the larvae were feeding.

2. White suckers and walleyes were not harmed at an oxygen tension of 50% saturation. Development of white suckers was inhibited at 25% saturation, and walleye survival dropped sharply at 35% saturation.

3. Developmental delay and mortality of coho salmon increased progressively with each reduced oxygen concentration tested including the highest reduced tension, 50% saturation.

4. Some slight delays of brook trout development occurred at all reduced oxygen tensions tested, but marked effects on development and poor survival occurred only at 20% saturation and below.

ACKNOWLEDGEMENT

We thank Roll Syrett, Lawrence Herman, Anthony Carlson, Charles Kleiner and Robert Drummond, who aided in solving the technical problems and assisted in the maintenance of the tests during the 2 years of study. Minnesota Department of Natural Resources personnel provided the walleyes and coho salmon.

REFERENCES

Allbaugh, C.A. and Manz, J.V., 1964. Preliminary study of the effects of temperature fluctuations on developing walleye eggs and fry. Progve Fish Cult., 26, 175-180.
American Public Health Association, 1965. Standard methods for the examination of water and waste-water. 12th ed. New York, 169 pp.
Anonymous, 1967. Temperatures for hatching walleye eggs. Progve Fish Cult., 29, 20.
Drummond, R.A. and Dawson, W.F., 1970. An inexpensive method for simulating diel patterns of lighting in the laboratory. Trans. Amer. Fish. Soc., 99, 434-435.
Garside, E.T., 1966. Effects of oxygen in relation to temperature on the development of embryos of brook trout and rainbow trout. J. Fish. Res. Bd Can., 23, 1121-1134.
Mason, J.C., 1969. Hypoxial stress prior to emergence and competition among coho salmon fry. J. Fish. Res. Bd Can., 26, 63-91.
McAfee, W.A., 1966. Eastern brook trout. In: A Calhoun (ed.), Inland Fisheries Management; California Department of Fish and Game, 242-260 pp.
Mount, D.I., 1964. Additional information on a system for controlling the dissolved oxygen content of water. Trans. Amer. Fish. Soc., 93, 100-103.
Oseid, D.M. and Smith, L.L., 1971a. Survival and hatching of walleye eggs at various dissolved oxygen levels. Progve Fish Cult., 33, 81-85.
Oseid, D.M. and Smith, L.L., 1971b. Survival and hatching of white sucker eggs at various dissolved oxygen levels. Progve Fish Cult., 33, 158-159.
Phillips, R.W., Campbell, H.J., Hug, W.L. and Claire, E.W., 1966. A study of the effect of logging on aquatic resources, a progress report, 1960-1966. Progress Memorandum, Fish. No. 3, Oregon State Game Comm., 28 pp.
Shumway, D.I., Warren, C.E. and Doudoroff, P., 1964. Influence of oxygen concentration and water movement on the growth of steelhead trout and coho salmon embryos. Trans. Amer. Fish. Soc., 93, 342-356.
Siefert, R.E., Spoor, W.A. and Syrett, R.F., 1973. Effects of reduced oxygen concentrations on northern pike (*Esox lucius* Linnaeus) embryos and larvae. J. Fish. Res. Bd Can., 30, (in press).
Sorenson, L., Buss, K. and Bradford, A.D., 1966. The artificial propagation of esocid fishes in Pennsylvania. Progve Fish Cult., 28, 133-141.
Van Horn, W.M. and Balch, R., 1956. The reaction of walleyed pike eggs to reduced dissolved oxygen concentrations. Purdue Univ. Engineering Ext. Dept., Series No. 91, 319-341.

R.E. Siefert
U.S. Environmental Protection Agency
National Water Quality Laboratory
6201 Congdon Boulevard
Duluth, Minn. 55804 / USA

W.A. Spoor
U.S. Environmental Protection Agency
National Water Quality Laboratory
6201 Congdon Boulevard
Duluth, Minn. 55804 / USA

Developmental Events

Brain Growth of Young Herring and Trout

A. Packard and A. W. Wainwright

INTRODUCTION

The job of the brain is to process information coming to it from the inside and outside worlds in such a way as to maintain an adequate representation of those worlds and to build and store programmes for appropriate action. In this sense the brain provides a representation of the animal's world - what J.Z. Young (1964) has called 'the model in the brain' - and we can say the *brain is the organ that represents the animal*. Unfortunately, the brain supplies no convenient two-dimensional plan of the life history of a fish as a scale does, but we can weigh the organ, and we can count the number of cells, total DNA quantities being used as an indicator of total cell numbers, and the total RNA as a measure of their activity.

In this paper we report changes in the total amount of DNA present in young herring and trout brains and in their wet weight. The study was undertaken out of basic curiosity as part of a project to find out how many cells there are at different stages of growth and what their relationship is to behaviour. Fish were chosen because their brains grow continuously (Geiger, 1956); and there was every reason to expect that they do so by the addition of cells. We also present evidence that linear eye dimensions can be a direct indicator of brain growth, at least during the early part of the life history. The only comparable studies we know of are those on the octopus (Packard and Albergoni, 1970; Giuditta et al., 1971), which gave rise to the present one.

MATERIAL AND METHODS

Herring (*Clupea harengus*) samples were from fish fertilized, hatched, and reared at the Dunstaffnage Marine Research Laboratory in 1972. Other herring samples included fish taken by beach seining in Loch Etive and from Oban trawler landings. Rainbow trout (*Salmo gairdneri*) were obtained in the "eyed" stage from New Zealand. One-year-old and 2-year-old *S. salar* were obtained from Faskally, Scotland. The following measures were made on the fresh specimens (usually immediately after death following anaesthesia in MS222): body weight (wet), body length (fork length), eye diameter (mean of longest and shortest diameters), brain weight (wet). Brains were lifted after sectioning the olfactory tracts and medulla at the narrowest point. Brains for nucleic acid estimation were kept on ice or stored at -20°C. Free nucleotides were removed in cold perchloric acid. The concentrations of RNA and DNA were measured by the ultraviolet absorption method following alkali extraction (RNA) and then acid extraction (DNA).

We have occasionally sampled freeze-dried material passed on to us by Dr. Ehrlich. The material is extremely easy to dissect and handle when one has trained oneself to recognize a freeze-dried brain, assuming the drying procedure has not shattered the internal organs.

RESULTS

a) Herring. The herring brain increases in absolute size from less than 50 µg at hatching to about 400 mg in large 3-year-olds. The brain weight-body weight relationship during a substantial part of larval and postlarval life is shown in Fig. 1 (upper curve). The straight

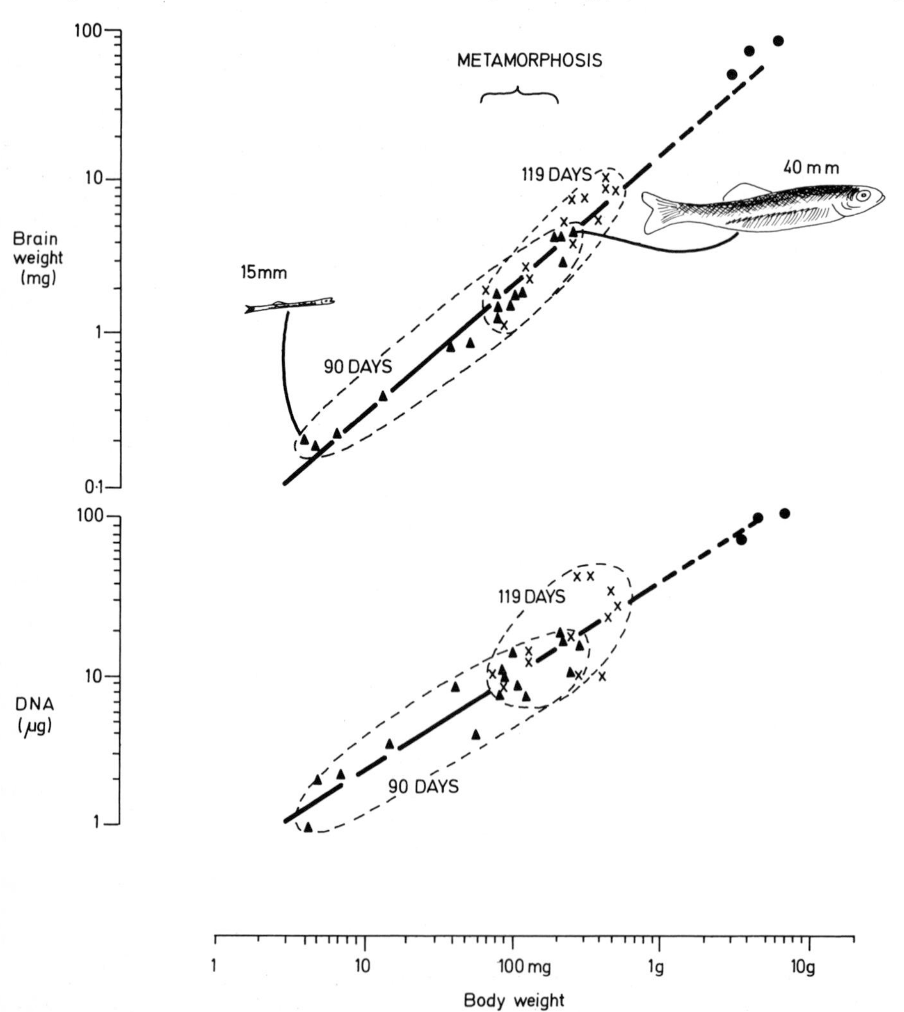

Fig. 1. Log/log representation of the brain wet weight (upper curve) and brain DNA content (lower curve) of young herring (*Clupea harengus*) sampled on two separate days - at 90 days (▲) and 119 days (x) of age. The values of α in the equation $y = bx^{\alpha}$ are the slopes of the regression lines. Drawings are of small and large specimens of the same day-age. Though quite different in size and growth stage, both fish hatched within two days of each other and were reared under the same aquarium conditions with abundant food (*Artemia* nauplii and fresh plankton). (●) Other reared herring sampled in 1971 are shown

line obtained when individual points are plotted on log/log paper reveal that brain growth is allometric. The points in the upper right of the figures suggest that the rate alters little up to 10 g body weight. Although increasing in size at a slower rate than the body (negative allometry), the slope of 0.88 is steeper than for any of the species reported by Geiger (1956). His values - representing overall brain growth rates of a variety of mainly freshwater species belonging to widely separated systematic groups - ranged from α = 0.28 to α = 0.76. No microbalance was available when we were making measurements on newly-hatched larvae in 1971, and no plots of larvae below 4 mg (15 mm) are included. The brain weight varies from 0.2 mg to 100 mg as body weight increases from 4 mg to 8 g, i.e. there is a 500-fold increase in the weight of the brain during a relatively restricted section of the total growth range.

The amount of DNA in the brains of these same fish is shown in Fig. 1 (lower curve). The increase with body weight is again negatively allometric, but the slope of the straight line fit is shallower than for brain wet weight showing that brain DNA increases at a slower overall rate than brain wet weight. For a 500-fold increase in brain weight, brain DNA increases 100-fold.

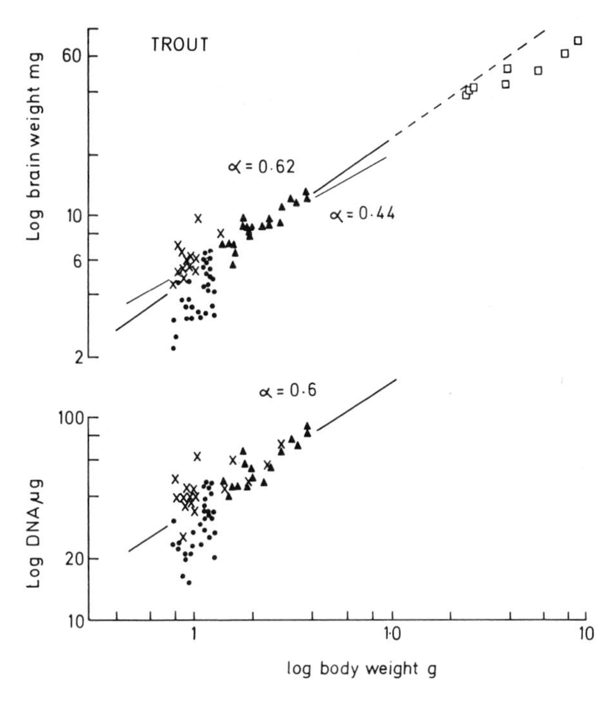

Fig. 2. Log/log representation of the brain wet weight (upper curve) and brain DNA content (lower curve) of young rainbow trout (*Salmo gairdneri*) hatched in the laboratory. The values of α in the equation $y = bx^{\alpha}$ are the slopes of the regression line fitted for alevins and early fingerlings beyond the yolk-sac stage. (From fertilized eggs of a single female supplied by the New Zealand Department of Internal Affairs). (●) Alevins to the end of the yolk-sac stage. (▲) Batch 1 fingerlings moved into feeding tank after yolk resorption. (x) Batch 2 fingerlings kept in the hatching tray after yolk resorption, including non-feeders and late-feeders. (□) 1-year-old and 2-year-old sea trout (*Salmo salar*)

b) Trout. Similar log/log plots of young trout up to and including the beginning of the fingerling stage have been made. The picture is confused by the presence of large amounts of yolk in the alevins, but straight lines have been fitted for the "post-larval" fingerling stage (Fig. 2). Two slopes are shown, one for fingerlings that began feeding at the end of yolk-sac resorption (α = 0.62), the other including fingerlings that started feeding late, or never fed (α = 0.44). Here brain weight and brain DNA of late alevins and normal fingerlings have similar slopes when plotted against body weight. The rates happen to be the same as the rate of DNA increase in the herring. The points in the upper right of Fig. 2 of 1-year-old and 2-year-old sea trout (*S. salar*) suggest that the brain weight-body weight rate does not change before 2 g body weight is reached.

c) Brain Dimensions as a Function of Other Parameters. The trout results concern a much narrower section of the growth range than the herring results, and they expose one of the difficulties inherent in relative growth studies: that of finding a meaningful rather than an arbitrary reference parameter. Except for the uptake of water and metabolic losses, the alevin hardly changes in weight from that of the fertilized egg. Use of body length as a parameter has a double disadvantage: it involves plots of linear dimensions against mass;

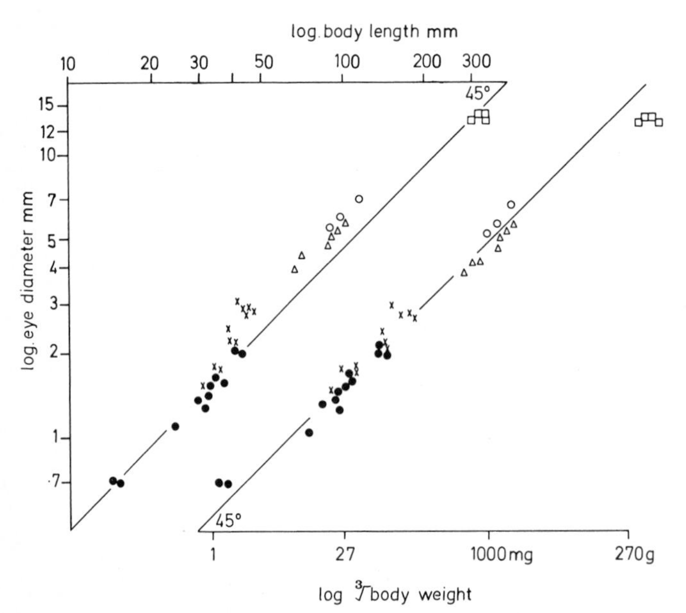

Fig. 3. Eye dimensions of herring (*Clupea harengus*) as a function of body length (left) and body weight (right). Logarithmic scales. The lines at 45° to the ordinate are the lines parallel to which points would fall if eye diameter increased in direct proportion to either body length or body weight. The fit between eye diameter and cube root of body weight for young herring passing through metamorphosis is a good one. Eye diameter toward the end of the growth range shows negative allometry both with respect to body length and cube root of body weight. (●) Laboratory-reared larval and metamorphosing herring sampled at 3 months of age. (x) Metamorphosing and post-metamorphic herring sampled at 4 months of age. (o) Other herring reared in 1971. (▲) Beach-seined herring, Loch Etive. (□) Commercially-fished adults landed at Oban

also, length measures do not account for changes of shape, and many fish, including the herring, change in shape radically before and during metamorphosis. The other candidates we have considered are cube root of body weight, eye diameter (and eye diameter cubed), and day-temperature age.

d) Mean-Eye Diameter and Cube Root of Body Weight. Mean-eye diameter increases allometrically with body length. In the trout (Lyall, 1957) the slope of the log/log curve is $\alpha = 0.88$ (negative allometry). In the young herring (Fig. 3) it is slightly above 1 (positive allometry). The right-hand set of points in Fig. 3 show, however, that when eye diameter is plotted against cube root of body weight on log/log paper they fall close to a straight line with a slope of 1 (45°), i.e. eye diameter of young herring increases as a cube root function of body weight.

The fact that mean eye diameter of the herring alters at the same rate as the cube root of body weight, at least up to 10 g body weight (100 mm), has encouraged us to adopt mean-eye diameter as a reference parameter. Figs 4 and 5 show log/log plots of brain weight of herring and trout against eye diameter cubed.

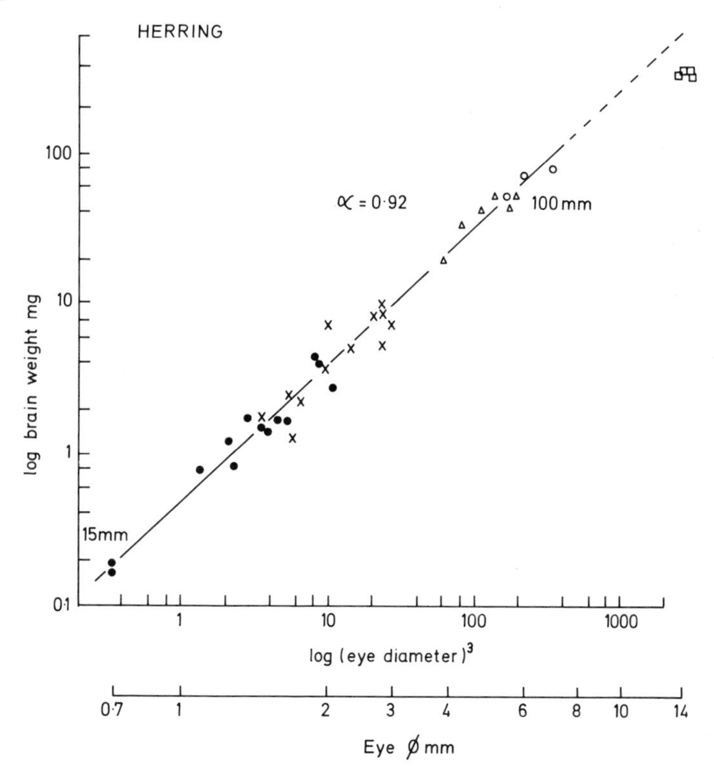

Fig. 4. Herring: Relative growths of eye and brain. Log/log representation of brain weight against mean eye diameter cubed for herring (*Clupea harengus*). The straight-line grouping of plots indicates that brain weight is allometrically related to eye size, their rates of change being constant for larval and post-larval stages between 15 mm and 100 mm body length. Laboratory-reared herring sampled at (●) 3 months, (x) 4 months, (o) 6 months of age. (△) Beach-seined herring, Loch Etive. (□) Adult 3-year-olds

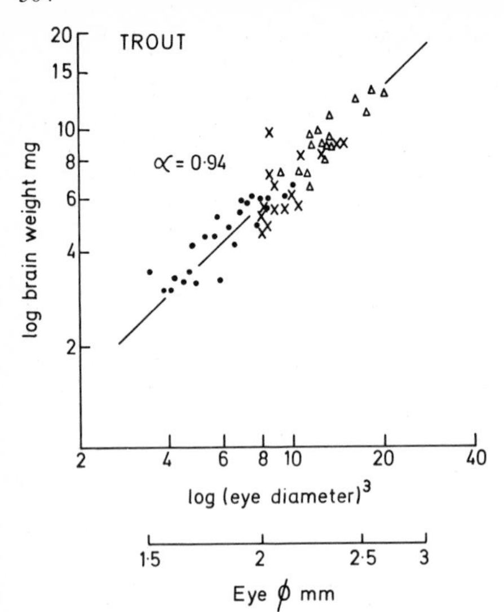

Fig. 5. Trout: Relative growths of eye and brain. Log/log representation of brain weight against eye diameter cubed for young trout (*Salmo gairdneri*) reared in the laboratory. Over a limited range of "larval" (alevin) and "post-larval" (fingerling) life brain weight is allometrically related to eye size. (●) Alevins. (△) Early fingerlings in feeding tank. (x) Non-feeding and late-feeding fingerlings

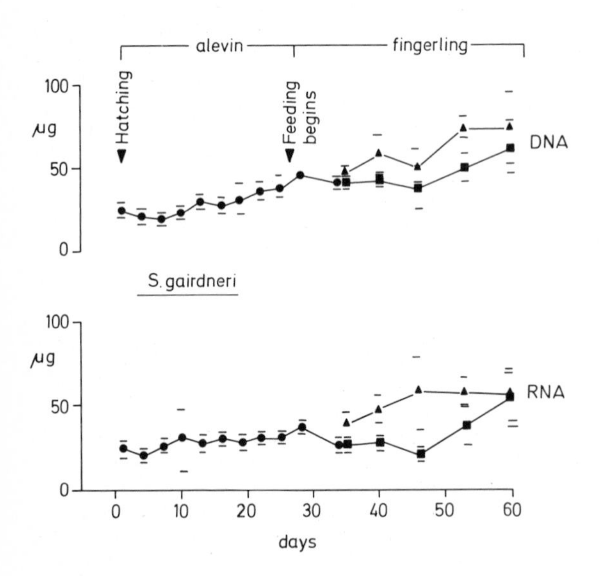

Fig. 6. Increases in brain DNA (upper plots) and RNA (lower plots) as a function of day-age in laboratory reared *Salmo gairdneri*. Sampling took place at 3-day intervals. Each plot is the mean from 3 fish. Standard deviations indicated by horizontal bars. (●) Alevins. (▲) fingerlings in feeding tank. (■) Non-feeding and late-feeding fingerlings

e) Day-Temperature Age. Most of the points from which the relative growth curves of Fig. 1 were established are from fish of only 2 day-temperature ages, sampled at 3 months and 4 months of age. They were the offspring of 5 females and 2 males; all had hatched between 8 and 10 March and been reared in 2 tanks, one held at 10°C, the other at 8°C rising to 14°C. All were feeding and actively shoaling when sampled. None were stunted and none apparently hypertrophied. They exemplify what has been called a "size-hierarchy" (Blaxter and Ehrlich, this symposium).

The rainbow trout that provided the plots in Fig. 2 were all the offspring of a single female and 3 males sampled at 3-day intervals from hatching onward. The early fingerling stages include equal numbers of 2 different batches: one batch was moved to a feeding tank at the end of the yolk-sac (alevin) stage; the other was kept on in the hatching tray where little or no feeding took place. It will be seen that when resorption of yolk stops and the fingerlings do not feed, brain growth (both wet weight and DNA) also stops. The kind of picture obtained when the brain weight or brain DNA of these same specimens is plotted as a function of day-age is shown in Fig. 6. There is considerable variation in the fingerlings by the end of the first month (60 days post-hatching). Note that the RNA content of the brain in the "non-feeders" drops down sharply.

DISCUSSION AND CONCLUSIONS

Size Hierarchies. The practice of sampling size hierarchies raises several points of interest and a number of problems. There is, of course, considerable advantage in being able to establish a relative growth curve by sampling a population on a limited number of occasions as we did for herring, rather than at regular intervals as was done for trout. As far as we can tell, the small animals sampled at 3- or 4 months-of-age are indistinguishable from the early stages of the large animals taken on the same day. The fish are at different stages of growth. The relative growth-curve - which we believe to be the normal curve for this particular population - has, in fact, been established from fish that were all growing at different rates! The size variance under the conditions of laboratory rearing appears to be greater than that of populations sampled at sea (Blaxter, 1968).

The Log/Log Curves and Their Relevance for Fisheries Studies. Despite the limitations of logarithmic presentation of data in allometry (Scholl, 1954), it remains the method of convenience for describing relative (i.e. differential) growth. The curves must not, however, be extrapolated beyond the particular section of the life history for which they are established. Over a wide range of larval and post-larval life the herring brain grows as approximately the 9/10 power of body weight and of eye dimensions cubed. Over a narrower range of early life the trout brain grows as the 0.62 power of body weight and as the 0.94 power of eye dimensions cubed. The amount of DNA in both brains appears to increase at slightly less than the 2/3 power of body weight.

Although the log/log data can be used to establish the equation $y = bx^{\alpha}$ - where y is brain weight or brain DNA, x the reference parameter (body weight, or eye diameter cubed, etc.) and α (a pure number representing the ratio of their specific growth rates) is the slope of the straight-line fit - we have limited ourselves to description of the data and not to their mathematical treatment. The question arises, however, how specific or general are the curves (and the equations that can be derived); could they be useful in fisheries studies? We have some evidence to suggest that the brain-growth curves are specific to the particular population and year class being studied. Our data for herring under the same conditions the previous year (1971) fall on a different line from the 1972-reared herrings shown here, the relative brain growth curve being slightly shallower and displaced above the 1972 curve. The different slope provides a different value of α; the displacement alone would change the level at which the straight line intercepts the ordinate on the log/log paper, thus changing the value of b. (N.B. b is the value of y when x is equal to unity and varies with the units of measurement used.)

Further data are needed before we can answer the question as to whether brain dimensions, or eye diameter, can be used to recognize fish of a given population or to pick out abnormal fish; there already is some evidence that they may be used for these purposes. The head dimensions of reared herring are larger than of wild specimens (Balbontin et al., 1973) and the brains of starved herrings (unpublished results) in rearing experiments fall above the straight-line fit established for non-starved reared fish, i.e. there is some brain sparing. Brain sparing is also evident in the non-feeding fingerlings of Fig. 2.

Cell Numbers. Assuming that brain cells show little polyploidy, or a constant proportion of polyploidy, DNA values will be a direct reflection of cell numbers in the brain. Two points seem worth noting: 1. the numbers of cells in the brains of fish increases during post-embryonic life by orders of magnitude unknown for any other vertebrate group. 2. Brain cell numbers are independent of day-temperature age and of the physical environment, but dependent on growth stage as reflected in other body dimensions.

We have not yet calibrated the herring and trout DNA figures against cell counts, but assuming that their diploid values are the same as those reported for *Clupea pallasii* and for *Salmo irideus* (Ohno and Atkin, 1966; Ohno et al., 1968) - respectively 2.0 - 2.8 pg and 5.6 pg i.e. 28-40% and 80% mammalian values - the maximum number of cells in the smallest brains would be between 350 and 500 thousand, and in the brains of the largest post-metamorphic herrings at 4 months between 11 and 15 million. The first of these two sets of figures is comparable to the number of cells present in the brain of a higher insect. It is probably not without significance that the herring larva's behaviour repertoire is more "insect-like" than "vertebrate-like", being characterized by fixed-action patterns in which experience plays a role limited to the switching on of stereotyped patterns of feeding, shoaling and light-dependent vertical movements, etc. (see, for instance, Rosenthal and Hempel, 1970; also Blaxter and Ehrlich, this symposium). However, the post-metamorphic fish, equipped with a larger number of brain cells comes to show a more varied repertoire in which exploration and learning play a part. Similar changes in behaviour are undergone by all vertebrates as they develop from egg to adult, but fish biologists are dealing with organisms that make the transition after the animal has already begun to interact with the external environment. Given the great differences in their brains, it is not surprising to find that larval characters are so different from adult ones and that the gradual transformations that take place at metamorphosis are accompanied by a gradual and not a step-wise increase in brain dimensions and DNA content.

ACKNOWLEDGEMENTS

We gratefully acknowledge the help of Drs. J.H.S. Blaxter, Karl Ehrlich, S.S. De Silva and Mr. F. Balbontin of the Dunstaffnage Marine Research Laboratory, Oban, for supplying herring larvae; Dr. D.H. Mills, Forestry and Natural Resources, Edinburgh, for trout-rearing facilities; the Director and staff of the Freshwater Fisheries Laboratory, Faskally, Pitlochry; and the New Zealand Department of Internal Affairs, Rotorua. The work was done with the aid of a grant from the Science Research Council of the United Kingdom for research on postembryonic development of the brain.

REFERENCES

Balbontin, F., De Silva, S.S. and Ehrlich, K.F., 1973. A comparative study of anatomical and chemical characteristics of reared and wild herring. Aquaculture in press.

Blaxter, J.H.S., 1968. Rearing herring larvae to metamorphosis and beyond. J. mar. biol. Ass. U.K., 48, 17-28.

Blaxter, J.H.S. and Ehrlich, K.F., 1973. Buoyancy and activity changes during growth and starvation of herring and plaice larvae. This symposium.

Ehrlich, K.F., 1973. Chemical changes during growth and starvation of herring larvae. This symposium.

Geiger, W., 1956. Quantitative Untersuchungen über das Gehirn der Knochenfische, mit besonderer Berücksichtigung seines relativen Wachstums. Acta. anat., 26, 121-163.

Giuditta, A., Libonati, M., Packard, A. and Prozzo, N., 1970. Nuclear counts on the brain lobes of *Octopus vulgaris* as a function of body size. Brain Res., 25, 55-62.

Lyall, A.H., 1957. The growth of the trout retina. Q. J. microscop. Sci. 98, 101-110.

Ohno, S. and Atkin, N.B., 1966. Comparative DNA values and chromosome complements of eight species of fishes. Chromosoma, 18, 455-466.

Ohno, S., Wolf, U. and Atkin, N.B., 1968. Evolution from fish to mammals by gene duplication. Hereditas, 59, 169-187.

Packard, A. and Albergoni, V., 1970. Relative growth, nucleic acid content and cell numbers of the brain in *Octopus vulgaris*. J. exp. Biol. 52, 539-553.

Rosenthal, H. and Hempel, G., 1970. Experimental studies in feeding and food requirements of herring larvae (*Clupea harengus*). In: Marine food chains (J.H. Steele, ed.). Edinburgh: Oliver and Boyd, 344-364.

Scholl, D.A., 1954. Regularities in growth curves, including rhythms and allometry. In: Dynamics of growth processes (11th Growth symposium). E.J. Boell (ed.).

Young, J.Z., 1964. A model of the brain. Oxford University Press.

A. Packard
Department of Physiology
University Medical School
Teviot Place
Edinburgh EH 8 9 AG /
GREAT BRITAIN

A.W. Wainwright
Department of Physiology
University Medical School
Teviot Place
Edinburgh EH 8 9 AG /
GREAT BRITAIN

Influence of Temperature and Salinity on Embryonic Development, Larval Growth and Number of Vertebrae of the Garfish, *Belone belone*

M. Fonds, H. Rosenthal, and D. F. Alderdice

INTRODUCTION

The garfish (*Belone belone* L.) commonly occurs in European coastal waters from Norway to the Mediterranean, including the Baltic Sea and the Black Sea (Ehrenbaum, 1904; Colette and Berry, 1965). Spawning takes place in the spring in coastal waters. Sexually mature garfish are caught regularly in the Dutch Wadden Sea in May and June at temperatures of 15-20°C and salinities of 20-30°/oo S. The 3 mm diameter eggs normally are found attached by their chorionic filaments to submerged vegetation. Observations on seasonal abundance of adult fish, artificial fertilization of the eggs, embryonic development, and growth of young fish were recorded earlier (Rosenthal, 1970; Rosenthal and Fonds, 1973).

The garfish is not considered to be an estuarine species. However, its coastal spawning habit, and the fact that it is known to enter estuaries frequently, leads to questions of possible adaptation of the species to estuarine conditions during early stages of development. Accordingly, its thermal plasticity and salinity tolerance were investigated by incubating eggs and rearing larvae over a range of combinations of temperature and salinity.

METHODS

Eggs and sperm were collected from live fish (see Table 1) caught in trap nets at the south point of the Isle of Texel. The eggs were fertilized in the laboratory, divided into subsamples, and assigned to the various incubation conditions within 1 h after fertilization.

Experimental Design. Ranges of salinity and temperature were selected on the basis of existing information on incubation of *Belone* eggs. The ranges selected, 10-45°/oo S and 9-24°C, are somewhat wider than those associated with the natural distribution of the species. Within those limits, a 13-point modified factorial design was employed with replicated tests at the centre of the design (Table 2). Three subsamples of eggs from different parent fish (Table 1) were incubated and reared at the constant salinity-temperature conditions of the 14 trials. Responses to the experimental conditions were determined for the following parameters: rate of development, survival of eggs to hatching, dimensions of newly-hatched larvae, growth of larvae and juveniles reared to about 35 mm total length, and meristic characters of the young fish reared in the same experimental conditions in which they had been incubated.

Table 1. Characteristics of the adult fish from which sex products were obtained for the 3 subsamples of fertilized eggs

Subsample	Date fertilized (1971)	Sex	Size (cm)		Weight (g)		Number of			Egg size (mm)
			Total Length	Depth at vent	Total	Gonads	Vertebrae	Anal rays	Dorsal rays	
1	2-6-15.00	♀	73.5	4.8	532.5	95.3	80	20	17	3.29 ± 0.06
2	3-6-17.00	♀	53.0	3.7	198.7	45.0	79	21	18	2.95 ± 0.06
		♂	50.0	2.8	120.0	5.3	79	21	18	
3	5-6-11.00	♀	66.5	4.5	388.0	65.0	80	21	18	3.06 ± 0.07
		♂	51.0	3.0	128.6	4.1	80	20	18	
		♂	52.0	3.0	128.4	3.9	80	21	17	
		♂	70.0	4.5	393.0	7.3	78	21	17	

Table 2. Salinity-temperature incubation conditions and resulting incubation periods from fertilization to median hatching time in the 14 trials examined

Trial	Design		Working levels[1]		Incubation time (h)[1] Subsample		
No.	S (‰)	Temp. (°C)	S (‰)	Temp. (°C)	1	2	3
1	12.5	10.06	12.58 ± 0.28	10.10 ± 0.28	-	-	-
2	12.5	22.94	12.83 ± 0.41	22.85 ± 0.25	263 ± 7	256 ± 9	254 ± 6
3	42.5	10.06	42.49 ± 0.60	10.10 ± 0.28	1704	-	-
4	42.5	22.94	42.82 ± 0.62	22.85 ± 0.25	267 ± 4	259 ± 6	274 ± 4
5A	27.5	16.50	27.52 ± 0.34	16.48 ± 0.23	516 ± 9	516 ± 8	504 ± 7
5B	27.5	16.50	27.54 ± 0.39	16.48 ± 0.23	503 ± 12	518 ± 7	499 ± 7
6	10.0	16.50	10.08 ± 0.29	16.48 ± 0.23	518 ± 11	518 ± 8	504 ± 9
7	45.0	16.50	45.09 ± 1.08	16.48 ± 0.23	512 ± 10	518 ± 8	510 ± 8
8	27.5	9.00	27.44 ± 0.35	9.08 ± 0.33	-	-	-
9	27.5	24.00	28.02 ± 0.61	23.96 ± 0.28	251 ± 4	240 ± 3	240 ± 5
10	20.0	13.20	20.10 ± 0.42	13.21 ± 0.20	866 ± 19	861 ± 16	863 ± 11
11	20.0	19.72	20.34 ± 0.36	19.72 ± 0.20	345 ± 6	342 ± 7	338 ± 3
12	35.0	13.20	35.06 ± 0.47	13.21 ± 0.20	865 ± 16	873 ± 12	860 ± 22
13	35.0	19.72	35.20 ± 0.42	19.72 ± 0.20	348 ± 6	342 ± 6	336 ± 8

[1] ± 1 standard deviation.

Incubation Methods. The required temperatures were obtained in 14 plastic containers of salt water suspended in a series of constant temperature baths. The desired salinities were provided by addition of distilled water or sea salt to filtered sea water; the volume of water in the containers was maintained at 10 l by daily adjustment with distilled water to compensate for evaporation. Salt water in the containers was renewed weekly, at which time 25 IU penicillin and 25 μg streptomycin per ml of salt water were added. Temperatures were checked every 3 h, and salinities every 2 days. Variations in temperature and salinity generally were kept within 2-3% of design levels (Table 2).

In each of the 14 containers the 3 subsamples of eggs were incubated in plastic cylinders (dia. 10 cm) provided with a plankton gauze bottom (500 μ mesh). Aeration at the centre of the containers, and horizontal slits in the cylinder walls, resulted in water circulation through the egg samples. Individual egg samples were examined daily to record state of development and rate of survival. At such times the adhesive filaments were cut to prevent clumping of eggs, and dead eggs were removed.

The incubating eggs were subjected to natural indirect daylight and, in addition, to continuous illumination from overhead fluorescent lights.

Newly-hatched larvae were recorded at 4-h intervals in order to estima the median hatching time in each trial. The larvae were narcotized with MS 222 and measured immediately by microscope and ocular micrometer. They were returned to separate rearing baskets in the appropria incubation containers, and reared to 3-4 cm total length within the following 20 days. The young fish were fed natural zooplankton (mainly copepods), *Artemia* nauplii, and occasionally dried aquarium fish food (Tetramin). They were finally killed with MS 222, measured and weighed (wet weight) and fixed in 4% formaldehyde. Vertebrae and fin rays were counted by microscopic examination after treatment of the young fish with alizarin red S in KOH-glycerine (Hollister, 1934; see also Taylor 1967).

Analysis of the Data. Data obtained for 10 of the measured variables were subjected to response surface analysis (Lindsey et al., 1970; Alderdice, 1972) to estimate the relation between response and the various salinity-temperature levels examined. Regression surfaces were fitted to each set of data using either a linear polynomial

$$Y = a + b_1S + b_2T + b_{11}S^2 + b_{22}T^2 + b_{12}ST$$

or the corresponding nonlinear form

$$Y^{\gamma} = a + b_1S^{\alpha 1} + b_2T^{\alpha 2} + b_{11}S^{2\alpha 1} + b_{22}T^{2\alpha 2} + b_{12}S^{\alpha 1}T^{\alpha 2}$$

where Y = response; S, T = salinity ($^{o}/oo$) and temperature (^{o}C), respectively; α_1, α_2, γ = power parameters derived by maximum likelihood procedures (Lindsey and Sandnes, 1970). If the nonlinear case provided a more plausible fit of the surface to the data, computations were based on that model; otherwise, they were completed on the basis of the linear model.

The data for each response were subjected to analysis of variance to ascertain if any of the coefficients could be omitted from the polynomials without seriously disturbing the adequacy of the model employe Finally, response surfaces were generated from the fitted polynomials. These are described in terms on contours of response around the centre of each surface, Y_S, which estimates the salinity-temperature conditions associated with maximum or minimum response on each surface.

RESULTS

A total of 13 attributes were investigated. Those analyzed by response surface techniques provide an initial appreciation of their relation to conditions of incubation and rearing (Table 3). The relationships examined are presented in the sections to follow.

Rate of Development. Incubation time from fertilization to several stages of embryonic development is shown in Fig. 1. Incubation time to hatching showed no relation to salinity (Table 2), but approaches an asymptote of about 7.1 days with increasing temperature. The rate of egg development (% per day) to the neurula stage and to hatching, describes an apparent sigmoidal relationship with temperature (Fig. 2), in a manner similar to that observed by Kinne and Kinne (1962) for

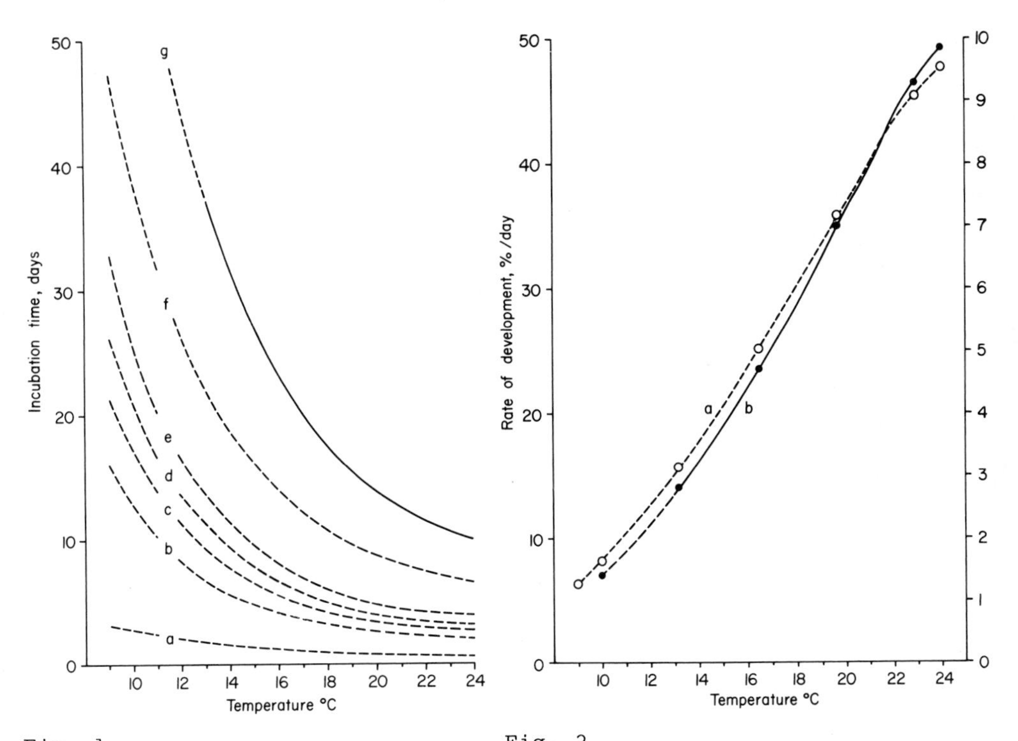

Fig. 1 Fig. 2

Fig. 1. Incubation time in days to various stages of egg development in relation to temperature. (a) blastodisc; (b) closure of blastopore, neurula; (c) tail bud; (d) heart function; (e) blood circulation, ventilation with caudal and pectoral fins; (f) dark pigmented embryo bent in a circle, gill ventilation; (g) hatching

Fig. 2. Rate of development in % per day, in relation to temperature. (a) Early development from fertilization to the neurula stage, left-hand scale; (b) from fertilization to hatching, right-hand scale

Table 3. Summary of response surface analysis on 10 measured attributes during development of eggs and young of *Belone belone*. The symbols in col. 2 refer to the corresponding elements of the polynomial. The linear (S, T) and quadratic (S^2, T^2) terms are associated with salinity and temperature, respectively. The last 4 columns show the calculated response, Y_s, at the centre (°/oo S, T°C) of the surface, and show whether the response Y_s is a maximum or a minimum

Measured attribute	Coefficients of the model explaining variance associated with regression	Centre of the response surface			Response at centre
		Y_s	‰ S	°C	
Eggs					
Percent total hatch	T, T^2	54.8%	34.4	17.4	max.
Percent hatch of neurulae	T, T^2	106.1%[1]	37.1	17.8	max.
Newly hatched larvae					
Body size, L × H mm	S, T^2	11.1	~0	14.0	max.
Yolk-sac size, L × H mm	S, T, T^2	0.26	~0	14.0	min.
Reared larvae					
Percent reared to 30-40 mm total length	S^2, T, T^2	107.2[1]	31.4	19.4	max.
Total length, mm	S^2, T, T^2	35.2	31.6	19.2	max.
Standard length, mm	S, S^2, T, T^2	27.6	17.7[2]	17.7	max.
Wet weight, mg	T^2	60	31.2	19.1	max.
Condition factor, W/(L × H^2)	S, S^2, T, T^2	74.0	33.5	15.1	min.
Number of vertebrae	T, T^2	79.79	41.1	21.2	min.

[1]Predicted without error, response at the centre would not be greater than 100%.

[2]Near-maximum values for standard length are calculated to range from about 14-32‰ S.

Table 4. Survival of eggs, larvae, and young fish in the 3 subsamples of eggs in relation to incubation and rearing conditions (see Table 2)

Trial	Initial number of eggs subsample			Surviving neuralae subsample			Pre-hatching survival subsample			Number of hatched larvae subsample			Number of larvae reared subsample		
	1	2	3	1	2	3	1	2	3	1	2	3	1	2	3
1	110	60	105	15	15	15	12	0	6	0	0	1	0	0	0
2	120	61	92	22	11	36	21	8	36	21	5	30	18	1	28
3	119	60	42	16	0	12	11	0	1	3	0	0	0	0	0
4	138	68	51	28	38	23	28	38	19	27	34	17	25	29	17
5A	134	77	53	31	58	23	31	54	23	30	54	23	29	49	22
5B	135	57	40	32	40	17	32	37	17	30	32	17	28	32	16
6	159	72	58	34	33	31	27	32	31	25	31	31	20	19	28
7	156	58	41	33	48	16	33	47	16	32	46	16	31	38	15
8	160	77	57	43	0	5	40	0	0	0	0	0	0	0	0
9	152	99	93	28	26	45	28	15	45	27	9	39	25	7	38
10	153	60	62	37	40	35	37	39	34	36	30	30	27	29	28
11	145	59	37	36	38	17	33	35	17	30	34	17	28	32	16
12	126	79	41	37	30	18	37	27	18	31	27	17	31	19	5
13	159	46	44	32	34	21	30	34	19	30	34	18	27	32	16

eggs of *Cyprinodon*. Satisfactory trend lines were fitted to the data for rate of development to hatching using either a second order expression:

$$Y = -4.3092 + 0.4628T + 0.0053T^2$$

or the logistic $Y = k/(1 + \exp(a + bX))$:

$$Y = 13.21/(1 + \exp 0.226\ (19.26 - T)$$

where Y = rate (% per day), T = temperature (°C), k = estimated maximum rate, equivalent to an incubation time of 7.6 days, the power 0.226 defines the slope, and 19.26°C is the centre of the trend where exp 0 = 1 and the rate is 13.21/2 = 6.6%. Extrapolation sets the developmental zero at 7-8°C approximately.

Survival of Eggs and Larvae. It was not possible to measure the rate of fertilization as the eggs initially were clumped together. After the sensitive period of gastrular overgrowth was complete the eggs were separated by cutting the adhesive filaments of the chorion.

Survival was determined at several stages of development: after blastopore closure (neurula), prior to hatching, after hatching (% initial eggs, and neurulae, which hatched), and after rearing to a total length of 30-40 mm (Table 4). Statistically, percent survival to hatching was related only to temperature over the broad range of salinities employed. However, both salinity and temperature influenced survival among hatched larvae reared to 30-40 mm total length. Maximum survival, both to hatching and in the rearing phase, occurred at sali-

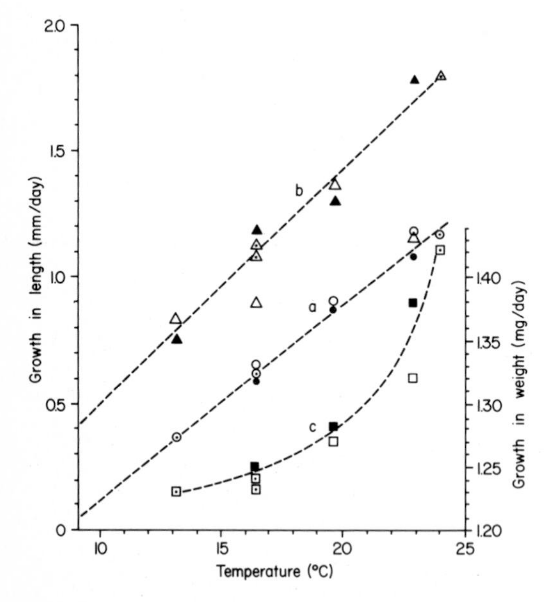

Fig. 3. Growth of (a) embryos during incubation, and (b,c) young fish during the rearing period, in relation to temperature (T, °C). (a,o) Increase in length, mm/day = 0.077 T - 0.656; (b,△) increase in length, mm/day = 0.093 T - 0.428; (c,□) increase in weight, mg/day = 0.656 $T^{0.236}$. Open symbols indicate low salinity trials (10-12.5°/oo), closed symbols high salinities (42-45°/oo), and dotted symbols intermediate salinities (20-35°/oo)

nities and temperatures calculated as 31.4 - 37.1°/oo and 17.4 - 19.4°C. Development of garfish eggs failed at 5°/oo S.

Growth Rate. Increase in length (L, mm) of embryos, and increase in length and weight (L, mm; W, mg) of young fish in the various test conditions was estimated in relation to temperature (T, °C) only (Fig. 3). Growth rate, approximated as increase in length in mm per day, showed a linear correlation with temperature, both for embryonic growth to hatch as well as for growth of larvae after hatch (Fig. 3a,b). Growth in mg per day was calculated from log wet weight/rearing time, assuming that increase in weight with time is exponential (Winberg, 1971). Growth of the reared larvae, in mg per day, showed an exponential relation with temperature (Fig. 3c). Growth rates of reared fish in some of the low salinity trials (10 and 12.5°/oo S) were influenced by a feeding problem. The copepods used as food died rapidly and sank to the bottom of the rearing cages. Hence, less food was available to the fish in the low salinity trials.

Body Dimensions. Body size (length x height, mm) and yolk-sac size (length x height, mm) of newly-hatched larvae were inversely related, larger larvae possessing smaller yolk sac (see Table 5). Both dimensions are associated with a linear salinity effect and a quadratic temperature effect. Body size is maximized and yolk-sac size is minimized at salinities and temperatures near 0°/oo S and 14°C (Table 3, Fig. 4).

Total length and wet weight of reared larvae (Table 5) appear to be maximized at 31°/oo S and 19°C. The condition of the reared fish, calculated as wet weight in mg per standard length x height^2 in mm^3, was minimized at 33.5°/oo S and 15°C (Table 3).

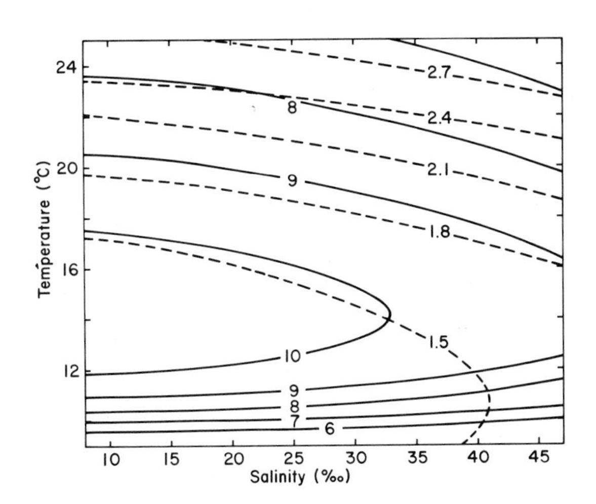

Fig. 4. Isopleths of body size (total length x height at vent, mm——) and yolk-sac size (length x height, mm---) of newly-hatched larvae, in relation to salinity-temperature incubation conditions

Table 5. Measurements of newly-hatched *Belone* larvae, and of young fish at the end of the rearing period (averages of the 3 subsamples)

	Newly hatched larvae			Reared larvae				
Trial	Total length (mm)	Body size (L × H, mm)	Yolk-sac size (L × H, mm)	Total length (mm)	Standard length (mm)	Wet weight (mg)	Condition factor (W/LH^2)	Number of vertebrae
1	-	-	-	-	-	-	-	-
2	12.6	8.00	2.27	29.1	23.6	42.9	96.2	80.77
3	-	6.39	1.55	-	-	-	-	-
4	11.7	7.24	2.68	31.1	24.6	44.4	86.9	80.00
5A	13.0	9.72	1.53	34.1	26.2	55.9	79.0	80.55
5B	13.1	9.69	1.48	34.2	26.8	57.3	77.0	80.66
6	13.4	10.45	1.42	29.8	23.2	45.6	85.2	80.98
7	12.7	9.01	1.73	33.9	26.5	56.7	75.3	80.50
8	-	-	-	-	-	-	-	-
9	11.9	7.59	2.58	31.4	25.3	49.9	87.0	80.14
10	13.3	10.20	1.35	27.9	22.4	38.9	76.8	81.97
11	12.9	9.19	2.07	35.0	27.6	53.4	77.6	80.19
12	13.3	10.00	1.53	26.0	21.2	33.9	75.0	81.88
13	12.5	8.59	2.19	33.5	26.6	54.9	77.1	79.85

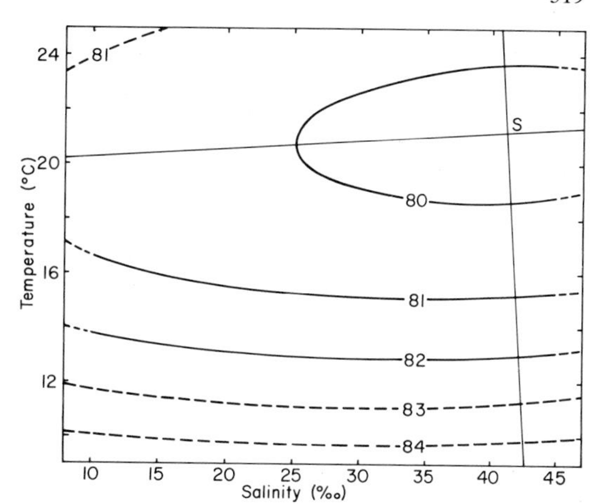

Fig. 5. Isopleths of vertebral number in relation to salinity-temperature incubation conditions (see text)

Vertebral Number. The relation between number of vertebrae and salinity-temperature incubation conditions (Table 6) may be expressed as the polynomial:

$$Y = 94.464 - 0.028\ S - 1.330\ T + 0.0008\ S^2 + 0.333\ T^2 - 0.0019\ ST$$

where Y = number of vertebrae; S and T = salinity (^{o}/oo S) and temperature (^{o}C) respectively. The linear and quadratic terms associated with salinity (S, S^2) and the interaction term (S X T) may be rejected, however, without seriously disturbing the adequacy of the model. That is, of the two factors considered, temperature has the greatest influence on vertebral number. A minimum vertebral number of 79.79 is estimated to occur at the centre of the salinity-temperature surface near 41°/oo S and 21°C (Table 3). Vertebral number is estimated to increase at higher and lower salinities and temperatures (Fig. 5), and sections through the surface at selected constant levels of salinity or temperature illustrate predicted changes in number of vertebrae (Fig. 6). Mean vertebral numbers of the different subsamples, in all trials grouped together, are more correlated with the size of the eggs than with the vertebral numbers of the parent fish (see Table 6).

Numbers of anal and dorsal fin rays showed little variation with salinity-temperature conditions of incubation. Analysis of variance in-

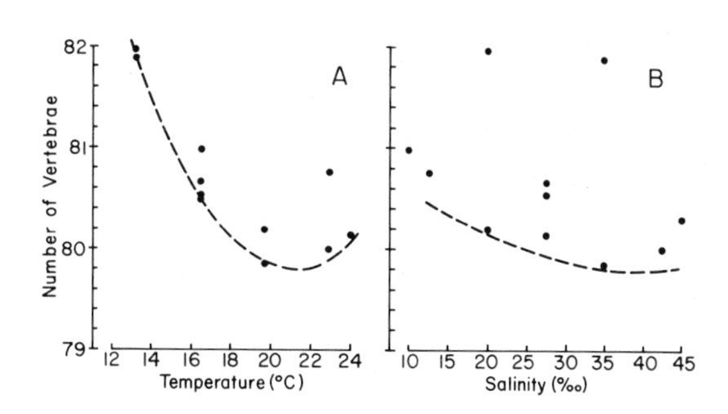

Fig. 6. Profiles of Fig. 5 through the surface centre at constant levels of salinity and temperature. The dotted lines show predicted changes of the vertebral number with (A) increasing incubation temperatures at 41.1^o/oo and (B) increasing incubation salinities at 21.1^{o}C

Table 6. Mean numbers of vertebrae, dorsal and anal finrays at the end of the rearing period. Results of all trials grouped into their appropriate subsamples

	Subsample 1	Subsample 2	Subsample 3
Vertebral number			
Mean	81.54	80.12	80.33
Stand. dev.	1.29	0.86	0.90
Range	79–85	78–82	77–83
Dorsal fin rays			
Mean	17.48	17.44	17.52
Stand. dev.	0.73	0.62	0.60
Range	15–19	15–19	16–19
Anal fin rays			
Mean	20.36	20.03	20.30
Stand. dev.	0.83	0.71	0.65
Range	18–22	18–22	18–22
Parent fish			
Vertebrae	80 x -	79 x 79	80 x 80
Dorsal rays	17 x -	18 x 18	18 x 17/18
Anal rays	20 x -	21 x 21	21 x 20/21
Egg size, mm	3.29	2.95	3.06

dicated that only temperature had a significant effect on numbers of rays. There was a slight tendency for both anal and dorsal rays to follow the reduction in number of vertebrae which occurred with increasing temperature.

DISCUSSION

Of the two environmental factors examined, temperature appears to have a predominant influence on eggs of the garfish during incubation. Over the range considered, rate of development and survival of eggs during incubation are largely independent of salinity level (10-45°/oo). Moreover, independent action of salinity and temperature is supported by the fact that for each of the attributes of development investigated, the salinity-temperature interaction terms were of the same order as the error terms.

Rate of development of fish eggs may be approximated by a second order polynomial (Alderdice and Forrester, 1968, 1971a, b). However, at higher temperatures, rate of transfer processes across the egg membrane may become limiting with respect to the metabolism of the embryo. It has been demonstrated, for example, that limitations in developmental rate associated with oxygen transfer may occur in species with large eggs and thick egg membranes, such as salmonids (Hayes et al., 1951; Smith, 1957; Alderdice et al., 1959; Hamdorf, 1961; Daykin, 1965).

Salinity and temperature limits for survival of garfish eggs may be estimated as approximately: salinity - somewhat less than 10°/oo to somewhat greater than 45°/oo; temperature - approximately 12-24°C. During other experiments, not mentioned here, it appeared that development of garfish eggs failed at 5°/oo S. According to Holliday and Blaxter (1960) eggs and larvae of the herring (*Clupea harengus*) tolerate a similar range of salinities. They found 100% fertilization of the eggs at 23-53°/oo, 50% hatch at 10-46°/oo, and more than 50% survival of the larvae at 3-53°/oo. Euryhalinity of eggs and larvae may be related to spawning in coastal and estuarine areas. Other marine species generally show a narrower range of tolerance in egg development with respect to salinity and temperature levels during incubation. Often there is also a low-low/high-high interaction between salinity and temperature such that maximum viability during development is maintained at higher salinities by a corresponding increase in temperature (Alderdice and Forrester, 1968, 1971a, b; Holliday, 1965; Westernhagen, 1970; Alderdice and Velsen, 1971).

Even though the effect of salinity on incubation success of garfish eggs cannot be demonstrated as significantly different from the error term, it is still concluded that salinity-temperature conditions associated with maximum hatching success are those at the centre of the calculated salinity-temperature surface, namely 34-37°/oo S and 17-18°C.

The suspected influence of salinity during incubation is supported by the recognition of salinity effects on body size and yolk size of newly-hatched larvae (Table 3, Fig. 4). The inverse relation between body size and yolk size at hatching may be due simply to a temperature effect on the stage at hatching, since body size steadily increases during development while yolk is decreasing (Hamdorf, 1961; Blaxter, 1969). However, if the rate of embryonic development increases more rapidly with incubation temperature than the embryonic growth rate (see Trifonova et al., 1939; Gabriel, 1944; Barlow, 1961), this may also result in hatching of smaller larvae at higher incubation temperatures. Salinity-temperature conditions associated with maximum larval size at hatching may not coincide with those conditions at which maximum hatching success is found (Alderdice and Velsen, 1971). In the garfish, largest larvae were produced at low salinities and a temperature near 14°C, while highest net production of tissue may be found in the larger number of somewhat smaller larvae produced at salinities and temperatures of 34-37°/oo S and 17-18°C (Table 3).

The calculated salinity-temperature conditions associated with maximum survival of young *Belone* after hatching (31.4°/oo S, 19.4°C) were not appreciably different from those resulting in maximum survival throughout incubation (Table 3). Wet weight and total length of reared fish also appeared to be maximized within the same salinity and temperature ranges (31-32°/oo S, 19.1°C). As indicated earlier, however, this relationship may have been biased by lower availability of food during rearing in some of the lower salinity tests. The broad range of salinity tolerance exhibited for incubating eggs of the garfish is probably exceeded by the larvae, juveniles, and adults. Adult garfish are commonly found at salinities as low as 7-9°/oo S. Young garfish, gradually adapted to increasing or decreasing salinities, reach a lethal limit below 2°/oo S and above 60°/oo S (Rosenthal and Fonds, 1973).

Vertebral number in fish is generally correlated with rate of development, which in turn is related to metabolic rate. Any environmental factor which can alter the metabolic rate and rate of development of fish eggs may also influence the number of vertebrae (for reviews, see Vladykov, 1934; Barlow, 1961; Fowler, 1970; Garside, 1966). Rate of development increases, and vertebral number generally decreases, with increasing incubation temperatures. However, a V-shaped relation between incubation temperature and vertebral number is also often found (Tåning, 1952; Lindsey, 1954; Seymour, 1959; Garside, 1966). A minimum vertebral number may be found at those conditions where the rate of histo-differentiation is highest in comparison with the embryonic growth rate (Gabriel, 1944). At each temperature level, mean vertebral numbers of the reared garfish were slightly higher at the lower salinity levels (Table 5, Fig. 5). This indicates that salinity may influence the rate of histo-differentiation in garfish embryos. Mean vertebral numbers showed a V-shaped relation with incubation temperatures, with a minimum vertebral number at 21°C, 41°/oo S (Figs 5, 6). This indicates that in garfish eggs the rate of histo-differentiation, relative to the embryonic growth rate, reaches a maximum at these conditions. The maximum may be due to limitations on diffusion of oxygen to the embryo at higher temperatures and salinities, imposed by the egg capsule (see Kinne and Kinne, 1962). Total mean vertebral numbers of the reared garfish in the 3 different subsamples (Table 6) appears to be more related to the size of the eggs than to the vertebral number of the parent fish (Table 1). A similar relation between yolk size and myomere number has been found in hybrid trout embryos (Garside and Fry, 1959; see also Lindsey and Ali, 1971).

The results lead to the following speculative arguments: although the garfish is not considered primarily to be an estuarine species, the euryhalinity of its early life history stages would make it highly adaptable to coastal and estuarine spawning conditions. The garfish spawns on algae or *Zostera*, which are generally found in coastal areas. Spawning success in more northerly waters (North Sea) appears to be strongly associated with the availability of warm coastal waters during the spawning season, such as those found in the Wadden Sea.

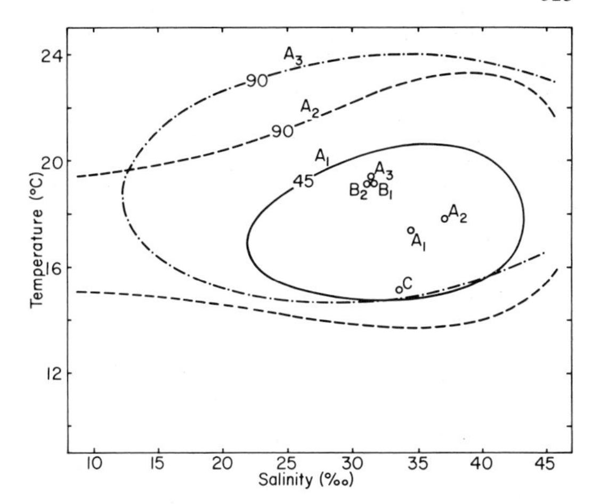

Fig. 7. Survival, size, and condition factor of eggs and young garfish in relation to salinity-temperature incubation and rearing conditions, showing calculated isopleths of near-maximum percent survival and the surface centres. A - survival: (A1) total hatch, %; (A2) hatch of neurulae, %; (A3) reared larvae, % of hatched larvae. B - size of larvae at end of rearing period: (B1) total length, mm; (B2) wet weight, mg. C - condition factor of reared larvae

Coupled with these higher temperature requirements during egg and larval development, and notwithstanding the euryhalinity of those stages, higher salinities than those found in estuarine conditions appear to be associated with maximum developmental success and survival of the early stages. For these reasons the suggestion arises that *Belone belone* may have extended its original range from more southern Atlantic waters.

SUMMARY

It appears that the attributes of early development of *Belone belone* which were examined resolve into three different groups, on the basis of salinity and temperature levels to which eggs and larvae were exposed during early development.

1. Survival of eggs during incubation and of larvae reared to 30-40 mm total length appears to be maximized at salinity-temperature conditions of approximately 31-37°/oo S and 17.5 - 19.5°C. Length and weight of reared larvae, and larval condition, appear to be optimized in the same approximate range, namely about 31-34°/oo S and 17.5 - 19.5°C, see Fig. 7.

2. Body size of newly-hatched larvae is maximized, and yolk-sac size is minimized, by exposure during incubation to low salinities below 10°/oo S at temperatures near 14°C. Although these conditions result in largest larvae at hatching, they do not appear to be advantageous to larvae during the subsequent rearing period.

3. Number of vertebrae is estimated to be a minimum after incubation to salinity-temperature levels at or near 41°/oo S and 21°C.

REFERENCES

Alderdice, D.F., 1972. Factor combinations. In: O. Kinne (ed.), Marine Ecology, Vol. I. London: Wiley-Interscience, 1659-1722 pp.
Alderdice, D.F. and Forrester, C.R., 1968. Some effects of salinity and temperature on early development and survival of the English Sole (*Parophrys vetulus*). J. Fish. Res. Bd Can., 25, 495-521.
Alderdice, D.F. and Forrester, C.R., 1971a. Effects of salinity, temperature and dissolved oxygen on early development of the Pacific Cod (*Gadus macrocephalus*). J. Fish. Res. Bd Can., 28, 883-902.
Alderdice, D.F. and Forrester, C.R., 1971b. Effects of salinity and temperature on embryonic development of the petrale sole (*Eopsetta jordani*). J. Fish. Res. Bd Can., 28, 727-744.
Alderdice, D.F. and Velsen, F.P.J., 1971. Some effects of salinity and temperature on early development of Pacific herring (*Clupea pallasi*). J. Fish. Res. Bd Can., 28, 1545-1562.
Alderdice, D.F., Wickett, W.P. and Brett, J.R., 1959. Some effects of exposure to low dissolved oxygen levels on Pacific salmon eggs. J. Fish. Res. Bd Can., 15, 229-250.
Barlow, G.W., 1961. Causes and significances of morphological variation in fishes. Syst. Zool., 10, 105-117.
Blaxter, J.H.S., 1969. Development: eggs and larvae. In: W.S. Hoar and D.J. Randall (Eds.): Fish Physiology, Vol. III, 177-252 pp. New York: Academic Press.
Colette, B.B. and Berry, F.H., 1965. Recent studies on the needle fishes (Belonidae): an evaluation. Copeia, 1965, 386-392.
Daykin, P.N., 1965. Application of mass transfer theory to the problem of respiration in fish eggs. J. Fish. Res. Bd Can., 22, 159-171.
Ehrenbaum, E., 1904. Eier und Larven von Fischen der Deutschen Bucht III. Fische mit festsitzenden Eier. *Ramphistoma belone* (L.) Wiss. Meeresunters. Helgoland N.F., 6, 127-200.
Fowler, J.A., 1970. Control of vertebral number in Teleosts - an embryological problem. Quart. Rev. Biol., 45, 148-167.
Gabriel, M.L., 1944. Factors affecting the number and form of vertebrae in *Fundulus heteroclitus*. J. Exp. Zool., 95, 105-147.
Garside, E.T., 1966. Developmental rate and vertebral number in Salmonids. J. Fish. Res. Bd Can., 23, 1537-1553.
Garside, E.T. and Fry, F.E.J., 1959. A possible relationship between yolk size and differentiation in trout embryos. Can. J. Zool., 37, 383-386.
Hamdorf, K., 1961. Die Beeinflussung der Embryonal- und Larvalentwicklung der Regenbogenforelle (*Salmo irideus* Gibb.) durch die Umweltfaktoren O_2-partialdruck und Temperatur. Z. vergl. Physiol., 44, 523-549.
Hayes, F.R., Wilmot, J.R. and Livingstone, D.A., 1951. The oxygen consumption of the salmon egg in relation to development and activity. J. exp. Zool. 116, 377-395.
Holliday, F.G.T., 1965. Osmoregulation in marine teleost eggs and larvae. Calif. Coop. Oceanic Fish. Invest. Rep., 10, 89-95.
Holliday, F.G.T. and Blaxter, J.H.S., 1960. The effects of salinity on the developing eggs and larvae of the herring. J. mar. biol. Ass. U.K., 39, 591-603.
Hollister, G., 1934. Clearing and dying fish for bone study. Zoologic 12, 89-101.
Kinne, O. and Kinne, E.M., 1962. Rates of development in embryos of a cyprinodont fish exposed to different temperature-salinity-oxyg combinations. Can. J. Zool., 40, 231-253.
Lindsey, C.C., 1954. Temperature controlled meristic variation in the paradise fish *Macropodus opercularis* (L.). Can. J. Zool., 30, 87-98.

Lindsey, C.C. and Ali, M.Y., 1971. An experiment with medaka, *Oryzias latipes*, and a critique of the hypothesis that teleost egg size controls vertebral count. J. Fish. Res. Bd Can., 28, 1236-1240.
Lindsey, J.K., Alderdice, D.F. and Pienaar, L.V., 1970. Analysis of nonlinear models - the nonlinear response surface. J. Fish. Res. Bd Can., 27, 765-791.
Lindsey, J.K. and Sandnes, A.M., 1970. Program for the analysis of non-linear response surface (extended version). Fish. Res. Bd Can. Tech. Rep., 173, 131 pp.
Rosenthal, H., 1970. Anfütterung und Wachstum der Larven und Jungfische des Hornhechtes *Belone belone*. Helgol. wiss. Meeresunters., 21, 320-332.
Rosenthal, H. and Fonds, M., 1973. Biological observations during rearing of garfish *Belone belone*. Mar. Biol., 21, 203-218.
Seymour, A., 1959. Effects of temperature upon the formation of vertebrae and finrays in young Chinook Salmon. Trans. Amer. Fish. Soc., 88, 58-69.
Smith, S., 1957. Early development and hatching. In: M.E. Brown (ed.), The physiology of fishes, Vol. I. New York: Academic Press, 323-369 pp.
Tåning, A.V., 1952. Experimental study of meristic characters in fishes. Biol. Rev., 27, 169-193.
Taylor, W.R., 1967. An enzyme method for clearing and staining small vertebrates. Proc. U.S. Nat. Mus. (Smithson. Inst.), 122 (3596), 1-17.
Trifonova, A.N., Vernidoube, M.F. and Philippov, N.D., 1939. La physiologie de la differenciation et de la croissance II. Les périodes critiques dans le dévelopment des Salmonidés et leur base physiologique. Acta Zoöl., (Stockholm), 20, 239-267.
Vladykov, V.D., 1934. Environment and taxonomic characters in fishes. Trans. Royal Can. Inst., 20, 99-140.
Westernhagen, H. von, 1970. Erbrütung der Eier von Dorsch (*Gadus morhua*), Flunder (*Pleuronectus flesus*) und Scholle (*Pleuronectes platessa*) unter kombinierten Temperatur- und Salzgehaltbedingungen. Helgol. wiss. Meeresunters., 21, 21-102.
Winberg, G.G., 1971. Methods for the estimation of production of aquatic animals (transl. A. Duncan). London: Academic Press, 175 p.

M. Fonds
Netherlands Institute for Sea Research
P.O. Box 59
Texel / NETHERLANDS

H. Rosenthal
Biologische Anstalt Helgoland
Zentrale Hamburg
2000 Hamburg - 50 / FEDERAL REPUBLIC OF GERMANY
Palmaille 9

D.F. Alderdice
Fisheries Research Board of Canada
Pacific Biological Station
Nanaimo, B.C. / CANADA

Artificial Gynogenesis and Its Application in Genetics and Selective Breeding of Fishes

J. G. Stanley and K. E. Sneed

INTRODUCTION

Gynogenesis, a process related to parthenogenesis, is the development of an ovum after penetration by a spermatozoan, but without syngamy. Gynogenesis is a natural mode of reproduction in a few fishes. It can be achieved artificially using various manipulations of events in fertilization and early development. The objectives of these manipulations are: first, to eliminate male inheritance by destroying DNA in the spermatozoa; and second, to restore diploidy to the ovum by interfering with either meiosis or cell cleavage.

Because inheritance patterns are altered, gynogenesis can be used as a tool in genetics studies and in selective breeding. Since male inheritance is excluded, offspring are exclusively female except when the female has heterogametic sex chromosomes or undergoes sex inversions, both of which are rare. Unisex populations can be used in fisheries management in situations where reproduction is undesirable or must be limited.

METHODS FOR PRODUCING GYNOGENETIC FISHES

Development of an embryo can be initiated by a number of methods. A classical procedure is to prick each ovum with a needle dipped in serum or blood. Weak electrical currents passed through the ovum are also effective (Lestage, 1933). In artificial gynogenesis, embryonic growth results from penetration of an ovum by a spermatozoan containing denatured DNA. High doses of radiation destroy DNA without seriously altering cytoplasmic components of spermatozoa and the irradiated spermatozoa retain their ability to initiate development, although motility is lost more quickly than in non-irradiated spermatozoa. Doses of 100 kiloroentgens of x-rays are recommended (Romashov et al., 1963; Golovinskaya et al., 1963), although gamma radiation from cobalt-60 also is effective (Purdom, 1969). Ultraviolet radiation from a 15 W sterilization lamp has been used to inactivate DNA of frog spermatozoa (Nace et al., 1970). Several dyes, such as trypaflavine, toluidine blue, and thiazine, denature DNA, as does nitrogen mustard (Beatty, 1964).

Embryonic development also can be initiated using sperm from distantly-related species. Loeb (1912) obtained pure *Fundulus* from *Fundulus* eggs treated with sperm from *Menidia*. Kasansky (1935) obtained development of carp (*Cyprinus carpio*) eggs treated with sperm from *Abramis brama*, and Buss and Wright (1956) found only pure brook trout (*Salvelinus fontinalis*) from brook trout eggs treated with sperm from brown trout (*Salmo trutta*). Development of 8 of 14 crosses between species from different orders followed the maternal type (Kryzhanovskii, 1968). Paternal genes apparently are not strongly expressed in hybrids of remote crosses.

Methods using sperm with denatured DNA to initiate development result in embryos which are haploid and fail to develop beyond the larval stage. In rare cases the embryo may be diploid because only one division occurred during meiosis. Hence, physical or chemical manipulations that interfere with meiosis increase the yield of diploid embryos.

In amphibians, hydrostatic pressure of 5000 lbs/sq in. prevents expulsion of the second polar body (Dasgupta, 1962). However, temperature shocks will sometimes prevent the second meiotic division. High temperatures appear to denature the spindle apparatus, while cold temperatures slow anaphase II, both of which inhibit the second meiotic division. The time of application of the temperature shock is critically important. In fish, hot and cold temperature shocks have been used to produce gynogenetic fish (Romashov and Belyaeva, 1965; Purdom, 1969).

Another method for increasing the number of chromosome sets is to block the second meiotic division or the first mitotic cleavage with colchicine, a drug frequently used in plants to induce polyploidy. Colchicine interferes with the spindle and prevents anaphase and cleavage. Colchicine has been used to treat fish eggs but without success at producing polyploidy (Lieder, 1964).

OCCURRENCE OF GYNOGENETIC FISHES

Several populations of fish reproduce by natural gynogenesis. These unisex fish may be triploid, as in the case of some populations of silver crucian carp, (Prussian crucian carp or European goldfish), *Carassius auratus gibelio* (Cherfas, 1966a) and in three forms of *Poeciliopsis* (Schultz, 1967), or diploid as in other silver crucian carp (Lieder 1955) and in *Poecilia formosa* (Hubbs and Hubbs, 1946; Kallman, 1962).

Gynogenesis can be produced in species that do not rely upon this process to reproduce in nature. Artificial gynogenesis or parthenogenesis resulting from experimental manipulation has been observed in 17 species of bisexual fish (Table 1), and has probably gone unrecognized in many other species. Apparently, artificial gynogenesis can occur in any species under suitable experimental conditions. Not included in Table 1 are numerous cases of gynogenesis or parthenogenesis in which embryos were haploid.

CYTOLOGICAL MECHANISMS OF GYNOGENESIS

In natural gynogenesis, the offspring receive only maternal genetic material and, therefore, might be expected to be haploid. This is not the case, however, since the polyploid number of chromosomes in these fish remains constant from one generation to the next. For constancy to occur, some alterations in either oogenesis or development are required. In artificial gynogenesis employing presently known methods, many haploid but only a few diploid individuals occur.

Table 1. Species of fishes in which artificial gynogenesis or parthenogenesis resulted in viable offspring

Species	Investigator
Sturgeon (*Acipenser güldenstädti*)	Romashov et al., 1963; Golovinskaya et al., 1963
Sterlet (*Acipenser ruthenus*)	Romashov et al., 1963; Golovinskaya et al., 1963
Beluga (*Huso huso*)	Romashov et al., 1963; Golovinskaya et al., 1963
Shad (*Alosa fallax*)	Lestage, 1933
Peled (*Coregonus peled*)	Tsoy, 1972
Arctic Cisco (*Coregonus lavaretus*)	Melander and Monten, 1950
Steelhead trout (*Salmo gairdneri*)	Tsoy, 1972; Vassileva-Dryanovska and Belcheva, 1965
Brown trout (*Salmo trutta*)	Oppermann, 1913; Purdom, 1969
Brook trout (*Salvelinus fontinalis*)	Buss and Wright, 1956
Common carp (*Cyprinus carpio*)	Kasansky, 1935; Golovinskaya, 1969
Goldfish (*Carassius auratus*)	Stanley (unpublished)
Grass carp (*Ctenopharyngodon idella*)	This study
Chebakhok (*Leuciscus bergi*)	Turdakov and Turdakov, 1959
Loach (*Misgurnus fossilis*)	Neifakh, 1959; Romashov and Belyaeva, 1964, 1965
Chinese bass (*Siniperca chua-tsi*)	Kryzhanovskii et al., 1953
Plaice (*Pleuronectes platessa*)	Purdom, 1969
Flounder (*Platichthys flesus*)	Purdom, 1969

Four mechanisms for maintaining chromosome constancy in gynogenetic fish have been described and are summarized in Fig. 1. In natural gyno genetic *Poeciliopsis*, chromosome replication without cleavage, called endomitosis, precedes meiosis (Cimino, 1972). This mechanism increases the chromosome number to 6n, which subsequently is reduced to 3n in meiosis. In the triploid gynogenetic silver crucian carp, the first meiotic division does not occur, so meiosis does not reduce the chromo some number (Cherfas, 1966b). The third mechanism for restituting chromosome number is the recombination of the second polar body with the female pronucleus, which is equivalent to failure of the second meiotic division. This mechanism has been observed during artificial gynogenesis in loach (Romashov and Belyaeva, 1964) and is suggested by the work of Purdom (1969). In the fourth mechanism, meiosis occurs, but replication of chromosomes without cleavage during the first mitosis restores diploidy, as observed in the silver crucian carp (Lieder, 1955, 1959).

The mechanism utilized for restoring diploidy affects the degree of heterozygosity in offspring. The level of heterozygosity remains constant from parent to offspring if diploidization is a result of endomitosis in oogenesis as in *Poeciliopsis* (Cimino, 1972) or elimination of meiosis I, as in triploid silver crucian carp (Cherfas, 1969). Fusion of the female pronucleus and second polar body produces a zygote with sister chromosomes and thus results in a homozygous condition for all gene pairs except those which might have been involved in earlier crossovers between homologous chromosomes. Because artificial gynogenesis involves such an alteration in the second meiotic division, the offspring would be expected to be homozygous for most gene pairs. Endomitosis in place of the first cleavage produces complete homozygosity.

GYNOGENESIS IN GENETIC STUDIES

Gynogenesis can be useful in genetic studies and selective breeding. If diploidy is restored by fusion of the female pronucleus with the second polar body, the gynogenetic offspring would be expected to be homozygous for every gene. This is because the second polar body contains the sister chromosome of each chromosome in the female pronucleu and, therefore, has identical genes. However, not all gynogenetic offspring are homozygous. As chromosomes pair in early meiosis, homologou chromosomes may twist around one another and break. If the pieces then fuse with the opposite chromosome, crossover results. Heterozygosity of the gynogenetic offspring occurs whenever a female pronucleus containing a crossover chromosome combines with the polar body containing the daughter chromosome without a crossover (or vice versa). The cross over-frequency can be calculated from the ratio of heterozygous to hom zygous progeny. The frequency of crossover is related to the distance between the gene and the centromere of the chromosome. Thus, these frequencies can be used to construct chromosome maps depicting the relative location of genes on chromosomes (Volpe, 1970; Nace et al., 1970). Because the technique of gynogenesis gives precise estimates of crossover frequencies, it allows construction of good maps.

Genetic studies on gynogenetic *Cyprinus carpio* (Golovinskaya, 1969) suggest that the frequency of crossover is high. Female carp having a genotype Ss produced a total of 61 gynogenetic offspring, of which 18 were genotype ss and 43 were either SS or Ss. Crossover frequency can be calculated from these data using ratios derived from Nace et al

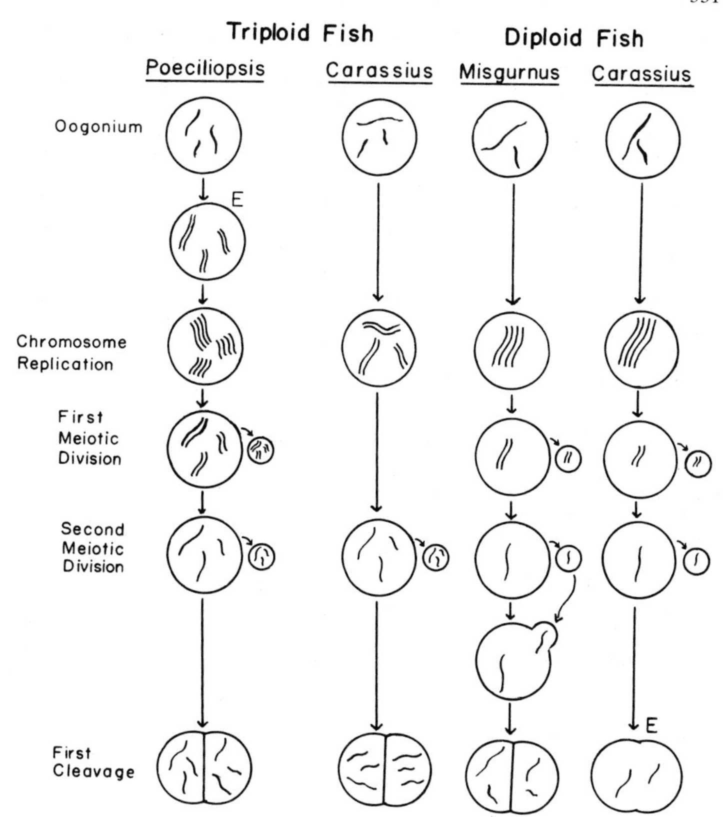

Fig. 1. 4 mechanisms whereby constancy in chromosome number is maintained in gynogenetic fishes. For simplicity only 1 chromosome from each set is shown. In unisex *Poeciliopsis* endomitosis (E) precedes meiosis and results in hexaploidy, which is returned to triploidy in meiosis (Cimino, 1972). Cherfas (1966b) reported that only 1 division of meiosis occurs in triploid forms of unisex *Carassius auratus gibelio*. In artificial gynogenesis as described in *Misgurnus fossilis* by Romashov and Belyaeva (1964) the 2nd polar body fuses with the ovum to restore diploidy. In diploid unisex *Carassius auratus gibelio* meiosis is normal with diploidy being maintained by endomitosis (E) after the 2nd meiotic division (Lieder, 1959)

(1970). These ratios are (1 + y)/2 (SS and Ss):(1 - y)/2 (ss), where y is the crossover frequency. Solving for y gives a value of approximately 42% crossovers. For another gene, all 17 gynogenetic offspring were heterozygous, suggesting a crossover frequency of 100%. It appears that autosomal crossovers occur frequently. Such crossovers may occur less frequently between sex chromosomes. Gordon (1937) and Kallman (1965) observed a crossover frequency of less than 1% between sex chromosomes in *Xiphophorus maculatus*, and Winge and Ditlevsen (1947) found crossovers of from 0 to 10% in *Poecilia reticulata*.

GYNOGENESIS IN SELECTIVE BREEDING

Gynogenetic offspring are useful in selecting for uncommon recessive traits or new mutants. The expression of a recessive trait in most populations occurs at a frequency equal to the square of the gene frequency, whereas gynogenetic offspring express a recessive trait with a frequency equal to the gene frequency in the population (except for crossovers). For example, if a recessive gene were present in 1% of the parents the gene will be expressed in only $(1/100)^2$ = 1/10 000 of the offspring from natural reproduction, whereas nearly 1% of gynogenetic broods would yield homozygous recessive individuals (depending on the degree of crossing over).

Combinations of traits can be more readily selected using gynogenesis than in usual breeding. For example, in a cross between a male and female, both heterozygous for three different genes, the chances of obtaining homozygosity for all three pairs in one offspring is 1/64. In artificial gynogenesis approximately one in eight have this combination (more precisely $(1-7y)(1-y')(1-y'')/8$, where y, y', and y" represent the respective crossover frequencies).

The fish breeder often does not know the genotype of his fish. Gynogenesis offers advantages in selective breeding since it does not require the identification of brood fish carrying recessive traits, nor are specific matings required between male and female fish. In the example given above involving the fish with the 3 gene pairs, conventional breeding requires that both parents carry the genes to be selected. In gynogenesis, only females need be considered. Thus, gynogenesis greatly increases the efficiency of selection for desirable traits in selective breeding programs.

Artificial gynogenesis is useful in producing inbred lines which can subsequently be crossed to produce hybrid vigor. This is the principal objective of the studies being conducted by Purdom in the United Kingdom. Except for crossovers, gynogenetic offspring are homozygous for every gene pair. Purdom (1969) calculated that gynogenetic offspring have a degree of homozygosity equivalent to 14 generations of sib-mating, assuming a mean crossover rate of 10% per generation. Until the mean crossover rate in fishes is known more precisely, the effectiveness of gynogenesis in producing homozygosity cannot be predicted, but it is probable that inbred lines can be obtained much more rapidly than with conventional methods of sib-crosses or back-crosses.

Because gynogenesis produces homozygosity, different lines of gynogenetic fish might be crossed to produce heterosis in the offspring. The problem is that gynogenetic fish are exclusively female and a technique such as sex-reversal must be used to obtain male fish. Sex-reversal has been successful in fish, if treatment with male hormone begins before gonadal differentiation (Yamamoto, 1958; Clemens and Inslee, 1968; Yamamoto and Kajishima, 1968).

GYNOGENESIS IN CONTROL OF REPRODUCTION

A persistent problem in fisheries management is the lack of control over reproduction in natural populations. Overpopulation due to excessive spawning leads to stunted fish. Gynogenesis offers a means of producing fish of one sex, thus permitting complete elimination of reproduction and regulation of population size.

Complete control of reproduction would be highly desirable for experimental stocking of exotic fishes. By using gynogenetic offspring, research on ecological effects of a proposed introduction could be accomplished without danger of permanent establishment. Should a candidate exotic fish prove undesirable, experimental populations could be destroyed or the fish simply be allowed to die of old age.

The principal objective of our research was to develop methods for producing unisex populations of grass carp, *Ctenopharyngodon idella* Val. Gynogenesis was one method that succeeded.

PRODUCTION OF GYNOGENETIC GRASS CARP

Gynogenetic development of grass carp eggs was induced using x-irradiated milt and cold shocks. Milt was irradiated with 75 kiloroentgens of x-rays applied over a 50-min period. Use of irradiated goldfish milt on grass carp eggs proved that male inheritance was excluded since no hybrids were observed. Cold shocks were applied to the eggs by adding ice cubes to the water until a temperature of 2°C was reached. This temperature was maintained by periodically adding ice. The cold shock was administered beginning 1 min after water was added to the egg-milt mixture and continued for 5, 10, or 15 min. After treatment, the eggs were quickly warmed to 26°C by placing them in screen baskets suspended in a large volume of water in a trough. Unfortunately, fry from the various treatment lots escaped through the 0.7 mm-mesh screen of which the baskets were made and became mixed in the trough. It therefore was impossible to determine the optimum duration for the cold shock, but of 40 000 eggs treated, 34 diploid grass carp hatched. An additional 151 embryos were found in which development was arrested and blood circulation had not become established. Their appearance fitted the classical description of the haploid syndrome. No embryos were found in control lots of eggs that received irradiated milt without cold shock and incubated in an adjacent trough.

Of the 34 presumed diploid gynogenetic grass carp, 24 fry survived and were stocked in a 0.04-ha pond. Six months later these fish averaged 280 mm in length and 290 g in weight. This growth was at least equal to that expected based on previous experience with the species. If gynogenetic fish lack vigor due to homozygosity (Romashov et al., 1963; Purdom, 1969) it was not obvious from the growth of these grass carp. Because the female grass carp grows more rapidly than the male, gynogenetic offspring might have better growth than the average for a normal bisexual population. These fish differed from hybrids of goldfish X grass carp which were characterized by a deep body, small eyes and fins similar to those of goldfish. These grass carp were identical to those produced using grass carp milt and we conclude that they were gynogenetic.

Although we succeeded in producing gynogenetic grass carp, it has not yet been confirmed that these are unisexual. Sexual maturity is not reached until age 3 to 5 years and sexing of immature fish is unreliable even by dissection and histological examination. Two mechanisms are known in fishes that could result in bisexual broods. A species with female heterogamety for sex chromosomes could produce gynogenetic offspring of both sexes. *Xiphophorus maculatus* from Mexico have this combination of sex chromosomes (Gordon, 1947), and Hickling (1960) suggested a similar situation for one species of *Tilapia*.

A second mechanism that might result in bisexual broods of gynogenetic fish is the presence of autosomal genes that modify the expression of sex chromosomes (Kallman, 1968). Inheritance of such sex-modifying genes results in sex differentiation independent of sex chromosomes. Homozygosity of sex-modifying genes is more likely in gynogenetic offspring than in normal individuals, thus enhancing the chance for sex-modification and production of fish of both sexes.

Despite remote possibilities for reproduction by gynogenetic fish, gynogenesis offers a means whereby reproduction can be controlled. Gynogenetic broods should prove particularly useful for initial stocking of exotic species for their evaluation prior to introduction. Although yields were low, each female grass carp lays several hundred thousand eggs and it should be possible to produce sufficient numbers of gynogenetic offspring for preliminary testing of this species in open waters of the United States.

SUMMARY

Gynogenesis is potentially valuable for providing unisex populations, in genetic studies and selective breeding for locating and selecting recessive genes and in studies of chromosome linkage and crossovers. Gynogenesis is especially useful in selective breeding, since homozygosity may be achieved in fewer generations than with inbreeding.

The grass carp (*Ctenopharyngodon idella*) was used as a test animal. Gynogenetic fry were produced by treating grass carp eggs with x-irradiated goldfish (*Carassius auratus*) sperm. A total of 185 embryos hatched of which 34 were presumed diploid out of 40 000 eggs which had received a 2°C temperature shock. No embryos were found in controls.

Offspring produced by gynogenesis could theoretically reproduce if the female of the species has heterogametic sex chromosomes or if spontaneous sex inversions occur. We conclude that production of gynogenetic fishes offers a means for stocking exotic species without the likelihood of reproduction.

ACKNOWLEDGEMENTS

We would like to thank Drs. F.E.J. Fry and Robert J. Valenti for critically evaluating an early version of this paper, and Mrs. Catherine Wilkerson at the Veterans Administration Hospital, Little Rock, Arkansas, USA for irradiating the milt used in this study. Financial support was provided in part by the U.S. Army Corp of Engineers.

REFERENCES

Beatty, R.A., 1964. Gynogenesis in vertebrates: fertilization by genetically inactivated spermatozoa. In: W.D. Carlson and F.X. Gassner (eds), Effects of ionizing radiation on the reproductive system. Oxford: Pergamon Press.

Buss, K.W. and Wright, J.E., Jr., 1956. The results of species hybridization within the family salmonidae. Prog. Fish Cult., 18, 149-158.

Cherfas, N.B., 1966a. Natural triploidy in females of the unisex form of the goldfish (*Carassius auratus gibelio* Bloch.). Genetika, 5, 16-24. (Transl. from Russ.)

Cherfas, N.B., 1966b. Meiotic analysis of unisexual and bisexual forms of crucian carp. Trud. vsesoiuz. nauch.-issled. Inst. ozer.-rech. ryb. Khoz., Tomsk., 14, 63-82. (In Russ.)

Cherfas, N.B., 1969. Results of a cytological analysis of unisexual and bisexual forms of silver crucian carp, pp. 79-89. In: B.I. Cherfas (ed.), Genetics, selection, and hybridization of fish. Ministry of Fish. RSFSR. Transl. from Russ. for Nat. Mar. Fish. Serv., Washington, D.C., 1972.

Cimino, M.C., 1972. Meiosis in triploid all-female fish (*Poeciliopsis*, Poeciliidae). Science, 175, 1484-1486.

Clemens, H.P. and Inslee, T., 1968. The production of unisexual broods by *Tilapia mossambica* sex-reversed with methyl testosterone. Trans. Amer. Fish. Soc., 97, 18-21.

Dasgupta, S., 1962. Induction of triploidy by hydrostatic pressure in the leopard frog, *Rana pipens*. J. exp. Zool., 151, 105-121.

Golovinskaya, K.A., 1969. Artificial gynogenesis in carp, pp. 74-78. In: B.I. Cherfas (ed.), Genetics, selection, and hybridization of fish. Ministry of Fish. RSFSR. Transl. from Russ. for Nat. Mar. Fish. Serv., Washington, D.C., 1972.

Golovinskaya, K.A., Romashov, D.D. and Cherfas, N.B., 1963. On the gynogenesis in carp caused by radiation. Trud. vsesoiuz. nauch.-issled. Inst. ozer.-rech. ryb. Khoz., Tomsk., 12, 149-168 (In Russ.).

Gordon, M., 1937. Genetics of *Platypoecilus*. III. Inheritance of sex and crossing over of the sex chromosomes in the platyfish. Genetics, 22, 376-392.

Gordon, M., 1947. Genetics of *Platypoecilus maculatus*. IV. The sex determining mechanism of two wild populations of Mexican platyfish. Genetics, 32, 8-17.

Hickling, C.F., 1960. The Malacca *Tilapia* hybrids. J. Genet. 57, 1-10.

Hubbs, C.L. and Hubbs, L.C., 1946. Breeding experiments with the invariable female, strictly matroclinous fish, *Molliensia formosa*. Genetics, 31, 218 (Abstract).

Kallman, K.D., 1962. Gynogenesis in the teleost, *Molliensia formosa* (Girard), with a discussion of the detection of parthenogenesis in vertebrates by tissue transplantation. J. Genet., 58, 7-24.

Kallman, K.D., 1965. Genetics and geography of sex determination in the poeciliid fish, *Xiphophorus maculatus*. Zoologica, 50, 151-190.

Kallman, K.D., 1968. Evidence for the existance of transformer genes for sex in the teleost *Xiphophorus maculatus*. Genetics, 60, 811-828.

Kasansky, W.J., 1935. Parthenogenetisch entwickelte junge Karpfen (*Cyprinus carpio* L.). Zool. Anz., 110, 191-193.

Kryzhanovskii, S.G., 1968. Aspects of the development of fish hybrids between various taxa. Nauka Press, Moscow, 220 pp. (In Russ. Rev. by Kosmider, P.N., 1969. J. Fish Biol., 1, 183-184).

Kryzhanovskii, S.G., Disler, N.N. and Smirnova, E.A., 1953. Ecologo-morphological regularities in the development of percoids. Trud. Inst. Morf. Zhiv., 10, 3-138 (In Russ.).

Lestage, J.A., 1933. L'industrialisation de la parthénogénèse artificielle chez l'agone (Clupeidae: *Paralosa lacustris lariana* Pir.). Ann. Soc. Zool. Belg., 64, 69-74.

Lieder, U., 1955. Männchenmangel und natürliche Parthenogenese bei der Silberkarausche *Carassius auratus gibelio* (Vertebrata, Pices). Naturwissenschaften, 42, 590.

Lieder, U., 1959. Über die Eientwicklung bei männchenlosen Stämmen der Silberkarausche *Carassius auratus gibelio* (Block) (Vertebrata, Pices). Biol. Zbl., 78, 284-291.
Lieder, U., 1964. Polyploidisierungsversuche bei Fischen mittels Temperaturschock und Colchizinbehandlung. Z. Fisch., 12, 247-257.
Loeb, J., 1912. Heredity in heterogeneous hybrids. J. Morph., 23, 1-15
Melander, Y. and Montén, E., 1950. Probable parthenogenesis in *Coregonus*. Hereditas, Lund, 36, 105-106.
Nace, G.W., Richards, C.M. and Asher, J.H., Jr., 1970. Parthenogenesis and genetic variability. I. Linkage and inbreeding estimations in the frogs, *Rana pipiens*. Genetics, 66, 349-368.
Neifakh, A.A., 1959. Effects of ionizing radiation on early development of fish. Trud. Inst. Morf. Zhiv., 24, 135-159 (In Russ.).
Oppermann, K., 1913. Die Entwicklung von Forelleneiern nach Befruchtung mit radiumbestrahlten Samenfäden. Arch. mikroskop. Anat., 83, 141-189.
Purdom, C.E., 1969. Radiation-induced gynogenesis and androgenesis in fish. Heredity, 24, 431-444.
Romashov, D.D. and Belyaeva, V.N., 1964. Cytology of radiation gynogenesis and androgenesis in the loach (*Misgurnus fossilis* L.). Dokl. Akad. Nauk SSSR, 157, 964-967. Transl. from Russ., In: Dokl. Biol. Sci., 157, 503-506.
Romashov, D.D. and Belyaeva, V.N., 1965. Increased yield of diploid gynogenetic loach larvae (*Misgurnus fossilis* L.) induced by temperature shock. Byulleten Moskovskogo Obschestva Ispytatelei Prirody, Biol. Ser., 70, 93-109 (In Russ.).
Romashov, D.D., Nikolyukin, N.I., Belyaeva, V.N. and Timofeeva, N.A., 1963. Possibilities of producing diploid radiation-induced gynogenesis in sturgeons. Radiobiologiya, 3, 104-109. Transl. from Russ., Radiobiology, 3, 145-154.
Schultz, R.J., 1967. Gynogenesis and triploidy in the viviparous fish *Poeciliopsis*. Science, 157, 1564-1567.
Tsoy, R.M., 1972. Chemical gynogenesis of *Salmo irideus* and *Coregonus peled*. Genetika, 8, 185-188.
Turdakov, F.A. and Turdakov, A.F., 1959. Parthenogenesis and other characteristics of issyk-kul chebachok development. Izv. Akad. Nauk. kirgiz. SSR, 1, 3-44 (In Russ.).
Vassileva-Dryanovska, O. and Belcheva, R., 1965. Radiation gynogenesis in *Salmo irideus* Gibb. Compte Rendus de l'Académie bulgare des Sciences, 18, 359-362.
Volpe, E.P., 1970. Chromosome mapping in the leopard frog. Genetics, 64, 11-21.
Winge, O. and Ditlevsen, E., 1947. Colour inheritance and sex determination in *Lebistes*. Heredity, 1, 65-83.
Yamamoto, T., 1958. Artificial induction of functional sex-reversal in the genotypic females of the medaka (*Oryzias latipes*). J. exp. Zool. 137, 227-262.
Yamamoto, T. and Kajishima, T., 1968. Sex-hormonic induction of reversal of sex differentiation in the goldfish and evidence for male heterogamety. J. exp. Zool., 168, 215-222.

J.G. Stanley
Fish Farming Experimental Station
P.O. Box 860
Stuttgart, Ark. 72160 / USA

K.E. Sneed
Bureau of Sports Fisheries and Wildlife
Department of the Interior
Washington, D.C. 20240 / USA

Gynogenesis in Hybrids within the Pleuronectidae

C. E. Purdom and R. F. Lincoln

INTRODUCTION

Hybrids between different genera within the flatfish family Pleuronectidae are commonplace. Naturally-occurring hybrids between the plaice (*Pleuronectes platessa*) and the flounder (*Platichthys flesus*) are common in the Baltic, and can be produced with ease by artificial fertilization of plaice eggs with flounder spermatozoa (Pape, 1935; von Ubisch, 1953). Natural flounder hybrids have also been reported in Japan (Hubbs and Kuronuma, 1942). Other pleuronectid hybrids, which have been produced artificially, include dab (*Limanda limanda*) x flounder (Riley and Thacker, 1969), plaice x dab and plaice x lemon sole (*Microstomus kitt*) and flounder x lemon sole (Thacker, personal communication).

Many hybrids are produced in attempts to create superior strains of fish for the purpose of fish cultivation. In this respect, the most interesting of the pleuronectids would be the halibut (*Hippoglossus hippoglossus*) which has a very fast growth rate in cold water and also has flesh of very high quality. Since hybrid characteristics are, in general, intermediate between those of the parents, the possible plaice x halibut hybrid would have great potential by virtue of the ease with which plaice eggs can be reared - experimental work with halibut eggs must await the successful maturation of female fish in captivity or the predictable capture of ripe fish from the sea.

Attempts were made to produce plaice x halibut and flounder x halibut hybrids by fertilizing eggs with halibut spermatozoa. In both cases the fertilization rates were good and early cleavage was normal. After gastrulation, however, the embryos of both hybrids developed very abnormally with short and thickened bodies. Survival to hatching was very poor and post-hatching mortality was complete within 24-48 h. The embryos closely resembled gynogenetic haploids which can be produced in plaice eggs by fertilization with radiation-inactivated spermatozoa (Purdom, 1969). Thus it seemed possible that the plaice x halibut and flounder x halibut hybrids were 'false hybrids' in the sense that the halibut spermatozoa had activated the eggs but had not contributed genetic material to the developing embryo.

Previous experience had suggested that a similar situation of false hybridization may have existed in the flounder x plaice hybrid, i.e. the reciprocal of the very successful plaice x flounder cross. When flounder eggs were fertilized with plaice spermatozoa the embryos were frequently very abnormal (Purdom, unpublished) although successful hatching has been reported (Pape, 1935).

The present experiments were performed to determine whether or not the abnormal hybrids were haploid. This was assessed by the use of post-fertilization cold shocks which can produce diploids from haploids, or triploids from diploid plaice zygotes respectively (Purdom, 1972), and by the comparison of embryos derived from eggs fertilized by normal spermatozoa, and by spermatozoa made genetically inert with ionizing radiation (Purdom, 1969). Some investigations were also made on the structure and size of sperm heads in halibut, plaice, and flounder. The eggs of these three species measure approximately 4, 2, and 1 mm in diameter, respectively, and the possible exclusion of spermatozoa on mechanical grounds was suspected.

MATERIALS AND METHODS

Halibut were caught by trawling on the Faroe Bank in May 1971. At the laboratory they were maintained in tanks of 65 m^3 capacity in which sea water was recirculated approximately 12 times per day through a gravel-filter bed. Temperature was controlled at about 12°C during the period June to October and allowed to adjust the ambient levels during the winter.

Ripe plaice and flounders were collected from the North Sea during the 1972-1973 spawning season and maintained in temporary holding tanks. Ripe eggs were stripped by hand and artificially fertilized in a small amount of sea water in 2 l crystallizing bowls. Incubation was carried out at 7°C.

Milt samples held on ice were irradiated with ^{60}Co gamma rays at a dose of 100 000 rad delivered in 160 min. Cold shocks were given from 10 to 20 min after fertilization by transferring eggs into bowls of sea water maintained at approximately -0.5°C where they were held for approximately 4 h. After this period the bowls were removed from the cold chamber and returned to the 7°C room and allowed to regain this temperature. Fertility counts were made on floating eggs at the two cell stage approximately 6 h after fertilization. Plaice eggs were scored for haploid or diploid embryos 5 days after fertilization, when closure of the gastral ring was complete. Flounder eggs, which are smaller than plaice eggs and have a shorter developmental time, were scored 4 days after fertilization. Daily counts of dead plaice eggs were made up to 20 days after fertilization. Mortality records were not kept for flounder eggs.

Measurement of sperm heads was carried out on electron micrographs of shadow-cast material fixed in osmic acid and magnified 9 200 times. Measurements of the major and minor diameters of sperm heads were made and the results expressed as the mean of the products of the two diameters. The structure of sperm heads was also studied from scanning electron micrographs after treatment of spermatozoa with ovarian fluid from plaice.

RESULTS

Plaice x Halibut Cross. All combinations of the plaice x plaice and plaice x halibut crosses were made using irradiated and non-irradiated spermatozoa with and without cold shocks. Table 1 lists these crosses and the treatments given to each, together with the percentage fertilit frequency of diploids at day 5 and survival to day 10 and day 20. Percentage fertilization was high in all the crosses and halibut milt was just as potent as plaice milt. Clear differences in the appearance

Table 1. Results of crossing plaice eggs (P) with plaice (P) or halibut (H) spermatozoa

Cross No.	Cross	Treatment	No. of eggs	Percentage fertility (200 eggs)	Percentage diploids (200 eggs) day 5	Percentage survival, day 10	Percentage survival, day 20
1	P x P	-	977	94.0	100.0	86.0	75.8
2	P x P	Sperm. irradiated	1 249	92.0	4.0	18.6	0.8
3	P x P	Sperm. irradiated, eggs cold-shocked	813	-	93.0	70.2	56.7
4	P x H	-	1 417	85.0	0.0	36.9	3.9
5	P x H	Eggs cold-shocked	1 027	-	94.5	70.2	55.0
6	P x H	Sperm. irradiated	1 132	90.2	3.5	39.8	5.0
7	P x H	Sperm. irradiated, eggs cold-shocked	1 013	-	93.5	78.6	64.4

of the embryos were observed at 5 days after fertilization. Embryos from cross numbers 2, 4, and 6 were all similar in appearance, showing the typical haploid syndrome already described. Some diploid embryos were observed in crosses 2 and 6; this is a common feature of diploid gynogenesis (Purdom, 1969). No spontaneous diploids were found in cross 4 but they have been observed in other experiments involving the same parents. All three cold-shocked groups (crosses 3, 5, and 7) gave consistently high frequencies of embryos identical in appearance to the normal diploid controls (cross 1), and all three showed similar survival patterns which were slightly lower than in the controls. Some variation was present in the haploid groups; the lowest survival occurred in cross 2 (0.8%) at day 20, but crosses 4 and 6 both showed considerably higher survivals (3.9 and 5.0%, respectively).

Flounder x Halibut and Flounder x Plaice Crosses. A shortage of good quality flounder eggs required for these experiments precluded a full investigation of these crosses to be made as for the plaice x halibut system. Five crosses were made including flounder controls. These are listed in Table 2, together with the appearance of the embryos at 4 days and the appearance and melanophore distribution in larvae 4 days after hatching.

In the flounder x plaice cross (cross 2) there was a high frequency of normal diploid-type larvae at day 4 with only about 10% abnormals of the possible haploid type. Following cold treatment (cross 3) there was no change in the frequency of haploid types. The larvae in cross 2 had a different pigment pattern from the pure flounder of cross 1, and in cross 3 the pigment pattern was approximately intermediate between that of 1 and 2. Thus there was no evidence that the abnormal embryos in cross 2 were haploid, and the normal embryos were clearly diploid.

A completely different result emerged for the flounder x halibut cross. Here also, all the embryos at day 4 in cross 4 were of the haploid type but after the cold shock (cross 5) the majority were of the diploid type. The larvae which hatched in crosses 4 and 5 were all of the pure flounder type. Thus fertilization of plaice or flounder eggs with halibut spermatozoa produced gynogenetic embryos but flounder eggs produced genuine hybrids after fertilization with plaice spermatozoa.

Sperm Head Measurement. Sperm heads were measured from electron micrographs at a magnification of x 9 200. Table 3 lists the results, which are expressed as areas of ellipses defined by the long and short diameters of the heads. There was significant variation between the 3 species of pleuronectids, with plaice having the larger spermatozoa and flounder the smallest.

Scanning electron micrographs of spermatozoa showed no evidence of surface structural differences between plaice, flounder and halibut. Spermatozoa from all three species showed rough, convoluted surface membranes. No evidence of acrosomal filaments was observed.

Table 2. Fertilization of flounder eggs with plaice or halibut spermatozoa, and the effect of cold shocks

Cross	Cross	Treatment	Appearance of embryo at 4 days after fertilization	Appearance and melanophore distribution in 4-day-old larvae
1	F x F	-	Normal diploid embryo	Larvae normal. Two bands of melanophores, one concentrated half-way along tail region in body and fin margin, the other anterior to anus in body and fin margins
2	F x P	-	90% normal diploid-looking embryos. 10% thickened embryos of haploid type	Majority of larvae bent. Melanophores throughout body region, none in fin margins
3	F x P	Cold-shocked	90% normal diploid-looking embryos. 10% haploid types present	Majority of larvae bent and runted. Few straight larvae showing melanophore distribution. As for F x F but reduced number in fin margins
4	F x H	-	Majority of embryos showing thickened and shortened bodies of haploid type. One or two of normal diploid type	Less than 1% hatch as straight larvae. Majority do not hatch. Pigment cell distribution similar to F x F spontaneous gynogenetic diploids
5	F x H	Cold-shocked	High percentage of normal-looking embryos of diploid type. Few haploid types present	Poor survival but majority of larvae straight. Distribution of melanophores similar to F x F

Table 3. Area of sperm heads in plaice, halibut, and flounder

Species	Number of sperm heads measured	Mean area in sq. μ	S.E	Values of P (t test)	
1 Plaice	9	1.92	0.11		
				0.02	
2 Halibut	10	1.59	0.04		0.001
				0.001	
3 Flounder	7	1.33	0.03		

DISCUSSION

The abnormal embryos which developed in plaice or flounder eggs after fertilization with halibut spermatozoa were shown to be similar to the haploid embryos which can be produced in these species by radiation gynogenesis (Purdom, 1969). The abnormal 'hybrids' showed a typical haploid syndrome and when zygotes were cold-shocked within 10 to 20 min of fertilization, the great majority of embryos were of the normal diploid type and resembled plaice or flounder, depending on the eggs used. This demonstrates conclusively that the embryos produced without cold shock were haploid. The effect of the cold shock is to suppress the metaphase of meiosis II and from a haploid this will produce a diploid, whereas from a diploid a triploid condition would be induced (Purdom, 1969, 1972). Since a triploid of hybrid origin would be intermediate in appearance between a pure-bred and the FI hybrid (Purdom, 1972), the plaice-like and flounder-like embryos produced after cold-shocking eggs fertilized with halibut spermatozoa must be gynogenetic diploids with no genetic material of halibut origin. The lack of involvement of halibut chromosomes in the 'hybrid' zygote was confirmed by the fact that radiation destruction of the genetic material of the halibut spermatozoa had no effect whatsoever on the nature of the embryos produced by apparent fertilization of plaice eggs.

The pattern of events in the flounder x plaice cross was quite different from that involving halibut spermatozoa. Only about 10% of embryos looked abnormal and haploid-like but the frequency of these did not differ between cold-shocked and non-cold-shocked groups of eggs. Thus the abnormalities observed in the flounder x plaice cross, although superficially like the haploid syndrome, were not caused by haploidy. These flounder x plaice embryos are genuine hybrids, and the abnormalities must represent ontogenetic accidents which may have no direct genetic determination.

Positive evidence is lacking as to the mode of action of halibut spermatozoa in activating the egg without functional involvement of the genetic material in the zygotes. The possibility that a mechanical factor was involved and that halibut spermatozoa might be too large to penetrate fully the micropyle of the plaice or flounder egg was ruled out by the measurement of sperm heads. The halibut spermatozoa appeared smaller than the plaice spermatozoa. Similarly, there was no apparent difference in the surface appearance of the spermatozoa of the three species and no evidence of acrosomal organelles - these are not thought to be involved in fertilization in teleosts (Ginzburg, 1968).

Two possibilities remain: first, that the halibut chromosomes do enter the egg but are rejected, and secondly that activation and penetration are distinct physiological events which can occur independently. The former seems unlikely, since intra-family hybridization is very common in fish and some inter-family hybrids have also been produced (Russell, 1939).

If the genetic material of the halibut was incompatible with that of plaice or flounder, it would seem likely that the hybrid might manifest this by developmental anomalies but not by complete rejection of the spermatozoa chromosomes. The morphology of the chromosomes of the Pleuronectidae is remarkably constant (Barker, 1972) and the halibut fits into this scheme, with a diploid complement of 48 telocentric chromosomes.

The possibility of a physiological block to spermatozoan penetration remains. Some attempts were made to investigate temperature effects, but fertilization at $3^{o}C$, which is the temperature at which halibut probably spawn, was no different from fertilization at $7^{o}C$. Similarly, the investigation of possible timing anomalies by treating eggs with nicotine prior to fertilization (Rothschild, 1953) also proved negative.

Although true hybrids cannot be produced between plaice and halibut at present, the gynogenetic offspring can be useful in the production of inbred lines and in genetic analysis in fish (Purdom, 1969). In this context, the gynogenomes produced by fertilizing plaice eggs with halibut spermatozoa appear more viable than those produced by fertilizing eggs with irradiated plaice spermatozoa. Several hundred gynogenomes have been reared through metamorphosis this year from plaice x halibut crosses which were cold-shocked. This compares very favourably with previous attempts to rear diploid radiation-induced gynogenomes.

Some genetic analysis has been performed with halibut spermatozoa-induced gynogenesis in plaice, and cross-over frequencies between the centromere and locus have been determined for two enzyme genes. These were glucose phosphate isomerase-b (3%) and phosphoglucomutase (45%). Full details of these crosses, and other genetic analyses on enzyme loci will be presented in a separate publication.

SUMMARY

1. Hybrids between plaice (*Pleuronectes platessa*) or flounder (*Platichthys flesus*) and halibut (*Hippoglossus hippoglossus*), using spermatozoa of the latter, were shown to be haploids arising by gynogenesis. Cold shocks given within 20 min after fertilization produced diploid embryos and larvae of maternal appearance. Irradiation of halibut spermatozoa with ^{60}Co gamma rays at a dose of 100 000 rad did not affect the patterns of fertilization and subsequent embryonic development.

2. Abnormalities in hybrids between female flounder and male plaice were not due to haploid gynogenesis. Cold treatments did not produce normal diploid offspring.

3. Physical attributes of halibut spermatozoa could not be invoked to explain the lack of fusion of halibut spermatozoan genetic material.

REFERENCES

Barker, C.J., 1972. A method for the display of chromosomes of plaice *Pleuronectes platessa*, and other marine fishes. Copeia, 1972, No. 2, 365-368.

Ginzburg, A.S., 1968. Fertilization in fishes and the problem of polyspermy. Moscow, Nauka Publ. Hse., 1968. Trans. from Russ. for the U.S. Nat. Mar. Fish. Serv., 1972, 366 pp. (TT 71-50111).

Hubbs, C.L. and Kuronuma, K., 1942. Hybridization in nature between two genera of flounders in Japan. Pap. Mich. Acad. Sci., Letters 27 (1942), 267-306.

Pape, A., 1935. Beiträge zur Naturgeschichte von *Platessa pseudoflesus*, ein Bastard zwischen Scholle und Flunder. Wiss. Meeresunters. Abt. Kiel, NF22, 5, 53-88.

Purdom, C.E., 1969. Radiation-induced gynogenesis and androgenesis in fish. Heredity, Lond., 24, 431-444.

Purdom, C.E., 1972. Induced polyploidy in plaice (*Pleuronectes platessa*) and its hybrid with the flounder (*Platichthys flesus*). Heredity, Lond., 29, 11-24.

Riley, J.D. and Thacker, G.T., 1969. New intergeneric cross within the Pleuronectidae, dab x flounder. Nature, 221 (5179), 484-486.

Rothschild, Lord, 1953. The fertilization reaction in the sea urchin. The induction of polyspermy by nicotine. J. Exp. Biol., 30, 56-67.

Russell, A., 1939. Pigment inheritance in the *Fundulus-Scomber* hybrid. Biol. Bull., 77, 423-431.

Von Ubisch, L., 1953. Über die Zahl der Flossenstrahlen bei *Pleuronectes platessa*, *Pl. flesus* und der Bastard und Rückkreuzungen zwischen beiden Arten. Zool. Anz., 151, 75-86.

C.E. Purdom
Ministry of Agriculture, Fisheries and Food
Fisheries Laboratory
Lowestoft / GREAT BRITAIN

R.F. Lincoln
Ministry of Agriculture, Fisheries and Food
Fisheries Laboratory
Lowestoft / GREAT BRITAIN

Behaviour

Vital Activity Parameters as Related to the Early Life History of Larval and Post-Larval Lake Whitefish (*Coregonus clupeaformis*)[1,2]

W. J. Hoagman

INTRODUCTION

The lake whitefish (*Coregonus clupeaformis* Mitchill) occurs in all of the Laurentian Great Lakes but is most abundant in Lakes Michigan, Huron, and Superior. It prefers cool, clean water and usually inhabits the shoal and shore zones to depths of 50 m; it makes a fall spawning migration to shallow water, and another limited shore movement in spring. The lake whitefish has been the most valuable commercial species in the Great Lakes since the early fishery. It usually brings a greater return than lake trout and yields 1.5 - 4.5 million kg/year.

Lake whitefish larvae have been studied in the field by Hart (1930), Bajkov (1930), Faber (1970), Reckahn (1970), and Hoagman (1973) but very little attention has been directed to behaviour in the laboratory. This study presents the first data on swimming speeds, temperature preferences, light preferences, feeding success, and related laboratory behaviour. The experiments were performed as the larvae grew from 12.6 to 24.9 mm total length. The specimens were captured from naturally-hatched eggs in Lake Michigan, of which a full account can be found in Hoagman (1973).

METHODS AND MATERIALS

The larvae were held in a series of 76 l aquaria in a large environmental chamber. The water from Lake Michigan was filtered and recycled in each tank. Each week the tanks were cleaned and half the volume replaced as the filters were renewed. Live plankton (mainly copepods and *Daphnia*) were added in generous quantities three to five times a day. At practically every feeding some organisms were still present from the previous feeding. Food was never used in any experimental tanks. The temperature and light in the environmental chamber were adjusted to follow the normal increase at their hatching zone in Lake Michigan, 45°N latitude. The chamber was equipped with fluorescent lights controlled by automatic timers. A small light was kept on at night to simulate natural conditions close to full moonlight.

Swimming speed, while feeding, was measured by timing the larvae as they swam below a plexiglass board marked with a cm grid suspended across the upper edges on the tank. The larvae were fed in their own holding tank and speed tests were made 15 min after food introduction.

---

[1]Contribution No. 96 of the Center for Great Lakes Studies, The University of Wisconsin - Milwaukee, Milwaukee, Wisconsin, 53201.

[2]Contribution No. 568 of the Virginia Institute of Marine Science, Gloucester Point, Virginia, 23062.

When 50 timings were made the entire distance divided by the entire time gave the average "searching" swimming speed for the tank conditions with abundant food.

Temperature preferences were determined by using a plexiglass cylinder 9 cm wide by 121 cm tall. This was marked externally by divisions every 15 cm. Water at the temperature of the environmental chamber was introduced and one or two small aquarium heaters started in the surface water, just after 20 to 25 larvae were introduced. Alongside the experimental cylinder, a control was run without heating. A vertical temperature gradient was established over the upper 45 cm by occasional mixing. Counts were made of the larvae in each section every 10 min (later 5 min) for 4 h and the temperature measured with a thermistor probe lowered in the cylinder.

Light preference was determined within a rectangular plexiglass tank 1.8 m long by 0.6 m high by 0.2 m wide, which was divided by a series of baffles. Each baffle had a single split in the middle 1.5 cm wide by 37 cm high. The entire tank was shielded from external light and one end was brightly illuminated with six 100 W bulbs. These lights were in a separate box mounted adjacent to the tank end and their light was directed into the plexiglass tank through a window in the box so there was no local heating. The baffles provided 6 sections (A-F) in which the larvae could swim freely either to a darker section or a brighter section until it reached either end section. Since all the baffles had a slit in the middle and all were lined up, this produced a shaft of light down through the tank. In effect, if a larva was near the edge of one section, it had to cross this vertical wall of bright light if it swam to the other side of the section. At 15-min intervals the larvae were counted in each section.

Sustained swimming speed was measured by allowing the larvae to swim against a flow of water circulated in a shallow cylindrical tank by means of 4 blowers 90° apart. The tank was 61 cm in diameter by 9 cm high, the water level being kept at 6 cm. The criterion for sustained speed was that the larvae could hold position for 60 sec. The voltage on the blowers could be varied, thus altering the velocity gradient from the centre to the edge of the tank. The larvae were watched under a triangular plexiglass board to establish at what point in the gradient they could stem the current. The current was calibrated by neutrally buoyant floats. Groups of 25 larvae were introduced with a low voltage setting on the blower. Observations were made as the power was raised and then lowered in 10 v steps. Burst speeds were measured by poking the larvae as they stemmed the current to see if they could advance when frightened.

Digestion rate was determined by holding a large quantity of newly-fed larvae in tanks without food and sacrificing them at predetermined intervals. Their stomach contents were then examined and compared with a sample taken at the start. The digestion experiments were made at the temperatures of the environmental chamber for that date.

RESULTS

Swimming and Feeding Speeds. The overall average swimming speed while feeding was 1.5 cm/sec. The searching speed of the larvae did not increase as they grew from 12.7 to 24.9 mm, or as the temperature increased from 7.2 to 14.8^{o}C (Table 1). The larvae would stop to snap at a copepod every 5 to 15 cm and it is the speed between snaps that is recorded here. Faber (1970) observed newly-hatched lake whitefish swimming at 2 to 3 cm/sec in South Bay, Lake Huron. The average swimming speed of *C. wartmanni* larvae held in aquaria by Braum (1967) was 2.5 cm/sec at all temperatures.

Table 1. Swimming speed of whitefish larvae while actively feeding in the laboratory

Water temperature C	Time (CDT)	Average speed (cm/sec)	Average total length (mm)
14.1	1900	1.82	12.8
7.2	1100	1.29	12.7
7.5	2145	0.80[a]	13.1
7.6	2115	1.70	15.2
12.0	2100	1.40	17.8
13.6	2000	1.15	19.3
14.2	1945	2.03	22.1
14.8	2100	1.21	24.9

[a]Food very heavy in tanks.

The larval whitefish could react to a food organism at a distance of approximately one cm to the side or front. Organisms below a line extended outward from their longitudinal axis were usually ignored. Since they swam continually at a slight angle, the zone in front and below the path where their head would cross was not important to the feeding space of the larvae. Therefore, in 1 h a larva would search 10.8 l (1.5 cm/sec x 2 cm x 3600 sec), or if the true feeding speed in nature is 2.5 cm/sec, 18 l/h would be searched. Braum (1964) found that *Coregonus wartmanni* could search 14.6 l/h in aquaria when they were approximately 10 mm. Blaxter and Staines (1971) give search volumes for herring, pilchard, and plaice and all were well below whitefish but herring of 13-14 mm could search 6-8 l/h. The calculations do not allow for the snapping time which from laboratory observations represent 1/5 to 1/3 of the total swimming time at all sizes. Allowing 1/4, the larvae in the aquaria would search 8.1 l/h, and in nature perhaps 13.5 l/h. Expressed on a 24-h basis this becomes 194 and 324 l of water searched. Since the larval whitefish were 100% active over the entire period and fed at all hours regardless of light intensity, the extrapolation to a 24 h day seems justified. Einsele (1963) found larvae *Coregonus* spp considerably reduced their food intake when light was reduced below 20 lux. From 1 to 20 lux in my experiments, the larvae did not slow their feeding activity when food was abundant (ca. 100-400 copepods/l).

Other information is required before one can speculate about a larva's chance of success in the environment. The feeding effectiveness was measured by observing hungry larvae on four separate occasions and counting the total number of encounters and the effective snaps which captured an organism. These observations were made in separate 7.6 l aquaria. The mean rates for 20 larvae were 62, 22, 38, and 29%. Braum (1967) found a rate of 21.6% for his larvae 9 to 16 days old. Feeding success increased with age in herring but not for plaice (Blaxter and

Staines, 1971). The over-all average of 37.8% may be slightly high but it indicates a whitefish larva 14-18 mm could capture approximately 1/3 of all the suitable organisms it encounters from the water it searches.

The sustained swimming speed of the larval whitefish increased linearly with total length and no significant differences were found at 11.5 and 14.5°C (Table 2). In general, the sustained swimming speed averaged 3.7 times the body length (BL) and the maximum burst speed average 7.6 times the body length. A sustained swimming speed of 3-4 BL/sec is common for a wide range of fishes (Houde, 1969) and burst speeds above 8 BL/sec were found for herring (Blaxter, 1962), plaice (Ryland, 1963), and goldfish (Radakov, 1964).

Table 2. Sustained swimming speed of whitefish larvae in moving water in the laboratory

Average length (mm)	Water temperature (C)	Maximum sustained speed		Maximum short distance speed	
		cm/sec	body len/sec	cm/sec	body len/sec
15.2[a]	7.5	3.0- 4.4[b]	2.4	5.4- 8.1[b]	4.4
15.8	11.5	4.7- 6.9	3.7	10.0-13.2	7.3
19.7	11.5	6.0- 8.3	3.6	14.4- 7.7	8.2
21.3	14.5	6.1- 8.9	3.5	13.1-17.5	7.2
28.8	14.5	10.0-12.9	4.0	18.2-25.2	7.5

[a]Using square tank.

[b]Expressed as range one standard deviation from either sie of mean

The relationship between speed and length varied as $L^{0.68}$ for sustained speed and $L^{0.91}$ for burst speed. Fry and Cox (1970) found $L^{0.5}$ for juvenile rainbow trout but they mention other studies using non-salmonids with exponents near unity.

Information on swimming speed can be used to calculate the theoretical rate and the size of the organisms which can escape towed nets. My towed net travelled at 67-90 cm/sec and had a 50 cm diameter opening with a 45 cm bridle (height of cone formed by bridle). If one assumes a slight shock wave ahead of the shackle (at tip of bridle), an "alert" zone exists 55 cm from the net mouth. Using a velocity of 80 cm/sec a larva in the alert zone would be reached by the net in 0.68 sec. A larvae therefore has 0.68 sec to swim the radius of the net (the maximum distance) to escape capture. But the apex and shackle should probably be eliminated from the radius because they would alert larvae to the approaching net. Using an effective radius of 17.5 cm one can calculate from swimming speed data that all larvae 33 mm or larger could swim fast enough to avoid capture. Larvae smaller than 33 mm could escape the net in proportion to how far they were from the net when they first noticed it and how far they were from the nearest edge of the cylinder of water which the net would strain. Field observations show the young whitefish become increasingly difficult to capture after 20 mm total length. Once a young whitefish is actually within the cone of the towed net it could not escape unless it could swim 80 cm/sec (the speed of the net). This speed could only be reached by fish about 100 mm.

Response to Light. The larval and post-larval whitefish showed no particular affinity for bright light. When placed in middle sections of the light chamber they gradually dispersed inversely to the light intensity but did not necessarily congregate in the darkest section except on one occasion (Table 3). The tendency to "seek" darker sections remained until a fish length of 23 to 30 mm.

Shkorbatov (1966) used a similar light chamber and found that many different species and hybrids of Russian coregonids, 10 days old, had a positive phototaxis. The reaction was most pronounced in the first several days of life, but gradually disappeared as age increased.

The larvae that I tested were approximately 2 weeks old at the first trial. These fish were placed in the darkest section first and at the end of 11 h had practically all moved to the brighter sections. Two weeks later all the fish were put in the brightest section first and had not moved at the end of 4 h. The next day another sample was placed in the middle section between the brightest and darkest. These fish gradually moved to the darker sections. Two identical trials spaced two weeks apart gave the same results. In every experiment some fish would find their way to the brightest section within a short time and thereafter the numbers in the brighter sections would remain constant (Table 4).

Several times in the darkened environmental chamber a bright light was brought to the side of the holding tanks. The larvae would not congregate in the beam or move to the side with the bright light. They would merely swim through the beam and back into the dark water.

In the laboratory the larval whitefish were always very close to the water surface regardless of the brightness of the room. In nature, surface tows during the day always yielded more larvae than subsurface tows in the same area (Hoagman, 1973). Surface tows at night yielded more larvae than surface tows in the same area during the day. A part of the difference probably was the greater visibility of the net during daylight but another part may be that some larvae were near bottom during the day, which would indicate a negative phototaxis.

Temperature Response. The lake whitefish is a stenothermic fish usually found in water below 20°C. This preference is manifested very early in life. The tendency for whitefish larvae to rise to the highest levels within a column of constant temperature is shown clearly by examination of the control cylinder counts (Table 4). Invariably, the majority of the young whitefish selected the upper waters. The response was the same at all fish lengths and control temperatures.

The tendency to rise to the surface was also a dominant reaction with fish in the experimental cylinders. They would continually penetrate the warm to very warm water, become immobilized, and drift motionless down to the cooler water. On reaching the cool water, they usually recovered, swam slowly until regaining full equilibrium, then would penetrate the warm water again. This stage of immobilization and series of events was termed heat narcosis by McCauley (1968). A similar behaviour was found with herring larvae in gradients of oil dispersants (Wilson, this symposium).

The grand average of all the 5- and 10-min counts provide some clues to temperature preference and the maximum temperature tolerated. The acclimization temperature did not seem to influence the maximum toler-

Table 3. Movement of larval whitefish in response to a light gradient. Counts of larvae represent average of 4 counts made to that hour. Foot candle measurements given below table for each section

Number of h after introducing fish	A	B	C	D	E	F	Average total length in mm	Water temperature °C
0	0	0	0	0	0	50	13.1	7.5
2	0	0	0	2	8	40		
3	0	0	0	3	8	39		
4	0	0	0	3	11	36		
11[a]	0	0	6	11	25	8		
0	60	0	0	0	0	0	15.2	7.5
1	60	0	0	0	0	0		
2	60	0	0	0	0	0		
4	59	1	0	0	0	0		
0	0	0	66	0	0	0	15.2	7.5
1	3	2	53	6	1	1		
2	4	1	43	9	6	3		
3	5	2	39	11	6	3		
4	5	0	35	14	6	6		
6	6	1	27	15	8	14		
7	6	0	24	12	12	14		
8	5	0	23	14	10	13		
9	5	1	23	13	14	13		
0	0	0	60	0	0	0	17.8	12.0
1	5	4	38	8	5	0		
2	5	4	33	10	5	2		
3	4	4	25	12	9	4		
4	5	2	24	6	11	10		
6	4	2	20	7	11	13		
7	4	1	22	6	10	14		
8	3	1	20	6	12	15		
9	3	1	17	5	13	17		
10	3	1	17	8	12	16		
0	0	0	64	1	0	0	19.3	13.5
1	6	6	20	18	13	0		
2	5	4	17	18	14	2		
3	4	5	19	18	13	2		
5	4	4	16	21	13	3		
6	4	3	16	20	14	5		
8	4	2	14	16	17	6		
9	4	3	12	16	18	6		
0	0	0	50	0	0	0	24.9	14.5
1	0	1	49	0	0	0		
2	0	2	48	0	0	0		
3	0	1	48	1	0	0		
4	0	1	47	0	1	1		
5	0	2	46	0	2	0		
6	1	3	42	0	3	1		
7	1	5	39	0	4	1		
10	2	6	35	1	5	1		
11	2	7	34	1	5	1		
lux measurements	2475	118	19	5	2	1		

[a]All larvae dead in sections at 11 hours. Counts shown for hour 11 are of dead larvae.

Table 4. Distribution of larval and postlarval whitefish in relation to a vertical temperature gradient in the laboratory. Counts are grand average of all readings for entire experiment. Time of experiment 3 - 4.5 hours

Type of Chamber	Section, temperature, and counts								Average fish length in mm
	A (top)	B	C	D	E	F	G	H	
Experimental-°C	26.5	15.7	10.3	7.9	7.0	6.4	6.1	6.1	12.6
Number	3	6	5	3	1	0	2	0	
Control, 6.0°C Number	14	2	1	1	0	1	1	0	
Experimental-°C	29.3	22.0	17.3	11.3	9.2	8.0	7.3	7.1	13.1
Number	1	1	7	6	2	4	2	2	
Control, 7.0°C Number	11	3	2	1	2	1	1	1	
Experimental-°C	27.2	24.8	19.0	14.7	11.9	9.4	7.7	7.0	13.1
Number	1	2	5	7	5	2	2	2	
Control, 7.1°C Number	19	3	3	2	2	2	2	0	
Experimental-°C	22.6	13.3	11.9	11.2	11.0	11.1	11.0	11.0	17.8
Number	4	9	5	2	1	2	1	1	
Control, 11.0°C Number	19	3	1	1	0	0	1	0	
Experimental-°C	24.7	18.1	13.3	12.0	12.1	12.0	11.9	11.9	17.8
Number	0	3	11	4	2	3	1	1	
Control, 11.9°C Number	15	5	1	2	0	1	0	0	
Experimental-°C	26.1	18.0	15.7	15.2	15.1	14.8	14.7	14.7	21.4
Number	1	4	8	5	2	1	1	2	
Control, 14.6°C Number	10	3	2	1	1	1	1	5	
Experimental-°C	24.6	23.7	19.6	17.6	16.2	15.3	15.1	14.9	26.5
Number	3	1	2	2	3	4	5	5	
Control, 14.7°C Number	14	2	1	1	1	2	2	2	

able temperature (Fig. 1). All three curves (data grouped by acclimization temperature) suggest a preferred temperature of 12° to 17°C. Coming from cool water, the larvae were usually thermally shocked at all temperatures above 20° to 22°C. The counts, however, only show the location of the larvae at a particular temperature and give no indication of the length of time spent in the sections. For the warmer sections the larvae could only spend a short time in the warm water, and thus the counts of 2 or 4 in the warm water are really an overestimate of their preference for that temperature.

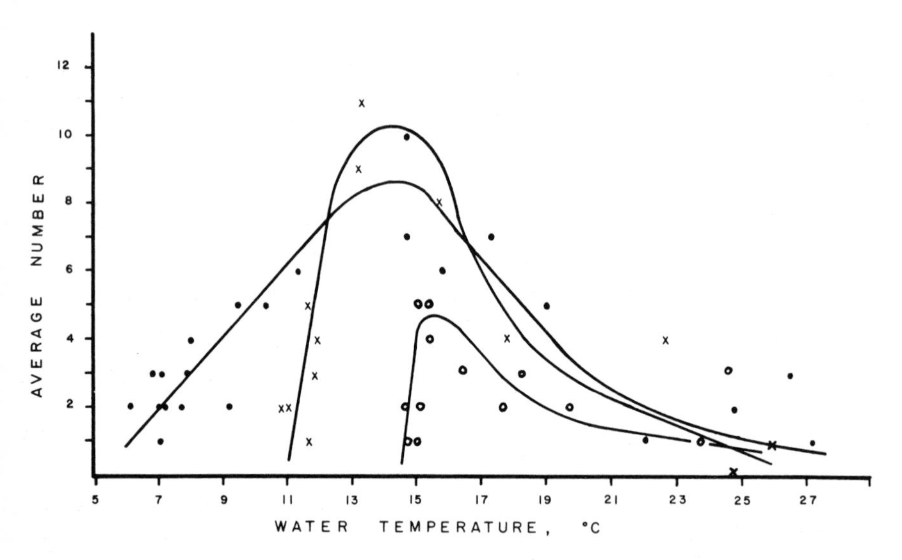

Fig. 1. Average numbers of larval whitefish within a vertical temperature gradient. Solid dots are for fish length of 12.9 mm average and acclimization temperature of 6.8°C average; crosses are for fish of 17.8 mm average length and acclimization temperature of 11.4°C; open circles are for fish of 23.1 mm average length and acclimization temperature of 14.6°C

The young lake whitefish investigated by Reckahn (1970) in South Bay, Lake Huron, preferred 17°C water throughout the summer. *Coregonus lavaretus* preferred 11.0° to 15.4°C when 60-days-old and their preference lowered to 8° to 12°C at 100-days-old after having been acclimated to 14°C in both cases (Mantelman, 1956). Shkorbatov (1966) studied many species of larval ciscoes and whitefish common to the Soviet Union and found the acclimization temperature influenced their eventual preferred temperature but in all cases they preferred several degrees higher than the holding temperature they came from.

Although this study was not intended to measure temperature tolerance of larval whitefish, the observations of behaviour and subsequent thermal shock suggest that when acclimated to 11° and 14°C, their resistance time is quite short at temperatures above 21°C. Studies on two other North American coregonids have shown that when the juveniles of *C. artedii* and *C. hoyi* are acclimated to 10°C, at least 50% will die when placed in water 24°C (Edsall and Colby, 1970; Edsall et al., 1970).

Digestion. The evacuation of the stomach and intestine was assumed equivalent to the rate of digestion. In fish of such small size no differentiation can be made between the stomach and intestine. The copepods were the only significant ingested food; thus rate of digestion was expressed as numbers of copepods remaining (partially digested or whole).

After 24 h the entire stomach contents of the whitefish fry had been digested (Table 5). Compared to the average number of copepods in the reference specimens, the test specimens had 68 and 99% fewer organisms in their stomach after 8 and 24 h respectively. From 0 h - 16 h, the rate of utilization was almost linear.

Table 5. Average number of copepods per stomach of larval whitefish held without food. 10 experimental fish sacrificed at each hour indicated. 39 and 31 reference specimens used respectively. Water temperature 14.4°C for both experiments. Average fish length in mm

	Reference specimens	Hour specimens taken						
		1	2	4	8	16	24	48
Copepods in stomach	9.7	8.1	8.5	5.3	2.8	0.7	0.0	--
Fish length	19.9	19.7	19.9	20.3	20.0	20.9	18.0	
Copepods in stomach	13.6	12.0	10.1	12.3	4.7	0.4	0.2	0.0
Fish length	25.1	25.5	24.7	24.3	23.5	25.9	25.6	23.2
Average percent decrease		14	19	27	68	95	99	100

This extremely rapid rate of food utilization has important consequences for the young whitefish. At this size their yolk sac is completely absorbed. They are frail, slender, and without large amounts of muscular flesh (compared to adult proportions). It is not difficult to understand, why the young whitefish 18-26 mm long weaken between 30 and 45 h when held without food, considering 95% of their ingested food is utilized after 16 h at 14.4°C. The smaller specimens digested their stomach contents sooner (Table 5).

The rates of food utilization for the young whitefish do not seem unusually fast in general. Windell (1967) found a 98% decrease of digestible organic material in bluegill (*Lepomis macrochirus*) after only 18 h. Other studies reported by Windell for *Gambusia* spp, and *Macropterus* spp showed complete digestion times of less than 24 h. Herring larvae 12-days-old (ca. 9 mm) evacuate their guts in 12-19 h at 9°C (Kurata, 1959); whereas Blaxter (1965) found complete digestion at 9 and 4 h at 7° and 15°C respectively.

By using the slope of the evacuation curve (rate of digestion), some idea of daily ration can be computed. The laboratory whitefish grew well, but somewhat slower than wild fish. This indicates they were getting more than the minimum daily ration, but less than the maximum. Growth rate and stomach contents show they were taking approximately 1/2 to 2/3 as much food as wild fry would be taking at these sizes (Hoagman, 1973).

From Table 5, 80% of the food is utilized in the first 12 h. Since the average number of copepods per stomach of the reference specimens was 11.7, the requirement for one-half day is 9.5 organisms (0.80 x 11.7).

If one assumes they eat at all periods of the day and night in the laboratory (all observations support this), the laboratory daily meal equals 18.6 copepods at these fish sizes and temperatures. Larvae in the wild had 11 to 17 copepods per stomach at fish sizes of 15 to 17 mm. The wild larvae, which averaged 20.8 mm long on 24 May, 1970, had 42.6 copepods per stomach and 68.6 *Nothalca* spp per stomach, which is much above the reference specimens from the laboratory at near the same size.

If laboratory larvae can live and grow well on only 19 copepods per day at 14.5°C, they might be able to live in nature on this ration. The laboratory larvae had abundant food, thus the number of organisms probably corresponds to their preferred intake. The preferred intake of wild whitefish larvae seems to be greater than the laboratory fish, and this was expressed in greater growth. Braum (1967) has calculated from feeding rates that *C. wartmanni* 31-days-old need 265 copepods/10 h at 15.6°C to satisfy maximal ration, and 14 copepods are necessary as a minimum ration at 11.0°C. Braum's method is somewhat obscure and he did not use digestion analysis; thus his data do not allow independent determination of the various rations.

Whatever the maximum or average ration for wild larval whitefish, it is clear from the results presented here that they need a fairly constant food input; what food they do ingest is rapidly digested and they require approximately 10-20 copepods per day as a minimum. The amount of food required for normal growth in the wild, if the same digestion rates are assumed (temperatures are similar), can be calculated from the average number of copepods per stomach of the preserved wild specimens. The number per day range from 11 to 70 for fish 15 to 23 mm from Lake Michigan and 10 to 18 for Green Bay fish at larvae sizes of 12.0 to 14.1 mm (Hoagman, 1973). It must be emphasized that this is not the minimum daily ration, i.e. that ration which just balances energy requirements and allows none for growth. Rather, it is somewhat above the minimum. The young whitefish should be able to live (stay alive) on something less, but the amount cannot be specified at this time.

From information on swimming speed and feeding effectiveness presented earlier, it is exceedingly doubtful that wild larvae would be unable to find and capture sufficient food for growth and maintenance in nature. Thus natural starvation seems remote, especially since they are continuously swimming and feeding at all light levels.

SUMMARY

Larval lake whitefish hatched in nature were transported to constant environmental conditions and experimented on over fish sizes of 12.6 to 24.9 mm. Swimming speed while feeding on live plankton was 1.5 cm/sec and did not increase proportional to size; but sustained swimming ability increased from 3.7 to 11.4 cm/sec and maximum short distance speed increased from 6.7 to 22.6 cm/sec. Body lengths per second remained nearly constant, averaging 3.7 times the body length. The overall relationship between speed and length was $L^{.68}$ for sustained and $L^{.91}$ for maximum speed.

The young whitefish were non-schooling and fed independently. The effective feeding rate on copepods was 37.8% of all encounters. Feeding ceased during absolute darkness while continuing at moonlight levels. Food intake and activity was continuous, enabling the larvae to search between 8.1 l/h and 14.5 l/h. From other tests and observations including digestion rate, death due to starvation in nature is extremely remote.

Whitefish larvae preferred temperatures between 12-17°C when acclimated at 6°, 11°, and 14.5°C before testing. Direct avoidance occurred at 20-23°C, with heat narcosis and death above 24°C. No change in temperature preference was apparent with increase in size. Light gradients from 2475 to 1 lux had no concentrating effects except at the smallest sizes. Beyond 14 mm total length the larvae ignored differences in light intensity.

Vertical distribution in the laboratory was epipelagic, confirming catches of 95% in the top metre in nature. Negative buoyance required constant swimming to remain neutral until filling of the air bladder at larger sizes.

REFERENCES

Bajkov, A., 1930. A study of the whitefish (*Coregonus clupeaformis*) in Manitoban Lakes. Contrib. Can. Biol. Fish. N.S., 5 (4), 433-455.

Blaxter, J.H.S., 1962. Herring rearing. IV. Rearing beyond the yolk-sac stage. Mar. Res. Scotland, No. 1, 1-18.

Blaxter, J.H.S., 1965. The feeding of herring larvae and their ecology in relation to feeding. Rep. Calif. Coop. Oceanogr. Fish. Invest., 10, 79-88.

Blaxter, J.H.S. and Staines, M.E., 1971. Food searching potential in marine fish larvae. 4th European Mar. Biol. Symp. D.J. Crisp, ed. Cambridge Univ., 467-485.

Braum, E., 1964. Experimentelle Untersuchungen zur ersten Nahrungsaufnahme und Biologie an Jungfischen von Blaufelchen (*Coregonus wartmanni* Bloch), Weissfelchen (*Coregonus fera* Jurine) und Hechten (*Esox lucius*). Arch. Hydrobiol. Suppl., 28, 183-244.

Braum, E., 1967. The survival of fish larvae in reference to their feeding behavior and the food supply. The Biological Basis of Fish Production, S.D. Gerking (ed.). John Wiley and Sons, 113-131.

Edsall, T. and Colby, P.J., 1970. Temperature tolerance of young-of-the-year cisco, *Coregonus artedii*. Trans. Amer. Fish. Soc., 99 (3), 526-531.

Edsall, T., Rottiers, D.V. and Brown, E.H., 1970. Temperature tolerance of bloater (*Coregonus hoyi*). J. Fish. Res. Bd Can. 27 (11), 2047-2052.

Einsele, W., 1963. Problems of fish-larvae survival in nature and the rearing of economically important middle European freshwater fishes. Calif. Coop. Oceanogr. Fish. Invest., 10, 24-30 (1965),

Faber, D.J., 1970. Ecological observations on newly-hatched lake whitefish in South Bay, Lake Huron. In: Biology of Coregonid Fishes, C.C. Lindsey and C.S. Woods (eds.). Univ. Manitoba Press, Winnipeg, Manitoba, 481-500 pp.

Fry, F.E.J. and Cox, E.T., 1970. A relation of size to swimming speed in rainbow trout. J. Fish. Res. Bd Can., 27 (5), 976-978.

Hart, J.L., 1930. The spawning and early life history of the whitefish, *Coregonus clupeaformis* (Mitchill), in the Bay of Quinte, Ontario. Contrib. Can. Biol. Fish., 6 (7), 167-214.

Hoagman, W.J., 1973. The hatching, distribution, abundance and nutrition of larval lake whitefish (*Coregonus clupeaformis* Mitchill) of Central Green Bay, Lake Michigan. Inst. Freshw. Res. Drottingholm, Sweden. Ann. Rep. No. 53 (in press).

Houde, E.D., 1969. Sustained swimming ability of larvae of walleye (*Stizostedion vitreum*) and yellow perch (*Perca flavescens*). J. Fish. Res. Bd Can., 26 (6), 1647-1659.
Kurata, H., 1959. Preliminary report on the rearing of the herring larvae. Bull. Hokkaido Reg. Fish. Res. Lab., 20, 117-138.
Mantelman, I.I., 1956. Izbiraemie temperaturi u molodi nekotopikh vidov promislovikh ryb. (Preferred temperatures of the young of some species of commercial fish). Trans. from Russ. Proc. Conf. Fish Physiol., Acad. of Science, Ichthyology Comm., Moscow, USSR (Pub. 1958).
McCauley, R.W., 1968. Suggested physiological interaction among rainbow trout fingerlings undergoing thermal stress. J. Fish. Res. Bd Can., 25 (9), 1983-1986.
Radakov, D.V., 1964. Velocities of fish swimming. Severtow Inst. of Animal Morph., Moscow, 4-28.
Reckahn, J.A., 1970. Ecology of young lake whitefish (*Coregonus clupeaformis*) in South Bay, Manitoulin Island, Lake Huron. In: Biology of Coregonid Fishes, C.C. Lindsey and C.S. Woods (eds.). Univ. of Manitoba Press, Winnipeg, Manitoba. 437-460.
Ryland, J.S., 1963. The swimming speeds of plaice larvae. J. Exptl. Biol., 40, 285-299.
Shkorbatov, G.L., 1966. Isbiraemaya temperatura i fototasis lichinok sigov. (The preferred temperatures and phototaxis of larval whitefish). Trans. from Russ., Zoological J. 14 (10), 1515-1525.
Windell, J.T., 1967. Rates of digestion in fishes. In: Biological Basis of Freshwater Fish Production. S. Gerking (ed.), John Wiley and Sons, New York.

W.J. Hoagman
Virginia Institute of Marine Science
Gloucester Point, Va. 23062 / USA

Effect of Prey Distribution and Density on the Searching and Feeding Behaviour of Larval Anchovy *Engraulis mordax* Girard

J. R. Hunter and G. L. Thomas

INTRODUCTION

Laboratory estimates of the minimum concentration of food required for survival of marine fish larvae usually are much higher than the average concentrations of food in the sea (O'Connell and Raymond, 1970; Hunter, 1972). A common explanation for the fact that laboratory food requirements exceed natural food densities is that larvae are able to find and remain in patches of food in the sea which are considerably above the average food density estimated from plankton net catches. This explanation is supported in part by Ivlev (1961) who demonstrated with carp fry that an increase in the degree of aggregation of prey had the same effect on food consumed as an increase in the overall density of food material. A patchy distribution of larval food occurs under natural conditions. Thus, the effect of prey distribution on feeding behaviour of larval fish and the scale of "patchiness" of food items in the sea must be known to estimate the impact of food distribution on the feeding and searching behaviour of larval anchovy. This paper described some aspects of the effect of prey distribution and density on the feeding and searching behaviour of larval anchovy *Engraulis mordax* Girard.

The prey used in most of the experiments was the dinoflagellate, *Gymnodinium splendens*, which forms dense, stable, and easily-recognizable aggregations in rearing containers. *Gymnodinium* is readily cultured (Thomas et al., 1973), and promotes growth in larval anchovy equivalent to a wild plankton diet during the first week in the life of an anchovy larvae (Lasker et al., 1970). For comparative purposes we also used the rotifer, *Brachionus plicatilis*.

Larvae were reared from the egg in air conditioned rooms in static sea water which varied in temperature from 17° to 19°C. Anchovy eggs hatch in about 2 days when kept at these temperatures. At hatching the yolk-sac larvae have functional olfactory organs and naked neuromasts but they do not have a functional eye (O'Connell, pers. comm.). About 50 h after hatching the mouth is functional, the gut expanded, and occasionally food is present in the gut (D. Kramer, unpubl.), although we have never seen larvae actively feeding at this stage. About 72 h after hatching only a few granules of yolk remain (D. Kramer, unpubl.), the eye becomes functional and active feeding begins.

During the first few days after yolk absorption anchovy larvae have a higher food density requirement than at any other time in the larval stage (Hunter, 1972). The larvae have a low level of feeding success, capturing only 11% of the prey at which they strike, and the volume of water searched for prey is much less than that for older larvae. In addition, at this stage, larvae die of starvation if they do not find food within 1.5 days (Lasker et al., 1970). These laboratory findings suggested that effect of food distribution was most important during the first days of feeding. For this reason we restricted our observations to larvae of ages 1 to 8 days.

METHODS OF FORMING PATCHES OF *GYMNODINIUM*

We consider here our observations of the behaviour of *Gymnodinium* aggregations and our techniques for formation of the aggregations. These observations are important because the aggregations played a critical role in our experiments; if similar aggregations form under natural conditions, they could be an important factor in the survival of marine fish larvae. *Gymnodinium splendens* is an unarmored, marine, halophytic dinoflagellate about 53 μ in diameter. Reproduction appears to be sexual and occurs under natural conditions only at night (Sweeney, 1959).

In the laboratory distinct, visible aggregations or patches of *Gymnodinium* were produced under a variety of conditions. The patches were yellow-green, and occurred at the surface in daylight and in darkness. The aggregations had a dense central mass with striae of cells extending from it. Cells were most dense at the water surface; the average density for 8 aggregations measured at the water surface was 30,000 cells/ml, whereas 10 cm below the water surface the mean density was 250 cells/ml. The patches varied in diameter from a few mm to ones greater than 10 cm. Occasionally more than one patch formed in a larval-rearing container.

Our procedure for establishing an aggregation of *Gymnodinium* in a rearing container was simply to add an inoculum sufficient to bring the average density of the cells in the container to 100 to 150 cells/ml. The density of inoculum ranged from 1500 to 2000 cells/ml. On occasion less-dense cultures were used, but these caused greater variability in the frequency of occurrence of aggregations and in their longevity. The cells were cultured by William Thomas and staff (Univ. of Calif., San Diego); the technique is described by Thomas et al. (1973). After inoculating the container, we covered it with either a black opaque or a transparent top whereupon aggregations, 5 to 10 cm diameter, formed within 24 h. The container contained only filtered sea water, *Gymnodinium*, and the medium in which the cells were originally cultured. The number of cells in the initial inoculum had a direct effect on the time necessary for the formation of the aggregation and its size. Inoculations that brought the initial cell density in the container to 20 to 80 cells/ml required 24 to 72 h to form aggregations and the aggregations were small, 0.5 to 5.0 cm diameter. When the initial density in the container was only 10 cells/ml, no patches formed after 6 days.

At the initial density in the container of 100 to 150 cells/ml patches formed equally well in constant dark (illumination $\leq 1 \times 10^{-6}$ ft-c) as they did under the fluorescent lamps located above the rearing containers. Distinct aggregations also formed in containers placed outdoors and exposed to direct sunlight. On occasion we used light to establish an aggregation of *Gymnodinium* in a particular position in a container. Once an aggregation had formed, however, it could not be moved to a different position in the container by illuminating a different section of the container. This was the case regardless of whether or not light had been used to establish the patch initially. The only way an established patch could be relocated was by dispersing the patch by stirring and allowing it to re-form. It would re-form in the light or in the dark and usually in a new region of the container. It could also be re-established in a particular region by illuminating one section of the tank. We occasionally used light to relocate an aggregation because we preferred to have the aggregations located near the center of the container. Patches were relocated by stirring the container, and placing an opaque cover over the container with a hole cut in it so that light entered in only one section of the tank. Aggregations, 5 to 10 cm diameter, re-formed after stirring within 3 to 4 h. Relocation of the aggregation was successful 87% of the time (N = 31).

The longevity of *Gymnodinium* aggregations was not measured because the ending of a larval fish experiment required the removal of the patch. During a period of low availability of anchovy eggs, however, of 18 containers inoculated with *Gymnodinium*, all retained aggregations for 10 days and 6 for 16 days.

In summary, the most important characteristics of *Gymnodinium* aggregations were that they were stable, easily detected visually, and formed spontaneously independent of light. The only difficulty encountered was that wind-driven currents sometimes caused the patch to disperse temporarily. This problem was avoided by keeping covers on the containers and turning the air conditioner off before uncovering them.

ATTRACTION OF LARVAE TO PATCHES OF *GYMNODINIUM*

We noticed at the beginning of this study that yolk-sac larvae appeared to be aggregated in patches of *Gymnodinium*. We performed a series of tests to confirm this observation and determine the role that size and density of the patch and the density of cells outside the patch played in the behaviour.

Patches of *Gymnodinium* of various sizes and densities were established in 10 l rearing containers using the methods described in the previous section. We then placed 100 anchovy eggs in each container. The containers in this and all other experiments described in this paper were kept in air conditioned rooms at 17° to 19°C and under light cycles of 14 h light, 10 h dark. On the day the larvae reached the age of 1 day we measured the maximum diameter of the patch and sampled it by selecting an acrylic cylinder similar in size, lowering it over the patch and sealing it with an acrylic plate that had a circular groove cut to fit the cylinder. In such a sample and in control samples we counted the larvae, measured the volume of the water, and measured the density of *Gymnodinium* by counting the cells in ten 0.03, or 0.01 ml samples; the number of larvae remaining in the container were also counted.

Our estimate of the density of the patch was the mean number of cells/ml in the cylinder containing the patch. These estimates were less than the actual density because the patch was most concentrated at the surface and usually did not extend to the bottom of the container. The estimate of the background concentration of *Gymnodinium* was the average of the 4 mean densities taken from the 4 control samples.

Effect of the Volume and Density. We took 27 sets of samples of 1-day-old yolk-sac larvae in containers having patches of different density and volume. A comparison of the percent of larvae captured per unit volume of the sample of the patch to that of samples taken outside the patch showed a difference at $P < 0.00006$ Mann-Whitney U Test (Siegel, 1956). To evaluate the importance of the various characteristics of the patch we plotted the log of the number of larvae captured in the patch against the log of patch density (cells/ml), density outside the patch (cells/ml), patch volume (ml), and number of larvae in the container. The number of 1-day-old larvae in the container varied among tests because of larval and egg mortality and infertile eggs. Consequently, N (the number of larvae present) had to be considered as a variable. Log transformations were used because a better fit was obtained with a multiplicative model than with an additive model.

Partial regression coefficients for number of larvae, patch density, and patch volume were significant ($P < 0.01$), whereas that for the cell density outside the patch was not. That the number of larvae present and patch volume are related to catch is obvious. On the other hand, these variables had to be considered in order to evaluate the effect of patch density. The final multiple regression of larvae caught (C) on the three independent variables - number of larvae present (N), volume of patch (V), and density of patch (D) - yielded the equation

$$C = N^{0.97}\ V^{0.51}\ D^{0.58} - 3.969$$

where coefficient of multiple correlation = 0.8493 and the standard error of the estimate = 0.1780. Since the coefficient for N was nearly unity and those for V and D were close to 0.5, the data were fitted to the general equation $C/N = K\sqrt{VD}$, using Marquardt's Algorithm for fitting nonlinear models (Conway, Glass and Wilcox, 1970) giving K = 0.000208 with 95% support plane confidence intervals for K of $0.000181 < K < 0.000235$ (Fig. 1). Thus, 1-day-old yolk-sac larvae aggregate in dense patches of *Gymnodinium* and the proportion of larvae attracted is correlated with the density and volume of the patch.

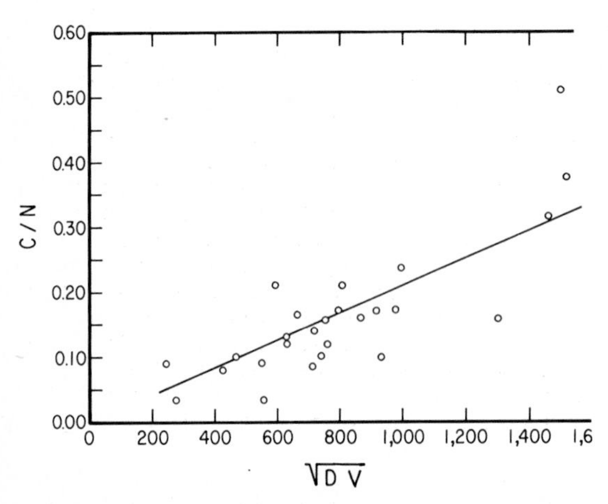

Fig. 1. Relationship between the proportion of 1-day-old anchovy larvae captured in a patch of *Gymnodinium spendens* and the square root of the volume and density of the patch; C/N is the number of larvae in the patch over the total larvae present in the tank; V is the volume of the patch in ml (range, 60 to 850 ml); and D is the density of the patch in cells/ml (range, 540 to 3240 cells/ml). Equation for line is, $C/N = 0.0002\sqrt{DV}$.

We also wished to establish that post-yolk-sac larvae that were actively feeding were also attracted to patches of *Gymnodinium*. For this purpose we made a short series of 11 tests of the same design on 4-day old larvae. The results, summarized in Table 1, indicate that more 4-day-old larvae were present in the patch than in the first control sample ($P < 0.01$, Wilcoxon matched-pairs, signed-ranks test; Siebel, 19

Table 1. Percent of 4-day-old anchovy larvae captured in patch of *Gymnodinium splendens* and in the first of 4 control samples; samples taken during day

Percent of larvae in				
Patch of *Gymnodinium*	First control sample	Total larvae	Patch density cells/ml	Patch volume ml
6	8	47	2,085	630
10	6	68	375	580
12	8	72	1,314	710
15	6	52	1,620	680
15	8	52	1,600	610
21	8	77	2,000	600
23	7	87	1,200	750
26	3	92	1,950	600
27	6	67	1,500	600
40	1	183	3,300	800
43	7	56	900	1,000

Effect of Light. The object of this series of experiments was to determine if yolk-sac anchovy larvae continue to aggregate in patches of *Gymnodinium* in the dark and if they could find patches in the dark. In the first of three experiments, we reared larvae from eggs in 10 l containers in which a patch of *Gymnodinium* had been established and sampled the larvae in and out of the patch 6 h after the onset of darkness on the day they reached 2 days of age. In the second experiment, we reared larvae in 1000 ml beakers, added them to 10 l containers in which a patch of *Gymnodinium* was established at the onset of darkness on the day they reached 2 days of age, and sampled the container in the dark 6 h later. In the third experiment we used the same procedure as in the second, except the larvae were placed in the dark for 24 h before taking the samples. In the second and third experiments we added the larvae by submerging the beaker in the 10 l container, a procedure that always resulted in the dispersion of the patch of *Gymnodinium*. Previous tests had shown, however, that the patch would re-form within 3 to 4 h.

All containers were covered with black plastic lids and were kept in a darkened room. At the end of an experiment the lids were removed and samples in and outside the patch taken simultaneously with 9.5 cm diameter cylinders. All containers were stocked with 100 eggs; because hatching rates varied, the number of larvae sampled was expressed as a percentage.

When larvae were associated with the patch of *Gymnodinium* up to the time of capture on the night of age 2 days (experiment 1) more larvae were present in the patch in the dark than in the control sample taken in the dark ($P < 0.01$, Wilcoxon matched-pairs, signed-ranks test). On the average, 24% of the larvae in the container were captured in the patch sample whereas only 4% were taken in the control sample (Table 2). Thus, larvae remained aggregated in the patches of *Gymnodinium* in the dark.

Table 2. Percent of yolk-sac larvae (age 2 days) captured in dark in a sample of a patch of *Gymnodinium splendens* and in a control sample taken simultaneously outside the patch; when the patch was not dispersed, 6 h after dispersion of the patch; and 24 h after dispersion

Percent of larvae in			
Patch of *Gymnodinium*	Control sample	Total larvae	Patch density cells/ml
Patch not dispersed			
8	3	58	550
9	0	72	1,830
15	2	57	780
21	1	70	2,380
22	5	43	600
27	0	41	2,560
30	1	89	3,480
31	12	68	2,020
37	10	93	1,630
42	2	103	910
52	8	63	740
6 h after dispersion			
3	7	59	900
3	6	67	450
4	1	103	1,000
5	5	125	1,240
6	1	151	1,000
10	4	67	1,380
16	12	69	1,200
17	3	66	450
19	2	63	1,820
21	12	65	2,280
24 h after dispersion			
13	1	78	1,350
23	8	78	1,110
24	1	95	1,500
26	0	92	1,170
27	3	89	1,410
27	1	85	600
27	0	93	1,770
28	2	82	1,440
28	3	79	1,350
29	4	78	1,620
29	1	106	2,050
30	2	91	1,830
37	2	86	2,220

In the two experiments in which larvae were added at the onset of dark, more larvae were taken in the sample from the newly re-formed patch of *Gymnodinium* than were taken in the control samples ($P < 0.05$ for experiment 2 and $P < 0.01$ for experiment 3 (Wilcoxon matched-pairs, signed-rank test; Siegel, 1956). After 24 h in the dark the percent of larvae captured in the patch was greater than after 6 h in the dark and about the same as in the first experiment. Presence of fewer larvae in the patch after 6 h might be caused by the shorter recruitment period, or because the patches were less dense on the average after 6 h than after 24 h. On the other hand, a long period may be required for the larvae to recover from the disturbance caused by being introduced into the container. It appears that search for concentrations of *Gymnodinium* can proceed on a 24-h basis regardless of the daily changes in light level.

These experiments also show that visual stimuli are not required for finding patches of food. One-day-old yolk-sac larvae have well-developed olfactory organs, and naked neuromasts; but the eyes are not developed. Thus, chemical, acoustic or tactile stimuli could be used by yolk-sac larvae to find concentrations of *Gymnodinium*. A chemical stimulus seems the most likely because algae can produce considerable quantities of extracellular substances (Fogg, 1962; Hellbust, 1965). On the other hand, in post-yolk-sac larvae the search pattern for food could also bring about an aggregation of larvae in areas of high food concentration.

STRUCTURE OF SEARCH PATTERNS

The object of these experiments was to determine how food type and distribution affected the structure of the search pattern of larval anchovy. For this study a large, circular, black fiberglass tank of 122 cm diameter with a 1 cm grid inscribed on the entire bottom was used. We filled the tank to a depth of 18 cm and inoculated it with a prey organism. When the desired food density and distribution was obtained we added eggs or larvae.

We recorded on a key board each time a larva completed a feeding act, when it swam in and out of a patch of *Gymnodinium*, and when it crossed a grid line. A different key was used for each of the four grid directions. If a larva did not move, no event was recorded. Each time a key was depressed the data and the elapsed time to the nearest 0.1 sec were entered on a 8 channel paper tape. The computer output included a plot of movements, swimming speed, frequency of feeding and directional probabilities. Directional probabilities were the proportion of grid intercepts made by a larva that were ahead, to the right, to the left, and backward (a 180° change in direction). Swimming speed was the sum of grid intercepts made by a larva divided by the observation period which was 5 min. It was expressed in cm/sec because the distance between grid lines was 1 cm.

To evaluate how differences in directional probabilities (the proportion of movements made in each grid direction) could affect the area covered by a larva during search we used a computer model of bounded random walks used by Cody (1971) to analyze the movements of bird flocks. The bounds of the walk are boundaries of an 11 X 11 unit grid. The larva is started in the centre and moves only along the sides of 1 X 1 unit squares, one side per step. Directional probabilities were fixed relative to the axis of motion of the larva. For each set of directional probabilities (data from one larva), 100 walks of 200 steps each were made. The results of the simulation included the percentage of possible points which had been visited, the proportion that were visited zero, once, twice, up to 10-plus visits.

Of the statistics generated by the random walk program we used only the percent area covered, the percentage of possible points visited one or more times during the walk for each larva. The percent area covered was considered to be a relative measure of the effect of the directional probabilities on the area searched by a larva. Some may consider its procedure too artificial; thus, in some cases we also compared the directional probabilities of groups directly. This was done by summing the frequencies each larva moved, ahead, right, left, and behind; the sum of the frequencies for each direction for each group of larvae were entered in a contingency table and the Chi-square test was used to determine differences between groups. This procedure inflated the number of observations because the total for each group was the total number of directional movements for the group rather than the total number of larvae observed. We preferred the values from the random walk program because the observations were independent and it indicated the possible effect of directional probabilities on the area searched.

We recorded search patterns and feeding rates of anchovy larvae in patches of *Gymnodinium*, out of patches, and at low densities of *Gymnodinium* where no patches were present. For comparative purposes, search patterns for larvae feeding on the rotifer *Brachionus plicatilis* over more limited density ranges were also studied. In all experiments larvae ranged in age from 4 to 8 days. No difference in behaviour associated with age was detected, so we combined the data for all ages.

Searching Behaviour of Larvae Fed on *Gymnodinium*. Anchovy larvae swam more slowly in patches of *Gymnodinium* ($P < 0.0006$) and fed more frequently than when not in such a patch ($P < 0.0006$ Mann-Whitney U test, Siegel 1956). Larvae inside a patch swam less frequently directly ahead and more frequently reversed their direction than did ones outside a patch ($P < 0.001$ Chi-square test for two independent samples). The percent area covered, computed by the random walk program from directional probabilities was less for larvae inside a patch than for ones outside ($P < 0.00006$ Mann-Whitney U test). Thus, on the basis of the directional characteristics of the search pattern alone, larvae inside patches of *Gymnodinium* would be expected to cover less area per unit time than ones outside of a patch. The combined effect of the reduction in speed and change in directional probability for larvae in a patch was that ones inside remained within a small area whereas those outside swam over a much greater area.

The change in speed of larvae when they entered or left patches of *Gymnodinium* was abrupt. In 7 observations, the larvae swam in and out of a patch during the 5-min observation period. In all 7, the speed of the larva was higher (median 0.30 cm/sec) when it was outside the patch than when it was inside (median 0.15 cm/sec). Directional probabilities and feeding rates followed the trends described above, but no statistical differences existed - probably because of the small sample size.

The behaviour of larvae in patches of *Gymnodinium* could be a discrete pattern that occurs only when food is at a very high density and sharp boundaries exist; on the other hand, larval searching behaviour may be modified continuously with changes in density and food distribution.

To determine which alternative was the more likely, we divided the data into 3 density classes: 1 to 21 cells/ml; 24 to 260 cells/ml and patch, $\geq 1{,}000$ cells/ml. A plot of the data segregated into the three classes indicated that the speeds and percent area covered for larvae in the 24 to 260 cells/ml class fell between those for larvae in the other two density classes (Fig. 2). This trend is also apparent in the medians (Table 3). The 95% confidence ellipse for these distributions plotted on a log scale also suggest a continuous change in characteristics of the search pattern (Fig. 3). A log transformation was required to normalize the distributions because speed distribution

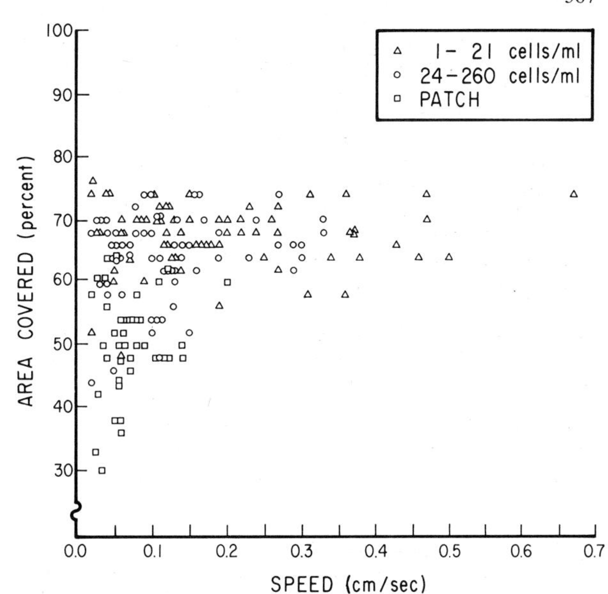

Fig. 2. Percent area covered from random walk program and speed for anchovy larvae 4 to 8 days old fed on *Gymnodinium splendens* at different densities

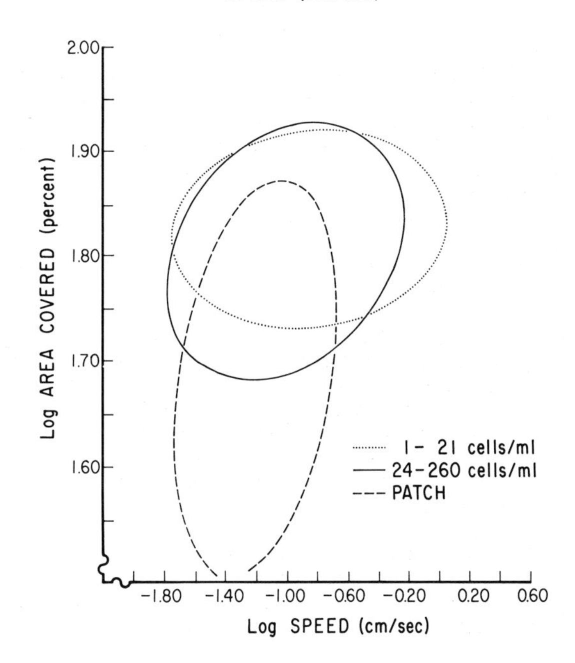

Fig. 3. 95% confidence ellipses for distribution of log of percent area visited on log of speed for larvae fed on *Gymnodinium splendens* at different densities

was strongly skewed. Comparisons of the speed, percent area covered, and feeding rates among the three classes of food density indicated that the distribution of each of these variables in each class was different from that in every other ($P < 0.05$ Mann-Whitney U test). We also compared the directional frequencies directly using the Chi-square test. These tests indicated that the frequency larvae moved in each grid direction differed among the three density classes ($P < 0.001$). From the above evidence we conclude that anchovy larvae responded to the density and distribution of *Gymnodinium* by continual modification

Table 3. Characteristics of search pattern of larval anchovy 4 to 8 days old in different concentrations of *Gymnodinium splendens*, and *Brachionus plicatilis*

Food type	Density range no/ml	N	Median cm/sec	Median feeding strikes/min	Median percent area covered	Mean directional probabilities				Total moves
						Ahead	Right	Left	Behind	
Gymnodinium splendens	1-21	72	0.151	0.00	68.5	0.355	0.311	0.280	0.054	3,821
"	24-260	65	0.105	0.21	65.0	0.319	0.304	0.306	0.071	2,416
"	patch	41	0.058	0.79	49.7	0.161	0.282	0.323	0.234	888
Brachionus plicatilis	10-15	38	0.212	0.86	68.3	0.341	0.305	0.315	0.039	2,465
"	42-72	23	0.162	0.79	65.3	0.324	0.305	0.302	0.068	1,247

of the speed and directional components of their searching behaviour. The result of these modifications was an expansion of searching area at low density and a reduction of the area searched at higher food densities.

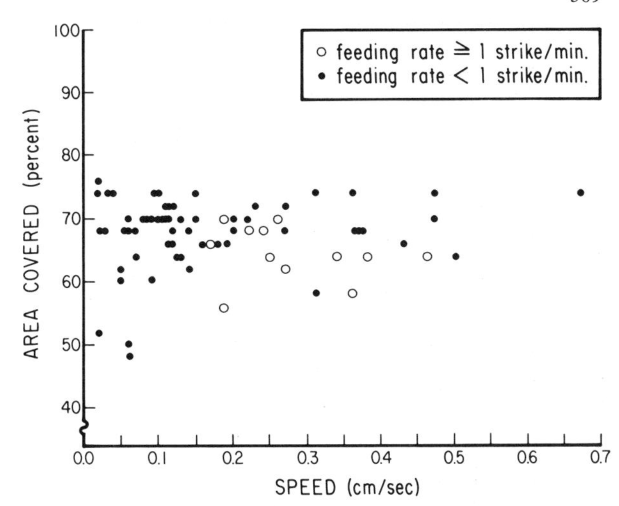

Fig. 4. Feeding rate, speed, and percent area covered for anchovy larvae 4- to 8-days-old fed on 1 to 21 *Gymnodinium splendens*/ml

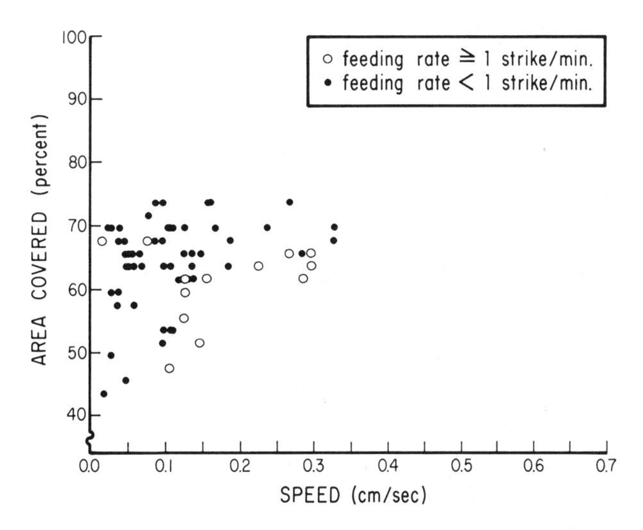

Fig. 5. Feeding rate, speed, and percent area covered for anchovy larvae 4- to 8-days-old, fed on 24 to 260 *Gymnodinium splendens*/ml

Within these classes a relationship existed between speed and feeding rate. Larvae that fed at higher rates tended to swim faster than those that fed less often or did not feed (Fig. 4 and 5). We measured the degree of correlation between feeding rate (completed feeding acts/min) and swimming speed by calculation of Spearman r_s and Students t associated with that value (Siegel, 1956). In each of the groups a strong positive correlation existed between feeding rate and swimming speed ($P < 0.005$). Thus, within each level the larvae that swam the fastest tended to feed more often, but larvae at lower densities swam faster and fed less often than ones at higher densities.

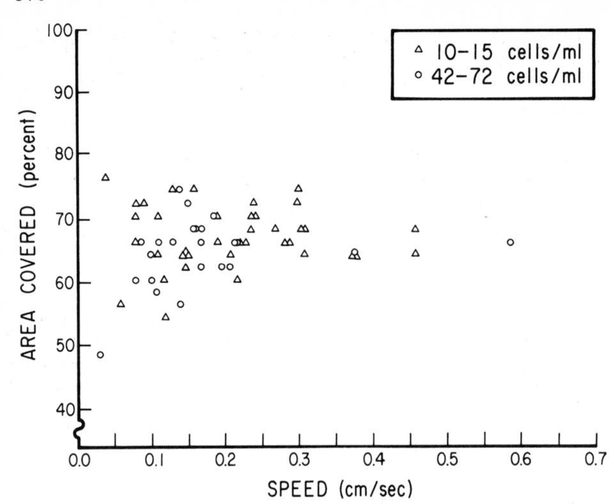

Fig. 6. Percent area covered calculated from random walk program and speed, cm/sec for anchovy larvae 4- to 8-days-old, fed on *Brachionus plicatilis* at different densities

Searching Behaviour of Larvae Fed on *Brachionus*. We used the same techniques and procedures to study the searching behaviour of larvae fed on the rotifer, *Brachionus plicatilis*, as we did for those fed on *Gymnodinium*; however, *Brachionus* was studied over a more limited density range. The data were divided into two density classes, 10 to 15 *Brachionus*/ml and 42 to 72/ml, to determine if density affected any characteristics of the search pattern. The distribution of speeds and percent area covered resembled those for *Gymnodinium* at comparable density levels; the medians were also similar (Fig. 6, Table 3). Comparisons of speed, percent area covered, and feeding rate between the

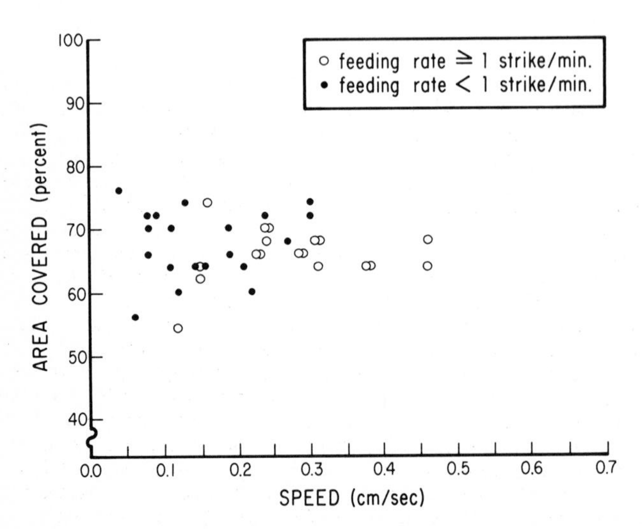

Fig. 7. Feeding rate, percent area covered and speed for anchovy larvae 4- to 8-days-old, fed on 10 to 15 *Brachionus plicatilis*/ml

two density classes of *Brachionus* showed a significant difference only in the case of percent area covered ($P < 0.05$ Mann-Whitney U test). The directional frequencies also differed between the two density classes ($P < 0.01$, Chi-square test).

A tendency existed in both density classes for the larvae that fed at higher rates to swim faster than those that fed at lower rates as was the case for *Gymnodinium* (Fig. 7 and 8). In both density classes of *Brachionus* the association was significant at $P < 0.005$ (Spearman rank correlation, Siegel, 1956).

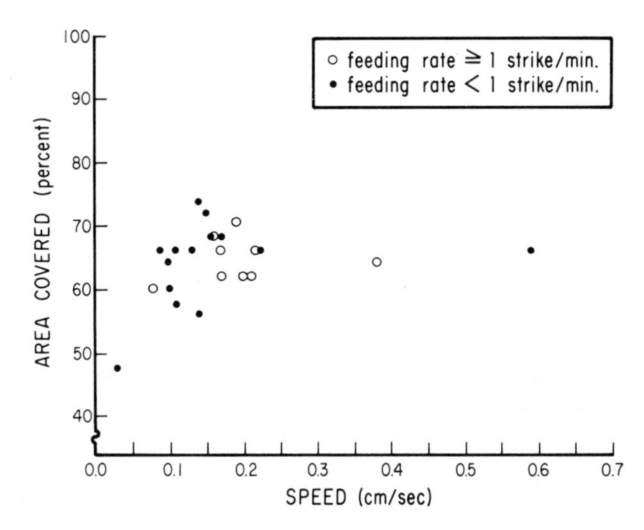

Fig. 8. Feeding rate, percent area covered and speed for anchovy larvae 4- to 8-days-old, fed on 42 to 72 *Brachionus plicatilis*/ml

DISCUSSION

The "non-randomness" of a search pattern of anchovy larvae is in agreement with other findings. Kleerekoper has shown that fish do not move at random in experimental environments that are void of directional cues (Kleerekoper, 1967; Kleerekoper et al., 1970). He also found that the presence of an odor without directional cues caused drastic changes in directional parameters (Kleerekoper, 1967). Beukema (1968) showed that sticklebacks found food in a maze about two times more efficiently than predicted from a random model. The food in the maze was at a lower density than would be expected under natural conditions and the fish improved their efficiency by making fewer reversals of direction than would be expected from the random model. The chance of an immediate occurrence of a reversal in direction was more than doubled when a prey was found. He also pointed out the adaptive advantage of this behaviour for a highly aggregated prey. Wyatt (1972) showed that the time plaice larvae spent swimming increased with a decrease in food density. In our experiments with *Gymnodinium* speed also increased with a decrease in food density and the probability of reversals in direction increased with density when either *Gymnodinium* or *Brachionus* was the prey. The increase in probability of direction reversals, 180° turns, was the principle reason for the decrease in area covered in the random walk at high food densities. We were unable to establish a direct link between reversals of direction and frequency of feeding; but since feeding and reversals of direction both increased with an increase in density they may be related.

The non-randomness of larval anchovy search patterns and the ability of yolk-sac and older larvae to find and remain in concentrations of the dinoflagellate, *Gymnodinium*, are adaptations that should equip them to take advantage of the contagiousness of the distribution of food in the sea. This could be of considerable adaptive significance, depending upon the extent of scale of patchiness of food in the sea. It could explain, for example, why laboratory estimates of food-density requirements of larval anchovy appear to be consistently higher than the average concentrations of food in the sea (O'Connell and Raymond, 1970; Hunter, 1972); and why no difference in mortality of anchovy larvae exists at the onset of feeding in the sea, although their food-density requirement determined in the laboratory is much higher at this time (Lenarz, 1972; Hunter, 1972). Laboratory estimates of food requirements for first-feeding larvae are higher because of their low level of feeding success, and they search a much smaller volume per unit time owing to their small size. If larvae can search for food and remain in concentrations while in the yolk-sac stage, the chance of finding a significant concentration of food before they starve is increased. Anchovy larvae can survive without food for 1.5 days after yolk absorption at temperatures of 22° to 15°C; but if the time to starvation is calculated from hatching, larvae can survive from 3 days at 22°C to 5.5 days at 15°C (Lasker et al., 1970). The variation in times is caused by temperature-dependent differences in the rate of yolk absorption. Since most anchovy eggs are spawned at 13° to 14°C, the time available to find food before starvation is increased 3 to 4 times and this conceivably could counteract the effect of a higher food density requirement.

Also pertinent is the extent that phytoplankton is eaten by larval anchovy under natural conditions. The importance of phytoplankton in the diet of anchovy larvae depends on the availability of patches and on whether or not larvae feed on these organisms in the sea. Phyto-plankton form dense patches in the sea. A remarkable example of this is a very extensive bloom observed in the Weddell Sea by El-Sayed (197

He recorded densities of phytoplankton of over 2000 organisms/ml for combined species and densities as high as 1700/ml for individual species. These densities are somewhat higher than the ones we typically measured in patches in our containers. Not all species of phytoplankton will sustain life in anchovy larvae, however. To survive, anchovy larvae appear to require that the cells be unarmoured and larger than 30 μ (Lasker et al., 1970). The abundance of phytoplanktonic organisms in the California Current meeting these requirements at the time of spawning of the anchovy is not known. On the other hand, stomach contents of anchovy larvae analyzed by Arthur (1956) indicate that phytoplankton may be of considerable importance in the diet of first-feeding anchovy. 11% of the items he identified in the guts of anchovy larvae 4.5 mm and smaller were phytoplanktonic organisms. If a group Arthur classed as unidentified spheres, presumably unicellular algae 20 μ in diameter is included, then phytoplankton comprised 32% of the diet. We consider 32% a conservative estimate because all organisms Arthur identified had hard parts that could withstand digestion; whereas the cells of a preferred phytoplankter, such as *Gymnodinium*, are broken down very rapidly in the gut and leave no identifiable elements. It should be noted, however, that phytoplankton is probably important for only about the first week of feeding. Lasker et al. (1970) showed that growth of larvae fed on *Gymnodinium* became retarded after about 1 week; Arthur (1956) showed a decline in the percent of phytoplankton in the guts of anchovy larvae with an increase in size. Only 10% of the items in guts of larvae 5.0 to 6.5 mm were phytoplankton and only copepods were present in the guts of larvae larger than 6.5 mm (Arthur, 1956).

SUMMARY

Searching and feeding behaviour of larval anchovy was studied when prey were highly aggregated and at various densities when prey were not highly aggregated. The prey used in most experiments was the dinoflagellate, *Gymnodinium splendens*; the rotifer, *Brachionus plicatilis*, was used in comparative studies.

Yolk-sac and post-yolk-sac larvae were found and remained in patches of *Gymnodinium* in light and in darkness. The number of larvae attracted to a patch depended upon the density and volume of the patch. The structure of searching behaviour at various prey densities was analyzed by recording movements of larvae as they swam over a grid. Speed and direction of search patterns of larval anchovy were density-dependent: larvae swam faster at low prey densities than they did at higher ones; and the expected area covered by a larva on the basis of directional components alone was greater at low prey densities than at higher ones.

ACKNOWLEDGEMENTS

We thank William Rand, Massachusetts Institute of Technology, for permission to use his computer model of bounded random walks, and Martin Cody, University of California at Los Angeles, for loaning a copy of the program deck. Alan Good wrote the program for processing the paper-tape data and James Zweifel fit various models to the data. Ms. Gail Theilacker developed the cylindrical sampler.

REFERENCES

Arthur, D.K., 1956. The particulate food and the food resources of the larvae of three pelagic fishes, especially the Pacific sardine, *Sardinops caerulea* (Girard). Thesis, Univ. Calif., San Diego, 231 p.

Beukema, J.J., 1968. Predation by the three-spined stickleback (*Gasterosteus aculeatus* L.): The influence of hunger and experience. Behaviour, 31 (1), 126 p.

Cody, M.L., 1971. Finch flocks in the Mohave Desert. Theor. Pop. Biol., 2 (2), 142-158.

Conway, G.R., Glass, N.R. and Wilcox, J.C., 1970. Fitting nonlinear models to biological data by Marquardt's algorithm. Ecology, 51 (3), 503-507.

El-Sayed, S.Z., 1971. Observations on phytoplankton bloom in the Weddell Sea. Antarct. Res. Ser., 17, 301-312.

Fogg, G.E., 1962. Extracellular products. In: Physiology and biochemistry of algae, R.A. Lewin (ed.). New York - London: Academic Press, 475-489 pp.

Hellebust, J.A., 1965. Excretion of some organic compounds by marine phytoplankton. Limnol. Oceanogr., 10 (2), 192-206.
Hunter, J.R., 1972. Swimming and feeding behavior of larval anchovy *Engraulis mordax*. Fish. Bull., U.S., 70 (3), 821-838.
Ivlev, V.S., 1961. Experimental ecology of the feeding of fishes. Yale University Press, 302 p.
Kleerekoper, H., 1967. Some aspects of olfaction in fishes, with special reference to orientation. Am. Zool., 7 (3), 385-395.
Kleerekoper, H., Timms, A.M., Westlake, G.F., Davy, F.B., Malar, T. and Anderson, V.M., 1970. An analysis of locomotor behaviour of goldfish (*Crassius auratus*). Anim. Behav., 18 (2), 317-330.
Lasker, R., Feder, H.M., Theilacker, G.H. and May, R.C., 1970. Feeding, growth, and survival of *Engraulis mordax* larvae reared in the laboratory. Mar. Biol., 5 (4), 345-353.
Lenarz, W.H., 1972. Mesh retention of larvae of *Sardinops caerulea* and *Engraulis mordax* by plankton nets. Fish. Bull., U.S., 70 (3), 839-848.
O'Connell, C.P. and Raymond, L.P., 1970. The effect of food density on survival and growth of early post-yolk-sac larvae of the northern anchovy (*Engraulis mordax* Girard) in the laboratory. J. Exp. Mar. Biol. Ecol., 5 (2), 187-197.
Siegel, S., 1956. Nonparametric statistics for the behavioral science New York: McGraw-Hill, 312 p.
Sweeney, B.M., 1959. Endogenous diurnal rhythms in marine dinoflagellates. Preprints Intern. Oceanogr. Congr., Aug.-Sept. 1959 (M. Sears, ed.), 204-207 p. Washington, D.C.: A. Assoc. Advancement Sci.
Thomas, W.H., Dodson, N. and Linden, C.A., 1973. Optimum light and temperature requirements for *Gymnodinium splendens*; a larval fish food organism. Fish. Bull., U.S., 71, 599-601.
Wyatt, T., 1972. Some effects of food density on the growth and behaviour of plaice larvae. Mar. Biol., 14 (3), 210-216.

J.R. Hunter and G.L. Thomas
National Marine Fisheries Service
Southwest Fisheries Center
La Jolla, Calif. 92037 / USA

Changes in Behaviour during Starvation of Herring and Plaice Larvae

J. H. S. Blaxter and K. F. Ehrlich

INTRODUCTION

The changes in weight and body chemistry of fish, which take place during starvation, have been fully reviewed by Ivlev (1961) and Love (1970). There are also substantial reductions in the tolerance of fish like carp and roach to low oxygen, salinity extremes, toxic substances such as phenol and to infection by pathogens (Ivlev, 1961). In terms of behaviour little is known. Certainly standard oxygen consumption falls in adult fish during starvation (Hickman, 1959; Beamish, 1964). Ivlev also described how feeding activity and stamina decrease, and susceptibility to predation increases, during starvation in some freshwater fish. Walker (1971) showed that starvation in cod caused a reduction in the diameter of both red and white muscle fibres.

The effect of starvation on marine fish larvae is known from the point-of-view of changes in body size, condition factor, and chemistry of the tissues (see Blaxter, 1969). In terms of behaviour Holliday et al. (1964) described small decreases in respiration rate in unfed herring larvae at the end of the yolk sac stage and Blaxter (1963) showed, also in herring larvae, how feeding intensity dropped after varying periods without food. It is perhaps surprising how little is known of the ecology of starvation in fish larvae since it is a period of high and often sustained mortality (see May, this symposium), which may have considerable effects on brood survival and the success of a year class. Furthermore, in ichthyoplankton surveys there may be dangers of selectively catching emaciated larvae.

The present study describes experiments on feeding success, buoyancy and activity of the larvae of herring *Clupea harengus* L. and plaice, *Pleuronectes platessa* (L.) during starvation. Buoyancy and activity are closely linked in terms of the problem of maintaining vertical position in the water column or in vertical migration. The experiments were partly designed to measure and qualify the point-of-no-return (PNR), a term used by Blaxter and Hempel (1963) to describe the point at which 50% of a larval population are too weak to feed, if food becomes available. Larvae after the PNR have reached a point of irreversible starvation though they may not be overtly moribund until some days later. Ecologically, the PNR is a more important point to establish than the time to death.

METHODS

<u>Material</u>. The herring larvae were reared from gametes obtained in February 1970-1972 from Firth of Clyde spawners. Using the method of Blaxter (1968) modified by Ehrlich (1972) it was possible to rear substantial numbers of larvae to and beyond metamorphosis. The plaice larvae were obtained from artificially-fertilized eggs supplied in

March 1970-1972 by staff of the White Fish Authority and reared according to the methods of Shelbourne (1964). The rearing temperatures commenced at 7-8°C and rose to 11-12°C by the end of the experiment. The sea-water salinity fluctuated somewhat but was usually of the order of 32-33°/oo. The various experiments were terminated after 90 days post-hatching in the herring when the larvae were about 25 mm long and well before the commencement of metamorphosis, and about 30 days post-hatching in plaice when they were about 10 mm long (Ryland's (1966) stage 4) and starting to metamorphose.

Feeding. The PNR was determined simply in some larvae by visual observations of the percentage of larvae hanging head-down in the water and obviously too inactive to feed. In two batches of older herring 30- and 50-days-old a sample was drawn at intervals after the commencement of starvation from a population and offered *Artemia* nauplii for a period of 1 h. The percentage feeding could be assessed as the guts are transparent and the food opaque.

Buoyancy. The larvae were anaesthetized in 1:20,000 MS222 and the average time was measured for 10-15 larvae to sink 10 cm in a 1 l glass cylinder. The temperature of the water was carefully controlled by placing the cylinders in a water bath within a constant temperature room run at 10°C. The main salinity used was 33°/oo, but lower and higher salinities were made up by adding distilled water or NaCl to sea-water. In all instances the larvae were rinsed in the experimental salinity before the experiment started. The fresh length was then measured under anaesthetic and the wet and dry weights determined on an electrobalance (Beckman EMB1) to an accuracy of about 1 μg (see Ehrlich, this symposium, for details of the drying and weighing procedure).

Activity. The apparatus used has been fully described by Blaxter (1973 It consists of a vertical perspex tube 120 cm high and 4.5 cm in diameter. At five levels pairs of matched thermistors are inserted. The thermistors are linked to a simple bridge circuit and 90 V DC. power supply. The circuit is balanced first by potentiometer. If a larva swims past one of the pairs of thermistors the circuit becomes imbalanced, this imbalance being recorded as a deviation (spike) from the base line on a pen recorder. The vertical tube is contained in a light-proof box over which an artificial light source can be placed. The intensity of this source, a 12V 48W tungsten iodide lamp within a light-proof housing can be varied by neutral density filters for which slots are provided in the lamp housing. Alternatively natural light from an outside window can be deflected to the top of the tube by a mirror.

A "normal" pattern of vertical movement of 50 larvae in response to natural light was obtained by recording from the thermistors through the dusk-night-dawn cycle. The surface record proved most useful in this respect and all results refer to the numbers of larvae at the surface of the tube. A vertical migration could also be imposed on the larvae in artificial light by the use of neutral density filters. In this case the light was reduced by a factor of ten in each of eight or nine steps - that is from the uninterrupted artificial light to the light with neutral density filters to the aggregate density of 9.0 interposed. (The density of the filter is given as a logarithm; thus a 1.0 filter reduces the light ten times, a 2.0 filter one hundred times and so on, the densities being additive when more than one is used). In one of these vertical migration experiments the light was reduced 100 million times.

Having established a normal pattern, experiments were repeated on larvae that had been starved to see when this pattern started to change. Two criteria on the pen record were used. The number of spikes per unit time was measured (usually for a period of 10 min) to get the number of larvae present at the surface. The height of the spikes gave a measure of the activity. Usually 20 spike heights in any one situation wer measured, transformed to logarithms (see Blaxter, 1973) and the mean and standard deviation calculated.

RESULTS

Feeding. The number of days to the PNR, based on visual assessments of the number of larvae hanging head down in the water, is shown in Table 1. Experiments on feeding rates after varying periods of starvation are shown in Fig. 1 for herring larvae 30- and 50-day-old together with some earlier results from Blaxter (1963). The two methods are in good agreement. Clearly the time to reach the PNR increases with age and is greater in plaice. This seems to be a reasonable result, because it is known that adult fish can survive without food for very long periods of weeks or months (see Love, 1970). Less active species, like the plaice, can survive longer than herring without food as fewer reserves are being used up in movement.

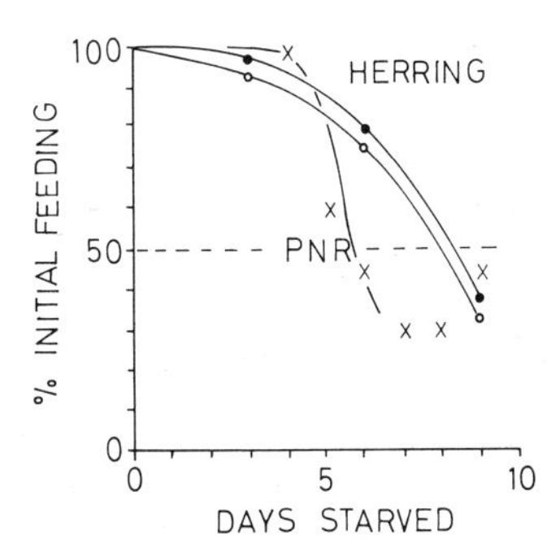

Fig. 1. Decrease in feeding rate of herring larvae after starvation. PNR - 'point-of-no-return'. o 30 days old, ● 50 days old, x 48 days old (from Blaxter, 1963)

Table 1. Days to point-of-no-return (PNR)

Species	Age, days from hatching or stage	Days to PNR
Herring	6 end of yolk-sac stage	6
	30	8
	50	8
	74	12
	88	15
Plaice	End yolk-sac stage	6
	Stage 3[a]	15
	Stage 4[a]	23

[a]See Ryland (1966)

The observations on the "head-down" position led to the buoyancy measurements. The head is likely to be the heaviest part of the larva, at least after the yolk is resorbed. According to Packard (pers. comm.), herring larvae appear to have a fatty substance within spaces of the head which may have a buoyancy function. If this is metabolized during starvation the larvae may show early signs of becoming "head-heavy"; such fat deposits in the head certainly influence balance in other fish species (Bone, 1972).

Buoyancy. Changes in the sinking rate and water content of herring and plaice larvae from hatching through the yolk-sac stage to subsequent starvation (no food being offered) are shown in Fig. 2 for sea-water (33°/oo) at 10°C. In herring throughout this period there is a steady decrease in sinking rate from about 0.4 cm/sec to neutral buoyancy past the PNR, the water content increasing from 80% to 91%. Larvae which were fed did not show this effect. There is a significant change when the larvae become truly moribund, in that they start to sink again at about 0.1 cm/sec. It is known (Holliday and Blaxter, 1960) that herring larvae are hypotonic to the sea-water medium, thus giving them some lift. As the larvae die osmoregulation probably fails causing a loss of hypotonicity; the specific gravity of the body then increases, resulting in a greater sinking rate.

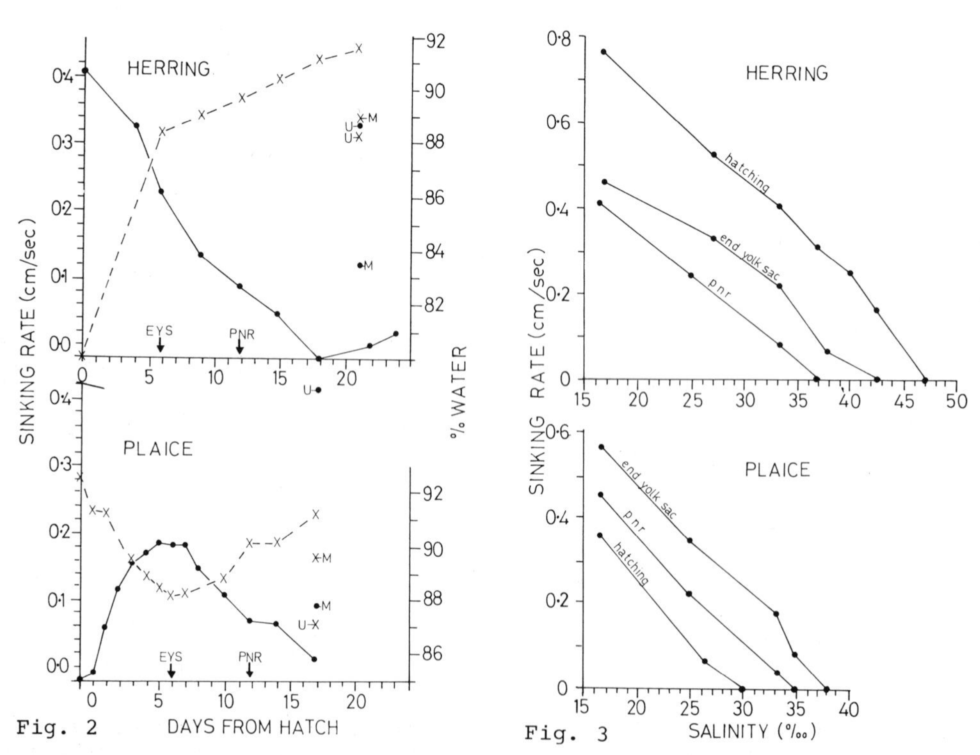

Fig. 2. Change in sinking rate and water content of herring and plaice larvae after hatching at 10°C in 33°/oo salinity. ● sinking rate, x water content; EYS - end of yolk-sac stage; PNR - 'point-of-no-return' M - moribund; U - unstarved

Fig. 3. The effect of salinity on sinking rate in herring and plaice larvae at three different stages. pnr - 'point-of-no-return'

In plaice the eggs were buoyant, but the sinking rate of the larvae steadily increased during yolk resorption. Later, in starving larvae, the sinking rate decreased until the larvae were about neutrally buoyant past the PNR. Fed larvae showed a steadily increasing sinking rate to 0.42 cm/sec by 17 days, 0.95 cm/sec by 31 days and 1.12 cm/sec by 38 days post-hatching (the latter two not shown in Fig. 3). As with herring the body fluids of plaice larvae are also hypotonic (Holliday and Jones, 1967) and osmoregulatory failure in moribund larvae probably causes an increase in sinking rate.

The effect of salinity of the medium on sinking rates of both species is shown in Fig. 3. The results follow closely those at 33°/oo salinity. Obviously, sinking rates decrease in high salinities, but the points of neutral buoyancy (where the lines intercept the x - axis) are interesting. In general the plaice larvae tend to be nearer neutral buoyancy in normal salinities than the herring.

Further experiments on older herring are shown in Fig. 4. The sinking rate increases with growth as the skeleton develops and water content

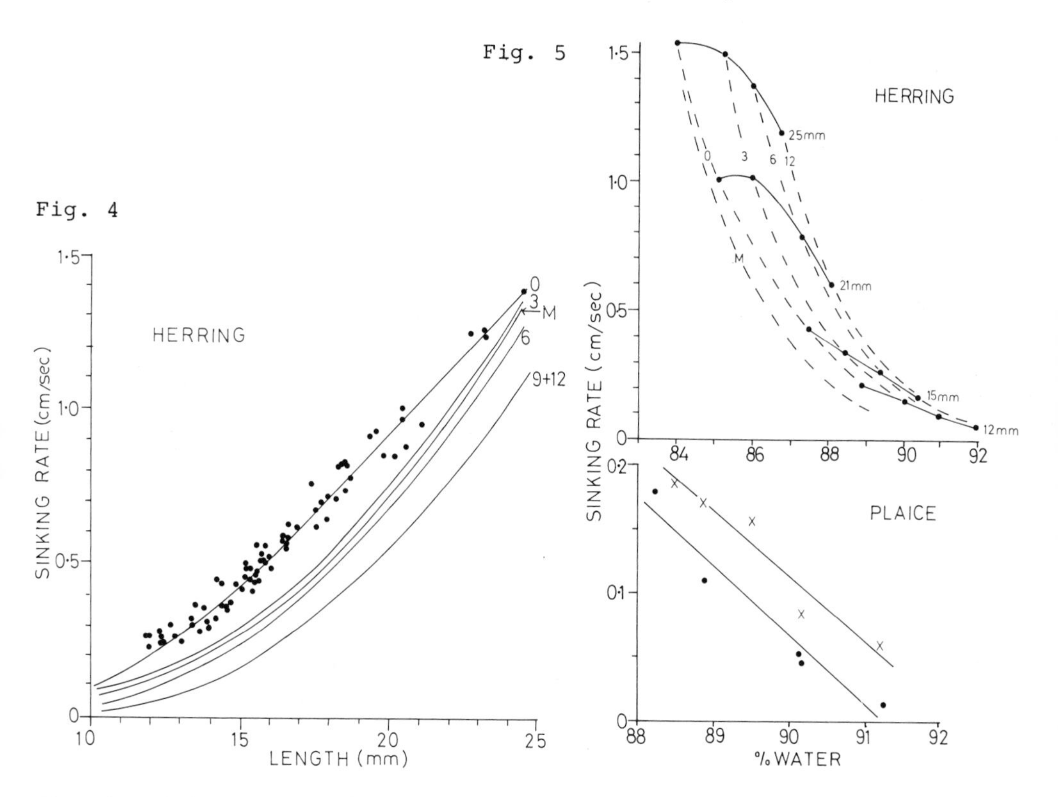

Fig. 4. Sinking rate of herring larvae of different length after varying periods of starvation (shown on right of each curve in days). M - moribund. Individual variation is shown for unstarved fish only

Fig. 5. The effect of water content on sinking rate in herring and plaice larvae. In the upper graph the full lines show length, the dashed lines days starved; M - moribund. In the lower graph x is from hatching to the end of the yolk-sac stage, ● from the end of the yolk-sac stage to starvation

falls, but regardless of length, starvation reduces the sinking rate with a parallel increase of water content. Water content is an important feature in the relationship, as shown in Fig. 5, for herring and plaice, but the changes in it do not entirely explain the alterations in sinking rate.

Activity. It was found that records of the numbers of larvae at the surface (top thermistor channel) were most useful in assessing the changes of activity during starvation. Fig. 6A, C shows the decline in numbers at the surface during starvation when herring and plaice larvae at the end of the yolk-sac stage were subjected to an artificial light cycle. Fig. 6B, D shows the results when older herring and plaice were used. Fig. 7A-D shows the changes in numbers of herring and plaice at the surface during the dusk-night period when a natural light cycle was used. Fig. 8 shows the mean number of larvae at the surface at 2300, 2400, and 0100 h after varying periods of starvation. Although records were also taken later during the night and at dawn there was a good deal of variability in the surface larval density and the dusk-early night results present the most consistent picture. Based on Figs 6, 7, and 8, Table 2 shows the number

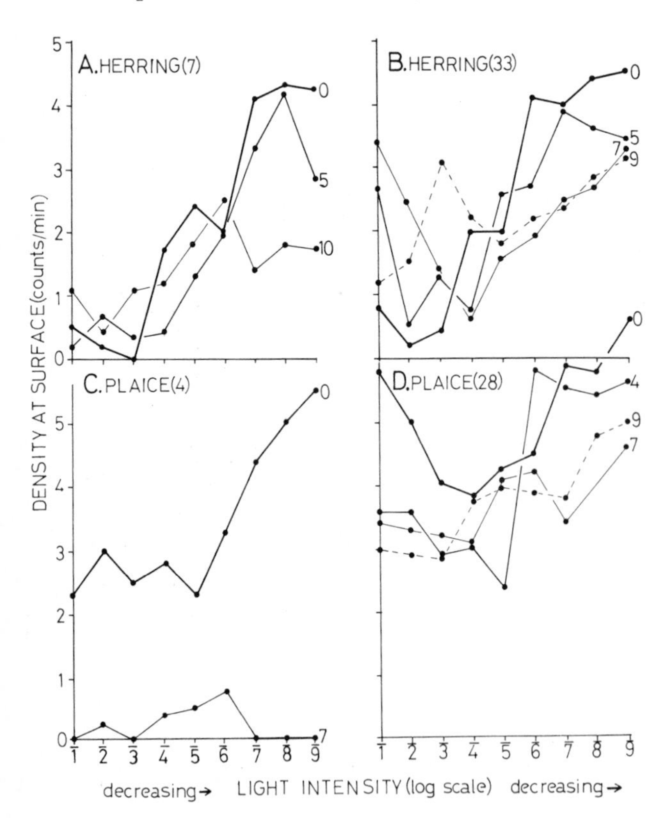

Fig. 6. Density of herring and plaice larvae at the surface of the experimental tube during an artificial light cycle. Each division on the abscissae denotes a reduction of 10 times in the light intensity. Figures at right of curves show number of days starved at $10^{o}C$. Figures at right of species show age in days post-hatching at the start of starvation

of days starved at 10°C before vertical migratory activity deteriorates markedly, within the limits of the data available. One of the problems to be faced was the tendency to experiment, towards the end of a period of starvation, with larvae which had survived best, in fact with the larvae which were most resistant to starvation at the beginning of the experiments. This experimental sampling problem will tend to give an optimistic picture of the ability of the average larva to survive in nature.

Table 2. Days to change of vertical migration

Species	Age, days from hatching	Days to behaviour change Natural light cycle	Artificial light cycle
Herring	7	10	10
	33	9	7
Plaice	4	5	?7
	28	7	7

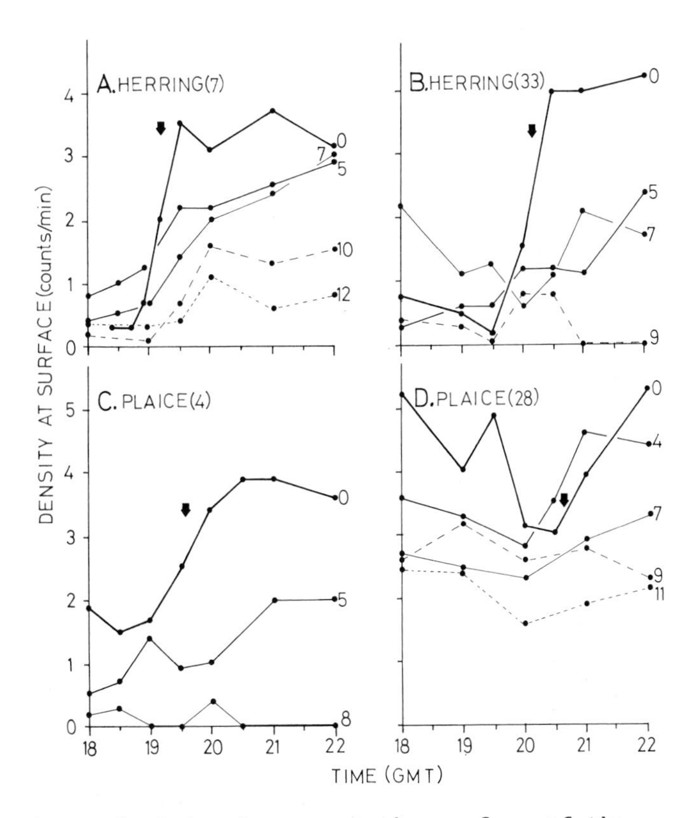

Fig. 7. Density of herring and plaice larvae at the surface of the experimental tube during dusk and early night. Figures at right of curves show number of days starved at 10°C. Figures at right of species give age in days post-hatching at start of starvation. Arrows show civil twilight

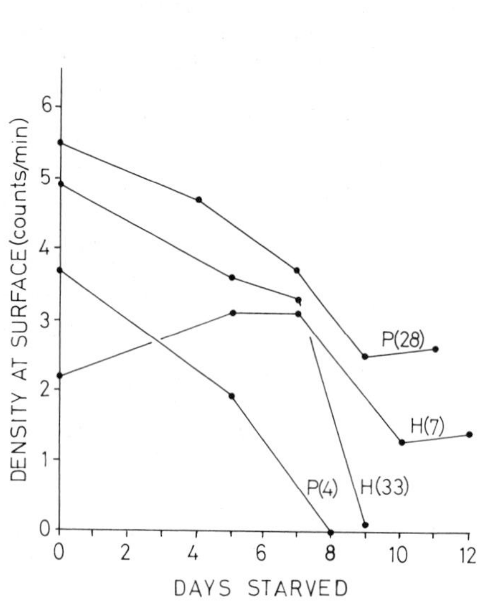

Fig. 8

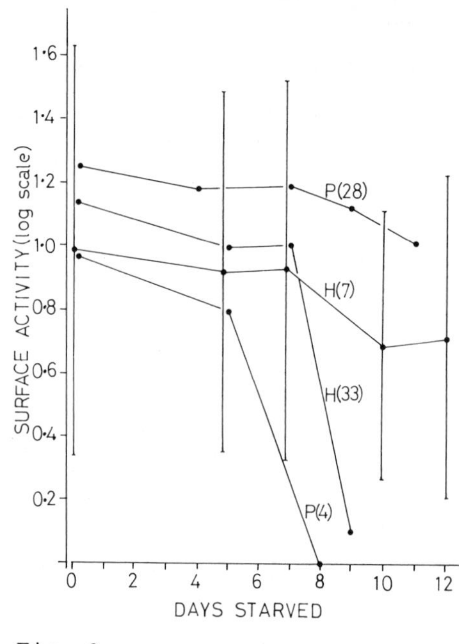

Fig. 9

Fig. 8. The mean density of herring (H) and plaice (P) larvae at the surface of the experimental tube at 2300, 2400, and 0100 h depending on number of days starved. The figure by the species shows the age in days post-hatching at the beginning of starvation

Fig. 9. The mean activity (see text for explanation) for herring (H) and plaice (P) larvae at the surface of the experimental tube at 2300, 2400, and 0100 h depending on days starved. The figure by the species shows the age in days post-hatching at the beginning of starvation. Vertical lines denote 2 x S.E. for herring 7 days old at start of experiment

The surface activity was measured on the surface channel during the early night, the values in Fig. 9 representing the mean log spike height of 20 observations at 2300, 2400, and 0100 h of a natural light cycle.

Both surface density and activity show that the maintenance of at least some semblance of a behaviour pattern is retained for a period of several days, at least as long as to the PNR, as determined by other criteria. Older plaice larvae appear to be most resistent to starvation. In general the larvae do not seem to conserve energy during early starvation by reducing their activity. The maintenance of activity over relatively long periods of starvation in the experiments appears more marked than it is actually likely to be in the sea. This is due to the tendency, mentioned above, to use those larvae surviving at the end of the experiment which were in the best nutritional condition at its beginning.

DISCUSSION

There are interesting differences between the sinking rate of herring and plaice larvae while living on their yolk reserves. Herring, hatching from demersal eggs, have a high initial sinking rate while plaice eggs and larvae at hatching are positively buoyant. Herring larvae must rise off the seabed towards the pelagic zone after hatching. As the yolk is resorbed and the sinking rate decreases they will require progressively less energy to maintain position in the water column. In plaice larvae the increasing sinking rate from hatching will help them to keep off the surface initially and eventually aid them in settling.

Based on behavioural criteria the time taken for well-fed herring and plaice larvae to reach the PNR at prevailing sea temperatures of about 9-10°C appears to be 5-9 days. There is evidence, however, that beyond a length of 20-25 mm in herring and Stage 3 in plaice the time may at least double. It is also likely that plaice will normally survive longer than herring without food as they become older, because their activity drops as they settle to the bottom. Clearly there will be individual differences based on individual body reserves and the time will be modulated by this factor as well as temperature. A likely situation in the sea is that larvae will not be totally deprived of food but find enough to extend the time to the PNR well beyond the experimental limits given here. Wyatt (1972) found slightly longer periods to the PNR in plaice larvae but his samples were small and criteria somewhat different.

In later stages still herring can survive at least 129 days at 6 to 12°C (Wilkins, 1967). The marked influence of age is probably partly the effect of a greater metabolic rate per unit weight in small organisms and partly limited storage capacity in larvae, both in organs like the liver and in the general muscle and body tissues (a main area for fat reserves in herring).

During early starvation there is very little immediate change in behaviour as expressed by feeding intensity or vertical activity. There is, however, an immediate increase in water content and reduction in sinking rate which may conserve energy, at least if the larvae under normal conditions are trying to maintain a certain depth. On the other hand, the dawn sinking into deeper water may be more difficult if the larvae become more buoyant as they starve.

The tendency to neutral buoyancy and the fall in activity during starvation may both influence sampling by plankton nets, larvae in poor condition being more vulnerable to capture. There is certainly no question of larvae well advanced in starvation sinking to the bottom where they would be inaccessible to nets. Some further work on the nutritional status of larvae caught by slow and fast plankton hauls might show whether there was a serious selection of inviable larvae by slow hauls, the larvae in good condition being able to escape the net.

The tendency to "head-heaviness" during starvation will also impair feeding behaviour which requires a careful manoeuvring near the prey. The younger larvae are the least efficient feeders so incipient starvation might be more serious for them (Rosenthal, 1969; Blaxter and Staines, 1971).

Table 3. Components of larval buoyancy (end of yolk sac (EY), point-of-no-return (PNR))

Herring

	Weight (mg) Wet	Dry
	1.5327	0.1689
	1.4200	0.1278

	ρ component	ρ sw−ρc		mg	% wet wt.	Rel % = % / (100−% Ash)	Partia Densit Rel% / 100 x
Water	1.0094	0.0161	EY	1.3638	88.98	89.72	0.9056
			PNR	1.2922	91.00	91.78	0.9264
Fat	0.926	0.0995	EY	0.0444	2.90	2.92	0.027C
			PNR	0.0317	2.23	2.25	0.0208
Protein plus carbohydrate	1.379	−0.3535	EY	0.1118	7.29	7.35	0.1014
			PNR	0.0841	5.92	5.97	0.0823
						Σ EY	1.0340
						Σ PNR	1.0295
Ash			EY	0.0127	0.83		
			PNR	0.0121	0.85		

Plaice

	Weight (mg) Wet	Dry
EY	1.0761	0.1271
PNR	0.8337	0.0822

	ρ component	ρ sw−ρc		mg	% wet wt.	Rel %	Partial density
Water	1.0114	0.0141	EY	0.9489	88.18	89.20	0.9021
			PNR	0.7515	90.14	91.22	0.9226
Fat	0.926	0.0995	EY	0.0244	2.27	2.30	0.0213
			PNR	0.0206	2.47	2.50	0.0231
Protein plus carbohydrate	1.379	−0.3535	EY	0.0905	8.41	8.51	0.1173
			PNR	0.0518	6.21	6.28	0.0867
						Σ EY	1.0407
						Σ PNR	1.0324
Ash			EY	0.0123	1.14		
			PNR	0.0980	1.18		

Table 3 (continued)

Vol.(ml)= Wt.(g)/ρc	Net force (dynes)= (ρ sw-ρc)x(vol)x(981 cm/sec^2)	Δdynes		dynes / wet wt.	Δ dynes / wet wt.
0.0013511	+0.0213			+0.0139	
0.0012802	+0.0202	-0.0011		+0.0142	+0.0003
0.0000479	+0.0047			+0.0031	
0.0000342	+0.0033	-0.0014		+0.0023	-0.0008
0.0000811	-0.281			-0.0183	
0.0000610	-0.0212	+0.0069		-0.0149	+0.0034
			Σ EY	-0.0013	
			Σ PNR	+0.0016	
			Δ	+0.0029	
0.0009382	+0.0130			+0.0121	
0.0007430	+0.0103	-0.0027		+0.0124	+0.0003
0.0000263	+0.0026			+0.0024	
0.0000222	+0.0022	-0.0004		+0.0026	+0.0002
0.0000656	-0.0228	+0.0098		-0.0211	
0.0000376	-0.0130	+0.0098		-0.0156	+0.0055
			Σ EY	-0.0066	
			Σ PNR	-0.0006	
			Δ	+0.0060	

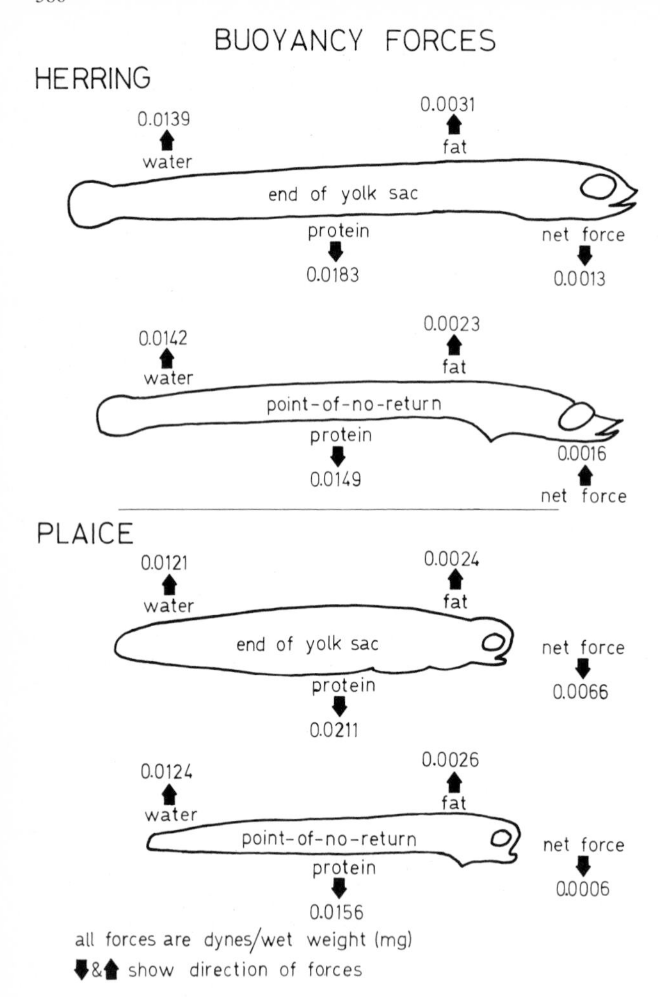

Fig. 10. Buoyancy forces on herring and plaice larvae at the end of the yolk-sac stage and at the 'point-of-no-return'. Note degree of emaciation of the body and slight shrinkage in length

It is of interest to consider the factors contributing to buoyancy in both herring and plaice larvae. This is summarized in Fig. 10 and Table 3 for larvae starved from the end of the yolk-sac stage. The density of the body water was based on salinity equivalents at 10°C of 12.3°/oo for herring (Holliday and Blaxter, 1960) and 15°/oo for plaice (Holliday and Jones, 1967) on the assumption that the body fluid concentration was maintained at this level until osmoregulation broke down when the larvae became moribund. Thus percent water increased during starvation, but it was assumed that its ionic concentration remained constant. Actual amounts of protein, carbohydrate, and ash were converted to percent wet weight and total lipid was taken as 100% minus the sum of these. Lipid density was taken from adult herring (Brawn, 1969); protein was calculated from total nitrogen X 6.025 (see Love, 1970). Protein plus carbohydrate density was taken as that for protein which was by far the major component. Protein density is calculated as the inverse of its partial specific volume (White et al., 1964). Ash was not directly considered as the larvae had no skeleton and it was likely that the main ash comprised the salts contributing to osmotic equilibrium already allowed for. The computed densities for both species at the end of the yolk-sac stage and at the PNR are only approximations because actual component densities were not known and because total fat and protein were not themselves chemically determined. There may also have been chemical and density changes in the lipids during starvation. The most interesting aspects are the relative amounts shown in Fig. 10. Water and lipid give lift while protein and carbohydrate cause sinking. Water is the dominant component giving, for example, five times the lift of lipid at the end of the yolk-sac stage in herring. Although the major lift is from water, which increases during starvation, it is decreasing protein which causes the much reduced sinking rate during starvation. The ability to catabolize protein from the start of food deprivation may be an adaptation to maintain buoyancy during pelagic life.

SUMMARY

1. The behaviour of herring and plaice larvae changes during starvation, the critical stage being called the point-of-no-return (PNR). This is the point where only 50% of the larvae are still able to feed if food becomes available. The larvae survive for some days after the PNR before they become moribund, but the effects of starvation have become irreversible.

2. The time taken to reach the PNR varies from 6 days at the end of the yolk-sac stage in both species to about 15 days in older larvae. Later still, especially in the plaice, it takes 3-4 weeks to reach the PNR.

3. During starvation there is a progressive decrease in sinking rate due to an increase in hypotonic body water and a decrease in body protein. Only when the larvae become moribund does osmoregulation fail and the larvae "dehydrate" and start to sink out of the planktonic zone.

4. Vertical migration in response to changes of light intensity continues beyond the PNR but there is a reduction in the number migrating and in their activity levels after a few days of starvation. There is no evidence for activity falling to compensate for lack of food.

5. There is a strong possibility that larvae during starvation will be selectively sampled by plankton nets. Particularly in an advanced state of starvation they will tend to float in midwater, their activity will become reduced and their ability to escape nets impaired.

REFERENCES

Beamish, F.W.H., 1964. Influence of starvation on standard and routine oxygen consumption. Trans. Amer. Fish. Soc., 93, 103-107.

Blaxter, J.H.S., 1963. The feeding of herring larvae and their ecology in relation to feeding. Calif. coop. ocean. Fish. Invest., 10, 79-88.

Blaxter, J.H.S., 1968. Rearing herring larvae to metamorphosis and beyond. J. mar. biol. Ass. U.K., 48, 17-28.

Blaxter, J.H.S., 1969. Development: Eggs and Larvae. In: Fish Physiology 3, 177-252. W.S. Hoar and D.J. Randall (eds.). New York - London: Academic Press, 485 pp.
Blaxter, J.H.S., 1973. Monitoring the vertical movements and light responses of herring and plaice larvae. J. mar. biol. Ass. U.K. 53, 635-647.
Blaxter, J.H.S. and Staines, M., 1971. Food searching potential in marine fish larvae. In: IV European Marine Biology Symposium, 467-485 pp. D.J. Crisp (ed.), Cambridge University Press, 599 pp.
Bone, Q., 1972. Buoyancy and hydrodynamic functions of the integument in castor oil fish, *Ruvettus pretiosus*. Copeia, 1972 (1), 78-87.
Brawn, V., 1969. Buoyancy of Atlantic and Pacific herring. J. Fish. Res. Bd Can., 26, 2077-2091.
Ehrlich, K.F., 1972. Morphometrical, behavioural and chemical changes during growth and starvation of herring and plaice larvae. Thesis, Stirling Univ., Scotland.
Hickman, C.P., Jr., 1959. The osmoregulatory role of the thyroid gland in the starry flounder *Platichthys stellatus*. Can. J. Zool., 37, 997-1060.
Holliday, F.G.T. and Blaxter, J.H.S., 1960. The effects of salinity on the developing eggs and larvae of the herring. J. mar. biol. Ass. U.K., 39, 591-603.
Holliday, F.G.T. and Jones, M.P., 1967. Some effects of salinity on the developing eggs and larvae of the plaice *Pleuronectes platessa*. J. mar. biol. Ass. U.K., 47, 39-48.
Holliday, F.G.T., Blaxter, J.H.S. and Lasker, R., 1964. Oxygen uptake of developing eggs and larvae of the herring (*Clupea harengus*). J. mar. biol. Ass. U.K., 44, 711-723.
Ivlev, V.S., 1961. Experimental ecology of the feeding of fishes. New Haven: Yale University Press, 302 pp.
Love, M., 1970. The chemical biology of fishes. London - New York: Academic Press, 547 pp.
Rosenthal, H., 1969. Untersuchungen über das Beutefangverhalten bei Larven des Herings *Clupea harengus*. Mar. Biol., 3, 208-221.
Ryland, J.S., 1966. Observations on the development of larvae of the plaice (*Pleuronectes platessa*) in aquaria. J. Cons. perm. int. Explor. Mer, 30, 177-195.
Shelbourne, J.E., 1964. The artificial propagation of marine fish. Adv. mar. Biol., 2, 1-83.
Walker, M.G., 1971. Effect of starvation and exercise on the skeletal muscle fibres of the cod (*Gadus morhua* L.) and coalfish (*Gadus virens* L.) respectively. J. Cons. perm. int. Explor. Mer, 33, 421-427.
White, A., Handler, P. and Smith, E.L., 1964. Principles of biochemistry. 3rd Ed. New York: McGraw-Hill, 1106 pp.
Wilkins, N.P., 1967. Starvation of the herring, *Clupea harengus* L.: survival and some gross biochemical changes. Comp. Biochem. Physiol. 23, 506-518.
Wyatt, T., 1972. Some effects of food density on the growth and behaviour of plaice larvae. Mar. Biol., 14, 210-216.

J.H.S. Blaxter and K.F. Ehrlich
Scottish Marine Biological Association
Dunstaffnage Marine Research Laboratory
P.O. Box 3
Oban, PA 34 4 AD, Argyll / GREAT BRITAIN

K.F. Ehrlich
Present address:
Occidental College
Biology Department
Los Angeles, Calif. 90041 / USA

The Ability of Herring and Plaice Larvae to Avoid Concentrations of Oil Dispersants

K. W. Wilson

INTRODUCTION

The pelagic habit of the eggs and larvae of many species of marine fish renders them especially susceptible to pollution in the surface waters. For example, this can occur from marine dumping operations or from accidental oil spillages and their subsequent treatment with chemical dispersants. Oil poses a particular threat because tanker movements, and therefore the chances of collisions involving tankers, are greatest in the coastal areas which are the most important spawning and nursery grounds.

The acute toxicity of oil (Kühnhold, 1969, 1972) and oil dispersants (Wilson, 1972) to the eggs and larvae of some marine fish has been demonstrated in the laboratory but information on their effects at sea is sparse. At the time of the "Torrey Canyon" operations (March 1967) some small flatfish were reported washed-up dead in areas adjacent to where shore spraying of oil with dispersants was being carried out (Simpson, 1968). Further offshore, Simpson did not find any adverse effects on size and composition of commercial catches of fin-fish. Smith (1968), however, reported that in samples taken in a high-speed plankton sampler 90% of the pilchard eggs were dead in areas of heaviest dispersant spraying, compared with 50% in other areas; young fish (3-20 mm length) were scarce or absent in the treated areas. Since the lack of young fish bore no relationship to the type of plankton found, it seemed, to Smith, "...an inescapable conclusion that the absence of young fish... was the effect of detergent (= dispersant) spraying carried out one or two days before the plankton samples were taken". However, it was not clear whether the low number of young fish was due to mortality or whether it had occurred through an active movement by the fish away from the affected area.

In the present investigation, the responses of the young stages of herring and plaice to concentration gradients of an oil dispersant have been studied and it has been possible to determine the ability of larvae to avoid potentially lethal concentrations of dispersants at sea.

METHODS

General. Larvae of the herring (*Clupea harengus* L.) were reared from Clyde spawners. The methods of fertilization of the ripe eggs, of incubation, and rearing described by Blaxter (1962, 1968a) were followed. Only larvae (10-12 mm) which had established feeding for about 1 week on *Balanus* nauplii were used. Larvae of plaice (*Pleuronectes platessa* L.) were reared from pond-spawned eggs according to the standard procedures described by Shelbourne (1964). Larvae (6-7 mm, stage

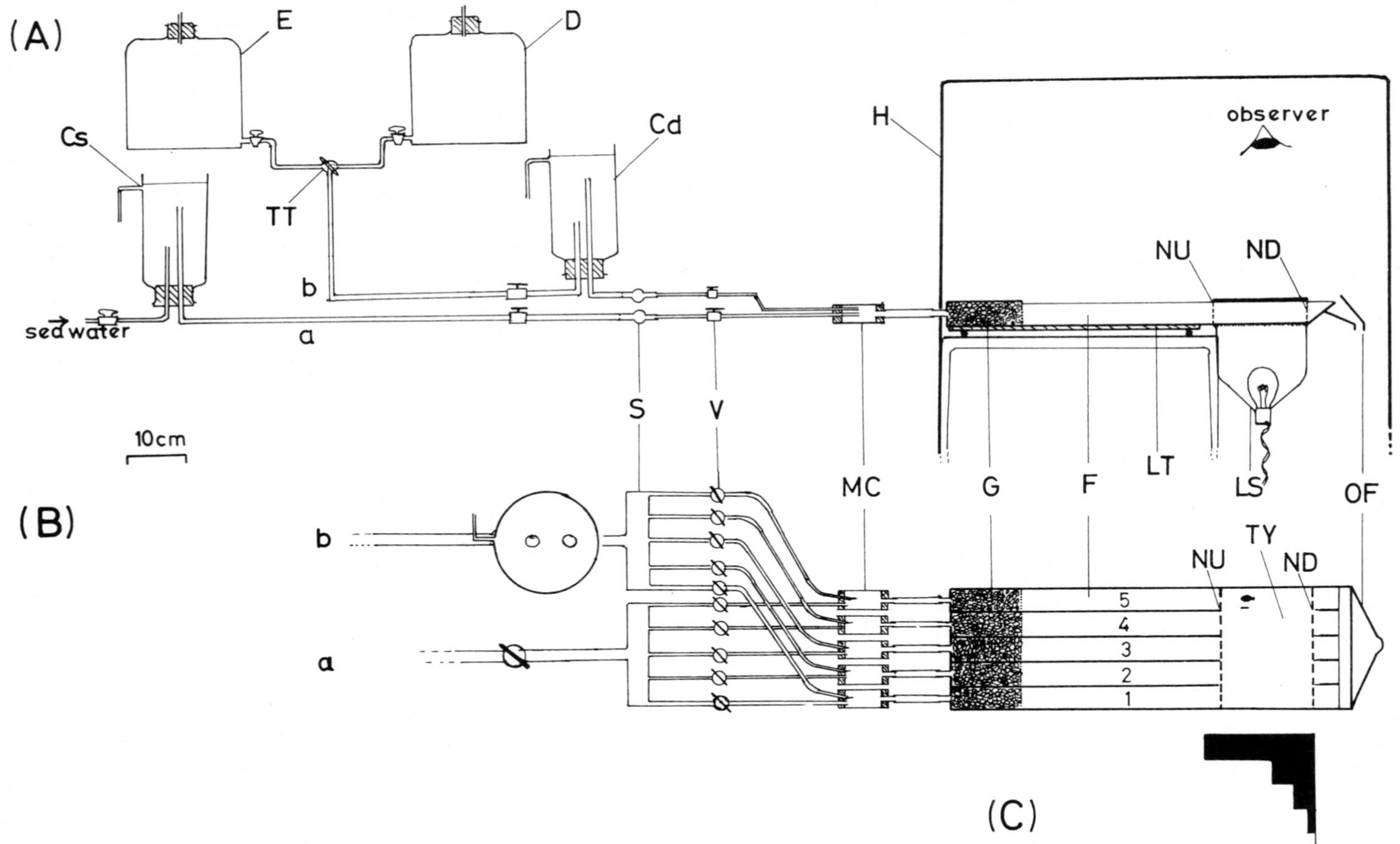

Fig. 1. (A) The fluvarium apparatus in vertical section, (B) the same in plan view, (C) the shape of the dispersant gradient. a, sea-water system; b, dispersant system; Cd, Cs, constant head reservoirs for dispersant and sea-water respectively; D, dispersant stock solution; E, sea-water stock solution; F, fluvarium; G, baffle of glass beads; H, light-proof canopy; LS, substage light source; LT, levelling table; MC, mixing chamber; ND, NU, downstream and upstream nets respectively; OF, overflow; S, channel separators; TT, three-way tap; TY, test yard; V, valve control. The 2 fish in the test yard, representing herring larvae and metamorphosed plaice, are drawn to scale

2a-b, Ryland, 1966) which had established feeding on *Artemia* nauplii and later newly metamorphosed larvae (12-14 mm, stage 5) were used.

All experiments were made with the dispersant BP 1002.

Horizontal Gradients. Horizontal gradients of dispersants were established using a fluvarium technique which generated a stepwise gradient at right angles to the direction of flow. The apparatus shown in Fig. 1 is modified after Höglund (1961). The fluvarium consisted of a shallow trough (600 x 200 x 45 mm deep) supported on a levelling table. The trough had a clear Perspex base and sides of black Perspex, and was divided for two-thirds of its length by vertical strips of black Perspex into five equal channels. Water entering each channel passed through a baffle of small glass beads which produced a uniform flow in each channel. When the rates of flow in the five channels were matched the water flowed through the open part of the trough, the "test yard", without mixing. The test yard, which was delimited at its upstream and downstream ends by fine-mesh netting, was covered with a sheet of clear Perspex. This removed the free water surface and enhanced the stability of the dispersant gradient.

The limits of the five channels in the test yard were marked on a sheet of white paper glued to its underside. This also served to diffuse the sub-stage light source. The fluvarium and observer were positioned in a light-proof housing to prevent extraneous factors affecting the orientation of the animals within the apparatus.

Sea-water (7-8°C, 33-34°/oo) and dispersant solution flowed in similar but separate systems through a constant head reservoir and a channel separator before meeting in the mixing chambers. These chambers were connected directly to the fluvarium. The rates of flow of sea-water and dispersant to each chamber were adjusted by the height of the reservoirs and by the flow control valves to achieve 1. the same total flow in each channel, and 2. a different dispersant concentration in each channel, thereby establishing a gradient in the test yard. Mean flows of 7.4 mm/sec for herring and 10.8 mm/sec for plaice were used. The shape of the resultant dispersant gradient (inset, Fig. 1) was calculated from the flows of sea-water and dispersant to each channel and checked by spectrophotometric measurements of dye solution.

Changes in the concentrations of the gradient were effected by changing the concentration of the dispersant stock solution. The shape of the gradient remained the same for all experiments.

In each experiment, sea-water from the mains and stock solution flowed through the apparatus for 10 min before 10 larvae were selected from holding tanks and transferred by pipette to the fluvarium. The larvae were added to the centre of the test yard by slightly raising the upstream net. After a further 10 min the flow of sea-water from the stock bottle was replaced by dispersant solution by turning the three-way tap. In control experiments, the sea-water stock solution was used throughout. The dispersant gradient quickly formed within the test yard and recording was started as soon as the dispersant flow was started.

The numbers of larvae in each channel were noted every 30 sec for a total test time of 30 min. The larvae were removed and the apparatus flushed through with sea-water for 30 min to clean it, ready for the next experiment. The results were analyzed according to Höglund (1961).

For each treatment, the distribution of the larvae in the test yard was represented as a frequency histogram in which the total number of observed visits to each channel during the experimental period was calculated as a percentage of the total number of observations. The distribution of the larvae at any instant was described by the mean position value, mpv. The number of observed visits to each channel was multiplied by the corresponding channel number and the sum of the five products divided by the total number of observations, i.e. $mpv = \Sigma(nc)/n$, where n is the number of visits to each channel and c the channel number. To simplify the description of the responses mean position values were based on four consecutive observations, i.e. mpv (2 min).

Vertical Gradients. Vertical gradients were established in 500 ml graduated measuring cylinders (250 x 52 mm diameter). 20 herring larvae were added to 400 ml of clean sea-water ($8^{o}C$, $33\text{-}34^{o}/oo$) in the cylinder and were attracted to the base with a bright light. 100 ml of dispersant solution (or diluent sea-water for controls) were carefully run on to the surface to ensure that a uniform layer of dispersant and a narrow mixed layer resulted above the sea-water column. The light source was moved and held at a distance vertically above the cylinder.

The numbers of larvae in each 100 ml division of the cylinder were recorded every 60 sec for a test period of 1 h. The cylinders were left in a regime of 12 h light/12 h dark for 48 h in a constant temperature room at $8^{o}C$ after which the distributions of the larvae were again recorded. The larvae were transferred to tanks of clean water and fed after a further 48 h.

Frequency histograms and mean position values were calculated as described for horizontal gradients.

RESULTS

Horizontal Gradients. In control situations both herring and plaice larvae orientated themselves in the direction of flow and swam freely around the test yard. Only rarely did they swim in the same direction as the current. Occasionally the larvae would cease swimming and, still orientated to the current, drift passively downstream. After drifting for a few cm or on touching the downstream net the larvae restarted swimming upstream.

Metamorphosed plaice were much less active than the larval stage, often remaining in the same position on the floor or sides of the test yard for many minutes. The fish swam slowly close to the surface

---

Fig. 2. The effect of dispersant gradients of different concentrations (given in parts/10^{6} above each histogram) on the distribution of fish within the test yard for (a) herring larvae, (b) plaice larvae and (c) metamorphosed plaice. The concentrations refer to the highest concentrations of the gradient; these occurred in channel 5. The white histogram shows the frequency distribution of visits to the 5 channels. The hatched area shows the distribution of mpv (2 min). The position of the mean mpv is indicated by the broken line, and its value (together with its estimated standard error) is given at the right of the histograms; n is total number of observations

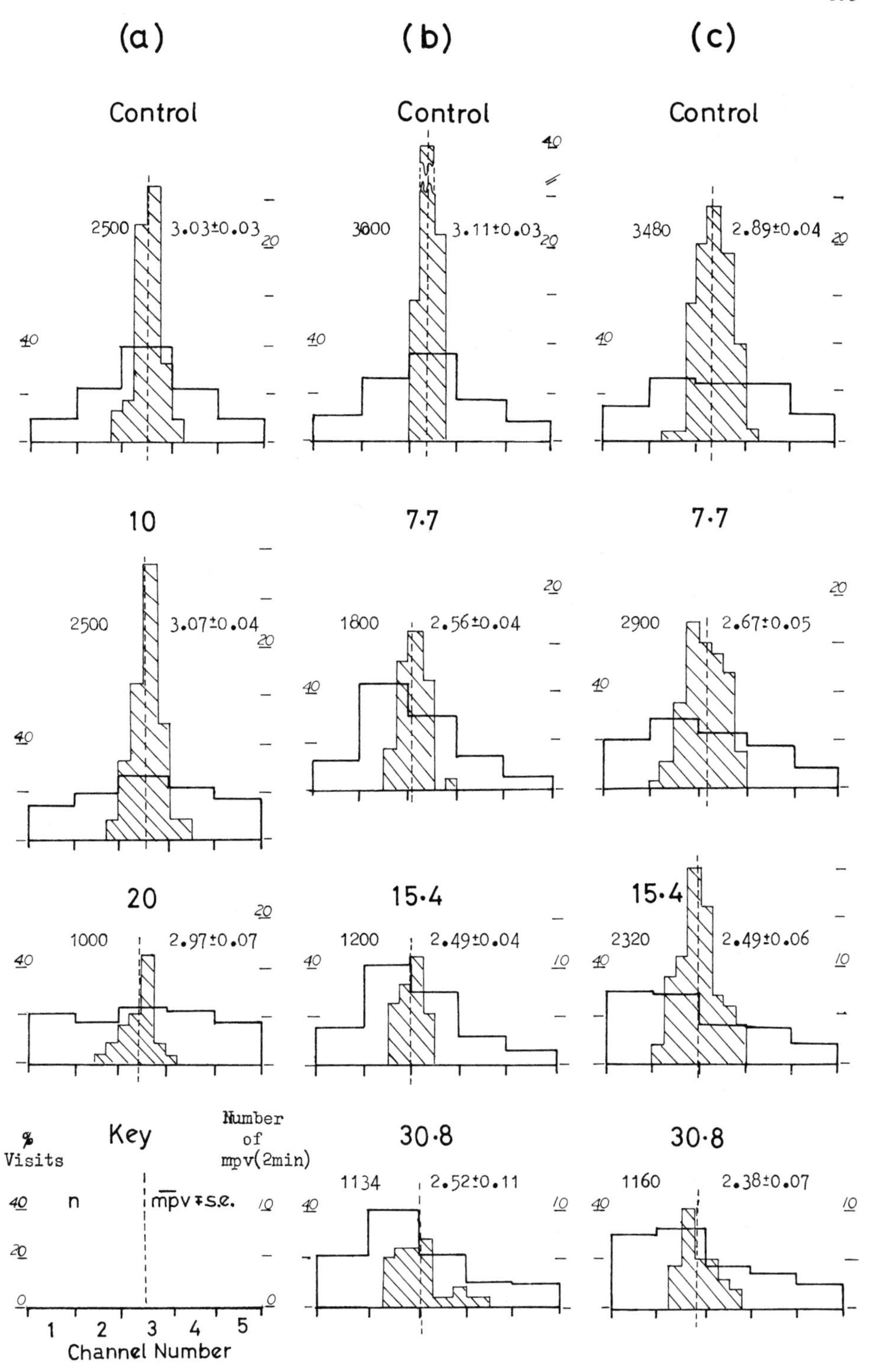
(a)
(b)
(c)
Control
Control
Control
2500
3.03±0.03
3600
3.11±0.03
3480
2.89±0.04
10
7·7
7·7
2500
3.07±0.04
1800
2.56±0.04
2900
2.67±0.05
20
15·4
15·4
1000
2.97±0.07
1200
2.49±0.04
2320
2.49±0.06
% Visits
Key
Number of mpv(2min)
n
mpv ∓ s.e.
30·8
30·8
1134
2.52±0.11
1160
2.38±0.07
Channel Number

using their lateral fins and only occasionally did they swim in mid-water. These differences in behaviour between the larvae and metamorphosed fish are shown in the frequency histograms (top row, Fig. 2). The distributions shown by the larvae are similar to those described by Lindahl and Marcström (1958) and Höglund (1961) for the "free-swimming" roach, in showing an aversion to the edges. For metamorphosed plaice there was a positive edge effect, for they settled on the sides of the test yard, and it was surprising that the resultant distributions were only slightly more uniform than those given by the larvae rather than showing a bias to the channels 1 and 5.

The distributions of the mpv (2 min) were approximately normal, with the mean coinciding with the median line of the test yard.

The test fish did not show a marked avoidance response to the gradients of BP 1002 but swam into dispersant solutions even at concentrations which caused instantaneous responses in the larvae - usually a vigorous shaking of the head. However, the frequency histograms showed that over an extended period the two species differed markedly in their response to the gradients (Fig. 2). After experiencing dispersants, herring larvae lost the aversion to the edges and indeed at the highest concentrations they frequently swam into the edges and often remained vigorously swimming for several minutes with their heads hard pressed in the angle formed by the sides and bottom of the test yard. This behaviour occurred equally at the dispersant and clean water edges.

Plaice larvae did not behave in this way and although they occurred more frequently in the cleaner channels the negative edge effect was

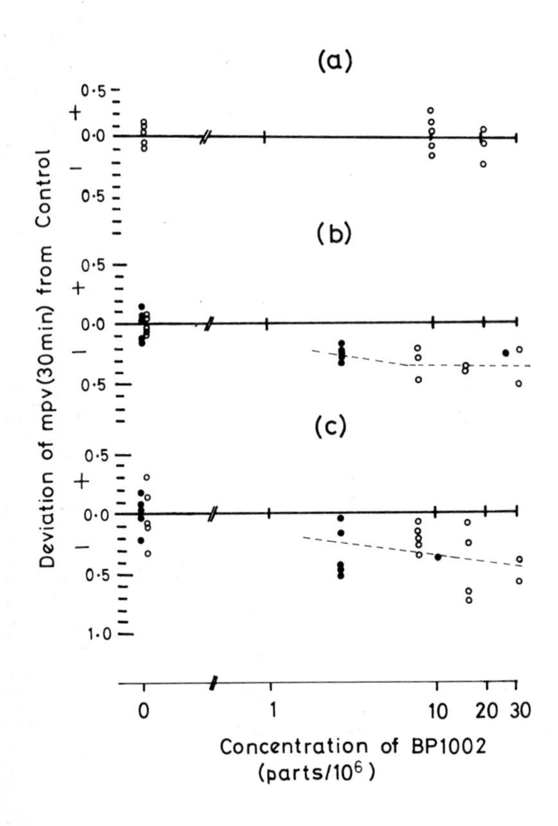

Fig. 3. Reaction diagrams for (a) herring larvae, (b) plaice larvae, and (c) metamorphosed plaice in gradients of dispersant. Solid points, 1969; open circles, 1970

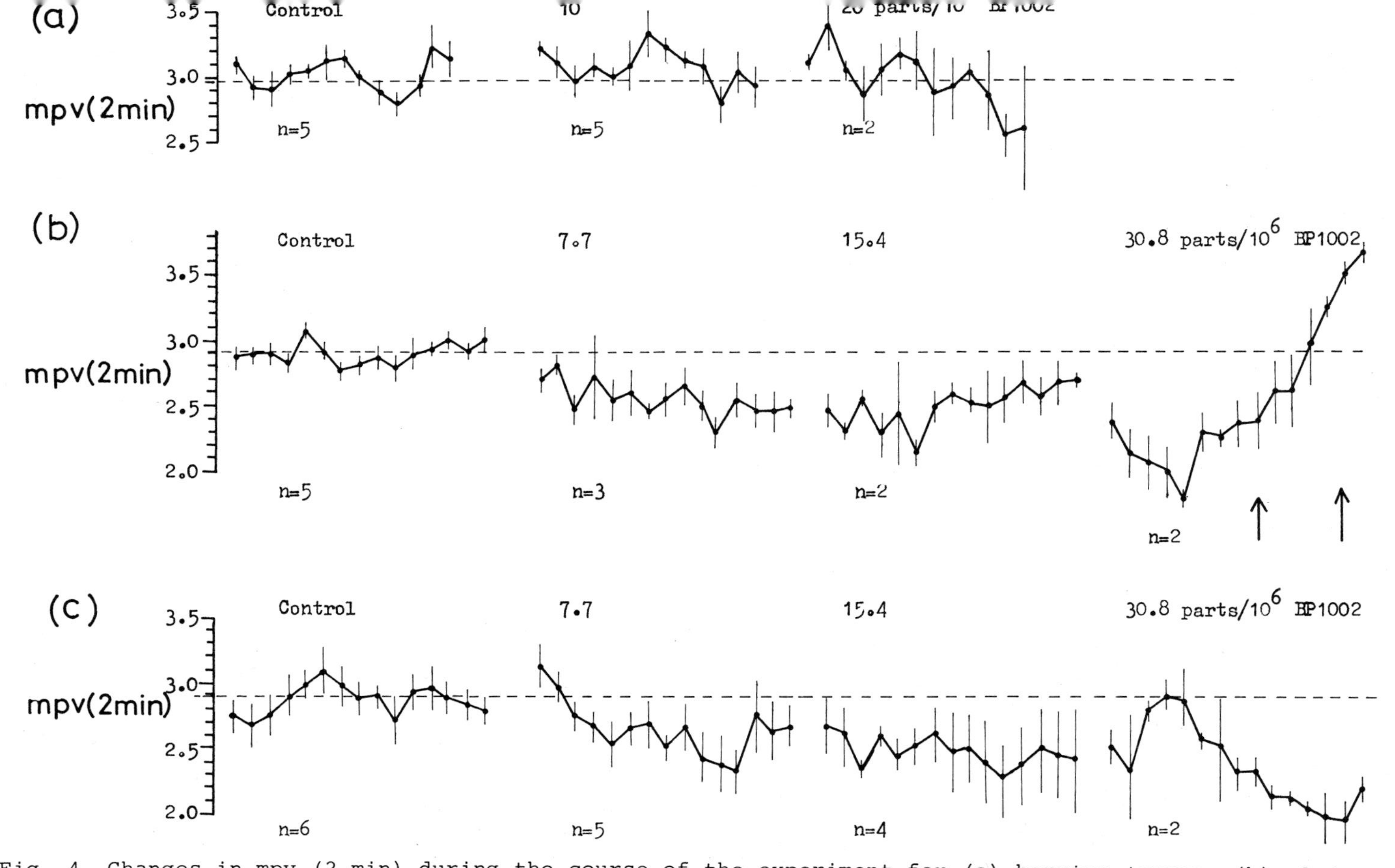

Fig. 4. Changes in mpv (2 min) during the course of the experiment for (a) herring larvae, (b) plaice larvae, and (c) metamorphosed plaice at different concentrations of dispersant. The mean and standard error of each mpv (2 min) is shown for the number of trials, n, for the test period of 30 min. Arrowed positions indicate loss of larvae on to the downstream net

still evident even at very high concentration gradients, and this behaviour restricted the magnitude of the lateral displacement. With metamorphosed plaice, which did not show a large negative edge effect, the frequency distributions tended further to the clean water channel with increasing concentration of the gradient (Fig. 2). The changes in frequency distributions were reflected by changes in the mean position values, and the magnitude of the response can be seen in Fig. 3 where the displacement of each mpv (30 min) from the mean value of the mpv (30 min) of the controls is plotted against the top concentration of the gradient. Positive values constitute an attraction and negative values an avoidance. On this basis, herring showed no response to the dispersants but plaice showed an avoidance. For metamorphosed plaice, the degree of avoidance increased with increasing concentration of the dispersant, whereas for plaice larvae a threshold of avoidance appeared to be reached at a top concentration of about 8 parts/10^6.

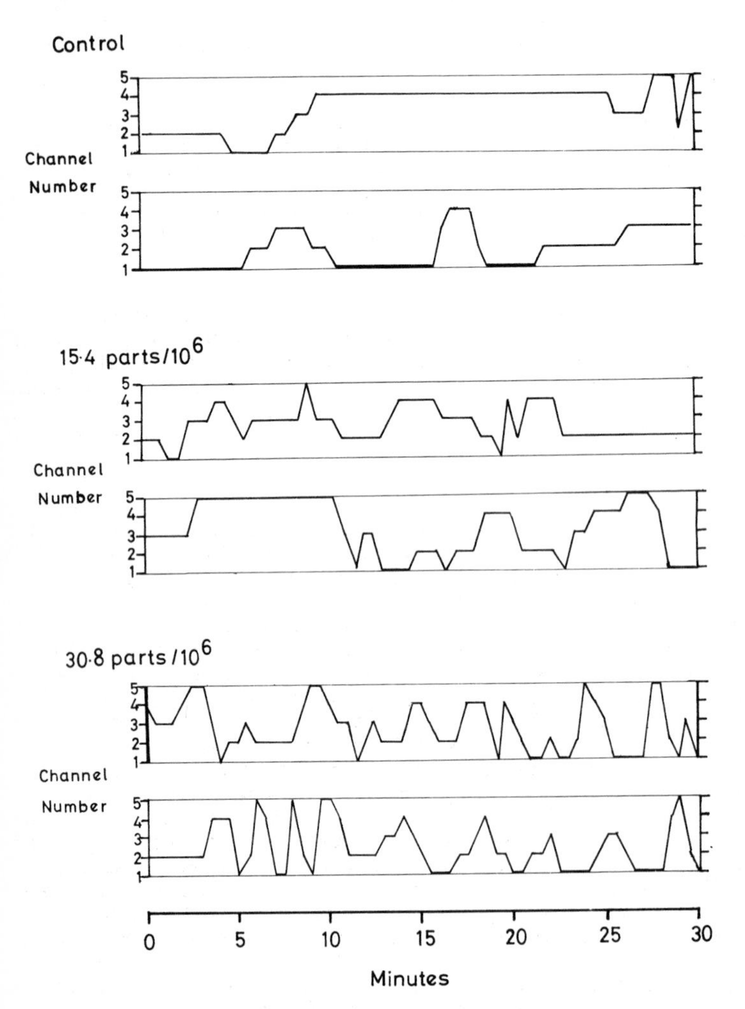

Fig. 5. The behaviour of individual metamorphosed plaice to different concentrations of dispersant. The concentration given is that in channel 5

The mean position values of the fish during the course of the experiments are shown in Fig. 4. The lack of an avoidance response in herring is confirmed, while for plaice larvae it is clear that the response was established quickly in all concentrations. Thereafter the response was dependent on the concentration of the gradient. At a top concentration of 7.7 parts/10^6 the avoidance increased throughout the test period. At 15 parts/10^6 the avoidance response was initially much greater but declined after about 15 min. At 30 parts/10^6 this pattern was repeated but the decline was more marked and led, after about 20 min, to an apparent attraction of the larvae to the dispersant. In fact, this change was due to an increased narcosis of the larvae by the high concentrations. This resulted in an accumulation of weak larvae in these channels, and after a time they were only capable of maintaining their position by sporadic short bursts of swimming; eventually they drifted on to the downstream net and in some cases were unable to swim off again.

Metamorphosed plaice did not respond as quickly as the larvae. There was no evidence of any degree of narcosis even at the highest concentrations, and a marked avoidance of these was recorded.

It was possible to identify individual metamorphosed fish by abnormalities of pigmentation (see Shelbourne, 1964) and to follow their movements during the normal experiments. From the typical examples shown in Fig. 5 there was a clear overall increase in the activity of the fish, related to the dispersant concentrations. This appeared to be generally true for larvae also.

Vertical Gradients. Under normal conditions the larvae showed the characteristic positive phototactic response at high light intensities (Blaxter, 1968b) and rapidly moved to the water surface where they remained actively swimming. Occasionally they would leave the surface and swim quickly down the cylinder, but after a short time at the bottom they returned to the surface. In the dark the larvae dispersed throughout the column.

In experimental situations, the larvae swam into the dispersant layer; however, at 100 parts/10^6 they did not go directly to the surface but showed some reaction at the sea-water/dispersant interface. The changes in the mean position values and frequency histograms were dependent on the concentration of the dispersant layer (Fig. 6), and on the time of exposure to the layer (Fig. 7). At 1.0 parts/10^6 the larvae appeared to behave normally but an analysis of the mpv (5 min) showed that they spent less time at the surface than the controls. This tendency was more clearly observed at 2.5 parts/10^6 in which the larvae also swam more frequently down the cylinder. At 10 parts/10^6 the toxic effects of the dispersant became noticeable and at 100 parts/10^6 a pattern of response was clearly established. Initially, the larvae became more active, swimming vigorously at the surface, but after 5-10 min the swimming became progressively more sporadic and of shorter duration, and since the larvae sank passively through the water column between these bursts of activity they eventually congregated on the bottom of the cylinder. After a period of inactivity here, the larvae recovered and exhibited the normal phototactic response. This pattern was confirmed by following the movements of single larvae (Fig. 7).

The pattern of narcosis and recovery persisted for some time, so that after 48 h the only differences in the distributions were the increased effects at 2.5 parts/10^6 (Fig. 6). After 96 h, however, many of the larvae in the highest concentrations were irreversibly affected, and only 20% recovered following transfer to clean water. At all other concentrations the larvae recovered, but the numbers that re-started feeding were small even in controls.

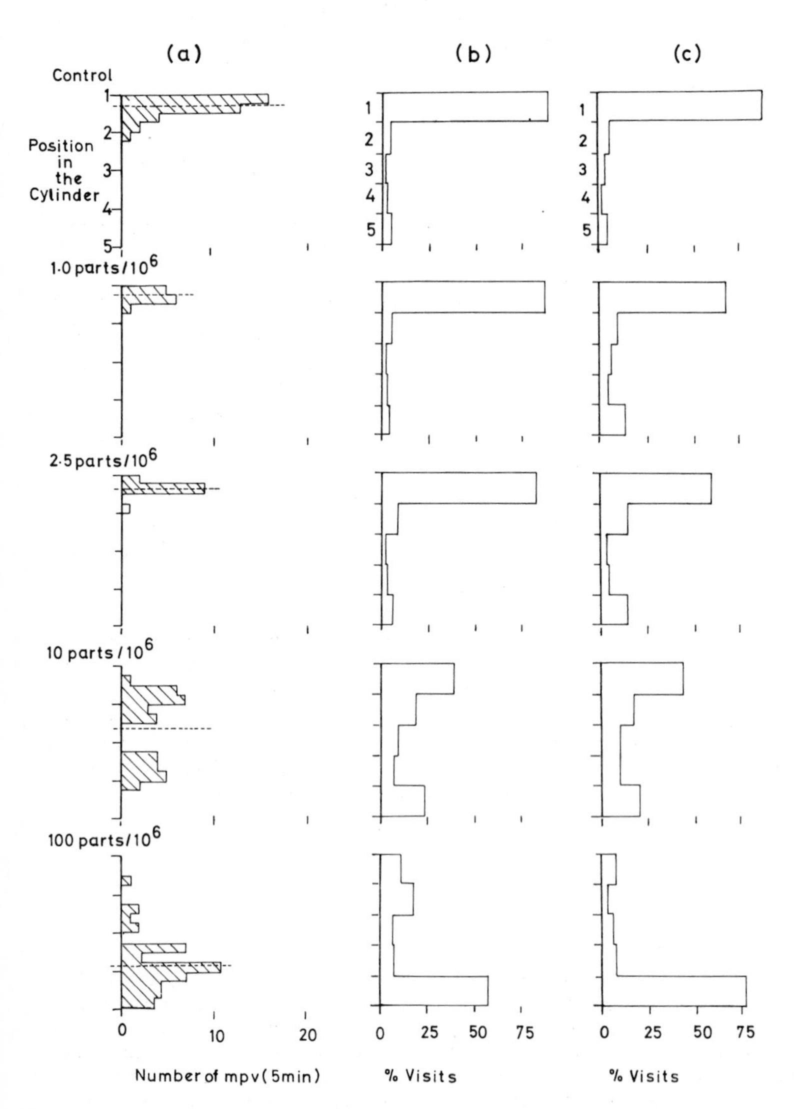

Fig. 6. Changes in the distributions of herring larvae in vertical gradients of dispersant of different concentrations: (a) mpv (5 min), (b) frequency histograms, (c) frequency histograms after 48 h

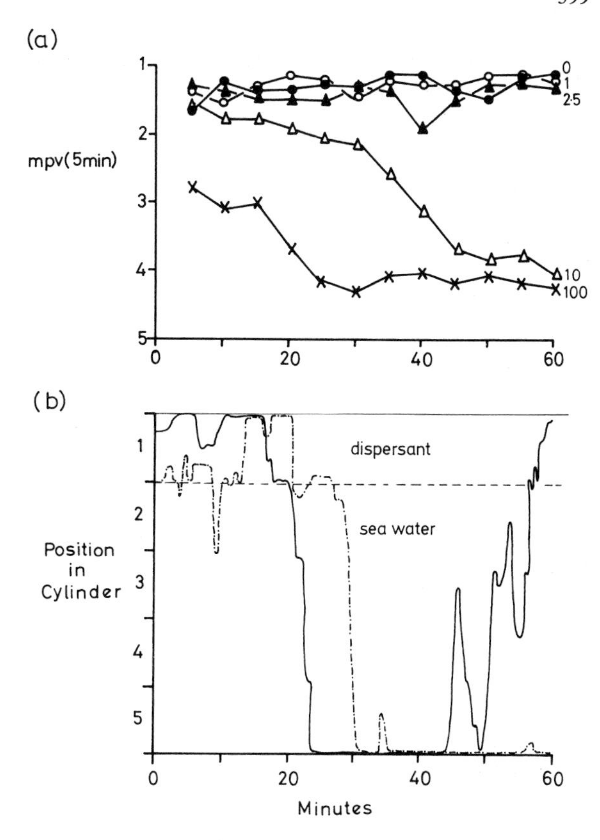

Fig. 7a and b. (a) Changes in the mean position of herring larvae with time in vertical gradients of different concentrations (parts/10^6). (b) Movement of individual larvae under a dispersant layer of 100 parts/10^6

DISCUSSION

The responses of animals to changes in the chemistry of the water are often a function of the type of apparatus used to generate the chemical changes and the reactions of test organisms to choice situations should be judged against the background of the technique employed. Thus, in the type of choice apparatus where clean and modified water meet (e.g. Jones, 1947; Sprague, 1964) it is not clear whether the response of the animals at the interface is a rheotactic or chemotactic response. Gamble (1971) concluded that the decisive responses of animals to changes of dissolved oxygen, observed by many authors using choice situations, resulted from rheotactic interference. Similarly, in the tank described by Ishio (1965) the concentration gradient is accompanied by a gradient in current speed. The fluvarium technique was designed to eliminate rheotactic interference by maintaining the chemical gradient perpendicular to any changes in flow characteristics, e.g. edge effects.

In the present study, the fish did not avoid the dispersants by directed swimming movements and it must be concluded that a chemotactic response was lacking or was, at best, readily sublimated to the drive of positive phototaxis in the vertical gradients. Nevertheless, both developmental stages of plaice showed an avoidance of the dispersants. This was no doubt due to orthokinesis. BP 1002 produced an almost instantaneous increase in activity and the animals were recorded less frequently in those sections where swimming was more vigorous. This pattern is similar to those described for *Astacus* (Höglund, 1961), *Gammarus* (Cook and Boyd, 1965) and *Corophium* (Gamble, 1971) in gradients of deoxygenated water. Changes in activity of larvae induced by dispersants have already been described (Wilson, 1972), and the more rapid establishment of the avoidance response in the free-swimming stages of plaice larvae and the consolidation of the avoidance response with time further substantiates the orthokinetic nature of the response. It has not been possible to assess the importance of klinokinesis.

The lack of avoidance shown by herring larvae can be explained by their greater sensitivity to dispersants than plaice larvae at a similar stage of development, and by behavioural differences between the two species. In plaice, experience of the high concentrations resulted in increased swimming speeds in short bursts, while in herring similar concentrations resulted in more or less continuous "frantic" swimming back and forth across the test yard.

Kühnhold (1972) observed that larvae of herring, cod, and plaice did not avoid high concentrations of oil dispersants and he concluded that the chemoreceptors were "blocked or destroyed rather quickly". A similar conclusion could be advanced for the behaviour of larvae to high concentrations of dispersants, but apart from the speed at which the larvae reacted there is no evidence to suggest mediation by chemoreceptors. It may be that the dispersants affect the central nervous system more directly and that their mode of action is a physical toxicity operating through space occlusion within the larvae as described by Crisp et al. (1967).

It now seems likely that larvae will not avoid potentially lethal concentrations of dispersant at sea by directional swimming but will, through increased activity, move by random movements away from these concentrations. However, in extensive areas of dispersants, or in very high concentrations, narcosis of the larvae is more likely to occur and under these conditions the buoyancy (or lack of it) of the larvae will determine their chances of survival. Thus, newly-hatched plaice larvae will float and remain in the dispersant slick but older larvae will sink into deeper and cleaner water where they will recover. Differences in the current velocities between the surface and the deeper water may carry the larvae to areas more suitable for their own survival and, equally important, for the survival of food items.

SUMMARY

1. The responses of larvae of herring and plaice to horizontal gradients of oil dispersants in sea-water have been studied using a five-channel fluvarium which maintained a stable gradient at right angles to the direction of flow.

2. Direct avoidance responses were not observed, but the mean distributions of plaice, though not of herring, tended to the clean water channel at all concentration gradients.

3. This tendency was consolidated with time and it appears that the changes in distributions were the result of an orthokinetic response and not a chemotaxis.

4. Herring larvae did not show any avoidance of vertical gradients but remained in the dispersant layer until they became narcotized.

5. Narcotized larvae sank into clean water and recovered, only to swim upward again into the dispersant layer.

6. It is concluded that larvae would not avoid areas of dispersant at sea but because of the effects of dispersants would sink or swim away from lethal concentrations.

ACKNOWLEDGEMENTS

I would like to thank Dr. J.H.S. Blaxter for his advice and encouragement throughout this work. The work was carried out during the tenure of a NERC research studentship.

REFERENCES

Blaxter, J.H.S., 1962. Herring rearing. IV. Rearing beyond the yolk-sac stage. Mar. Res., No. 1, 18 pp.
Blaxter, J.H.S., 1968a. Rearing herring larvae to metamorphosis and beyond. J. mar. biol. Ass. U.K., 48, 17-28.
Blaxter, J.H.S., 1968b. Visual thresholds and spectral sensitivity of herring larvae. J. exp. Biol., 48, 39-53.
Cook, R.H. and Boyd, C.M., 1965. The avoidance by *Gammarus oceanicus* Segerstråle (Amphipoda, Crustacea) of anoxic regions. Can. J. Zool., 43, 971-975.
Crisp, D.J., Christie, A.O. and Ghobashy, A.F.A., 1967. Narcotic and toxic action of organix compounds on barnacle larvae. Comp. Biochem. Physiol., 22, 629-649.
Gamble, J.C., 1971. The responses of the marine amphipods *Corophium arenarium* and *C. volutator* to gradients and to choices of different oxygen concentrations. J. exp. Biol., 54, 275-290.
Höglund, L.B., 1961. The reactions of fish in concentration gradients. Rep. Inst. Freshwat. Res. Drottningholm, No. 43, 147 pp.
Ishio, S., 1965. Behavior of fish exposed to toxic substances. Advances in Water Pollution Research, Proc. 2nd Int. Conf. Tokyo (O. Jaag, ed.), 1964, 1, 19-33.
Jones, J.R.E., 1947. The reactions of *Pygosteus pungitius* (L.) to toxic solutions. J. exp. Biol., 24, 110-122.
Kühnhold, W.W., 1969. Der Einfluß wasserlöslicher Bestandteile von Rohölen und Rohölfraktionen auf die Entwicklung von Heringsbrut. Ber. dtsch. wiss. Komm. Meeresforsch., 20, 165-171.
Kühnhold, W.W., 1972. The influence of crude oils on fish fry. In: M. Ruivo (ed.) Marine pollution and sea life, 315-318 pp. London: Fishing News (Books), 624 pp.

Lindahl, P.E. and Marcström, A., 1958. On the preference of roaches (*Leuciscus rutilus*) for trinitrophenol, studied with the fluvarium technique. J. Fish. Res. Bd Can., 15, 685-694.
Ryland, J.S., 1966. Observations on the development of larvae of the plaice, *Pleuronectes platessa* L., in aquaria. J. Cons. perm. int. Explo Mer, 30, 177-195.
Shelbourne, J.E., 1964. The artificial propagation of marine fish. Adv. mar. Biol., 2, 1-83.
Simpson, A.C., 1968. The Torrey Canyon disaster and fisheries. Lab. Leafl. Fish. Lab., Lowestoft, N.S. No. 18.
Smith, J.E. (ed.), 1968. Torrey Canyon: pollution and marine life. Cambridge University Press, 196 pp.
Sprague, J.B., 1964. Avoidance of copper-zinc solutions by young salmon in the laboratory. J. Wat. Pollut. Control Fed., 36, 990-1004
Wilson, K.W., 1972. Toxicity of oil-spill dispersants to the embryos and larvae of some marine fish. In: M. Ruivo (ed.) Marine pollution and sea life, 318-322 pp. London: Fishing News (Books), 624 pp.

K.W. Wilson
Department of Natural History
University of Aberdeen
Aberdeen / GREAT BRITAIN

Present address:
Ministry of Agriculture, Fisheries and Food
Fisheries Laboratory
Burnham on Crouch, Essex / GREAT BRITAIN

Taxonomy

The Role of Larval Stages in Systematic Investigations of Marine Teleosts: The Myctophidae, A Case Study[1]

H. G. Moser and E. H. Ahlstrom

ABSTRACT

One group of well-known taxonomic characteristics, those of the embryonic and larval stages, have received scant attention from systematic ichthyologists. Using the family Myctophidae as a case study, evidence was gathered to show that larval characters can aid significantly in differentiating taxa and defining evolutionary lineages within a fish group.

The lantern fish family, Myctophidae, is the most speciose and widespread family of mid-water fishes in the world ocean. As presently recognized it contains about 30 genera and 300 nominal species. Myctophid larvae are highly prominent in the plankton, comprising about 50% of all fish larvae taken in open-ocean plankton.

Our studies of larvae of this family have included material from all oceans. We have developmental series for 29 myctophid genera, and for many genera we have series of all known species. This has afforded a more comprehensive view of the range and variability of larval characteristics, and we are increasingly impressed with the functional independence of larval and adult characters. It is apparent that the planktonic world of the larvae and the nektonic world of the adults are two quite separate evolutionary theatres. Our studies of larval lantern fishes have disclosed a full range of characteristics - from generalized to specialized and conservative to labile - equal in scope to those of the adults. These characters fall into several categories. An important group is the shape of various structures such as the eye, head, trunk, gut, and fins, especially the pectoral fins. Another group is the sequence of appearance and the position of fins, photophores, and bony elements. Another is the size of the larvae when they transform into juveniles. Pigmentation provides an important group of characteristics based on the position, number, and shape of melanophores. Finally, these are the highly specialized larval characters such as voluminous finfolds, elongated and modified fin rays, chin barbels, preopercular spines, etc.

The most trenchant characteristic of larval myctophids is eye shape. Our study shows that lantern fish larvae fall naturally into two groups on the basis of eye shape. Larvae of the sub-family Myctophinae have narrow elliptical eyes; some species have ventral prolongations of choroid tissue and some have the eyes on stalks (Moser and Ahlstrom, 1970); those of the Lampanyctinae have round or nearly round eyes (Moser and Ahlstrom, 1972). The development of a group of photophores during the larval period occurs in members of both sub-families, but much more commonly among Lympanyctinae; the sequence and number of photophores developed during the larval period is an especially useful character in revealing phylogenetic affinities when used in conjunction with other larval features such as body shape and pigment pattern (Moser and Ahlstrom, 1972).

[1] Published in Fish. Bull., U.S. 72 (2): 391-413, April 1974

The larvae illustrated in this paper comprise 55 species representing 24 genera. Emphasis is placed on larvae of the sub-family Myctophinae, in order to indicate the value of larval characters in showing intrageneric relationships. For example, larval characters substantiate that larvae of the two recognized subgenera of *Protomyctophum*, i.e. *Protomyctophum* sensu strictu and *Hierops*, are closely related but can be separated on the basis of eye shape. Three developmental lines are shown for *Electrona* on the basis of eye shape and amount of choroid tissue developed under the eye, with other characters. The uniqueness of the larvae of *Metelectrona* strongly suggests that *Metelectrona* is a valid genus. The genus *Benthosema* has four types of larvae that comprise two highly divergent species pairs.

The genus *Hygophum* affords an excellent example of the taxonomic utility of larval stages. The juveniles and adults of some species are notoriously difficult to identify but the larvae are readily identified. We have 11 such distinct larval types, whereas only 9 species are currently known from the adults. Search for adults of the two remaining larval types has led to the discovery of two undescribed species. A study of the larvae, but not the adults, shows that there are 3 highly distinct sub-generic groups, each containing from 2 to 6 closely related species.

The genus *Myctophum* has a diversity of larval form unmatched in the family, including the most aberrant of all lantern fish larvae, that of *Myctophum aurolaternatum*. The majority of larvae fall into distinct species groups, but there are common characters of pigmentation and eye shape which appear in all groups. Our work in the larvae should help to define the number of species in the genus, and provide some insight into phyletic lines within the genus.

The concluding section of the paper deals with myctophid evolution, to which larval studies can contribute significantly - especially in regard to the evolution of photophore patterns. According to our theory (see Moser and Ahlstrom, 1972), ancestral myctophids had a generalized arrangement of unspecialized photophores - one at the posterior margin of each scale pocket and a group of similar photophores on the head. We proposed that the specific photophore patterns of contemporary myctophids evolved through progressive enlargement and specialization of certain photophores of the generalized pattern and concurrent diminution or loss of the unspecialized photophores. Myctophinae would thus be considered highly specialized, since it is here that diminution of secondary photophores has reached the highest degree. Both larvae and adults of Myctophinae are considered highly advanced and conspicuous adult characters such as low photophores, prominent gas bladders, short jaws, and often narrow caudal peduncles as specialized adaptations of active, near-surface dwelling fishes. The presence of small secondary photophore and dorsal positioning of primary photophores in many Lampanyctinae genera indicate a retention of the ancestral condition. The Lampanyctinae are generally deeper living than the Myctophines, and the specializations of adults, such as long jaws and fat-invested swim bladders are basically related to their habitat.

REFERENCES

Moser, H.G. and Ahlstrom, E.H., 1970. Development of lantern fishes (family Myctophidae) in the California Current. Part I. Species with narrow-eyed larvae. Bull. Los Angeles Co. Mus. Nat. History., Sci., No. 7, 145 p.

Moser, H.G. and Ahlstrom, E.H., 1972. Development of the lantern fish, *Scopelopsis multipunctatus* Brauer 1906, with a discussion of its phylogenetic position in the family Myctophidae and its role in a proposed mechanism for the evolution of photophore patterns in lantern fishes. Fish. Bull., U.S., 70 (3), 541-564.

H.G. Moser and E.H. Ahlstrom
National Marine Fisheries Service
Southwest Fisheries Center
La Jolla, Calif. 92037 / USA

The Larval Taxonomy of the Primitive Myctophiform Fishes

M. Okiyama

INTRODUCTION

The order Myctophiformes is comprised of about 15 families, the relationships of which have hitherto not been very clear. However, the order is frequently divided into two suborders, Myctophoidea and Alepisauroidea; the family Aulopidae is generally considered to be the most primitive (Gosline et al., 1966). The definition of the "primitive myctophiform fishes" in the present paper is provisional; the families treated here are exclusively referable to Myctophoidea, but special emphasis will be placed on Aulopidae and its close relatives.

As Gosline et al. (1966) stated, the information on reproduction and development of the Myctophiformes is scanty, especially for Myctophoidea. Recent study, including scrutiny of the literature, however, provided further information on the larval stages of this suborder covering 9 out of 10 families. The purposes of this paper are threefold: 1. to describe the unknown larvae, 2. to review the larval stages of Myctophoidea, and 3. to speculate on the possible familial relationships from the larval standpoint. Subordinal concept follows Gosline et al. (1966), Rosen and Patterson (1969), and Paxton (1972).

DEVELOPMENT OF A JAPANESE AULOPID, *HIME JAPONICA*

Literature: Larvae of this species have not been described previously, but the egg and newly hatched larva illustrated by Mito (1960, Pl. 11) as II.A, No. 1, seems to be referred to this species, see Fig. 1 and Table 1.

Morphology: Early larvae are extremely slender. Body depth at pectoral fin base is 8% of standard length (unless otherwise noted, larval size is given by SL in this paper) in the smallest specimen available (3.69 mm). This increases to about 12% by the end of the larval stage. Snout to anus distance slightly decreases from 59-65% of SL in the larvae less than 10 mm, to an average of about 57% in the advanced specimens. Head length increases also from 21-23% of SL in the early larvae to about 30% in the early juveniles. Its width shows similar growth. The dorsal profile of head is slightly concave and the tip of the snout is somewhat pointed in the early larvae, but the head becomes massive and rounded in lateral profile as growth proceeds. The mouth is terminal, slightly oblique in position and moderately large. Posterior tip of the maxillary reaches below the vertical of anterior border of pupil. Supramaxillary is not formed during the larval stages. Eye is nearly rounded but generally the vertical measurement is slightly greater than the horizontal in the larvae. No choroid tissue is found. Gut of the early larvae is a long cylindrical tube with three distinct sections. The anteriormost section of the

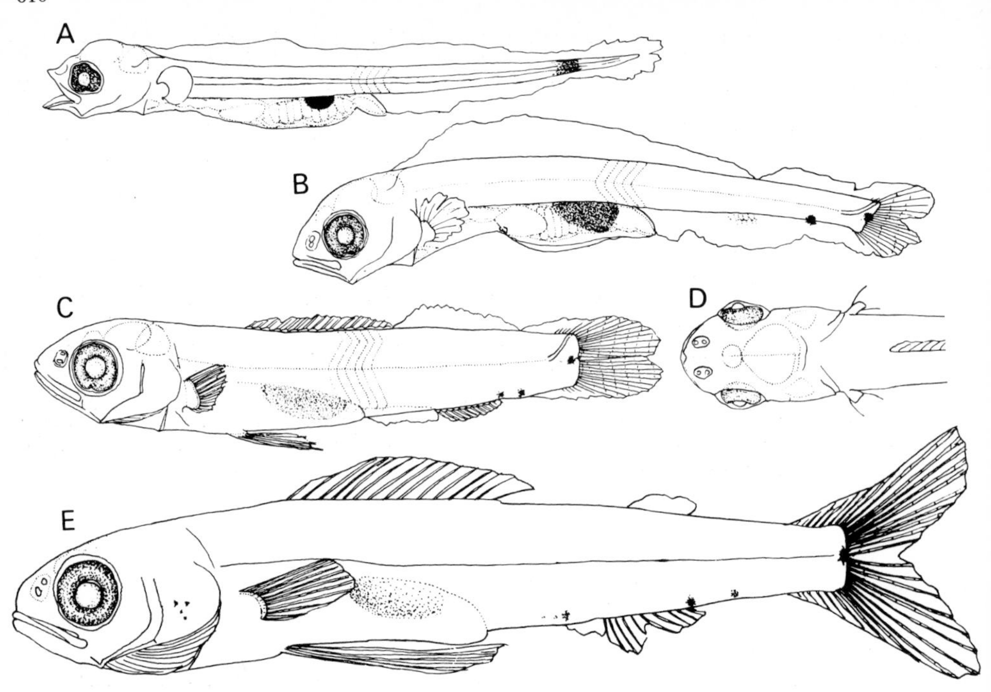

Fig. 1. Developmental stages of *Hime japonica*. A - 3.69 mm; B - 6.18; C-D - 13.3; E - 21.8

Table 1. Measurements and counts of early developmental stages of *Hime japonica* collected from the Japan Sea during 1964-1967

Total length (mm)	3.80	6.35	9.0	13.1	16.6	24.0
Standard length	3.69	6.15	8.1	11.9	13.2	21.8
Snout to anus	2.15	3.75	5.0	6.9	8.5	12.2
Snout to anal fin	-	-	-	8.3	9.5	15.4
Head length	0.80	1.54	1.87	3.44	4.13	5.[illegible]
Eye diameter	0.29	0.50	0.65	0.89	1.13	1.6
Snout length	0.15	0.37	0.54	0.67	1.00	1.2
Body length	0.52	1.00	1.44	2.06	2.69	3.5
Dorsal fin rays	-	-	-	-	ca. 15	15
Anal fin rays	-	-	-	10	10	10
Pectoral fin rays	-	-	ca. 5	ca. 8	11	11
Pelvic fin rays	-	-	-	-	ca. 8	9
Myotomes	41	43	42	42	43	

oesophagus is exceptionally long, attaining a maximum of 50% of gut length. In the advanced larva of 13.2 mm, the intestinal section occupies by far the greater portion than the remaining parts. Anus opens far in advance of, but slightly closer to, the anal fin than the pelvic throughout all stages. Gas bladder is absent.

All fins are small. Full complements of the dorsal, anal, and pectoral fins are formed simultaneously between sizes of 11.9 and 13.3 mm. Pelvic fin appears in 6.2 mm larva but its complete numbers of rays are first present in 17.5 mm. Notochord begins to flex at about 7.0 mm and is fully flexed before larvae are 9.0 mm. A total of 19 principal caudal rays are ossifying at about 8.0 mm. Adipose fin is well formed in a juvenile of 20 mm.

Larvae and juveniles are poorly pigmented. In the smallest larva of 3.69 mm, pigmentation is restricted to the eyes and caudal part near the tip of notochord, besides a single pigment spot in the peritoneum. A pigment band in the caudal region is a temporal one and its dorsal element disappears in a larva of 3.81 mm. There are two additional melanophores in much advanced larvae; one on the base of the caudal fin first appearing at 6.19 mm, and the other just behind the anal fin base becoming distinct at 8.55 mm. The peritoneal pigment spot is an oval thick membrane of darkly pigmented structure and seemingly has no direct association with the surrounding organs. Its longitudinal length increases from 6% to 15% of SL through the larval stages.

REVIEW OF THE LARVAL FORMS OF THE OTHER MYCTOPHOIDEA

Aulopidae (Fig. 2)

A fine series of larval stages of *Aulopus filamentosus* ranging from 6.80 to 35.0 mm was described by Sanzo (1938). Larvae are moderately slender and transparent with large fan-shaped pectoral fins and a depressed snout. A total of 12 to 13 peritoneal pigment spots appear in about 10 mm larva, whereas small pigment spots of the peripheral areas of the pectoral fins are distinct in the smallest larva.

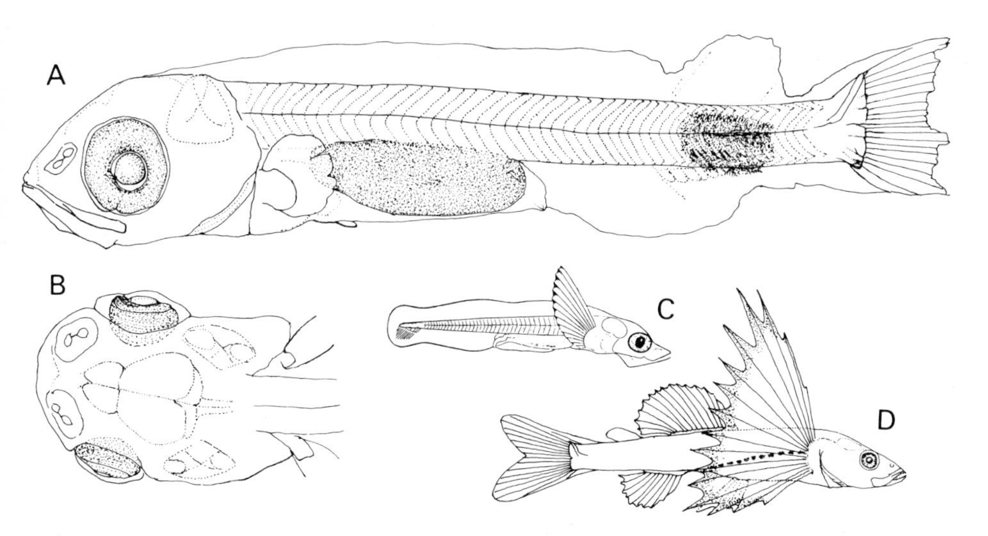

Fig. 2. Developmental stages of Aulopidae. A-B - *Hime*-type larva of 12.3 mm; collected at the mouth of the Gulf of California, on 17 February 1972. C - *Aulopus filamentosus*, 9.44 mm; D - the same, 30.9 (C-D, after Sanzo, 1938)

Through the courtesy of Dr. Ahlstrom, an apparently undescribed larva probably referable to this family has been put at my disposal. Head is massive, and broader than its depth. Its length is 30% of SL. Body is moderately deep and laterally flattened. Its depth at the pectoral fin insertion is the same as the head length. Anus opens slightly behind the middle point of body with wide space from anal fin. Gut is a simple tube, two thirds of its length being occupied by intestine of larger diameter. Notochord is completely flexed. Except for the dorsal fin, all fins including the adipose are developing but small. A single peritoneal pigment spot is extremely large and spreading over the full length of the intestine. A prominent melanistic blotch is present in the transparent tail just above the posterior half of the anal fin base. SL, 12.3 mm; head length, 3.75; snout to anus, 7.19; eye diameter, 1.16 x 1.28; myotomes, ca. 49; Af rays, ca. 9; Cf rays, 10 + 9 (Fig. 2A-B).

Its specific identification is impossible, since none of the aulopid fishes have been recorded from around the locality. In view of its close resemblance to larvae of *Hime japonica*, it seems that there are at least two distinct types of larval forms in the Aulopidae, which can be named *Aulopus*- type and *Hime*- type. Table 2 shows the differences between these two types.

Table 2. Comparison of the early developmental stages of 2 aulopid genera

	Hime	*Aulopus*
Eggs:	(Spawned eggs in formalin)[1]	(Matured ovarian eg
Diameter	1.18 - 1.34 (mean 1.27)	1.36 - 1.44
Oil globule	absent	numerous
Sculpture on membrane	anomalous hexagonal meshes	absent
Larvae and juveniles:		
Peritoneal pigment spots	single; large	12-13; small
Pectoral fin	small; transparent	large; pigmented
Adipose fin	opposite to anal fin	behind of anal fin

[1] *Hime japonica*

Synodontidae (Fig. 3)

This is one of the best known families in the early life stages (Zvjagina, 1965). The known larvae, including those of the 3 genera, bear a striking resemblance to each other, as can be seen in Fig. 3. The number, position and size of the peritoneal pigment spots lying in pairs along the ventro-lateral sides of gut, are of great aid in identifying the protracted larval stages. The intraspecific consistenc of this character is sometimes obscure in *Synodus variegatus* (Fig. 3C-E); it has a maximum of 13 spots but often loses the bilateral symmetry of the pairs. These pigment spots are evenly spaced at considerable distance, usually more than the pertinent spot sizes. The presence of a fulcral scale-like structure in the caudal region of *Saurida* sp. of 37 mm is worth mentioning (Okiyama, unpubl.).

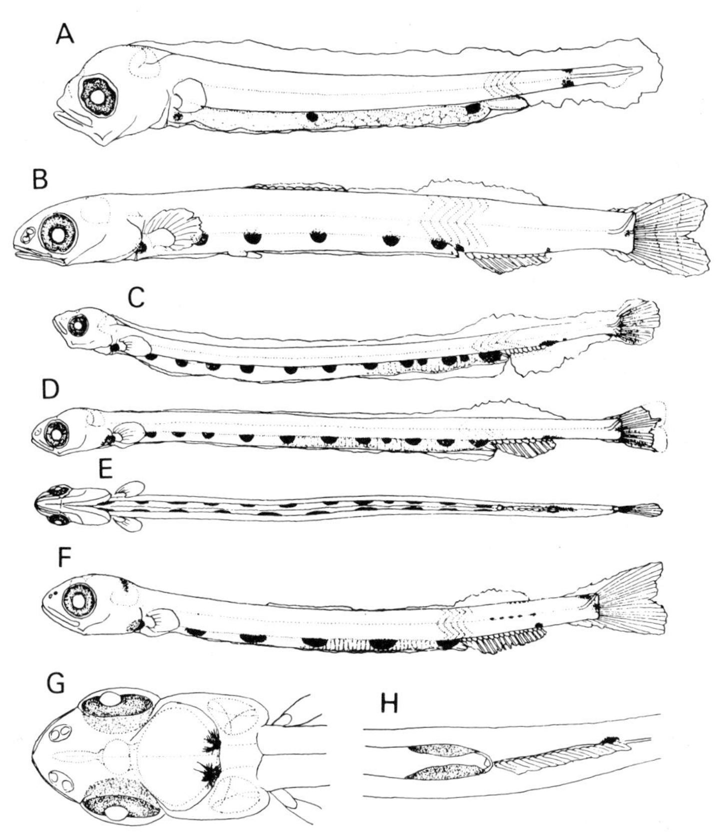

Fig. 3. Developmental stages of Synodontidae collected from the Japan Sea. A - *Saurida* sp., 4.75 mm; B - *Saurida undosquamis*, 15.6; C - *Synodus variegatus*, 12.9; D-E, the same, 13.2; F-H - *Trachinocephalus myopus*, 19.8

Bathysauridae

Larval stages of this family have not been known until recently when Rosen (1971) concluded that *Macristium chavesi* of the family Macristiidae is a juvenile form of a bathysaurid fish. Only three specimens, 20 to 110 mm long, have been described in the literature, while Ida and Tominaga (1971) recorded another *Macristium* specimen of considerable size from the Indian Ocean. In this Indian specimen, 6 peritoneal pigment spots in a saddle-shape are clearly developed with moderate spacing (Tominaga, pers. comm.). None of the material has been identified to species, but several species may be involved in view of the differences of meristic characters and conditions of ossification.

Chlorophthalmidae

Our knowledge on the larval stages of this family are restricted to those of *Chlorophthalmus agassizi* (Tåning, 1918; and others) and *C. mento* (Ahlstrom, 1971). They are extremely similar in appearance and essentially identical.

Bathypteroidae (Fig. 4)

A sketch of a 14.2 mm specimen, probably referable to this family, was kindly provided to me by Dr. Ahlstrom. This individual is slender and transparent with an enlarged fan-like pectoral fin. When depressed its posterior end reaches over the anus, but no rays are deformed into filaments. Eye is still distinct. Posterior end of small jaw gape is just below the anterior rim of eye. It has a full complement of all fin rays, but uniquely lacks the peritoneal pigment spot. Anus opens closely before the anal fin. Despite some differences, such as slightly posterior position of the dorsal and pelvic fins, fin disposition as well as meristics of this specimen fit best those of the family Bathypteroidae, genus *Bathypterois* (Ahlstrom, pers. comm.). SL, 14.2 mm; Df, 11; Af, 9; Pf, 17; Pel, 8.

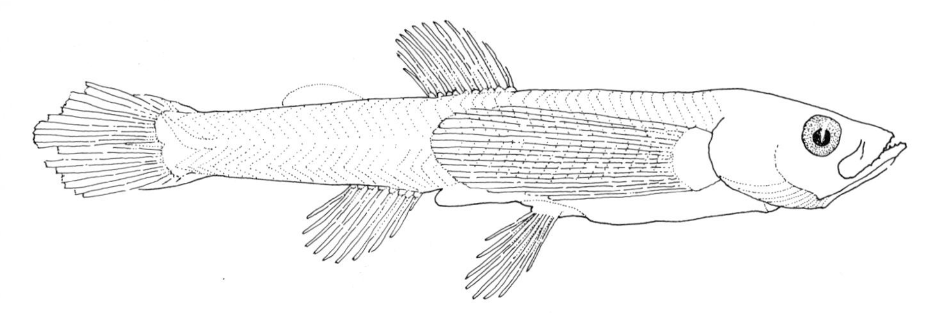

Fig. 4. Developmental stage of Bathypteroidae. *Bathypterois* sp. 14.2 mm

Ipnopidae

Larval study of this family involves a similar history to the Bathysauridae, since the clue to explain the *Macristiella*-ipnopid linkage was first suggested by Rosen (1971). His assumption was immediately confirmed by Okiyama (1972) who concluded that *Macristiella* specimens are the larval forms of the genus *Bathytyphlops* of the Ipnopidae. In a report on a new form of *Macristiella*, Parin and Belyanina (1972) also suggested its close relation with *Bathytyphlops* among the 3 ipnopid genera. A total of 19 specimens of the so-called *Macristiella*, ranging from 7.7 to 47 mm in total length, are described in the literature. Their specific identification still remains to be determined.

Neoscopelidae (Fig. 5, Table 3)

No information is available on neoscopelid larvae except a brief comment on the superficial resemblance of larval *Scopelengys* and *Lampanyctus* (Moser and Ahlstrom, 1970). Recently, several larval specimens referable to *Neoscopelus macrolepidotus* and *Scopelengys dispar* were sent to me for study through the courtesy of Drs. Ida and Ahlstrom.

They are structurally similar; advanced larvae are rather deep-bodied, large-headed and laterally compressed with large fan-shaped pectoral fins of about head length. Eye is rounded and not so large. No choroid tissue is present. Anus opens just in front of anal fin. Gas bladder is distinct in all larvae of both genera but it seems less apparent in the larger specimen of *Scopelengys* which reportedly lacks this organ (Ebeling, 1966). Fin differentiation is rapid, particularly in the pectoral fin. Full complements of all fin rays may be attained at

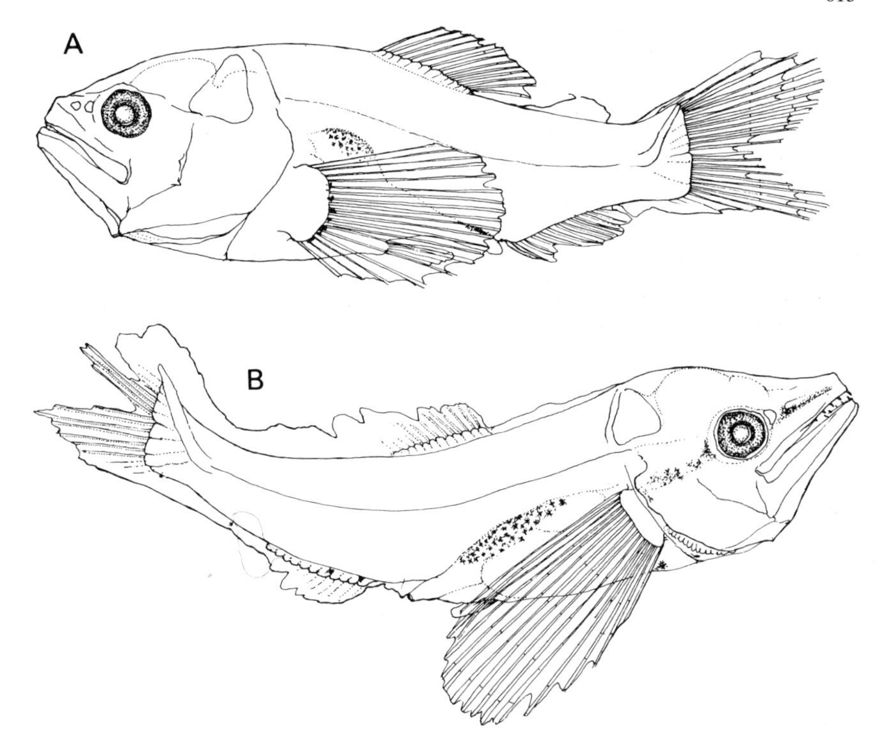

Fig. 5. Developmental stages of Neoscopelidae. A - *Neoscopelus macrolepidotus*, 6.56 mm; B - *Scopelengys dispar*, 6.25 mm

less than 10 mm length. Pigmentation is scanty in both genera but sharply contrasts between them. Most definitive is the pigment on the side of the head both in front of and behind the eye in *Scopelengys*, since no corresponding melanophores are seen in *Neoscopelus*. Peritoneal pigments are often coalesced into a patch around the gas bladder in both genera, but clearly differ from the peritoneal pigment spots common in other myctophoid larvae. Small pigment spots along the pectoral fin base are unique in *Neoscopelus* whereas a few spots along the ventral edge of body are restricted in *Scopelengys*. Likewise, a long snout is peculiar to the latter and preopercular spines are developed only in the former.

Other Families

In contrast to myctophid larvae which are well studied (Moser and Ahlstrom, 1970), larval forms of the remaining two families, Harpadontidae and Notosudidae, are poorly known. Quite recently Ahlstrom (1972) included a brief account of the larval characteristics of the Notosudidae.

Table 3. Measurements and counts of the early developmental stages of Neoscopelidae

	N. macrolepidotus		*S. dispar*	
Total length (mm)	11.6	8.13	7.19	7.03
Standard length	9.69	6.56	6.25	6.41
Snout to anus	6.80	4.62	4.18	3.69
Head length	3.57	2.50	2.17	2.22
Eye diameter	0.68	0.61	0.48	0.44
Snout length	1.14	0.61	0.48	0.44
Body depth(at Pf base)	3.38	2.15	1.36	1.38
Pectoral fin length	2.40	1.72	2.28	2.28
Dorsal fin rays	11	13	ca. 12	ca. 12
Anal fin rays	11	12	ca. 10	--
Pectoral fin rays	18	18	15	-
Pelvic fin rays	8	9	-	-
Branchiostegals	2+7	-	-	-
Myotomes	30	-	31	-
Date of collection	17 July 1970		18 August 1967	1 March 1967
Locality	8°21.7' S		13°41' N	05°03' S
	151°06.0' E		103°23' W	97°57' W

Table 4. Comparisons of some meristic characters and larval structures of Myctophoidea. (Compiled chiefly on the basis of McAllister (1968); Rosen (1972); Paxton (1972) and present materials)

	VN	Br.	Position of anus	Gas bladder	Peritoneal pigment spots	Pf structure
Aulopidae	41-53	13-16	with moderate space from anal fin	absent	1 or 12-13	small or large
Synodontidae	44-58	11-18	just in front of anal	absent	6-13	small
Harpadontidae	57[1]	17-26	just in front of anal	absent	?	?
Bathysauridae	50-63	8-13	just in front of anal	absent	6	very large
Chlorophthalmidae	38-49	7-10	near pelvic	absent	1	small
Bathypteroidae	49-60	11-14	variable	absent	0	large
Ipnopidae	51-80	8-17	near pelvic	absent	1	large
Neoscopelidae	30-31[2]	8-11	just in front of anal	present	0	large
Myctophidae	28-44	7-11	just in front of anal	present	0	small to moderately large
Notosudidae	45-66	8-10	near pelvic	absent	0	small

[1]Data on *Harpadon microchir*, 61 cm.

[2]Data on adults of *Neoscopelus macrolepidotus* and *Scopelengys dispar*.

LARVAL CHARACTERISTICS AND THE FAMILIAL RELATIONSHIPS

Despite incomplete material, previous reviews showed that there are several larval characteristics which are shared by many families and are likely to contribute to the generic or higher classification of the Myctophoidea (Table 4). On the basis of these factors, the larval forms now under consideration are tentatively grouped as follows:

A. Peritoneal pigment spot single
   a) Pectoral small; anus with wide space from anal - Aulopidae (*Hime*-type), Chlorophthalmidae.
   b) Pectoral large; anus with wide space from anal - Ipnopidae.

B. Peritoneal pigment spots numerous
   a) Pectoral small; anus just in front of anal - Synodontidae.
   b) Pectoral large; anus with wide space from anal - Aulopidae (*Aulopus*- type).
   c) Pectoral large; anus just in front of anal - Bathysauridae.

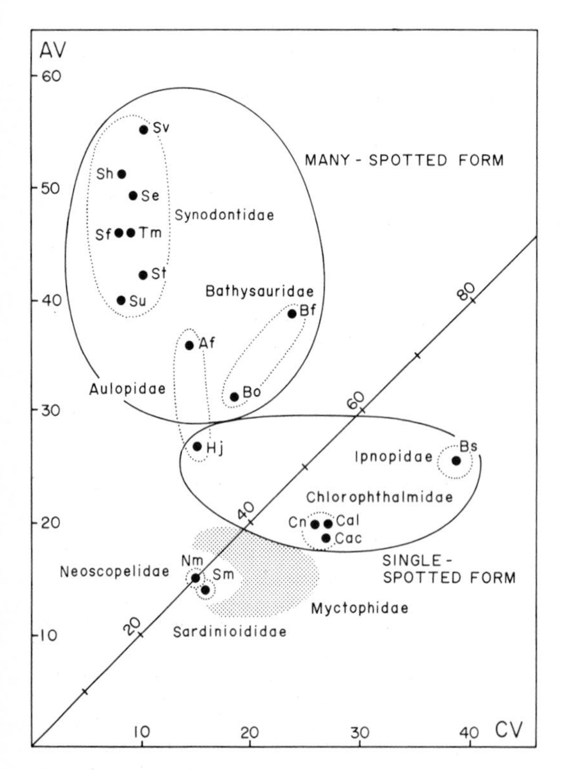

Fig. 6. Relationships between the numerical characters of vertebrae (AV, abdominal vertebrae; CV, caudal vertebrae) and main larval groups of Myctophoidea. Af, *Aulopus filamentosus*; Bf, *Bathysaurus feris*; Bo, *B. obtusirostris*; Bs, *Bathytyphlops* sp.; Cac, *Chlorophthalmus acutifrons*; Cal, *C. albatrossis*; Cn, *C. nigromarginatus*; Hj, *Hime japonica*; Nm, *Neoscopelus macrolepidotus*; Se, *Saurida elongata*; Sf, *Synodus foetens*; Sh, *S. hosinonis*; Sm, *Sardinioides minimus*; St, *Saurida tumbil*; Su, *S. undosquamis*; Sv, *Synodus variegatus*; Tm, *Trachinocephalus myopus*. Position of Myctophidae is approximate (after Paxton, 1972). Other spots are based on various sources including original

C. Peritoneal pigment spots absent

a) Pectoral small; anus with wide space from anal - Notosudidae.

b) Pectoral small to moderately large; anus just in front of anal - Myctophidae (exclusive of *Protomyctophum*).

c) Pectoral large; anus just in front of anal - Neoscopelidae, Bathypteroidae (*Bathypterois*).

Here, special emphasis is placed on the peritoneal pigment spots, in view of their integrity within Myctophiformes. As to the origin or probable function of these structures, no reasonable explanation has been given; but evidence that the occurrence of these spots is correlated with the lack of a gas bladder is significant. Fin elongation is especially frequent in the pectorals, and long pectoral rays are more advanced in ossification than the other rays in *Aulopus* and Neoscopelidae. Thus, it is of interest that a pair of elongate rays are exceptionally occurring in the pectoral of larval *Loweina* and *Tarletonbeania* (Moser and Ahlstrom, 1970). Differences in the anus location and gut structure are also likely to have certain indication of familial separation.

Fig. 6 shows the relationships between the above mentioned three groups and the numerical characteristics of the vertebrae. This diagram includes one remarkable feature; although these vertebral characteristics do not always represent those of the species whose larval stages have been studied, there is a significant separation between the single and many spotted groups. Many spotted forms have decidedly higher counts in the abdominal vertebrae, but in the reverse regard to the caudal elements. Among non-spotted forms, Neoscopelidae and Myctophidae are clearly distinguished from the single and many spotted groups, whereas the remaining two families cannot be defined strictly.

Apart from the dichotomy of the Aulopidae, this scheme illustrates the two main lines of evolution of the Myctophoidea currently recognized (cf. Paxton, 1972). It is quite puzzling that the striking contrast between the two types of the aulopid larvae (Table 2) is not reflected in the systematics of the adults (Mead, 1966). I feel, however, that certain distinctive features in *Aulopus japonica* of Mead (1966) including the short snout, head and body, coupled with small number of vertebrae, probably warrant at least the distinct generic status of *Hime*.

Judging from the many similarities between the *Hime*-type aulopid and the chlorophthalmid larvae, *Hime*-type larvae seem to be less specialized than the *Aulopus*-type. The close resemblance of the vertebral characteristics of the *Sardinioides* and *Hime* rather than of the *Sardinioides* and *Aulopus* (Fig. 6) may support the above assumption, as *Sardinioides* is currently considered to be a distinct myctophiform member very close to the ancestral stock of the Myctophiformes (Goody, 1969). Perhaps key characters of the Aa Group are responsible for the most generalized features of the primitive myctophiform larvae. On the other hand, dissimilarity of their gut structure should be considered an indication of the significant evolutionary divergence associated with the two main lines mentioned. So far as the gut structure is concerned, *Hime* is indistinguishable from Group B. The uniquely elongate oesophagus of larval *Hime japonica* thus seems to represent an odd structure related to the evolutionary jump which possibly took place in the early phase of aulopid evolution along the line to increase the abdominal elements. There are a number of additional lines of evidence that suggest the possible linkage between Aulopidae, particularly *Hime*, and Synodontidae. These include the resemblance of the egg and newly hatched larva, presence of a fulcral scale-like structure and similar feature of the epurals. On this last point, Rosen and Patterson (1969) stated that loss of one epural occurring in Synodontidae is foreshadowed in some specimens of *Hime japonica*.

The increase in the number of peritoneal pigment spots also accords with this trend. For example, the highest counts of abdominal vertebrae and peritoneal pigment spots are duplicated in the larvae of *Synodus variegatus*, among known Japanese synodontid forms. I believe that the Bathysauridae is a possible offshoot of the ancestral stock in common with the Synodontidae, because of the six peritoneal pigment spots and similar anus location.

Relationships among the remaining families are too complex to reveal any reasonable associations. Within these loose assemblages, however, the similarities of larval stages of Chlorophthalmidae and Ipnopidae as well as *Protomyctophum* (of Myctophidae, as suggested by Moser and Ahlstrom, 1970) are conspicuous. If these are not the results of convergence, the above evidence is in favor of the recent arrangement of families with monophyletic origin from chlorophthalmid-like ancestral stock (Paxton, 1972). This standpoint on larvae, however, cannot elucidate the admitted bathypteroid lineage between Bathypteroidae and Ipnopidae (Rosen and Patterson, 1969). Problems in this connection apparently lie in the rather sporadic appearances of the non-spotted forms. In such circumstances, a familial pair of Myctophidae and Neoscopelidae may be secured by the similar number of vertebrae, presence of gas bladder and rapid ossification. This myctophid lineage is unique because it has a smaller number of vertebrae than its ancester like Chlorophthalmidae. Evolutionary aspects of the remaining two families of non-spotted groups are as yet beyond the scope of this larval study.

Finally, I wish to emphasize the need to study advanced juveniles along with larval stages for myctophoid taxonomy, since the majority of forms in question are known to show drastic transformation, especially with respect to the eye and mouth features during the protracted early life stages.

ACKNOWLEDGEMENTS

My grateful thanks go to Dr. E.H. Ahlstrom, NOAA, USA, who not only provided precious material and unpublished information indispensable to this work, but gave me encouragement throughout; also to Drs. Y. Tominaga and H. Ida of Tokyo University, for the specimens, literature, and useful advice. Thanks are also do to Dr. S. Ueyanagai of the Far Seas Fisheries Research Laboratory, for his continued encouragement and help, to Dr. O. Tabeta of Shimonoseki University of Fisheries, Mr. K. Maruyama of Iwate Prefectural Fisheries Experimental Station, and Mr. T. Yamakawa of Kochi High School, who kindly provided the material for this study. I extend my thanks to many other scientists who assisted me in covering the literature. I am also indebted to Dr. A. Furukawa and the staff of Japan Sea Regional Fisheries Research Laboratory in various phases of the study.

REFERENCES

Ahlstrom, E.H., 1971. Kinds and abundance of fish larvae in the eastern tropical Pacific, based on collections made on Eastropac I. Fish. Bull., 69, 3-77.
Ahlstrom, E.H., 1972. Kinds and abundance of fish larvae in the eastern tropical Pacific on the second multivessel Eastropac survey, and observations on the annual cycle of larval abundance. Fish. Bull., 70, 1153-1242.
Ebeling, A.W., 1966. Review of Fishes of the Western North Atlantic, order Iniomi and Lyomeri. Copeia, 1966, 895-898.
Goody, P.C., 1969. The relationships of certain Upper Cretaceous teleosts with special references to the myctophoids. Bull. Brit. Mus. (Nat. Hist.), Geol., suppl., 7, 1-255.
Gosline, W.A., Marshall, N.B. and Mead, G.W., 1966. Order Iniomi. Characters and synopsis of families. In: Fishes of the Western North Atlantic. Mem. Sears Found. Mar. Res., Mem. 1, 5, 1-29.
Ida, H. and Tominaga, Y., 1971. On a *Macristium* specimen from the Indian Ocean. Japan. J. Ichthyol., 18, 103. (In Japanese).
Mito, S., 1960. Keys to the pelagic fish eggs and hatched larvae found in the adjacent waters of Japan. Sci. Bull. Fac. Agr., Kyushu Univ., 18, 71-94. Pls. 2-17. (In Japanese).
Moser, H.G. and Ahlstrom, E.H., 1970. Development of lantern fishes (family Myctophidae) in the California Current. Part I. Species with narrow-eyed larvae. Bull. Los Angeles Co. Mus. Nat. Hist., Sci., 7, 1-145.
Okiyama, M., 1972. Morphology and identification of the young ipnopid, "*Macristiella*" from the tropical western Pacific. Japan. J. Ichthyol., 19, 145-153.
Parin, N.V. and Belyanina, T.N., 1972. Novye dannye o rasprostranenii i morfologii ryb roda *Macristiella*. Vopr. Ihtiol., 12, 1120-1123.
Paxton, J.R., 1972. Osteology and relationships of the lantern fishes (family Myctophidae). Bull. Los Angeles Co. Mus. Nat. Hist., Sci., 13, 1-81.
Rosen, D.E., 1971. The Macristiidae, a Ctenothrissiform family based on juvenile and larval scopelomorph fishes. Amer. Mus. Nov., 2452, 1-22.
Rosen, D.E. and Patterson, C., 1969. The structure and relationships of the paracanthopterygian fishes. Bull. Amer. Mus. Nat. Hist., 141, 357-474.
Sanzo, L., 1938. Uova ovariche e svilluppo larvale di *Aulopus filamentosus* Cuv. Mem. Com. talassogr. Ital., 254, 3-7.
Tåning, A.V., 1918. Mediterranean Scopelidae (*Saurus*, *Aulopus*, *Chlorophthalmus* and *Myctophum*). Rep. Danish Oceanogr. Exped. 1908-10 Mediter. Adj. Seas, Vol. 2, 7, 1-154.
Zvjagina, O.A., 1965. Materialy o razvitii jashcherogolobykh ryb (Pisces, Synodontidae). Trudy Inst. Okeanol., 80, 147-161.

M. Okiyama
Japan Sea Regional Fisheries
Research Laboratory
Nishi-Funamicho
Niigata / JAPAN

Analysis of the Taxonomic Characters of Young Scombrid Fishes, Genus *Thunnus*[1]

W. J. Richards and T. Potthoff

INTRODUCTION

The young stages of fishes of the family Scombridae are among the most difficult to identify to generic levels and particularly to the species level. These stages are, for the most part, easily identified to the proper family, but the eggs are unknown except for a few species This paper reports on our studies which attempt to solve some of these identification problems for the larvae and transforming stages. A great deal of research has been done in this area by us and others; but our recent findings cast grave doubts on the identification methods that have been used in the past. To give some indication of the amount of work already directed to these problems, a recent bibliography of young scombrids covering the years 1880-1970 lists 170 papers dealing with aspects of the identification of eggs, larvae, and juveniles (Richards and Klawe, 1972).

To achieve accurate identifications of young stages, adult taxonomy must be thoroughly understood and morphometric and meristic data on adults must be available. This family contains about 40 species allocated to 14 genera. Most of the taxonomic problems of the adults are fairly well understood; the only striking confusions exist with the species of *Scomberomorus* in the Indo-Pacific Ocean and the exact nature of the species of *Auxis* (see Collette and Gibbs, 1963; Fraser-Brunner, 1950; Gibbs and Collette, 1967; and Matsui, 1967). A large amount of anatomical data has been published for this family which is useful to the larval taxonomist. As a result, there seems to be little disagreement among researchers as to the generic identification of the young stages of *Thunnus, Katsuwonus, Euthynnus, Auxis, Scomber, Rastrelliger, Acanthocybium, Allothunnus, Sarda, Scomberomorus*, and *Grammatorcynus*. Three genera - *Cybiosarda, Gymnosarda*, and *Orcynopis* - are poorly understood due mainly to inadequate material. Within all these genera, many of the young stages of the species are well understood because of adequate work or their monotypicity or allopatricity. The major difficulties lie in the species of *Scomberomorus* because of confused adult taxonomy, and hence little attention to the larvae, and in the species of *Thunnus* because of their similarity. Our work has been confined to *Thunnus*, but what we have learned about this genus may also be applicable to the other scombrids.

A larval tuna identification workshop in 1970 (see Matsumoto et al., 1972) summarizes the problems in larval *Thunnus* identification and forms the base for our studies. The prime reason for our investigation was the point raised by Richards on page 9 of that report "... (he) suspects that *T. atlanticus* larvae are very similar to larvae of *T. obesus*. This suspicion is based on the great abundance of larvae

[1]Contribution No. 227, Southeast Fisheries Center, National Marine Fisheries Service, NOAA, Miami, Fl 33149.

Table 1. Characters to separate larvae of *Thunnus* in size range of 3-10 mm SL (adapted from Matsumoto et al., 1972)

Characters:	T. albacares	T. alalunga	T. thynnus (Atlantic)
Black pigmentation:			
upper jaw	Appears at 5.8 mm SL, mostly after 6.0 mm SL	Appears at about 5 mm SL	No observation
lower jaw	Appears at 4.5-6.0 mm SL	Appears at 9-10 mm SL	
	At tip on inner edge; migrate to outer edge with further growth	At tip on outer edge	2 on inner edge
Dorsal edge trunk	None	None	1 or 2
Lateral line	None	None	0-2 near mid-trunk
Ventral edge trunk	None	None	1-4
Red pigmentation (number of cells)			
Dorsal edge trunk	0, 1, 2, (3) [mean = 0.6]; near caudal peduncle	2, 3, (4) [mean = 2.6] from caudal peduncle to mid-second dorsal fin base	Streak on caudal peduncle[1]
Lateral line	(0), 1, 2, 3, 4, (5) [mean = 2.4]	(2), 3, 4, (5) [mean = 3.5]	Indistinct[1]
Ventral edge trunk	3-12 [mean = 7.0]	5-12 [mean = 8.0]	Streak anus to caudal peduncle[1]
Lower jaw ventral view	No observation	No observation	Streak along margin anterior half of jaw and midline[1]

[1]Only one larva taken in a day tow was examined.

Table 1. (continued)

T. thynnus (Pacific)	*T. tonggol*	*T. maccogii*	*T. obesus*
Appears above 6 mm SL	No observation	Appears above 5 mm SL	Appears above 5 mm SL
2 on inner edge above 4 mm SL	No observation	Appears above 4 mm SL	0-2 cells on inner edge below 4 mm SL
1 or 2	1, 2 or more	1 or 2, very small	None
None	None	0 or 1 near mid-trunk	None
2 or more	2 or more	1-3	1 or more
1-5, mostly 3	No observation	No observation	0, 1, (2)
Number not available	No observation	No observation	0, 1, 2, 3, 4
Number not available	No observation	No observation	1-8 [mean = 5.3]
2 well spaced on anterior half	No observation	No observation	1 on each side near tip

resembling those of *T. obesus* in this area (tropical western Atlantic), particularly at times and places where *T. obesus* adults are rarely found or absent".

Larval *Thunnus* are identified to species solely on the distribution of melanophores and erythrophores found on the tip of the lower jaw and on the trunk - both the dorsal and ventral edges of the trunk and laterally on the trunk, exclusive of the pigment over the gut. Table 1 summarizes these characteristics. Larvae greater than 10 mm standard length (SL) lose most of these pigment characteristics primarily through development of additional melanophores, which obscure the larval melanophores. For these internal characteristics, principally features of the axial skeleton, number of gillrakers, fin rays, fin supports, and shape of the lateral line, must be depended upon for accurate identification (Potthoff and Richards, 1970; Matsumoto et al., 1972; and Potthoff, 1974).

Basically, larval *Thunnus* types are assigned by taxonomists to particular species by indirect means, since the larval pigment characteristic do not persist into the later stages where other methods can be applie These larval types are subjectively determined by the variability foun in the various pigments, in conjunction with the known distribution of adults. Errors can develop using this methods. An apparent error was made by Matsumoto (1962) when he attributed one type of larvae to *T. alalunga*. Yabe and Ueyanagi (1962) disputed Matsumoto since they felt that *T. alalunga* was represented by an entirely different type of larvae. These arguments are continued and explained by Matsumoto et al. (1972) for the species of *Thunnus*. It is most difficult to determine whether or not the variation observed in pigment types really represent specific or individual variations. The melanophores and erythrophores that are used vary greatly in size and shape. At times they are greatly expanded and quite easily seen. Erythrophores from day-caught larvae are expanded to the point where they appear as faint pink strips; those from night-caught larvae are contracted to round, bright orange spots. Some melanophores in their contracted state are only 10 μ in diameter and may also be confused with particles of dirt.

Confirmation of identification for larvae can be achieved best through rearing products from known parents in the laboratory. This is a difficult process, but it has been achieved with many fishes and most recently for *T. albacares* (Harada et al., 1971). Mori et al. (1971) confirmed that *T. albacares* had been identified correctly through study of the reared material. The identity of *T. atlanticus* larvae remains the enigma for the western Atlantic Ocean. As mentioned earlier, Richards (in Matsumoto et al., 1972) suspected that *T. atlanticus* larvae were similar to *T. obesus* because of the abundance of the *T. obesus* type, the lack of any other type despite the abundance of *T. atlanticus* adults and the rarity of *T. obesus* adults in the tropical western Atlantic Ocean. Wise and Davis (1973), in their analyses of distribution of tunas based on the Japanese long line fishery, show very few records of *T. obesus* in the Caribbean Sea and Gulf of Mexico. Rather, the major concentration of *T. obesus* is throughout the central areas of the North and South Atlantic. Because the Japanese long liners do not record *T. atlanticus* in their fishery catch data, it is plausible that some of their western Atlantic *T. obesus* records are, in fact, large *T. atlanticus*.

Juárez (1972) described the larval form of *T. atlanticus*. We believe that this identification is correct, although several statements made by Juárez indicate that her concept of larval tuna identification differs from that expressed by Matsumoto et al. (1972). No mention is made by Juárez of melanophores occurring on the trunk, and the pigmentation she describes would be appropriate for *T. albacares*. Juárez's largest specimen (7.3 mm SL) was cleared and stained, revealing the characteristic 19 precaudal vertebral count and first closed haemal arch on the 11th vertebrae which is diagnostic for *T. atlanticus*.

METHODS

To study this problem, we followed two courses. First, we made a detailed study of the number and position of the diagnostic melanophores found on these larvae, particularly the number and position of melanophores on the tip of the lower jaw and the number and location (myomere on which they occurred) of melanophores on the trunk. Most specimens were collected in the Straits of Florida; others in the Gulf of Mexico and in the Caribbean Sea. A series of *Thunnus* larvae from the Gulf of Guinea were examined, for this area is well outside of the known range of *T. atlanticus*. Second, specimens greater than 6.0 mm SL (when the axial skeleton develops), from which this pigmentation data had been collected, were cleared and stained for study of the axial skeleton, using the method described by Taylor (1967). A further problem which prevents observation of bone structures is improper original fixing and subsequent preservation which prevents staining or renders the specimens opaque. Complete data were not available for all specimens; therefore numbers in the tables may not coincide with the number of specimens mentioned in the text. A total of 421 specimens was used in this study.

RESULTS

Description of the Melanophore Pattern on *T. albacares* and *T. alalunga* Types

The *T. albacares* and *T. alalunga* larval types have been distinguished from the remaining *Thunnus* types by the lack of melanophores along the dorsal and ventral body margins. We examined 89 larvae, ranging in length from 2.8 to 12.3 mm SL, that had been collected from the eastern and western Atlantic oceans. These larvae were identified as *T. albacares* - *T. alalunga* types by the lack of dorsal and ventral body pigment. *T. albacares* larvae are distinguished from *T. alalunga* larvae by the presence or absence of pigment on the lower jaw tip (see Table 1). We investigated this as well as the occurrence of melanophores on the upper jaw and in the caudal region. These last two areas were chosen because variation was observed there. The data are summarized in Table 2. Upper jaw pigment does not occur in this group until 4.5 mm SL, and most larvae larger than 5.0 mm SL have upper jaw pigment. Contrary to Matsumoto et al. (1972), lower jaw pigment is sometimes present in larvae less than 3.0 mm SL; that paper states that lower jaw pigment appears in *T. albacares* at 4.5 to 6.0 mm SL and at 9 to 10 mm SL in *T. alalunga*. One can assume then that all of our specimens that lack lower jaw pigment are *T. alalunga*, except for those less than 6.0 mm SL. These small specimens then could be either species. Using the presence of upper jaw pigment coupled with the lack of lower jaw pigment in specimens between 5.0 and 5.8 mm SL would indicate *T. alalunga*. We had 5 specimens from 4.0 to 6.0 mm SL that lacked lower jaw pigment and, of these, 3 had upper jaw pigment and 2 had none.

There appeared to be no relation to the presence or absence of melanophores occurring on the caudal fin with occurrence on the jaw tips, since most specimens had caudal melanophores. These caudal melanophores occurred on the fin rays or in the area of the hypural bones. 80% of the specimens with caudal pigment had 1 melanophore, while 20% had 2.

Table 2. Frequency of occurrence of melanophores on the tips of the upper and lower jaws and on the tail in *T. albacares* - *T. alalunga* type larvae. Data presented by size groups. Numbers do not agree because many specimens are damaged in at least 1 of the 3 areas examined

Size	Upper Jaw		Lower Jaw		Caudal Fin	
(SL in mm)	present	absent	present	absent	present	absent
> 3.0	-	2	2	-	2	-
3.0-3.49	-	3	-	3	3	-
3.5-3.99	-	14	10	4	12	-
4.0-4.49	-	9	8	1	8	1
4.5-4.99	1	8	9	1	6	-
5.0-5.49	4	2	5	1	4	-
5.5-5.99	4	-	2	2	3	-
6.0-6.49	8	-	7	1	3	1
6.5-6.99	11	1	10	2	4	5
7.0-7.49	2	2	4	1	-	-
7.5-7.99	-	-	1	-	-	-
8.0-8.49	2	-	2	-	1	-
8.5-8.99	-	-	1	-	-	-
9.0- >	1	-	1	2	-	1

Description of the Melanophore Pattern on *T. thynnus* Type Larvae

T. thynnus type larvae are characterized by melanophores occurring on both the ventral and the dorsal margins of the body (Matsumoto et al., 1972). In addition to dorsal and ventral melanophores, we noted several specimens that had melanophores occurring laterally on the body and several with distinct internal melanophores near the vertebral column. Therefore, we included in this analysis those specimens lacking dorsal melanophores but did have lateral melanophores and/or internal melanophores in addition to ventral melanophores. This was done because 1. many typical *T. thynnus* larvae had these lateral or internal melanophore 2. lateral melanophores had been previously noted on *T. thynnus* larvae (Matsumoto et al., 1972), and 3. internal melanophores have not been previously reported for any *Thunnus* larvae but were common in typical *T. thynnus* types.

Pigment patterns were examined on 83 specimens ranging in SL from 3.0 mm to 10.2 mm: 22 had dorsal, lateral, ventral, and internal melanophores; 19 had dorsal, lateral, and ventral; 19 had dorsal and ventral; 18 had dorsal, ventral, and internal; 4 had lateral and ventral; and 1 had only internal and ventral. The number and location of each melanophore was recorded for each specimen. These numbers and locations were tabulated (Tables 3 to 6) to see if a particular pattern emerged which would be useful for defining the larvae. Table 3 shows the locations of the ventral melanophores. The first occurring ventral melanophore, regardless of the number present, generally is found between the 21st and 27th myomere. The second melanophore generally occurs between the 25th and 33rd myomere, and the third melanophore generally between the 28th and 34th myomere. The fourth, fifth, and sixth melanophores were found too infrequently to be of any analytical value. From inspection of the data there is an anterior displacement of the melanophores, which increases with the number of melanophores present. Also with increasing numbers of melanophores there seems to be a posterior displacement of the last melanophore. This is shown by comparing the means of the last melanophore when two occur (32.6), the last when three occur (34.4), and the last when four occur (35.0). For all ventrally-occurring melanophores the mean myomere number was 29.0. Likewise, the dorsal melanophores were tabulated and a pattern emerged, similar to the ventral melanophore location (Table 4). The mean myomere number was 27.4.

Tabulation of the lateral melanophores is shown in Table 5. Both sides of each specimen were counted. The melanophores occurred over a wide range of myomeres and showed a pattern similar to that of the ventral chromatophores. On the left side 36 melanophores occurred and 41 occurred on the right side. The mean myomere number of those occurring on the left side was 23.3 and on the right it was 23.4.

Tabulation for the internal melanophores is shown in Table 6. These melanophores varied in their position within the myomere where they occurred. For analysis, nine designations of position were made. Some melanophores occurred immediately above and below the notochord on the vertical axis of the body and are designated A and B respectively in Table 6. Some occurred immediately to the left or right of the notochord on the horizontal body axis and these are designated OL and OR respectively. Some occurred in the muscle mass diagonally from the horizontal or vertical planes of the body axis and these are designated BL (below the notochord on the left side), BR (below the notochord on the right side), AL (above the notochord on the left side), and AR (above the notochord on the right side). A few melanophores appeared to be directly on the notochord and these are designated O. As with the other melanophores, a similar pattern was noticed with regard to myomere position. The mean myomere number of the 60 internal melanophores examined was 20.0.

The number of melanophores from each area (dorsal, ventral, lateral, and internal) were compared with one another to see if any pattern would emerge that would indicate if more than one type or possible species might exist. These comparisons are presented in Tables 7 to 11. No pattern emerged which would indicate anything but that a single species was involved.

With regard to body melanophores, no differences were noted as to the size of the specimens except for those specimens greater than 9 mm SL (in these larger specimens the larval pigment was obscured by the development of the juvenile pigmentation on the skin).

Table 3. Frequency of occurrence of the location of the ventral melanophores in *Thunnus thynnus*. Each melanophore is recorded for the myomere on which it occurs. The number of specimens examined and the mean number of the myomere are shown

	Myomeres																													
Melanophores	12	13	14	15	16	17	18	19	20	21	22	23	24	25	26	27	28	29	30	31	32	33	34	35	36	37	38	39	No.	Mean
1 melanophore present	-	-	-	-	-	-	-	-	-	-	-	1	2	-	-	2	1	2	2	-	-	-	-	-	-	-	-	-	10	27.1
2 melanophores present																														
first melanophore	-	1	-	-	-	-	1	-	-	3	-	-	1	2	2	2	2	4	3	2	-	-	-	-	-	-	-	-	23	26.0
last melanophore	-	-	-	-	-	-	-	-	-	-	-	-	-	-	-	-	2	2	3	4	1	3	-	2	3	1	2	-	23	32.6
3 melanophores present																														
first melanophore	-	-	-	-	1	-	1	-	-	2	-	2	1	4	3	2	3	1	-	-	-	-	-	-	-	-	-	-	20	24.6
second melanophore	-	-	-	-	-	-	1	-	-	-	-	-	-	-	2	2	1	3	4	5	2	-	-	-	-	-	-	-	20	28.9
last melanophore	-	-	-	-	-	-	-	-	-	-	-	1	-	-	-	-	-	1	1	1	-	1	2	1	6	5	1	-	20	34.4
4 melanophores present																														
first melanophore	-	-	-	-	-	-	-	-	1	1	-	1	2	1	1	1	-	-	-	-	-	-	-	-	-	-	-	-	8	23.8
second melanophore	-	-	-	-	-	-	-	-	-	-	-	1	-	1	-	2	2	1	1	-	-	-	-	-	-	-	-	-	8	27.1
third melanophore	-	-	-	-	-	-	-	-	-	-	-	-	-	-	-	-	-	2	1	4	-	-	1	-	-	-	-	-	8	30.8
last melanophore	-	-	-	-	-	-	-	-	-	-	-	-	-	-	-	-	-	-	-	-	2	-	-	2	3	-	1	-	8	35.0
5 melanophores present																														
first melanophore	-	-	-	-	-	-	-	-	1	1	1	-	-	-	-	-	-	-	-	-	-	-	-	-	-	-	-	-	3	21.0
second melanophore	-	-	-	-	-	-	-	-	-	-	-	1	1	-	-	-	1	-	-	-	-	-	-	-	-	-	-	-	3	25.0
third melanophore	-	-	-	-	-	-	-	-	-	-	-	-	-	-	-	1	1	-	1	-	-	-	-	-	-	-	-	-	3	28.3
fourth melanophore	-	-	-	-	-	-	-	-	-	-	-	-	-	-	-	-	-	-	-	1	2	-	-	-	-	-	-	-	3	31.7
last melanophore	-	-	-	-	-	-	-	-	-	-	-	-	-	-	-	-	-	-	-	-	-	-	-	1	1	-	1	-	3	36.3
6 melanophores present																														
first melanophore	-	1	-	-	-	-	-	-	-	-	-	-	-	-	-	-	-	-	-	-	-	-	-	-	-	-	-	-	1	12.0
second melanophore	-	-	-	-	-	-	-	-	-	-	1	-	-	-	-	-	-	-	-	-	-	-	-	-	-	-	-	-	1	22.0
third melanophore	-	-	-	-	-	-	-	-	-	-	-	-	-	-	-	-	-	1	-	-	-	-	-	-	-	-	-	-	1	29.0
fourth melanophore	-	-	-	-	-	-	-	-	-	-	-	-	-	-	-	-	-	-	-	1	-	-	-	-	-	-	-	-	1	31.0
fifth melanophore	-	-	-	-	-	-	-	-	-	-	-	-	-	-	-	-	-	-	-	-	-	1	-	-	-	-	-	-	1	33.0
last melanophore	-	-	-	-	-	-	-	-	-	-	-	-	-	-	-	-	-	-	-	-	-	-	-	-	-	-	-	1	1	39.0

Table 4. Frequency of occurrence of the dorsal melanophores in *Thunnus thynnus*. Each melanophore is recorded for the myomere on which it occurs. The number of specimens examined and the mean number of the myomere are shown

Melanophores	3	6	12	13	14	15	16	17	18	19	20	21	22	23	24	25	26	27	28	29	30	31	32	33	34	35	36	37	38	No.	Mean
1 melanophore present	1	-	1	-	-	-	-	1	-	-	-	-	1	2	-	-	1	2	4	3	4	1	2	-	-	-	-	-	-	23	25.8
2 melanophores present																															
first melanophore	1	-	-	-	-	-	-	1	-	-	-	2	2	1	1	3	1	5	1	5	2	-	1	-	-	-	-	-	-	26	25.1
last melanophore	-	-	-	-	-	-	-	-	-	1	-	-	-	-	-	-	-	-	3	1	-	2	5	4	5	2	1	2	-	26	32.1
3 melanophores present																															
first melanophore	-	1	-	-	-	1	-	1	1	1	1	3	-	1	-	-	1	-	-	-	-	-	-	-	-	-	-	-	-	11	18.8
second melanophore	-	-	1	-	-	-	-	-	-	-	-	-	-	-	-	-	1	1	2	3	3	-	-	-	-	-	-	-	-	11	27.1
last melanophore	-	-	-	-	-	-	-	-	-	-	-	-	-	-	-	-	-	-	1	-	2	-	-	3	3	2	-	-	-	11	32.6
4 melanophores present																															
first melanophoere	-	-	-	-	-	-	-	1	-	-	-	-	-	-	1	-	-	-	-	-	-	-	-	-	-	-	-	-	-	2	20.5
second melanophore	-	-	-	-	-	-	-	-	-	-	-	-	-	-	1	-	-	-	-	-	-	1	-	-	-	-	-	-	-	2	27.5
third melanophore	-	-	-	-	-	-	-	-	-	-	-	-	-	-	-	-	-	-	-	-	-	1	-	-	1	-	-	-	-	2	32.5
last melanophore	-	-	-	-	-	-	-	-	-	-	-	-	-	-	-	-	-	-	-	-	-	-	-	-	-	1	1	-	-	2	35.5

Table 5. Frequency of occurrence of lateral melanophores in *Thunnus thynnus*. Both sides of each larva were counted and each melanophore is recorded for the myomere on which it occurs and is designated R for occurring on the right side, L for occurring on the left side. The number of specimens examined and the mean number of the myomere are shown

Melanophores	Myomeres 10	11	12	13	14	15	16	17	18	19	20	21	22	23	24	25	26	27	28	29	30	31	32	33	34	35	36	37	38	Right No.	Right Mean	Left No.	Left Mean
1 melanophore present	-	-	-	-	LL	-	-	LR	R	-	R	R	L	-	R	L	-	R	-	L	RR	R	L	RR	-	L	-	-	-	11	25.8	8	23.5
2 melanophores present																																	
first melanophore	-	-	-	RR	R	R	-	L	-	L	LL	-	-	L	-	-	L	L	-	-	-	-	-	-	-	-	-	-	-	4	13.8	7	21.7
last melanophore	-	-	-	-	-	-	-	-	R	R	R	-	-	R	-	-	-	-	R	RR	LL	-	-	R	-	-	-	-	R	9	26.3	2	30.0
3 melanophores present																																	
first melanophore	-	R	L	-	-	-	-	-	-	-	-	-	R	-	-	L	-	-	-	-	-	-	-	-	-	-	-	-	-	2	16.5	2	18.5
second melanophore	-	-	-	L	-	-	-	-	-	R	-	-	-	-	-	-	-	-	-	LR	-	-	-	-	-	-	-	-	-	2	24.0	2	21.0
last melanophore	-	-	-	-	-	-	-	-	-	-	-	-	-	-	-	L	-	-	-	-	-	R	-	RL	-	-	-	-	-	2	32.0	2	29.0
4 melanophores present																																	
first melanophore	-	R	-	-	-	-	-	-	L	-	-	-	-	-	-	-	-	-	-	-	-	-	-	-	-	-	-	-	-	1	14.0	1	18.0
second melanophore	-	R	-	-	-	-	-	-	-	-	-	-	-	-	-	-	L	-	-	-	-	-	-	-	-	-	-	-	-	1	14.0	1	26.0
third melanophore	-	-	-	-	-	-	-	-	-	-	-	-	-	-	L	-	-	R	-	-	-	-	-	-	-	-	-	-	-	1	24.0	1	27.0
last melanophore	-	-	-	-	-	-	-	-	-	-	-	-	-	-	-	-	-	L	R	-	-	-	-	-	-	-	-	-	-	1	28.0	1	27.0
5 melanophores present																																	
first melanophore	L	-	R	-	-	-	-	-	-	-	-	-	-	-	-	-	-	-	-	-	-	-	-	-	-	-	-	-	-	1	12.0	1	10.0
second melanophore	-	-	-	-	-	-	-	-	-	R	R	-	-	-	-	-	-	-	-	-	-	-	-	-	-	-	-	-	-	2	19.5	0	
third melanophore	-	-	-	-	-	-	-	-	-	-	-	L	-	-	-	-	L	-	-	-	-	-	-	-	-	-	-	-	-	0		2	23.5
fourth melanophore	-	-	-	-	-	-	-	-	-	-	-	-	-	-	-	-	-	-	-	-	RL	-	-	-	-	-	-	-	-	1	30.0	1	30.0
last melanophore	-	-	-	-	-	-	-	-	-	-	-	-	-	-	-	-	-	-	-	-	-	-	-	L	L	-	-	-	-	0		2	33.5
6 melanophores present																																	
first melanophore	-	-	-	R	-	-	-	-	-	-	-	-	-	-	-	-	-	-	-	-	-	-	-	-	-	-	-	-	-	1	13.0	0	
second melanophore	-	-	-	-	L	-	-	-	-	-	-	-	-	-	-	-	-	-	-	-	-	-	-	-	-	-	-	-	-	0		1	14.0
third melanophore	-	-	-	-	-	-	-	-	L	-	-	-	-	-	-	-	-	-	-	-	-	-	-	-	-	-	-	-	-	0		1	18.0
fourth melanophore	-	-	-	-	-	-	-	-	-	-	-	-	L	-	-	-	-	-	-	-	-	-	-	-	-	-	-	-	-	0		1	22.0
fifth melanophore	-	-	-	-	-	-	-	-	-	-	-	-	-	-	-	-	-	-	-	R	-	-	-	-	-	-	-	-	-	1	29.0	0	
sixth melanophore	-	-	-	-	-	-	-	-	-	-	-	-	-	-	-	-	-	-	-	-	-	R	-	-	-	-	-	-	-	1	31.0	0	

Table 6. Frequency of occurrence of internal melanophores in *Thunnus thynnus*. Each melanophore is recorded for the myomere on which it occurs. The number of specimens examined and the mean number of the myomere are shown. See the text for explanation of the location of the letters A, B, OR, OL, BL, BR, AL, AR, and O

Melanophores	2	7	9	10	11	12	13	14	15	16	17	18	19	20	21	22	23	24	25	26	27	28	29	30	31	32	33	34	No.	Mean
1 melanophore present	-	-	-	-	-	-	-	A	B	-	-	B	AL	OL	OL	-	OL	-	A	A	-	-	-	-	BR	AR	OR	-	18	22.4
	-	-	-	-	-	-	-	-	B	-	-	-	AL	-	BL	-	-	-	-	?	-	-	-	-	A	-	-	-		
	-	-	-	-	-	-	-	-	B	-	-	-	-	-	-	-	-	-	-	-	-	-	-	-	-	-	-	-		
2 melanophores present																														
first melanophore	-	OL	OL	-	B	-	BL	OL	B	BR	-	-	-	-	BR	OR	-	-	A	-	B	-	-	-	-	-	-	-	15	16.6
	-	-	-	-	-	-	-	OR	-	B	-	-	-	-	-	-	-	-	B	-	-	-	-	-	-	-	-	-		
	-	-	-	-	-	-	-	B	-	-	-	-	-	-	-	-	-	-	-	-	-	-	-	-	-	-	-	-		
last melanophore	-	-	-	-	-	OL	-	B	-	B	BL	-	AL	-	-	B	-	A	-	-	-	B	A	O	OL	-	-	-	15	22.0
	-	-	-	-	-	A	-	-	-	OL	-	-	-	-	-	-	-	-	-	-	-	-	OR	-	AR	-	-	-		
4 melanophores present																														
first melanophore	A	-	-	-	-	BR	-	-	-	-	-	-	-	-	-	-	-	-	-	-	-	-	-	-	-	-	-	-	3	8.7
	-	-	-	-	-	BL	-	-	-	-	-	-	-	-	-	-	-	-	-	-	-	-	-	-	-	-	-	-		
second melanophore	-	-	-	-	-	-	A	AR	-	-	-	-	-	-	-	-	-	-	-	-	-	-	-	-	-	-	-	-	3	13.3
	-	-	-	-	-	-	OL	-	-	-	-	-	-	-	-	-	-	-	-	-	-	-	-	-	-	-	-	-		
third melanophore	-	-	-	-	-	-	-	B	-	-	-	-	-	-	OR	-	-	-	-	-	-	-	-	-	-	B	-	-	3	22.3
last melanophore	-	-	-	-	-	-	-	-	-	O	-	-	-	-	-	-	-	-	-	-	-	-	-	-	-	-	-	OR	3	28.0
	-	-	-	-	-	-	-	-	-	-	-	-	-	-	-	-	-	-	-	-	-	-	-	-	-	-	-	B		

Table 7. Comparison of the number of dorsal and ventral melanophores occurring in *Thunnus thynnus* specimens

Number of Ventral Melanophores	Number of Dorsal Melanophores 0	1	2	3	4
1	1	5	1	3	-
2	1	12	11	1	1
3	3	5	8	5	-
4	-	3	4	2	1
5	-	2	1	-	-
6	-	-	-	1	1

Table 8. Comparison of the number of lateral and internal melanophores occurring in *Thunnus thynnus* specimens

Number of Internal Melanophores	Number of Lateral Melanophores 0	1	2	3	4	5	6
0	11	9	3	1	-	1	-
1	6	4	3	2	1	1	1
2	5	4	4	1	-	-	-
3	-	-	-	-	-	-	-
4	2	1	-	-	-	-	-

Table 9. Comparison of the dorsal/ventral melanophore combinations and the number of lateral melanophores in *Thunnus thynnus* specimens

Number of Lateral Melanophores	Dorsal/Ventral Melanophore Combinations																
	0/1	1/1	2/1	3/1	1/2	2/2	3/2	4/2	0/3	1/3	2/3	3/3	1/4	2/4	1/5	2/5	3/6
0	-	5	-	-	5	2	-	-	1	-	5	-	1	3	1	1	1
1	1	-	1	1	2	3	1	-	2	2	-	3	1	1	-	-	-
2	-	-	-	1	3	4	-	-	-	1	1	1	-	-	-	-	-
3	-	-	-	-	1	1	-	1	-	-	-	-	-	-	1	-	-
4	-	-	-	-	-	1	-	-	-	-	-	-	-	1	-	-	-
5	-	-	-	1	-	-	-	-	-	1	-	-	-	-	-	-	-
6	-	-	-	-	-	-	-	-	-	-	-	1	-	-	-	-	-

Table 10. Comparison of the dorsal/ventral melanophore combinations and the number of internal melanophores in *Thunnus thynnus* specimens

Number of Internal Melanophores	Dorsal/Ventral Melanophore Combinations																	
	0/1	1/1	2/1	3/1	1/2	2/2	3/2	4/2	0/3	1/3	2/3	3/3	1/4	2/4	3/4	1/5	2/5	3/6
0	1	3	-	2	5	4	1	-	2	1	4	-	1	1	-	1	-	-
1	-	-	-	1	2	4	-	-	1	1	1	3	2	1	1	1	-	-
2	-	-	1	-	2	3	-	1	-	2	1	2	-	2	-	-	-	1
3	-	-	-	-	-	-	-	-	-	-	-	-	-	-	-	-	-	-
4	-	1	-	-	-	-	-	-	-	1	-	-	-	-	-	-	1	-

Table 11. Comparison of the dorsal/ventral melanophore combinations with the lateral/internal melanophore combinations in *Thunnus thynnus* specimens

Lateral/Internal Melanophore Combinations	Dorsal/Ventral Melanophore Combinations																	
	0/1	1/1	2/1	3/1	1/2	2/2	3/2	4/2	0/3	1/3	2/3	3/3	1/4	2/4	3/4	1/5	2/5	3/6
0/0	-	3	-	-	3	1	-	-	-	-	2	-	-	1	-	1	-	-
1/0	1	-	-	-	1	1	1	-	2	1	-	-	1	-	-	-	-	-
2/0	-	-	-	1	-	2	-	-	-	-	-	-	-	-	-	-	-	-
3/0	-	-	-	-	1	-	-	-	-	-	-	-	-	-	-	-	-	-
5/0	-	-	-	1	-	-	-	-	-	-	-	-	-	-	-	-	-	-
0/1	-	-	-	-	-	-	-	-	1	-	1	-	2	1	1	-	-	-
1/1	-	-	-	1	-	1	-	-	-	-	-	2	-	-	-	-	-	-
2/1	-	-	-	-	2	1	-	-	-	-	-	-	-	-	-	-	-	-
3/1	-	-	-	-	-	1	-	-	-	-	-	-	-	-	-	1	-	-
4/1	-	-	-	-	-	1	-	-	-	-	-	-	-	-	-	-	-	-
5/1	-	-	-	-	-	-	-	-	-	1	-	-	-	-	-	-	-	-
6/1	-	-	-	-	-	-	-	-	-	-	-	1	-	-	-	-	-	-
3/2	-	-	-	-	-	-	-	1	-	-	-	-	-	-	-	-	-	-
0/2	-	-	-	-	2	1	-	-	-	1	-	-	-	-	-	-	-	1
0/4	-	1	-	-	-	-	-	-	-	-	-	-	-	-	-	-	1	-
1/2	-	-	1	-	-	1	-	-	-	-	-	1	-	-	-	-	-	-
1/4	-	-	-	-	-	-	-	-	-	1	-	-	-	-	-	-	-	-
2/2	-	-	-	-	-	1	-	-	-	1	1	1	-	-	-	-	-	-

The presence or absence of melanophores on the tips of the jaws and on the caudal fin is shown in Table 12. All but one specimen had melanophores on the lower jaw. Upper jaw melanophores were acquired at about 5 mm SL and were consistently present on specimens longer than 6.0 mm SL. Caudal pigment occurred on only 14 of the 61 specimens examined for this feature.

Table 12. Frequency of occurrence of melanophores on the tips of the upper and lower jaw and on the tail in *Thunnus thynnus* larvae. Data presented by size groups. Numbers do not agree because many specimens are damaged in at least 1 of the 3 areas examined

Size	Upper Jaw		Lower Jaw		Caudal Fin	
(SL in mm)	present	absent	present	absent	present	absent
3.0-3.49	-	2	1	1	2	-
3.5-3.99	-	2	2	-	1	1
4.0-4.49	-	8	8	-	4	4
4.5-4.99	1	12	13	-	2	8
5.0-5.49	4	12	17	-	1	14
5.5-5.99	9	6	15	-	4	8
6.0-6.49	5	-	5	-	-	4
6.5-6.99	no specimens					
7.0-7.49	1	-	1	-	-	1
7.5-7.99	2	-	2	-	-	2
8.0-8.49	no specimens					
8.5-8.99	1	-	1	-	-	1
9.0- <	2	-	2	-	-	2

Description of the Melanophore Pattern on *T. obesus* Type Larvae

T. obesus type larvae are characterized by melanophores on the ventral margin of the body and lack of melanophores on the dorsal margin (Matsumoto et al., 1972). As noted earlier, it is this type of larvae that Richards suspected as including *T. atlanticus*. Comparisons of melanophores were made between specimens occurring in the eastern Atlantic Ocean, which is outside the known range of *T. atlanticus*, with specimens from the western Atlantic Ocean, which could include both *T. obesus* and *T. atlanticus*. Tables 13 and 14 depict the distribution of ventral melanophores for each group. Although the sample size from the western Atlantic is twice as large, the distributions show only slight differences. Differences were noted in the number of melanophores present between the two areas. Of the 162 specimens studied from the western Atlantic, 79 had one melanophore, 55 had two melanophores, 22 had three melanophores, 4 had four melanophores, and 2 had

Table 13. Frequency of occurrence of the location of ventral melanophores in *Thunnus obesus* type larvae from the western Atlantic Ocean. Each melanophore is recorded for the myomere on which it occurs. The number of specimens examined and the mean number of the myomere are shown

Melanophore	Myomere 14	15	16	17	18	19	20	21	22	23	24	25	26	27	28	29	30	31	32	33	34	35	36	37	38	No.	Mean
1 melanophore present	1	-	-	-	-	-	1	6	2	1	2	6	11	15	12	10	7	1	4	-	-	-	-	-	-	79	26.7
2 melanophores present																											
first melanophore	-	1	1	2	3	-	2	1	3	7	3	7	9	5	3	4	2	1	1	-	-	-	-	-	-	55	22.9
last melanophore	-	-	-	-	-	-	-	-	-	-	-	1	3	7	4	9	9	4	1	5	5	1	3	2	1	55	30.6
3 melanophores present																											
first melanophore	1	-	-	2	2	-	3	4	3	1	2	2	-	1	1	-	-	-	-	-	-	-	-	-	-	22	21.4
second melanophore	-	-	-	-	1	-	-	-	1	1	2	4	4	3	1	3	-	1	-	1	-	-	-	-	-	22	26.1
last melanophore	-	-	-	-	-	-	-	-	-	-	-	-	1	5	2	2	5	2	2	-	1	1	1	-	-	22	29.8
4 melanophores present																											
first melanophore	-	-	-	-	-	-	-	-	-	1	-	-	1	-	-	-	-	-	-	-	-	-	-	-	-	2	24.5
second melanophore	-	-	-	-	-	-	-	-	-	-	-	-	-	-	1	1	-	-	-	-	-	-	-	-	-	2	28.5
third melanophore	-	-	-	-	-	-	-	-	-	-	-	-	-	-	-	-	-	-	1	1	-	-	-	-	-	2	32.5
last melanophore	-	-	-	-	-	-	-	-	-	-	-	-	-	-	-	-	-	-	-	1	-	1	-	-	-	2	34.0
6 melanophores present																											
first melanophore	-	1	-	-	-	-	-	-	-	-	-	-	-	-	-	-	-	-	-	-	-	-	-	-	-	1	15.0
second melanophore	-	-	1	-	-	-	-	-	-	-	-	-	-	-	-	-	-	-	-	-	-	-	-	-	-	1	16.0
third melanophore	-	-	-	-	-	-	-	1	-	-	-	-	-	-	-	-	-	-	-	-	-	-	-	-	-	1	21.0
fourth melanophore	-	-	-	-	-	-	-	-	-	-	-	-	-	1	-	-	-	-	-	-	-	-	-	-	-	1	27.0
fifth melanophore	-	-	-	-	-	-	-	-	-	-	-	-	-	-	-	-	-	1	-	-	-	-	-	-	-	1	31.0
last melanophore	-	-	-	-	-	-	-	-	-	-	-	-	-	-	-	-	-	-	-	-	-	1	-	-	-	1	35.0

Table 14. Frequency of occurrence of the location of ventral melanophores in *Thunnus obesus* type larvae from the eastern Atlantic Ocean. Each melanophore is recorded for the myomere on which it occurs. The number of specimens examined and the mean number of the myomere are shown

Melanophores	Myomere 17	18	19	20	21	22	23	24	25	26	27	28	29	30	31	32	33	34	35	36	37	No.	Mean
1 melanophore present	1	-	-	5	4	7	7	4	8	5	2	3	2	6	4	4	2	2	-	1	1	69	27.8
2 melanophores present																							
first melanophore	-	-	-	-	2	-	-	-	-	3	3	2	2	2	1	-	-	-	-	-	-	15	27.1
last melanophore	-	-	-	-	-	-	-	-	-	-	2	1	1	-	-	2	5	-	1	2	1	15	32.3
3 melanophores present																							
first melanophore	-	-	-	-	-	-	-	-	-	1	-	-	-	-	-	-	-	-	-	-	-	1	26.0
second melanophore	-	-	-	-	-	-	-	-	-	-	-	-	1	-	-	-	-	-	-	-	-	1	29.0
last melanophore	-	-	-	-	-	-	-	-	-	-	-	-	-	-	1	-	-	-	-	-	-	1	31.0

six melanophores; whereas of the 85 eastern Atlantic specimens, 69 had one melanophore, 15 had two melanophores, 1 had three melanophores, while no specimens were found with more than three melanophores. In the event that these melanophores migrate as the fish grows, we also compared the position of the ventral melanophore, in those cases when only one occurs, to see if the position changed with size of the fish. This comparison was done with both eastern and western Atlantic materia and we found no indication that these melanophores changed position with age.

We also studied melanophores that occur on the tips of the jaws and on the tail. Table 15 compares jaw melanophores for the eastern and the western Atlantic specimens. Upper jaw pigment is acquired between about 5 and 6.5 mm SL on specimens from both regions. The majority of specimens from both areas had lower jaw pigment by about 4 mm SL. Eastern Atlantic specimens all had this pigment at 5 mm SL, but some of the western Atlantic specimens had not acquired it until about 7.5 mm SL.

Table 15. Frequency of occurrence of melanophores on the tips of the upper jaw and lower jaw for the *T. obesus* type larvae from the eastern and western Atlantic Oceans. Data presented by size groups. Numbers do not agree because many specimens are damaged in at least 1 of the areas examined

Size	Eastern Atlantic				Western Atlantic			
	Upper Jaw		Lower Jaw		Upper Jaw		Lower Jaw	
(SL in mm)	present	absent	present	absent	present	absent	present	absent
< 3.0	-	2	-	2	-	-	-	-
3.0-3.49	-	7	1	6	-	-	-	-
3.5-3.99	-	18	16	2	-	3	1	2
4.0-4.49	-	20	18	2	-	23	21	2
4.5-4.99	-	12	11	1	1	37	29	9
5.0-5.49	3	2	5	-	8	18	25	1
5.5-5.99	8	2	10	-	10	7	14	3
6.0-6.49	8	1	9	-	19	3	16	6
6.5-6.99	1	-	1	-	14	-	12	2
7.0-7.49	1	-	1	-	7	-	6	1
7.5-7.99	-	-	-	-	6	-	6	-
8.0-8.49	-	-	-	-	2	-	2	-
8.5-8.99	-	-	-	-	3	-	3	-
9.0-9.49	-	-	-	-	2	-	1	1

Caudal fin melanophores were seen in about two-thirds of both the eastern and the western Atlantic specimens. They consisted generally of one, rarely two, melanophores in the area of the forming hypural plate and principal caudal rays. The remaining specimens lacked caudal melanophores.

Comment on the Diagnostic Value of the Red Pigment Patterns on *Thunnus* Type Larvae

Our data on the occurrence and distribution of erythrophores on *Thunnus* larvae was included in Matsumoto et al. (1972). It suffices to say that day-caught larvae (which most of our specimens were) are very difficult to analyze because of the expanded nature of the erythrophores. Suitable fixatives for erythrophores need to be developed to allow sufficient time for detailed studies. We have used some antioxidants for this, but have had little success.

Identification of Cleared and Stained Specimens

Upon completion of the pigment analyses, representative specimens were selected, cleared and stained, and examined for diagnostic osteological studies. 27 specimens of the *T. albacares* - *T. alalunga* larval types, 12 *T. thynnus* larval types, 2 eastern Atlantic Ocean *T. obesus* larval types, and 63 western Atlantic Ocean *T. obesus* larval types were prepared and identified from their oesteological characters. An additional 60 specimens were cleared and stained, but we were unable to identify them because of lack of development or poor preservation. The osteological characters we used were principally the number of trunk and caudal vertebrae, the position of the first closed haemal arch, the pattern of dorsal fin and anal fin pterygiophores, and the number of gill-rakers on the first ceratobranchial gill arch. These characters were found to be diagnostic for juvenile *Thunnus* (Potthoff and Richards, 1970; Matsumoto et al., 1972; Potthoff, 1974). The only exception is that *T. albacares* cannot be separated from *T. obesus*, but together they are separable from the others. These characters are summarized in Tables 16 and 17.

The 27 *T. albacares* - *T. alalunga* larval types, were identified from osteological characters as follows: *T. atlanticus* 20 (5.8 to 12.3 mm SL); *T. albacares/T. obesus* 5 (6.1 to 12.0 mm SL); *T. alalunga* 1 (7.2 mm SL); and 1 specimen (6.5 mm SL), which could be *T. alalunga* or *T. albacares/ T. obesus*. The one specimen (7.2 mm SL) that was identified as *T. alalunga* from the vertebral count and pterygiophore pattern was not positively considered *T. alalunga* based on pigmentation because of a slight indication of a melanophore on the lower jaw. Two specimens (6.9 and 7.1 mm SL), which perfectly fit the pigmentation pattern of *T. alalunga*, proved to be *T. atlanticus* based on vertebral counts. It must be remembered, though, that Potthoff (1974) in his examination of 118 juvenile *T. alalunga* found 4 specimens with the same vertebral counts as *T. atlanticus* (Table 16). Therefore, the possibility is not ruled out that these are aberrant *T. alalunga*. Based on pigmentation, the remaining 24 specimens were all considered to be *T. albacares*, except for the 6.5 mm SL specimen which was considered to be *T. alalunga* based on pigment, and *T. alalunga* or *T. albacares/T. obesus* based on osteology.

Of the 12 *T. thynnus* larval types that were cleared and stained, two were large larvae (9.6 and 10.1 mm SL) and were not included in the pigmentation analysis because juvenile pigment had obscured much of

Table 16. Precaudal and caudal arrangement of the vertebrae and total vertebral number in juvenile *Thunnus*. (After Potthoff, 1974)

	% total variability from mode	Precaudal, caudal, and total number of vertebrae							
		16+22=38	18+20=38	19+19=38	17+22=39	18+21=39	19+20=39	18+22=40	19+21=40
T. thynnus	5.1	1	2	-	1	149	2	1	1
T. alalunga	3.4	-	-	-	-	114	4	-	-
T. atlanticus	1.9	-	-	1	-	1	105	-	-
T. albacares/obesus complex	14.6	-	-	-	5	35	-	-	1

the larval pigment. These two specimens were identified as *T. atlanticus* from their osteological characters. We had identified them as *T. thynnus* because of what appeared to be remnants of dorsal melanophores. An 8.7 mm SL specimen, which had lateral and ventral pigment and thus considered to be *T. thynnus*, was identified as a *T. atlanticus* from its osteological characters (19 + 20 vertebrae, first haemal arch on 11th vertebrae, and number of gillrakers). The remaining 9 specimens had typical *T. thynnus* pigment and were identified as *T. thynnus* from their osteological characters (18 + 21 vertebrae and haemal arch on 10th vertebrae).

Each of the 2 eastern Atlantic *T. obesus* type larvae (6.4 and 7.2 mm SL) had 21 caudal vertebrae. One can only assume, then, that they are not *T. atlanticus*.

60 of the 63 western Atlantic *T. obesus* types were identified as *T. atlanticus*. One specimen (7.5 mm SL) had two distinct ventral melanophores and was identified as *T. alalunga* from its vertebral count, location of first closed haemal arch, and pterygiophore pattern. Each of the remaining 2 specimens (6.6 and 7.4 mm SL) had 21 caudal vertebrae and therefore probably are not *T. atlanticus*. Pigment characters of all authenticated *T. atlanticus* larvae are given in Table 18.

CONCLUSIONS

Our results indicate to us, at least for the larvae of *Thunnus* from the Atlantic, that pigmentation characteristics are not necessarily reliable indicators of specific differences. Using pigmentation characters, most of the larvae of *T. atlanticus* resemble the descriptions of *T. obesus* and consequently are inseparable from *T. obesus*. Additionally, our evidence indicates that some *T. atlanticus* larvae lack ventral pigment and therefore can be confused with either *T. albacares* or *T. alalunga*. Also, some *T. albacares* and *T. alalunga* larvae may possess ventral melanophores. Based on pigment characters, we had provisionally characterized as *T. thynnus* larvae which lacked dorsal melanophores but which possessed lateral or internal and ventral melanophores. Our data are limited, in this case, because we were able only to verify osteologically that one of these specimens was *T. atlanticus*. The other 4 specimens of this type were too small or too poorly preserved for us to obtain reliable results. However, all authenticated larvae of *T. thynnus* had both dorsal and ventral melanophores and usually lateral and internal melanophores as well.

To identify *Thunnus* larvae from osteological features is a time-consuming task that is limited by specimen size and may be limited by preservation. We had only four specimens (5.8 - 5.9 mm SL) less than 6.0 mm SL with sufficiently developed osteological features by which we could verify the number of caudal vertebrae. For accurate identifications, however, there is no other alternative than to follow this process.

In light of our results, we believe that the status of the larvae of *T. tonggol* of the Indo-Pacific area should be defined because there is a possibility that it may be similar to or indistinguishable from other species of *Thunnus*, as is the case for *T. atlanticus*. It also would be desirable to make such detailed studies on the two species of *Thunnus (alalunga* and *thynnus)* that spawn in the Mediterranean Sea. In the southern oceans that status of *T. maccoyi* larvae should also be studied. A continued effort on authenticating larval types through osteological methods is called for.

Table 17. Comparison of diagnostic characters for the juvenile *Thunnus* species. Parentheses indicate rare occurrence. (After Potthoff, 1974)

Character	T. thynnus	No.	%	T. alalunga
Number of vertebrae, precaudal and caudal	18 + 21 = 39	149	95	18 + 21 = 39
	16-19 + 20-22 = 38-40	8	5	19 + 20 = 39
First haemal arch on vertebra number	10	137	88	10
	11, (9)	19	12	(9)
Pattern of single second dorsal fin pterygiophores for interneural spaces[1]	1,1 — — — — — 1,1	140	95	1,1,1, — — — — — 1
	1,1 — — — — — — 1	8	5	
Pattern of single anal fin pterygiophores for interneural spaces	— — — — — 1,1	116	84	— — — — — 1,1
	— — 1 — — 1,1 } — — — — 1,1,1 }	23	16	— — 1 — — 1,1
Gillraker number over ceratobranchial bone	17,18,(19,20)	102		14,15,(16)

[1]The pattern of 2nd dorsal fin pterygiophores for interneural spaces differs from that expressed in Matsumoto et al. (1972) because a slightly different method of counting was used. For a description of the method consult Potthoff (1974). He positions the 2nd dorsal fin 1 interneural space anterior to that expressed by Matsumoto et al. (1972).

Table 17. (continued)

No.	%	T. atlanticus	No.	%	Thunnus spp.	No.	%
114	97	19 + 20 = 39	105	98	18 + 21 = 39	35	85
4	3	18, 19 + 19,21 = 38,39	2	2	17, 19 + 21,22 = 39,40	6	15
115	99	11	101	94	11	38	93
1	1	12, (10)	6	6	10,12	3	7
116	100	1,1,1,——————1	42	46	1,1 —————1,1	40	98
		1,1 ———————1	32	35	1,1,1,—————1	1	2
		1,1 ——————1,1	18	19			
74	66	————— 1	55	61	————1,1,1	31	78
38	34	————1,1	34	39	—————1,1	9	22
81		11,12,(13)	92		14,15,16	27	

Table 18. Summary of pigment types of *T. atlanticus* larvae based on specimens authenticated by osteological methods. The number of specimens and size range for each body pigment type are given, and the number of specimens are given for presence or absence of jaw and caudal pigment. Damaged specimens account for those where no data are given

Body pigment type	No.	Size range (mm SL)	Upper jaw pigment			Lower jaw pigment			Caudal pigment		
			Present	Absent	No data	Present	Absent	No data	Present	Absent	No data
Ventral pigment only:											
melanophore number unknown	4	6.0-11.0	-	-	4	-	-	4	-	-	4
1 melanophore present	29	5.9-11.8	8	-	21	19	3	7	6	2	21
2 melanophores present	15	6.4-12.1	3	1	11	8	1	6	3	1	11
3 melanophores present	9	5.9- 7.9	-	-	9	6	3	-	-	-	9
4 melanophores present	2	7.5- 8.5	-	-	2	1	-	1	-	-	2
5 melanophores present	1	6.0	-	-	1	-	-	1	-	-	1
No ventral pigment	20	5.8-12.3	6	-	14	9	2	9	2	2	16
Ventral and dorsal pigment	2	9.6-10.1	2	-	-	2	-	-	2	-	-
Ventral and lateral pigment	1	8.7	1	-	-	1	-	-	-	-	1

SUMMARY

The specific identification of larval scombrids has been based primarily on the distribution of melanophores on various parts of the body. Examination of these pigment distributions has revealed a great amount of variability as to the exact location and number of melanophores on larvae in the genus *Thunnus* from the Atlantic Ocean. Specimens from which melanophore data were obtained were cleared and stained for study of their osteology and for accurate identification. Results have shown that melanophore distributions are unreliable characters for specific identification. This is particularly evident where western Atlantic larvae identified as *T. albacares*, *T. alalunga*, or *T. obesus*, using the traditional melanophore character method, were found to be larvae of *T. atlanticus* based on osteological features.

ACKNOWLEDGEMENTS

We wish to express our appreciation to E.H. Ahlstrom and W.M. Matsumoto for kindly reviewing this manuscript.

REFERENCES

Collette, B.B. and Gibbs, R.H., Jr., 1963. A preliminary review of the fishes of the family Scombridae. FAO Fish. Rep., (6) 1, 23-32.
Fraser-Brunner, A., 1950. The fishes of the family Scombridae. Ann. Mag. Nat. Hist., 12 (3), 131-163.
Gibbs, R.H., Jr. and Collette, B.B., 1967. Comparative anatomy and systematics of the tunas, genus *Thunnus*. U.S. Fish Wildl. Serv. Fish. Bull., 66 (1), 65-130.
Harada, T., Mizuno, K., Murata, O., Miyashita, S. and Hurutani, H., 1971. On the artificial fertilization and rearing larvae in yellowfin tuna. Mem. Fac. Agric. Kinki Univ., (4), 145-151.
Juárez, M., 1972. Las formas larvarias del *Thunnus atlanticus*. Mar y Pesca (Inst. Nac. Pesc., Cuba), (78), 26-29.
Matsui, T., 1967. Review of the mackerel genera *Scomber* and *Rastrelliger*, with description of a new species of *Rastrelliger*. Copeia, (1), 71-83.
Matsumoto, W.M., 1962. Identification of larvae of four species of tuna from the Indo-Pacific region I. Dana Rep., (55), 16 pp.
Matsumoto, W.M., Ahlstrom, E.H., Jones, S., Klawe, W.L., Richards, W.J. and Ueyanagi, S., 1972. On the clarification of larval tuna identification particularly in the genus *Thunnus*. Fish. Bull., 70 (1), 1-12.
Mori, K., Ueyanagi, S. and Nishikawa, Y., 1971. The development of artificially fertilized and reared larvae of the yellowfin tuna, *Thunnus albacares*. Bull. Far Seas Fish. Res. Lab., (5), 219-232.
Potthoff, T., 1974. Osteological development and variation in young tunas, genus *Thunnus* (Pisces, Scombridae), from the Atlantic Ocean. Fish. Bull., 72 (2).
Potthoff, T. and Richards, W.J., 1970. Juvenile bluefin tuna, *Thunnus thynnus* (Linnaeus), and other scombrids taken by terns in the Dry Tortugas, Florida. Bull. Mar. Sci., 20 (2), 389-413.
Richards, W.J. and Klawe, W.L., 1972. Indexed bibliography of the eggs and young of tunas and other scombrids (Pisces, Scombridae), 1880-1970. NOAA Tech. Rep. NMFS SSRF, (652), 107 pp.

Taylor, W.R., 1967. An enzyme method of clearing and staining small vertebrates. Proc. U.S. Nat. Mus., 122 (3596), 17 pp.
Wise, J.P. and Davis, C.W., 1973. Seasonal distribution of tunas and billfishes in the Atlantic. NOAA Tech. Rep. NMFS SSRF, (662), 24 pp.
Yabe, H. and Ueyanagi, S., 1962. Contribution to the study of the early life history of the tunas. Occ. Rep. Nankai Reg. Fish. Res. Lab., (1), 57-72.

W.J. Richards and T. Potthoff
National Marine Fisheries Service
Southeast Fisheries Center
Miami Laboratory
75 Virginia Beach Drive
Miami, Fla. 33149 / USA

Present State of Billfish Larval Taxonomy

S. Ueyanagi

INTRODUCTION

Billfishes are important both in the sport and tuna longline fisheries. Despite some research on the early life history many problems still remain to be examined. This report summarizes recent research on the identification of the larvae.

Of the two families the Xiphiidae is monotypic while the Istiophoridae consists of 3 genera and 11 species (Nakamura et al., 1968; Robins, in press). All species are listed in Table 1.

Of particular interest in billfish taxonomy is that billfish resemble tuna. The adults are large, surface-swimming fish whose habits are very similar to the tuna's; however, their distribution is markedly different. Most of the tuna species show "cosmopolitan" distribution, whereas among the billfishes only the swordfish can make this claim. Cognate species of the istiophorids are found in the Atlantic and the Indo-Pacific Oceans. Furthermore, while Indo-Pacific black marlin also occur in the Atlantic, it is believed that they have strayed there from the Indian Ocean (Ueyanagi et al., 1970).

IDENTIFICATION OF INDO-PACIFIC BILLFISH LARVAE

Identification of six Indo-Pacific billfishes including swordfish is based on characteristics in the largest of a developmental series of larvae. The identification of the larvae of Atlantic species has not progressed as far (Table 2).

Meristic factors such as fin-ray counts and vertebral formula are not particularly useful for distinguishing the various istiophorid species from one another (Richards, in press). Probably the most useful factor is head morphology (Ueyanagi, 1963a). The snout is short in all istiophorid larvae under about 5 mm in body length, but in larger specimens the snout lengthens greatly in *Istiophorus* and *Tetrapturus*. At lengths greater than about 12 mm the elongated snouts of *Istiophorus* and *Tetrapturus* readily distinguish them from the shorter-snouted *Makaira*.

Fig. 1 shows the ratio of snout length to orbit diameter in relation to standard length for 5 Indo-Pacific species. Sailfish larvae, with long snout and small eyes, differ most from the blue and black marlin larvae with their short snouts and relatively large eyes. The larvae of shortbill spearfish and striped marlin fall between these two extremes. The variation in growth of snout relative to growth of the larvae is greatest in shortbill spearfish. In larvae smaller than about 7 mm, the ratio of snout length to diameter of orbit differs very little among blue marlin, black marlin, striped marlin, and shortbill spearfish; identification by this method is thus difficult.

Table 1. Billfish species, their English common names and distribution

Species	Distribution	English name
Xiphias gladius Linnaeus, 1758	Cosmopolitan	Swordfish
Istiophorus platypterus (Shaw and Nodder, 1792)	Indo-Pacific Ocean	Pacific sailfish
Istiophorus albicans (Latreille, 1804)	Atlantic Ocean	Atlantic sailfish
Tetrapturus angustirostris Tanaka, 1914	Indo-Pacific Ocean	Shortbill spearfish
Tetrapturus belone Rafinesque, 1810	Mediterranean Sea	Mediterranean spearfish
Tetrapturus georgei (Lowe, 1840)	Mediterranean Sea and northeastern Atlantic Ocean	Roundscale spearfish
Tetrapturus pfluegeri Robins and de Sylva, 1963	Atlantic Ocean	Longbill spearfish
Tetrapturus albidus Poey, 1860	Atlantic Ocean	White marlin
Tetrapturus audax (Philippi, 1887)	Indo-Pacific Ocean	Striped marlin
Makaira mazara (Jordan and Snyder, 1901)	Indo-Pacific Ocean	Blue marlin
Makaira nigricans Lacépède, 1803	Atlantic Ocean	Atlantic blue marlin
Makaira indica (Cuvier, 1831)	Indo-Pacific Ocean, possibly Atlantic Ocean	Black marlin

Table 2. List of the species and stages of billfishes which have been described to date

Species	Occurrence	Post-Larvae	Juveniles
Xiphias gladius	I, P, A, M	X	X
Istiophorus platypterus	I and P	X	X
Istiophorus albicans	A	X	X
Tetrapturus angustirostris	I and P	X	X
Tetrapturus audax	I and P	X	X
Makaira indica	I and P	X	-
Makaira mazara	I and P	X	X
Makaira nigricans	A	X	X
Tetrapturus albidus	A	-	X
Tetrapturus pfluegeri	A	-	X
Tetrapturus belone	M	-	X
Tetrapturus georgei	A	-	-

I = Indian Ocean; P = Pacific Ocean; A = Atlantic Ocean; M = Mediterranean Sea

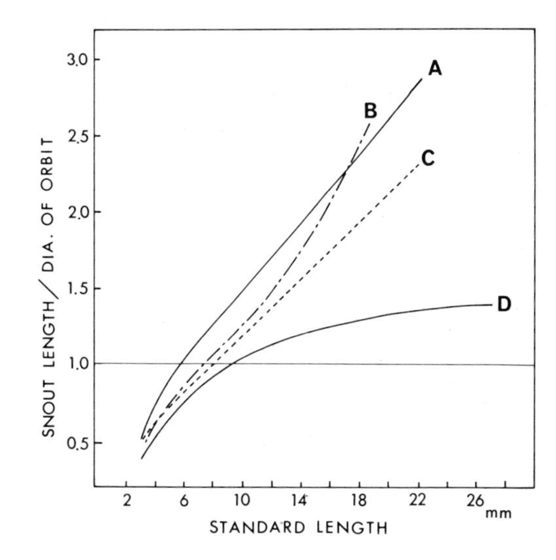

Fig. 1. Ratio of snout length to diameter of orbit in relation to standard length in 5 species of Indo-Pacific billfishes. Curves identified by A, B, C, and D represent sailfish, shortbill spearfish, striped marlin, and blue and black marlins, respectively. (Adapted from Ueyanagi, 1963a)

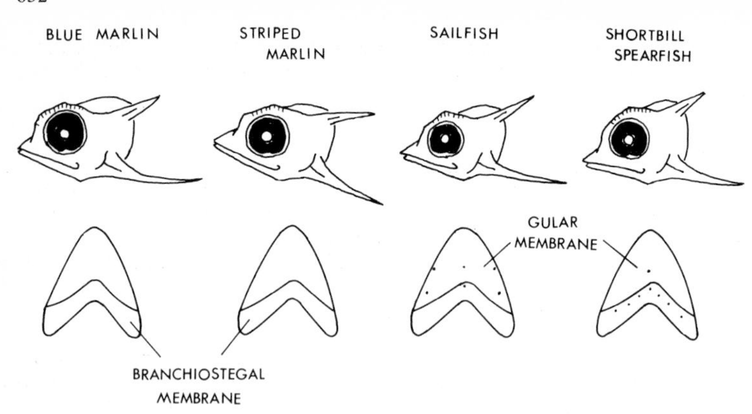

Fig. 2. The head profile and the pigmentation pattern on the lower side of the head of larvae (around 6 mm in standard length) of 4 species of Indo-Pacific billfishes

As seen in Fig. 2, however, the head profile and pigmentation on the lower part of the head are specific characteristics that make identification of the various istiophorid larvae possible (Ueyanagi, 1963a, in press). In the figure, the pigment which appears on the tip of the lower jaw in all larvae has been omitted. Because of the lack of a sufficient number of specimens, the black marlin is not included here.

HEAD PROFILE AND PIGMENTATION PATTERN ON THE LOWER SIDE OF THE HEAD

Blue Marlin. The larvae of blue marlin are characterized by a relatively short snout, large eyes, and forward placement of the anterior edge of orbit; the center of eye is located above the level of the tip of snout. The pterotic spine rises obliquely from its base and the preopercular spine runs nearly parallel to the ventral profile of the body. There is no pigment on the lower part of the head with the exception of the tip of the lower jaw. In rare cases pigment has been found to occur just above the midline of the gular membrane.

Striped Marlin. In the striped marlin the anterior edge of the orbit does not project forward as in the blue marlin; the center of the eye is located at about the same level as the tip of snout; the pterotic spine runs nearly parallel to the body axis, and the preopercular spine is inclined sharply downward, forming a large angle with the body axis. As in the case of blue marlin larvae, there is generally no pigmentation on the lower part of the head. Occasionally, however, pigment occurs just above the midline of the gular membrane or mid-anterior portion of the branchiostegal membrane.

Sailfish. Larval sailfish are characterized by an elongated snout, small eyes, and a relatively small head depth. The anterior edge of orbit is not angular and the tip of the snout is at a lower level than the center of the eye. Both the pterotic and preopercular spines are similar in shape to those in blue marlin. Pigment spots appear characteristically on the posterior peripheral area of the gular membrane and also just above the midline of the gular membrane. However, there are also some non-pigmented types of sailfish larvae, which will be discussed later.

Shortbill Spearfish. As in the blue marlin, the center of eye is located at a higher level than the snout tip; the anterior edge of the orbit does not, however, project forward. The pterotic and preopercular spines are similar in shape to those in the blue marlin. The preopercular spine is shorter than in blue marlin and is also inclined further downward. Pigment characteristically occurs on the branchiostegal membrane and also above the midline of the gular membrane.

There are problems in using head profile or spine shape as diagnostic characters in distorted specimens. In working with distorted specimens, profile characters such as snout-eye level must be considered by "estimating" the original shape of the specimen, and this can only be done by amassing considerable experience. Nevertheless, there is no way of escaping subjectivity. Richards (in press) has also pointed out similar problems in the use of such diagnostic characters. Obviously there is need to find additional diagnostic characters in order to substantiate identification of these larvae.

Pigmentation on the lower portion of the head is a good characteristic for shortbill spearfish and sailfish larvae, but presents some problems in the case of the blue marlin and striped marlin larvae. In larval sailfish, there is also the non-pigmented type of larvae in addition to the pigmented type. These two types may represent two separate populations of sailfish (Ueyanagi, in press).

EXAMINATION OF ATLANTIC BILLFISH LARVAE

The description of the developmental series of larvae of Atlantic billfishes is limited to that of sailfish. Only preliminary work is in progress on the larvae of other species, because there are inadequate specimens to examine a complete developmental series. Consequently, it is still difficult to definitely identify larvae of the various species. The following is a tentative report on Atlantic species based on specimens collected from the Atlantic Ocean by Japanese research vessels from 1965 to 1971. The collection comprises 66 specimens of 3.5 - 52 mm standard length (only 3 specimens were larger than 20 mm). The areas of collection extended between lat 23°N and 27°S and included the probable spawning areas of the various species. Among the specimens, the probable species included *Makaira nigricans*, *Istiophorus albicans*, *Tetrapturus albidus*, *T. pfluegeri*, and *T. georgei*.

Head morphology, which was useful in separating larvae of Indo-Pacific billfishes, was used in identifying these species. This was done on the assumption that the various Atlantic species would show similarity in head morphology, as to their cognate Indo-Pacific species. The assumption was based on the similarity in the growth pattern of the snout in the Atlantic and Indo-Pacific billfishes (Fig. 3).

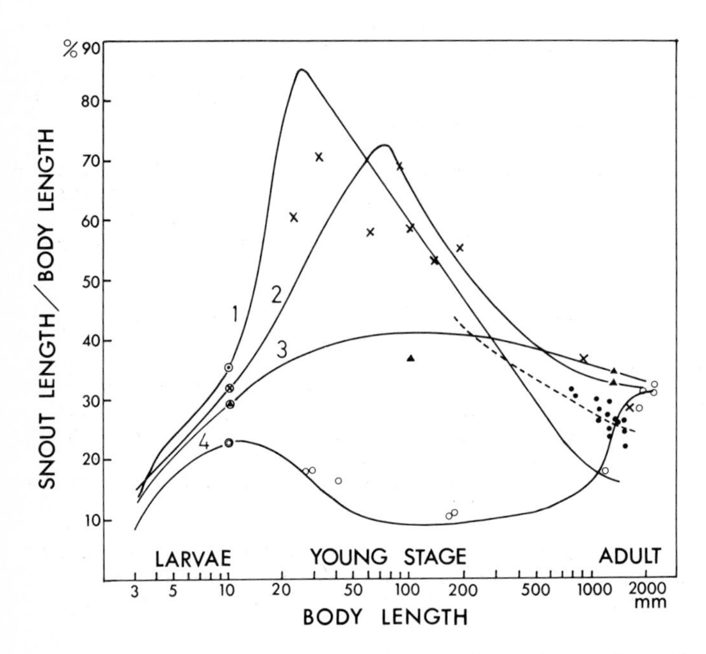

Fig. 3. Relation of snout length to body length during growth of the istiophorid species. Curve 1 represents shortbill spearfish, 2 - Indo-Pacific sailfish, 3 - striped marlin and 4 - Indo-Pacific blue marlin; data points are ● for longbill spearfish, X for Atlantic sailfish, ▲ white marlin and o Atlantic blue marlin, respectively. Body length is the distance from the posterior margin of the orbit to the tip of the center rays of the caudal fin. (Adapted from Ueyanagi, 1963b)

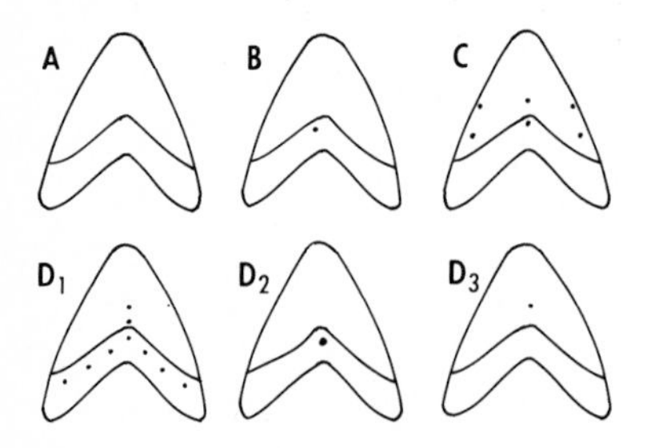

Fig. 4. The pigmentation pattern on the lower side of the head of Atlantic billfish larvae. A, B, C, and D (1-3) correspond to the head-profile types of blue marlin, striped marlin, sailfish, and shortbill spearfish of the Indo-Pacific, respectively.

Of the 66 Atlantic specimens examined, 4 were so distorted that their head profiles could not be determined. The remaining 62 specimens were separated into 4 head-profile types: blue marlin, striped marlin, sailfish, and shortbill spearfish types. The pigmentation pattern on the lower part of the head of these 4 types of larvae is as shown in Fig.4. In the figure, the pigment which appears on the tip of the lower jaw in all larvae has been omitted.

Of the group of larvae classed as blue marlin type (A), the majority lacked pigment on either the gular membrane or the branchiostegal membrane. About 10% of the specimens were found with a single spot of pigment above the midline of the gular membrane.

The larvae with the striped marlin type of head profile (B) lacked pigmentation on the gular membrane but had one spot on the mid-anterior portion of the branchiostegal membrane. The sailfish type larvae (C) exhibited pigment on the postero-peripheral and postero-medial parts of the gular membrane.

Larvae with a head profile of the shortbill spearfish type (D) had three different pigmentation patterns. Type D_1 had pigmentation on the branchiostegal membrane and on the midline of the gular membrane. Type D_2 had a single, distinct bit of pigment on the mid-anterior portion of the branchiostegal membrane. Most lacked any pigment on the gular membrane but there was an occasional specimen with a single spot. Type D_3 was pigmented on the midline of the gular membrane, thus differing from the other two types; but this occurred on only two specimens, so no conclusive statements can be made.

In both the sailfish type (C) and spearfish type D_1 larvae, the spots of pigment appearing on the midline of the gular membrane varied in numbers from one to three but this is believed attributable to individual variations.

Types A, B, C, D_1, classified on the basis of head profile and pigmentation pattern, are tentatively identifiable as Atlantic blue marlin, white marlin, Atlantic sailfish, and longbill spearfish, respectively. The shortbill spearfish types D_2 and D_3 larvae are more difficult to identify. However, since type D_2 larvae include large specimens in which the anus is located considerably anterior to the origin of the first anal fin (a characteristic of adult longbill spearfish), it appears probable that these are also larvae of longbill spearfish. Type D_3 larvae have been collected only from waters of the northeastern Atlantic corresponding to the distribution of *T. georgei* (Fig.5), suggesting that these may be larvae of that species.

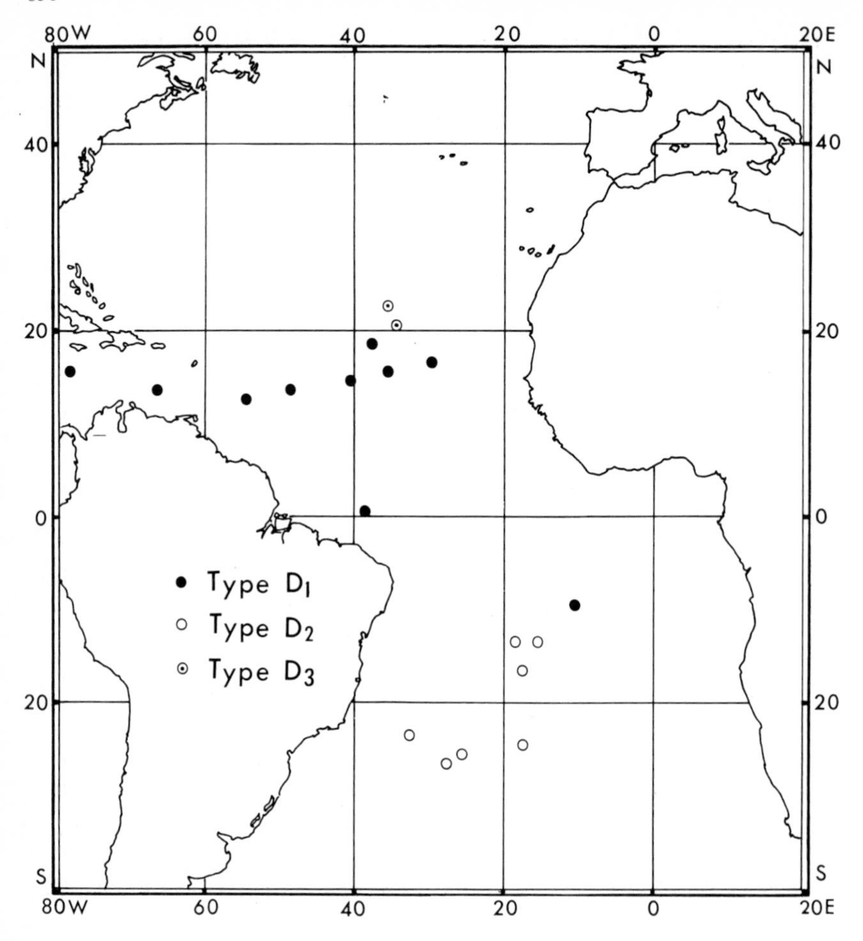

Fig. 5. The occurrence of the 3 types of spearfish larvae (on the basis of pigmentation patterns in Fig. 4) in the Atlantic Ocean

VARIATIONS IN PIGMENTATION

Although the pigmentation patterns on the lower part of the head of larval billfishes are thought to be species-specific, there were two different patterns among the larvae believed to be longbill spearfish. There were also the non-pigmented as well as the pigmented types of Indo-Pacific sailfish larvae. Regarding the latter, Ueyanagi (in press) found a rather clear separation in their distribution (Fig. 6). Sailfish larvae are generally found in waters close to land masses. While both types of larvae were so distributed, the non-pigmented larvae were confined to waters south of about lat 10°S. The non-pigmented type of sailfish larvae occurred exclusively in the Coral Sea. There is little reason to believe that this difference between the two types is due to individual variations in pigmentation. Rather, it appears to be genetic in nature, indicating that the two types belong to separate subpopulations of the Indo-Pacific sailfish.

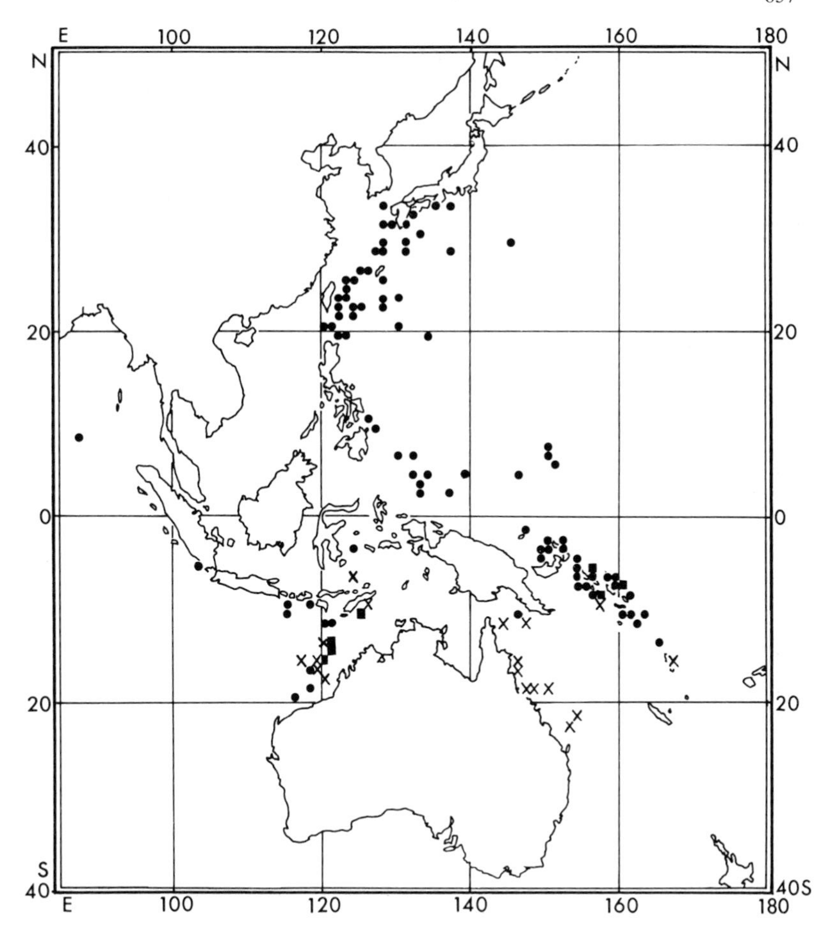

Fig. 6. The occurrence of the pigmented and non-pigmented types of sailfish larvae in the Indian and Pacific Oceans. ● - pigmented larvae (A); X - larvae without pigment (B); ■ - both A and B. (Adapted from Ueyanagi, in press)

The two patterns of pigmentation in larvae identified as longbill spearfish also probably represent genetic differences. It is clear that the distribution of all three types is discrete (Fig. 5): Type D_1 appeared between lat 20°N and 10°S, type D_2 in waters south of lat 10°S with no overlap between the two. Type D_3, possibly the larvae of *T. georgei*, occurred farthest north.

The areas of occurrence of types D_1 and D_2 larvae correspond to the probable spawning grounds of longbill spearfish, and the time of collection to the probable spawning seasons, based on gonad maturity data (Ueyanagi et al., 1970). If these two types of larvae differ in pigmentation due to individual variations, then both types might be expected to occur randomly in the distributional areas. However, the localized distribution strongly suggests genetic variations, with these two types representing separate subpopulations of the longbill spearfish.

In the tropical Atlantic, the surface currents have their boundaries at around lat 10°S. The currents are directed in a northwesterly direction to the north of lat 10°S, and southwesterly to the south of lat 10°S. This current boundary, as habitat for the early life stages of fish spawned in tropical waters, probably has a great influence and may be a splitting mechanism. In relation to this, it is interesting to note that the two types of longbill spearfish larvae were found separated at just about lat 10°S.

If the morphological differences in larvae of Indo-Pacific sailfish and the longbill spearfish are indeed due to genetic differences as surmised, then the study of larval morphology is not only useful for species identification but may contribute to the study of population identification.

ACKNOWLEDGEMENTS

I am grateful to Dr. Elbert H. Ahlstrom of the National Marine Fisheries Service, La Jolla, who encouraged me to prepare this paper. I wish to sincerely thank Mr. Tamio Otsu of the National Marine Fisheries Service, Honolulu, who helped me with the English translation of the manuscript.

REFERENCES

Nakamura, I., Iwai, T. and Matsubara, K., 1968. A review of the sailfish, spearfish, marlin and swordfish of the world. Misaki Mar. Biol. Inst., Kyoto Univ. Spec. Rep., 4, 1-95.

Richards, W.J., in press. Evaluation of identification methods for young billfishes. U.S. Dept. Commer., NOAA Tech. Rep. NMFS SSRF.

Robins, C.R., in press. The validity and status of the roundscale spearfish, *Tetrapturus georgei*. U.S. Dept. Commer., NOAA Tech. Rep. NMFS SSRF.

Ueyanagi, S., 1963a. Methods for identification and discrimination of the larvae of five istiophorid species distributing in the Indo-Pacific. Rep. Nankai Reg. Fish. Res. Lab., 17, 137-150.

Ueyanagi, S., 1963b. A study of the relationships of the Indo-Pacific istiophorids. Rep. Nankai Reg. Fish. Res. Lab., 17, 151-165.

Ueyanagi, S., in press. On an additional diagnostic character for the identification of billfish larvae with some notes on the variations in pigmentation. U.S. Dept. Commer., NOAA Tech. Rep. NMFS SSRF

Ueyanagi, S., Kikawa, S., Uto, M. and Nishikawa, Y., 1970. Distribution, spawning, and relative abundance of billfishes in the Atlantic Ocean. Bull. Far Seas Fish. Res. Lab., 3, 15-55.

S. Ueyanagi
Far Seas Fisheries Research Laboratory
1000 Orido
Shimizu 424 / JAPAN

The Diverse Patterns of Metamorphosis in Gonostomatid Fishes – An Aid to Classification

E. H. Ahlstrom

The gonostomatid lightfishes rank second only to myctophid lanternfishes in abundance in the sea. The gonostomatid genus *Cyclothone* has been singled out as the most abundant group of fishes in the ocean. These observations were based on abundance of adults in mid-water trawl hauls. If abundance is based on larvae, the gonostomatid genus *Vinciguerria* could be the most numerous group of fishes in the ocean. Without attempting to decide whether *Cyclothone* or *Vinciguerria* ranks first in abundance, we can assume that the gonostomatids obviously are an important group of oceanic fishes.

Grey (1964), in her impressive contribution on gonostomatid fishes, recognized 21 genera as valid, with between 52 to 55 valid species. A few species subsequently have been added and a few synonymized, bringing the number of species close to 60. Weitzman (in press) will reduce the number of genera to 20 (pers. comm.). The Gonostomatidae are closely related to the hatchetfishes, family Sternoptychidae; the most recent review of the latter family was made by Baird (1971).

Life histories have been published for 12 of the 20 genera of gonostomatid fishes and for slightly more than a third of the species. In addition to the published record I have larval series for an additional 3 genera and 9 species.

The larvae of Gonostomatidae constitute a moderately homogeneous group. By this I mean that it is not difficult to identify larvae to the family level despite differences noted below. About half the known gonostomatid larvae have moderate to fairly heavy pigmentation that aids in identification. The gut (digestive tract) is more variable in length between larvae of different genera of gonostomatids than perhaps any other family of fishes. It ranges from fairly short, as in *Danaphos*, to a full 95% of the standard length, as visable in an interesting *Maurolicine* larva I will discuss later; occasionally the gut can be trailing, as in *Ichthyococcus*. Gut length is even shorter in the sternoptychid genus *Sternoptyx*. Eyes in larvae are either round or narrowed; but when narrowed they lack the specializations found in the eyes of various myctophid larvae. The most strikingly narrowed larval eyes are found in the sternoptychid genus *Argyropelecus*.

The anal fin in gonostomatids is usually long, occupying most of the tail portion of the body. The shorter dorsal fin, however, is variously placed in relation to the anal. It may completely precede the anal, as in *Danaphos* or *Woodsia*; it may partly precede the anal as in *Vinciguerria* or *Maurolicus*; the origin of the dorsal may be opposite the origin of the anal as in *Cyclothone*; or the anal origin may precede the dorsal by several rays, as in *Gonostoma* or *Araiophos*. In the majority of gonostomatid larvae the dorsal and the anal fins form in about the same relative positions they will retain in later life - but there are interesting exceptions. These are associated with larval forms with relatively long intestines in which the anus shifts anteriorward during metamorphosis, as does the anal fin relative to the dorsal.

Grey (1964) illustrated such a shift in *Pollichthys mauli*, and a marked shift must occur in an interesting *maurolicine* larva that is discussed later.

The caudal fin of gonostomatids is interesting in several respects. The principal caudal rays are invariably 19, a character shared with most salmoniform fishes. The bones that support the principal rays, however, can retain the primitive complement of 4 superior and 3 inferior hypurals as in *Gonostoma* or *Diplophos*, be variously reduced as in *Vinciguerria* (3 + 2) or *Danaphos* (3 + 1) to only 1 superior and 1 inferior hypural as in *Cyclothone* or *Araiophos*, or achieve the ultimate reduction to a single plate in the sternoptychid genus *Sternoptyx*. Similarly epurals can retain the primitive complement of 3, or be variously reduced to none. In no other group of fishes has such a variable pattern of reduction of caudal supporting bones been observed. In the 6 other stomiatoid families of fishes, for example, hypurals are stabilized at 3 superior and 3 inferior plates. In most fishes the procurrent C rays are about equal in number, both in the dorsal group and in the ventral, but in various gonostomatids-sternoptychids the ventral group is reduced in number of rays relative to the dorsal - obviously to accommodate the posteriormost photophores of the AC group. Among the genera with a reduced ventral complement of procurrent caudal rays are *Vinciguerria, Polymetme, Maurolicus, Danaphos*, and *Argyropelecus*; among the genera in which no such reduction occurs are *Cyclothone, Diplophos, Araiophos*, and *Sternoptyx*.

Although larvae of gonostomatids and sternoptychids exhibit marked variability in certain characteristics such as gut length, eye shape, pigmentation and fin position, none of these is of primary value in tracing relationships among genera and species. The patterns of photophore development, during the larval and metamorphic stages and specialization in photophore groups in adults, constitute the most trenchant characteristics for showing such relationships. The notations used to designate photophores follow Grey (1964), with the exception that her IV group is divided into IP + PV groups.

Before discussing photophore information in larval and metamorphic stages, I will briefly characterize photophore groups in adults. Most photophores in gonostomatids are in a ventral series usually extending along the length of the body from symphysis to base of caudal. In addition, photophores are present between branchiostegal rays (6 or more pairs per side), on the head and operculum (3 to 5 pairs), and usually a lateral body series (occasionally several lateral series). In the majority of gonostomatid genera (13) all photophores remain individually separate, but in the other 7 genera some or most of the photophores can be variously clustered into groups with common bases. A similar pattern of clustered photophores is found in the 3 sternoptychid genera.

In most families of stomiatoid fishes, photophores form simultaneously during a relatively brief metamorphic period. They form initially as unpigmented organs and subsequently acquire pigment and structure. This is the pattern of formation, for example, in the families Stomiatidae, Chauliodontidae, Melanostomiatidae, etc.; it is also found in some genera of gonostomatids.

In several genera of gonostomatids, for example *Vinciguerria* and *Cyclothone*, most photophores are laid down initially during a white photophore stage, and only a few photophores are late forming. More commonly, most ventral photophores form simultaneously during a white photophore stage; the lateral photophores form later and usually gradually.

This is the situation in *Ichthyococcus*, *Pollichthys*, and *Diplophos*. All genera with clustered photophores and at least three genera with single photophores (*Gonostoma*, *Margrethia*, and *Bonapartia*) have a protracted metamorphosis with gradual formation of photophores. Even when photophore formation is gradual, each addition is initially unpigmented and then becomes pigmented. Recent additions among pigmented photophores are usually smaller than those formed earlier, hence easily recognized.

The sequence of photophore formation has been as well documented for *Vinciguerria* as for any gonostomatid genus. Developmental series have been described for the 4 recognized species. Sanzo (1913b) showed from life-history studies that 2 species of *Vinciguerria* in the Mediterranean could be distinguished as larvae - *V. attenuata* and *V. poweriae*. Jespersen and Taning (1926) provided additional life-history information for these species and for *V. nimbaria* (as *V. sanzoi*). Ahlstrom and Counts (1958) described the life history of *V. lucetia* and provided information on *V. nimbaria* and *V. poweriae* from the eastern North Pacific; Silas and George (1969) provided additional information concerning development of *V. nimbaria*. The developmental pattern of all 4 species is strikingly similar.

At first formation, photophores are colorless in *Vinciguerria*, but they soon become pigmented. Most photophores appear simultaneously except for 3 to 7 pairs (Fig. 1A).

In *V. lucetia* the late-forming photophores are invariably the following:

A. The upper opercular pair of photophores.
B. The symphysial pair under the lower jaw.
C. A median pair of photophores of the AC group.
D. Two to 4 pairs of lateral photophores of the posterior lateral group.

Only 2 of the 4 recognized species of *Vinciguerria* develop a symphysial pair of photophores; except for this, the late-forming photophores in all 4 species are as listed above, the number of late-forming lateral photophores is reduced to 1 or 2 pairs in *V. poweriae*.

Even though the metamorphic period is relatively short in *Vinciguerria*, Ahlstrom and Counts (1958) indicated a natural division into 3 stages: 1. pro-metamorphosis, the white photophore stage, 2. mid-metamorphosis, when marked changes in body form occur in addition to the development of photophores into functional organs and the formation of several late-forming photophores, and 3. post-metamorphosis during which photophores formation is completed and juvenile body form attained. All stages are commonly taken together in plankton haul sampling no deeper than 200 m.

In *Cyclothone*, all ventral photophores are laid down simultaneously as white stage photophores, and on some species most lateral photophores as well. Developmental series have been described for two species of *Cyclothone*, *C. braueri*, and *C. pygmaea* by Jespersen and Taning (1926) and possibly *C. atraria* (Mukhacheva, 1964 as *C. microdon*). In addition we have developmental series for *C. acclinidens*, *C. signata*, and *C. alba*. Body form and pigmentation are strikingly similar in *Cyclothone* larvae of the various species, which complicates establishing life-history series in this most speciose of gonostomatid genera. Adult taxonomy has recently been clarified by Mukhacheva (1964) and Kobayashi (1973); 12 species are now recognized. Larvae of *Cyclothone* are rather shallowly distributed - i.e. they occur principally in the upper mixed

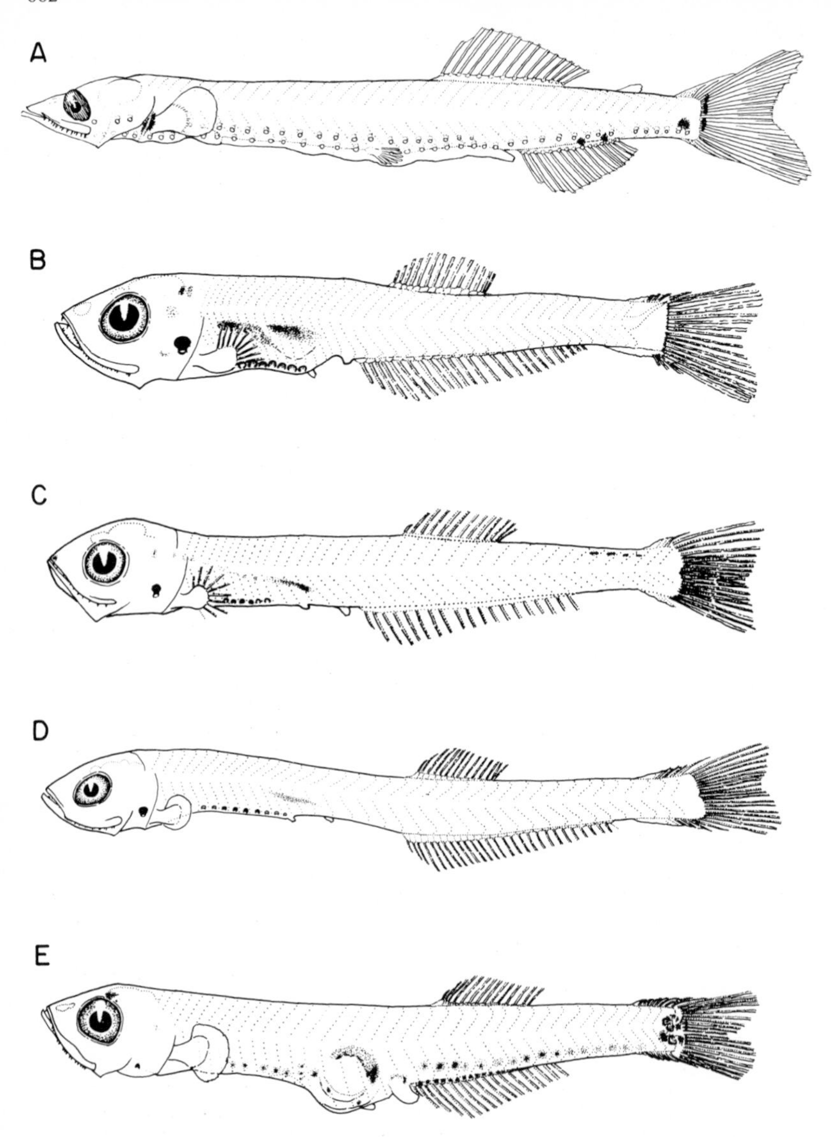

Fig. 1. Larvae and metamorphosing specimens of gonostomatids. (A) *Vinciguerria lucetia*, 15.0 mm, pro-metamorphic stage (from Ahlstrom and Counts, 1958). (B) *Gonostoma elongatum*, 9.8 mm, early metamorphosis. (C) *G. bathyphilum*, 11.0 mm, early metamorphosis. (D) *G. ebelingi*, 15.0 mm, early metamorphosis. (E) *G. atlanticum*, 12.0 mm, late larva

layer (Ahlstrom, 1959) and remain in this depth zone through the pro-metamorphic or white stage of photophore formation; they move to deeper levels to complete photophore formation. Later metamorphic stage specimens of *Cyclothone* are never taken in plankton collections made within the upper 200 m. It should be emphasized that the white photophore stage in *Cyclothone* is strikingly similar to that observed in *Vinciguerria*.

Larvae of *Ichthyococcus* develop more specialized larval characteristics than other gonostomatids, including an elongated ray on the lower part of the pectoral, reminiscent of similar pectoral development on larvae of the myctophids *Loweina* and *Tarletonbeania* (Moser and Ahlstrom, 1970), and a trailing gut. Information on the number of photophore formed initially in *Ichthyococcus* is somewhat inconsistent among authors (Sanzo, 1913a, 1930; Jespersen and Taning, 1926; Grey, 1964) but most photophores of the ventral group form simultaneously as unpigmented "white" photophores, whereas the lateral series forms latter. A 20.3 mm pro-metamorphic specimen of *I. ovatus* collected near Hawaii had 12 BR (definitive), 8 IP (definitive), 17 PV (definitive), 9 VAV (definitive), and 8 + 3 AC (1 or 2 lacking) and no lateral photophores. The AC photophores were divided into 2 groups, a larger anterior group and a posterior group of 3, with 1 or 2 photophores lacking between the 2 groups. A mid-metamorphic specimen illustrated in Jespersen and Taning (1926) has a less complete series of AC (7 + 3), although 13 pairs were developed in the lateral series. Jespersen and Taning indicated that the last photophores to form were the upper OP, posterior photophores of the lateral series, and AC photophores between the anterior and posterior groups. Marked changes in body proportions, particularly a striking increase in body depth, occur during metamorphosis of *Ichthyococcus*. Although metamorphosis is more gradual than in *Vinciguerria* or *Cyclothone*, it is basically similar.

Grey (1964, Fig. 28) illustrates 2 metamorphic-stage specimens of *Pollichthys mauli* of about equal size (16-17 mm), but at different stages of development. At first, I found these illustrations disturbing. The termination of the gut and origin of the anal fin appeared to be too far back on the body on the larger but less advanced specimen. I now know the reason for this: in various gonostomatids the gut length (i.e. snout – anus length) shortens during metamorphosis and in some the origin of the anal fin moves forward relative to the origin of the dorsal fin. This was happening in *Pollichthys* between the 2 metamorphic stages illustrated by Grey.

A less complete complement of photophores is laid down initially than in the preceding genera. Not only is the lateral group of photophores late forming, but the ventral AC group is less complete. The pattern of AC development is strikingly similar to the patterns to be discussed for *Gonostoma* and *Margrethia*. The two posterior AC photophores develop independently of the main group of AC photophores; in the latter the middle photophores develop initially and are added to both anteriorly and posteriorly until the complete complement is formed. The more advanced stage illustrated by Grey has the ventral and head photophores complete, but has only 6 of the lateral group of photophores which will number 19-21 when complete. I have examined post-metamorphic stage specimens of *Pollichthys*, but not the earlier metamorphic stages.

Diplophos has the most elongate larva among gonostomatids, and attains the largest size before transformation (ca. 45-50 mm). Developmental stages of *D. taenia* were described by Jespersen and Taning (1919). Although the majority of ventral photophores form simultaneously in *Diplophos taenia*, few series are actually complete. For example, on a 44.2 mm transforming specimen, the isthmal photophores were incomplete anteriorly, the anterior VAV photophore was small and barely formed; posterior AC group was incomplete. However, the two posteriormost photophores of the AC group were well formed and separate from the main group of AC photophores. The lateral series was partly formed, 25 + 8 photophores, but several anterior photophores were lacking as well as most photophores above the VAV and AC ventral groups. One photophore of the lateral midline group was formed - the posteriormost photophore of this group, which forms far out on the caudal fin (about 2 mm from the caudal base). A 46 mm specimen had midline photophores formed (87 + 1) but still lacked posterior photophores in the lateral group. Even on juveniles this lateral series is often incomplete.

A developmental feature shared by the above genera is that most or all ventral photophores form simultaneously as white stage photophores. In none of these genera are all photophores formed initially, but the number of late-forming photophores are usually the following: OA serie of lateral photophores of which some or all are late-forming; AC (anal caudal ventral series) which usually form as 2 separate groups with 1 to several photophores lacking (*Cyclothone*, an exception); OP photophores, of which the upper pair is invariably late-forming; and SO photophores are often the last to form on species possessing this pair

Larvae and pro-metamorphic specimens of Atlantic material tentatively identified as *Polymetme*, have most of the ventral photophores formed, much in *Ichthyococcus*. Larval stages, up to ca. 16 mm are known for *Woodsia nonsuchae* and *Yarrella argenteola*, but not metamorphosing specimens For both, metamorphosis is assumed to be similar to *Ichthyococcus*. Developmental stages are unknown for *Triplophos* and *Photichthys*.

The genera of gonostomatids-sternoptychids with a protracted metamorphosis and gradual formation of all body photophores can be broken down into 3 clusters of genera:

1. Genera with all photophores remaining individually separate, but with a protracted metamorphosis: *Gonostoma*, *Bonapartia*, and *Margrethia*.

2. *Maurolicus* and 6 related genera having photophores in clusters with common bases: *Danaphos*, *Valenciennellus*, *Araiophos*, *Argyripnus*, *Sonoda*, *Thorophos*, and *Neophos*.

3. The 3 genera of the family Sternoptychidae: *Sternoptyx*, *Argyropelecus*, and *Polyipnus*.

I will deal first with the group that has the least specialized photophores:

Gonostoma, *Bonapartia*, and *Margrethia*. All 3 genera lack photophores on the isthmus, and the lateral (OA) series of photophores is lacking on *Bonapartia* and *Margrethia*.

Developmental series have been described for 3 species of *Gonostoma*: *G. denudatum* (Sanzo, 1912), *G. elongatum* (Jespersen and Taning, 1919 as sternoptychid larva B; Grey, 1964 as *G. elongatum*) and *G. gracile* (Kawaguchi and Marumo, 1967), and for *Bonapartia pedaliota* (Jespersen and Taning, 1919; Grey, 1964), and *Margrethia obtusirostra* (Jespersen and Taning, 1919; Grey, 1964). I have been fortunate in obtaining larvae and/or metamorphosing specimens of the 3 remaining species of *Gonostoma*: *G. atlanticum*, *G. bathyphilum* and *G. ebelingi*, hence developmental stages are known for all species of this group.

Gonostoma appears to be the pivotal genus with respect to photophore development. Among species of *Gonostoma*, photophores may be formed gradually over a rather extended size range as in *G. elongatum* (Fig. 1B) *G. denudatum*, *G. bathyphilum* (Fig. 1C), and probably *G. ebelingi* (Fig. 1D) or somewhat more rapidly as in *G. gracilis* and *G. atlanticum* (Fig. 1E). The forms with gradual development are closely allied in development to *Bonapartia* and *Margrethia*, as well as to the Maurolicine and sternoptychid genera. The gradual formation of photophores during a protracted metamorphosis is a major evolutionary trend first evidenced in *Gonostoma*.

The OP_3 pair of photophores is the first to form in all 6 species of *Gonostoma* as well as in *Margrethia* and *Bonapartia*. The next photophores to form (1 or several) are in the PV series, but otherwise the sequence is various, as among the above species (Table 1).

The sequence of photophore formation is contrastingly different in two species of *Gonostoma* with a protracted metamorphosis - *G. denudatum* and *G. elongatum*. In the latter the metamorphosis stage extends from about 6.0 mm to 22.5 mm. Although photophores form gradually in all groups, the sequence among photophore groups is as follows in *G. elongatum*: OP, PV, BR, VAV, ORB and AC, OA and SO. The addition of photophores to the AC group occurs only after most PV and VAV photophores are formed; the initial photophore to form is an inner photophore of the anterior AC group. In contrast, photophores form much later on *G. denudatum* (between ca. 18-34+ mm) and in a different sequence: OP, PV, posterior AC group, BR and VAV, anterior AC group, ORB and OA, SO. Photophores begin forming in the posterior AC group immediately after the first PV photophore is laid down. When photophores are formed in the anterior AC group the 3rd or 4th is laid down first and additional photophores are added both anteriorly and posteriorly.

Metamorphosis in *Margrethia obtusirostra* occurs gradually between about 5.8 to 19.0 mm; photophores are added in the following sequence: OP, PV, VAV and AC (both groups), BR, ORB, and SO. Metamorphosis in *Bonapartia pedaliota* according to Grey (1964) occurs between ca. 9 to 25 mm and is rather similar in sequence to *G. elongatum*.

Although the first few photophores are laid down gradually in *Gonostoma atlanticum*, most of the ventral photophores are laid down as a group. The same general pattern of photophore formation has been described for *Gonostoma gracile* (Kawaguchi and Marumo, 1967). The development of ventral photophores in these two species contrasts sharply with the patterns described above. The lateral series of photophores are laid down after the ventral, much as in *Pollichthys* or *Ichthyococcus*.

Among Maurolicine genera and species rather complete developmental series are known for *Maurolicus mulleri* (Holt and Byrne, 1913; Jespersen and Taning, 1926; Sanzo, 1931; Okiyama, 1971), *Valenciennellus tripunctulatus* (Jespersen and Taning, 1919; Grey, 1964), and *Danaphos oculatus* (original) and less complete series for *Araiophos eastropas* (Ahlstrom and Moser, 1969) and *Argyripnus atlanticus* (Badcock and Merritt, 1972). Information on sequence of photophore formation is contained in Table 2.

The life-history stages are best known for *Maurolicus mulleri* (Fig. 2A) - a fascinating species from egg to adult. Photophores are precociously laid down in this species on larvae as small as 5.5 mm. The first to form are branchiostegal photophores, followed soon by PV photophores. The OP_3 pair form as early as 6.7 mm, the ORB pair soon after (6.9 mm), IP and AC photophores begin to form by 7.5 mm, VAV by 8.6, and OA by 9.0 (see Table 2). Photophore formation is complete by 19.0 mm or sooner. The AC photophores in *Maurolicus* are divided into 3 groups consisting of: an anterior single photophore, a large middle cluster (13-14 photophores), and a posterior cluster (ca. 8 photophores). The first AC photophores to form are in the large middle cluster, but soon thereafter they also form in the posterior cluster. The early forming photophores are in the middle of each cluster and photophores are added in both directions. This pattern appears to be similar to that described for *Margrethia* and *Bonapartia*.

Table 1. Sequence of photophore formation in *Gonostoma*, *Bonapartia*, and *Margrethia*

		ORB	OP	SO	BR	PV	VAV
Gonostoma elongatum	adult	1	3	1	9	15	(4)-5
	6.0	0	1	0	0	0	0
	7.5	0	1	0	0	5	0
	10.2	0	1	0	2/1	10	2
	13.0	0	1	0	2	11	3
	14.0	1	1	0	2	11	2/3
	16.7	1	1	0	3	11	4
	22.5	1	3	1	9	15	5
Gonostoma denudatum	adult	1	3	1	9	15-16	5
	18.25	0	1	0	0	1	0
	19.0	0	1	0	0	2	0
	20.75	0	1	0	1	3	1
	24.75	0	1	0	3	6	3
	29.65	0	2	0	5	14	5
	34.0	1	3	0	9	16	5
	39.0	1	3	1	9	16	5
Gonostoma gracile	adult	1	2	1	9	13-15	4-5
	15.5-17.0	0	1	0	0	0	0
	20.0	1	2	1	2	13	5
	22.0	1	2	1	9	14	4
Gonostoma ebelingi	adult	1	2	1	9	15	10
	13.8	0	1	0	0	7	0
	15.0	0	1	0	0	9	0
Gonostoma bathyphilum	adult	1	2		9	11-12	4-5
	11.0	0	1	0	0	5	0
	14.8	1	1	0	4	10	2
Gonostoma atlanticum	adult	1	2	1	9	15-16	5
	12.0		1				
	13.0	0	1	0	0	1	0
	14.5	0	1	0	0	2	0
	18.8	1	2	0	9	16	5
Margrethia obtusirostra	adult	1	3	0	9-12	13-15	4
	5.8	0	1	0	0	2	0
	6.4	0	1	0	0	6	2
	8.0	0	1	0	2	10	4
	11.3		2	0	6	14	4
	15.0	1	3	0	9	14	4
Bonapartia pedaliota	adult	1	3	1	11-13	14-15	5-(6)
	9.5	0	1	0	2	3	0
	12.0	0	1	0	4	5	2
	14.0	1	1	0	5	10	4
	16.0	1	1	0	6	11	5
	23.0	1	3	0	11	14	5

Table 1. (continued)

AC	OA + ODM	Source
21-23	13-15	Grey 64
0	0	Orig
0	0	Orig
0	0	Orig
0	0	Orig
1+	0	Grey
1+	0	J & T 19
22	13	Grey 64
17-20	13-15	Grey 64
0	0	Sanzo 12
+2	0	Sanzo 12
+3	0	Sanzo 12
3 + 3	0	Sanzo 12
11 + 3	0	Sanzo 12
15 + 5	13	Sanzo 12
15 + 5	13	Sanzo 12
17-19	11-12 + 6-7	K & M 67
0	0	K & M 67
17	0	K & M 67
18	12 + 4	K & M 67
19	21	Grey 64
0	0	Orig
0	0	Orig
20-21	14	Grey 64
0	0	Orig
0	0	Orig
19	13	Grey 64
		Orig
0	0	Orig
0	0	Orig
19	0	Orig
13-14 + 3-4	0	Grey 64
0	0	Orig
1 + 2	0	Orig
1 + 2	0	Orig
5 + 3	0	Orig
11 + 4	0	Orig
16-18 + 2-3	0	Grey 64
0	0	Grey 64
0	0	Grey 64
3 + 1	0	Grey 64
5 + 2	0	J & T 19
14 + 2	0	Grey 64

Table 2. Sequence of photophore formation in *Maurolicus muelleri* and allied maurolicine genera

		ORB	OP	SO	BR	IP	PV	VAV	AC	OA	Source
Maurolicus muelleri	adult	1	3	1	(6)	(6)	(12)	(6)	1 + (13/14) + (8)	(2) + 7	Orig
	5.5	0	0	0	(1/2)	0	0	0	-0-	-0-	Orig
	6.2	0	0	0	(2)	0	(2)	0	-0-	-0-	Orig
	6.5	0	0	0	(2)	0	(4)	0	-0-	-0-	Orig
	6.7	0	1	0	(3)	0	(5)	0	-0-	-0-	Orig
	6.9	1	1	0	(4)	0	(8)	0	-0-	-0-	Orig
	7.5	1	1	0	(4)	1	(9)	0	0 + (2) + 0	-0-	Orig
	8.6	1	2	0	(5)	(3)	(12)	(2)	0 + (3) + (3)	-0-	Orig
	9.0	1	2	0	(5)	(3)	(11)	(2)	0 + (3) + (3)	1	Orig
	9.7	1	3	0	(5)	(5)	(11)	(3)	0 + (4) + (6)	(2) + 1	Orig
	10.8	1	3	0	(6)	(5)	(12)	(4)	0 + (5) + (6)	(2) + 2	Orig
	13.5	1	3	0	(6)	(6)	(12)	(6)	0 + (9) + (7)	(2) + 6	Orig
Danaphos oculatus	adult	1	3	0	(6)	(3) + (4)	(11)	(5)	(3) + 16 + (4) + 1	6	Orig
	16.5	0	0	0	(2)	-0-	0	0	-0-	0	Orig
	16.5	0	0	0	(3)	-0-	(3)	0	-0-	0	Orig
	19.2	0	0	0	(4)	-0-	(10)	0	-0-	0	Orig
	21.0	1	1	0	(5)	(2) + (4)	(10/11)	0	(2) + 0 + 0 + 0	0	Orig
	21.3	1	1	0	(4/5)	(3) + (4)	(10)	0	(3) + 0 + (2) + 0	0	Orig
	21.8	1	2	0	(5)	(3) + (4)	(11)	(2)	(3) + 8 + (4) + 0	2	Orig
	24.2	1	2	0	(6)	(3) + (4)	(11)	(2)	(3) + 9 + (4) + 0	2	Orig
Valenciennellus tripunculatus	adult	1	3	0	(6)	(3) + (4)	(16-17)	(4-5)	(3) + (3) + (3) + (2) + (4)	(2) + 3	Grey 64
	8.6	0	0	0	(3)	-0-	(3)	0	-0-	-0-	Orig
	9.5	0	0	0	(4)	-0-	(6)	0	-0-	-0-	Orig
	12.0	0	0	0	(4)	-0-	(13)	(2)	-0-	-0-	Orig
	13.2	0	0	0	(4)	-0-	(14)	(3)	-0-	-0-	Orig
	17.0	1	2	0	(4-5)	(3) + (4)	(15)	(5)	(3) + (3) + 0 + (3) + (4)	(2)	Grey 64
Araiophas eastropas	adult	1	1	0	(6)	(2)	(3) + 3-4 + (2)	(3)	(2) + 2 + (2)	no	A & M 69
	11.2	0	0	0	(3)	0	(2)	0	-0-	-	A & M 69

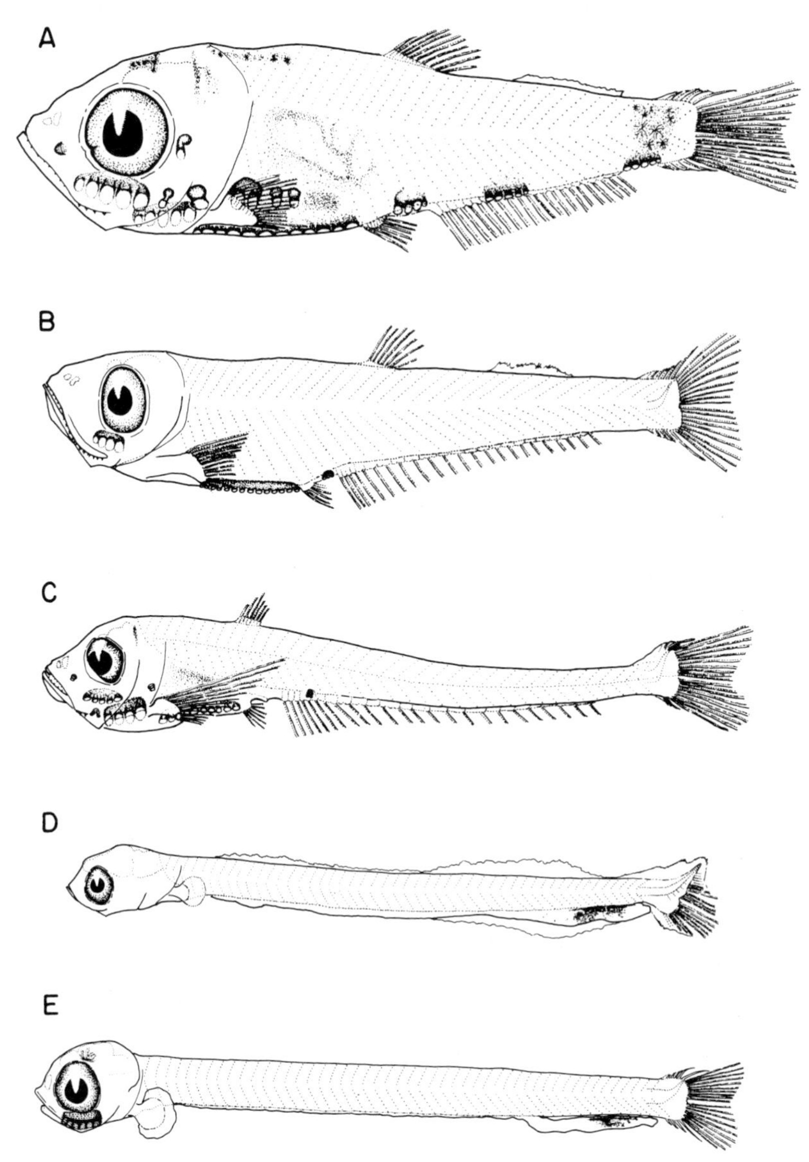

Fig. 2. Larvae and metamorphosing specimens of gonostomatids. (A) *Maurolicus muelleri*, 10.8 mm, middle metamorphosis. (B) *Valenciennellus tripunctalatus*, middle metamorphosis. (C) *Danaphos oculatus*, middle metamorphosis. (D) Maurolicine alpha - 7.5 mm larvae. (E) Same, 16.0 mm, early metamorphosis

The initiation of photophore formation on *Valenciennellus tripunctulatus* is not quite as precocious as in *Maurolicus*: BR and PV photophores were first observed on 8.6 mm larvae. The sequence also is different: the VAV photophores form relatively much sooner and the AC group much later (Fig. 2B). These differences in sequence facilitate identification.

Larvae of *Danaphos* and *Valenciennellus* occur together in the eastern Pacific, and small specimens are somewhat similar in appearance. *Danaphos* has a shorter gut; both lack pigmentation (except peritoneal). Hence it is fortunate that *Valenciennellus* begins photophore formation at a relatively small size, whereas *Danaphos* delays photophore formation until about 16.5 mm. As in the preceding 2 genera, photophore formation begins in the BR and PV groups. Thereafter the sequence differs from that in *Maurolicus* and *Valenciennellus*, although it is closer to *Maurolicus* (Fig. 2C).

We have observed only larvae and early transformation-stage specimens of *Araiophos eastropas*. The larvae have a substantially longer gut than the preceding 3 genera (snout-anus in *Araiophos* larvae - 72-79% SL). Metamorphosis beings at about 11.0 mm, with the initial formation of photophores in the BR and PV groups.

The developmental pattern, shared in common by all Maurolicine genera for which metamorphic stages are known, is a gradual metamorphosis with the initial formation of photophores in the BR and PV groups.

I am illustrating two sizes of a fascinating Maurolicine larvae that I call simply "alpha" (Fig. 2D, E). The eye is large, to the point of dominating the head; the gut is exceptionally long for a gonostomatid larva (ca. 95% SL). The posterior portion of the gut is becoming detached from the body in the larger specimen, undoubtedly to accommodate the anal fin which hasn't yet begun to develop. However, the branchiostegal photophore group is complete - a cluster of 6 branchiostegal photophores with a common base. No other photophores have formed. The number of myomeres is in the mid-40's. The larva is obviously of a primitive Maurolicine gonostomatid. Among described species only *Neophos nexilis* Myers has this number of vertebrae (44-45, S. Weitzman, pers. comm.). If not this, it represents an undescribed form.

Inasmuch as the Maurolicine genera are as closely related to the Sternoptychidae as to the gonostomatid with separate photophores, I will give a little information on the sequence of photophore formation in *Sternoptyx diaphana* and *Argyropelecus lychnus*.

The first photophore pair to form in *Sternoptyx* is the subopercular pair (OP_3). This is soon followed by BR, and then both PV and IP photophores. The lateral (suprapectoral) photophores begin forming before the AC photophores. Marked changes occur in body depth and form during transformation which occurs over a relatively short size range, as measured by SL.

I have not been as successful with transforming specimens of *Argyropelecus lychnus* in tracing the sequence in which photophores form. When I have found transforming specimens, a number of photophores were already developed. The BR and IP groups form their full complement of photophores before the PV group and the AC photophores begin to form before the VAV group.

In my attempt to coordinate what is known concerning metamorphic patterns, I have separated the gonostomatids-sternoptychids into 4 groups of genera.

Group A. Genera in which most or all ventral photophores are laid down initially during a pro-metamorphic (white photophore) stage and which have all photophores separate - *Vinciguerria, Cyclothone, Ichthyococcus, Pollichthys, Diplophos*, and presumably *Woodsia, Yarrella*, and *Polymetme*.

Group B. Genera with a gradual, protracted metamorphosis, with all photophores remaining individually separate and initial photophore formation in OP and PV groups - *Gonostoma, Bonapartia*, and *Margrethia*.

Group C. Genera having some or most photophores in clusters with common bases, a gradual, protracted metamorphosis, with initial photophore formation in BR and PV groups - *Maurolicus, Valenciennellus, Danaphos, Araiophos, Argyripnus*, and presumably *Thorophos, Neophos*, and *Sonoda*.

Group D. More highly specialized fishes having most photophores in clusters with common bases; they have a gradual, protracted metamorphosis with striking changes in body form - sternoptychid genera *Sternoptyx, Argyropelecus*, and *Polyipnus*.

The triad of genera in group B is a closely knit assemblage, with *Gonostoma* the pivotal genus. Metamorphosis among the 6 species of *Gonostoma* shows much greater diversity than among species of other genera, as for example *Vinciguerria* or *Cyclothone*. As pointed out earlier, all species of *Gonostoma* first form the OP_3 pair and 1 to several PV pairs before forming other photophores; thereafter the patterns diverge. Most of the photophores of the ventral series are laid down fairly rapidly in *G. gracile* and *G. atlanticum* - features that ally this pair of species to the genera included in group A. The species of *Gonostoma* with a protracted metamorphosis, as for example *G. elongatum* and *G. denudatum*, share developmental features with the Maurolicine genera (group C).

Certain developmental patterns are common to all 4 groups, and to me the most striking of these is the pattern of photophore formation within the AC group. During formation of these photophores there is usually a sharp separation into 2 groups - a posterior group of 2 or more AC photophores separate from the main AC group. In the anterior AC group, 1 or several of the inner photophores develop initially and are added to both anteriorly and posteriorly. When the posterior group contains 3 or more photophores, they form similarly; eventually the 2 groups may unite. This pattern of AC development is found in *Pollichthys, Ichthyococcus*, and *Diplophos* in group A, in several species of *Gonostoma* (example *G. denudatum*), as well as *Bonapartia* and *Margrethia* in group B, and in a modified form among Maurolicine (group C) and Sternoptychidae (group D) genera, i.e. separation in 3 or more AC groups, each with an inner to outer sequence of photophore formation.

Although *Gonostoma* is considered the pivotal genus, with developmental patterns that show relationships to both the Maurolicine line and the genera in group A; I consider it closer to the genera in group A. There are objections to considering it as basic stock from which the other lines could be derived.

First of all there is the problem of isthmal photophores; these are lacking in *Gonostoma-Margrethia-Bonapartia* but developed in all Maurolicine genera, and in all genera of group A except *Cyclothone*. It is more logical to derive both lines from an ancestor that possessed this group of photophores than from one that lacked it. There is the additional problem of separate photophores vs. clustered photophores when relationships with Maurolicine genera are considered. We are also confronted with the fact that in some larval characteristics, such as gut length, several Maurolicine genera appear to be more primitive than *Gonostoma*. Undoubtedly the divergence between species with single vs. clustered photophores came early in the evolution of gonostomatid-like fishes.

Bassot (1966, 1970) used the structure of light organs to trace evolutionary lineages among stomiatoid fishes. He described 3 types of light organs, the most primitive type was found in *Gonostoma, Bonapartia, Cyclothone*, and *Diplophos (Manducus)*; a distinctively different type in Maurolicine and sternoptychid genera; and third type in *Vinciguerria, Ichthyococcus, Yarrella* and in most stomiatoids (*Chauliodus, Stomias*, etc.). His findings based on light organs are in general agreement with the groups discussed above, except for placement of 2 genera - *Cyclothone* and *Diplophos*.

The question arises as to the relation of the Maurolicine genera to the three genera of sternoptychids. The latter show more marked specializations in both larval and adult characters: they are a more specialized group than the Maurolicine genera. The two assemblages have several important characters in common: clustered photophores, a protracted metamorphic stage, and similar structure of light organs. The Maurolicine genera certainly are as closely related to the sternoptychids as to other genera of gonostomatids. The present separation of the two groups into separate families is artificial. If the sternoptychids are combined with the gonostomatids, the former would have priority as the family name. I personally favor the inclusion of all gonostomatid-sternoptychid fishes in a single family. If divisions are to be made within the family, the genera with clustered photophores would constitute one subfamily, those with separate photophores the other. I find the genus *Gonostoma* too ambivalent in developmental patterns to separate it sharply from other gonostomatid genera with separate photophores.

SUMMARY

The gonostomatid light fishes rank second only to myctophid lantern fishes in abundance in the sea. The family is made up of 20 genera and approximately 60 species, all possessing photophores. Photophore patterns have been used as a primary character in adult taxonomy. The diversity of patterns of photophore acquisition - ranging from most photophores being formed simultaneously to a gradual formation of photophores during a protracted metamorphosis - is used to trace relationships among gonostomatid genera. On the basis of metamorphic patterns, the gonostomatid genera fall into three groups. One group includes genera in which most or all ventral photophores are laid down initially during a "white" photophore stage and which have all photophores individually separate (*Vinciguerria, Cyclothone, Ichthyococcus*, etc.). The second group includes genera with a gradual, protracted metamorphosis, but with all photophores individually separate (*Gonostoma, Bonapartia*, and *Margrethia*). The third group includes genera having some or most photophores in clusters with common bases which are laid down gradually during a protracted metamorphosis (*Maurolicus, Valenciennellus, Danaphos, Araiophos*, etc.). *Gonostoma* is considered the pivotal genus with developmental patterns that show relationships to the other two groups. It is pointed out that the third group is as closely related in developmental pattern to the sternoptychids as to other genera of gonostomatids, and that the two families should be combined into a single family.

ACKNOWLEDGEMENTS

I am indebted to a number of persons for assistance and specimens during the preparation of this manuscript. Particular thanks are due to Henry Orr for preparing the 2 plates of illustrations. H. Geoffrey Moser and John Butler provided help in many facets of the research. Many workers provided specimens including Wm. Richards and Thomas Potthoff, National Marine Fisheries Service, Miami, Florida; Thomas Clarke, Hawaii Institute of Marine Biology; Walter Matsumoto, National Marine Fisheries Service, Honolulu; Shirley Imsand and Michael Barnett, Scripps Institution of Oceanography. Technical assistance during the study was provided by Elizabeth Stevens, Elaine Sandknop, and Patricia Lowery of the National Marine Fisheries Service, La Jolla. I also wish to thank S. Weitzman and N. Merritt for their helpful reviews of the manuscript.

REFERENCES

Ahlstrom, E.H., 1959. Vertical distribution of pelagic fish eggs and larvae off California and Baja California. Fish. Bull., U.S., 60, 107-146.

Ahlstrom, E.H. and Counts, R.C., 1958. Development and distribution of *Vinciguerria lucetia* and related species in the eastern Pacific. Fish. Bull., U.S., 58, 363-416.

Ahlstrom, E.H. and Moser, H.G., 1969. A new gonostomatid from the tropical eastern Pacific. Copeia (3), 493-500.

Badcock, J.R. and Merritt, N.R., 1972. On *Argyripnus atlanticus*, Maul 1952 (Pisces, Stomiatoidei), with a description of post-larval forms. J. Fish. Biol., 4, 277-287.

Baird, R.C., 1971. The systematics, distribution, and zoogeography of the marine hatchetfishes (family Sternoptychidae). Bull. Mus. Comp. Zool., Harvard Univ., 142 (1), 128 p.

Bassot, Jean-Marie, 1966. On the comparative morphology of some luminous organs. In: Frank H. Johnson and Yata Haneda (eds.) Bioluminescence in Progress. Princeton University Press, 557-610 pp.

Bassot, Jean-Marie, 1970. Structure and evolution of the light organs of stomiatoid fish. In: The world ocean. Joint Oceanography Assembly, Tokyo, 13-25 September 1970. Contribution in Biological Oceanography, mimeo, 4 pp.

Grey, Marion, 1964. Family Gonostomatidae. In: Fishes of the western North Atlantic, 78-240. Mem. Sears Found. Mar. Res. 1, part 4.

Holt, E.W.L. and Byrne, L.W., 1913. Sixth report on the fishes of the Irish Atlantic Slope. Fisheries, Ireland, Sci. Invest. (1912), I, 27 pp.

Jespersen, P. and Taning, A. Vedel, 1919. Some Mediterranean and Atlantic Sternoptychidae. Preliminary note. Saertryk af Vidensk. Medd. fra Dansk naturh. Foren., 70, 215-226.

Jespersen, P. and Taning, A. Vedel, 1926. Mediterranean Sternoptychidae. Rep. Danish Oceanogr. Exped. 1908-1910, col. 2 (Biology) A, 12, 1-59.

Kawaguchi, K. and Marumo, R., 1967. Biology of *Gonostoma gracile* (Gonostomatidae) 1. Morphology, life history and sex reversal. Inform. Bull. Planktol. Japan, Commem. Dr. Y. Matsue, 53-69.

Kobayashi, B.N., 1973. Systematics, zoogeography, and aspects of the biology of the bathypelagic fish genus *Cyclothone* in the Pacific Ocean. Thesis, Univ. of Calif., Scripps Inst. Oceanogr.

Moser, H.G. and Ahlstrom, E.H., 1970. Development of lantern fishes (family Myctophidae) in the California Current. Part 1. Species with narrow-eyed larvae. Bull. Los Angeles County Mus. Nat. Hist. Sci. 7, 145 pp.

Mukhacheva, V.A., 1964. The composition of species of the genus *Cyclothone* (Pisces, Gonostomatidae) in the Pacific Ocean. Trudy Inst. Okeanol. Akad. Nauk SSSR 73: 93-138 (Eng. trans. from Russ. in Isrea program for scient. trans., 1966, IpST Cat. No. 141).

Okiyama, Muneo, 1971. Early life history of the gonostomatid fish, *Maurolicus muelleri* (Gmelin), in the Japan Sea. Bull. Jap. Sea. Reg. Fish. Res. Lab., 23, 21-52.

Sanzo, Luigi, 1912. Comparsa degli organi lummosi in una serie di larve di *Gonostoma denudatum* Raf. R. Comitato Talassografico Italiano Mem., 9, 22 pp.

Sanzo, Luigi, 1913a. Larva di *Ichthyococcus ovatus*. R. Comitato Talassografico Italiano Mem., 27, 6 pp.

Sanzo, Luigi, 1913b. Stadi post-embrionali di *Vinciguerria attenuata* (Cocco) e *V. poweriae* (Cocco) Jordan ed Evermann. R. Comitato Talassografico Italiano Mem., 35, 7 pp.

Sanzo, Luigi, 1930. Uova, sviluppo embrionale, stadi larvali, post-larvali e giovanili de Sternoptychidae e Stomiatidae. 2 *Ichthyococcus ovatus*. R. Comitato Tallasografico Italiano Monogr., 2, 69-119.

Sanzo, Luigi, 1931. Uova, larvae e stadi giovanili di Teleostei. Sottordine: Stomiatoidei. Fauna e Flora del Golfo di Napoli, Monogr. 38: 42-92.

Silas, E.G. and George, K.C., 1969. On the larval and postlarval development and distribution of the mesopelagic fish *Vinciguerria nimbaria* (Jordan and Williams - family Gonostomatidae) off the west coast of India and the Laccadive Sea. J. Mar. Biol. Ass. India 11 (1, 2), 218-250.

E.H. Ahlstrom
National Marine Fisheries Service
Southwest Fisheries Center
La Jolla, Calif. 92037 / USA

Early Life History of *Limanda yokohamae* (Günther)

T. Yusa

INTRODUCTION

Aquaculture is becoming an increasingly important industry in Japan. The Japanese government is interested in developing the culture of flatfishes of which *Limanda yokohamae*, the mud dab, is one of the most important species taken by gillnet in coastal waters. Its potential use in coastal aquaculture led to the present investigations on egg and larval development. Natural spawning grounds of the mud dab are found in the coarse sand among rocky reefs, where the demersal eggs adhere to each other and to the coarse sand.

METHODS

Samples of eggs and milt were obtained from mature fish, and the eggs were fertilized by the ordinary dry method. The adhesive eggs were brushed onto microscope slides in single layers and fertilized. The slides were transferred to seawater in constant temperature baths. Six constant temperatures were used: 2.5, 5, 10, 15, 20, and 25°C. Average seawater density (at 15°C) was near 1.025.

Foods used in the experiment to feed hatched larvae were young stages of the following organisms: the rotifer *Brachionus plicatilis*, larvae of saltwater mussels and sea urchins, and early nauplii of brine shrimps.

RESULTS

Egg Development. Hatching occurred in temperatures of 5, 10, and 15°C, but not at 2.5, 20, and 25°C (Yusa et al., 1971). Egg survival to hatching was 50% at 5°C, 70% at 10°C, and 10% at 15°C. The duration of the hatching period varied inversely with temperature. The relation between temperature (X) and incubation period (days) is expressed by the formula

$$y = 33.9e^{-0.18X}$$

Larval Development. The total body length of newly hatched larvae averaged 3.6 mm. Pigmentation of the larvae provides important taxonomic characters. There are groupings of yellow and black chromatophores in six regions of the body; the pectoral fin fold, the dorsal surface of the gut, and four locations associated with the trunk musculature.

Prolarvae of the mud dab (from hatching until 15 days after hatching at 8°C) showed strong positive phototaxis about one week after hatching.

Metamorphosis. Metamorphosis starts about 32 days after hatching at 8°C, and is completed after about 18 days. About 70 days after hatching (8~11°C) the larvae were 13-15 mm in total length and they became negatively phototactic.

Synonomy. The adult of *L. yokohamae* closely resembles that of *L. schrenki*. Yamamoto (1951) suggested that *L. yokohamae* is a synonym of *L. schrenki*. However, the following differences indicate that these are in fact two separate species:

Adhesive layer. The adhesive layer on the outside of the chorion in eggs of *L. yokohamae* is 20-30 μ in thickness in early stages of development. However, in similar eggs of *L. schrenki*, it is 40-60 μ in thickness. This adhesive layer in eggs of *L. yokohamae* does not remain constant in thickness. When germinal ring formation is about at the equatorial plane, the adhesive coat of most eggs becomes concentrated at points of attachment of the egg with other surrounding eggs or substrate. In comparison, the adhesive layer in eggs of *L. schrenki* was observed to be of uniform thickness until hatching occurred.

Incubation period. The incubation period for eggs of *L. yokohamae* (about 16 days) is longer than that for *L. schrenki* (about 12 days) at 4-6°C.

Pigmentation. In most flatfish larvae, pigmentation patterns of the body and fin folds remain fairly constant until the early post-larval stage (50 days after hatching in *L. yokohamae* at 8-11°C). Shortly after hatching there are 4 chromatophore groups in the trunk region of *L. yokohamae*. In *L. schrenki* there are only 2 groups of these chromatophore There are also stellate chromatophores near the margin of the dorsal fin fold in both *L. yokohamae* and *L. schrenki*. These chromatophores tend to be concentrated above the groups of chromatophores located in the trunk region. Although fin fold chromatophore patterns are less well defined, they tend to be located in four regions in *L. yokohamae*, and in two regions in *L. schrenki*.

REFERENCES

Yamamoto, K., 1951. On the egg and larvae of *Limanda schrenki* synonymatic problem between *Limanda schrenki* and *Limanda yokohamae*. Sci. Rep. Hokkaido Fish Hatchery, 6 (1,2), 173-179 (Japanese).

Yusa, T., Forrester, C.R. and Iioka, C., 1971. Egg and larvae of *Limanda yokohamae* (Günther). Fish. Res. Bd Can. Tech. Rep. No. 236, 1-21.

T. Yusa
Tohoku Regional Fisheries Research Laboratory
Shiogama City
Miyagi Prefecture 985 / JAPAN

Larvae of Some Flat Fishes from a Tropical Estuary

K. P. Balakrishnan and C. B. L. Devi

INTRODUCTION

Many species of fishes occur in Cochin Backwater (a tropical estuary); among flat fishes (Heterosomata) only the genus *Solea* Quensel 1806 and the genus *Cynoglossus* Hamilton-Buchanan 1822 are represented. Even though adults of these species do not contribute to a major commercial estuarine fishery, the presence of their eggs and larvae in the plankton cannot be overlooked.

Scattered references are available dealing with eggs and larvae of flat fishes of the Indian waters. Most of them are fragmentary and hence it is difficult to rely on them to asses the parentage. An attempt has been made to identify the larvae of flat fishes occurring in the Cochin Backwater.

MATERIAL AND METHODS

Larval stages were sorted out from formalin preserved plankton samples taken from different Stations and in different seasons using the H-T net (Tranter et al., 1972). Length (standard length) groups preferred for description have been selected based on their morphological differentiation. Alizarin red S stain was used to determine the sequence of ossification. Body measurements were taken using stage and ocular micrometers. All drawings were made using Projectina Optik, Switzerland.

RESULTS

1. Larvae of *Solea heinii* Steindachner 1902.

2.2 mm (Fig. 1): Eyes are symmetrical and black. Alimentary canal consists of a single circular coil with rectal portion directed obliquely backward. Swim bladder is present. Myotomes are indistinct. Median fin folds are continuous and confluent with the caudal. A hump is present on the posterior dorsal aspect of the skull. Pectoral girdle is developed but the fin is not rayed. Ventral half of the cleithrum is obliquely bent forward. Irregular pigment patches and spots are distributed as shown in the figure.

2.4 mm (Fig. 2): Intestinal coil becomes oval by its forward extension and the space between the cleithrum and the intestine is reduced. The hump is more prominent.

2.5 mm (Fig. 3): Rectal portion of the intestine is directed vertically down. 32 vertebral segments are discernible; urostyle is straight. Interspine complexes differentiate from the anterior end which are better developed in the anal fin fold.

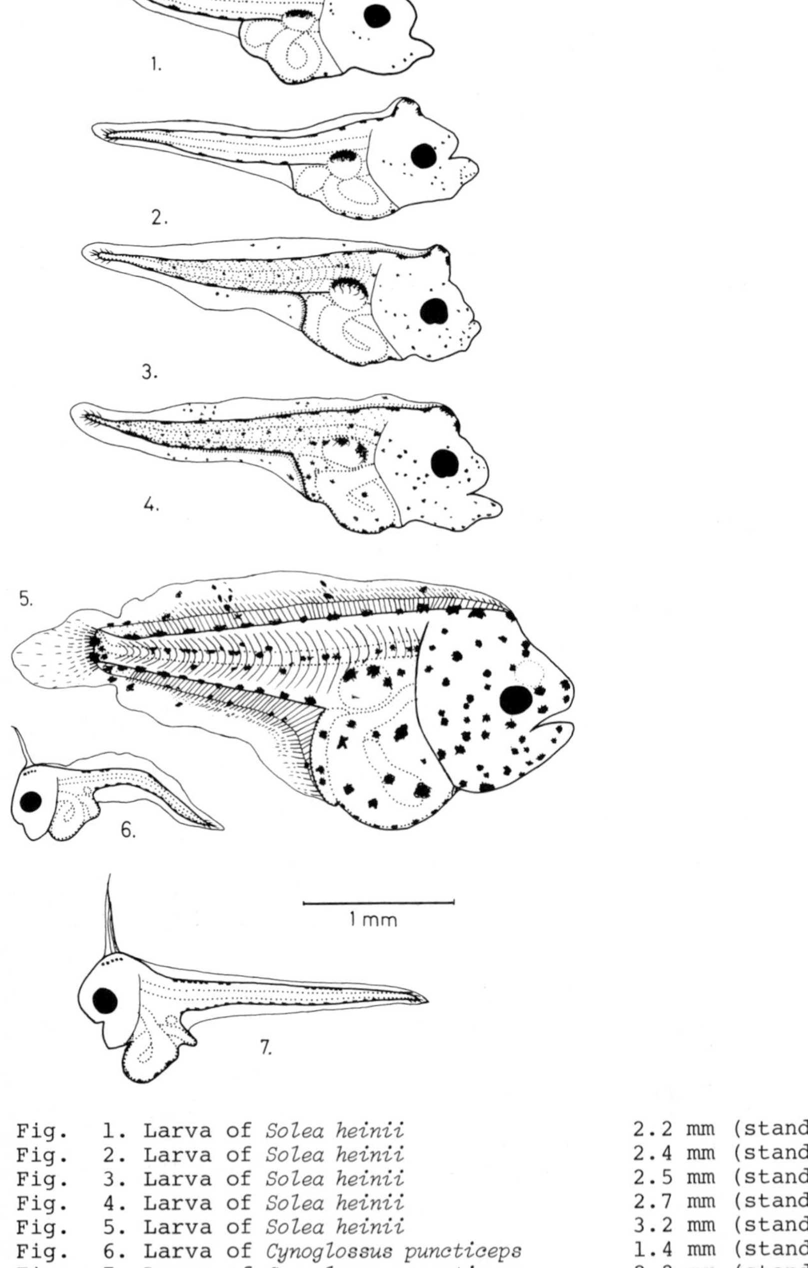

Fig. 1. Larva of *Solea heinii* 2.2 mm (standard length)
Fig. 2. Larva of *Solea heinii* 2.4 mm (standard length)
Fig. 3. Larva of *Solea heinii* 2.5 mm (standard length)
Fig. 4. Larva of *Solea heinii* 2.7 mm (standard length)
Fig. 5. Larva of *Solea heinii* 3.2 mm (standard length)
Fig. 6. Larva of *Cynoglossus puncticeps* 1.4 mm (standard length)
Fig. 7. Larva of *Cynoglossus puncticeps* 2.2 mm (standard length)
Fig. 8. Larva of *Cynoglossus puncticeps* 2.5 mm (standard length)
Fig. 9. Larva of *Cynoglossus puncticeps* 3.4 mm (standard length)
Fig. 10. Larva of *Cynoglossus puncticeps* 4.2 mm (standard length)
Fig. 11. Metamorphosed larva of *C. puncticeps* 4.9 mm (standard length)
Fig. 12. Metamorphosed larva of *C. puncticeps* 4.3 mm (standard length)
Fig. 13. Metamorphosed larva of *C. brevis* 4.0 mm (standard length)

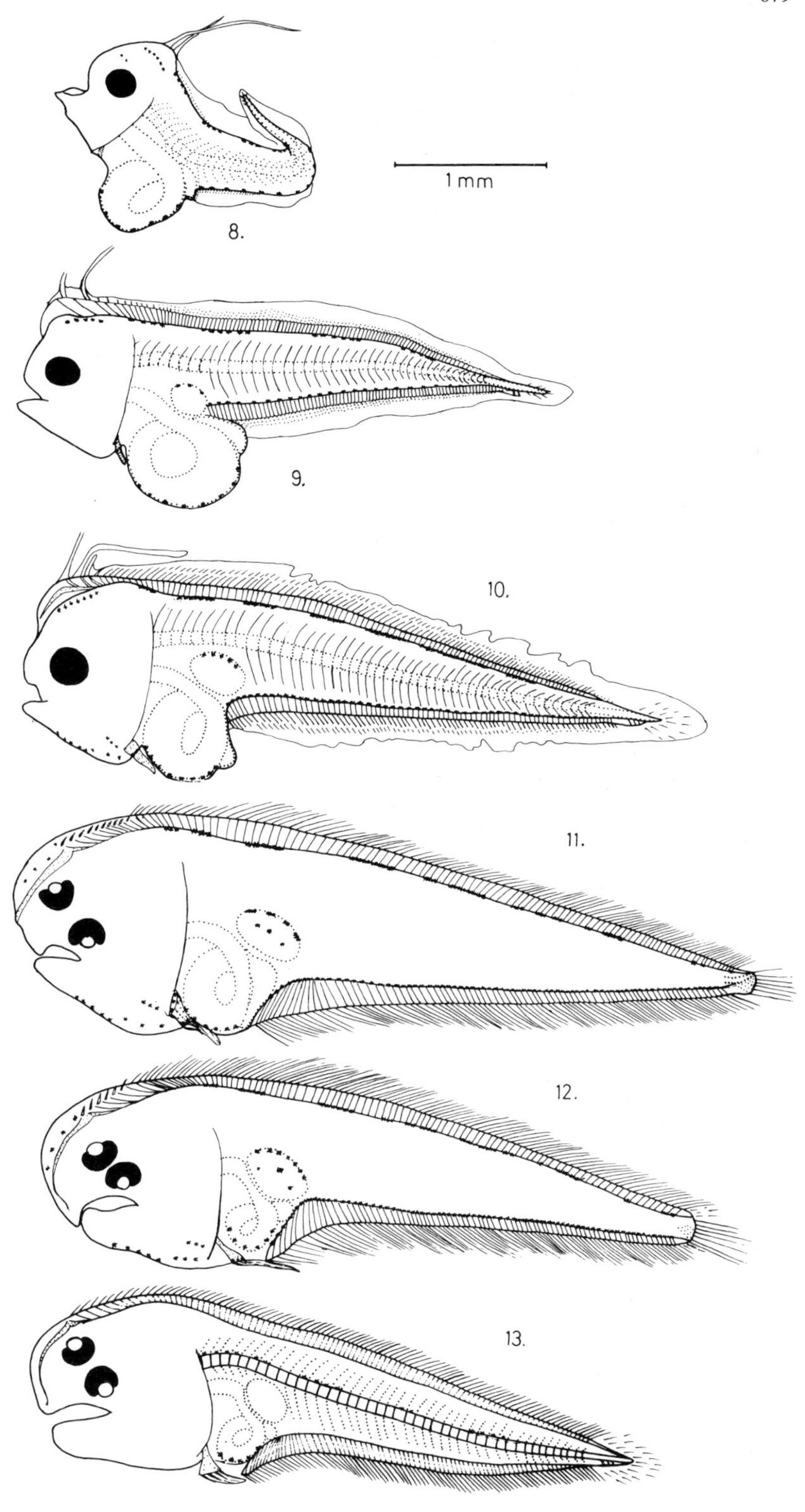
1 mm
8.
9.
10.
11.
12.
13.

2.7 mm (Fig. 4): 33 vertebral segments can be distinguished. The hump gets slightly flattened. Number and intensity of pigments increase.

3.2 mm (Fig. 5): The left eye is found near the dorsal margin. Abdomen bulges out at the ventral profile. Rectal portion is pushed forward by the forward extension of the anal fin. 37 vertebrae, including urostyle, are visible. The centra are not ossified, but the neural and haemal spines have become bony. Median fins are continuous, but remain separate from the caudal. 68 dorsal and 55 anal rays could be deciphered from the bases of fin rays. Hypurals are well developed and the urostyle is strongly deflected dorsalward. Bases of 14 caudal rays are visible. Epurals are not differentiated. Ventral fin rudiments are found at the lower end of the cleithrum.

2. Larvae of *Cynoglossus puncticeps* Richardson 1846.

1.4 mm (Fig. 6): Eyes are black and symmetrical. Mouth and anus present. Intestinal coil is circular. Anus opens on the right side. Rectal portion is directed obliquely backward. Intestine contains a large number of copepod remnants. Abdomen projects beyond the ventral body wall. A tiny swim bladder is present. Dorsal fin fold commences from the level of the eye where a tentacular process supported by the first interneural spine and ray is present. The larvae are very delicate and in many cases the abdomen is torn apart.

2.2 mm (Fig. 7): The second elongated ray supported by the second inter neural spine is differentiated. Ventral fin rudiments are seen at the posterior distal end of the cleithra.

2.5 mm (Fig. 8): Interspines begin to differentiate, but hypurals and epurals do not. Neural and haemal spines are faintly marked in 3.4 mm larvae (Fig. 9). Bases of 98 dorsal and 79 anal rays are discernible. Anterior end of the dorsal fin fold continues to grow beyond the elongated rays, extends over to the snout and is supported by the prolongation of the first interneural spine. Ventral fin rays are in the process of formation and appear as a tiny bud.

4.2 mm (Fig. 10): Swim bladder occupies the space between 5th and 10th vertebrae. Centra are not clearly marked but 48 segments could be deciphered from the neural and haemal spines. About 102 interneural and 82 interhaemal spines are visible. Epural elements are not developed. Embryonic rays are present in the caudal fin fold.

4.9 mm (Fig. 11): (Metamorphosed): Right eye lies in front of the left one. Interorbital space 0.12 mm. A tubular nostril is found in front of the left eye. Cleft of the mouth is asymmetrical, that of the left being oblique and of the right side curved. The diameter of the intestinal coil is reduced. The abdomen is protected by the forward extension of the anal fin and the backward prolongation of the ventral fins. Anus opens on the right side. 49 centra, including the urostyle, are discernible. The forward extension of the dorsal fin fold becomes closely applied to the snout and the first interneural spine extends to its tip. This corresponds to the rostral hook of the adult. 107 dorsal and 83 anal rays are present. Elongated dorsal rays are lost. In the caudal fin 7 rays borne on hypurals are seen. Epural is developed but does not carry rays. Pectoral fin is absent; 4 rays are seen in the ventral fin.

Some of the larvae appear to undergo metamorphosis when they are about 4.3 mm (Fig. 12) in which the interorbital space is 0.045 mm, 48 vertebrae including urostyle 98 dorsal, 78 anal, and 7 caudal rays are present. Though less in numerical count the 4.3 mm larvae are more advanced than the 4.9 mm larvae.

3. Larvae of *Cynoglossus brevis* Günther 1862.

4.0 mm (Fig. 13): (Metamorphosed): Right eye lies a little in front of the left one and the interorbital space is only 0.015 mm. Cleft of the mouth is asymmetrical that of the left side being oblique and extends up to the mid level of the eye. Anus opens on the right side. An oval swim bladder lies between 6th and 10th vertebrae. The abdomen is protected by the forward extension of the anal and backward prolongation of the ventral fins. Anterior margins of 43 centra are differentiated. Urostyle is straight. Interspines are not fully differentiated in the median fin folds. Nevertheless, 94 dorsal and 76 anal rays are discernible. Elongated rays of the dorsal fin are absent. Rostral hook reaches the tip of the premaxilla. Hypurals are not developed and caudal fin remains in the embryonic stage. Pectoral fin is absent. Ventral fin has 4 rays.

4. Larvae of *Cynoglossus cynoglossus* Hamilton Buchanan 1822.

1.6 mm (Fig. 14): Eyes are unpigmented, mouth absent. Intestine runs almost parallel to the body wall. At about its middle the intestine is flexed preparatory to form a coil. Yolk remnants are seen in front of the flexed portion. The median fin folds are well preserved unlike in other larvae and the distortion of the body is comparatively less. Rudiments of the first elongated dorsal ray supported by the first interneural spine is seen in a small tentacular organ at the anterior end of the dorsal fin fold. Pectoral fin present. Irregular dark brown pigment patches and spots are distributed as shown in the figure.

1.9 mm (Fig. 15): Mouth is developed. Intestine consists of a circular coil. Abdomen does not project beyond the ventral body wall. The tentacular organ gains in length.

3.5 mm (Fig. 16): 47 vertebrae including a straight urostyle are present. Swim bladder occupies the space between 5th and 10th vertebrae. First two neural spines are ossified. Two elongated rays are present in place of the dorsal tentacle. Rostral hook extends over the snout. 89 interneural and 67 interhaemal spines are visible. Bases of fin rays are also discernible in the median fin folds. In the caudal fin only embryonic rays are found. Ventral fin radials are developed.

4.0 mm: 98 interneural and 80 interhaemal spines are discernible. Fin rays are only faintly marked. Hypurals and epurals are not differentiated. Ventral fin rudiments are present. In 4.4 mm larvae 4 hypurals are developed in the caudal fin. 4 ventral rays are also differentiated. In the dorsal fin fold 3 elongated rays are present at the anterior end which are well separated from each other, while the remaining rays are crowded together. In 4.6 mm larvae about 101 dorsal, 78 anal and 4 caudal rays are seen.

4.1 mm (Fig. 17): 46 vertebrae including urostyle, 97 dorsal, 78 anal, and 7 caudal rays are present. The rostral hook remains separate from the ethmoidal region of the cranium indicating advancement in development over the 4.6 mm larvae.

4.7 mm (Fig. 18): (Metamorphosed): Right eye lies a little in front of the left eye. Interorbital space is 0.023 mm. Swim bladder occupies the space between 5th and 10th vertebrae. 47 vertebrae including a urostyle, 102 dorsal, 76 anal, and 7 caudal rays are present. Epural is not developed. Elongated dorsal rays are absent at the anterior end. Rostral hook reaches the tip of the premaxilla and is closely applied to the ethmoidal region of the cranium. Pectoral fin is absent. Ventral fin has 4 rays which extends backward to meet the forward extension of the anal fin. Food remnants are not seen in any of the larvae.

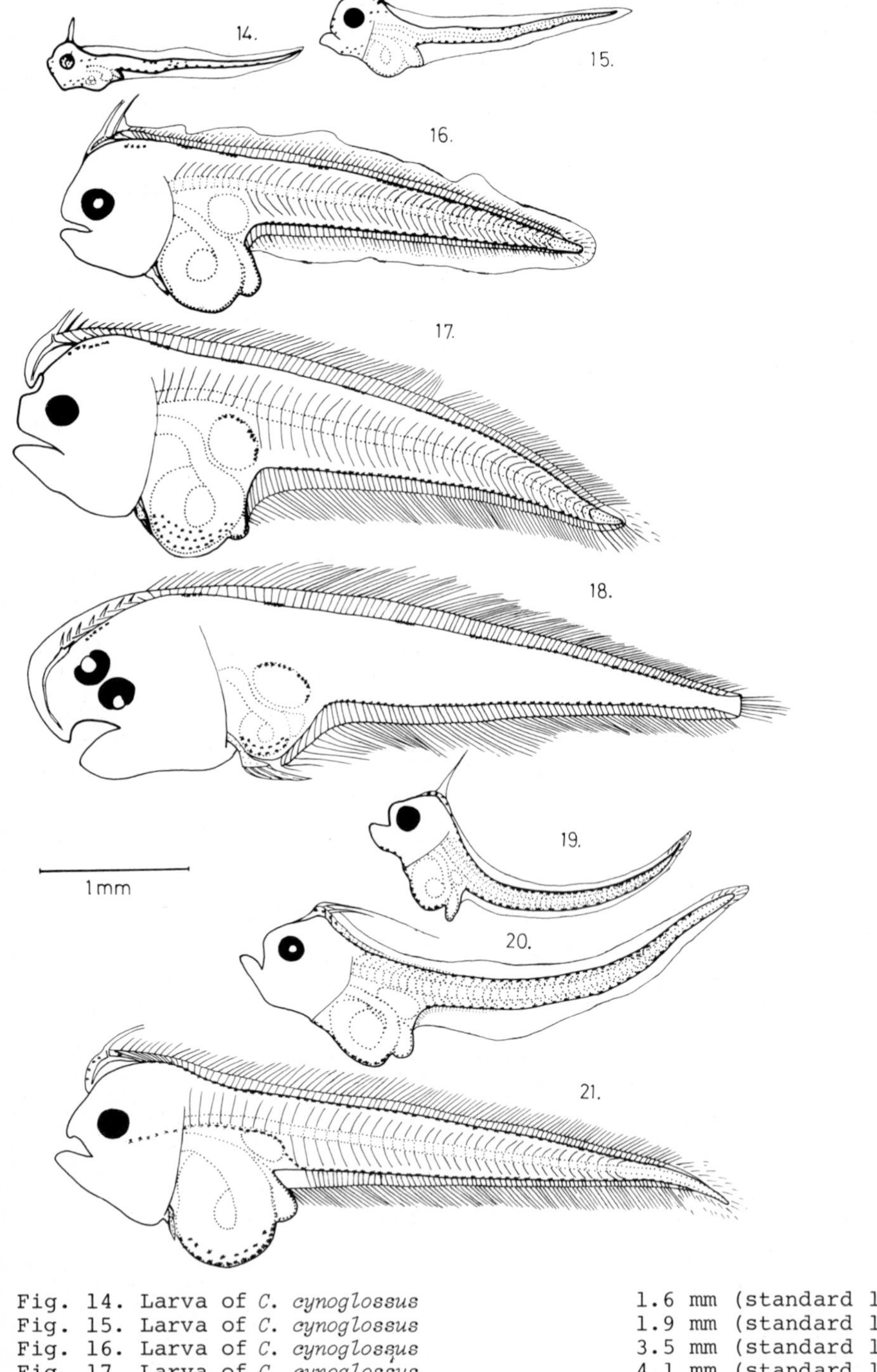

Fig. 14. Larva of *C. cynoglossus*	1.6 mm (standard length)
Fig. 15. Larva of *C. cynoglossus*	1.9 mm (standard length)
Fig. 16. Larva of *C. cynoglossus*	3.5 mm (standard length)
Fig. 17. Larva of *C. cynoglossus*	4.1 mm (standard length)
Fig. 18. Metamorphosed larva of *C. cynoglossus*	4.7 mm (standard length)
Fig. 19. Larva of *C. lida*	2.1 mm (standard length)
Fig. 20. Larva of *C. lida*	3.3 mm (standard length)
Fig. 21. Larva of *C. lida*	4.6 mm (standard length)

5. Larvae of *Cynoglossus lida* (Bleeker) 1852.

2.1 mm (Fig. 19): Eyes are black and symmetrical. Mouth and anus well developed. Intestine consists of a circular coil with rectal portion remaining separate from the rest. A tiny swim bladder is present. Two elongated rays are developed at the anterior end of the dorsal fin fold. Pigmentation is as shown in the figure.

3.3 mm (Fig. 20): 44 vertebral segments are discernible with difficulty. Interspine complexes are differentiated. Ventral fin rudiments are also visible.

4.6 mm (Fig. 21): 45 vertebral segments including a straight urostyle are discernible. Interspines are developed. These are comparatively smaller than those of other species studied thus far. Bases of 100 dorsal and 82 anal rays are discernible. Hypurals and epurals are not yet developed. Rostral hook reaches only half way over the snout. Ventral fin rays are not present.

DISCUSSION

Larvae of *Solea heinii* possess a hump at the posterior dorsal aspect of the skull (at least in early stages), a circular intestinal coil which soon transforms into an oval one and fills the space between the cleithrum and intestine, a much forwardly bent distal half of the cleithra, a snout which is almost equal to or exceeds the horizontal diameter of the eye, and dense dark brown pigment patches or spots all over the body. The position of the left eye in 3.2 mm larvae is an indication that metamorphosis has started. The fin formula of advanced larvae agrees with those of the adults of *Solea heinii* as described by Norman (1928). It differs from *Solea ovata* (Richardson) in the presence of a hump, number of fin rays, pattern of pigmentation and length of larvae at different stages (Balakrishnan, 1963).

The larvae of *Cynoglossus puncticeps* are comparatively small. The presence of elongated rays at the anterior end of the dorsal fin fold, development of ventral fin at a very early stage, intestine loaded with copepods, flimsy abdomen and body parts that are easily dismembered, the number of vertebrae, fin rays and shape of the body in advanced stages and metamorphosis taking place when the larvae attain a length of 4.3 to 4.9 mm are all helpful characteristics in distinguishing the larvae of *C. puncticeps* from other cynoglossids.

Larvae of *C. brevis* appears to metamorphose earlier than those of *C. puncticeps* because the metamorphosed larvae of *C. brevis* has only 4.0 mm length. The head length is more but the number of vertebrae is fewer than *C. puncticeps*. Early ossification of centra, incomplete differentiation of interspines (even in metamorphosed larvae), better development of all the dorsal and anal fin rays, delay in the formation of hypural and epural elements and in the presence of embryonic fin rays in the caudal fin could be considered distinctive characteristics of *C. brevis*.

The larvae of *C. cynoglossus* appear to have a robust body. Their fin folds retain good shape and their bodies undergo very little distortion on preservation. Hence the larvae could be easily picked out from the plankton at all stages without much difficulty. The larvae are apparently longer than those of *C. puncticeps* because even at 1.6 mm length the functional mouth and eye are not differentiated. The intestine contains no food remnants at any stage. The meristic characters are distinct from the larvae of other species and the number of fin rays agrees with that of the adult *C. cynoglossus*.

In *C. lida* the ratio between standard length and depth at cleithra are comparatively more than *C. puncticeps*, *C. brevis*, and *C. cynoglossus*. The interspines are shorter than other species. *C. lida* appears to be the largest of the four species described. When the other three species metamorphose at a length of 4.0 to 4.9 mm, larvae of *C. lida* remain only as a symmetrical postlarva even at 4.6 mm. The rostral hook, reaching only half way over the snout in 4.6 mm larva, indicates that when the larvae reach metamorphosing stage they would have attained a length greater than that of other species. The number of fin rays almost agrees with that of the adult.

From the above it is apparent that most of the larvae of the tropical estuarine waters metamorphose before they attain a standard length of 4 to 5 mm.

REFERENCES

Balakrishnan, K.P., 1963. Fish eggs and larvae collected by the research vessel 'Conch'. 2. Larvae of *Arnoglossus tapaenosoma* Blkr., *Bothus ocellatus* Agassiz, *Laeops güntheri* Alc., *Solea ovata* Rich. and *Cynoglossus monopus* Blkr. Bull. Dept. Mar. Biol. Oceanogr. Univ. Kerala, 1, 81-96.

Norman, J.R., 1928. The flat fishes (Heterosomata) of India with a list of specimens in the Indian Museum Part II. Rec. Ind. Mus., 30, 173-215.

Tranter, D.J., Devi, C.B.L. and Balakrishnan, K.P., 1972. Heron-Tranter Net. In IOBC Handbook. Proceedings of the Workshop on Plankton Methods, 3, 6-7.

K.P. Balakrishnan
Department of Marine Sciences
University of Cochin
Cochin 682016 / INDIA

C.B.L. Devi
Indian Ocean Biological Centre
National Institute of Oceanography
Cochin 682018 / INDIA

Morphological Studies of Two Pristigasterinae Larvae from Southern Brazil

Y. Matsuura

INTRODUCTION

This study is a part of the Sardine Project SOL to investigate the spawning and early life history of the Brazilian sardine and its relatives, and includes beam-trawl samplings in Southern Brazil, from Cabo de São Tomé (22°00'S) to Cabo de Sta. Marta Grande (28°35'S). The purpose is to demonstrate differences among very similar larvae and juveniles of clupeoid fishes. Even though the species studied may not be of direct commercial value, morphological and ecological studies of these fishes are necessary for an understanding of the interrelationships of the different forms and their place in the ecosystem. In this paper the larval stages of *Pellona harroweri* (Fowler) and *Chirocentrodon bleekerianus* (Poey) are described and compared, with a discussion of the systematic status of the latter. Comments are also made about the position of *Chirocentrodon* within the family Clupeidae.

MATERIAL AND METHODS

The material used in this work was collected with a beam-trawl net (1.5 m mouth opening) from the sea bottom by the R/V "Prof. W. Besnard" and "Emilia" of the Instituto Oceanográfico da USP during 1970 and 1971. Beam-trawl and otter-trawl samplings were made over the continental shelf (from the coast to 200 m depth line) of the region studied. Sampling methods and station data are given by Iwai (1973).

The fish were fixed and preserved in a solution of 10% formalin. All drawings were made with the help of a universal projector, using magnification of X 10 for larvae, and X 20 and X 50 for the maxillary bone and caudal skeleton. The fish for osteological studies were cleared and stained according to the method of Clothier (1950). Counts and measurements were made in stained specimens. We used 150 specimens of *P. harroweri* (10.3 - 58.0 mm standard length) and 167 of *C. bleekerianus* (12.6 - 58.5 mm S.L.). The terminology of the caudal skeleton is based on Nybelin (1963) as modified by Monod (1967 - use of the term 'Parhypural' = hypural 1 of authors). Measurements of larvae and juveniles less than 20 mm were taken with a micrometer eyepiece under the stereomicroscope. The larger fish were measured with dial calipers. Fin-ray counts were made under microscope magnification.

The measurements and counts are made as follows:

Standard length: Distance from tip of snout to posterior end of hypural plate, or tip of urostyle in smaller specimens. The total length was not taken due to caudal fin damage in many specimens.

Body depth: Vertical distance from origin of dorsal fin to ventral edge. The greatest body depth was used for adults of *C. bleekerianus* to compare with the data in the literature.

Head length: Distance from tip of snout to posterior edge of operculum, or to the shoulder girdle in small larvae.

Predorsal, preanal, and preventral distances: Distance from the tip of snout to anterior edge of the base of first fin-ray of each fin.

Fin-ray counts: The terminal divided ray of the dorsal and anal fins were counted as one when these were connected to the last pterygiophore All rudimentary anterior rays of the dorsal and anal fins were included in the counts. Normally 1 unbranched anterior soft ray was observed in the dorsal fin and 2 rays in the anal fin, but 2 rays in the dorsal fin and 3 rays in the anal fin were also observed.

In the study of body proportions we used regression analysis. Regressions were computed for all variables measured, but only those showing allometric growth are presented. In this case the independent variables were separated at the point of inflection into two parts and the linear regressions calculated for each part. To show the variation in position of fin origins and the changes in certain body parts, body proportions as a percentage of standard length were plotted against standard length, using the data from the linear regressions.

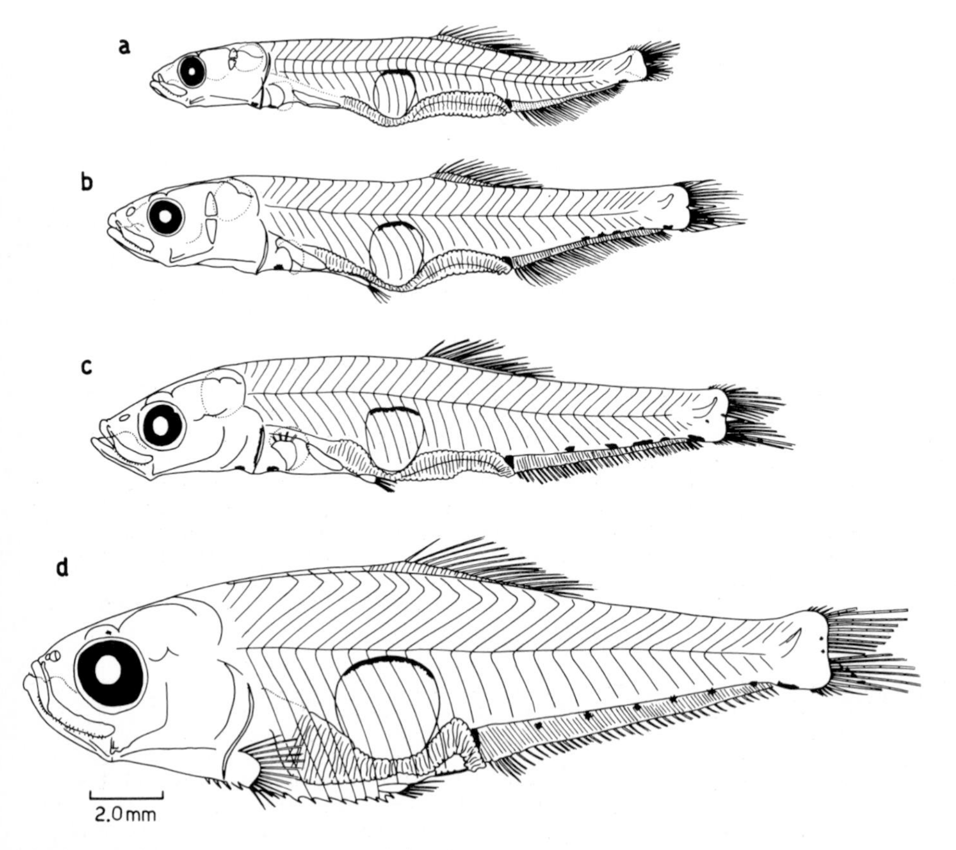

Fig. 1a-d. Development of the post-larvae of *Pellona harroweri*. (a) 13.5 mm; (b) 15.5 mm; (c) 17.5 mm; (d) 22.0 mm (in standard length)

I. *Pellona harroweri* (Fowler, 1917)

1. Body Proportions. Fig. 1 shows the development of the post-larvae of *P. harroweri*. The regression data are shown in Table 1. The regression lines of predorsal and preanal distances to standard length show an inflection at 25 mm (Fig. 2).

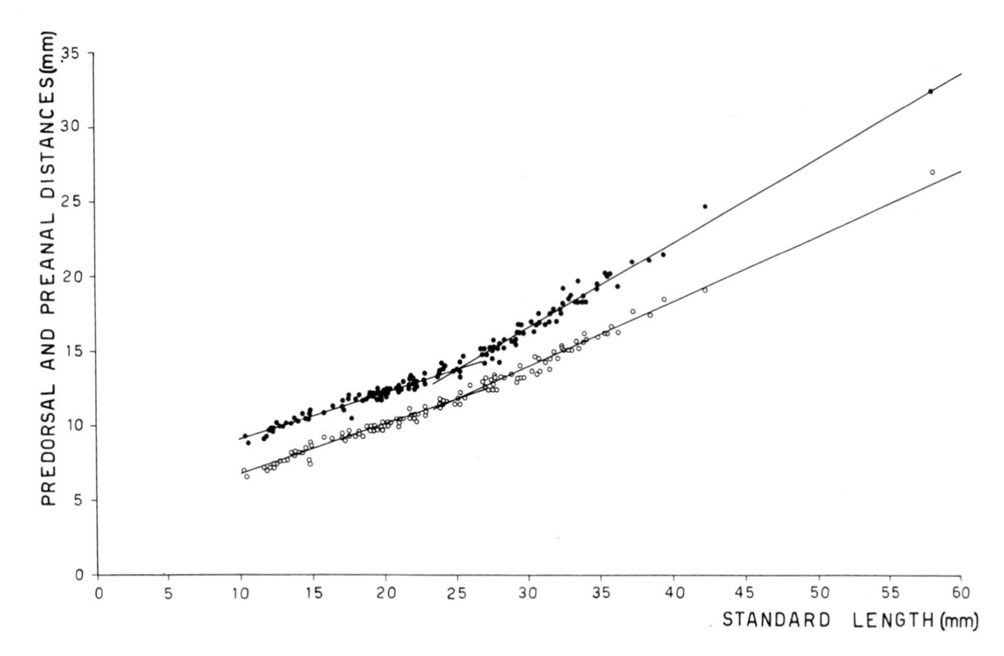

Fig. 2. Regressions of predorsal and preanal distances on standard length of *P. harroweri*. o predorsal distance, • preanal distance

The position of the anal and dorsal fin origins varies with growth. In general we can say that when compared to most other clupeoid fishes, adult *P. harroweri* have the origin of the anal fin considerably forward and the base of this fin rather long. At the post-larval stage of 13.5 mm S.L. (Fig. 1), the anal fin base is short and the origin of the fin considerably backward. From this size the anal fin origin advances to the stage of 25 mm length. This is shown in Fig. 3a. It can be seen that after the intersection corresponding to the standard length of 25 mm, the preanal distance remains stable. The origin of the dorsal fin shows a similar pattern (Fig. 3a). The body depth increases with size. In general, small specimens have a more elongated body than larger specimens (Figs 1 and 3a).

2. Pigmentation. The development of pigment cells is very poor in the post-larval and juvenile stages of *Pellona harroweri*. Therefore the termination of metamorphosis is not clear. The pigmentation in the smallest specimen examined (10.3 mm) consists only of some chromatophores on the antero-ventral part of the pectoral fin membrane and on the dorsal part of the air bladder. The 13.5 mm specimen has one chromatophore on the antero-ventral part of both the shoulder girdle and the pectoral fin membrane. A characteristic and noteworthy melanophore lies above and behind the anus (Fig. 1a). The 17.5 mm specimen shows the melanophore at the shoulder girdle becoming larger and 4 pronounced

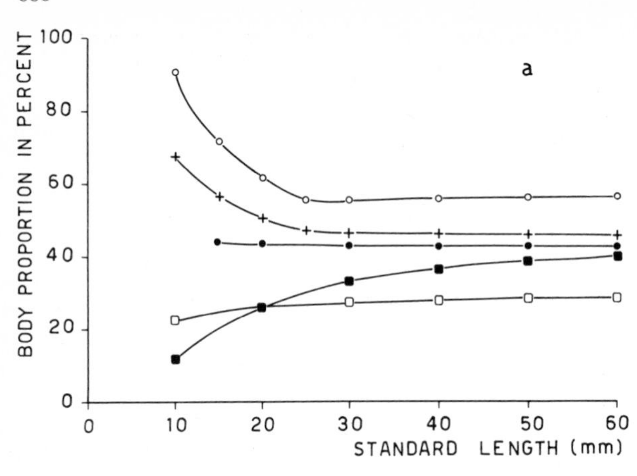

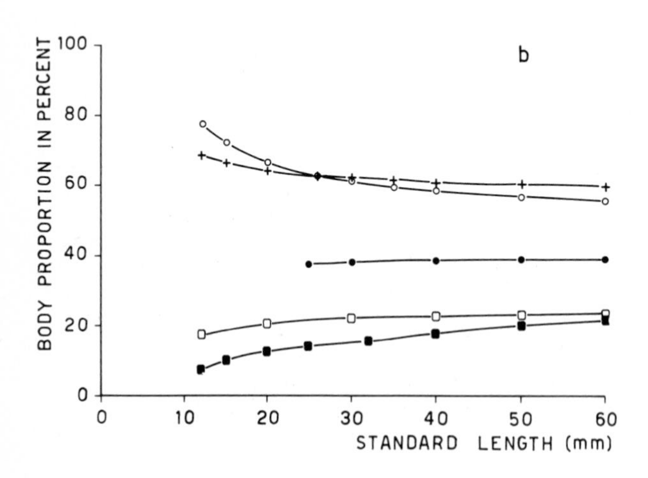

Fig. 3a and b. Changes in certain body parts in percent to standard length in the larvae and juveniles of *P. harroweri* (a) and *Chirocentrodon bleekerianus* (b). o preanal distance, + predorsal distance, ● pre-ventral distance, □ head length, ■ body depth

melanophores at the anal fin base. There is also one melanophore on the parhypural and others lie on the ventral lobe of caudal fin (Fig. 1c). The 22.0 mm specimen (Fig. 1d) shows the melanophores becoming larger and denser on the anal fin base. Their size and position vary. From this size onward the pigmentation on the brain case starts to develop and become intense. The melanophores of the antero-ventral part of the body, both of the pectoral fin and the shoulder girdle, disappear.

Table 1. Regression data for *P. harroweri* at post-larval and juvenile stages

Dependent variables (y)	Specimen size range	n	a	b	s^2	F
Body depth[a]	10.3-58 mm	150	-4.2575	0.4719	0.1691	11443
Head length[a]	10.3-58 mm	150	-0.6895	0.2951	0.1043	7257
Predorsal distance	10.3-25 mm	88	3.4114	0.3352	0.0572	2623
Do	25 -58 mm	61	0.5353	0.4475	0.1563	1981
Preanal distance[a]	10.3-25 mm	88	5.8695	0.3214	0.0695	1978
Do	25 -58 mm	61	-0.4241	0.5689	0.2325	2152
Preventral distance[a]	15.7-58 mm	127	0.2140	0.4220	0.1186	8255
Eye diameter[b]	11.6-58 mm	35	-0.5538	0.3893	0.0447	1726
Snout length[b]	11.6-58 mm	35	-0.3349	0.3420	0.0413	1442
Caudal peduncle[c]	11.6-58 mm	35	0.4119	0.2724	0.0537	1652

Independent variables (x): [a]standard length; [b]head length; [c]body depth. n = total number of specimens, a = y-intercept of regression line, b = slope of regression line, s^2 = mean square deviation, F = variance ratio.

3. Meristic Characters. All fin rays are complete in the 21 mm specimen. The first ossification of the fin rays is observed in the caudal fin. In the smallest specimen (10.3 mm) the principal caudal fin rays show complete ossification with 19 principal rays. This is expected because the caudal fin is more important for swimming than any other fin. Later, at a length of about 14 mm, the dorsal and anal fins become completely ossified. The ossification of the ventral fin may be delayed until the larvae reach a length of 16 mm. The pectoral fin membrane is present in the smallest larvae; but the ossification of the fin rays starts only at a standard length of about 17 mm and is complete by about 21 mm. Table 2 shows the complete fin ray counts.

Table 2. Meristic counts of *P. harroweri* larvae and juveniles

	n	$\bar{x}$	s	Range
Dorsal fin ray	132	15.99	0.2619	15 - 17
Pectoral fin ray	30	14.23	0.6260	13 - 16
Anal fin ray	128	39.38	1.1162	37 - 42
Caudal fin ray	149	18.96	0.3047	18 - 19
Ventral scute, anterior	77	18.43	0.5238	18 - 20
posterior	65	6.00	0.1766	5 - 7
Vertebrae	136	39.10	0.3706	38 - 40

n = number of specimens, $\bar{x}$ = average of meristic counts, s = standard deviation.

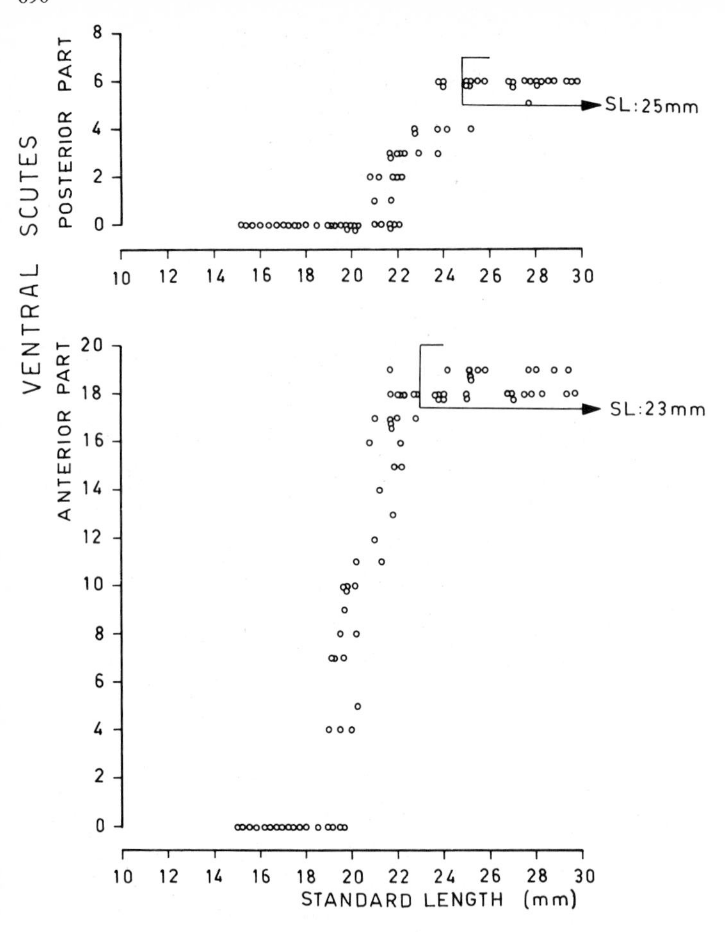

Fig. 4. Development of ventral scutes in the post-larvae of *P. harroweri*

At 21 mm all fin rays are complete, but the ventral scutes are not. Their ossification starts at about 19 mm and the full number is completed at a larval size of 25 mm (Fig. 4). The number of ventral scutes is important in the identification of species of the genus *Pellona*. Table 2 shows the average number obtained for 77 specimens examined.

4. Osteology of the Caudal Skeleton and Maxillary Bone. Ossification of the vertebrae starts at the middle of the body trunk and proceeds anteriad and posteriad connecting the vertebral column with the penultimate centrum which has ossified somewhat earlier. The anteriormost centra complete ossification at 13 mm length. The range and mean vertebral number for 136 specimens is given in Table 2.

Fig. 5 shows the development of the caudal skeleton. The urostyle is formed through the union of the last three ural centra with the first pre-ural centrum. In large specimens, the third centrum cannot be seen under the uroneurals. Along with the urostyle, 1 parhypural and 6 hypurals are formed. The third hypural has a single base but is split

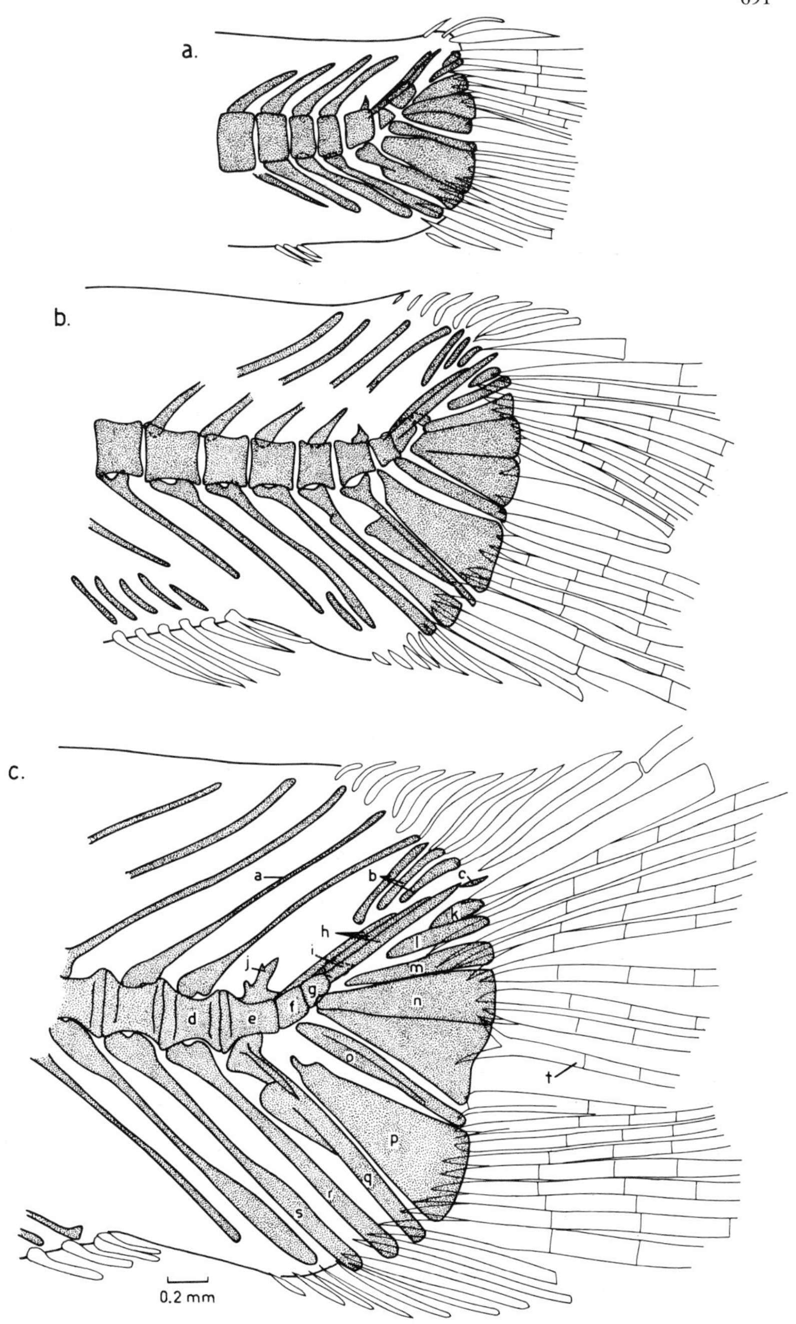

Fig. 5a-c. Caudal skeleton of *P. harroweri*. Stippling indicates alizarin stained bones. (a) 13.5 mm; (b) 17.5 mm; (c) 22.0 mm (in standard length). a - neural spine, b - epurals, c - 4th uroneural, d - 2nd preural centra, e - 1st preural centra, f - 1st ural centra, g - 2nd ural centra, h - uroneural, i - 3rd ural centra, j - specialized neural process, k~p - hypurals, q - parhypural (= haemal arch 1), r~s - haemal spines, t - caudal fin rays

longitudinally in certain stages of development. The lack of a basal articulation between the 1st hypural (2nd hypural of authors) and the urostyle is an important characteristic of the Clupeomorpha (Gosline, 1960).

The 10 principal rays of the upper caudal lobe are associated with 4 hypurals (3rd to 6th) and the 9 principal rays of the lower caudal lobe with the parhypural and 2 hypurals (1st to 2nd). Middle rays have short bases, not overlapping the hypurals, whereas most clupeids have long bases over the hypurals. Twelve procurrent caudal rays on the dorsal side and nine on the ventral side were observed. The specialized neural process is already present in the specimen of 13.5 mm S.L., but the epurals are absent. At 17.5 mm three epurals are present. At this size, one melanophore appears on the parhypural and some melanophores are also present on the lower lobe of the caudal fin. At 22.0 mm, the fourth uroneural appears on the top of the first uroneural and a lateral spine is present on the base of the parhypural. This spine is about one third as long as the parhypural.

The only unequivocal character separating *Pellona* from *Ilisha* is the presence of a toothed hypomaxilla in the former (Berry, 1964a, b). Fig. 6 shows the development of the maxillary bone. At 13.5 mm the

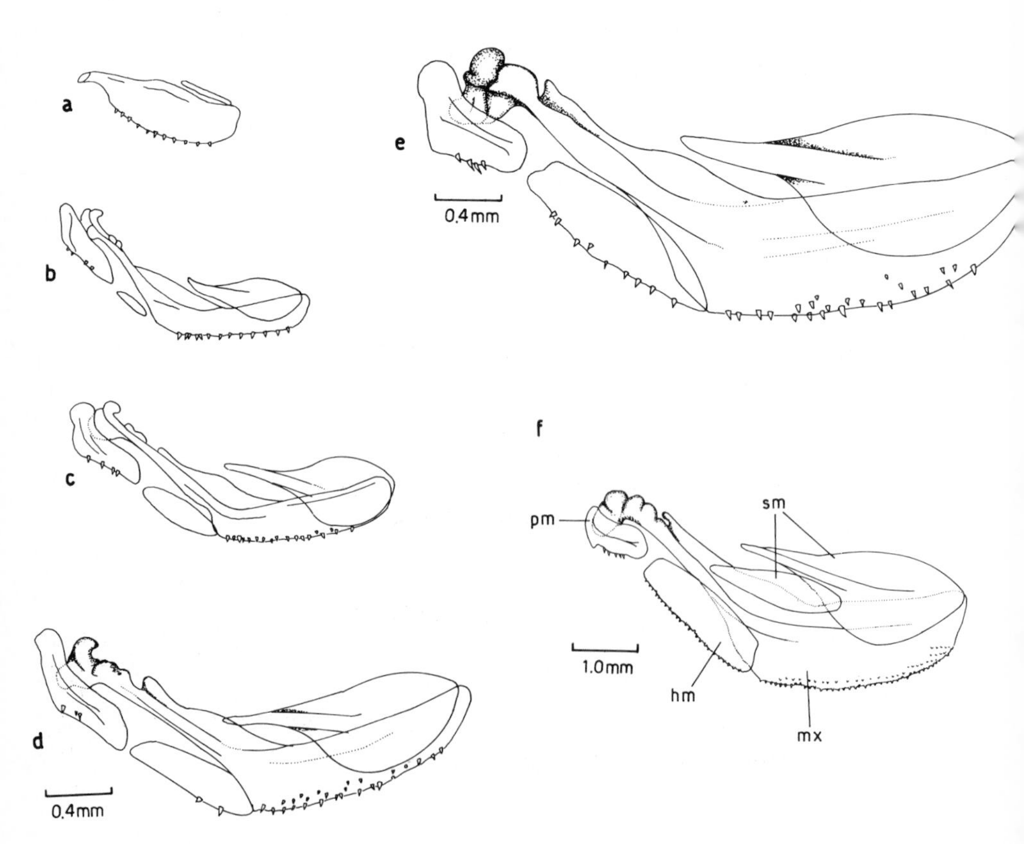

Fig. 6a-f. Development of the maxillary and associated bones of *P. harroweri*. (a) 13.5 mm; (b) 17.5 mm; (c) 19.0 mm; (d) 22.0 mm; (e) 27.3 mm; (f) 44.3 mm (in standard length). pm - premaxilla, sm - supramaxilla, hm - hypomaxilla, mx - maxilla

hypomaxilla is not present (Fig. 6a), but at 17.5 mm it appears (Fig. 6b); at this size the premaxilla does not. At 22.0 mm two conical teeth are found on the hypomaxilla (Fig. 6d). At 27.5 mm the premaxilla, maxilla, hypomaxilla and 2nd (posterior) supramaxilla are well developed, but the 1st (anterior) supramaxilla is not yet ossified (Fig. 6e). At 44 mm the anterior supramaxilla is developed between the maxilla and the anterior arm of the 2nd supramaxilla. The posterior tip of the maxilla reaches one third but not more than one half eye diameter beyond the anterior eye border.

The post-larval stage of this species may be considered complete at 25 mm standard length. All fin rays and ventral scutes are completed, the vertebrae are completely ossified, and the advance of the anal and dorsal fin origins has come to a halt.

II. *Chirocentrodon bleekerianus* (Poey, 1867)

1. Body Proportions. Fig. 7 shows the development of the post-larvae of *C. bleekerianus*. The regression data are presented in Table 3. The regression line of body depth to standard length shows an inflection at 32 mm standard length. The body depth does not show a pronounced increase as does that of *P. harroweri* (Fig. 3). Development of the two species is similar up to the stage of 15 mm. From this size on, the depth of the body varies differently. The regression line of body depth to standard length is shown in Fig. 8.

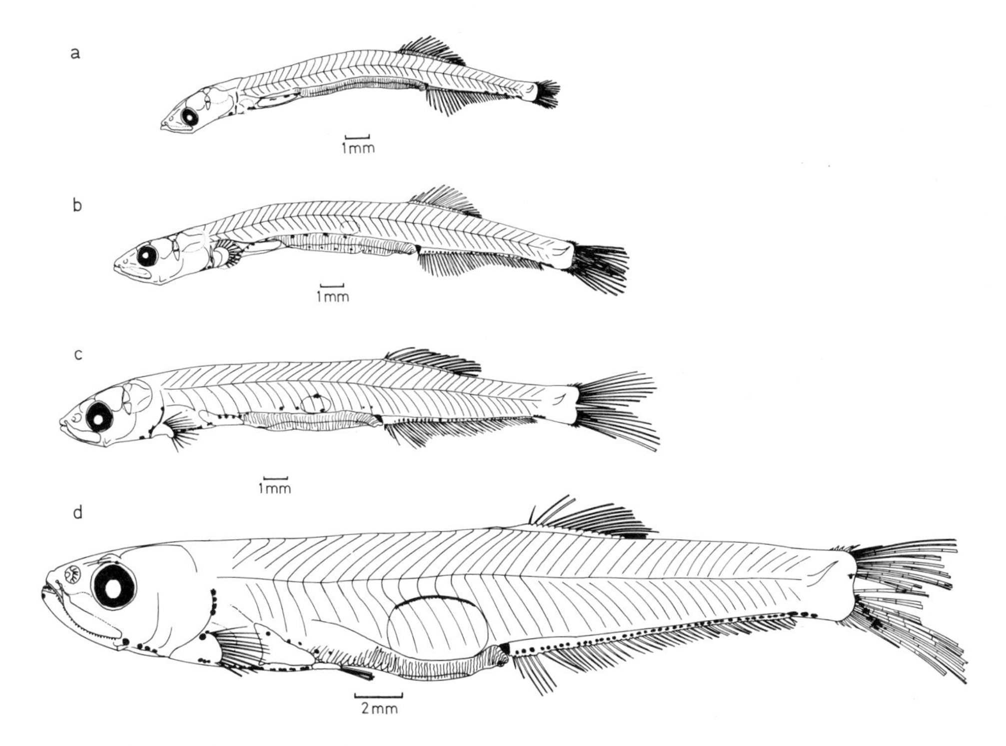

Fig. 7a-d. Development of the post-larvae of *C. bleekerianus*. (a) 16.0 mm; (b) 19.5 mm; (c) 22.0 mm; (d) 34.6 mm (in standard length)

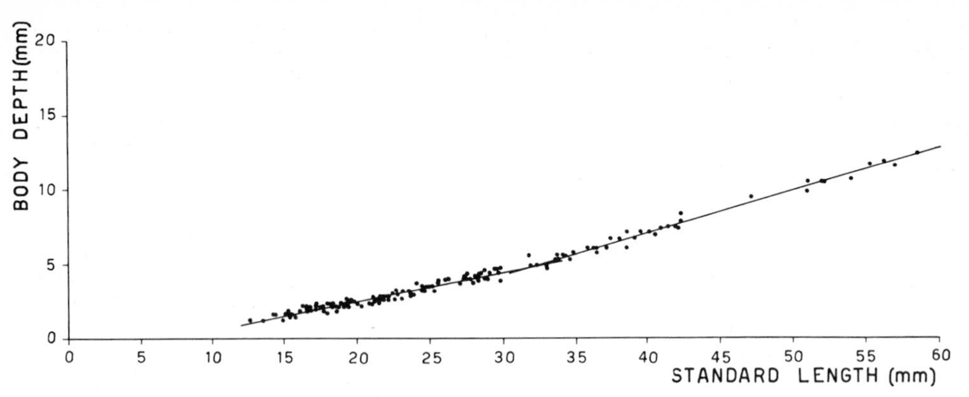

Fig. 8. Regression of body depth on standard length of *C. bleekerianus*

Table 3. Regression data for *C. bleekerianus* at post-larval and juvenile stages

Dependent variables (y)	Specimen size range	n	a	b	s^2	F
Body depth[a]	12-32 mm	124	-1.5309	0.1998	0.0558	1980
Do	32-58 mm	41	-4.2098	0.2831	0.0721	2745
Head length[a]	12-58 mm	167	-0.8980	0.2478	0.1203	8505
Predorsal distance[a]	12-58 mm	164	1.3427	0.5760	0.2645	20467
Preanal distance[a]	12-58 mm	165	3.2849	0.5012	0.2181	18812
Preventral distance[a]	25-58 mm	87	-0.7322	0.4027	0.1470	7616
Eye diameter[b]	13-57 mm	30	-0.2271	0.2590	0.0128	1766
Snouth length[b]	13-57 mm	30	-0.2209	0.2959	0.0223	1325
Caudal peduncle[c]	13-57 mm	30	-0.0174	0.5145	0.0219	3382

Independent variables (x): [a] = standard length; [b] = head length; [c] = body depth. n = total number of specimens; a = y-intercept of regression line; b = slope of regression line; s^2 = mean square deviation; F = variance ratio.

The origin of the anal fin in adult *C. bleekerianus* is relatively advanced and the anal fin base is long (33 to 37% in standard length). The advancement of the anal fin origin starts at the beginning of the post-larval stage and continues up to the length of 40 mm (Fig. 3b). The ratio of the predorsal distance to standard length decreases more slowly than that of the preanal distance. The two linear regressions cross at about 26 mm (Fig. 3). This shows that the origin of the dorsal fin is in front of the anal fin in smaller specimens, while in specimens larger than 26 mm S.L., this relationship is reversed. The formation of the ventral fin is observed at about 25 mm length. The ratio of preventral distance to standard length remains almost constant during development.

2. Pigmentation. Early post-larval stages of *C. bleekerianus* show little pigmentation. At about 12 mm the fish has one melanophore, triangular in shape, lying above and behind the anus. One melanophore lies on the antero-ventral part of shoulder girdle and another is behind the pectoral fin. A series of melanophores are also present on the anal fin base and there is a patch on the ventral part of the caudal peduncle.

Fig. 7b shows the distribution of melanophores at the anal fin base of a 19.5 mm specimen. Three melanophores are found on the antero-ventral part of the shoulder girdle and another group appears above this. The pigmentation behind the pectoral fin develops posteriad. At this size, a small melanophore appears on the lower lobe of the caudal fin. The 22 mm specimen has one group of melanophore on the dorsal surface of the air bladder and the pigmentation behind the pectoral fin shows further development along the digestive canal (Fig. 7c). The 36.5 mm specimen (Fig. 7d) presents well developed pigmentation in the ventral parts of the body: 4 melanophores on the pharynx, a patch of melanophores on the pectoral fin base, 1 series on the belly from the pectoral fin to the ventral fin, 1 melanophore on the 4th hypural and 1 on the tip of the mandible.

At this size the general appearance of this species is rather similar to that of an engraulid larvae of the same size. It possesses a large mouth with a long slender maxilla and an advanced anal fin origin. But it can be distinguished from engraulid larva of this area by the pigmentation along the anal fin base, the triangular melanophore near the anus and the series of melanophores on the ventral edge of pectoral region.

3. Meristic Characters. The complete number of fin rays is present in the 26 mm specimen. The first ossification of the fin rays occurs in the caudal and dorsal fins; the smallest specimen (12.6 mm) shows a perfect ossification of these fin rays. Later, at about 24 mm, the ventral fin becomes completely ossified. At 26 mm the anal and pectoral fin rays are completely ossified.

The ossification of the ventral scutes in this species is delayed and is completed only at a length of 52 mm. The ossification starts at about 35 mm and the part anterior to the ventral fin completes its ossification at about 50 mm; the posterior part ossifies completely at about 52 mm (Fig. 9). The anterior part has 15 to 17 scutes and the posterior part 9 to 10. The total number of ventral scutes is 24 to 27 with a mean of 25.65 ($s = 0.7451$).

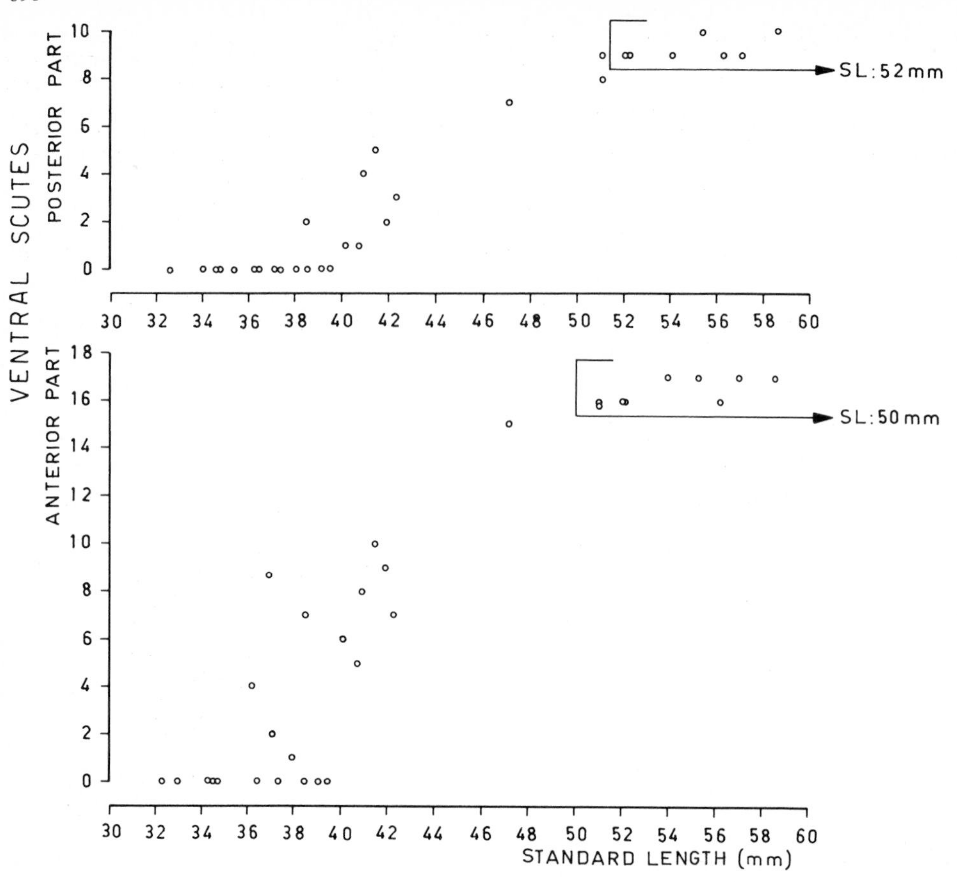

Fig. 9. Development of ventral scutes in the post-larvae of *C. bleekerianus*

The ossification of the vertebrae is completed at about 13 mm. Table 4 shows the ranges and means for fin rays, vertebrae, and ventral scutes.

Table 4. Meristic counts of *C. bleekerianus* larvae and juveniles

	n	$\overline{x}$	s	Range
Dorsal fin ray	135	15.83	0.5670	14 - 17
Pectoral fin ray	49	13.53	0.5041	13 - 14
Anal fin ray	50	42.06	1.7427	39 - 47
Ventral scute, anterior	9	16.44	0.5270	16 - 17
posterior	8	9.25	0.4628	9 - 10
Vertebrae	145	45.14	0.7166	43 - 47

n = number of specimens, $\overline{x}$ = average of meristic counts, s = standard deviation.

4. Osteology of the Caudal Skeleton and Maxillary Bone. Fig. 10 shows the development of the caudal skeleton. The formation of the urostyle and hypurals is similar to that of *P. harroweri*. Along with the parhypural, 6 hypurals are formed. The 10 principal rays of the upper caudal fin lobe are associated with 4 hypurals (3rd to 6th) and the lower 9 with the parhypural and 2 hypurals (1st to 2nd). The number of procurrent caudal rays was 10 or 11 on the dorsal side and 9 to 11 on the ventral side. The structure of the caudal skeleton of the 12.6 mm specimen is not clear although it is well stained. At a length of 19.8 mm (Fig. 10a), the parhypural and 6 hypurals can be distinguished, but the specialized neural process was not observed. A uroneural can be seen along the urostyle, but the third uroneural does not appear at this size. At 22.4 mm (Fig. 10b), the specialized neural process and the third uroneural are present. At 28 mm (Fig. 10c), 3 epurals can

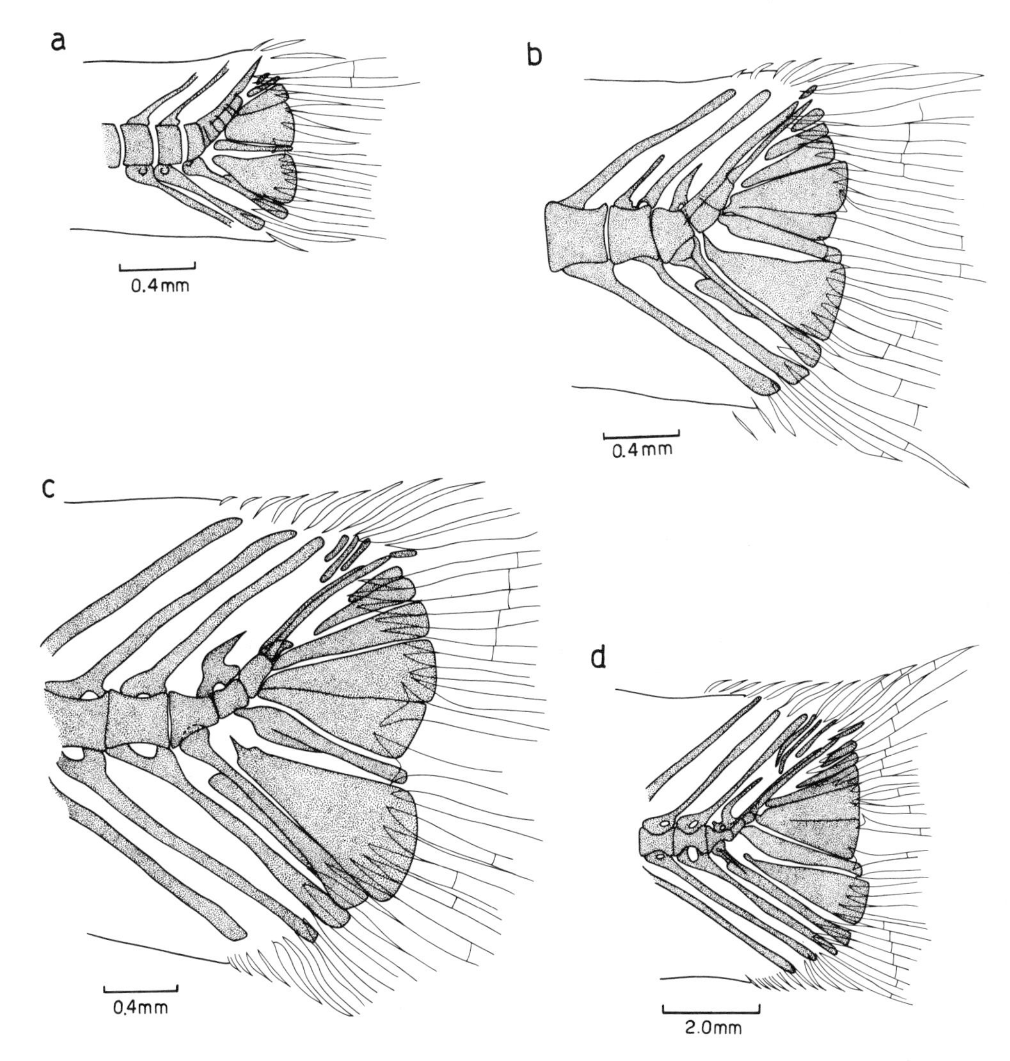

Fig. 10a-d. Caudal skeleton of *C. bleekerianus*. Stippling indicates alizarin stained bones. (a) 19.8 mm; (b) 22.4 mm; (c) 28.0 mm; (d) 56.0 mm (in standard length)

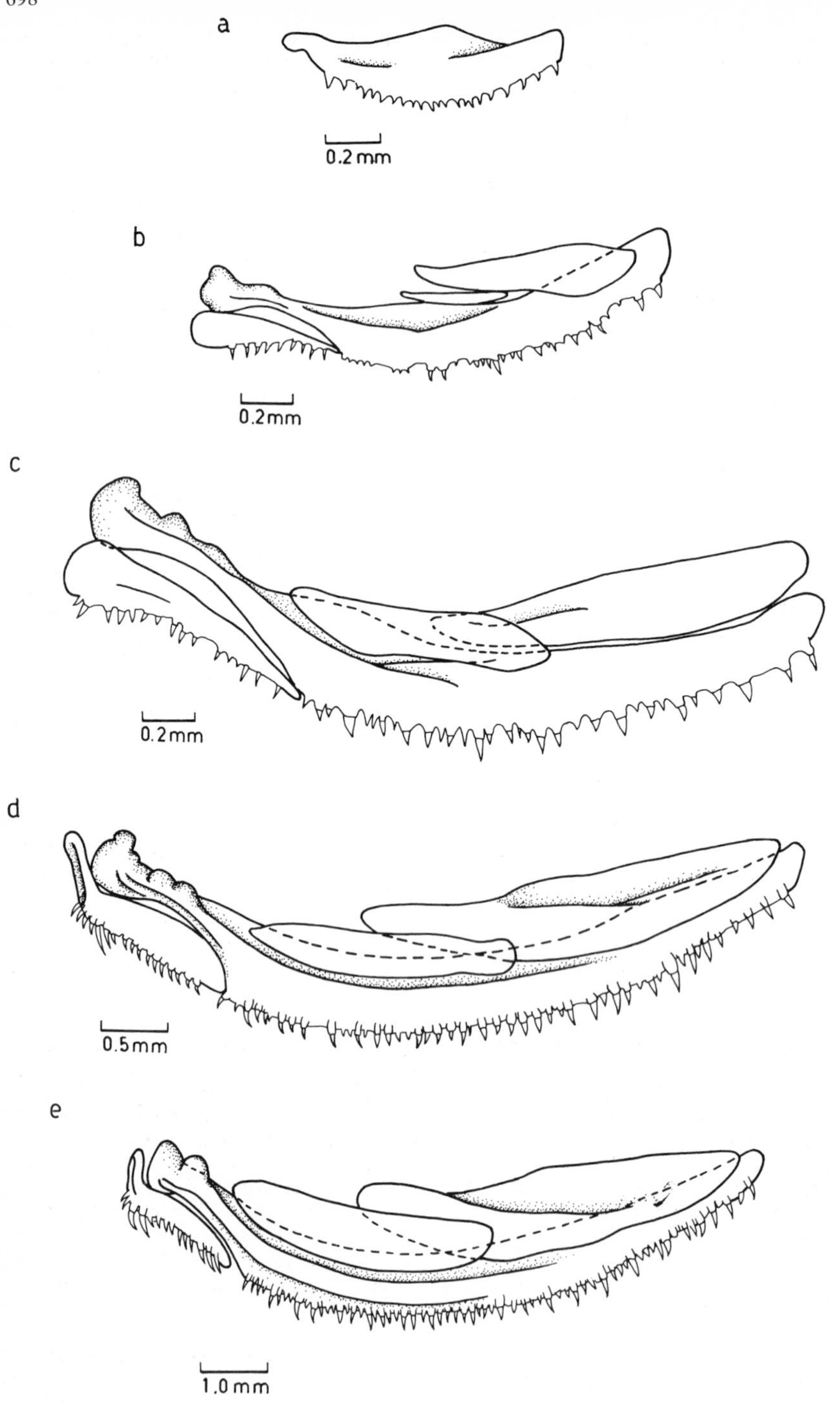

Fig. 11a-e. Development of the maxillary and associated bones of *C. bleekerianus*. (a) 19.8 mm; (b) 21.6 mm; (c) 28.0 mm; (d) 38.0 mm; (e) 56.0 mm (in standard length)

be seen above the uroneurals. At 56.0 mm (Fig. 10d), the specialized neural process and epurals are well developed and many melanophores appear in the caudal region. One lateral spine is present at the base of the parhypural.

Fig. 11 shows the development of the maxillary bone. The genus *Chirocentrodon* has no hypomaxilla, differing in this character from the genus *Pellona*. At 19.8 mm (Fig. 11a), the maxilla is present and the conical teeth are evident. Other maxillary bones are not yet ossified at this size. At 21.6 mm (Fig. 11b), the well-developed maxilla, toothed premaxilla and two supramaxillae are present. The conical teeth on the maxilla and premaxilla are irregular in size. At 28.0 mm (Fig. 11c), the well-developed 1st (anterior) supramaxilla is visible and at 38.0 mm (Fig. 11d), the premaxilla has well-developed conical teeth. At 56.0 mm (Fig. 11e), two large canine teeth can be seen on the premaxilla and this characteristic is important in distinguishing the *Chirocentrodon* from *Ilisha*. The dentition on the maxilla and premaxilla continues developing and forms an irregular pattern of long and short conical teeth. All teeth develop on the maxilla's edge in one row. The posterior tip of the maxilla fails to reach the posterior eye border at 58 mm S.L., but in the adult (80 mm S.L.) it extends beyond the eye to give an engraulid-like appearance.

5. The Synonymy of *Chirocentrodon bleekerianus* (Poey). On the basis of specimens from Cuba, Poey gave the first description of this species as *Pellona bleekeriana* in 1867. Günther, independently of Poey's work, described *Chirocentrodon taeniatus* in 1868. *Chirocentrodon* has, generally, been placed in the family Clupeidae, but Jordan and Seale (1926), who evidently considered *C. taeniatus* Günther distinct from *P. bleekerianus* (Poey), established a new genus, *Medipellona*, for *P. bleekeriana* and doubtfully assigned it to the family Engraulidae.

Myers (1929) referred *Medipellona* to *Chirocentrodon* and included *Ilisha caribbaea* Meek and Hildebrand in the genus *Chirocentrodon*. Breder (1942) identified the specimens he collected in Cuba as *C. bleekerianus*; comparing them with *P. bleekeriana*, *C. taeniatus*, and *I. caribbaea*. He found no significant differences and decided they all belonged to the same species. He concluded that "the synonymy of the thus monotypic genus *Chirocentrodon*, as here understood, should stand as *Chirocentrodon bleekerianus* (Poey)".

Hildebrand (1963) examined and compared specimens of *C. bleekerianus* from Cuba, Jamaica, Puerto Rico, Trinidad, Panama, Venezuela, and Brazil (São Paulo) and gave the first detailed description of the species and its geographical range. On the basis of specimens collected at Santos, state of São Paulo, Tommasi (1964) described a new species, *Chirocentrodon cladileokae*. The differences cited for this new species were as follows: "*C. cladileokae* sp. n., difiere de *C. bleekerianus* por presentar mayor número de escudos ventrales (18 posteriormente a las ventrales en *cladileokae* y 10-11 en *bleekerianus*), por la dorsal (15-17 en *cladileokae*, 14-15 en *bleekerianus*), por la anal 41-45 en *cladileokae*, 39-42 en *bleekerianus*) y por el hocico (3,08 a 3,72 en *cladileokae*, 2,75 a 3,4 en *bleekerianus*)".

The type specimens of *C. cladileokae* are lost. I have examined some topotypes and find that the morphometric characteristics agree almost entirely with those of *C. bleekerianus* (Table 5). The only significant difference between the Santos fishes of Tommasi and *C. bleekerianus* was said to be the number of the ventral scutes (posterior to the ventral fin). Tommasi claimed for his species 18 posterior ventral scutes, but all the topotypes examined have never had more than 10 scutes in the posterior part; I believe Tommasi's counts were not accurate. The specimens from Santos, therefore, should also be referred to as *C. bleekerianus*, *C. cladileokae* being a synonym.

Table 5. Measurements of *Chirocentrodon bleekerianus* (Poey)

	Poey 1867	Günther 1868	Breder 1942	Hildebrand 1963	Tommasi 1964	Matsuura
SL	100	89	82	42 -84	67 -107.5	80 -90
No. specimens	7	5	1	22	15	12
DF	15	15	15	14 -16	15 - 17	14 -17[a]
AF	43	41	39	39 -43	41 - 45	39 -47[a]
HL	-	4.3+	4.3	3.85- 4.55	3.87- 4.21	4.1 - 4.4
BD	-	4.6+	4.7	3.85- 5.55	3.84- 4.21	4.0 - 4.5
Eye	3.5	-	3.3	3.0 - 4.0	3.17- 3.80	3.23- 4.20
Snout	-	-	3.3	3.25- 4.0	3.08- 3.73	2.87- 3.70
V.S.	25	?+11	15+10	24 -28	?+ 18	16,17+ 9,10
G.R.[b]	15	17	17	14 -18	14 - 16	15 -17
A.F. base	-	-	3.1	2.5 - 2.95	-	2.7 - 3.0
Maxillary	1.2	-	1.3	1.25- 1.55	-	1.22- 1.48
Vert.	-	-	-	44 -45	-	43 -47[a]
PD	-	-	-	1.6 - 1.7	1.59- 1.69	1.60- 1.76
PA	-	-	-	-	1.66- 1.81	1.73- 1.88

[a]Used smaller specimens for meristic counts. [b]Gill rakers on lower limb of the first gill arch.
Note: All measurements refer to standard length, except the maxillary length which was tested against head length.
SL: standard length; DF: dorsal fin rays; AF: anal fin rays; HL: head length; BD: body depth; Eye: eye diameter; Snout: snout length; V.S.: ventral scutes; AF base: anal fin basement; Vert.: vertebrate number; PD: predorsal distance; PA: preanal distance.

SUMMARY

Larval developments of two clupeid fishes, *Pellona harroweri* and *Chirocentrodon bleekerianus*, belonging to the subfamily Pristigasterinae are described and figured. The samples were collected in southern Brazil with beam-trawl net. Morphometrical and osteological studies are made and a systematic status of *C. bleekerianus* is presented. *Chirocentrodon cladileokae* Tommasi 1964 is considered a synonym of *C. bleekerianus*.

ACKNOWLEDGEMENTS

I gratefully acknowledge the assistance given by various staff members of the Instituto Oceanográfico da USP, especially in sampling cruises and in laborarory work.

Special thanks are extended to Drs. Naercio A. Menezes and Plínio S. Moreira for their critical reading of the manuscript and Prof. Frank R. Shaffer for revision of the English. Special thanks are also due to Dr. Gelso Vazzoler and Admir. Alberto S. Franco for their support and encouragement on the project. I would like to thank Dr. P.J.P. Whitehead of the British Museum for revision of the manuscript and Miss Leko Kanno, who prepared most of the figures.

This study was accomplished under a research contract at the Instituto Oceanográfico da U.S.P., on a fellowship of the Fundação de Amparo à Pesquisa do Estado de São Paulo (Processo: biológicas 70/578).

REFERENCES

Berry, F.H., 1964a. A hypomaxillary bone in *Harengula* (Pisces: Clupeidae). Pacific Science, 18 (4), 373-377.
Berry, F.H., 1964b. Aspects of the development of the upper jaw bones in Teleosts. Copeia, 2, 375-384.
Breder, C.M., Jr., 1942. The reappearance of *Chirocentrodon bleekerianus* (Poey) in Cuba. Copeia, 3, 133-138.
Clothier, C.R., 1950. A key to some Southern California fishes based on vertebral characters. Calif. Fish and Game, Fish. Bull., 79, 83.
Gosline, W.A., 1960. Contributions toward a classification of modern Isospondylous fishes. Bull. Br. Mus. (Nat. Hist.) Zool., 16 (6), 327-365.
Günther, A.K.L., 1868. Catalogue of the fishes in the collection of the British Museum, 7, 435.
Hildebrand, S.F., 1963. Fishes of the Western North Atlantic. Family Clupeidae. Sears Found. Mar. Res., Yale Univ., 438-442 pp.
Iwai, M., 1973. Pesca exploratória e estudo biológico sobre camarão na costa centro-sul do Brasil do N/O Prof. W. Besnard em 1969-1971. SUDELPA e Inst. Oceanogr. Univ. S. Paulo, 71 pp.
Jordan, D.S. and Seale, A.S., 1926. Review of the Engraulidae, with descriptions of new and rare species. Bull. Mus. Comp. Zool., 67 (11), 355-418.
Monod, T., 1967. Le complexe urophore des Téléostéens: typologie et évolution (note préliminaire). Colloques int. Cent. natn. Rech. scient., Paris, 163, 111-131.
Myers, G.S., 1929. Mutanda ichthyologica II. *Heringia* vs. *Rhinosardinia* (Clupeidae), and *Entonanthias* vs. *Mirolabrichthys* (Anthiidae). Copeia, 170 p.
Nybelin, O., 1963. Zur Morphologie und Terminologie des Schwanzskelettes der Actinopterygier. Ark. Zool., Stockholm, (2) 15, 485-516.
Poey, F., 1867. Repertorio fisico-natural de la isla de Cuba, 2-242.
Tommasi, L.R., 1964. Sôbre tres especies de peces marinos del litoral San Pablo (Brasil). Neotropica, 10 (31), 29-35.

Y. Matsuura
Instituto Oceanográfico da Universidade de São Paulo
C.P. 9075
São Paulo / BRAZIL

A Comparison of Larvae of the Deepwater and Fourhorn Sculpin, *Myoxocephalus quadricornis* L. from North America. I. Morphological Development

N. Y. Khan and D. J. Faber

INTRODUCTION

Sculpins of the genus *Myoxocephalus* live in freshwater lakes in northern North America and in coastal estuarine waters extending from Alaska eastward to Labrador including Hudson Bay and adjoining waters. There is controversy regarding the taxonomic status and origin of the freshwater population (McAllister, 1959; Hutchinson, 1967; McPhail and Lindsey, 1970; and others) and this study was undertaken to determine the morphological and ecological differences between the larvae of freshwater and estuarine forms. Owing to page limitations set up for the Symposium on the Early Life History of Fish, this study has been divided into three parts: I. Morphological development, II. Seasonal Occurrence and Distribution (Faber, Khan and Wells, MS.), and III. Growth Characteristics (Khan and Faber, MS.). Parts II and III will be published elsewhere.

A number of larval Cottidae have been known since early 1900 (Ehrenbaum, 1905) but the genus *Myoxocephalus* was revised by Berg and Popov (1932) who placed *M. quadricornis* together with other species that some other authors had placed in the genera *Cottus* and *Oncocottus*. Most North American authors (for example, Walters, 1955 and McAllister, 1959) concur with this placement. The most recent work (McPhail and Lindsey, 1970) treats the freshwater North American form as a subspecies, *M. quadricornis thompsoni* (Girard), of the marine form, *M. quadricornis* (L.). The other species placed within the genus *Myoxocephalus* that have their larvae described are *M. scorpius* (L.)(Holt, 1893 and Khan, 1971), *M. aeneus* (Mitchill)(Khan, 1971), and *M. octodecemspinosus* (Mitchill)(Khan, 1971).

The free-swimming larvae of *M. quadricornis* were found to be vulnerable to plankton nets and this paper describes the general morphology of *M. quadricornis* larvae collected in this manner. Previously published illustrations of larval *M. quadricornis* are discussed in light of our present findings. It is shown that larvae from freshwater populations differ in several distinctive ways from larvae of estuarine populations.

This study suggests that valuable knowledge can be gained through larval fish collections and, indeed, the information can be used to help solve taxonomical and ecological problems.

MORPHOLOGICAL DEVELOPMENT

The specimens of sculpin larvae examined in this study were collected from wild indigenous sculpin populations by means of fine-mesh, conical townets (mesh openings ranged from 0.2 - 0.6 mm). Larval deepwater sculpins were collected in shallow waters of Lake Huron (0-2 m) and in deep offshore waters of Lake Michigan (0-91.5 m); larval fourhorn sculpins were collected in shallow estuarine waters of the Beaufort Sea (0-2 m). The specimens were immediately preserved in 5-10% formalin solution and were examined later in the lab. Details about the sampling gear and habitats will be presented in a later paper (see Faber, Khan and Wells, MS). The shrinkage of our formalin preserved larvae was not taken into account in the presentation of length data. Drawings were made with the aid of a camera lucida.

The terminology proposed by Balon (1970) is being used in this study because of the inconsistent terminology that North American workers have used in the past. In summary, Balon has proposed an "Embryonic Period", divided into cleavage, embryonic, and eleuterembryonic phases; a "Larval Period" divided into protopterygiolarval and pterygiolarval phases; and a "Juvenile Period". We have subdivided our descriptions into 2-mm length intervals for convenience and compatibility with the statistical presentation of meristic characters. Although ranges of meristic characters are presented here, a later paper (Khan and Faber, MS.) critically compares variations of meristic counts and of relative growth rates of body parts.

Deepwater Sculpin Larvae

Freshwater *M. quadricornis*, which are well-known as deepwater sculpins in North America, live landlocked in deep, cold lakes in northern North America. A similar freshwater form lives in Eurasia and although it is not known as the deepwater sculpin there, we have included information on its description here. McPhail and Lindsey (1970) have summarized the latest information about their biology in North America. They state that spawning occurs in summer but our results (Faber, Khan and Wells, MS.) suggest that deepwater sculpins spawn through the winter, spring, and early summer months. In addition, we have determined that deepwater sculpin larvae disperse widely into the open water shortly after hatching. Thus juveniles and adults, which live on the lake bottom, inhabit a different habitat than do their larvae.

Little information is available regarding the morphological development of the deepwater sculpin from the egg through the Larval Period and into the Juvenile Period. Jacoby (1953) reported ovarian eggs of deepwater sculpins from Lake Superior which ranged from 1.5 - 2.2 mm (mean 1.8 mm) in diameter. Since cottid eggs swell on contact with water, the actual diameter of the eggs during incubation would be larger but their actual size has not been reported. It is supposed that the eggs of the deepwater sculpins are adhesive and are deposited in nests. Westin (1968a) recorded eggs ranging from green to bluish-green and yellow to reddish yellow in several Swedish lakes. We never observed eggs during our study and the spawning information we have arrived at was obtained indirectly through the occurrence of free-swimming larvae in the lakes.

Three previous studies concerning the larval development of the freshwater forms have been published, one in North America (Fish, 1932) and two in Europe (Sundevall, 1855 and Nordqvist, 1915). Fish (1932) illustrated 7 stages of larvae and juveniles collected from Lake Erie. She identified 5 stages as *Cottus bairdi kumlieni* (Figs 128-132), one stage as *Cottus bairdii bairdii* (Fig. 133), and one juvenile as *Triglopsis thompsoni* (= *M. q. thompsoni*) (Fig. 127). The morphometric characteristics associated with each figure leave no doubt that all these figures represent various sizes in the free-swimming larval development of the deepwater sculpin in Lake Erie. All other species of sculpins in the Great Lakes belong to the genus *Cottus* and their early larval development is very different from those in the genus *Myoxocephalus*. Sundevall (1855) illustrated two specimens (Figs 5-8) which Ehrenbaum (1905) later reproduced. The first specimen (Figs 5 and 6) does not look like a *M. quadricornis* larva because of the enlarged parietal spines and general details in the head and intestinal region. Ehrenbaum (1905) states that this specimen looks very much like *Cottus scorpius* (= *M. scorpius*). The second specimen (Figs 7 and 8) is morphologically similar to our freshwater specimens. Nordqvist (1915) described seven stages and illustrated three of these (Figs 22-25) from Lake Vättern, Sweden. They are morphologically similar, except for pigmentation, to larvae of the North American deepwater sculpin.

Description of Larvae

The 57 specimens that were examined ranged from 7.7 - 17.5 mm and this range covered the late portion of the Embryonic Period (eleuterembryonic phase) and most of the Larval Period (protopterygiolarval and

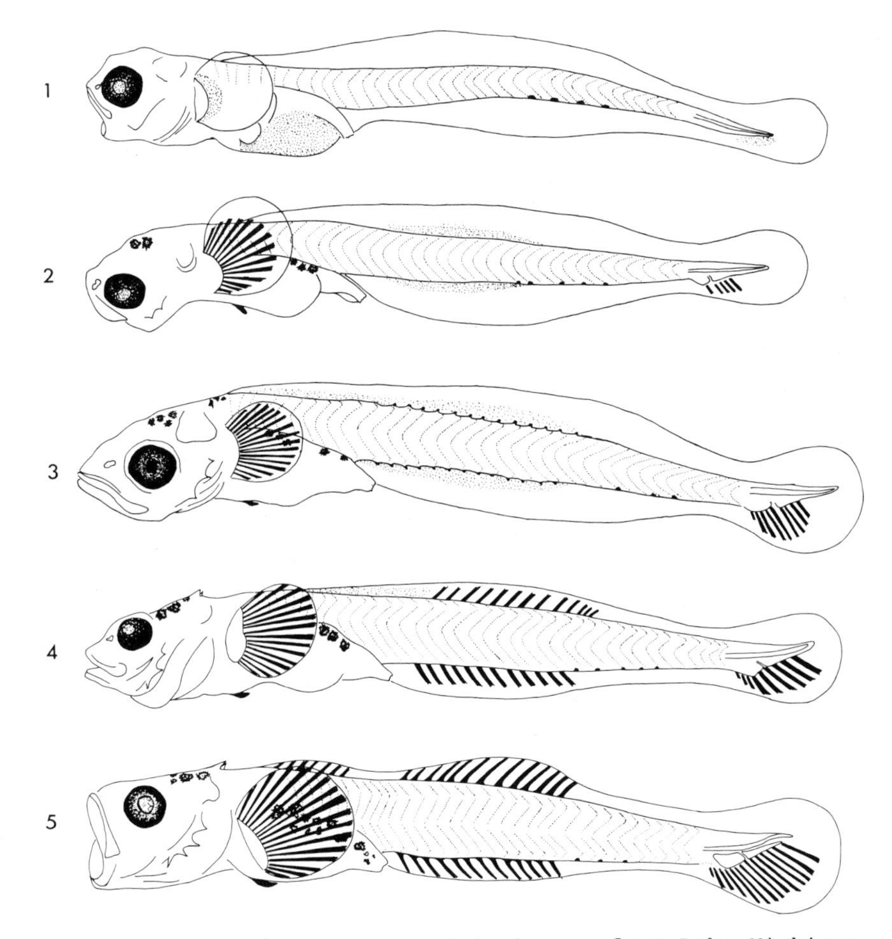

Figs 1-5. Representative deepwater sculpin larvae from Lake Michigan. Fig. 1. 9 mm T.L., collected on 3 May 1964. Fig. 2. 10.5 mm T.L., 1 May 1964. Fig. 3. 12.5 mm, 1 May 1964. Fig. 4. 14.9 mm, 3 May 1964. Fig. 5. 17.1 mm, 10 May 1964

pterygiolarval phases). The illustrations (Figs 1-5) are representations of individual specimens picked at random from the available Lake Michigan specimens in May near the mean of each length-interval.

8-10 mm Length Interval (Fig. 1): Larvae examined ranged from 8.7 - 10.0 mm (mean 9.4 mm). This length interval lies within the eleuterembryonic phase of the Embryonic Period. All larvae examined had some yolk material and the smaller sized had most. An oil globule was never observed. At a length of about 8.5 mm an accumulation of mesenchyme appears on the finfold at the site of the developing hypurals and ventral caudal fin rays. The vent is located anterior to the ventral section of the finfold on the midventral axis and it opens at the margin of the finfold. The notochord remains straight. The pectoral fin is fan-shaped with a rather broad base. Smaller specimens lack pigmentation of any kind but at about a length of 9 mm, a number of melanophores appear at the site of the peritoneal concentration. At about the same length, melanophores appear in the anisomeric medio-ventral row. Myomeres are chevron-shaped and range in number from 37-43. Nostrils are single.

10-12 mm Length Interval (Fig. 2); Larvae examined ranged from 10.3 - 11.9 mm (mean 10.9 mm). This length interval lies within the proto-pterygiolarval phase of the Larval Period. Finfold is still complete; mesenchyme appears in the regions of the second dorsal and anal fins. A small section of the finfold is present anterior to the anus. The hypurals and caudal fin rays become visible at about 10.5 - 10.9 mm. Pectoral fin rays and pelvic fin buds appear early. Some early larvae are still without pigmentation but the cephalic and peritoneal concentrations begin to develop. Myomeres are still chevron-shaped and range from 36-41. The preopercular spines begin to appear and number up to two.

12-14 mm Length Interval (Fig. 3): Larvae examined ranged from 12.2 - 14.0 mm (mean 13.0 mm). This length interval lies at the beginning of the pterygiolarval phase. The finfold remains complete while the paired and unpaired fin rays differentiate. The beginning of fin rays in both second dorsal and anal fins appear at about 12 mm. Second dorsal fin-ray counts range from 7-13; anal fin-ray counts range from 10-13. Mesenchyme appears at the site of first dorsal fin. Caudal fin-ray counts range from 9-13. Pectoral fin-ray counts range from 16-17 while pelvic fin rays remain undifferentiated. Most larvae show the peritoneal concentration, cephalic concentration and anisomeric medio-ventral row of pigmentation but wide variations occur in numbers and individual shapes of melanophores. Melanophores in the anisomeric medio-ventral row vary from 0-6. Myomeres develop a piscine shape and counts range from 33-42. Preopercular spines range from 2-4 and a parietal spine appears at about 13 mm. The nostrils are completely constricted by 14 mm forming the deepwater sculpin's characteristic double nostril.

14-16 mm Length Interval (Fig. 4): Larvae examined ranged from 15.0 - 15.9 mm (mean 15.6 mm). This length interval also lies within the pterygiolarval phase. Finfold gradually narrows. Fin-ray counts for both second dorsal and anal fins range from 12-14. First dorsal fin rays appear very late with counts ranging from 3-8. Homocercal caudal fin continues to develop with fin-ray counts ranging from 9-15. Pectoral fin-ray counts range from 15-18; pelvic fin rays are still not visible. Melanophores increase in both peritoneal and cephalic concentrations while the number of melanophores in the anisomeric medio-ventral row continues to vary from 0-6. Counts of piscine-shaped myomeres vary from 35-39. Larvae now show four preopercular and two parietal spines.

16-18 mm Length Interval (Fig. 5): Larvae examined ranged from 16.0 - 17.5 mm (mean 16.7 mm). This length interval also lies within the pterygiolarval phase. Fin-ray counts of pectoral, first dorsal, second dorsal, anal, and caudal fins range from 17-18, 5-8, 13-17, 12-16, and 12-18 respectively. Pelvic fin rays begin to appear at about 17 mm. Peritoneal and cephalic concentrations continue to enlarge but the anisomeric medio-ventral row still averages only four melanophores with a maximum of six. Counts of piscine-shaped myomeres varies from 37-42. Preopercular and parietal spines are 4 and 2 respectively.

Fourhorn Sculpin Larvae

Fourhorn sculpins live littorally in coastal arctic waters in North America from Alaska, eastward throughout the Canadian Arctic Archipelago and Hudson's Bay, and along the Atlantic coast as far south as Labrador. They are always found near shore, especially in estuarine areas, and sometimes they move into fresh water by ascending rivers for considerable distances. Most of our knowledge about the biology of the fourhorn sculpin is due to workers in Sweden and the U.S.S.R. Walters (1955), Backus (1957), and McPhail and Lindsey (1970) present the latest information on the North American form. The estuarine form differs from the landlocked freshwater form mainly in size and development of parietal spines, i.e. the estuarine form grows larger and has better developed parietal spines than those in freshwater (Hutchinson, 1967). Spawning takes place in winter (usually December and January) and hatching occurs during May and June (Faber, Khan and Wells, MS.). As with deepwater sculpins, the spawning information we have arrived at was obtained indirectly through the occurrence of free swimming larvae in Kugmallit Bay, N.W.T. It was found that their larvae disperse widely and can be found along shore lines and in shallow pelagic regions.

Information about the early morphological development of the marine fourhorn sculpin from the egg through the larval period is scattered through the literature and some is incorrect. The embryology has not been described but Westin (1968a, 1969) has studied the variation in egg colour, time of egg development and related subjects. Andriyashev (1964, from Pokrovskaya) states that fertilized eggs range from 2.8 - 3.0 mm in arctic seas and Westin (1968b) states that they range from 2.4 - 2.9 mm in the Baltic. Westin (1968a) showed that these eggs were coloured by environmental conditions in various shades of green, turquoise, brown, and reddish brown. Later, Westin (1969) found that fourhorn sculpin eggs are externally fertilized, expelled all at once, adhesive, laid in small clumps in nests, and tended by males during incubation which lasts for 97 days at 1.5^{o}C and 74 days at 2.0^{o}C. We never observed eggs during our study.

A number of previous studies have been published that show illustrations of larval fourhorn sculpins (the marine form). Zviagina (1963) has produced the most accurate and useful descriptions. She drew six larvae (Figs 1a, b, c, d, e, f) ranging from 12.3 - 32 mm. McIntosh and Prince (1890) published one stage (Fig. 11) which they thought was a fourhorn sculpin and Holt (1893) illustrated a larva (Fig. 45) which he thought could have been a fourhorn sculpin. Bruun (1925), however, has shown that the two latter descriptions of larvae from the British Isles belong to the Norway Bullhead, *Taurulus (Cottus) lilljeborgi* (Collett). Ehrenbaum's (1905) illustrations (Fig. 24) were redrawn from Sundevall (1855) who drew them from a lake in Sweden, thus neither Ehrenbaum nor Sundevall show descriptions of marine fourhorn sculpin larvae.

Johansen (1912) shows three illustrations (Figs 11, 12, 13) which he has identified as larvae of the fourhorn sculpin from coastal Greenland. These specimens do not show the general shape nor pigment pattern of juvenile fourhorn sculpins. In addition, Johansen (1912, p. 647) found the development of spines in these larvae take place in the following sequence: 1. occipital (parietal), 2. preopercular, 3. frontal, and 4. nasal. Our specimens (and Zviagina's, 1963) showed the appearance of preopercular spines before the pair of parietals. Alekseeva (1949) shows an illustration of a larva measuring 7.1 mm from Pechova Bay but this larva is probably not a larval stage of *M. quadricornis*. The length is too short and the number of melanophores in the anisomeric medio-ventral pigment line is too few. The 7.1 mm larva illustrated in Alekseeva (1949) is similar to an 8.4 mm larvae of *Triglops murrayii* (Günther) illustrated by Khan (1971). Knipowitsch (1907) also illustrated a larva which he thought was *M. quadricornis*.

The 84 specimens that were examined for this study covered only the Larval Period (protopterygiolarval and pterygiolarval phases). The illustrations (Figs 6-8) are representations of individual specimens picked at random from the available Kugmallit Bay specimens near the mean of each length interval.

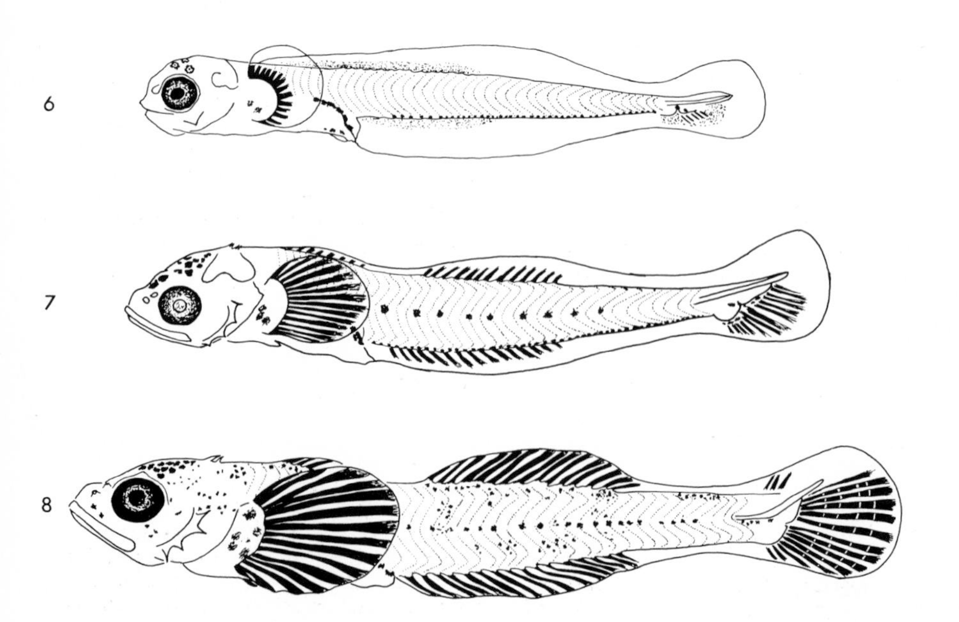

Figs 6-8. Representative fourhorn sculpin larvae from Tuktoyaktuk Harbour. Fig. 6. 12.8 mm T.L., all collected during July 1970. Fig. 7. 14.4 mm T.L. Fig. 8. 17.0 mm T.L.

12-14 mm Length Interval (Fig. 6): Larvae examined ranged from 12.3 - 14.0 mm (mean 13.24 mm). This length interval lies at the end of the protopterygiolarval and the beginning of the pterygiolarval phase. The unpaired fins begin to differentiate late and second dorsal fin-ray counts range from 8-13; and anal fin-ray counts range from 10-15. First dorsal fin only shows mesenchyme. Caudal fin-ray counts range from 4-7. Pectoral fin rays begin to show but pelvic fins remain as buds. All larvae show peritoneal concentration, cephalic concentration and anisomeric medio-ventral row of pigmentation; but variations occur in numbers and shapes of melanophores. Large stellate melanophores are consistent in the cephalic concentration. Small melanophores in the anisomeric medio-ventral row range in number from 40-65. At about 13 mm 6 or 7 medio-lateral melanophores appear. Other areas where 2 or 3 melanophores occur are base of pectoral fins, behind nostrils, on premaxilla and just anterior to the anus. Myomeres are all chevron-shaped at beginning but develop a piscine pattern by 14 mm. Myomeres range from 41-45. Late larvae show preopercular spines ranging from 2-4 and a pair of parietal spines. Nostrils are single.

14-16 mm Length Interval (Fig. 7): Larvae examined ranged from 14.2 - 15.9 mm (mean 15.5 mm). This length interval also lies within the pterygiolarval phase. Finfold remains complete but is gradually narrowing. Fin-ray counts for second dorsal and anal fins are 13-15 and 14-17 respectively. First dorsal fin rays appear late with counts ranging from 6-9. Caudal fin developes considerably during this length interval and fin-ray counts range from 4-17; pectoral fin rays range from 15-17. Pelvic fin rays are not visible. Melanophores increase in number in positions mentioned in 12-14 mm length interval. Small surface melanophores appear along the dorsal fins. Counts of piscine-shaped myomeres range from 43-46. Larvae show four preopercular and two parietal spines. Nostrils are constricted early into an anterior and posterior nostril.

16-18 mm Length Interval (Fig. 8): Larvae examined ranged from 16.0 - 17.4 mm (mean 16.4 mm). This length interval also lies within the pterygiolarval phase. Fin-ray counts of second dorsal, anal, and caudal fins range from 13-16, 14-16, 11-14 respectively. All pectoral fin-ray counts were 16. Pelvic fin rays appear late. Numerous surface melanophores appear at the beginning of adult pigmentation. The number of melanophores along medio-lateral row increases to 18-35. The numbers of melanophores along anisomeric medio-ventral row remains high but sink into the dermis and become obscured. Counts of piscine-shaped myomeres vary from 42-45. Preopercular and parietal spines are 4 and 2, respectively.

DISCUSSION

This comparative morphological study of the larvae of *Myoxocephalus quadricornis* from estuarine and freshwater habitats of North America showed that the fourhorn sculpin larvae differ from the deepwater sculpin larvae in a number of characters. Fourhorn larvae possess a greater number of myomeres than deepwater sculpin larvae throughout larval development and this is reflected in the number of vertebrae in the adults (Khan, 1971). As a result the former larvae become free-swimming at a longer total length than do deepwater sculpin larvae. In addition, the two forms have striking differences in pattern of pigmentation. Generally, melanophores are much reduced in numbers in deepwater sculpin larvae throughout larval development. A distinctive mediolateral row of melanophores is characteristic of fourhorn sculpin larvae while it is lacking in the larvae of deepwater sculpin. The tendency towards reduced pigmentation in the freshwater form is also reflected in the anisomeric medio-ventral row of melanophores (i.e. 0-6 melanophores in the freshwater form as opposed to 40-65 in the estuarine form). The larvae of the freshwater form of *M. quadricornis* described and illustrated by Nordqvist (1915) also lacked mediolateral melanophores and the medio-ventral melanophores were about 20 in number. Our results, therefore, suggest that freshwater *Myoxocephalus* larvae from North America are morphologically similar to European freshwater *Myoxocephalus* larvae, while at the same time low salinity *Myoxocephalus* larvae from North America are morphologically similar to those living in low salinity waters in Europe.

The morphological characteristics of larvae within the genus *Myoxocephalus* are extremely variable. The larvae of *M. quadricornis* become free-swimming at lengths (8-10 mm) that are longer than *M. aeneus*, *M. octodecemspinosus*, and *M. scorpius* (5-7 mm)(Khan, 1971). The pattern of pigmentation also varies between species. For example, *M. scorpius* has abundant melanophores in the mid-lateral region while these melanophores are lacking in other species of the genus, i.e. *M. aeneus*, *M. octodecemspinosus*. The present study shows that even larvae within the same species can vary in their pigmentation pattern. Ehrenbaum (1905, p. 65) stated that the characteristic pigment pattern of Cottidae larvae consists of a concentration of the peritoneum and a row of melanophores along the base of the anal fin but this pattern of pigmentation is not confined solely to Cottidae larvae. Certain species within the families Scorpaenidae (Dannevig, 1919) and Liparidae (Aoyama, 1959) and others also exhibit the same pattern. Because of these variations in pigment pattern, it is not possible to assign any basic pattern of pigmentation to larvae within the family Cottidae or even to those within the genus *Myoxocephalus*. Although differences in patterns of pigmentation on developing teleost and amphibian embryos have been studied (Parker, 1958 and DuShane, 1953), the causes have not been identified.

SUMMARY

Morphologically rather similar sculpins known as *Myoxocephalus quadricornis* L. live in freshwater lakes and in coastal arctic waters in northern North America. This morphological study, which is part of a larger, comprehensive study of larval Cottidae, was carried out in order to identify differences and similarities among larvae collected from different habitats. Free-swimming larvae of deepwater sculpins from the Laurentian Great Lakes and of fourhorn sculpins from the Beaufort Sea were collected with conical plankton nets. The morphological development of larvae collected from these bodies of water are described and illustrated. It is shown that larval deepwater sculpins from lakes differ, in several distinctive ways, from larval fourhorn sculpins living in low salinity arctic waters in their pattern of pigmentation, number of myomeres and size. These results suggest that freshwater *Myoxocephalus* larvae from North America are morphologically quite similar to the same freshwater larvae in Europe while, at the same time, the low salinity *Myoxocephalus* larvae from North America are similar to those living in low salinity waters in Europe.

ACKNOWLEDGEMENTS

We wish to thank Tex Wells, National Sport Fisheries Service, Ann Arbor, Michigan, for providing specimens of larval deepwater sculpins from Lake Michigan and Len Marhue, National Musuem of Natural Sciences, Ottawa, for collecting larval fourhorn sculpins in Kugmallit Bay, N.W.T. In addition we would like to express our appreciation for boats and logistics support provided by the Polar Continental Shelf Project, Marine Sciences Branch, Department of Environment, Ottawa; the Biological Laboratory, National Sport Fisheries Service, Ann Arbor, Michigan, and the Fisheries Research Station, Ontario Department of Lands and Forests, South Baymouth, Ontario.

REFERENCES

Alekseeva, S.P., 1949. Ikrinki i mal'ki ryb iz Pechorskogo zaliva (Eggs and fingerlings of fish from Pechora Bay). Trudy Vses. nauchno-issled rybn. khoz-va i okean., 17, 175-188.

Andriyashev, A.P., 1964. Fishes of the northern seas of the USSR. Israel program for scientific translations, Jerusalem, U.S. Dept. of Commerce, Wash., D.C. 617 pp.

Aoyama, T., 1959. On the egg and larval stages of *Liparis tanakae* (Gilbert et Burke). Bull. Seikai Reg. Fish. Res. Lab., Nagasaki, No. 18, 69-73.

Backus, R.H., 1957. The fishes of Labrador. Bull. Amer. Mus. Nat. Hist., 113 (4), 273-337.

Balon, E.K., 1971. The intervals of early fish development and their terminology. Vest. Cesk. Spolec. Tool., 35 (1), 1-8.

Berg, L.S. and Popov, A., 1932. A review of the forms of *Myoxocephalus quadricornis* (L.). C.R. Acad. Sci. USSR, 6, 152-160.

Bruun, A., 1925. On the development and distribution of the Norway Bullhead (*Cottus lilljeborgi* Collett). Cons. perm. int. Explor. Mer, Publ. de Constance, No. 88, 3-15.

Dannevig, A., 1919. Canadian fish eggs and larvae. In: J. Hjort (ed) Canadian Fisheries Expedition, 1914-15. Dept. Nav. Serv. Can. King's Printer, Ottawa, 1-74.

DuShane, G., 1943. The embryology of vertebrate pigment cells. Part I. Amphibia. Quart. Rev. Biol., 18, 109-127.

Ehrenbaum, E., 1905. Nordisches Plankton, Eier und Larven von Fischen. Kiel-Leipzig, reprinted 1964, 414 p.

Faber, D.J., Khan, N.Y. and Wells, L. (MS.). A comparison of the fresh-water and estuarine sculpin, *Myoxocephalus quadricornis* L. from North America. II. Seasonal occurrence and distribution.

Fish, M.P., 1932. Contributions to the early life histories of sixty-two species of fishes from Lake Erie and its tributary waters. Bull. U.S. Bur. Fish, 67, 293-398.

Holt, E.W.L., 1893. Survey of fishing grounds, West coast of Ireland 1890-1891: On the eggs and larval and postlarval stages of tele-osteans. Scient. transact. Royal Dublin Society, 5 (Series II), 20-26.

Hutchinson, G.E., 1967. A treatise on Limnology Vol. II. New York: John Wiley and Sons, 1115 p.

Jacoby, C., 1953 (MS.). Notes on the life history of the deepwater sculpin, *Myoxocephalus quadricornis* L. in Lake Superior, M. Sc. Thesis, Univ. of Michigan, Ann Arbor, 21 p.

Johansen, F., 1912. The fishes of the Danmark Expedition Meddel. om. Grønl., 45, 631-675.

Khan, N., 1971 (MS.). Comparative morphology and ecology of the pelagi larvae of nine Cottidae (Pisces) of the northwest Atlantic and St. Lawrence drainage. Thesis, Univ. of Ottawa, Ottawa, 234 p.

Khan, N. and Faber, D. (MS.). A comparison of the freshwater and estuarine sculpin, *Myoxocephalus quadricornis* L. from North America. III. Growth characteristics.

Knipowitsch, N., 1907. Zur Ichthyologie des Eismeeres (The ichthyology of the Polar Sea). Zapiski Akad. nauk, fiz.-mat. otd., 18 (5).

McAllister, D.E., 1959. The origin and status of the deepwater sculpin, *Myococephalus thompsonii*, a nearctic glacial relict. Bull. Nat. Mus. Can., 172, 44-65.

McIntosh, C.W. and Prince, E., 1890. On the development and life histories of the teleostean food and other fishes. Trans. Roy. Soc. Edin., 35, pt. III.

McPhail, J.D. and Lindsey, C.C., 1970. Freshwater fishes of northwestern Canada and Alaska. Bull. Fish. Res. Bd Can., No. 173, 381 p.

Nordqvist, H., 1915. Bidrag till Kännedomen om vära sötvattensfiskars larvstadier (Contributions to the knowledge of the larval stage in our freshwater fish)(Eng. trans. in Lib. Nat. Museums Can., Ottawa). Arkiv för Zoologi 9 (4), 1-49.

Parker, G., 1948. Animal colour changes and their neurohumours. Hafne Publishing Co., New York, 377 p.

Sundevall, C.J., 1855. Om Fiskyngels Utveckling (Development of fish larvae). Kungl. Vet. Akad. Handl., 1, 1-24.

Walters, V., 1955. Fishes of western arctic America and eastern arctic Siberia. Taxomony and zoogeography. Bull. Amer. Mus. Nat. Hist., 106 (5), 261-368.

Westin, L., 1968a. Environmentally determined variation in the roe colour in the fourhorn sculpin, *Myoxocephalus quadricornis* (L.). Oikos 19, 403-407.

Westin, L., 1968b. The fertility of fourhorn sculpin, *Myoxocephalus quadricornis* L. Inst. of Freshwater Res., Drottingholm, Rep. No. 48, 67-70.

Westin, L., 1969. The mode of fertilization, parental behaviour and time of egg development in the fourhorn sculpin, *Myoxocephalus quadricornis* (L.). Inst. of Freshwater Res. Drottingholm, Rep. No. 49, 175-182.

Zviagina, O.A., 1963. Materialy po razmnozheniyu i razvituju ryb morya Laptevykh 2. Ledovitomorskaya rogatka, i 3. Aziatskaya Koryushka. (Materials on the reproduction and development of fish of the Laptev Sea 2. Arctic sculpin and 3. Asiatic melt)(Eng. trans. in Lib. Nat. Museums Can., Ottawa). Trudy Instit. Okean., 62, 3-13.

N.Y. Khan
University of Ottawa
Ottawa / CANADA

Present address:

Aquatic Contaminants Unit
Fisheries and Marine Service
Environment Canada
Ottawa / CANADA

Daniel J. Faber
National Museums of Natural Sciences
Ottawa, KIA OM8 / CANADA

Rearing

Artifical Insemination in Trout Using a Sperm Diluant

R. Billard, J. Petit, B. Jalabert, and D. Szollosi

INTRODUCTION

In the practice of artificial insemination in salmonids, efficiency is usually poor since one male is used to fertilize only a few females. Some attempts have been made to improve the efficiency of artificial insemination, either by using minute quantities of sperm or by diluting sperm. The use of small quantities of sperm allows fertilization of 24 females by one male (Nursall and Hasler, 1952). Sperm dilution, which was performed in saline by Scheuring (1925) and Nomura (1964), has been reported to improve the percentage of fertilized eggs when compared with results obtained following fertilization by pure milt (Poon and Johnson, 1970; Plosila et al., 1972). Ginsburg (1968), however showed that the dry method was the most efficient. Lately, we showed that pH and osmotic pressure were the most important parameters for the diluant (Petit et al., 1973). In this study, effects of various parameters and components of the diluant on eggs, spermatozoa and percentage of fertilization are examined.

MATERIAL AND METHODS

Experiments were conducted in 1972-73 with 3- to 5-year-old male and female rainbow trout, *Salmo gairdneri* Richardson, obtained from several local fish farms. Inspection of the fish for readiness to spawn was carried out 2 or 3 times a week. Running males and females were anaesthetized in MS 222. Before stripping, the fish were dried with towels to prevent contamination of gametes by water. Excess coelomic fluid was discarded by straining eggs in a strainer. All experiments were carried out at about 10°C; eggs from several females were pooled, mixed, and divided at random in batches of 200 eggs each. In all experiments only 1 male was used; milt was stripped and sperm motility was determined by light microscopy. Sperms having low motility were discarded. Sperm concentration was measured with a Thomas haemocytometer; 10 ml of diluant was usually used per batch. Sperm dilution: 10^{-5}, 10^{-4}, 10^{-3}, 10^{-2}, corresponded to 0.1, 1, 10, 100 µl of sperm per batch of 200 eggs; minute quantities of sperm, such as 0.1 and 1 µl, were measured with 0.5 and 1 µl microsyringes (Scientific glass engineering PTY. Ltd.). In each experimental batch, 10 ml of various diluants were added, while control batches were fertilized according to the dry method used on fish farms: eggs without coelomic fluid were mixed with pure milt in excess, and then freshwater was added. Egg incubation was carried out at 13.0 ± 1°C in a special incubator divided into small compartments. Ten days after insemination, egg samples were cleared in Stockard's solution. Fertilized and dead eggs were counted; percentage of fertilization (which is, in fact, the percentage of embryos) was calculated with 95% confidence limits. The χ^2 test was used to compare the percentages of fertilized eggs. Five experiments were performed and specific details for each of them are given below.

Experiment No. 1: Comparison of Saline and Freshwater as Diluants for Fertilization. The saline solution used in this experiment was buffered at pH 9.5 with 0.02 M carbonate-bicarbonate buffer. Osmotic pressure was 250 mosmols. Freshwater, in which the animals were kept and the eggs incubated, was used as a diluant for controls. The mineral content of the water was 200 mg/l. Egg diluant and sperm of various concentrations were mixed together simultaneously and transferred into incubators 15 min later.

Experiment No. 2: Effects of Osmotic Pressure of Diluant on Gametes and Rate of Fertilization. Seven diluants with various osmotic pressures were used in this experiment (Π = 0, 50, 100, 150, 200, 250, 300 mosmols). Double distilled water, reoxygenated by agitation, was used as zero mosmol diluant. The other diluants were adjusted for required osmotic pressure with NaCl after pH adjustment at 9.5 with 0.02 carbonate-bicarbonate buffer. After fertilization, the pH of the mixture dropped to 9.35 - 9.28 with saline, and to about 8.00 with double distilled water. Effect of the diluant on rate of fertilization was measured by adding and mixing eggs, diluant, and sperm at various concentrations. This mixture was left without any agitation for 15 min, and then transferred to incubators. The direct effects of diluant on eggs were shown by adding in fertilized eggs and diluant first. Sperm was added every 0.5, 1, 2, 6, 10, 20, and 40 min afterwards. Fifteen minutes after insemination, eggs were transferred into incubators. Effects of various osmotic pressures on spermatozoa were demonstrated in the same way: addition of sperm in diluant first, and mixing with eggs afterwards, according to the following intervals: 15 sec, 30 sec, 1, 2, 4, 6, 8, 10 min.

Experiment No. 3: Effect of Diluant pH and Buffer on Fertilization Rate. The saline solution at Π = 250 mosmol was buffered with 3 buffers: carbonate-bicarbonate glycocolle and tris at various pHs ranging from 8.1 to 10.4. For each buffer, two molarities, 0.02 M and 0.1 M, were used. Sperm concentration was 10^{-3} (10 µl in 10 ml diluant). Eggs were transferred into incubators 15 min after insemination.

Experiment No. 4: Optimum Timing for Transferring Inseminated Eggs into Water. The experiment was carried out to determine how long inseminated eggs should stay in the diluant before transfer to incubators. After mixing eggs, sperm (at the dilution rate of 10^{-3} and 10^{-4}), and the diluant (saline solution Π = 250; pH = 9.5; buffer: carbonate-bicarbonate 0.02 M), we transferred the mixture into incubators every 1, 2, 4, 6, 8, 10, 20, and 40 min.

Experiment No. 5: Minimum Number of Spermatozoa Required for Fertilization of One Egg. In order to determine the minimum amount of spermatozoa necessary to fertilize one egg, insemination was performed on batches of about 100, 200, 400, and 800 eggs with 2 rates of sperm dilution: 10^{-3} and 10^{-4}. Diluant: saline solution, Π = 250 mosmol, pH 9.5 buffered with Tris buffer 0.02 M. In this experiment, 20 ml of diluant was used instead of 10 as in the previous experiments.

RESULTS

Experiment No. 1: Comparison of Saline and Water as Diluants for Fertilization. In results shown in Table 1, the superiority of the saline over freshwater for fertilization rate is obvious, whatever the sperm concentration is. Statistical analysis is presented in Table 2. In saline at 10^{-2} and 10^{-3} sperm concentration, percentage of fertilized eggs is not significantly different; at 10^{-4} and 10^{-5}, it drops significantly. Following sperm dilution in water, the percentage of fertilized eggs is always significantly lower than after dilution in saline. In controls when sperm in excess is mixed with eggs, the rate of fertilization is significantly lower than that obtained with smaller amounts of sperm (10^{-2} and 10^{-3}). Thus, large amounts of spermatozoa are not favorable for fertilization efficiency.

Table 1. Comparison of rate of fertilization following sperm dilution (either in saline or water)

Diluant / Dilution of sperm		Saline pH= 9.5 π= 250 mosmol.			Water		
		% ferti-lized eggs	lower limit	higher limit	% ferti-lized eggs	lower limit	higher limit
10^{-2}		91.8	87	95	77.9	72	83.5
10^{-3}		90.6	86.5	94.5	70.8	64.5	77
10^{-4}		77.9	72	83.5	35.5	28.9	42.5
10^{-5}		34.8	28.5	42	2.5	1.5	5.5
Contro (pure sperm)	BEGINNING				66.9	60	73.5
	END				64.1	57	70.5

Table 2. Statistical analysis of the results from Table 1

Comparison between	χ_2	P
Controls (pure sperm) beginning and end	0.414	
Diluted sperm in NaCl 10^{-2} and 10^{-3}	0.077	>0.05
Diluted sperm in water 10^{-2} and 10^{-3}	2.60	>0.05
Diluted sperm in NaCl 10^{-2}, 10^{-3} and 10^{-4}	21.257	<0.1
Diluted sperm in NaCl 10^{-2}, 10^{-3} and controls	82.604	<0.1
Diluted sperm in water 10^{-2}, 10^{-3} and controls	10.960	<0.1

Experiment No. 2: Effect of Osmotic Pressure of Diluant on Gamete Survival and Fertilization Rate. Effects of various osmotic pressures on fertilization (Fig. 1). High rate of fertilization was recorded following insemination at 10^{-2} and 10^{-3} sperm dilution with diluant with osmotic pressure ranging between 100 and 250 mosmol. The same pattern was obtained when sperm dilution was 10^{-4}, but percentage of fertilized eggs was significantly reduced. In distilled water, results were surprisingly low at 10^{-3} and 10^{-2} sperm concentration.

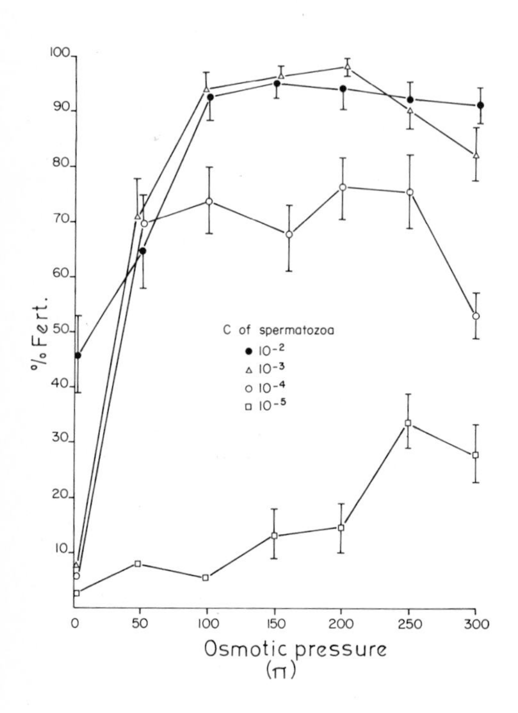

Fig. 1. Effects of various osmotic pressures of diluant on percentage of fertilization (measurement of embryonic development rate 10 days following insemination) in rainbow trout

Effects on eggs (Fig. 2). The ability of eggs to be fertilized, following exposure to diluants of various Π, was variable. Egg fertility dropped within 2 min exposure to distilled water, and 15 to 20 min exposure to diluant at 50 mosmol. In other diluants, egg fertility lasted more than 40 min. The best results were obtained when eggs were kept in diluants at 100, 150, and 200 mosmol. At 250 and 300 mosmol, egg fertility dropped drastically in 1 min.

Effects on spermatozoa (Fig. 3). Spermatozoa did not survive for a long time (not more than 2 min) in the various diluants. Within 15 sec, the ability of spermatozoa to fertilize eggs dropped, especially with low osmotic pressures (0 to 100 mosmol). Thirty seconds after mixing sperm with diluants at Π = 150, 200 and 300 mosmol, some spermatozoa still retained their fertilizing capacity.

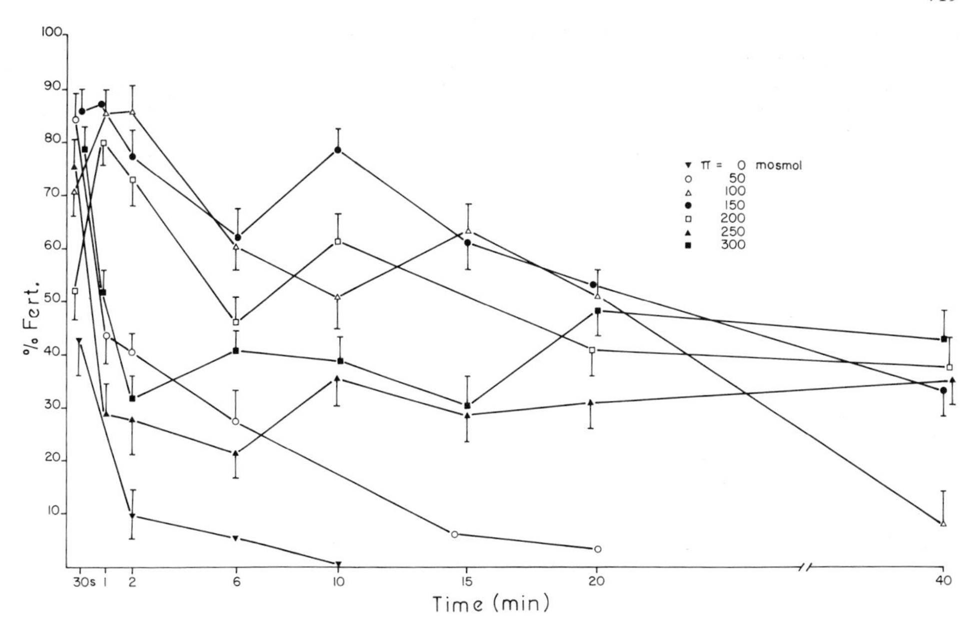

Fig. 2. Effects of various osmotic pressures of diluant on egg fertilizability. Time in minutes corresponds to duration of exposure of eggs to diluant before insemination

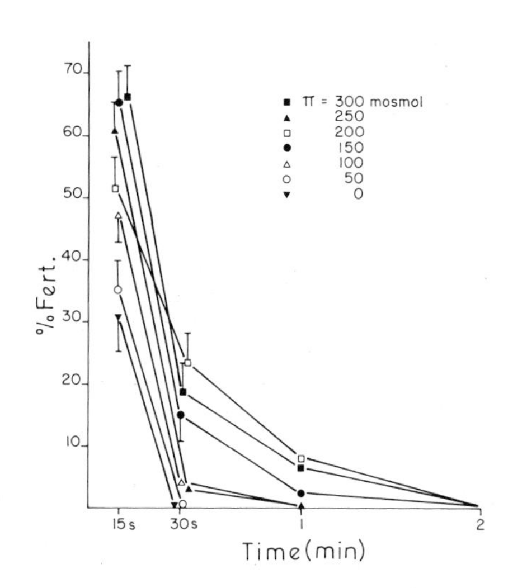

Fig. 3. Effects of various osmotic pressures of diluant on spermatozoa survival; abscissa: length of time of sperm exposure to diluant before eggs are added

<u>Experiment No. 3: Effect of pH and Buffer of Diluant on Fertilization Rate</u>. Whatever the buffer used in the diluant, percentage of fertilization was not significantly different at a molarity of 0.02 M (Fig. 4). Between pH 8.1 and 9.7, the fertilization rate stayed very high, but was lowered beyond 9.7. When the molarity of diluant was 0.1 M, the rate of fertilization was lower, especially beyond pH 7.9 with carbonate-bicarbonate and glycocolle. Thus, 0.1 M buffer concentration, which was better for maintaining medium pH after insemination, should not be used in the diluant.

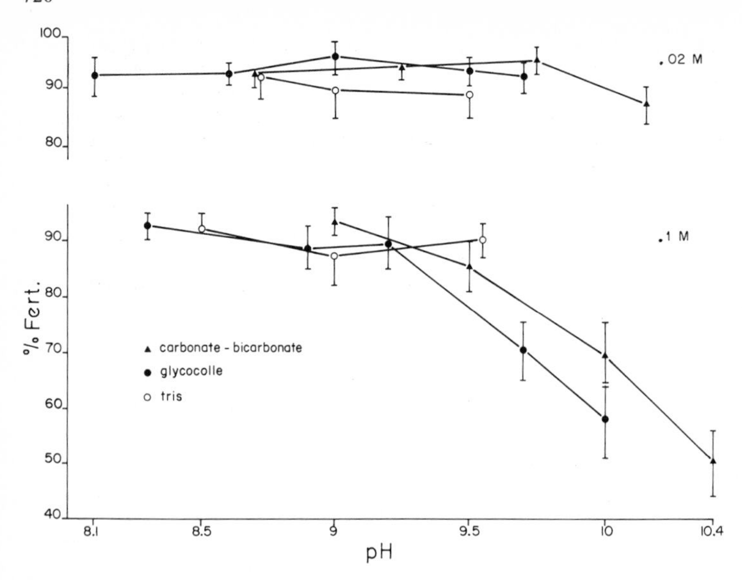

Fig. 4. Effects of various buffers in diluant on fertilization rate in rainbow trout

<u>Experiment No. 4: Optimum Timing for Transferring Inseminated Eggs into Water</u> (Fig. 5). When fertilized eggs were transferred immediately into water, fertilization rate was normal. Within 6-8 min, transfer affected the rate of fertilization, which was significantly lower than at 2 or 10 min, respectively. From 10 to 20 min, the percentage of fertilized eggs was back to normal. Beyond 20 min, keeping fertilized eggs in the fertilization medium was not very satisfactory since fertilization rate dropped significantly at 40 min exposure. In both sperm concentration 10^{-3} and 10^{-4}, the pattern was the same. Depression of fertility occurring in the first 10 min was even more marked at 10^{-4}.

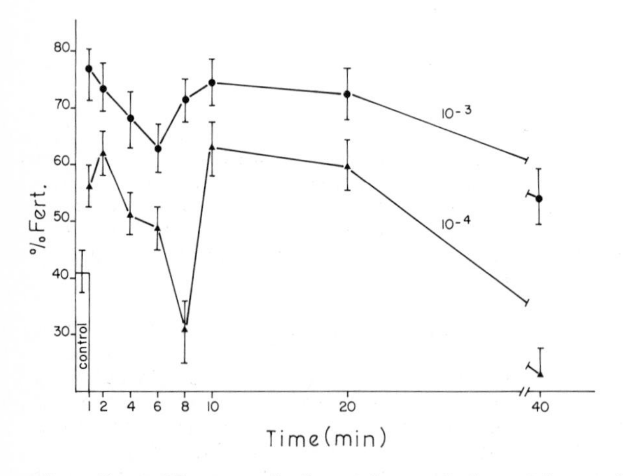

Fig. 5. Effects of duration of inseminated eggs in diluant on fertilization rate after insemination; abscissa: duration of exposure of eggs to the diluant after insemination

<u>Experiment No. 5: Minimum Number of Spermatozoa for Fertilization of One Egg</u> (Table 3). At the same sperm concentration, the percentage of fertilized eggs stays at the same level, whatever the egg concentration. When insemination is performed with 800 eggs at 10^{-3} sperm dilution, fertilization percentage is even higher. The rate of fertilization is significantly higher at 10^{-3} sperm concentration than at 10^{-4}. For optimum concentration of egg and sperm (771 eggs in 20 ml diluant, and 20 µl sperm = 10^{-3} sperm concentration), giving maximum fertilized eggs, the number of spermatozoa required to fertilize one egg was approximately 200 000. Lower sperm per egg ratios were recorded, especially at 10^{-4} sperm concentration, but percentage of fertilized eggs was not very good.

Table 3. Percentage of fertilized eggs and number of spermatozoa/egg following insemination with various concentration of sperm and eggs

Sperm dilution	No eggs Total	None fertilized	Fertilized	% of fertilized	Lower limit	Higher limit	No spz/egg
10^{-3}	88	26	62	70.46 (a)	64	78	2 237 903
	196	55	141	71.94 (a)	65	78	984 042
	321	84	237	73.83 (a)	70	78	585 443
	771	151	620	80.42 (a)	77	82.5	223 790
10^{-4}	139	45	94	67.63 (b)	57	76	147 606
	185	57	128	69.19 (b)	62	75	108 398
	418	164	254	60.77 (b)	56	65.5	47 193
	842	257	585	69.48 (b)	66	73	23 718

χ^2 test (a) $0.5\% < P < 1.0\%$
(b) $1.0\% < P < 2.5\%$

DISCUSSION

Saline solution used as a diluant for insemination increases fertilization rate and allows substantial reduction in the amount of spermatozoa used. As the minimum quantity of spermatozoa necessary for fertilization is approximately 200 000 per egg, and as a male can produce several hundred million spermatozoa (Bratanov and Dikov, 1961; Billard et al., 1971), it is theoretically possible to inseminate several hundred females with the sperm of only one male.

The diluant used in this technique should be a saline solution with a rather wide range in osmotic pressure (this point will be discussed later) and buffered with either organic buffer (glycocolle, Tris) or carbonate-bicarbonate, 0.02 M at pH between 8.5 and 9.5. After fertilization, eggs can be transferred immediately into freshwater in less than 2 min. If not, eggs should be left in the diluant for 15 min because of an unexplainable drop in fertility occurring when eggs are transferred into water between 2 and 10 min following insemination (Fig. 5).

Effects of the osmotic pressure of the diluant on retention of fertility were not exactly the same for sperm and eggs. Optimum osmotic pressure ranged from 150 to 250 mosmol for sperm, and from 100 to 200 for eggs. However, when both gametes were immediately mixed together with diluant, the percentage of fertilized eggs was maximum and constant from 100 to 300 mosmol at 10^{-2} sperm concentration. At lower sperm concentration, rate of fertilization dropped at 300 mosmol.

It is well known that gamete fertilizibility in salmonids declines very sharply in water. In this study, there was no indication of sperm viability after 30 sec, which is in agreement with Ginsburg (1963). Buss and Corl (1966) found traces of viability remaining at 2 min, but sperm concentration was higher than in the present experiment. Egg fertility drops progressively to zero in a few minutes in water (variation from 2 to 20 min), according to various authors (Ginsburg, 1961; Smirnov, 1963; Buss and Corl, 1966). Survival of spermatozoa was much better in saline (Π = 200 to 300 mosmol), as shown previously by Tourdakov (1970) and Billard and Breton (1970).

In conclusion, in the practice of artificial insemination it is therefore possible to use a saline solution with a definite composition as a diluant. This allows minute quantities of sperm to fertilize more eggs than with the usual dry method.

SUMMARY

A new technique of artificial insemination using a diluant for sperm is described. Ova and sperm are mixed together with a diluant simultaneously. Inseminated eggs should be transferred into water in less than 2 min, or in 15 min following insemination. Diluant is a saline solution buffer at pH 8.5 to 9.5, and with osmotic pressure varying from 150 to 250 mosmol. Minimum amount of spermatozoa required for fertilization is about 200 000 per egg.

ACKNOWLEDGEMENT

This work was supported by the Institut National de la Recherche Agronomique.

REFERENCES

Billard, R. and Breton, B., 1970. Modifications ultrastructurales et cytochimiques des spermatozoïdes après dilution chez les Poissons d'eau douce. Congr. int. Microsc. électron., 7, 637-638.

Billard, R., Breton, B. and Jalabert, B., 1971. La production spermatogénétique chez la Truite. Ann. Biol. anim. Biochem. Biophys., 11, 190-212.

Bratanov, C. and Dikov, V., 1961. Sur certaines particularités du sperme chez les Poissons. C.R. IVème Congr. int. Reprod. anim. (La Haye), 895-897.

Buss, K. and Corl, K.G., 1966. The viability of trout germ cells immersed in water. Prog. Fish Cult., 28, 152-153.

Ginsburg, A.S., 1961. The block to polyspermy in sturgeon and trout with special reference to the role of cortical granules (alveoli). J. Embryol. exp. Morph., 9, 173-190.
Ginsburg, A.S., 1963. Sperm-egg association and its relationship to the activation of the egg in salmonid fishes. J. Embryol. exp. Morph., 11, 13-33.
Ginsburg, A.S., 1968. Differences in morphology and properties of gametes in fishes and their importance for the elaboration of adequate methods of artificial insemination. VIème Congr. int. Reprod. anim. Insém. artif. (Paris), II, 1037-1040.
Nomura, N., 1964. Studies on reproduction of rainbow trout activities of spermatozoa in different diluents and preservation of semen. Bull. jap. Soc. Scient. Fish, 30, 723-733.
Nursall, J.R. and Hasler, A.D., 1952. The viability of gametes and the fertilization of eggs by minute quantities of sperm. Progr. Fish Cult., 14, 165-169.
Petit, J., Jalabert, B., Chevassus, B. and Billard, R., 1973. L'insémination artificielle de la truite. I - Effets du taux de dilution, du pH et de la pression osmotique du dilueur sur la fécondation. Ann. Hydrobiol. (in press).
Plosila, D.S., Keller, W.T. and McCartney, T.J., 1972. Effects of sperm storage and dilution on fertilization of brook trout eggs. Progr. Fish Cult., 34.
Poon, D.C. and Johnson, A.K., 1970. The effect of delayed fertilization on transported salmon eggs. Progr. Fish Cult., 32, 81-84.
Scheuring, L., 1925. Biologische und physiologische Untersuchungen und Forellensperma. Arch. Hydrob., suppl. IV, 181-318.
Smirnov, A.I., 1963. Conservation dans l'eau de l'aptitude des oeufs et des spermatozoïdes du saumon *Oncorhynchus gorbuscha* (W) à la fécondation. Nauchn. Dokl. Vyssh. Shk biol. Nauki SSSR, 3, 37-41.
Tourdakov, A.F., 1970. Effects of various concentration of the ringer solution on trout sperm behaviour. Biol. Nauki, 74, 20-24.

R. Billard, J. Petit, B. Jalabert and D. Szollosi
Laboratoire de Physiologie des Poissons, I.N.R.A.
78350 Jouy-en-Josas / FRANCE

Laboratory Rearing of Common Sole (*Solea solea* L.) under Controlled Conditions at High Density with Low Mortality

J. Flüchter

INTRODUCTION

The results of previous rearing experiments (Flüchter, 1966) gave the impression that the survival rate of sole larvae was closely connected with the presence of certain ciliates. It seemed that there was some necessity for ciliates at first feeding and that early larval mortality was due to a lack of ciliates in the diet. The results of the 1972 sole-rearing experiments did not support this hypothesis but showed that it is possible to manipulate actively the survival rates of larvae, thus opening the way to investigations on the life requirements of fish larvae in general.

METHODS

Obtaining viable sole eggs depends on the spontaneous spawning of the parent fish in captivity. Inadequate nutrition was found to be the main cause of failure of spawning. If nutrition was corrected, successful spawning occurred in small containers, allowing eggs to be collected from individual females. The eggs were incubated, as previously described (Flüchter, 1964), in aquaria with inside filtration. The hatched larvae were washed in Trypaflavin-0.1% seawater solution for 12 sec to remove ciliates, then transferred to jars with sterile seawater (filtered through poresize 0.22 μ) in groups of 100 per litre and in a few cases smaller groups of 50 per litre and kept under constant temperature at 15°C or 20°C with continual illumination (3000 - 5000 lux).

Different species of protozoans suspected of being the first food were then added from pure cultures. Some days after hatching freshly hatched *Artemia* nauplii (younger than 24 h post-hatching) were offered in all experiments and the percentage of feeders was noted. The *Artemia* nauplii had been washed previously with freshwater (40°C) for 30 min so that they did not carry unwanted organisms with them on introduction to the rearing containers. *Artemia* nauplii were first offered on the 5th to 6th day after hatching at 20°C or on the 6th to 8th day at 15°C. A special starvation experiment showed that *Artemia* should not be offered until the 4th day at 20°C and 6th day at 15°C. Food should not be offered later than the 6th day at 20°C and later than the 9th day at 15°C. Using these feeding criteria nearly 100% feeders were obtained. To prevent the undesirable effect of surplus uneaten *Artemia* nauplii in later experiments, the requirements per larvae of different size were estimated as shown in Table 1.

Small containers were used to rear the larvae where the presence or absence of microfauna could be most easily controlled. Sole larvae can be satisfactorily reared in small volumes and high densities since they are not very sensitive to metabolites or pollutants; they also react favourably to the confined conditions, are not phototactic, and are not aggressive. A high density of larvae was suggested to control the unwanted growth of microorganisms; it was assumed that the chance of ciliates surviving in a high predator density would be small and no balance of ciliate growth and sole feeding would be established. It was thus planned that the larvae would be dependent on the food offered to them and not on organisms growing in an uncontrolled way in the containers.

Table 1. Number of *Artemia* nauplii taken per feed (2 feeds/day at 20°C, 1 feed/day at 15°C

Larval body length mm	*Artemia* nauplii/larva
3.5	3 - 5
4.0 - 5.0	7 - 10
5.0 - 5.5	10 - 15
5.0 - 6.5	15 - 20
6.5 - 8.0	ca. 30

Some larvae were reared, as previously described, in aquaria with inside filtration as a control (Flüchter, 1964). The water in the jars without such filtration was changed generally when the sole were 4 to 6 weeks old, when they were measured. The sole were also measured at the start of metamorphosis. In all about 200 batches of larvae were used.

RESULTS

It should be stressed that in all experiments *Artemia* nauplii were added as the main food. Depending on the other organisms added, different types of mortality and survival were found. Three main types were reproduced several times. The first type, characterized by nearly 100% of larvae starting to feed and continuing to feed to a point where they developed into healthy metamorphosed fish, was produced when:

1. the red flagellate *Cryptomonas* sp. was added,
2. *Cryptomonas* together with the ciliate *Uronychia transfuga* were added,
3. the green flagellate *Dunaliella* sp. together with the ciliate *Holosticha* sp. were added,
4. pure sterile filtered 1972 seawater was used and the density of larvae was below 50/l.

See Fig. 1 for an example of the results.

The second type characterized by no feeding on freshly hatched *Artemia* nauplii and a complete mortality of the early larval stages at the first-feeding stage was caused when:

1. small harpacticoid copepods such as *Tisbe helgolandica* were added at concentrations higher than 5000/l,
2. young *Artemia* nauplii at the beginning of the phototactic stage were added,

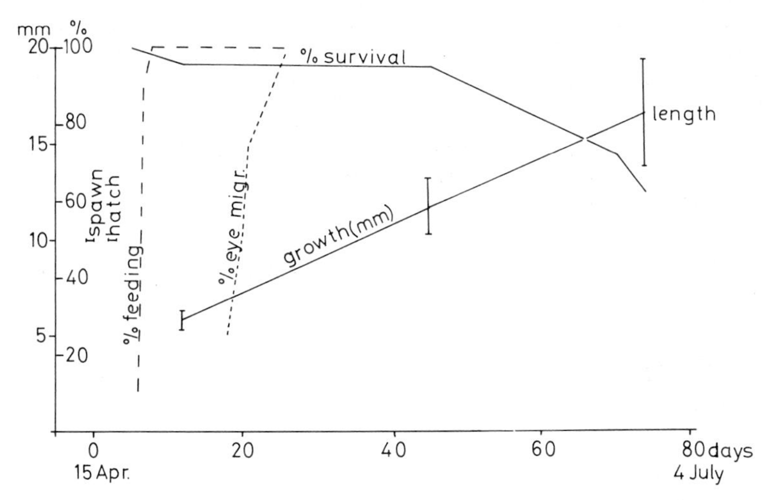

Fig. 1. First type: sterile filtered seawater + *Cryptomonas* at 20°C giving good survival to the healthy young fish stage. First length measurement made at metamorphosis (mean length and standard deviation given); % eye migration shows proportion of larvae starting to metamorphose

3. when the hypotrich ciliate *Euplotes* sp. (c. 55 μ x 25 μ in size) was added,
4. when the green flagellate *Tetraselmis tetrahele* was added,
5. when the blue-green alga *Phormidium* sp. was present.

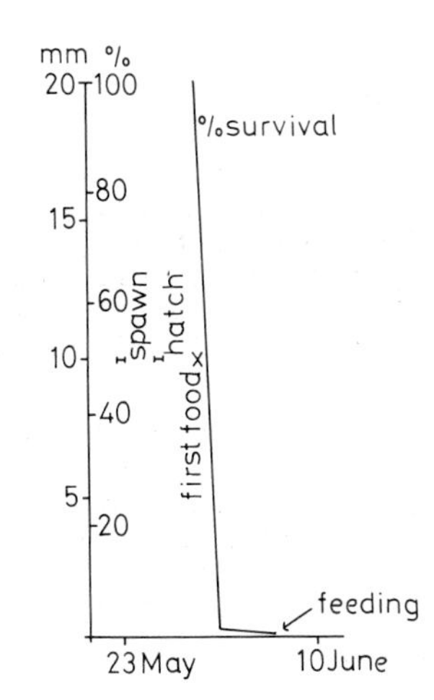

Fig. 2. Second type: sterile filtered seawater + harpacticids giving high larval mortality, only a few larvae started feeding

The same early high mortality but with a few survivors occurred when:

1. only about 1000 harpacticids were added,
2. the brown alga *Ectocarpus* sp. was present.

See Fig. 2 for an example of the results.

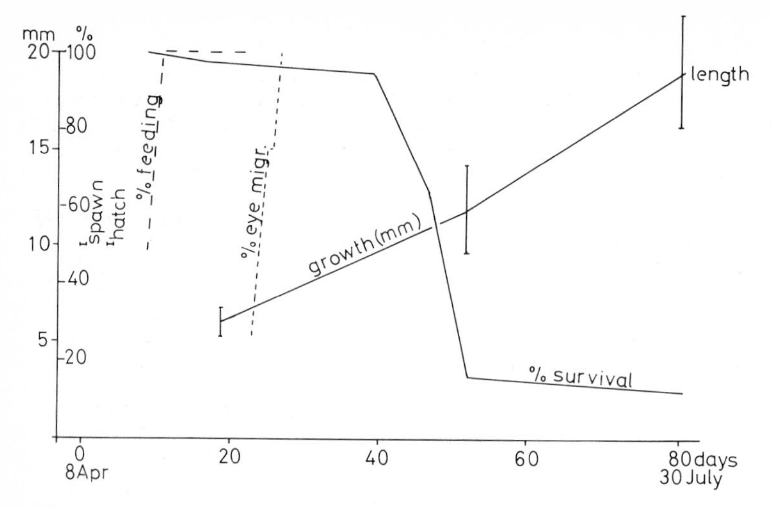

Fig. 3. Third type: pure sterile filtered seawater giving a good start to the larvae but sudden high mortality of the newly metamorphosed fish. Length given with standard deviation

The third main type, characterized by a good start and a high survival rate of the larvae, followed, however, by a sudden high mortality at the end of eye migration, was obtained when:

1. using pure sterile filtered water and adding more than 100 sole larvae/l,
2. using sterile filtered seawater in which before about 5000 *Artemia* nauplii has lived for 7 days at 15°C without feeding,
3. *Uronychia transfuga* alone (without *Cryptomonas*) was added,
4. *Dunaliella* or *Cryptomonas* were added autoclaved at 120°C,
5. the dinoflagellate *Prorocentrum micans* was added.

See Fig. 3 for an example of the results.

In the experiments chemical checks showed that larval viability was not reduced by ammonia up to 10 mg NH_4/l (0.5 mg free ammonia/l) at pH 8.2, nor by pH variations between 7.3 and 8.5. The range of salinity tolerated by feeding larvae was 22 to 53°/oo. The oxygen content of the water was generally above 80% saturation. It should be stressed that the patterns of survival were independent of parentage.

Lasker (1970) reported that anchoveta fed successfully on both the rotifer *Brachionus plicatilis* and on *Gymnodinium splendens*. When offered to sole these food species produced no better result than pure sterile filtered seawater.

DISCUSSION

The best interpretation of the results is, in my opinion, that a dissolved organic substance exists which is required by the larvae. This would explain the unpredictability of rearing experiments. Small crustaceans such as *Artemia* and *Tisbe*, also the ciliate *Euplotes*, the blue-green alga *Phormidium*, and the green flagellate *Tetraselmis* may be able to extract this special substance from the seawater; that means they would be better competitors for the substance and thus the feeding of the sole larvae might finally be inhibited. That *Tetraselmis* can extract dissolved carbohydrates and organic nitrogen compounds from the seawater (Gooday, 1970), and that *Artemia* can extract dissolved nucleic acid components and vitamins of the B-group from seawater (Provasoli and D'Agostino, 1969), supports this interpretation. It is rather unlikely that harpacticids or *Artemia* produce harmful metabolities which are responsible for the mortality, since sole larvae are very resistant to pollution. The rapid growth of the 'few survivors' could be explained by a second substance for which only the sole larvae compete. Though it is not known whether these suspected substances may at any time become the limiting factor for young sole in the natural environment, it is quite clear that they can influence the course of experiments. It may be said that sole larvae proved to be especially favourable test organisms for investigations aimed at the question of whether fish larvae have a requirement for dissolved organic substances in seawater. At present the problems of determining very low concentrations of dissolved organic substances in seawater by chemical methods have not been fully resolved (see Bohling, 1970).

SUMMARY

1. Sole larvae were reared in small containers at high density (100/l).

2. First feeding and the survival rate depend upon the presence of certain members of the microfauna.

3. Three types of survival were observed:
a) Good development to the healthy young-fish stage
b) No feeding on *Artemia* nauplii when first feeding should start, followed by a total mortality
c) High percentage feeding when *Artemia* first offered followed by a later high mortality

4. Pollution cannot have been responsible for the differences in survival rates.

It is proposed that there is a dissolved organic substance that sole larvae require. Small crustaceans, ciliates, and algae may extract this from the seawater more efficiently than sole, being better competitors for the substance.

REFERENCES

Bohling, H., 1970. Untersuchungen über freie gelöste Aminosäuren im Meerwasser. Mar. Biol., 6, 213-225.
Flüchter, J., 1964. Eine besonders wirksame Aquarienfilterung und die Messung ihrer Leistung. Helgolander wiss. Meeresunters., 11, 168-170.
Flüchter, J., 1966. Spawning, first feeding and larval behaviour of North Sea sole. ICES CM 1966 C, 3.
Gooday, G.W., 1970. A physiological comparison of the symbiotic alga *Platymonas convolutae* and its free-living relatives. J. mar. biol. Ass. U.K., 50, 1-9-208.
Lasker, R., Feder, H.M., Theilacker, G.H. and May, R.C., 1970. Feeding, growth and survival of *Engraulis mordax* larvae reared in the laboratory. Mar. Biol., 5, 345-353.
Provasoli, L. and D'Agostino, A., 1969. Development of artificial media for *Artemia* salina. Biol. Bull., 136, 434-453.
Spotte, S.H., 1970. Fish and invertebrate culture. New York - London: Wiley-Interscience, 145 pp.

J. Flüchter
Bayerische Landesanstalt für Fischerei
8130 Starnberg / FEDERAL REPUBLIC OF GERMANY
Weilheimer Str. 8a

Progress towards the Development of a Successful Rearing Technique for Larvae of the Turbot, *Scophthalmus maximus* L.

A. Jones, R. Alderson, and B. R. Howell

INTRODUCTION

Assessments based on economic and biological factors have demonstrated the potential of the turbot *Scophthalmus maximus* L. as a species for commercial cultivation (Jones, 1972a; Purdom et al., 1972). On economic grounds the turbot is an attractive proposition because of its high market price (which in 1972 averaged 62 pence per kg at first sale on fish markets in England and Wales). Its important biological factors include fast growth rate and efficient food conversion. Progress in the cultivation of turbot has, over the past few years, been impeded by difficulties in larval rearing. Work on turbot larval rearing began at the MAFF Fisheries Laboratory, Lowestoft in 1966 and the early results have already been published (Jones, 1972b). Although larvae reached late stages of development in these experiments, none succeeded in completing metamorphosis, and it was not until 1972 that larvae were reared beyond metamorphosis for the first time at the Lowestoft and Port Erin Laboratories. This paper describes these successful rearing experiments and discusses ways in which larval survival in future experiments might be improved.

METHODS AND RESULTS

Rearing Conditions. Turbot larvae were hatched from eggs obtained at sea from wild fish and from a spawning stock which had been held in the Lowestoft Laboratory for 3 years. At sea the incubation temperature could not be controlled and rose from 9 to 14°C; in the laboratory the eggs were incubated at 12°C. After hatching, rearing was carried out at temperatures of between 13.5 and 18°C.

Larvae were fed initially on the rotifer *Brachionus plicatilis* Müller, cultured in both laboratories on *Dunaliella tertiolecta* Butcher (Howell, 1973), followed by *Artemia* nauplii which were first offered between days 12 and 14 after hatching. As the larvae grew they were provided with *Artemia* metanauplii and finally adult *Artemia*, both cultured in the laboratory on *Dunaliella*. In addition to these basic foods the larvae reared in Lowestoft from eggs stripped at sea (Experiment 1) were also provided with natural plankton during the rotifer-feeding stage. This plankton was obtained from the coastal waters near the Lowestoft Laboratory by suspending a 200 m.p.i. (8 meshes/mm) International Fine Net in the tidal current. Organisms greater than 250 μm were removed by filtration, and the remaining material, consisting mainly of copepod nauplii, was offered to the larvae.

At Lowestoft, larvae were reared to metamorphosis in two separate experiments; experiment 1, using eggs fertilized at sea and experiment 2, using eggs from captive spawning stock. In both experiments 1400 l (152 x 152 x 61 cm) black butyl rubber tanks were stocked with approximately 5000 yolk-sac larvae per tank. At day 26 after hatching, when the larvae were actively feeding on *Artemia* nauplii, they were transferred to a larger tank of 3800 l capacity (249 x 183 x 84 cm). This transfer was an attempt to counteract an increase in larval mortality which appeared to be associated with contact of the larvae with the sides and bottom of the tank. A high concentration of ciliates present in the detritus on the tank bottom may have contributed to this mortality. A final transfer to smaller, 100 l (120 x 57 x 27 cm) tanks was made on day 42, to enable adequate concentrations of larger food organisms to be maintained. Aeration and a continuous inflow of fresh filtered sea water was provided in all tanks. The overflows left the tanks by standpipes, thus maintaining a surface run-off which prevented the accumulation of scum at the water surface. Continuous light was provided by single 65-80 watt 'warm white' fluorescent tubes suspended above the rearing tanks, giving a surface illumination of 900 lux.

At Port Erin only one rearing trial resulted in the production of metamorphosed turbot, the larvae being derived from the same egg stock used at Lowestoft for experiment 1. At day 7 after hatching, 1000 actively-feeding larvae were transferred from a 120 l (120 x 60 x 30 cm) stock tank to a 300 l (120 x 60 x 60 cm) tank in which a partially self-sustaining rotifer/alga culture had been established. Six days prior to the transfer of the turbot larvae 40 l of algal-rich sea water, taken from an outside seawater pond and filtered through a 60 μm mesh net, was added to clean sea water in this tank. Algal nutrient salts, as recommended by Walne (1966) for the culture of *Isochrysis galbana* and other algae, were added and rotifers were introduced. Natural daylight supplemented by fluorescent tubes provided illumination at the water surface of up to 1600 lux. Although the rotifers were initially sustained by the mixed algal species which developed, periodic additions of *Dunaliella* were necessary to maintain their reproduction. The tank, though provided with gentle aeration, was static until 11 days after the addition of the larvae. By this time the high density of food organisms had led to a deterioration in water conditions, and a flow of fresh sea water was commenced.

At day 26 after hatching the 232 surviving larvae were transferred to clean tanks - some to tanks of the same size (300 l) and 100 to a 1500 l tank (300 x 86 x 80 cm). This last tank was illuminated with tungsten halogen tubes to give a light intensity at the water surface of between 1600 and 2700 lux, sufficient to maintain the reproduction of the *Dunaliella* with which it had been inoculated. When the larvae were added the algal-cell density had reached 150 cells/μl and this supplied food for the *Artemia* nauplii, enabling them to grow and provide the larvae with a larger food organism. Development to metamorphosis was completed in this tank. None of the larvae transferred to the 300 l tank survived to metamorphosis.

In addition to turbot a small number of hybrids of the brill *Scophthalmus rhombus* L. female with turbot male, obtained from a fertilization carried out at sea using freshly-trawled fish, were also reared during 1972.

Growth and Survival. The mean lengths of larvae at stocking in experiments 1 and 2 at Lowestoft were 3.4 and 3.1 mm respectively. At metamorphosis the larvae ranged in size from 23.0 to 30.0 mm. A total of 40 larvae completed metamorphosis between days 45 and 72 in experiment 1; and 18 larvae between days 50 and 80 in experiment 2. The 46 larvae reaching metamorphosis in the one successful rearing trial at Port Erin ranged in size from 27 to 39 mm at day 70.

The general behaviour of the turbot larvae, observed during the rearing experiments, differed markedly from that of larvae of the Pleuronectidae and Soleidae previously reared in the laboratory. Initial orientation and first feeding, as with other fish larvae, took place in mid-water, but by day 20 the larvae were concentrated at the water surface where they remained until moving to the bottom at metamorphosis. They were aided in maintaining their position at the surface by the presence of a well-developed swimbladder, which was lost at metamorphosis.

Another striking feature in their development was the change in swimming position of the larvae while at the water surface. Although initially they swam with the dorso-ventral plane in a vertical direction, they later assumed a swimming position more nearly horizontal, with the left side of the body uppermost. This change was generally accompanied by a reduction in apparent pigmentation - the early larvae were very densely pigmented whereas the later stages, from about day 35 onward, were almost completely transparent, the internal organs becoming clearly visible through the body surface.

The increase in mean length of turbot reared in experiment 1 from hatching to day 248 is shown in Fig. 1, together with mean monthly temperature. The rapid fall in growth rate from day 150 may be related to the decrease in temperature. The growth rate of brill x turbot hybrids reared at Port Erin (Fig. 1) did not decline at this time, probably because these fish were maintained at a constant temperature of 18°C.

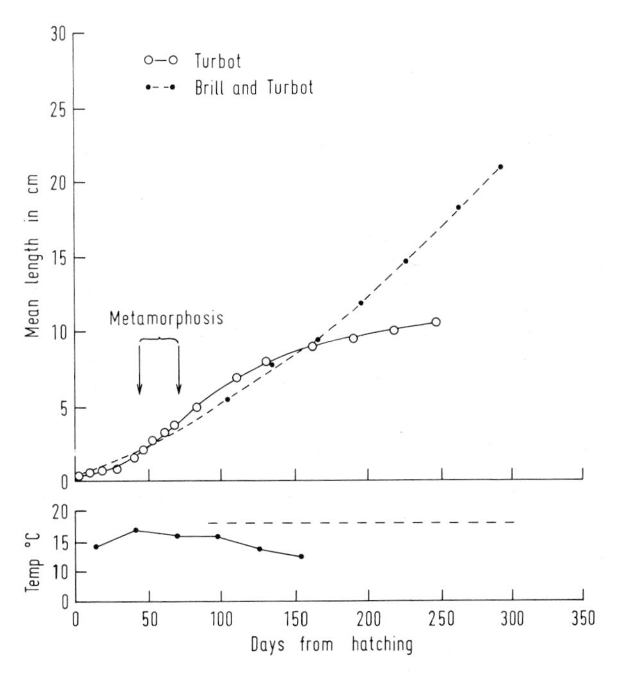

Fig. 1. The growth in length of turbot and brill x turbot hybrids reared in the laboratory. Mean monthly temperature is also shown

Larval survival from hatching to metamorphosis in experiments 1 and 2 at Lowestoft was 0.8 and 0.36% respectively. The pattern of survival is shown in Fig. 2. Data on the survival of early-stage larvae up to day 17 had been taken from a series of experiments conducted at Lowestoft using 40 l tanks. Survival of older larvae to metamorphosis is shown using data from experiment 1. It is clear from Fig. 2 that mortality rate was highest between days 15 and 30.

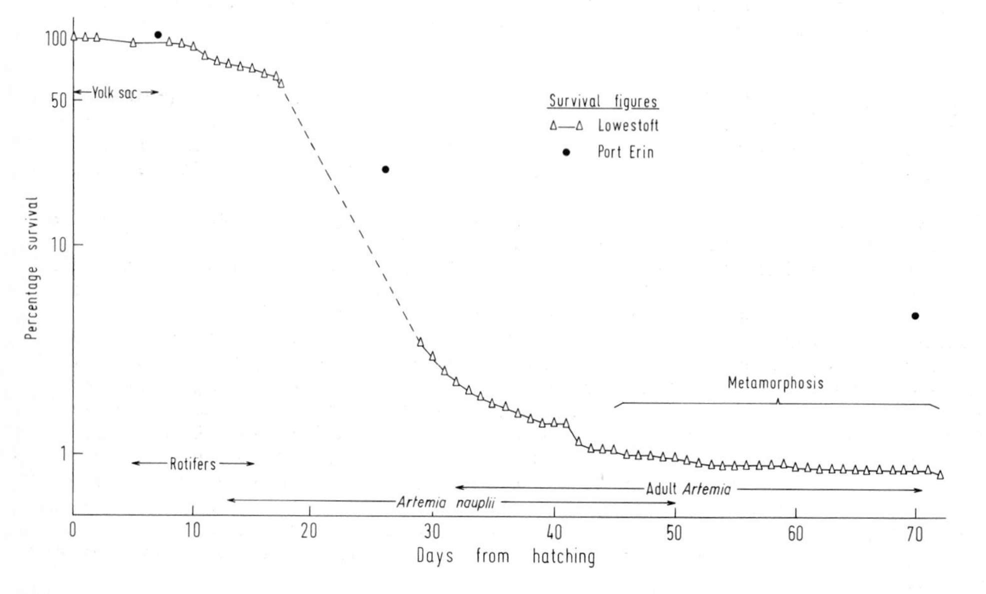

Fig. 2. The survival of turbot larvae during the rearing experiments

In the successful rearing trial at Port Erin, 232 of the 1000 actively feeding larvae selected at day 7 were alive at day 26; 100 of these were moved to a large 1500 l tank and 46 completed metamorphosis, giving a survival from day 7 to metamorphosis of 4.6% (Fig. 2). The remaining 132 larvae surviving at day 26 were transferred to small 300 l tanks and did not survive. If it can be assumed that the 100 larvae transferred to the large tank were representative of the 232 survivors at day 26, this would represent a survival of approximately 10% from day 7 to metamorphosis. Possible factors influencing survival are discussed in the next section.

DISCUSSION

Analysis of Treatments. Since these attempts to rear turbot larvae had to be of an empirical nature, few firm conclusions can be made regarding the special requirements of this species. However, the possible importance of certain factors are indicated by the results of the experiments.

Tank Size. One of the main differences between the successful rearing of turbot larvae reported here and earlier attempts (Jones, 1972b) is that we used tanks of a much greater capacity than those previously used. However, the significance of this factor is not clear. Tanks of 40 l capacity have in the past proved quite adequate for rearing the larvae of other flatfish at high density, and with good survival to metamorphosis (Shelbourne, 1964; Howell, 1972, 1973). It is, however, possible that turbot larvae, which in the later stages swim actively at the surface of the water, may in some way be more sensitive than other larvae to contact with solid surfaces.

Oxygen Tension. Unlike many flatfish, the larvae of the turbot have a functional swimbladder; it has been suggested (Purdom, 1973) that failure to adequately fill the swimbladder, due to low dissolved oxygen levels, may have contributed to previous rearing failures.

In the Lowestoft experiments the tanks were irrigated with fresh sea water throughout the rearing period and oxygen concentrations were maintained within 90% of saturation. These high levels were not maintained in the initially static rearing conditions in the Port Erin experiment; coupled with the development of a high density of food organisms, this resulted in the oxygen concentration falling to 64% of saturation, and the pH to 7.48, before a flow of fresh sea water through the tank was established on day 18 after hatching. The poor water quality during this period may indeed have been responsible for the large range in size of swimbladder noted in the transparent late-stage larvae, and also for the observation that some larvae were unable to maintain their position at the surface without continuous very active swimming. This did not however, appear to prevent these larvae from reaching metamorphosis, and there is no evidence to suggest that the low oxygen concentration had a significant effect on survival.

Food. Differences in the growth of larvae in their early stages were evident in some experiments carried out at Port Erin. When the 300 l tank (Tank A), with the rotifer/alga mixture, was established using algae from an outside pond, an identical tank (Tank B) adjacent to it was also stocked. The water for this tank was passed through an ultra-violet light sterilizing unit and, after algal nutrient salts had been added, the tank was inoculated from a pure culture of *Dunaliella*. Rotifers were then added and when the larvae were transferred to Tank A, the same number were also put into Tank B. The latter tank initially had a lower level of rotifers than the first, but further additions combined with their natural reproduction soon led to an excess level of food organisms in both tanks. The mean preserved wet weights of the larvae determined from samples taken at day 24 are given in Table 1. Survival in these two tanks, measured at day 26, was 23% in A and 22% in B.

Table 1 also shows the results of a later experiment using larvae hatched from eggs obtained from a captive spawning stock maintained by the White Fish Authority at Hunterston, in which a similar result was obtained. In this experiment two 220 l tanks were each stocked with 1000 late yolk-sac larvae. Both tanks were provided with gentle aeration and a continuous flow of fresh sea water; one tank also received a slow flow, approximately 16 l/day, of water taken from the outside seawater pond. An examination of the algae present on this occasion, carried out by Dr. M. Parke at the Plymouth Marine Laboratory revealed a mixture of organisms including diatoms in which colourless flagellates predominated. Rotifers reared on *Dunaliella* were added daily and an excess level of rotifers was always present in both tanks. Loss through the outflow and by predation, however, meant that the rotifers in the tank receiving pond water did not establish an expanding and reproducing population, as had occurred in the previous experiment.

Table 1. Wet weights of preserved turbot larvae from rearing trials at Port Erin

Tank	Treatment	Wet Weight (mg)		Day of Sample
		Mean	S. Deviation	
A	*Dunaliella* and Pondwater	4.85	1.48	24
B	*Dunaliella* alone	1.59	0.28	24
1	No algae present	0.66 1.28	0.13 0.65	11 19
2	Pond water added	0.75 4.27	0.14 1.30	11 19

These results show that in both cases the larvae in the tanks receiving pond water reached a larger size than those without this addition. Variations in the density of food organisms may have contributed to these differences, but it is also possible that the water from the outside pond may have indirectly supplied additional food. If the turbot larva feeds on a rotifer which is actively feeding on algae then it must also take in the partially digested algae in the rotifer gut. Species of algae cultured in the laboratory have been shown to differ in their nutritive value to certain bivalve molluscs (Walne, 1970); Helm (1969) demonstrated that mixtures of algal species were of greater value than single species when fed to oyster larvae. It is perhaps possible that the difference in growth of the turbot larvae may have been similarly related to differences in the algal species on which the rotifers were feeding. It would, however, be surprising if rotifers grown only on *Dunaliella* were to prove inadequate as a food for turbot larvae, since Howell (1973) showed that plaice and sole larvae could be reared to metamorphosis on rotifers alone. Moreover, a batch of Dover sole larvae was reared to metamorphosis during 1972 on rotifers from the same culture used to supply the turbot rearing experiments (Howell, unpubl.). The brill x turbot hybrids, too, developed normally even though they did not receive a pondwater supplement during their rotifer feeding stage. This problem clearly requires further, more detailed investigation before these results can be fully understood.

CONCLUSIONS

Although newly metamorphosed turbot may be taken on beaches at certain times of the year it is unlikely that they could be caught in sufficient numbers to support a commercial farming operation. Therefore, a successful rearing technique is essential if turbot farming is to become a reality. Although the results discussed in this paper show that turbot can be reared artificially, from the egg to metamorphosis, the optimum conditions for rearing still have to be defined. Further investigation of the effect of diet and food density on survival may be important in this context.

REFERENCES

Helm, M.M., 1969. The effect of diet on the culture of the larvae of the European flat oyster, *Ostrea edulis* L. ICES CM 1969/E, 7, 6 pp. (mimeo).

Howell, B.R., 1972. Preliminary experiments on the rearing of larval lemon sole *Microstomus kitt* (Walbaum) on cultured foods. Aquaculture, 1, 39-44.

Howell, B.R., 1973. Marine fish culture in Britain VIII. A marine rotifer, *Brachionus plicatilis* Müller, and the larvae of the mussel, *Mytilus edulis* L. as foods for larval flatfish. J. Cons. perm. int. Explor. Mer, 35 (1)(in press).

Jones, A., 1972a. Marine fish farming; an examination of the factors to be considered in the choice of species. Lab. Leafl. Fish. Lab. Lowestoft, N.S. (24), 16 pp.

Jones, A., 1972b. Studies on egg development and larval rearing of turbot, *Scophthalmus maximus* L. and brill, *Scophthalmus rhombus* L. in the laboratory. J. mar. biol. Ass. U.K., 52, 965-986.

Purdom, C.E., 1973. Turbot shows promise for marine farming. Fishing News int. 12 (2), 33.

Purdom, C.E., Jones, A. and Lincoln, R.F., 1972. Cultivation trials with turbot (*Scophthalmus maximus*). Aquaculture, 1 (2), 213-230.

Shelbourne, J.E., 1964. The artificial propagation of marine fish. Adv. mar. Biol., 2, 1-83.

Walne, P.R., 1966. Experiments in the large-scale culture of *Ostrea edulis* L. Fish. Invest., Lond., 25 (4), 53 pp.

Walne, P.R., 1970. Studies on the food value of 19 genera of algae to juvenile bivalves of the genera Ostrea, Crassostrea, Mercenaria and Mytilus. Fish. Invest., Lond., 26 (5), 62 pp.

A. Jones
Fisheries Laboratory
Lowestoft / GREAT BRITAIN

Present address:

The British Oxygen Company
Shearwater Equipment
55 London Road
Horsham, Sussex / GREAT BRITAIN

R. Alderson and B.R. Howell
Fisheries Laboratory
Port Erin, Isle of Man / GREAT BRITAIN